# MOLECULAR BIOPHYSICS

# MOLECULAR BIOPHYSICS

**M. V. Volkenstein**

*Institute of Molecular Biology*
*Academy of Sciences of the USSR*
*Moscow, USSR*



ACADEMIC PRESS New York • San Francisco • London 1977

A Subsidiary of Harcourt Brace Jovanovich, Publishers

ACADEMIC PRESS, INC.
111 Fifth Avenue, New York, New York 10003

*United Kingdom Edition published by*
ACADEMIC PRESS, INC. (LONDON) LTD.
24/28 Oval Road, London NW1

**Library of Congress Cataloging in Publication Data**

Volkenstein, Mikhail Vladimirovich, Date
Molecular biophysics.

Translation of Molekuliarnaia biofizika.
Includes bibliographies and index.
1. Molecular biology. 2. Biological physics.
I. Title. [DNLM: 1. Molecular biology. 2. Biophysics. QT34 V917ma]
QH506.V5813 574.8'8 73-18990
ISBN 0-12-723150-1

PRINTED IN THE UNITED STATES OF AMERICA

# Contents

## Chapter 4 **The Physics of Proteins**



## Chapter 5 **X-Ray Analysis, Optics, and Spectroscopy of Biopolymers**



## Chapter 6 **The Physics of Enzymes**



## Chapter 7 **Cooperative Properties of Enzymes**

## Chapter 8 **The Physics of Nucleic Acids**



## Chapter 9 **Biosynthesis of Protein**

# Preface

The science of biophysics, currently attracting many disciples, has counted among its practitioners long-established scientists such as Helmholtz, investigating the transmission of nerve impulses, and Maxwell, examining the phenomenon of color vision. The merger of biology and physics has attained consummation in the second half of this century concurrently with the development of sophisticated instrumentation and the molecular approach to the study of life phenomena.

In this volume, the fundamental principles of biophysics and their application to the study of the physical properties of biological macromolecules are presented. In a subsequent volume—"General Biophysics"—a thorough treatment of the thermodynamics of biological systems, an analysis of recognition, selection, and regulatory phenomena, as well as membrane, nerve excitation, mechanochemical, and photobiological processes will be examined.

Biophysics and biology are rapidly developing fields. New findings necessitate revision of old ideas and the introduction of new postulates. In the present book we have attempted to differentiate clearly between established principles and speculative ideas. It is our hope that the factual and the theoretical will be apparent to the reader.

I thank N. G. Esipova, G. V. Gursky, V. I. Ivanov, A. M. Jabotinsky, M. A. Livshits, A. I. Poletayev, Yu. A. Sharonov, and V. G. Tumanian for their valuable help in the preparation of this book.

# MOLECULAR BIOPHYSICS

# Chapter 1

# Biology and Physics

## 1.1 Introduction

Inanimate and animate bodies are both composed of atoms and molecules and thus obey the general laws governing the structure and properties of matter and field. Contemporary physics is turning to the study of life, and the problem of the relationship between physics and biology has assumed particular interest.

That physics and biology are closely interrelated was recognized early in the development of the natural sciences. Later, however, as biologists gained deeper insights into the complexities and pecularities of the life processes, the paths of physics and biology diverged more and more, and the fundamental biological laws, primarily the Darwinian law of natural selection, came to be regarded as totally incompatible with physics.

The interaction of biology and chemistry evolved in a different way. The chemistry of life, organic chemistry, was at first considered completely distinct from inorganic chemistry, for it appeared that the substances present in living organisms could not in principle be obtained *in vitro:* they were amenable to analysis but not to synthesis. The formation of an organic substance was believed to require the participation of a special agent--a vital force. Organic chemistry was looked upon as a reliable basis for vitalism.

Chemistry, however, overcame this dichotomy. In 1777 Lavoisier showed that respiration and combustion were essentially identical processes in that both oxidized organic substances to form water and carbon dioxide. In 1828 Wöhler synthesized, for the first time, an organic substance, urea $CO(NH_2)_2$, from inorganic compounds. From this time on organic chemistry ceased to be the chemistry of life and became the chemistry of carbon compounds. Inspired by this achievement, the more farsighted thinkers of the last century rejected vitalism in favor of materialistic natural sciences. Engels, for example, developed the following ideas [1]: Chemistry brings us close to organic life and it has advanced far enough to guarantee that it alone will explain to us the dialectical transition to an organism;...It is necessary to attain only one more goal: to explain the rise of life from inorganic nature; in the current stage of scientific development this means only the following: to prepare protein bodies from inorganic substances. These ideas led Engels to his formula of life, which says that: Life is the mode of existence of protein bodies, the essential element of which consists in continual metabolic interchange with the natural environment outside them. This formula indicated the fundamental features of life, the knowledge of which thus came down to the understanding of the artificial synthesis of proteins.

We know now that the actual situation is not as simple as that. Some proteins have already been synthesized, yet the problem of life is far from being solved. If an organism is a protein system, we must understand how that system works. As Engels correctly noted, a prerequisite for this understanding is the consideration of exchange with the environment. Therefore we have to deal with an open protein system.

Science has demonstrated that proteins are indeed responsible for the functioning of living organisms. Life, however, requires many other low- and high-molecular substances, primarily nucleic acids, and is impossible without a diversity of interacting substances without chemical heterogeneity. An individual chemical substance of any complexity--whether a protein or a nucleic acid--does not live. It makes no sense to speak of living molecules. A living organism and any functional part thereof invariably represents a complicated heterogeneous system of interacting components, such as large and small molecules, ions, and supramolecular structures.

The advanced ideas of the 19th century were not able to overcome vitalism, proponents of which kept finding new arguments in the development of science itself.

Two great evolutionary theories were developed in the 19th century. The second law of thermodynamics (Clausius, Gibbs, Boltzmann) is suggestive of the evolution of matter in an isolated system toward its most probable state, characterized by maximum disorder, that is, maximum entropy. In contrast to this concept, the theory of biological evolution (Darwin) posits the increased orderliness and complexity of living systems from primitive microorganisms to *Homo sapiens* with his thinking brain. These two theories conflicted with each other because biological evolution, phylogeny, and ontogeny, were quite inconsistent with the equilibrium thermodynamics of isolated systems.

On the other hand, biology had a powerful impact on physics in the 19th century. The law of conservation of energy, the first law of thermodynamics, was discovered by Mayer, Joule, and Helmholtz. It is well known that Mayer based his theory on observations of living human organisms, but it is not so well known that Helmholtz also based his theories on biological phenomena, on a clear cut antivitalistic concept. He wrote that "according to Stahl, the forces acting in a living body are in fact physical and chemical forces of organs and things, but there is some vital soul or force inherent in the body which can tie up or set free their activity. ...I have found that Stahl's theory ascribes to any living body properties of the so-called *perpetuum mobile*. ...So I am confronted with the question of what kind of relations should hold between different forces of nature if we assume that the *perpetuum mobile* is impossible..." [2].

The basic question we have to answer when attempting to develop and study biophysics (i.e., the physics of living nature) is that of the interrelationship between biological and physicochemical phenomena. Two alternatives are apparent: either biology contains something basically foreign to physics and chemistry, or else life is a manifestation of physical and chemical processes occurring in complicated open systems; *tertium non datur*. Either biology contradicts physics or the contradiction is only apparent and the vitalistic theory is untenable whatever its form.

Modern vitalism does not deny the applicability of physics to the study of life processes but, as we will see, urges the creation of a new physics--one that has not yet been formed. On the other hand, physical interpretations of fundamental biological processes are often viewed as impermissible reductionism, striving to reduce the complicated biological laws to simpler physical ones.

However, arguments concerning "reducibility" or "irreducibility" are meaningless, for this approach is not an attempt to subordinate biology to physics, but rather an effort to ascertain the unity of animate and inanimate nature. Physics, as the science of matter and field, is by no means simpler than biology; hence, the notion of reductionism is inappropriate. One should speak not of reductionism but rather of the integration of different domains of knowledge. Thus it is quite clear now that chemical transformations do not involve any phenomena other than physical, and in this sense chemistry is "reducible" to physics. But this awareness in no way jeopardizes the independence or significance of chemistry; rather it provides chemistry with a much more solid and generalized foundation.

Nonetheless, the question whether contemporary physics can assure the knowledge of life phenomena is not devoid of meaning, and in this connection it is appropriate to discuss the concepts put forward by some biologists and physicists.

Bertalanffy developed ideas concerning the so-called general theory of systems (see [3]). He believed that biological phenomena could be understood in terms of the exact sciences and that the imaginary conflict with thermodynamics vanished as soon as it was recognized that organisms are open systems which exchange matter and energy with the external environment. But canonical thermodynamics was concerned with isolated systems, and a thermodynamics of open systems (nonequilibrium thermodynamics) was necessary in order to interpret biological phenomena in physical terms. Bertalanffy considered systems theory as a basis for scientific biology. The system is a totality of interacting objects, and its properties are not reducible to the sum of the properties of its constituent elements. By considering systems it is possible to explore the problems of organization and the integrity of dynamic interaction. These problems are of critical importance for biology.

At a time when the thermodynamics of open systems had not yet been formulated, Bauer discussed the nonequilibrium properties of organisms and devised his fundamental law of biology, which stated that "all living systems are never in equilibrium and use their own free energy to perform continual work against the equilibrium as required by the laws of physics and chemistry under given external conditions" [4]. Bauer's ideas were not understood by his contemporaries, nor for that matter are they appreciated by some present-day commentators (see, e.g., [5]). Although Bauer approached present-day biophysics, his works are of only historical interest today. It is important, however, that he tried to prove that life could be interpreted in atomic-molecular terms, saying that "a nonequilibrium state of living matter and, consequently, its constantly maintained work capacity, are determined ... by the molecular structure of living matter, while the source of the work performed by living systems is eventually the free energy peculiar to that molecular structure, to that state of molecules" [4].

Niels Bohr tackled the problem of the relationship between physics and biology on the basis of the principle of complementarity [6]. He considered the biological laws to be complementary to those obeyed by inanimate bodies and held that it was impossible to determine at one time the physicochemical properties of an organism and life phenomena, because knowledge of the former precludes knowledge of the latter. Life should be considered as the "basic postulate of biology, not susceptible of further analysis, in the same way as the existence of the quantum of action ... forms the elementary basis of atomic physics" [6, p. 21]. Bohr thus considered biological and physicochemical studies to be complementary, that is, incompatible, though not contradicting each other. This concept has nothing in common with vitalism, for it rejects the existence of any limiting boundary in the application of physics and chemistry to the solution of biological problems. "... no result of biological investigation can be unambiguously described otherwise than in terms of physics and chemistry, just as any account of experience even in atomic physics must ultimately rest on the use of the concepts indispensable for a conscious recording of sense impressions" [6, p. 21].

Using the same complementarity principle, claims were made to the effect that knowledge of morphology and functionality, homology and analogy, environment and internal state, and heredity and adaptation were incompatible. It was thought that the investigation of some one aspect of a biological phenomenon so strongly affects the other aspects that the latter becomes unknowable in principle [7]. Because all noncommutative factors act concurrently in life, the latter is unknowable. One can study the atomic and molecular structure of an organism but one has to kill the organism in order to do so.

This viewpoint was not a new one. As Goethe's Mephistopheles said:

> Wer will was Lebendig's erkennen und beschreiben,
> Sucht erst den Geist herauszutreiben,
> Dann hat er die Teile in seiner Hand,
> Fehlt leider! nur das geistige Band.
> *Encheiresin naturae* nennt's die Chemie,
> Spottet ihrer selbst, und weiss nicht wie.

*Encheiresin naturae* is the way of nature, the mode of its action. Goethe thought that an organic creature is so many-sided externally and so diverse and inexhaustible internally that it is impossible to pick out a sufficient number of initial points to survey it or to develop a sufficient number of organs so as to disassemble it without killing it.

Bohr's views changed with the development of contemporary biology. He referred later to a complementarity between "arguments based on the full resources of physical and chemical science, and concepts directly referring to the integrity of the organism transcending the scope of these sciences" [6, p. 76]. The application of the complementarity principle in biology was argued to be justified not by the postulative character of the concept of life but rather by the extreme complexity of the organism as an integral system. In his last lecture on this subject [8] Bohr referred only to the practical rather than the fundamental complementarity involved in the inexhaustible complexity of life. In his letter to the present writer (reproduced in [9]) Bohr in fact abandoned the views contained in his earlier papers, saying "I am well aware that some of my early utterances have caused misconception of my general attitude."

In 1945 Schrödinger published a farsighted book devoted to the relationship between physics and biology [10]; in it he discussed three problems of fundamental importance for biophysics. The first problem concerned the thermodynamic bases of life. The distinction between an organism and an inanimate body consists, he said, in the high orderliness of the organism, which is similar in this respect to an "aperiodic crystal," and in the ability of this orderliness to sustain itself and to produce ordered phenomena. This is a matter of self-regulation and self-reproduction of organisms and cells. Schrödinger attributed this property to the fact that any organism is an open system in a nonequilibrium state owing to the outflow of entropy to the environment. Organisms continuously create "order from order," "extract order from the environment" in the form of "a well ordered state of matter in foodstuffs." He provided an answer to the question of the cause of macroscopicity, the multiatomicity of organisms. In a system consisting of a small number of atoms, fluctuations should destroy any order. It is precisely because of their multiatomicity that organisms can exist in accordance with the laws of thermodynamics.

The second problem is the molecular basis of life, regarding which Schrödinger argued in favor of a materialistic interpretation of the molecular nature of genes and raised the question of the structure of the hereditary substance and the reasons for its stable reproduction in a series of generations. This question was answered by molecular biology, the origin and development of which were much stimulated by the Schrödinger book.

A third problem was that of the quantum-mechanical laws, which are clearly manifested in radiobiological phenomena. In discussing the works of Timofeev-Ressovsky, Delbrück, and others, Schrödinger emphasized that biological processes are consistent with the laws of physics.

Schrödinger's book is very important because in it he not only showed that physics does not contradict biology, but he also correctly outlined the future of biophysics.

Elsässer opposed physics to biology [11]. The store of information contained in the original germ cell, the zygote, is much smaller than that in the adult multicellular organism. In his view, the increase in the amount of information is inexplicable in physical terms because this is a specific "biotonic" regularity. His ideas on this matter will be detailed later (p. 22).

Wigner believed that the self-reproduction of biological molecules and organisms contradicted quantum mechanics [12] and held that the probability of existence of self-reproducing states was practically zero. According to him, the Hamiltonian describing the behavior of a complex system can be represented by a random symmetrical matrix. The state of an organism can be described by a vector **v**, and an analogous vector **w** designated for nourishing products. The common vector for the organism plus food is then

$$\phi = v \times w \qquad (1.1)$$

After reproduction, we get the vector

$$\psi = v \times v \times r \qquad (1.2)$$

where r describes the results of metabolism and the coordinates of two organisms. We have an N-dimensional space for the organism and an R-dimensional space for r. If the "collision matrix" S which represents the final state resulting from interaction between organism and food is a random, stochastic matrix, then

$$v_k v_\lambda r_\mu = \sum_{k',\lambda',\mu'} S_{k\lambda\mu,k'\lambda'\mu'} v_{k'} w_{\lambda'\mu'} \qquad (1.3)$$

We get $N^2R$ equations. The number of unknowns (N values of v, R values of r, and NR values of w; i.e., N + R + NR) is much smaller than the number of equations. It would be a miracle if these unknowns were to satisfy the written expression. Wigner followed Elsässer in considering the reduplication of biological macromolecules to be a "biotonic" phenomenon.

In actual fact, as shown by Eigen [13], the matrix S is not a random one. Wigner did not take into account the instructive functions of informational macromolecules. The entire presentation given by Wigner contradicts reality and his conclusion that it is necessary to modify quantum mechanics to make it applicable to biology proved to be untenable. At the same time, the application of quantum mechanics to macroscopic systems requires special consideration.

An important paper by Eigen [13] devoted to the self-organization and evolution of biological macromolecules produced convincing arguments in favor of the sufficiency of contemporary physics for the explanation of biological phenomena.

A living organism is an open, self-regulating, and self-reproducing heterogeneous system, whose most important functional substances are biopolymers--proteins and nucleic acids. Such a system must be investigated physically and chemically. Our knowledge of it should rest on the elucidation of the physical features of life, that is, the physical considerations of the development of nonequilibrium, orderliness, and systematicity in the organism.

## 1.2 Finalism and Causality

Before discussing the physical basis of life phenomena, we will consider an important feature of life that is usually considered to contradict physics. Biology naturally makes use of a finalistic treatment of the phenomena under study. The development of a zygote into an adult organism can be described using the notion of the goal. The goal of development is the formation of an organism. Its structure is expedient, in that it corresponds to the conditions of existence. Even in the early stages of embryogenesis, definite groups of cells are predestined to develop into some definite organ, and this determines their functionality down to the molecular level. Phylogeny, or evolutionary development, can be described in the same way. This development is directed toward the greatest adaptation of the population (the elementary evolutionary system) to the external conditions.

In this sense an organism is like an engine designed according to some plan for the attainment of a definite goal. Scientific biology does not, of course, consider developmental processes in teleological terms. The attainment of a goal in ontogeny and phylogeny is a result of real causes (natural selection, etc.). Emphasizing the existence of some plan of development, Monod introduced the notion of teleonomy as opposed to teleology [14], having in mind the causality of development. The extraordinary complexity of a living organism (a "living engine") determines its finalistic description, which is not peculiar to conventional physics and chemistry. It is obvious that a statement like: "Sodium and chlorine ions interact for the purpose of forming a cubic crystal" is meaningless. On the contrary, the statement "because sodium and chlorine ions have such and such charges and radii, a NaCl crystal must belong to the cubical crystal system" possesses a clear meaning. Physicists usually ask "why?," whereas biologists often ask "what for?"

The notion of expediency is closely linked with that of optimality. Optimality means attaining some result (goal) through the smallest possible expenditure of energy, the formation of a system which would best execute certain functions, etc.

Biological finalism expresses, on the one hand, the complexity of biological phenomena and structures which prevents their causal explanation at the atomic-molecular level. On the other hand, it characterizes the irreversibility and "antientropicity" of development which implements a plan, a program, the instructing action of information (see p. 25). In actual fact, there is no contradiction between finalism and causality. Finalism arises in physics whenever its principles are formulated as variational. Here are some examples.

A very general formulation of the law of motion of mechanical systems is contained in Hamilton's principle of least action. The Lagrange function $L(q, \dot{q}, t)$, depending on time, coordinates, and velocities, satisfies the condition

$$S = \int_{t_1}^{t_2} L(q, \dot{q}, t)\, dt = \text{Minimum} \tag{1.4}$$

In other words, the variation $\delta S$ is equal to zero. The action S is minimal, that is, the system moves between two sets of coordinates $q^{(1)}$, $q^{(2)}$ and velocities $\dot{q}^{(1)}$, $\dot{q}^{(2)}$, corresponding to the times $t_1$, $t_2$, in such a way that S becomes minimal. *The goal of the mechanical system is its minimal action* and its motion is optimal in this sense.

But expression (1.4) is equivalent to the Lagrange equations of motion

$$\frac{d}{dt}\frac{\partial L}{\partial \dot{q}} - \frac{L}{q} = 0 \tag{1.5}$$

The Lagrange function L expresses the difference between kinetic and potential energy

$$L = \sum_a \frac{m_a v_a^2}{2} - U(r_1, r_2, \cdots) \tag{1.6}$$

where $m_a$ are masses, $r_a$ are radius vectors, and $v_a$ are velocities of material points. Equations (1.4) and (1.5) can be rewritten in terms of Newton's equations of motion

$$m_a \frac{d^2 r_a}{dt^2} \equiv m_a \frac{dv_a}{dt} = -\frac{\partial U}{\partial r_a} \tag{1.7}$$

The finalistic expression (1.4) comes down to the causal equations (1.7) describing motion as a result of the action of forces. Other examples of physical laws formulated in finalistic terms include Fermat's principle in optics, Le Chatelier's principle in thermodynamics, and Lenz's rule in electrodynamics. The number of such examples is in fact unlimited.

The equations of motion (1.8) of mechanics are reversible, since they contain only the second derivatives in time and therefore do not vary if the sign of time is reversed, $t \to -t$. However, the equations of mechanics have solutions corresponding to either stable or unstable equilibria and motions. Neither state of equilibrium of the pendulum shown in Fig. 1.1 contradicts statics, but state 1 is stable and state 2 is unstable. There always exist small forces and small deviations from the initial state of a material system which perturb equilibria and motions. These perturbations do not change

*FIG. 1.1 Stable (1) and unstable (2) equilibrium of a pendulum.*

state 1 but do strongly change state 2. Equilibria and motions which are slightly changed by small perturbations are stable and those strongly changed are unstable. But what do "slightly" and "strongly" mean? The general problem of stability of motion was solved in the classical work of Liapounov (1892), who formulated the criteria of stability [15,16]. If any perturbation however small (but not zero), alters the magnitude of some characteristic of motion in such a way that this magnitude deviates more and more from its value in unperturbed motion, then the unperturbed motion is unstable relative to this characteristic. The motions of the pendulum at small deviations from equilibrium state 1 are described by the equation

$$\ddot{\phi}+\omega^2\phi = 0 \tag{1.8}$$

where $\phi$ is the angle of deviation and $\omega$ is the cyclic frequency of oscillation, equal to

$$\omega = 2\pi(g/\ell)^{1/2} \tag{1.9}$$

where g is the gravitational acceleration and $\ell$ is the pendulum length. The solution of (1.8) is

$$\phi \equiv x = A\cos(\omega t+\alpha), \quad \dot{\phi} \equiv y = -A\omega\sin(\omega t+\alpha) \tag{1.10}$$

in which A is the amplitude of vibration and A cos $\alpha$ is the initial phase. Excluding t from expressions of x and y, we get a set of trajectories of motion at the phase plane x,y differing by values A:

$$(x^2/A^2)+(y^2/A^2\omega^2) = 1 \tag{1.11}$$

*FIG. 1.2 Phase trajectories of a pendulum.*

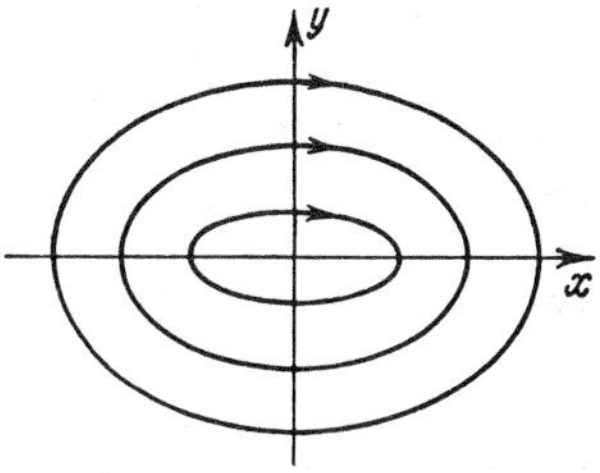


These are the equations of ellipses (Fig. 1.2). The equation of motion at the phase plane is

$$dy/dx = -\omega^2 (x/y) \tag{1.12}$$

Equation (1.12) is the equation of an integral curve possessing the singular point x = 0, y = 0, through which none of the integral curves pass. Near this point all integral curves are closed and possess no singularities. Such a point is called the center. The motions around the center are periodic.

The state of equilibrium is stable if, for any given region $\varepsilon$ of the permissible deviations from the equilibrium state, there exists a region $\delta(\varepsilon)$ surrounding this state and possessing the property that every motion beginning inside $\delta$ will never reach the boundaries of region $\varepsilon$. Conversely, the state of equilibrium is unstable if there exists a region $\varepsilon$ for which the region $\delta(\varepsilon)$ does not exist. In our example the stability of pendulum oscillations is characterized by the conditions: if for t = 0

$$|x(0)| < \delta \quad \text{and} \quad |y(0)| < \delta' \tag{1.13}$$

then for $0 < t < \infty$

$$|x(t)| < \varepsilon \quad \text{and} \quad |y(t)| < \varepsilon' \tag{1.14}$$

The singular point of the center type corresponds to the stable state of equilibrium [17]. The problems of stability are important in the theory of automatic regulation, in the theory of vibrational motions, etc. Obviously, the problem of the stability of the dynamic state of a biological system inevitably arises in biophysics.

In the general case the state of a physical system is characterized not only by mechanical but also by thermodynamic conditions. The second law of thermodynamics--the striving of the entropy of an isolated system toward the maximum (a finalistic formulation)--can be expressed as a variational principle:

$$(\delta S)_{eq} = 0 \tag{1.15}$$

Obviously the thermodynamic treatment of biological phenomena--developmental processes in the first place--must be based on a consideration of the irreversible transitions in open systems.

The main difficulties are due to the dynamic, "engine-like," rather than statistical, character of biological systems. However, this does not imply any contradiction between biology and physics.

## 1.3 Thermodynamics and Biology

A physical consideration of any kind of system, including living ones, starts with its phenomenological, thermodynamic description. Further study adds a molecular content to such a description.

The main thermodynamic features of living systems are that they are open and exist in a nonequilibrium state. Accordingly, what is needed is not thermostatics but thermodynamics in the real sense of the word, taking into account the variation of thermodynamic quantities with time. Nonequilibrium thermodynamics is a kind of kinetic treatment. Its detailed presentation will be given in "General Biophysics" and only some fundamentals will be presented here.

The total change of entropy in an open nonequilibrium system is the sum of the entropy production $d_iS$ inside the system and the flow of entropy $d_eS$ into the system from the environment and *vice versa*:

$$dS = d_iS + d_eS \tag{1.16}$$

Let us introduce an adiabatic cover around the system. Then $d_eS = 0$ and, according to the second law,

$$dS = d_iS > 0 \tag{1.17}$$

If the entropy production is due to chemical reactions, then

$$d_iS = (-1/T) \sum_k \mu_k \, dn_k \tag{1.18}$$

where $\mu_k$ is the chemical potential of the kth component and $dn_k$ is the change in its molar concentration.

We introduce now the chemical reaction coordinate $\xi$ expressing the extent of reaction

$$dn_k = \nu_k \, d\xi \tag{1.19}$$

where $\nu_k$ is the stoichiometric coefficient. For instance, in the case of glucose oxidation

$$C_6H_{12}O_6 + 6O_2 \rightarrow 6CO_2 + 6H_2O$$

$$\nu_{C_6H_{12}O_6} = -1, \quad \nu_{O_2} = -6, \quad \nu_{CO_2} = 6, \quad \nu_{H_2O} = 6$$

We then have

$$d_iS = (-d\xi/T) \sum_k \mu_k \, \nu_k = (A \, d\xi)/T \tag{1.20}$$

The quantity

$$A = -\sum_k \mu_k \nu_k \tag{1.21}$$

is called the affinity. According to (1.20), $A > 0$. In the equilibrium state $A = 0$, as does $d\xi$. Indeed

$$A = RT[\ln K - \ln \prod_k a_k \nu_k] \tag{1.22}$$

where K is the equilibrium constant and $a_k$ is the activity of the kth component.

The rate of entropy production is

$$d_iS/dt = \int \sigma \, dV \geq 0 \tag{1.23}$$

The entropy production in a unit volume is $\sigma \geq 0$, since the nonequality (1.23) is valid for every macroscopic volume.

In the case of a chemical reaction

$$\sigma = (-d\xi/dt)(1/T) \sum_k \mu_k \nu_k = Av/T \geq 0 \tag{1.24}$$

where v is the rate of reaction. If many reactions proceed simultaneously

$$\sigma = (1/T) \sum_i A_i v_i \tag{1.25}$$

A single reaction for which $A_1v_1 < 0$ is impossible. However, if another reaction occurs concurrently for which $A_2v_2 > 0$, the first reaction becomes possible if $A_1v_1 + A_2v_2 > 0$. The thermodynamic coupling in open systems makes possible processes not realizable in isolated systems.

In general, the dissipation function $\sigma$ is expressed as the sum of the products of certain flows J (in the chemical reactions, v) and their macroscopic causes, generalized forces X (in chemical reactions, A/T)

$$\sigma = \sum_i J_i X_i \tag{1.26}$$

The flows and forces are interdependent. Their connection can be readily established in the case of one flow arising from the action of one force. Such is the solution of the Fourier equation of heat conduction (flow-heat, force-difference in temperature), Fick's equation of diffusion (flow-matter, force-difference in concentration), etc. In a chemical reaction we are concerned with scalar flows and forces. Thus in the linear approximation

$$J_i = \sum_{k=1}^{n} L_{ik} X_k, \qquad i = 1,2,\ldots,n \tag{1.27}$$

where $L_{ik}$ are the so-called phenomenological coefficients. For chemical reactions

$$v_i = d\xi_i/dt = \sum_k L_{ik} (A_k/T) \tag{1.28}$$

The linear approximation is valid near equilibrium. As was shown by Onsager [18], we can always select flows and forces in this region so that the matrix $(L_{ik})$ becomes symmetrical

$$L_{ik} = L_{ki} \tag{1.29}$$

Condition (1.29) is based on the theory of fluctuations and on the principle of microscopic reversibility. This condition was generalized by Casimir [19].

Closeness to equilibrium for a chemical process means that

$$A_k << RT \tag{1.30}$$

Relation (1.28) can be expressed in the form

$$v_i = (1/T) \sum_k L_{ik} \sum_\ell (\partial A_k/\partial \xi_\ell)_{T,p} \, \delta\xi_\ell \tag{1.31}$$

Matrix $(L_{ik})$ and tensor $(\partial A_k/\partial \xi_\ell)_{T,p}$ can be diagonalized. Equation (1.31) is then transformed linearly to the form

$$v_i = d\xi_i'/dt = - \delta\xi_i'/\tau_i \tag{1.32}$$

where $\xi_i'$ is the normal mode, and $\tau_i$ is the relaxation time, having substantial and positive value. The solution of (1.32) describes the approach to equilibrium:

$$\delta\xi_i' = \delta\xi_{i0}' \exp(-t/\tau_i) \tag{1.33}$$

The condition of stability of the equilibrium state is given by $\sigma \geq 0$ or

$$\sigma = (1/T^2) \sum_{i,k} L_{ik} A_i A_k \geq 0 \tag{1.34}$$

Any fluctuation near the equilibrium state can only decrease the entropy

$$\delta_i S \leq 0 \tag{1.35}$$

or

$$\delta_i S = (1/T) \sum_k A_k \, \delta\xi_k \leq 0 \tag{1.36}$$

Expanding $A_k$ near the equilibrium state and diagonalizing the tensor, we get

$$A_k = A_k^{(e)} + (\partial A_k'/\partial \xi_k')_e \, \delta\xi_k' + \cdots = (\partial A_k'/\partial \xi_k')_e \, \delta\xi_k' + \cdots \tag{1.37}$$

Therefore

$$\delta_i S = (1/T) \sum_k (\partial A_k'/\partial \xi_k')_e (\delta\xi_k')^2 \leq 0 \tag{1.38}$$

This is true since $(\partial A_k'/\partial \xi_k') < 0$.

If the system is open and $d_e S \neq 0$, then at constant boundary conditions the system tends toward a nonequilibrium steady state in which

$$dS = d_iS+d_eS = 0 \qquad (1.39)$$

or

$$d_eS = -d_iS < 0$$

The steady state is maintained by the flow of entropy into the surrounding medium.

Let us consider an isolated system containing an organism and an external medium. The organism obtains nourishment, water, and oxygen from this medium and in turn excretes various substances. There is heat exchange between organism and medium. Such conditions actually describe the situation of an astronaut in a spaceship. The astronaut is an open system in relation to the ship, but the ship as a whole is well isolated. The total change in entropy of the whole system is, according to the second law of thermodynamics,

$$dS = dS_1+dS_2 > 0$$

where $dS_1$ is the change of the astronaut's entropy and $dS_2$ that of the surrounding medium. Since the astronaut is an open system,

$$dS_1 = d_iS_1 + d_eS_1$$

The change of entropy of the medium surrounding the astronaut occurs only as a result of the exchange of heat and matter with the astronaut. It can be assumed that the medium itself does not produce entropy. We then have

$$dS_2 = -d_eS_1$$

Therefore

$$dS = d_iS_1 > 0$$

If the state of the astronaut is a steady one, then

$$dS_1 = 0, \quad d_eS_1 = -d_iS_1 < 0, \quad \text{and} \quad dS_2 = dS = d_iS_1 > 0$$

Thus the steady state of the astronaut is maintained through the increase in entropy of the surrounding medium, determined by the outflow of entropy from the astronaut to the medium, which compensates for entropy production in the body. This is the meaning of Schrödinger's words that "an organism feeds with negative entropy" [10]. The value of $S_2$ increases as a result of heat production by the astronaut and because the entropy of the excreted substances is greater than that of the substances consumed by the astronaut. The steady state of the astronaut will be maintained until the nutrient substances in the surrounding medium are exhausted or until irreversible processes in the organism (aging) change it. In this sense the steady state is not perpetual. Its maintenance depends on the existence of two time scales--the time of entropy exchange with the surrounding medium, and a much longer time of the exhaustion of nutrient substances and/or of aging. Our simple calculation considers only the rapid exchange process and not the slow irreversible processes.

We illustrate this situation by another elementary example. Suppose we have two big heat reservoirs at temperatures $T_1$ and $T_2$ connected by a thin heat conductor. A steady heat flow is established in the conductor, every part of which is in a steady state. This state will persist until temperatures $T_1$ and $T_2$ change. Clearly, they tend to become equal. The larger the reservoirs, the smaller will be the time required to establish the steady state relative to the time taken to attain the equilibrium corresponding to $T_1 = T_2$. In calculating the properties of the steady state in the fast time scale, temperatures $T_1$ and $T_2$ are assumed to be constant, that is, to change in the slow time scale. Thus the steady state is reached in the presence of a large irreversible process.

Life itself, the existence of the biosphere, can be approximately treated as a steady process occurring against the background of the grandiose process of the cooling of the sun. The state of the biosphere is steady only in a limited time interval because the population of the species *Homo sapiens* constantly increases and destroys other plant and animal species (excepting the cultivated ones, whose populations also increase). *Homo sapiens* destroys biogeocenoses on the scale of the entire biosphere.

Let the flows and coefficients $L_{ik}$ be constant in the steady state. Entropy production (a function of dissipation) $\sigma$ depends on generalized forces $X_k$. As was shown by Prigogine and Glansdorff [20,21], $\sigma$ is a minimum in the steady state in relation to the variations of $X_k$. In other words

$$d_X\sigma = \sum_k J_k \, dX_k \leq 0 \tag{1.40}$$

If the steady state is approached, entropy production decreases toward a minimum, that is, $d_X\sigma = 0$. For a chemical reaction

$$T \, d_X\sigma = \sum_k v_k \, dA_k \leq 0 \tag{1.41}$$

The equality sign refers to the steady state. We can always form suitable linear combinations of $v_k$ and $A_k$ such that all $A_k$ are independent. In this case the equality $d_X\sigma = 0$ means that in a steady state all $V_k{}^{(0)} = 0$. If the steady state is perturbed

$$v_k = v_k{}^{(0)} + \delta v_k = \delta v_k, \quad A_k = A_k{}^{(0)} + \delta A_k$$

The variation of $\sigma$ produced by fluctuations of the affinity is

$$T\ \delta_X\sigma = \sum_k v_k\ \delta A_k = \sum_k \delta v_k\ \delta A_k \geq 0 \tag{1.42}$$

and in general

$$\delta_X\sigma = \sum_k \delta J_k\ \delta X_k \geq 0 \tag{1.43}$$

This is the condition for stability of the steady state. It is valid near equilibrium in the linear approximation of irreversible thermodynamics. In this case, a system shifted from the steady state returns to it without oscillations.

Consider as an example a simple autocatalytic reaction near equilibrium

$$B + C \underset{k_-}{\overset{k_+}{\rightleftarrows}} 2C$$

The kinetic equation is

$$v = k_+[B][C] - k_-[C]^2 \tag{1.44}$$

At constant [B]

$$\delta v = k_+[B]\ \delta[C] - 2k_-[C]\ \delta[C] \tag{1.45}$$

and since near equilibrium $k_+[B] \cong k_-[C]$,

$$\delta v = -k_-[C]\ \delta[C] \tag{1.46}$$

At the same time [cf. (1.22)]

$$A = RT\{\ln K - \ln [C]/[B]\} \tag{1.47}$$

Hence

$$\delta A = -RT\ \delta[C]/[C] \tag{1.48}$$

and

$$T\ \delta_X\sigma = \delta A\ \delta v = RT\ k_-(\delta[C])^2 \geq 0 \tag{1.49}$$

that is, condition (1.43) is fulfilled.

Linear irreversible thermodynamics, developed by Prigogine [20] (see also [22-24]), gives the general explanation of the "antientropicity" of biological processes, disclosing the possibility of the existence of an open system in a steady but nonequilibrium state. Study of a series of biophysical phenomena, including membrane transport, shows that Onsager's relations (1.29) are valid [25]. However, linear nonequilibrium thermodynamics is obviously inapplicable to the study of ontogeny and phylogeny, to processes of the formation of organized structures from nonorganized ones, or to periodic processes. Biology requires nonlinear thermodynamics. Situations far from equilibrium, for which the steady states can be unstable--that is, situations where (1.43) is not obeyed--are relevant here.

The inapplicability of linear thermodynamics to developmental processes is quite obvious. Near equilibrium every catalyst (enzyme) accelerates both forward and reverse reactions at an equal rate.

Consider the same autocatalytic reaction proceeding far from equilibrium. If $k_-[C] << k_+[B]$, then

$$v = k_+[B][C] \tag{1.50}$$

and if [B] is a constant

$$\delta v = k_+[B]\ \delta[C] \tag{1.51}$$

Again

$$\delta A = -RT\ \delta[C]/[C] \tag{1.48}$$

We get

$$T\ \delta_X\sigma = \delta A\ \delta v = -RTk_+([B]/[C])(\delta[C])^2 \leq 0 \tag{1.52}$$

The steady state is unstable.

Such nonstabilities have been shown to underlie growth and evolution. Prigogine and his co-workers have reported that combining autocatalytic reactions and transport processes may result, under conditions far from equilibrium, in the formation of ordered "dissipative structures" determined by energy dissipation rather than by conservative molecular forces [21,26]. On this basis, Eigen has developed a molecular theory of the self organization, natural selection, and evolution of biopolymers [13].

## 1.4 Information Theory and Biology

The phenomenological treatment of living systems quite naturally makes use of information theory concepts. The existence and development of the cell and organism, which are open and self-regulated systems, are based on the generation and transmission of information through channels of direct and reverse communication.

Communication theory employs a quantitative measure of the available information, that is, a measure of the quantity of information contained in a message. This theory is concerned with the modes of message coding, utilization of redundancy (discussed later), optimization of "noise" control, etc.

The amount of information in $P_1$ events selected from among $P_0$ events of equal probability is determined to be

$$I = K \log(P_0/P_1) \tag{1.53}$$

and if $P_1 = 1$, then

$$I = K \log P_0 \tag{1.54}$$

This definition gives the additivity of information when independent probabilities are multiplied. Thus the result of throwing a die gives the information K log 6, while the result of throwing two dice gives the double information 2K log 6 = K log 36. For quantitative calculations it is necessary to determine the base of logarithms and the constant K. Using base 2 and taking K = 1, we get

$$I = \log_2 P_0 \tag{1.55}$$

This definition of the quantity of information corresponds to problems with a certain number n of different independent choices, each choice being reduced to a binary alternative. In this case, and according to (1.55)

$$P_0 = 2^n$$

The unit of information content is the bit (binary digit), corresponding to a choice between two alternatives at $P_0 = 2$. When throwing a die, $I = \log_2 6 = 2.585$ bits.

How many bits does an arbitrary three-digit number contain? The first digit has 9 different values, from 1 to 9, while the second and third ones have 10 values, each from 0 to 9. We get

$$I = \log_2 9 + 2 \log_2 10 = 2.64 + 2(3.32) = 9.28 \text{ bits} \tag{1.56}$$

This generally accepted definition of information corresponds to the binary system of calculus, in which every number is written as the sum of powers of the number 2 by means of the figures 0 and 1. One decimal unit gives 3.32 bits, that is, the binary notation of a number requires on average 3.32 times as many digits as the decimal one.

If the initial events are not equally probable and the distribution of probabilities is given, then the information obtained in choosing one of several events is less than it would be in the case of equal probabilities. Knowledge of the distribution of probabilities implies that some preliminary information is available which is to be subtracted from that obtained.

Assume that we have a message containing G sequential cells --a text composed of G sequential letters. Every cell can contain one of M letters (for English, M = 26). The message contains $N_1, N_2, \ldots, N_M$ letters of a given type. We have

$$G = \sum_{j=1}^{M} N_j$$

The probability of appearance of a given letter is

$$P_j = N_j/G \quad \text{and} \quad \sum_{j=1}^{M} P_j = 1$$

The total number of messages P in a text of G letters written in the M-letter language is

$$P = G!/\prod_{j=1}^{M} N_j!$$

Introducing natural logarithms, we get the information content of one message

$$I = K \ln P = K[\ln(G!) - \sum_{j=1}^{M} \ln(N_j!)] \cong K(G \ln G - \sum_{j=1}^{M} N_j \ln N_j)$$

It is assumed that G, $N_j >> 1$ and Stirling's theorem is applicable. Hence

$$I \cong -KG \sum_{j=1}^{M} \frac{N_j}{G} \ln \frac{N_j}{G} = -KG \sum_{j=1}^{M} p_j \ln p_j$$

We have obtained Shannon's formula. Turning to the binary definition, we take

$$K = (\ln 2)^{-1} \quad \text{and} \quad I = -G \sum_{j=1}^{M} p_j \log_2 p_j$$

Then the information per letter is

$$i = \frac{I}{G} = - \sum_{j=1}^{M} p_j \log_2 p_j$$

Actually, any text written in a given language contains less information than is indicated here because it has additional constraints. The probabilities of the sequential appearance of letters are interdependent; they form a Markov chain. The probability of appearance of a given letter depends on the preceding letters; thus, in the Russian language, the appearance of a vowel after a vowel is much less likely than after a consonant. Any correlation of events decreases the information content in the communication about them.

The foregoing explanation is essential for the theory and practice of communication. However, information theory has direct physical significance as well. Like information, entropy is expressed as the logarithm of thermodynamic probability, that is, of the number of microstates contained in a given macrostate

$$S = k \ln P \tag{1.57}$$

where $k = 1.38 \times 10^{-16}$ erg-deg$^{-1}$ (Boltzmann's constant).

Let us compare formulas (1.54) and (1.57). By presenting information in the form

$$I = k \ln P \tag{1.58}$$

it is possible to express its content in entropy units. Does such an expression have any real meaning, however? A decrease in the number of microstates of a system means an increase in the information in this system and a decrease in its entropy. Therefore, information is equivalent to negentropy, that is, to entropy with a minus sign. Let the number of microstates decrease from $P_0$ to $P_1$. The information obtained is

$$I = k \ln (P_0/P_1) = S_0 - S_1$$

that is, I is equal to the decrease of entropy or the increase of negentropy; I is complementary to entropy. This treatment of information was proposed by Szillard [27].

If we mix $N_1$ molecules of gas 1 and $N_2$ molecules of gas 2, the entropy increases by the value of the entropy of mixing

$$\Delta S = S - S_0 = -kN(p_1 \ln p_1 + p_2 \ln p_2) > 0$$

where $p_1 = N_1/N$, $p_2 = N_2/N$, and $N_1 = N_1 + N_2$. It is equivalent to the decrease of information by the same amount. Therefore the initial information was

$$I_0 = -kN(p_1 \ln p_1 + p_2 \ln p_2)$$

This expression coincides with Shannon's formula. One bit of information is equivalent to k ln 2, it is $\sim 10^{-16}$ erg-deg$^{-1}$, which is a very small quantity. This definition is not subjective. Obtaining information about a physical system requires real physical action on the system. It is impossible to obtain information about events in an adiabatically isolated system. Any experiment producing information about a system increases that system's entropy or the entropy of its environment and, according to the second law, the increase in entropy always exceeds the amount of information obtained. We have to pay for information in entropy because any observation involves an increase in entropy, an irreversible process.

A liquid contains more entropy than a crystal, and its state is attainable by a greater number of modes of molecular distribution than the crystal state. By crystallizing a liquid, we obtain information about the positions of molecules in a crystalline lattice. We must cool the liquid, and it is impossible to do this by an adiabatic process. We need a refrigerator, the temperature and entropy of which will increase as the liquid cools. The increase in entropy of the refrigerator will exceed the obtained information expressed in entropy units.

In analyzing the second law, Maxwell suggested a fanciful mental experiment: a vessel, containing a gas, is divided into two parts by a partition in which there is a door. The door is controlled by a microscopic demon inhabiting one part of the vessel. He lets fast molecules in and slow ones out. As a result, a difference in temperature arises between the two halves of the vessel. The second law is thus violated. Brillouin investigated this paradox in terms of information theory [28]; he concluded that the demon has to illuminate a molecule in order to see it, that is, he must have a light source not in equilibrium with the surrounding medium. Such a source has lower entropy; it possesses negentropy, from which the demon gets information. This information is used to reduce the entropy of the gas, that is, to increase its negentropy. Calculation of the balance of entropy shows that the entropy of the entire system (gas and light source) increases and the second law holds.

As has been stated (p. 6), Elässer believed the development of an adult organism from a zygote to be incompatible with the laws of physics. He considered the increase in information content during morphogenesis and growth to be "biotonic." These ideas concerning epigenesis conflict with the concept of preformism, according to which all the information about a future organism is stored in its zygote.

It is difficult to estimate the information content of a cell or an organism. It may be thought that the information (i.e., orderliness) assuring biological functionality is completely determined by atomic structure [29,30]. Dankoff and Quastler estimated the quantity of information required for the selection of one atom from among the set of atoms in a cell and for the determination of its position with the precision determined by the temperature. Using their method, we get 24.5 bits per atom, $\sim 10^{10}$ per cell of *Escherichia coli*, $10^{15}$ per mammalian zygote, and $10^{25}$ per human organism (i.e., $10^9$ erg-deg$^{-1}$ = 24 cal-deg$^{-1}$-mole$^{-1}$ in terms of entropy). If these calculations are meaningful, then

$$I_2 \equiv I_{org} \gg I_{zyg} \equiv I_1$$

Elsässer considered the increase of information from $I_1$ to $I_2$ incomprehensible in physical terms, but there is actually no difficulty here.

We can write the change in information content in an open system, such as (1.16), that is,

$$\Delta I = \Delta_i I + \Delta_e I$$

The first term $\Delta_i I < 0$ represents the decrease in I resulting from internal processes for which $\Delta_i S > 0$. In the growing organism $\Delta_e I > 0$. We have

$$I_2 = I_1 + \Delta_e I + \Delta_i I$$

Since $I_2 \gg I_1$, it is obvious that $\Delta_e I \gg I_1 - |\Delta_i I|$.

The open system draws information from the environment, the entropy of which increases. The organism growing from a zygote is like a crystal growing from a seed placed in a saturated solution. In either case, the increase in orderliness, or I, is exceeded by the increase in the entropy of the surrounding medium (the refrigerator, in the case of crystallization). Elsässer's concept is vitalistic, for it implicitly assumes that the second law is not obeyed in animate nature.

Raven, who criticized Elsässer's ideas, thinks that all

the cells in an organism contain the same genotypic information, and that the development of an organism involves an increase not in the amount of information but rather in its redundancy [31]. He regards the zygote as a communication channel, with the parents serving as the source of information and the growing organism as the recipient of information. Development can be viewed as the decoding of information. Raven believes that it is feasible to make an absolute estimate of the information content in the zygote and the organism. In actual fact, as shown by Apter ([32], see also [33]), such an estimate will always be relative and conventional. The identity of genes in cells does not signify redundancy. Development results from interactions among the different parts of the embryo, and information is contained not only in the chromosomes but in every intra- and intercellular relationship. The concepts of epigenesis and those of preformism are largely unsuitable for the description of development, which cannot be reduced to an increase in, or the maintenance of, the amount of information in a cell. The problem is not one of devising such a description, but of elucidating the essence of development, its physical nature, its atomic-molecular basis.

Approaches that use the notion of quantity of information are important for theoretical biology and biophysics not because they resolve the epigenesis-preformism controversy. They are important, first, because the translating of biological concepts into the language of information theory makes those concepts more rigorous and clear cut and imparts to them a new content. Such work has been done by Schmalhausen, who developed the informational aspects of the theory of evolution [34,35]. Second, study of the cell and the organism as complicated self-regulating systems necessitates an informational approach because control or regulation are effected by means of the forward and reverse flow of information. The problem facing biophysics in this area is to ascertain the material nature of information storage, coding, transmission, and reception in living systems. The problem of the genetic code (Chapter 9) has a clear informational meaning. Questions about the amount of genetic information coded by DNA molecules, the noise resistance of the code, and the mechanisms by which the informational program is executed are all quite meaningful.

Information is transmitted in an organism by molecules and ions and is thus of a chemical nature. Information reception is essentially the recognition of molecules by molecules and is determined by their interactions. In this sense, the organism is a chemical engine, and this makes it different from man-made "robots." Present-day cybernetic devices operate by means of electric or magnetic signals, while in an organism the molecules themselves act as signals.

Quastler introduced the principle of signature into

molecular information theory [29]. Molecular interactions mean the transmittal of information. Generally, not all the information stored in a molecule is transmitted, but only a part of it, which Quastler called the signature. A molecule's signature represents a specific set of those of its characteristics owing to which the molecule gets involved in a particular chemical reaction. The problems of ambiguity and nonambiguity of molecular signatures and of their reliability have real physical meaning [36].

A genetic message's information can be written as

$$I = I_0 - I_n$$

where $I_0$ is ideal information and the $I_n$ are "noise" losses due to errors in the process of chromosome duplication and in the reception of genetic information, that is, in protein synthesis (see Chapter 9). The $I_n$ can be expressed through the probability of "reading" or the transition from a letter of kind i to one of kind j (coding), which we will denote by $p_i(j)$. The probability of appearance of the letter i will be denoted $p(i)$. Then

$$I = I_0 + \sum_{i,j} p(i)p_i(j) \log_2 p_i(j)$$

Suppose that some agent exerts an influence on the genetic system characterized by measure $\lambda$. For example, $\lambda$ is a dose of ionizing radiation. We have

$$\frac{dI}{d\lambda} = \log_2 e \sum_{i,j} \{p(i) \frac{dp_i(j)}{d\lambda} + p_i \ln p_i(j) \frac{dp_i(j)}{d\lambda}$$

$$+ p_i(j) \ln p_i(j) \frac{dp(i)}{d\lambda}\}$$

The derivative increases to infinity owing to the second term on the right-hand side of the equation if $p(i)$ and $dp_i(j)/d\lambda$ are different from zero and one of the probabilities $p_i(j)$ tends to zero. Since

$$\sum_j p_i(j) = 1$$

then, in the ideal case, when one of the quantities $p_i(j) = 1$, all the others are zero. The system becomes unstable, $dI/d\lambda \to \infty$. This means that a system formed by isogenous organisms is unstable. A real system must have noise; that is, genetic messages in different isogenous organisms are not identical. There must be mutation or, in terms of information theory,

$$p(i) \neq 0 \quad \text{and} \quad dp_i(j)/d\lambda \neq 0$$

According to Dankoff's principle [32], the optimum attained by natural selection does not mean an ideal system without noise. In information transmission, noise is controlled by means of redundant information. Thus a wrecked ship transmits the SOS signal repeatedly so as to increase the probability of the signal's reception. The Dankoff principle states that the optimal amount of redundant information in the evolution of a species and an individual is equal not to that amount which obliterates all errors but to that which minimizes the total cost of all errors as well as the cost of their correction. As many errors are made as is compatible with continued existence using the minimal amount of redundant information. Thus some stationary state is achieved. These notions can evidently be formulated in terms of thermodynamics if the entropic measure of information is used.

The foregoing considerations generally exhaust the capabilities of information theory operating only with the concept of quantity of information. It follows from the relation of quantity of information to entropy (with a minus sign) that the concept of amount of information can contribute nothing more to the interpretation of biological processes than the notion of entropy. The generation and instructive value of information, the qualities which determine a message's specific biological function, are ignored in the concept of amount of information. A concept of information value in relation to a particular process is essential when we deal with a developing, evolving system. The notion of information value is concerned with the sense of information, which may be evaluated only with reference to specific physical phenomena.

Eigen was the first to attempt a physical definition of the value of the information contained in biologically functional macromolecules in the process of their natural selection [13]. The selective value, according to Eigen, is a dimensionless kinetic parameter characterizing the autocatalytic synthesis of macromolecular chains occurring at the expense of chains having a smaller value, which undergo destruction. The process takes place in an open stationary system that is far from equilibrium. Eigen's theory agrees with the theory of dissipative systems of Prigogine and Glansdorff and describes a certain initial stage of prebiological evolution.

The theory of Eigen is very important for biophysics. It is based on actual properties of biological macromolecules. The presentation, analysis, and development of this theory are possible only when we are thoroughly acquainted with biological macromolecules. Consequently, Eigen's theory will be discussed in "General Biophysics."

## 1.5 Cooperativity

The formulation of a physics of living systems requires investigation of their microscopic molecular properties. These properties are specific.

The cell and the organism are heterogeneous condensed systems built up of quasicrystalline bodies (supramolecular structures) and liquids, of big and small molecules. The supra molecular structures in an organism are highly organized and are predominantly linear or two-dimensional systems. Examples of linear systems are the myofibril of a muscle and the axon of a nerve cell; examples of two-dimensional systems are various cellular and intracellular membranes, β forms of protein structures, etc. When we speak about the quasicrystallinity of such structures, we have in mind their high degree of organization, expressed in many cases as structural periodicity.

On the other hand, a high degree of order (in one, two, and three dimensions) also characterizes the molecular level of organization. The macromolecules of biopolymers--of proteins and nucleic acids--possess not only a given genetically determined sequence of monomeric links, but also a definite overall structure.

The notion of an aperiodic crystal, introduced by Schrödinger, is applicable to the molecular, supramolecular, and cellular levels of structure of living systems.

Condensed systems containing large numbers of particles are cooperative. Their properties are determined by the interaction of particles, and they cannot be understood by studying only single particles. A specific feature of the condensed system is its ability to change its state and overall structural organization, that is, to undergo phase transitions, particularly order-disorder transitions. The notion of cooperativity was introduced (by Fawler) in order to explain phase transitions. The cooperative transition is a self-amplifying transformation. Consider a simple example--the van der Waals equation of state of a real gas

$$(p + (N^2a/V^2))(V - Nb) = NkT \qquad (1.59)$$

where $N >> 1$ is the number of gas molecules.

Let us compress the gas at a temperature below the critical one. The decrease in volume V causes an increase in internal pressure $a/V^2$, which in turn diminishes the volume, etc. Transition proceeds until the transformation of gas into liquid is complete. It is a typical cooperative process occurring according to the all-or-none principle.

Thermodynamics formulates the conditions of phase equilibrium (see [37]): Two phases are in equilibrium if their temperatures, pressures, and chemical potentials are equal:

$$T_1 = T_2, \quad p_1 = p_2, \quad \mu_1(p,T) = \mu_2(p,T) \tag{1.60}$$

Since the chemical potentials depend on p and T, equilibrium is possible only at definite values of these quantities. The condition of equality of chemical potentials is equivalent to

$$H_2 - H_1 - T_{tr}(S_2 - S_1) = 0$$

where H is the enthalpy. Hence, the temperature of transition

$$T_{tr} = \Delta H/\Delta S \tag{1.61}$$

where $\Delta H$ is the amount of heat absorbed or liberated in the transition (e.g., the heat of melting or of crystallization). There is a discontinuity in the enthalpy in such a transition, that is, a phase transition of the first kind. The discontinuity in enthalpy, and also in entropy and volume, is determined by the jumplike change in the structure of the substance. There is an obvious discontinuity in symmetry in crystal-liquid transitions--a symmetry element is either present or lacking, it cannot vanish or arise gradually. Therefore there is no critical point for the crystal-liquid transition. A liquid-gas transition, on the other hand, can occur gradually throughout the critical region because the symmetry of both states is the same (see [37]). The usual liquid-gas transition at temperatures below the critical one is a phase transition of the first kind. Transitions between different crystalline modifications are also transitions of the first kind with a change in symmetry.

Cases in which the structure of a system changes gradually and its symmetry changes by a jump are also phase transitions, but with different features [37]. Consider a set of electronic spins in a ferromagnet in the form of a regular two-dimensional lattice whose points contain arrows which can be oriented upward and downward. Let the spins initially be oriented randomly, and the probabilities of their direction upward and downward be equal. Now reorient, one after another, the downward-directed arrows. When the last arrow has been rotated upward, the symmetry will change by a jump and the substance will turn from a paramagnetic into a ferromagnetic state.

Such a transition (like the order-disorder transitions in binary alloys) is a phase transition of the second kind [37,38]. It is characterized by monotonic change in properties such as enthalpy, specific volume, etc., but involves a break in their derivatives, such as heat capacity, coefficient of expansion, and compressibility.

Thus the phase transition implies the presence of singularities on the curves describing thermodynamic quantities as functions of external parameters, including the temperature. Thermodynamic quantities are expressed with the help of the partition function of the system

$$Z = \sum_r g_r \exp(-E_r/kT) \tag{1.62}$$

where $E_r$ is the energy of the rth state of units making up the ensemble and $g_r$ is the statistical weight of that state. We have the following expressions for free energy, internal energy, and heat capacity (per mole)

$$F = -RT \ln Z \tag{1.63}$$

$$E = Rt \frac{d \ln Z}{dT} \tag{1.64}$$

$$C_V = R \frac{d}{dT} (T \frac{d \ln Z}{dT}) \tag{1.65}$$

etc. Singularities can arise only as a result of the interactions of units, which determine the dependence of the state of each unit on the states of other ones [39].

Theoretical deduction of the van der Waals equation goes beyond the scope of thermodynamics, for it is based on a molecular model. This is precisely the concern of the methodology of theoretical investigation of a cooperative system: a molecular model is suggested and the partition function satisfying this model is calculated, and the thermodynamic functions are derived from this partition function and compared with experiment.

Consider some models of cooperative systems. The simplest approach in statistics is based on the molecular field approximation. A particle is acted upon by forces produced by other similar particles. It is in a force field, described by some average value, in a molecular field. The states of the particle are dealt with by statistical mechanical methods for determining the average field produced by this particle and affecting the neighboring particles. This average field must coincide with the molecular one--the theory should be self-consistent. The derivation of the van der Waals equation uses essentially this method [39].

Consider now the model of the ferromagnet suggested by Ising [40]--the spin lattice (p. 123). The state of arrow j (j = 1,2,3,...,N, N being the total number of arrows) can be defined by the parameter $\sigma_j$ equal to 1 and -1, depending on the direction of the arrow. The energy of interaction of the neighboring spins

$$E_{int} = - \sum_{i,j} \varepsilon \sigma_i \sigma_j \tag{1.66}$$

that is, the energy of a pair of parallel arrows, is lower by $2\varepsilon$ than that of a pair of antiparallel arrows.

Ferromagnetism is determined by the parallel arrangement of electronic spins. Energy $2\varepsilon$ characterizes the advantage of the parallel arrangement because of the quantum-mechanical exchange interaction. Let us describe the interaction by magnetic field H' acting on every spin. We take H' to be proportional to the magnetization M, that is, to the difference of the numbers of spins facing upward and downward

$$H' = \beta M \tag{1.67}$$

Magnetization is equal to the magnetic moment of the unit volume, that is, to the number of spins in this unit volume n multiplied by the average spin moment $\bar{\mu}$. The partition function of the spin is

$$Z = \exp(\mu H_{eff}/kT) + \exp(-\mu H_{eff}/kT) \equiv 2\cosh(\mu H_{eff}/kT) \tag{1.68}$$

since the spin (whose magnetic moment is $\mu$) has only two states--parallel and antiparallel to the field $H_{eff}$. Free energy (see (1.63))

$$F = -kT \ln Z = -kT \ln\{2\cosh(\mu H_{eff}/kT)\} \tag{1.69}$$

and

$$\mu = -\partial F/\partial H_{eff} = \mu \tanh(\mu H_{eff}/kT) \tag{1.70}$$

Hence

$$M = n\bar{\mu} = n\mu \tanh(\mu H_{eff}/kT) \tag{1.71}$$

If there is an external field H

$$H_{eff} = H+H' = H+\beta M$$

and

$$M = n\mu \tanh\{(\mu/kT)(H+\beta M)\} \tag{1.72}$$

This transcendental equation gives the self-consistent definition of M and H'.

Let the given spin possess z nearest neighbors. On average, $z_+$ of them are oriented upward and $z_-$ downward. We have

$$z_+ + z_- = z, \qquad z_+ - z_- = z(M/n\mu)$$

Assume that

$$\mu H' = \varepsilon(z_+ - z_-) = z\varepsilon(M/n\mu)$$

Hence

$$\beta = z\varepsilon/n\mu^2$$

and

$$M = n\mu \tanh\left(\frac{\mu H}{kT} + \frac{z\varepsilon}{kT}\frac{M}{n\mu}\right) \tag{1.73}$$

Solving this equation graphically, we find the intersection of the curve and straight line

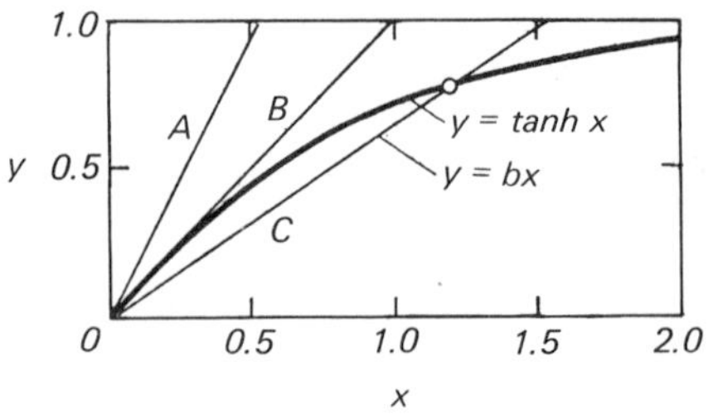


*FIG. 1.3 Magnetization curves. For straight-line curves: A, $T = 2T_C$; B, $T = T_C$; C, $T = 2T_C/3$.*

$$y = \tanh(a+x), \qquad y = bx$$

where

$$x = (z\varepsilon/kT)(M/n\mu), \qquad a = \mu H/kT, \qquad b = kT/z\varepsilon$$

If a = 0 (in the absence of external field H), we obtain the result shown in Fig. 1.3. The system is magnetized spontaneously, that is, it acquires M other than zero in the absence of an external field if

$$T < T_C = z\varepsilon/k \tag{1.74}$$

where $T_C$ is the Curie temperature, the temperature of the ferromagnetic-paramagnetic phase transition.

The molecular field approximation is substantiated in the Bragg-Williams method. Let N be the total number of spins, $N_+$ of them facing upward and $N_-$ downward. The number of possible microstates

$$W = N!/N_+!N_-! \tag{1.75}$$

The entropy is equal to (according to Stirling's formula)

$$S = k \ln W = -k\left(N_+ \ln \frac{N_+}{N} + N_- \ln \frac{N_-}{N}\right) \tag{1.76}$$

Introduce deviations of $N_+$ and $N_-$ from the average value, which in the absence of interaction is equal to N/2

$$N_+/N = (1/2)(1+x), \qquad N_-/N = (1/2)(1-x)$$

Then

$$S = -kN\{(1/2)(1+x)\ \ln[(1/2)(1+x)]$$
$$+(1/2)(1-x)\ \ln[(1/2)(1-x)]\} \tag{1.77}$$

The interaction energy [cf. (1.66)]

$$E = -\varepsilon(N_{++}+N_{--}-N_{+-}) \tag{1.78}$$

where $N_{++}$ is the number of pairs of ++ neighbors, $N_{--}$ the number of -- pairs, and $N_{+-}$ that of +- pairs. The total number

of pairs of neighboring spins is $\frac{1}{2}(zN)$.
The average values of $N_{++}$, $N_{--}$, $N_{+-}$ are

$$\bar{N}_{++} = \frac{1}{2} zN_{+}(N_{+}/N) = \frac{1}{8} zN(1+x)^{2}$$
$$\bar{N}_{+-} = zN_{+}(N_{-}/N) = \frac{1}{4} zN(1-x^{2}) \tag{1.79}$$
$$\bar{N}_{--} = \frac{1}{2} zN_{-}(N_{-}/N) = \frac{1}{8} zN(1-x)^{2}$$

Substituting these values in E, we get

$$E = -\frac{1}{2} z Nx^{2} \tag{1.80}$$

We now find the free energy $F = E - TS$ and from the condition of its minimum, the most probable value x. From $\partial F/\partial x = 0$ we get

$$\tanh[(z\varepsilon/kT)x] = x \tag{1.81}$$

which coincides with (1.73). The molecular field approximation is equivalent to the approximation that takes the entropy to be equal to the entropy of mixing, and the energy is obtained by averaging all possible configurations. A more exact solution of Ising's problem is given by Bethe-Pajerls' approximation (cf. [41]) or by the quasichemical approximation, which takes account of the local correlation of spins. A lattice point and z of its nearest neighbors are now discussed and the probabilities of spin distribution in these z points at the given direction of spin in the initial lattice point are determined. In Bethe-Pajerls' method the short-range order (i.e., the $N_{++}$ value) is determined by the long-range order,

$$\bar{N}_{++} = (z/2N)N_{+}^{2}$$

In the quasichemical approximation, however, these values are calculated independently but are expressed via the same parameter of the lattice. It is assumed that

$$\bar{N}_{++}\bar{N}_{--}/\bar{N}_{+-}^{2} = \frac{1}{4} \exp(-\varepsilon/kT) \tag{1.82}$$

In the Bragg-Williams approximation, according to (1.79)

$$\bar{N}_{++}\bar{N}_{--}/\bar{N}_{+-}^{2} = \frac{1}{4} \tag{1.83}$$

The factor $\frac{1}{4}$ appears because of the symmetry numbers of the states (++) and (--), equal to 2. Formula (1.82) is similar to the expression for the equilibrium constant for the "chemical reaction"

$$2(+-) \rightarrow (++) + (--)$$

This approximation is therefore called quasi-chemical.

A rigorous calculation of the partition function for the two-dimensional Ising problem was done by Onsager (cf. [41,42]).

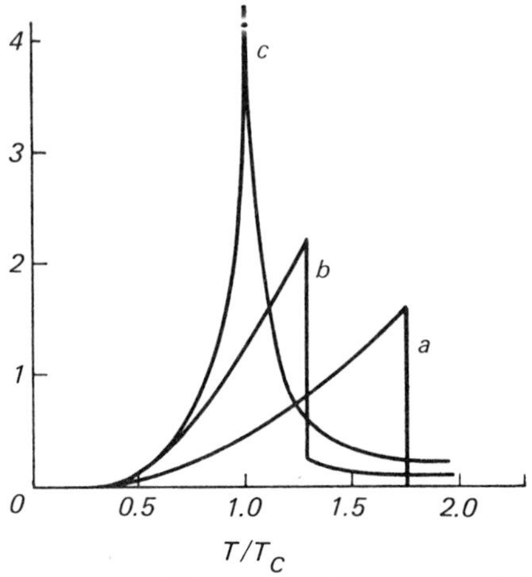


*FIG. 1.4 Phase transition of the second kind: a, calculation according to Bragg and Williams; b, quasichemical approximation; c, rigorous method.*

The analytical expression for a three-dimensional problem was not derived, but it can be solved numerically.

Figure 1.4 shows the heat capacity jump at the Curie point--the phase transition of the second kind--calculated according to Bragg and Williams, by the method of quasichemical approximation, and by the rigorous method.

In biology we encounter one-, two-, and three-dimensional cooperative systems, containing large numbers of statistical units. Cooperativity is a fundamental feature of molecular systems in biology and is responsible for many important phenomena (cf. [43]). Methods for investigating cooperative processes are very important in theoretical biophysics.

The foregoing discussion concerned the statistical physics of cooperative systems in states of thermodynamic equilibria. Despite the general nonequilibrium of biological systems, statistical techniques contribute greatly to our understanding of them. No less important, however, is the study of the kinetics of cooperative processes.

The thermodynamics of a phase transition does not mean that it occurs by an anomalously high rate. A liquid can crystallize very slowly at the phase transition temperature, the rate of crystallization depending on factors that do not figure in statistical thermodynamics, such as by the rate of formation of the crystal germs, that is, by the nucleation rate and by the crystal growth rate. The noncooperative kinetic process of a substance's transition from one state to another usually requires activation energy. The initial and final states are separated by an activation barrier. The rate constant (i.e., the number of transitions through the barrier in a given direction per unit time) is described by Arrhenius's formula

$$k = A \exp(-E_a/RT) \quad (1.84)$$

where $E_a$ is the activation energy and A is the preexponential factor. The meaning of this formula is spelled out later in the theory of absolute reaction rates (Chapter 6).

In some processes $E_a$ is a real constant, independent of temperature. In a self-reinforcing process the rate constant itself changes during the process; in other words, $E_a$ changes in the process of transition. Thus the viscous flow of a liquid is a cooperative process because the moving particles interact. In order to move, they have to "push aside" their neighbors and in this way overcome a certain energy barrier; that is, they must possess some activation energy. The situation is similar to that of a person who makes his way toward the exit in an overcrowded bus. In order for him to reach his destination, other passengers have to move back or aside. The activation energy of the cooperative process depends on the number of particles that have already passed the barrier. The greater the number of people who have left the bus, the easier it is to move toward the exit.

In the absence of cooperativity, the dependence of the logarithm of the rate of the process on the inverse temperature is linear--we get, from (1.84), $\ln k = \ln A - E_a(RT)^{-1}$. Cooperativity violates the linearity. Formally, cooperativity can be expressed by the dependence of $E_a$ on the temperature.

In contrast to thermodynamics, the general kinetics of cooperative processes has not been studied adequately. Physics in effect does not yet possess a complete kinetic theory of crystallization or melting. Some approximate approaches to one-dimensional problems, which are particularly important for biophysics (template synthesis of biopolymers) have been proposed, however (cf., e.g., [43,44]).

The notion of cooperativity introduced with reference to phase transformations in ensembles containing large numbers of particles is treated in modern biophysics in broader terms. Every phenomenon determined by interaction, even that of a small number of elements in a system, is called cooperative. In this sense a chemical reaction of the type

$$A + B \rightleftarrows AB$$

$$AB + B \rightleftarrows AB_2$$

is considered cooperative if the equilibrium constant of the second stage differs from that of the first one. In this case the binding isotherm of substance B by substance A may have a singularity, a flection. If, on the other hand, the equilibrium constants are equal, this isotherm goes smoothly to

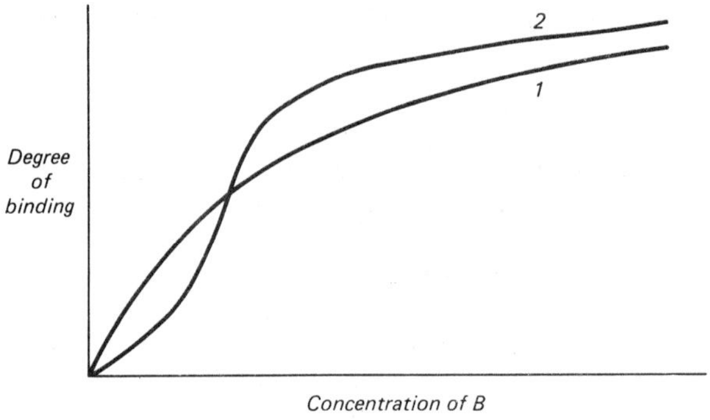


*FIG. 1.5 (1) simple and (2) cooperative binding isotherms.*

saturation, like Langmuir's isotherm (Fig. 1.5). The S-shaped character of the binding curve generally suggests the cooperativity of the system (see Chapter 7).

A heterogeneous system can also be considered cooperative if its properties depend essentially on the interactions between the homogeneous and heterogeneous elements. We thus pass from the notion of cooperativity to the more general notion of a system in the sense introduced by Bertalanffy [3].

Statistical physics is faced with great difficulties in interpreting the behavior of an aggregate of interacting heterogeneous elements containing small numbers of every kind of element. The behavior of such an aggregate is determined not by statistical but by dynamic laws. The interaction of its elements results in the organization of the system, that is, in a spatial and temporal ordering of the processes developing in it. In contrast to the statement of Bertalanffy, the notion of organization also arises in conventional physics in a quite natural way. Examples include the condensation of gases, the crystallization of liquids, and the spontaneous magnetization of a ferromagnet. Another example, one that does not concern a system containing large numbers of units, is the distribution of electrons in an atom. The organizing principle here is the Pauli principle. However, conventional physics does not take into account the individuality of heterogeneous elements. The main differences between an organized biological system and a nonliving system such as is usually studied by physicists are that the biological system is open, far from being in equilibrium, and heterogeneous; hence, the great problems, which are far from solution. (The problems of organization, systematicity, and development in living organisms will be discussed in "General Biophysics.")

## 1.6 Biophysics

Biophysics is the physics of living systems. Thermodynamic and informational analyses of life phenomena have removed the apparent contradictions between physics and biology. We have to agree with Eigen's statement that contemporary physics is adequate for explaining life phenomena, and for validating biology [13]. Such validation requires the modification of important physical principles and introduction of new notions (such as that of the selective value of information), but not the formation of a new physics or of new principles. A new physics, such as quantum mechanics or the theory of relativity, arises as a result of establishing the scope of applicability of earlier theories and ideas. Thus far we have not encountered any such limits for physics in biology.

A biophysical investigation begins with the formulation of a physical problem on the basis of general physical laws and atomic-molecular (i.e., quantum-mechanical) notions. Physics proceeds through phenomenology--first via thermodynamics and information theory--to the atomic-molecular investigation of living bodies. A living body is macroscopic in principle, and contains a large number of atoms and molecules, links of polymeric chains, all possessing, to a certain extent, independent degrees of freedom. The ordering of a biological system, its ability to develop, would not exist should this system be microscopic and therefore subject to fluctuations [10].

A biological problem can be solved by means of physics (e.g., with the aid of an electron microscope), but this would not make such an investigation biophysical. Conversely, a physical problem can be solved by biological methods. Thus, the formulation of the problem of the genetic code--the correlation between the sequence of amino-acid residues in a protein chain and the sequence of nucleotides in DNA--is a physical statement of the problem based on the physicochemical hypothesis suggesting the existence of a code. However, the solution to this physical problem was obtained by purely biological (p. 559) and chemical (p.584) methods.

The statement of a physical problem is always preceded by hard work in the fields of biology, physiology, biochemistry, cytology, etc. Biophysics is a broad field for new and major discoveries and the solution of genuine riddles of nature. It may seem that if a physicist approached a biological problem, he would make such discoveries relatively quickly because his ideas and methods are very powerful. The actual situation, however, is quite different. The complexity of biological objects and phenomena hinders the formulation of physical problems, which is possible only as a result of a deep biological investigation.

Therefore, work in the field of biophysics requires a great deal from an investigator; this is inevitable at the interface of sciences. A biophysicist is a physicist possessing broad biological erudition, who is also able to formulate and solve physical problems. Biological erudition implies more than knowledge of the special fields of biology bearing directly on the subject at hand, say, molecular biology or physiology. One who is alien to living nature, who is not familiar with zoology and botany, does not understand biology. The knowledge of those fundamentals of biology that are sometimes looked upon disparagingly by physicists forms the biological Weltanschauung without which no genuine biophysics can be constructed.

The ultimate objectives of biology and biophysics are the same--knowledge of the essence of life phenomena. The applied objectives in medicine and pharmacology, in agriculture and technology, are also alike. But although it is a part of physics biophysics should not be considered an auxiliary biological discipline. It should be emphasized that the application of physical and mathematical methods to the solution of biological problems does not imply biophysical investigation. No exact knowledge is possible without mathematical tools. The modern zoologist resorts to sophisticated mathematical procedures in studying population dynamics, but this procedure does not make him either a mathematician or a biophysicist.

What is essential are not methods but physical ideas, the formulation and solution of physical problems. Contemporary biophysics is usually divided into three areas: molecular biophysics, cell biophysics, and the biophysics of complex systems. This division is somewhat artificial although convenient.

Molecular biophysics lies at the interface of molecular physics and molecular biology. It is the molecular physics of biological processes, of biologically functioning molecules. Molecular physics and molecular biophysics are concerned with three groups of problems. They investigate the structure of molecules, their equilibrium interrelations, and their properties, as well as the kinetics of their interactions and transformations. Structural studies employ a number of physical and chemical methods.

The theory of the structure of the electronic shell of molecules and of the phenomena determined by that shell constitutes quantum mechanics and quantum chemistry. The whole of chemistry--the phenomenon of the chemical bond, the transformations of bonds through reactions--is determined by quantum-mechanical laws. Quantum mechanics plays the same role in biophysics as in chemistry and in the physics of molecules; it is the basis for our understanding of molecular structure, the nature of molecular interactions, and electronic (e.g., spectral) properties. In many cases, however, the problems associated with the electronic properties of molecules can be solved by means of a semiempirical classical theory using, in particular, the so-called valence-optical scheme (cf. [45,46]).

The essential feature of the chief biologically functional substances is their macromolecularity. Proteins and nucleic acids are big molecules, biopolymers. Therefore, molecular biophysics is mainly macromolecular biophysics, or the physics of biopolymers. It makes wide use of theoretical and experimental methods developed earlier for the study of macromolecules of nonbiological origin.

It is impossible to draw a boundary between molecular biophysics and biophysical chemistry, just as it is impossible to establish the boundary between molecular physics and physical chemistry. Any classification of the fields of knowledge is always historical and conventional in character. Molecular physics and physical chemistry differ not so much by the objects they study and the subject under investigation as by their respective approaches, which are determined to some extent by tradition.

In those cases in which biophysicists study biomolecules *in vitro* the concepts of equilibrium thermodynamics are applicable, and in this sense the investigation of protein denaturation does not differ from that of any physicochemical process in a nonliving system. Experimental study and theoretical calculations of equilibria are also very important for the investigation of open living systems, about which these methods provide basic information that is indispensable. Thus the *in vitro* reduplication of DNA realized in Kornberg's experiments (cf. p. 535) is essential for understanding the duplication of DNA in dividing cells, which are open systems. Quantitative characteristics of equilibria found in *in vitro* experiments are important also for the behavior of the systems *in vivo*.

The same considerations apply to kinetic studies. The kinetics of enzymic processes studied *in vitro* do not differ from the kinetics of any other chemical reactions, and serve as the basis for investigation of these processes in an organism.

Enzyme physics is becoming one of the central areas of molecular biophysics [47]. Its problems are the elaboration of experimental and theoretical methods of investigation of enzymes and enzymic processes, the unraveling of the physical and physicochemical mechanisms of enzymic catalysis. The problems of molecular kinetics in biophysics are inexhaustible, and include active transport, ion exchange, and mechanochemical processes.

The rise of molecular biophysics is associated with the development of molecular biology. It is a new field of biophysics, the emergence of which could not have been predicted several decades ago. Biophysics of the cell is the conventional title of the oldest traditional domain of biophysics.

The physics of isolated processes in the living organism, it comes into close contact with physiology. Isolation means mental (and experimental) severing of the links connecting these processes with the integral living system--a method that is indispensable at some level of investigation. The three most developed areas of cellular biophysics are the biophysics of contractile systems (primarily of muscular activity), the biophysics of neural conduction, and the biophysics of the sense organs. Helmholtz had subjected to strict physico-chemical analysis a series of problems pertinent to cellular biophysics. Sechenov stated that physiology was the physico-chemistry of the living organism. Later Lazarev formulated the problem of the scientific development of biophysics, and suggested an interesting ionic theory of neural excitation. Physiology has long applied physical methods, for which reason many physiological studies are often identified with biophysical ones. But we have already said that such identification is not justified so long as the problem is formulated as a physiological rather than a physical one. Thus the study of the flight of insects (the quantitative registration of nerve impulses and wing movements, etc.) remains purely physiological until the problem of the mechanism of the auto-oscillating process involved is posed.

By removing the object of investigation from the organism (isolating a muscle or nerve axon) it is possible to establish fundamental laws that remain valid in the intact system. A squid *Loligo* whose axon has been isolated is killed, but study of the axon enables us to understand processes that occur in the living squid and, in the final analysis, in any animal organism.

Today cellular biophysics is becoming integrated with molecular biophysics. The establishment of the molecular and supramolecular structure of muscle, nerve, etc., leads to a molecular interpretation of corresponding processes. Problems of cellular biophysics are much more complicated than those of molecular biophysics, since they concern heterogeneous supramolecular systems rather than single molecules and their interactions.

The present-day biophysics of complicated systems is concerned with the physical essence of the organism or of some of its functional subsystems as a whole. Most prominent are those features which are completely ignored in molecular biophysics and nearly completely in cellular biophysics. These include the organism's properties as an open system, its self-regulation and self-reproduction. Not only the organism, but the population, the biocenose, and the biosphere as a whole, are complex systems in this sense. In this respect the biophysics of complex systems becomes integrated with theoretical biology.

The central problem of theoretical biology and the

biophysics of complex systems is of development, or ontogeny and phylogeny. The differentiation of cells, the rise of a complex organism from a zygote, poses a host of unresolved problems. Their solution will have tremendous theoretical and practical (the problem of cancer!) implications.

Methods of investigation in the biophysics of complicated systems are specific. Today they consist in experimental and theoretical modeling, the elaboration of appropriate mathematical machinery with which to interpret complex phenomena of regulation, etc. A biological system is studied in its dynamics, its interrelations with the environment. A complex system exists by maintaining its nonequilibrium steady state or by changing irreversibly. It may be thought that the investigation of periodic processes, the study of "biological clocks," can aid our understanding of internal linkages in a complex system. Considering a certain radioelectronic device as a "black box," we test it in different oscillatory conditions and as a result we ascertain its inner structure.

Contemporary biophysics of complex systems is a phenomenological field of physics using cybernetic concepts and mathematical models. The mathematical biophysics of Rashevsky [48] belongs entirely to this field.

The interrelation between the biophysics of complex systems and the other areas of biophysics mentioned earlier is similar to the relationship between phenomenological thermodynamics and molecular physics. The behavior of a complex biological system has its basis in the properties of biological molecules and the structures they form. Further development of biophysics should lead to its integration in a general molecular treatment of complex systems. We already speak about the molecular foundations of evolution [13,49]. However, a number of biological phenomena arise only at the level of a complex system. Thus higher nervous activity, which originates in molecular processes, is realized only in a complex system such as the brain. That the behavior of a complex system can be interpreted in molecular terms is evidenced by studies of molecular regulatory systems, for example, the operon in molecular genetics or biochemical chains, catalyzed by allosteric enzymes (Chapter 7). Obviously a general molecular treatment of complex systems is necessary. The description and explanation of their concrete functioning must of course be based on physicomathematical modeling, cybernetics, and regulation theory. Thus, knowing the electronic basis of the work of a transistor, we do not resort to the quantum theory of solids in order to calculate a radioelectronic circuit.

We see that biophysics itself constitutes a complex system of knowledge. Like the physics of inanimate nature, it contains phenomenological and molecular areas of study. The complicated structure of biophysics reflects the complexity of life phenomena.

In the literature we encounter one-sided and therefore inadequate definitions of biophysics. Biophysics is identified with physical chemistry [50], physiology [51], and systems theory [52]. In a number of monographs and handbooks on biophysics certain problems are dealt with in depth [52-56]. However, these works lack any sequentially developed physical concept. We feel today an acute need for the formation of biophysics as an inextricable part of physics.

In conclusion, a few words concerning radiobiology. The impact of short-wave radiation on organisms, cells, supramolecular structures, and biological molecules is subject to physical treatment. However, radiobiology studies life under abnormal conditions (if one ignores the background of cosmic radiation). It is a special field which has developed intensively during the last three decades owing to its enormous practical importance, particularly for medicine. Therefore it is undesirable to present radiobiology in a monograph devoted to the biophysics of normal life processes.

The foundation of biophysics must ultimately be sought in atomic-molecular structure and functionality. This book is devoted to the physics of the substances most important for life--proteins and nucleic acids. The emergence and rapid development of molecular biology have proceeded in close contact with physics and have made molecular biophysics today a well-developed domain of science. Molecular biophysics should be used as the basis for constructing the physics of supramolecular systems, the physics of developmental processes, and the physics of life as a whole. These problems will be dealt with in "General Biophysics."

## References

References followed by (R) are in Russian.

1. F. Engels, "Dialektik der Natur." Gospolitizat, Moscow, 1955.
2. H. Helmholtz, "Über sich selbst." Ed. Teubner, Leipzig, 1966.
3. L. von Bertalanffy, "General System Theory." Braziller, New York, 1969.
4. E. Bauer, "Theoretical Biology." Ed. All-Union Inst. of Experimental Medicine, Moscow, 1935 (R).
5. B. Tokin, "Theoretical Biology and the Works of E. Bauer." Ed. of Leningrad Univ., Leningrad, 1965 (R).
6. N. Bohr, "Atomic Physics and Human Knowledge." Wiley, New York, 1958.
7. A. Meyer-Abich, Acta Biotheoret, 11, 2 (1955).
8. N. Bohr, in Symposia of the Society for Exp. Biol., N14, 1.

Cambridge Univ. Press, London and New York, 1960.
9. M. Volkenstein, "Cross-Roads of Science." Nauka, Moscow, 1972 (R).
10. E. Schrödinger, "What Is Life?" Cambridge Univ. Press, London and New York, 1945.
11. W. Elsässer, "The Physical Foundation of Biology." Pergamon, Oxford, 1958.
12. E. Wigner, "Symmetries and Reflexions." Indiana Univ. Press, Bloomington, Indiana, 1967.
13. M. Eigen, Naturwissenschaften, 58, 465 (1971); Quart. Rev. Biophys. 4, 149 (1971).
14. J. Monod, "Le hasard et la necéssité." Seuil, Paris, 1971.
15. A. Liapounov, "General Problem of Stability of Motion." Gostekhizdat, Moscow, 1950 (R).
16. N. Chetaiev, "Stability of Motion." Gostekhizdat, Moscow, 1946 (R).
17. A. Andronov, A. Witt, and S. Khaikin, "Theory of Vibrations." Fizmatgiz, Moscow, 1959 (R).
18. L. Onsager, Phys. Rev. 37, 405; 38, 2265 (1931).
19. H. Casimir, Rev. Mod. Phys. 17, 343 (1945).
20. I. Prigogine, "Introduction to Thermodynamics of Irreversible Processes." Wiley (Interscience), New York, 1962.
21. P. Glansdorff and J. Prigogine, "Thermodynamic Theory of Structure, Stability and Fluctuations." Wiley (Interscience), New York, 1971.
22. S. de Groot, "Thermodynamics of Irreversible Processes." North Holland Publ., Amsterdam and Wiley (Interscience), New York, 1952.
23. K. Denbigh, "Thermodynamics of the Steady State." Methuen, London, 1951.
24. R. Haase, "Thermodynamik der irreversiblen Processe." Dietrich Steinkopff Verlag, Darmstadt, 1963.
25. A. Katchalsky and P. Curran, "Non-equilibrium Thermodynamics in Biophysics." Harvard Univ. Press, Cambridge, Massachusetts, 1965.
26. I. Prigogine and G. Nicolis, Quart. Rev. Biophys. 4, 107 (1971).
27. L. Szillard, Z. Phys. 53, 840 (1929).
28. L. Brillonin, "Science and Information Theory." Academic Press, New York, 1962.
29. H. Quastler, "The Emergence of Biological Organization." Yale Univ. Press, New Haven, Connecticut, 1964.
30. H. Quastler, in Symp. Informat. Theory in Biol., October 1956, (H. Yockey, ed.), Pergamon, Oxford, 1957.
31. C. Raven, "Oogenesis." Pergamon, Oxford, 1961.
32. M. Apter, "Cybernetics and Development." Pergamon, Oxford, 1966.
33. M. Apter and L. Wolpert, J. Theor. Biol. 8, 244 (1965).
34. I. Schmalhausen, "Cybernetical Problems of Biology."

Nauka, Novosibirsk, 1968 (R).

35. I. Schmalhausen, "Factors of Evolution." Nauka, Moscow, 1968 (R).
36. H. Yockey, in Symp. Informat. Theory Biol., October 1956 (H. Yockey, ed.), Pergamon, Oxford, 1957.
37. L. Landau and E. Lifshitz, "Statistical Physics." 2nd ed., Addison-Wesley, Reading, Massachusetts, 1969.
38. M. Leontovitch, "Introduction to Thermodynamics." Gostekhizdat, Moscow, 1950 (R).
39. R. Kubo, "Statistical Mechanics." North-Holland Publ., Amsterdam, 1965.
40. E. Ising, Z. Phys. 31, 253 (1925).
41. K. Huang, "Statistical Mechanics." Wiley, New York, 1963.
42. T. Hill, "Statistical Mechanics." McGraw-Hill, New York, 1956.
43. M. Volkenstein, Biophys. J. 2, N 2, Part 2, 189 (1962).
44. M. Volkenstein, J. Gottlieb, and O. Ptitsyn, Phys. Solids 3, 420 (1961) (R).
45. M. Volkenstein, "Molecular Optics." Gostekhzdat, Moscow, 1951 (R).
46. M. Volkenstein, "Structure and Physical Properties of Molecules." Ed. Acad. Sci. USSR, Leningrad, 1955 (R).
47. M. Volkenstein, "Enzyme Physics." Plenum Press, New York, 1970.
48. N. Rashevsky, "Mathematical Biophysics." Univ. of Chicago Press, Chicago, Illinois, 1960.
49. C. Anfinsen, "The Molecular Basis of Evolution." Wiley, New York, 1959.
50. B. Tarusov (ed.), "Biophysics." Vysshaja Shkola, Moscow, 1968 (R).
51. P. Makarov (ed.), "Lectures on Biophysics." Ed. of Leningrad Univ., Leningrad, 1968 (R).
52. W. Beier, "Biophysik." Thieme, Leipzig, 1968.
53. W. Beier, "Biophysik." Thieme, Leipzig, 1960.
54. E. Ackerman, "Biophysical Science." Prentice-Hall, Englewood Cliffs, New Jersey, 1962.
55. Biophysical science. A study program, Rev. Mod. Phys. 31, N 1 (1959).
56. R. Setlow and E. Pollard, "Molecular Biophysics." Addison-Wesley, Reading, Massachusetts, 1962.

# Chapter 2

# Chemical Foundations of Biophysics

## 2.1 Chemistry and Biology

An organism exists as a result of chemical transformations of substances from without and the excretion of substances into the surrounding medium. Biology and biophysics are closely related to chemistry. Physical methods alone cannot reveal the structure of complex molecules. Chemical investigation, which has already disclosed the detailed structure of highly complicated bioorganic substances--steroids, porphyrin compounds, hormones, etc.--is also needed. Chemical and physicochemical methods have provided our basic knowledge of the structure of biopolymers (i.e., of proteins and nucleic acids).

Chemistry is the study of the structure of substances and their transformations during chemical reactions. The development of biochemistry has resulted in the decoding of the most important processes that occur in living organisms. Some fundamental principles of the chemistry of life have to be emphasized: the unity of the chemical mechanisms in nature; the heterogeneity of the living chemical system; the essential role of the minute individual features of molecules; the principle of chemical, molecular communication--all of which will be discussed in the following presentation. Here we limit ourselves to fundamental notions and some examples.

The enormous variety of biological species and individuals does not mean that an inexhaustible variety of chemical substances determines their existence. On the contrary, certain fundamental substances and chemical mechanisms are common to all living things. All proteins are made up of a limited number of amino acid residues, all nucleic acids of the still smaller number of nucleotides. The same atomic structures appear in the various organisms of plants and animals, not only in biopolymers but in the small molecules as well. The fundamental sources of energy, and they are chemical sources, are the same for all living matter. The most important energy source is adenosine triphosphate (ATP). The fundamental processes also belong to the same general type: respiration, which consists of the oxidation of the organic substances to form carbon dioxide and water, occurs in the overwhelming majority of organisms. The variety of organisms is due to variations in the combination of the same substances and atomic groups, variations in their relationships in space and time.

The chemical heterogeneity of a living system means more than just the presence of different molecules in the cell. Spatial heterogeneity is also necessary for life; the separation of substances in space by semipermeable walls (membranes) and the existence of concentration gradients.

In conventional chemistry small differences in molecular structure, say, the difference between the methyl and ethyl groups, do not have a pronounced influence on the properties of a complex molecule. In bioorganic chemistry, however, small differences between molecules are of great importance. Especially important are the conformational variations produced by rotations of atomic groups around single bonds. Chemists only recently began to study conformations, although conformational properties determine the biological function of biopolymers as well as of the small molecules.

Life is characterized by chemical individualization, which is not of great importance in conventional chemistry. Biological molecules and macromolecules have a definite composition and chemical structure, in contrast to synthetic polymers, which are always represented by the mixture of macromolecules of various lengths, with various "defects" in structure. The properties of a given protein are individual and definite; the properties of a synthetic polymer have the sense of averaged characteristics. The law of definite composition is not valid in the chemistry of synthetic polymers.

We now discuss some properties of the fundamental biopolymers--of proteins, nucleic acids, and carbohydrates; of their monomers; and of some low-molecular substances essential for biology (namely, bioregulators).

## 2.2 Amino Acids

The chemical structure of the amino acids, whose residues form proteins and polypeptides (protein chains with lengths of less than 100 links), is

$$
\begin{array}{c}
R \diagdown \quad \diagup NH_2 \\
C \\
H \diagup \quad \diagdown COOH
\end{array}
$$

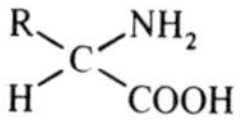

where R represents various hydrocarbon radicals, some of which may contain O, S, and N atoms. The formula shown here is not precise. A series of facts shows that in a neutral medium amino acids exist as dipolar ions (zwitterions).

$$
\begin{array}{c}
R \diagdown \quad \diagup \overset{+}{N}H_3 \\
C \\
H \diagup \quad \diagdown COO^-
\end{array}
$$

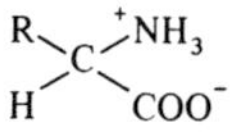

In building the protein chain, the amino acids are linked together via the peptide bond -NH-CO-, with consequent elimination of water (p. 104).

All natural proteins are formed from the 20 canonical amino acid residues listed in Table 2.1. Nineteen of them are true amino acid residues but Pro is an imino acid one, corresponding to the imino acid proline. The residues of cysteine (Cys-SH) in proteins are frequently bound in pairs by covalent disulfide cross link, forming Cys-S-S-Cys.

The 20 residues contain 20 different radicals R. In Table 2.1 the residues are classified according to their electrochemical properties. If we classify them according to chemical structure, we meet with a large variety of residues:

A. Aliphatic amino acid residues (15). Ala, Arg, Asn, Asp, Cys(Cys-SH), Gly, Gln, Glu, Ile, Leu, Lys, Met, Ser, Thr, Val, including:

   (a) Hydrocarbon residues (5): Ala, Gly, Ile, Leu, Val;

   (b) Residues containing the hydroxyl group (2): Ser, Thr;

   (c) Acid residues (containing COOH)(2): Asp, Glu;

   (d) Amides (containing $CONH_2$)(2): Asn, Gln;

   (e) Residues containing the basic amino group (2): Arg, Lys;

   (f) Residues containing sulfur (2): Cys(Cys-SH), Met.

B. Amino acid residues containing π-electron rings (4): His, Phe, Trp, Tyr.

C. Imino acid residues (1): Pro.

*TABLE 2.1*
*Canonical Amino Acid Residues*

| No. | Amino acid | Abbreviation | | Formula |
|---|---|---|---|---|
| I | Neutral residues | | | |
| 1 | Glycyl<br>Glycine | Gly | G | $-CO-CH_2-NH-$ |
| 2 | Alanyl<br>Alanine | Ala | A | $-CO-CH(NH-)-CH_3$ |
| 3 | Valyl<br>Valine | Val | V | $-CO-CH(NH-)-CH(CH_3)_2$ |
| 4 | Leucyl<br>Leucine | Leu | L | $-CO-CH(NH-)-CH_2-CH(CH_3)_2$ |
| 5 | Isoleucyl<br>Isoleucine | Ile | I | $-CO-CH(NH-)-CH(CH_2-CH_3)(CH_3)$ |
| 6 | Phenylalanyl<br>Phenylalanine | Phe | F | 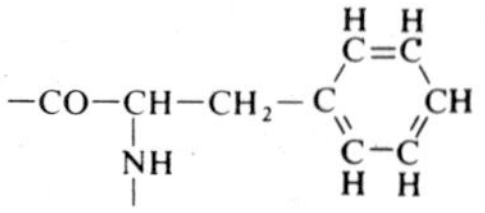 |
| 7 | Prolyl<br>Proline | Pro | P | 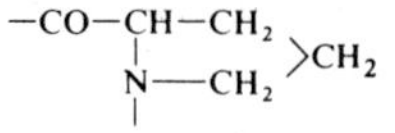 |
| 8 | Tryptophanyl<br>Tryptophan | Trp | W | 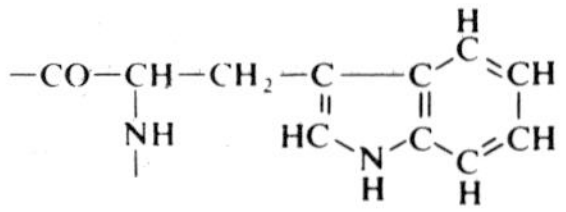 |
| 9 | Seryl<br>Serine | Ser | S | $-CO-CH(NH-)-CH_2-OH$ |
| 10 | Threonyl<br>Threonine | Thr | T | $-CO-CH(NH-)-CH(OH)(CH_3)$ |
| 11 | Methionyl<br>Methionine | Met | M | $-CO-CH(NH-)-CH_2-CH_2-S-CH_3$ |
| 12 | Asparaginyl<br>Asparagine | Asn<br>Asp-$NH_2$ | N | $-CO-CH(NH-)-CH_2-CO-NH_2$ |

*TABLE 2.1 (continued)*

| No. | Amino acid | Abbreviation | | Formula |
|---|---|---|---|---|
| 13 | Glutaminyl<br>Glutamine | Gln<br>$Glu-NH_2$ | Q | $-CO-\underset{\underset{\vert}{NH}}{CH}-CH_2-CH_2-CO-NH_2$ |
| 14 | Cystinyl<br>Cystine | Cys(-S-) | C | $-CO-\underset{\underset{\vert}{NH}}{CH}-CH_2-S-S-CH_2-\underset{\underset{\vert}{NH}}{CH}-CO-$ |
| 14a | Cysteinyl<br>Cysteine | Cys-SH | C | $-CO-\underset{\underset{\vert}{NH}}{CH}-CH_2-COO^-$ |
| II | Acid residues (as anions) | | | |
| 15 | Aspartyl<br>Aspartic acid | Asp | D | $-CO-\underset{\underset{\vert}{NH}}{CH}-CH_2-CH_2-COO^-$ |
| 16 | Glutamyl<br>Glutamic acid | Glu | E | $-CO-\underset{\underset{\vert}{NH}}{CH}-CH_2-S^-$ |
| 17 | Tyrosyl<br>Tyrosine | Tyr | Y | $-CO-\underset{\underset{\vert}{NH}}{CH}-CH_2-C\genfrac{}{}{0pt}{}{\diagup \overset{H}{C}=\overset{H}{C} \diagdown}{\diagdown \underset{H}{C}-\underset{H}{C} \diagup}C-O^-$ |
| III | Basic residues (as cations) | | | |
| 18 | Histidyl<br>Histidine | His | H | $-CO-\underset{\underset{\vert}{NH}}{CH}-CH_2-C=CH$<br>$H_2N^+ \quad N$<br>$CH$ |
| 19 | Lysyl<br>Lysine | Lys | K | $-CO-\underset{\underset{\vert}{NH}}{CH}-CH_2-CH_2-CH_2-CH_2-\overset{+}{N}H_3$ |
| 20 | Arginyl<br>Arginine | Arg | R | $-CO-\underset{\underset{\vert}{NH}}{CH}-CH_2-CH_2-CH_2-NH-\underset{\underset{}{NH_2}}{C}=\overset{+}{N}H_2$ |

Besides the 20 canonical residues, some proteins contain some of their derivatives. Of these minor residues oxyprolyl

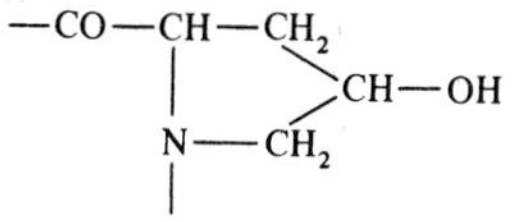


(Opro) has special importance because it is contained in large amounts in one of the universal animal proteins, collagen (p. 239).

A series of enzymes contains oxylysyl (Olys)

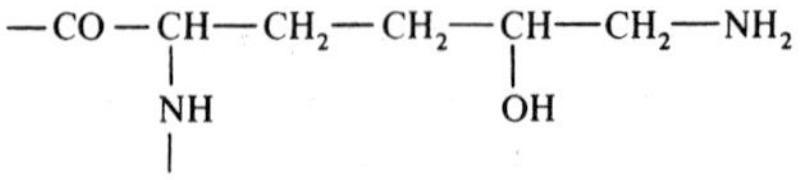
—CO—CH—CH₂—CH₂—CH—CH₂—NH₂
NH
OH

The proteins of some marine organisms contain diiodotyrosyl and dibromotyrosyl, etc. Many noncanonical amino acids (ornithine, α,γ-diaminovaleric acid, α,γ-diaminobutyric acid, etc.) are present not in proteins but in the low-molecular peptides taking part in metabolism.

All variations in the function of proteins are due to variations in the sequence of the 20 residues in the polypeptide chains. A limited number of types of atomic groups R is sufficient to ensure the molecular-biological processes.

## 2.3 The Properties of Electrolytes

Free amino acids and their basic and acidic residues in proteins are electrolytes and in water dissociate into ions. According to Brönsted's definition, an acid is a molecule from which a proton can be removed; a base, a molecule that can accept a proton. Water itself acts as an acid in the reaction

$$H_2O \rightleftarrows H^+ + OH^-$$

and as a base in the reaction

$$H_2O + H^+ \rightleftarrows H_3O^+ \text{ (oxonium ion)}$$

Water is a very weak electrolyte; its dissocation constant is very low. At 25°C in pure water,

$$[H^+][OH^-] = K[H_2O] = 10^{-14} \text{ (mole liter}^{-1})^2$$

Since $[H^+] = [OH^-]$ the concentration of hydrogen ions in water is $10^{-7}$. It is convenient to use the negative logarithm of this quantity, the pH. For water (i.e., for a neutral medium), the pH = 7. Correspondingly, for acids, pH < 7 and for bases, pH > 7.

Biological electrolytes are usually weak, that is, their degree of dissociation is small. The dissociation constant of an acid, the equilibrium constant of the reaction

$$HA \rightleftarrows H^+ + A^-$$

is equal to

$$K = \frac{a_{H^+} a_{A^-}}{a_{HA}} \tag{2.1}$$

where $a_{H^+}$, $a_{A^-}$, and $a_{HA}$ are activities of the corresponding substances in solution. In nearly ideal solutions, activities do not differ practically from molar concentrations. Hence

$$[H^+]^2 = K[HA]$$

and

$$pH = \frac{1}{2}(pK - \log [HA]) \qquad (2.2)$$

The degree of ionization of a weak acid in water is equal to

$$\alpha = \frac{[A^-]}{[A^-]+[HA]} \cong \frac{[H^+]}{[HA]} = \left(\frac{K}{[HA]}\right)^{1/2} \qquad (2.3)$$

The strength of an acid, its degree of dissociation, increases with K (i.e., decreases with an increase in pK); the strength of a base decreases with an increase in K (i.e., increases with pK). The condition $[A^-] = [H^+]$ is valid only in a neutral solution. In general

$$pH = pK \pm \log (a_A/a_{HA})$$

or in an ideal solution

$$pH = pK \pm \log[\alpha/(1-\alpha)] \qquad (2.4)$$

The plus sign corresponds to the acid, the minus sign to the base, for which

$$\alpha = \frac{[HA]}{[A^-]+[HA]}$$

This formula describes the reaction

$$A^- + H^+ \rightleftarrows HA$$

where $A^-$ is the base.

For a nonideal solution

$$pH = pK + \log(\gamma_{A^-}/\gamma_{AH}) \pm \log[\alpha/(1-\alpha)] \qquad (2.5)$$

where the $\gamma$s are the activity coefficients.

The nonideality of an electrolyte solution is due to the nonchemical interactions of molecules and ions. At a sufficiently high concentration of ions, every ion is surrounded by an ionic atmosphere of ions of opposite sign. The theory of Debye and Hückel makes possible the calculation of the thermodynamic characteristics, and hence of the activity coefficients, for nonideal solutions of electrolytes.

All of the ions under investigation form a cooperative system with strong interactions (p. 26). In contrast to the Ising lattice, the strength of the interactions here decreases slowly with the distance between particles, since the interactions are electrostatic, and follow Coulomb's law. However, the molecular field approximation can be also employed here.

Let the volume unit contain $n_i$ ions of the kind i (i = 1, 2, 3, ...) with charges $e_i = |e|z_i$, where e is the absolute value of the electron charge and $z_i$ is the number of + or − charges on the ions (the valence of the ion). Since the entire solution is electrically neutral

$$\sum_i e_i n_i = 0 \tag{2.6}$$

Every ion is under the influence of electrostatic forces produced by other ions. In the molecular field approximation the given ion is acted upon by an average field of surrounding particles. If the particles are identical, the same average field characterizes the action of the chosen ion on other particles. The field is self-consistent. The electrostatic field is a potential one. The average potential obeys the Poisson equation

$$\Delta\psi(r) = -\frac{4\pi}{\varepsilon}\rho(t) = -\frac{4\pi}{\varepsilon V}\sum_i e_i n_i \exp -\frac{e_i\psi(r)}{kT} \tag{2.7}$$

where $\varepsilon$ is the dielectric constant and $\rho$ is the numerical density of the ions, depending on the distance r from the given ion. In accordance with the Boltzmann statistics, $\rho(r)$ depends exponentially on the free energy of the ion $e_i\psi$. If $e_i\psi \ll kT$

$$\Delta\psi(r) = \kappa^2\psi(r) \tag{2.8}$$

where $\kappa$ is the Debye-Hückel parameter, the radius of the ionic atmosphere

$$\kappa^2 = \frac{4\pi}{\varepsilon VkT}\sum_i e_i^2 n_i \tag{2.9}$$

This expression for the density $\rho(r)$ in Eq. (2.7) corresponds to the average field determined by the distribution of ions. In its turn the field affects this distribution.

Using the boundary condition $\psi(\infty) = 0$, we find the solution to Eq. (2.7)

$$\psi(r) = (A/r)e^{-\kappa r} \tag{2.10}$$

For the evaluation of A let us assume that the ions are of a finite size and have a diameter a. The potential of the ion at a distance $r \le a$ is $\psi_i = e_i/\varepsilon r$ We can write the continuity condition for the electric induction at r = a as

$$-\varepsilon(\partial\psi/\partial r)\Big|_{r=a} = (\varepsilon A/r^2)e^{-\kappa r}(1 + \kappa r)\Big|_{r=a} = (e_i/a^2)$$

Hence the average potential is

$$\psi(r) = \frac{e_i}{\varepsilon r}\,\frac{e^{-\kappa(r-a)}}{1 + \kappa a} \tag{2.11}$$

We now calculate the free energy. A system containing a central ion and a surrounding ionic atmosphere is similar to a galvanic element which can be discharged infinitely slowly by a reversible and isothermic way. Let us gradually reduce the charges of the ions from $e_i$ to 0. At r = a it follows from

(2.11) that

$$\psi = \frac{e_i}{\varepsilon a(1+\kappa a)} \tag{2.12}$$

The part of the average potential depending on the charges of the ionic atmosphere and not on the charge of the ion is equal to

$$\psi_e = \frac{e_i}{\varepsilon a(1+\kappa a)} - \frac{e_i}{\varepsilon a} = - \frac{e\kappa}{\varepsilon(1+\kappa a)} \tag{2.13}$$

If the ion charges decrease from $e_i$ to $\eta e_i$ $(\eta<1)$, $\kappa$ also decreases to $\eta\kappa$ because of Eq. (2.9). Hence $\psi_e$ decreases to $\eta^2\psi_e$ and the corresponding free energy change is

$$F_e = - \sum_i \frac{n_i e_i^2 \kappa}{\varepsilon} \int_0^1 \frac{\eta^2 \, d\eta}{1+\eta\kappa a} = - \sum_i \frac{n_i e_i}{3\varepsilon} \kappa g(\kappa a) \tag{2.14}$$

where

$$g(x) = \left(\frac{3}{x^3}\right) \{\ln(1+x) - x + \frac{1}{2}x^2\} = 1 - \frac{3}{4}x + \frac{3}{5}x^2 + \cdots$$

and if $\kappa a \ll 1$

$$F_e = - \sum \frac{n_i e_i^2 \kappa}{3\varepsilon} = - \frac{1}{3} \left(\frac{\Sigma n_i e_i^2}{\varepsilon}\right)^{3/2} \left(\frac{4\pi}{kT}\right)^{1/2} = - \frac{\kappa^3 kT}{12\pi} \tag{2.15}$$

This is the free energy of interactions of ions. The change in the free energy if an ion of a given kind is removed from solution is

$$\frac{\partial F_e}{\partial n_i} = - \frac{e_i^2 \kappa}{2\varepsilon} = kT \ln \gamma_i \tag{2.16}$$

where $\gamma_i$ is the activity coefficient. We get

$$\ln \gamma_i = - \frac{1}{2} \frac{e^2 z_i^2}{\varepsilon kT} \kappa = - \frac{e^3 a_i^2 (2\pi)^{1/2}}{(\varepsilon kT)^{3/2}} I \tag{2.17}$$

$$I = \frac{1}{2} \sum_i n_i z_i^2 \tag{2.18}$$

is the ionic strength of the solution. Ionic strength and pH (pK) are the most important characteristics of an electrolyte solution.

We see that the activity coefficient of a given ion depends on the concentrations of all the other ions. The Debye-Hückel theory concerns strong electrolytes. Amino acids are weak electrolytes. However, if the ionic strengths are high, the activity coefficients of amino acids differ from unity and experiments give not the pK values but effective dissociation constants

$$pK' = pK + \log(\gamma_{A^-}/\gamma_{HA}) \tag{2.19}$$

For monovalent electrolytes the ratio $(\gamma_{A^-}/\gamma_{HA})$ does not differ much from unity and pK' is close enough to pK (the difference is 0.1-0.2). Therefore

$$pH \cong pK \pm \log\,[\alpha/(1-\alpha)] \tag{2.20}$$

If a weak acid is titrated by a strong base, we have reactions of the kind

$$HA + Na^+ + OH^- \rightarrow A^- + Na^+ + H_2O$$

Hence $[A^-] \cong [Na^+]$ and $[H^+]$ is small. We get

$$\alpha = \frac{[A^-]}{[A^-]+[HA]} \cong \frac{C_{NaOH}}{C_{HA+A^-}} \tag{2.21}$$

where α is equal to the ratio of the concentrations of the strong and weak electrolytes. The determination of α, pH, and hence pK can be made by means of titration of an acid by alkali or of a base by acid. Chemical indicators are used for approximate determinations. Much more precise results are given by the potentiometric method, which measures the relative potential of an electrode placed in a solution. This potential is directly related to the activity of the ion (say, proton) in solution. Titration changes the activity and electrode potential. A glass electrode is used to determine the pH. This electrode and the reference electrode (usually a calomel electrode) are standardized in a solution of known concentration of the substance under investigation or of known pH. The difference in potentials is measured directly in pH units. The pK is estimated from the pH value at the neutralization point according to formula (2.4). Methods of titration of biological substances are described in [1].

Amino acids are amphoteric electrolytes having two values of pK, corresponding to titration by alkali (neutralization of the $COO^-$ group) and by acid (neutralization of the $NH_3^+$ group). In Table 2.2 some values of $pK_1$, $pK_2$, and $pH_i$ are listed. The last value corresponds to the isoelectric point. If an amino acid is positively charged, it moves toward the cathode; if negatively, toward the anode. At the isoelectric point the molecule of an amphoteric electrolyte is neutral, it remains stationary and does not take part in electric conduction. We have

$$pH_i = \frac{1}{2}(pK_i + pK_2)$$

Study of the electrochemical properties of amino acids confirms their dipolar nature (p. 45). The heats of ionization of organic acids

*TABLE 2.2*
*Electrochemical Constants of Amino Acids*[a]

| Amino acid | $pK_1$ | $pK_2$ | $pK_3$ | $pH_i$ |
|---|---|---|---|---|
| Glycine | 2.35 | 9.78 | | 6.1 |
| Alanine | 2.34 | 9.87 | | 6.1 |
| Valine | 2.32 | 9.62 | | 6.0 |
| Leucine | 2.36 | 9.60 | | 6.0 |
| Serine | 2.21 | 9.15 | | 5.68 |
| Proline | 1.99 | 10.60 | | 6.30 |
| Tryptophan | 2.38 | 9.39 | | 5.89 |
| Aspartic acid | 2.09 | 3.87[b] | 9.82[c] | 3.0 |
| Glutamic acid | 2.19 | 4.28[b] | 9.66[c] | 3.2 |
| Tyrosine | 2.20 | 9.11[c] | 10.1[d] | 5.7 |
| Cysteine | 1.96 | 8.18[c] | 10.28[e] | 5.07 |
| Arginine | 2.02 | 9.04[c] | 12.48[f] | 10.8 |
| Lysine | 2.18 | 8.95[g] | 10.53[h] | 9.7 |
| Histidine | 1.77 | 6.10[i] | 9.18[c] | 7.6 |

[a] From [2]; $pK_1$ belongs to the COOH($COO^-$) group.
[b] $COO^-$.
[c] $NH_3^+$.
[d] OH.
[e] SH.
[f] Guanidine.
[g] $\alpha$-$NH_3^+$.
[h] $\varepsilon$-$NH_3^+$.
[i] Imidazole.

$$RCOOH \rightleftarrows RCOO^- + H^+$$

are small, of the order of 1 kcal/mole. On the other hand, the dissociation of the derivatives of the amonium ion

$$RN^+H_3 \rightleftarrows RNH_2 + H^+$$

is characterized by heats of the order of 12 kcal/mole. Amino acids in an acid solution have the heats of ionization of from 1.3 to 2.1 kcal/mole, in an alkaline solution of 10-13 kcal/mole. This corresponds to the reactions:

$$H_3N^+\cdot CHR\cdot COO^- + H^+ \rightleftarrows H_3N^+\cdot CHR\cdot COOH$$

$$H_3N^+\cdot CHR\cdot COO^- + OH^- \rightleftarrows H_2N\cdot CHR\cdot COO^- + H_2O$$

Other arguments in favor of the dipolar structure of amino acids are the high values of the dielectric constants of their solutions and the high melting points of solid amino acids, due to their molecules being bound in a crystal by strong electrostatic forces.

All this shows that it is necessary to work with biological substances at constant values of pH. Stabilization of the pH is achieved by means of buffer solutions. In the presence of neutral salts, the dissociation of weak acids and bases does not depend on dilution. Consider the solution of a weak acid HA and of its sodium salt NaA. For the acid

$$HA \rightleftarrows H^+ + A^-$$

The equilibrium constant is

$$K_a = \frac{[H^+][A^-]}{[HA]} \tag{2.22}$$

and since $[A^-] = [H^+]$,

$$[H^+] = (K_a[HA])^{\frac{1}{2}} \tag{2.23}$$

In the presence of NaA, which is dissociated much more completely than HA, $[A^-] \cong [NaA]$. Hence

$$K_a = \frac{[H^+][NaA]}{[HA]} \tag{2.24}$$

and

$$[H^+] = K_a \frac{[HA]}{[NaA]} \tag{2.25}$$

The concentration of $H^+$ ions depends on the ratio of concentrations of acid and salt but not on dilution. Thus, in a solution of 0.1 N acetic acid and 0.1 N sodium acetate the pH = 4.628. If we dilute the solution 10 times, the pH becomes 4.670, at a hundredfold dilution the pH = 4.73.

If we add 1 ml of 0.01 N HCl to 1 liter of water, the pH decreases from 7 to 5. Addition of the same quantity of HCl to a solution of 0.1 N $CH_3COOH$ and 0.1 N $CH_3COONa$ causes the pH to decrease from 4.628 to 4.540. Acetate, phosphate, and other buffers are used for the stabilization of pH (cf. [1,3]). [3]).

## 2.4 Amino Acid Composition of Proteins

Proteins are high-molecular compounds with exactly defined chemical structures and molecular weights. The protein molecule contains one or several polypeptide chains, built by the condensation of amino acids. Effective methods have been worked out for the extraction and purification of proteins [1,2]. Although proteins can be obtained in the crystalline form, many important proteins have only recently been obtained in a pure form (e.g., acetylcholinesterase). Methods of identification of characteristic groups and bonds are described in the biochemical literature (cf. in particular [2]).

Linked amino acids form the peptide bonds -NH-CO- in a protein chain. At one end of the chain is the COO group (C terminal) and at the other end is the $N^+H_3$ group (N terminal). Molecular weights of proteins vary over a broad range, from several tens of thousands to several millions (hemocyanins). Typical molecular weights of single polypeptide chains forming a protein molecule are of the order of 20,000 corresponding approximately to 150-180 amino acid residues (the average molecular weight of an amino acid residue is 177). It is customary to call molecules containing less than 100 residues not proteins but polypeptides. Examples are some hormones (e.g., insulin and adrenocorticotropin; cf. p. 60). Synthetic polyamino acids and their derivatives are also often called polypeptides.

Pauling and Corey discovered, by means of X-ray analysis of model low-molecular amides, $R-CO-NH_2$, that the peptide (amide) bond has the specific structure shown in Fig. 2.1 [4].

*FIG. 2.1 The peptide bond; angles in degrees, bond lengths in angstroms.*

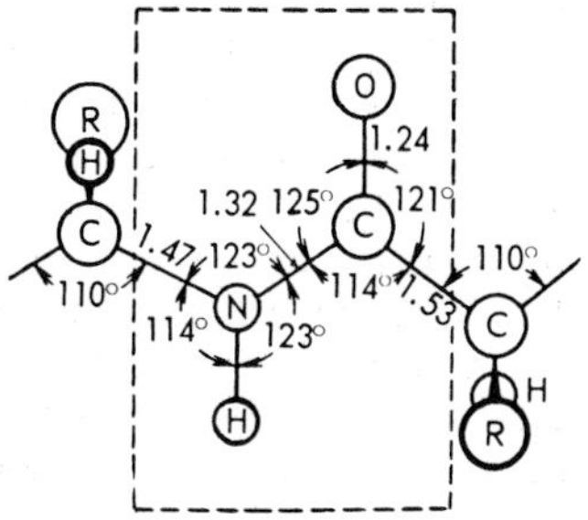


The four atoms C,O,N,H lie in one plane. The N-C bond length (1.32 Å) is much shorter than the similar bond in aliphatic amines, $R-NH_2$, where its length is 1.47 Å. These facts imply the conjugation of the N-C and C=O bonds, the overlapping of their electron clouds, and the shift of electron density from nitrogen to oxygen. The N-C bond has to be considered not as a single bond but as partly double. The coplanarity of the peptide bond determines essential features of a polypeptide chain as a whole (cf. Section 4.2).

The chemical, and therefore biological, behavior of a protein is determined by its amino acid composition and by the sequence of residues in the chain, that is, by its primary structure. Modern methods enable us to determine the composition of a protein without great difficulty. The protein is split into amino acids hydrolytically in a reaction that is the reverse of polycondensation

$$-CHR_1-NH-CO-CHR_2- + H_2O \rightarrow -CHR_1-NH_2 + HOOC-CHR_2-$$

Hydrolytic splitting occurs under the action of acids and

alkalis and proteolytic enzymes, proteases catalyzing the breakage of peptide bonds. The hydrolyzate is a mixture of amino acids. Qualitative determination of its composition is performed with the help of characteristic chemical reactions which in many cases give colored products (cf. [2]). Quantitative determination of the hydrolyzate composition is achieved by chromatographic separation into pure amino-acid fractions.

Chromatography is a method of separating the components of a gaseous mixture or a solution based on their passage through a porous substrate. This method was discovered in 1903 by Tswett [5]. Various modifications of the methods used for analysis of proteins and amino acids are described in [1].

Chromatography requires a column containing a porous substrate (e.g., the powdered $Al_2O_3$), at the top of which is put a quantity of the solution under investigation (it will be totally adsorbed on the column). The pure solvent is then passed through the column at a particular rate of flow. The solvent washes out, or elutes the adsorbed substances and they are displaced downward. Displacement of the substance that is less strongly adsorbed occurs faster than that of the substance that is more strongly adsorbed on the column. Every substance is adsorbed at the upper part of the column and desorbed at its lower part. Therefore, maximum adsorption is reached with the formation of zones of adsorbed substances, which move along the column at various rates because the sorption depends on their and the sorbent's chemical nature. If the differences in sorption are great, then after some time the zones containing the different components of the mixture are completely distinct and it is possible to extract these components. The essence of chromatography is the repetition of the sorption-desorption process, culminating in the separation of the components of the mixture.

Chromatography is clearly based on kinetic, dynamic phenomena. To understand it we have to consider the phenomena of adsorption and (or) ion exchange in heterogeneous systems and the distribution of a substance in two liquid phases. In the simplese case molecular adsorption is described by the Langmuir isotherm

$$m = kc/(1+ac) \tag{2.26}$$

where m is the quantity of adsorbed substance, c is its concentration in solution, and a and k are constants. In more complicated cases the empirical equation of Freundlich can be useful

$$m = bc^{\varepsilon} \tag{2.27}$$

where b and $\varepsilon$ are constants.

The distribution of a substance between two liquid phases

follows from the condition of equality of its chemical potentials in equilibrium

$$\mu_i^{(1)} = \mu_i^{(2)}$$

or

$$\mu_i^{(1)}(0) + RT \ln a_i^{(1)} = \mu_i^{(2)}(0) + RT \ln a_i^{(2)} \quad (2.28)$$

Hence the ratio of the activities a of the components in both phases is constant.

$$a_i^{(1)}/a_i^{(2)} = \text{const} \quad (2.29)$$

which means that the activity (concentration) a of the substance in one phase depends linearly on the activity (concentration) of the same substance in another phase.

Let us consider equilibrium chromatography, assuming that the equilibrium between the substance on the solid sorbent and in the solution is established instantly and longitudinal diffusion can be neglected. The equation of the material balance for the displacement of one component along the column has the form

$$-\partial c/\partial x = \alpha(\partial c/\partial V) + \rho(\partial m/\partial V) \quad (2.30)$$

where m is the amount of substance adsorbed by 1 g of sorbent, $\rho$ the number of grams of sorbent in 1 cm of column length, c the concentration of the substance, V the volume (in milliliters) of the passed solution, $\alpha$ the fraction of the pores in the column, and x the distance from the upper level of the column. Equation (2.30) can be rewritten in the form

$$-w\,(\partial c/\partial x) = \alpha(\partial c/\partial t) + \rho(\partial m/\partial t) \quad (2.31)$$

where w is the rate of solvent flow. To solve such equations, we have to evaluate the dependence of m and c on w, t, and x. This solution gives the laws of deformation of the boundaries of chromatographic zones.

If the solution is a mixture of two substances, they are adsorbed at different levels on the column. At dynamic conditions of solvent flow, the first, or more mobile, component emerges from the column first, and later both components emerge in the form of the original solution. The output curve (the dependence of the final concentration on the volume of eluent) looks like that shown in Fig. 2.2. This is a frontal analysis. The concentration of the first component at the first stage is denoted by $C_{11}$, the concentration of the second component at the first and second stages by $C_{12}$ and $C_{22}$, respectively. Volumes $V_1$ and $V_2$ are called the volumes of delay. If the adsorption isotherms are those of Langmuir where $m_1$ and $m_2$ are the amounts of the first and second components in the adsorbed state per gram of sorbent,

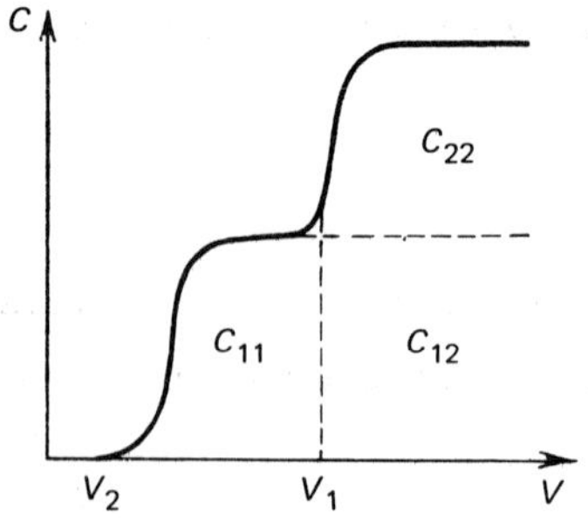


*FIG. 2.2 Output curve, concentration vs. eluent volume.*

$$m = \frac{k_2C_{22}}{1+\ell_1C_{12}+\ell_2C_{22}}, \qquad m_2 = \frac{k_2C_{22}}{1+\ell_1C_{12}+\ell_2C_{22}} \tag{2.32}$$

where $m_1$ and $m_2$ are the amounts of the first and second components in the adsorbed state per gram of sorbent, and k and $\ell$ are constants. Thus,

$$m_1 = V_2C_{12}-(V_2-V_1)C_{11}, \qquad m_2 = V_2C_{22} \tag{2.33}$$

From (2.32) and (2.33) it follows that

$$C_{12} = C_{11}\,\frac{1-V_1/V_2}{1-k_1/k_2} \tag{2.34}$$

The values $V_1$ and $V_2$ can be estimated from chromatographic experiments with single components. Thus it is possible to find $C_{12}$ if $C_{11}$ is known.

Actually the attainment of equilibrium is not instantaneous. Therefore, the zones are somewhat diffuse. The theory taking into account kinetic factors explains this phenomenon. The distribution of concentration in a zone is Gaussian. The theory and methods of chromatography have been described in many monographs (cf. e.g., [1,6-10]).

The two-dimensional separation of amino acids on paper, the method of "fingerprints," is visual. Dense filter paper serves as the sorbent. A drop of hydrolyzate is put on the edge of a rectangular sheet of paper, and eluted by a special solvent, starting at one edge and flowing toward the other. After some definite time the paper is dried, and the sheet is turned 90° and similarly eluted by another solvent. The sheet is dried, sprayed with solutions of substances that give colored reaction products with amino acids, and heated. Colored spots appear on the sheet corresponding to different amino acids. Their positions make it possible to determine the composition of the protein. By cutting out the spotted areas and extracting the amino acids with solvents, we can determine the weight content of the amino acid.

A particularly effective method of investigating of the amino acid composition of proteins is ion-exchange chromatography, particularly on the cation exchanger Dowex-50, which contains sulfo groups that bind the $NH_3^+$ group of amino acids. Elution is performed at different pHs, buffer concentrations, and temperatures. Amino acids are washed out in a definite sequence. Small portions of the eluate are collected automatically in test tubes. Every fraction is mixed with a reagent (ninhydrin) that causes coloration and is heated according to standard procedures, after which the intensity of coloration is measured. Calibration is performed for every amino acid, taking into account the dependence of light absorption on concentration. The absorption of single portions of eluate in test tubes is plotted on a graph. The area of each peak corresponds to the quantity of a specific amino acid in the hydrolyzate (Fig. 2.3)[11]. The whole procedure is done automatically, so that determination of the amino acid composition of a protein is now a routine process.

## 2.5 The Primary Structure of Proteins

The sequence of amino acid residues in a polypeptide chain is its primary structure. Primary structure is determined by means of the partial hydrolysis of proteins with the help of specific proteases which catalyze the splitting of the peptide bond only between definite residues. Thus, trypsin attacks only those peptide bonds which are formed by the CO groups of residues of basic amino acids--Arg or Lys. This results in a mixture of short polypeptide chains or oligomers. These short chains, which are called peptides, are studied by chemical and physicochemical methods (chromatography, mass spectrography). By reacting it with another enzyme, a protein can be cut at other bonds and another mixture of peptides can be formed. N= and C= terminal residues of the protein are determined by means of their chemical modification before hydrolysis. Knowing the structure of peptides obtained by specific splitting of a protein by different enzymes, we can establish its primary structure. Assume that the chain has the

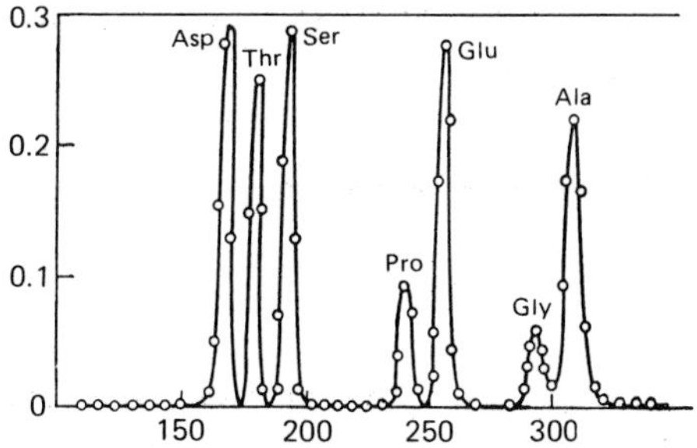


*FIG. 2.3 Part of the curve of the amino acid analysis of hydrolyzate of ribonuclease. Ordinate, intensity of color with ninhydrin; abscissa, eluate volume.*

structure

A→B→C→D→A→E→F→C→A→G→H→F→G→B→C→A

where the letters denote residues and the arrows the peptide bonds -NH-CO-. We subject the protein to enzyme splitting of the bond of the CO group with residue A. We obtain the peptides

A, B→C→D→A, E→F→C→A, G→H→F→G→B→C→A

Reacting it with an enzyme that splits the bond with residues D and F, we get the peptides

A→B→C→D→, A→E→F, C→A→G→H→F, G→B→C→A

The peptide sequence is not known but can be determined because peptides can be combined in only one way [2]

A B→C→D→A E→F→C→A G→H→F→G→B→C→A

A→B→C→D A→E→F C→A→G→H→F G→B→C→A

This is the method Sanger used to determine the structure of insulin for the first time [12]. Since then the primary structures of many protein chains have been established. The data on all such determinations made before 1968 are presented in [13].

Every protein chain has a definite primary structure and is in this sense "a text written in a 20-letter alphabet." In Figs. 2.4-2.6 are shown the primary structures of some peptidic hormones, corticotropin (Fig. 2.4a) and insulin

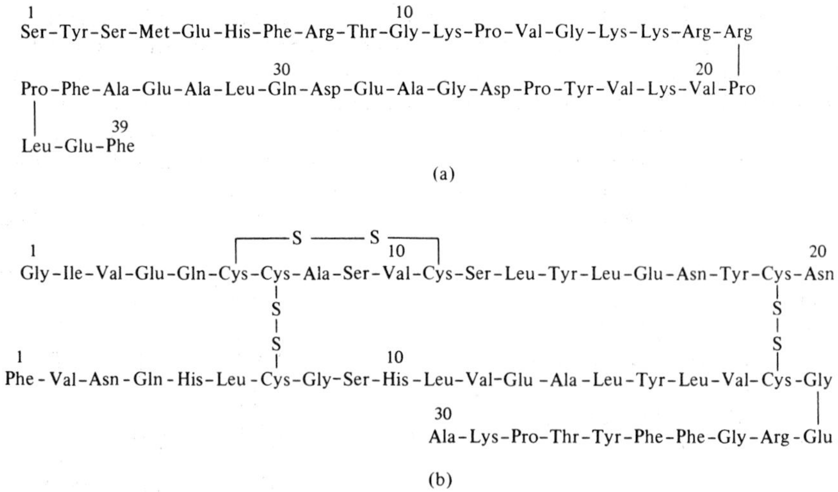


*FIG. 2.4 The primary structure of (a) β-corticotropin of pig and (b) insulin of ox.*

(Fig. 2.4b); and proteins, ribonuclease (Fig. 2.5) and myoglobin (Fig. 2.6). The protein text contains definite information, the amount of which we can estimate. At every protein site there can be one of 20 possible residues. The quantity of information in a chain containing N residues is (p. 19)

$$I = N \log_2 20 = 4.32 \text{ N bits}$$

In fact, I is smaller because the frequencies of appearance of different residues are not equal. At the same time, the analysis of sequences in proteins and their fragments has shown that no correlation exists between neighboring residues [14]. The question of correlation at great distances cannot be considered solved.

That primary structure of a protein determines its function can be demonstrated by two groups of facts: first, by the difference between and similarity of the structures of proteins of the same type belonging to different species; and second, by the pathological changes in protein functions that occur if some residues are substituted (mutations).

In the works of Margoliash *et al.* (cf. [15,16]) the cytochrome *c*, hemoglobin, and other proteins of vertebrates are

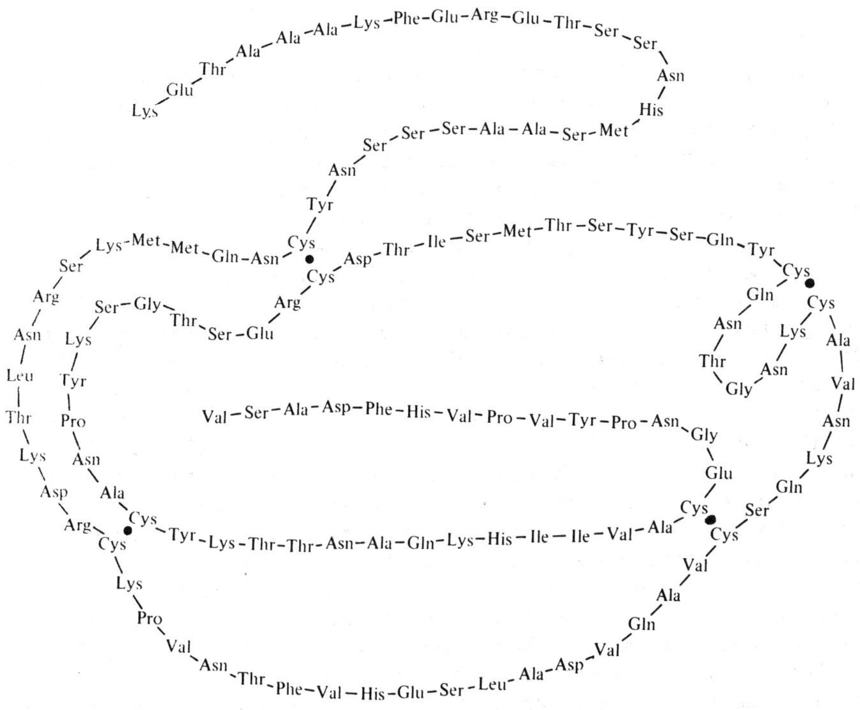


*FIG. 2.5 The primary structure of ribonuclease of ox.*

1 10
Val-Leu-Ser-Glu-Gly-Glu-Try-Gln-Leu-Val-Leu-His-Val-Try-Ala

30 20
Ile-Leu-Ile-Asp-Gln-Gly-His-Gly-Ala-Val-Asp-Ala-Glu-Val-Lys

40
Arg-Leu-Phe-Lys-Ser-His-Pro-Glu-Thr-Leu-Glu-Lys-Phe-Asp-Arg

60 50
Asp-Glu-Ser-Ala-Lys-Met-Glu-Ala-Glu-Thr-Lys-Leu-His-Lys-Phe

70
Leu-Lys-Lys-His-Gly-Val-Thr-Val-Leu-Thr-Ala-Leu-Gly-Ala-Ile

90 80
Ala-Leu-Pro-Lys-Leu-Glu-Ala-Glu-His-His-Gly-Lys-Lys-Lys-Leu

100
Gln-Ser-His-Ala-Thr-Lys-His-Lys-Ile-Pro-Ile-Lys-Tyr-Lys-Glu

120 110
Pro-His-Arg-Ser-His-Leu-Val-His-Ile-Ile-Ala-Glu-Ser-Ile-Phe

130
Gly-Asn-Phe-Gly-Ala-Asp-Ala-Gln-Gly-Ala-Met-Asn-Lys-Ala-Leu

150 140
Gly-Leu-Glu-Lys-Tyr-Lys-Ala-Ala-Ile-Asp-Lys-Arg-Phe-Leu-Glu

153
Tyr-Gln-Gly

*FIG. 2.6 The primary structure of myoglobin.*

compared. Table 2.3 contains the data for cytochrome *c*. The function of cytochrome *c*--its participation in the oxidative phosphorylation--is the same in various organisms. Variations in its primary structure characterize the animal species. The comparative study of primary structures makes it possible to follow biological evolution at the molecular level.

Mutations are expressed by a change in the primary structure of a protein (by the substitution of amino acid residues), that is, by distortion of the protein text. A "misprint" in the protein text always has serious biological implications. These phenomena have been studied in detail in the case of hemoglobin. A series of hereditary blood diseases (i.e., anemias) is known. In the case of so-called sickle-cell anemia, which is widespread in some regions of Africa, Southeast Asia, and the Mediterranean, erythrocytes have the form of a sickle. Sickle-cell hemoglobin (S-hemoglobin, different from the normal A-hemoglobin) forms crystal-like structures; the erythrocytes become stuck together and hemolyzed. Serious disruptions in the blood circulation ensue, with the result that the disease kills people, frequently at an early age. Mediterranean anemia (T-hemoglobin) is characterized by the disintegration of erythrocytes, the compensatory overproduction of marrow tissue (resulting in skeletal deformations), and swelling of the liver and spleen. Other anemias are also very dangerous. These diseases are inherited in a recessive way

TABLE 2.3[a]

*Primary Structure of Cytochrome c*

```
                1              5                   10
Horse   Acetyl Gly-Asp-Val-Glu-Lys-Gly-Lys-Lys-Ile-Phe-
                               15                  20
               Val-Fln-Lys-Cys-Ala-Glu-Cys-His-Thr-Val-
                           |___________|
                               Heme
                               25                  30
               Gln-Lys-Gly-Gly-Lys-His-Lys-Thr-Gly-Pro
                               35                  40
               Asn-Leu-His-Gly-Leu-Phe-Gly-Arg-Lys-Thr-
                               45                  50
               Gly-Gln-Ala-Pro-Gly-Phe-Thr-Tyr-Thr-Asp-
                               55                  60
               Ala-Asn-Lys-Asn-Lys-Gly-Ile-Thr-Trp-Lys-
                               65                  70
               Glu-Glu-Thr-Leu-Met-Glu-Try-Leu-Glu-Asn-
                               75                  80
               Pro-Lys-Lys-Tyr-Ile-Pro-Glu-Thr-Lys-Met-
                               85                  90
               Ile-Phe-Ala-Gly-Ile-Lys-Lys-Lys-Thr-Glu-
                               95                  100
               Arg-Glu-Asp-Leu-Ile-Ala-Tyr-Leu-Lys-Lys-
                           104
               Ala-Thr-Asn-Glu-COOH
```

Differences from horse cytochrome *c*

| | |
|---|---|
| Pig and cow (3 differences) | 47Ser, 60Gly, 89Gly |
| Kangaroo (6 differences) | 33Asn, 35Ile, 58Ile, 62Asp, 89Gly |
| Man (11 differences) | 11Ile, 12Met, 15Ser, 46Tyr, 47Ser, 58Ile, 60Gly, 62Asp, 83Val, 89Gly, 92Ala |
| Hen (11 differences) | 3Ile, 15Ser, 44Glu, 47Ser, 60Gly 62Asp, 89Ser, 92Val, 100Asp, 103Ser, 104Lys |
| Tuna (19 differences) | 4Ala, 9Thr, 22Asn, 28Val, 33Trp, 44Glu, 47Ser, 54Ser, 58Val, 60Asn, 61Asn, 62Asp, 89Gly, 92Gln, 95Val, 100Ser, 103Ser, 104 fails. |
| Butterfly (24 differences) | The sequence -4 to -1 is added: Gly-Val-Pro-Ala, 2Asn, 3Ala, 5Asn, 13Arg, 22Ala, 28Val, 54Ala, 60Gly, 61Asp, 62Asp, 65Phe, 82Val, 85Leu, 88Ala, 89Asn, 92Ala, 100Glu, 101Ser, 103 fails, 104Lys-COOH |

TABLE 2.3 (continued)

| | |
|---|---|
| Yeast (46 differences) | The sequence -5 to -1 is added: Thr-Glu-Phe-Lys-Ala, 2Ser, 3Ala, 4Lys, 7Ala, 8Thr, 9Leu, 11Lys, 12Thr, 13Arg, 15Glu, 16Leu, 25Pro, 28Val, 35Ile, 39His, 40Ser, 44Gln, 46Tyr, 47Ser, 53Ile, 54Lys, 56Asn, 57Val, 58Leu, 60Asp, 62Asn, 63Asn, 64Met, 65Ser, 69Thr, 81Ala, 83Gly, 85Leu, 88Glu, 89Lys, 90Asn, 92Asn, 96Thr, 102 fails, 103Cys-SH, 104Glu-COOH |

[a] From [15].

according to Mendel's law. In other words, the anemia is strongly exhibited by homozygous and not by the heterozygous individual. The maintenance of a high level of SA-heterozygotes in the geographic regions involved is associated with the spread of malaria there. Malaria is one of the chief causes of death in these areas. SA-heterozygotes have greater resistance to malaria, because the plasmodium of the malaria mosquito does not reproduce in sickle-cell blood as well as in normal blood.

Pauling and his co-workers [17] discovered that the motility of S- and A-hemoglobins in electrophoresis is different. This is explained by the difference in amino acid composition and therefore by the number of charged groups. Pauling defined the hemoglobin diseases as molecular ones. As was shown by Ingraham [18], the distinction between anomalous and normal hemoglobin is determined by substitution of only one residue in the protein chain. The sense of the text is drastically altered if only one letter is changed. Table 2.4 lists some of the pathogenic substitutions in human hemoglobin [13, 19].

Chemists have devised a method for synthesizing the peptide bond (cf. [20]). These procedures have nothing in common with the way protein is synthesized in a living cell (Chapter 9) and are used for the synthesis of polyamino acids, homopolymers that are similar to proteins in some respects. However, if the primary structure of a protein is known, its synthesis *in vitro* is also possible. The polypeptide hormones corticotropin and insulin have already been synthesized. Merrifield automated the method and for the first time obtained artificially a real protein having an enzymatic function--ribonuclease [21].

*TABLE 2.4*
*Mutational Substitutions in Human Hemoglobin*[a]

| α Chain | | | | β Chain | | | |
|---|---|---|---|---|---|---|---|
| | Residue | | | | Residue | | |
| Designation | No. | Substituted | Substituting | Designation | No. | Substituted | Substituting |
| J Toronto | 5 | Ala | Asp | S | 6 | Glu | Val |
| J Texas | 16 | Lys | Glu | C | 6 | Glu | Lys |
| G Audhali | 23 | Glu | Val | G San Jose | 7 | Glu | Gly |
| G Honolulu | 30 | Glu | Gln | E | 26 | Glu | Lys |
| Norfolk | 57 | Gly | Asp | K Ibadan | 46 | Gly | Glu |
| M Boston | 58 | His | Tyr | J Bangkok | 56 | Gly | Asp |
| G Philadelphia | 68 | Asn | Lys | M Saskatoon | 63 | His | Tyr |
| M Shibata | 87 | His | Tyr | Zürich | 63 | His | Arg |
| J Capetown | 92 | Arg | Glu | M Milwaukee | 67 | Val | Glu |
| Cheepsake | 92 | Arg | Leu | New York | 113 | Val | Glu |
| J Tongarekee | 115 | Ala | Asp | D Punjab | 121 | Glu | Gln |
| O Indonesia | 116 | Glu | Lys | O Arabia | 121 | Glu | Lys |

[a]More than 100 mutant human hemoglobins are known.

## 2.6 Asymmetry of Biological Molecules

Molecules lacking planes and a center of symmetry are dissymmetrical or chiral. The word chirality (from the ancient Greek χειρ, the hand; cf. chirurgy, chiromancy) means the distinction between the left and right side of such a molecule. Consequently, chiral substances exist in two forms, right and left. These two configurations cannot be matched by rotation of the molecule as a whole in space or by rotation of some molecular group around a single bond. They are related as the right and left hands are. In Fig. 2.7 are shown two configurations, two mirror images, of secondary butanol.

We meet chirality most frequently imposed by the so-called asymmetric carbon atom. In saturated (aliphatic) organic compounds, the four valences of the carbon atom are oriented tetrahedrally. If two of them are bound to similar groups, as in the $CR^1_2R^2R^3$ molecule, then the $CR^2R^3$ plane constitutes a plane of symmetry in the molecule and there is no chirality. The asymmetric carbon atom is bound to four different groups ($CR^1R^2R^3R^4$), and such a molecule lacks a plane and center of symmetry. The molecule of secondary butanol belongs to this class.

Chirality is clearly a property of the overwhelming majority of complex chemical compounds. A substance obtained by chemical synthesis *in vitro* occurs as a racemic mixture of two mirror images. According to the second law of thermodynamics, the amount of right and left enantiomers in a mixture is equal, since this corresponds to a maximum mixing entropy. If amino acids are obtained by synthesis, all of them except glycine ($CH_2NH_2COOH$, which does not contain an asymmetric carbon atom) occur in the form of racemic mixtures of right and left enantiomers.

In nature, however, definite configurations of all important substances are fixed, starting with amino acids. Amino acid residues in proteins always belong to the "left" or L, series. Their absolute configuration is represented by Fig. 2.8.

Asymmetric molecules, unlike their racemic mixtures, possess optical activity and circular dichroism. They rotate the plane of polarization of light and they absorb right and left circularly polarized light differently (cf. Chapter 5). The optical activity of the mirror images differs not in magnitude but only in sign of rotation; the same is true for circular dichroism. We designate L-amino acids L not because they rotate the plane of polarization of light to the left. The L series of organic compounds comes from levorotatory glyceraldehyde $HOC \cdot C^{*}H(OH) \cdot CH_2OH$. All L compounds can be obtained in principle from this substance via the substitution of corresponding atoms and groups linked by the asymmetric carbon atom $C^{*}$, without a change in molecular configuration. In this way both right- and left-rotating derivatives can be obtained.

Asymmetry is characteristic of proteins, carbohydrates, nucleic acids, and low-molecular compounds functioning in the cell. Consequently, in the processes of metabolism which proceed without racemization (i.e., without mutual transformation of the mirror images), the cell can assimilate only those enantiomers which correspond to the structure of its biological molecules. An organism assimilates L- but not D-amino acids. Finding itself in an "antiworld," where plants and animals contained molecules of opposite configuration, an organism would perish from starvation despite the abundance of food [22].

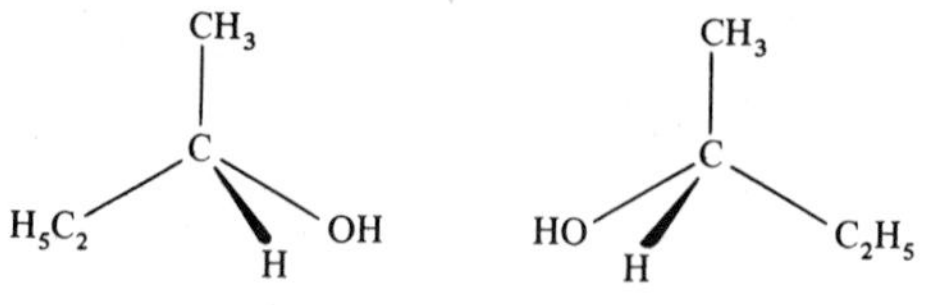


*FIG. 2.7 Mirror images of secondary butanol.*

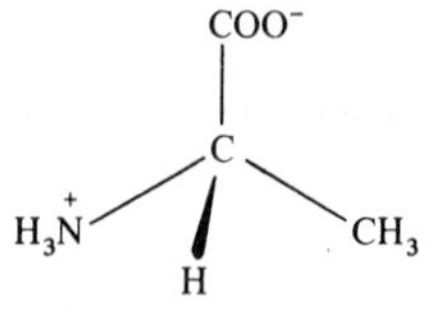


*FIG. 2.8 Absolute configuration of L-alanine.*

In Gause's book [23] many examples are quoted attesting to the biological distinctions between right and left configurations of colecules. Substances are known that are poisonous in the form of one enantiomer and harmless in its mirror form. D-aspartic acid is sweet, whereas L-aspartic acid is tasteless. Pasteur showed that bacteria feed mainly on one enantiomer of a given substance, etc.

Pure enantiomers are obtained *in vitro* from their racemic mixtures with the help of asymmetric substances of biological origin (usually alkaloids). Reacting the racemic mixture D,L with the substance L' we get

$$D,L + L' \rightarrow DL' + LL'$$

The substances DL' and LL' are not mirror images (as DL' and D'L would be). All the physiocochemical properties of enantiomers except optical activity are identical. These properties coincide for D and L but differ for DL' and LL'. Hence, the substances DL' and LL' can be separated, for instance, by means of crystallization.

For the separation of enantiomers an asymmetric influence is necessary: the substance or the being must be able to recognize the difference between the right and left forms. Mirror images were discovered in 1848 by Pasteur when, in the course of studying tartaric acid, he discovered that there are right and left forms of its crystals. Sorting the crystals, Pasteur obtained pure enantiomers of tartaric acid. Obviously Pasteur played the role of an asymmetric effector--being built asymmetrically, the human being knows the difference between right and left.

Not only biological molecules but whole organisms are asymmetric. Figure 2.9 shows a view of colonies of *Bacillus mycoides*, which usually form left-twisted structures. Right strains are very rare. The same figure shows left and right shells of the mollusk *Fructicola lantzi* from Middle Asia [24]. One of these forms is much more abundant than the other.

Chirality occurs also in crystals built up of symmetric molecules. Thus quartz ($SiO_2$) can be crystallized in right

*FIG. 2.9 Colonies of Bacillus mycoides and shells of the mollusk Fructicola lantzi.*

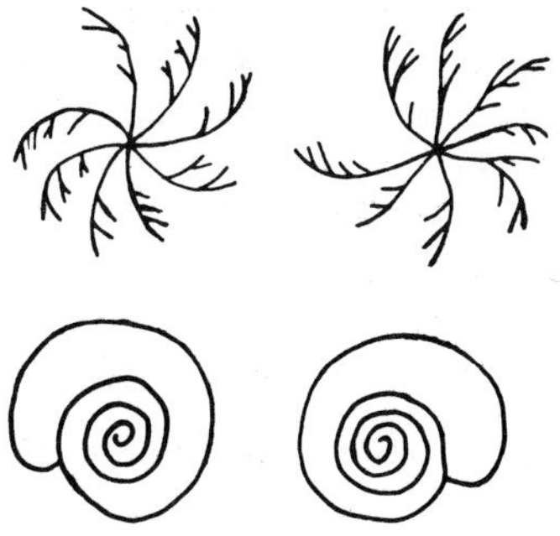

and left forms. Consequently, these crystals are optically active. Asymmetry is lost in melted crystals; melted vitreous quartz is not optically active. Every quartz deposit contains on the average equal amounts of right and left crystals. Pure enantiomers can be obtained by means of crystallization. By priming the racemic mixture with a right or a left crystal, we get preferential crystallization of that enantiomer.

The asymmetry of biological molecules is of great interest for biophysics in three respects. First, asymmetry determines the specificity of biochemical reactions, particularly the interactions of enzymes with different metabolites. Asymmetry has to be taken into account to understand these processes (cf. Chapter 6).

Second, the presence of asymmetry makes possible the investigation of the structure of biologically functional substances and their transformation by means of spectropolarimetry (Chapter 5). Physicists and chemists are grateful that nature has made its molecules optically active.

Finally, the asymmetry of biological systems down to the molecular level of structure is of great importance for biology in the broadest sense. The origin of asymmetry and its fixation in the process of biological evolution is mysterious in many respects. If asymmetry arose as the result of fluctuations, then its fixation in evolution has to be treated according to the general pattern of the "antientropicity" of life processes. Asymmetry means definite order; the separation of an enantiomer gives an amount of information equal to 1 bit per molecule.

## 2.7 Nucleic Acids

The second most important type of biopolymer is the nucleic acids. They serve as necessary participants in protein synthesis (Chapters 8 and 9). Nucleic acids were discovered much later than proteins. Miescher extracted them for the first time from salmon sperm in 1872.

The basic chain of a nucleic acid contains alternating links of phosphoric acid and sugar: ribose in ribonucleic acids (RNA's), deoxyribose in deoxyribonucleic acid (DNA). In this sense the chains of RNA and DNA lack any primary structure; they are a kind of consistent ornament, but not texts. However, every sugar is linked to a nitrogen base, and these bases are not repeated. The general pattern of the chain is

```
nitrogen base          nitrogen base
      |                      |
 — sugar — phosphate — sugar — phosphate —
```

Sugars are asymmetric carbohydrates belonging to the D series.

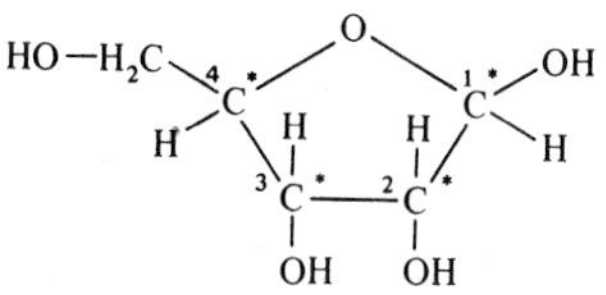

D-ribose

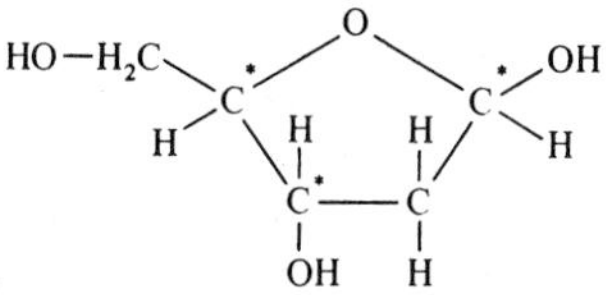

D-2-deoxyribose

Nitrogen bases are derivatives of pyrimidine and purine the heterocyclic compounds, containing nitrogen. Pyrimidine and purine possess the properties of a base, since their nitrogen atoms can bind protons, thereby obtaining positive charges.

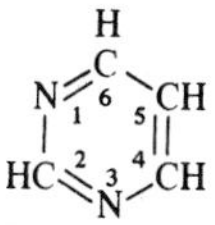

pyrimidine

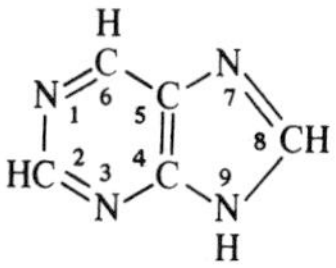

purine

Just as proteins containing 20 amino acid residues, DNA and RNA contain 4 nitrogen bases. However, this rule is not so rigorous; besides the canonical bases, DNA and RNA contain--in much smaller amounts--the so-called minor bases. The formulas for the canonical bases of DNA are:

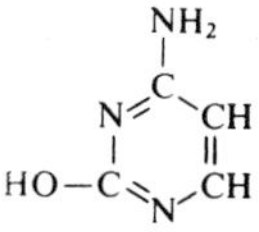

cytosine (C)

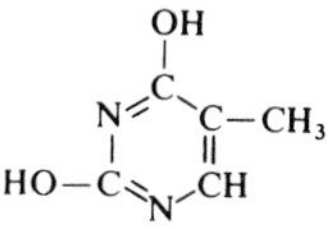

thymine (T)

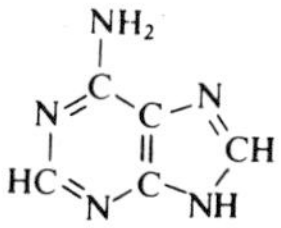

adenine (A)

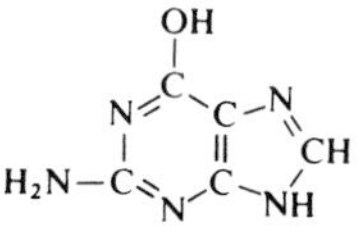

guanine (G)

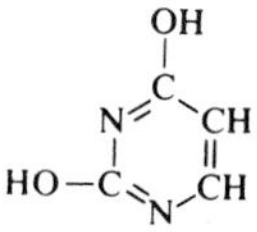

uracil (U)

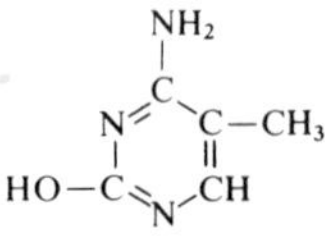 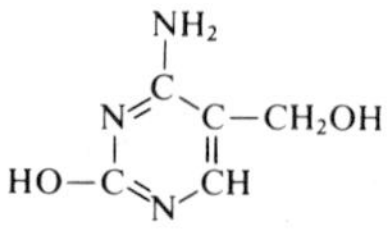 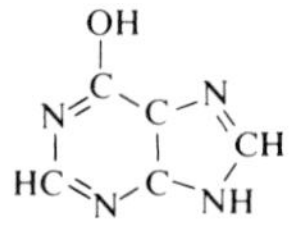 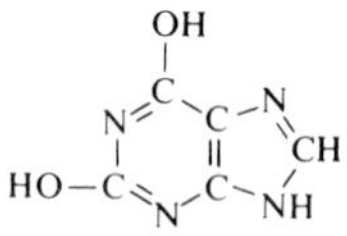

5-methylcytosine (t-MC) | 5-oxymethylcytosine (5-OMC) | hypoxanthine (I) | xanthine (X)

The letter I denotes hypoxanthine because the corresponding nucleoside (discussed later) is called inosine.

All these nitrogen bases can exist in several tautomeric forms produced by the transfer of hydrogen atoms from the hydroxyl group OH and the imino group NH to carbon and nitrogen atoms. For instance, the following tautomers of T and U are possible:

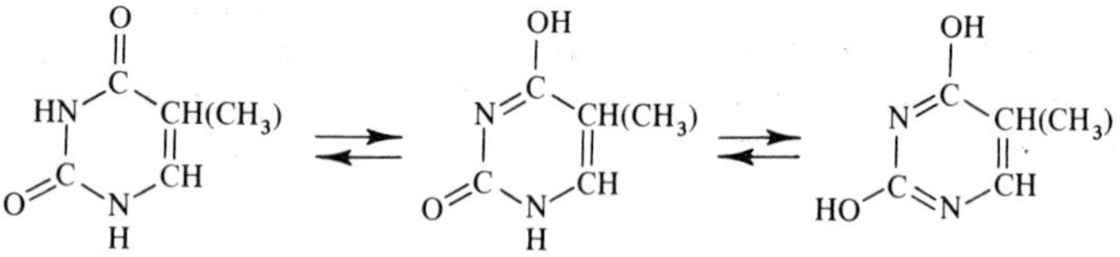

This is the keto-enol tautometry (keto group C=O, en is the designation of the double bond, ol that of the hydroxyl group OH). The equilibrium is usually shifted toward the keto form.

Tautomers are in a state of dynamic equilibrium depending on temperature. The amount of a tautomer in an equilibrium mixture is determined by its free energy in accordance with the Boltzmann distribution. The tautometry of the nitrogen bases may have biological implications (cf. p. 602).

Nucleosides are compounds of the nitrogen bases with ribose (ribonucleosides) and deoxyribose (deoxyribonucleosides). The 3-nitrogen of pyrimidine and 9-nitrogen of purine are linked with 1-carbon of sugar. The nucleosides analogous to G, T,

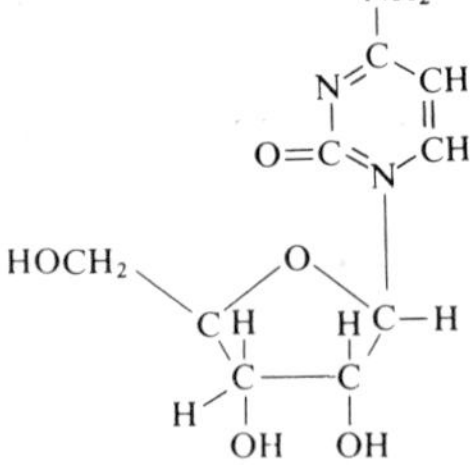

cytidine

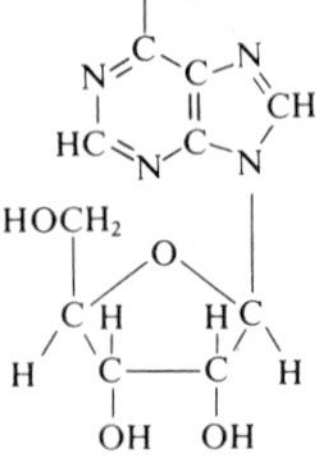

adenosine

and U are called, respectively, guanosine, thymidine, and uridine. Phosphorous esters of nucleosides are called nucleotides--ribonucleotides and deoxyribonucleotides. Evidently the esterification of ribose can occur in the positions of three hydroxyls--the 5', 3', and 2' positions. 2'-phosphorylation does not occur in nucleic acids; it is impossible in the case of deoxyribose. We have

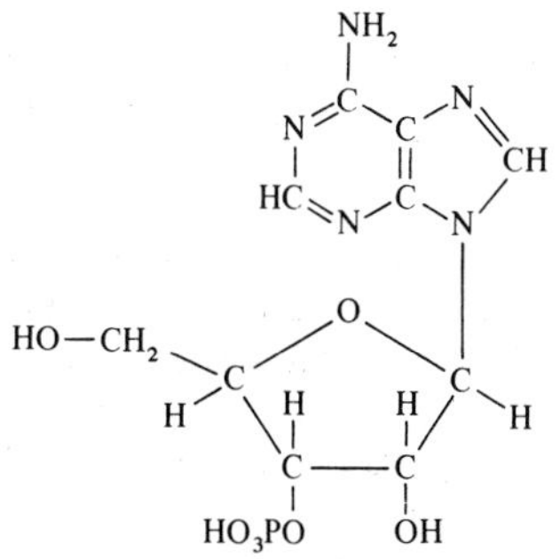

adenosine 3'-phosphate

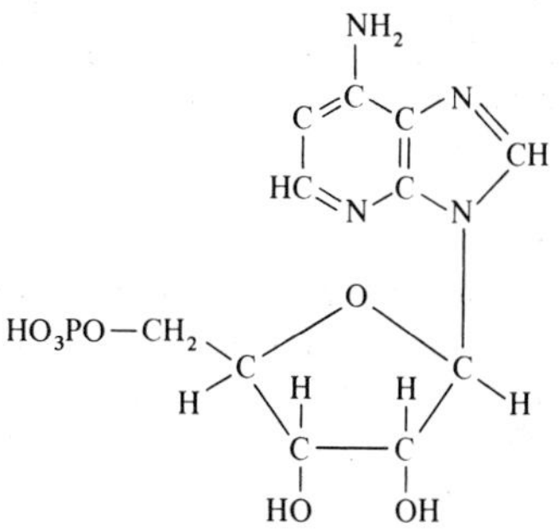

adenosine 5'-phosphate
(adenosine monophosphate, AMP)

Nucleoside 5'-phosphates can be phosphorylated further, forming di- and triphosphates.

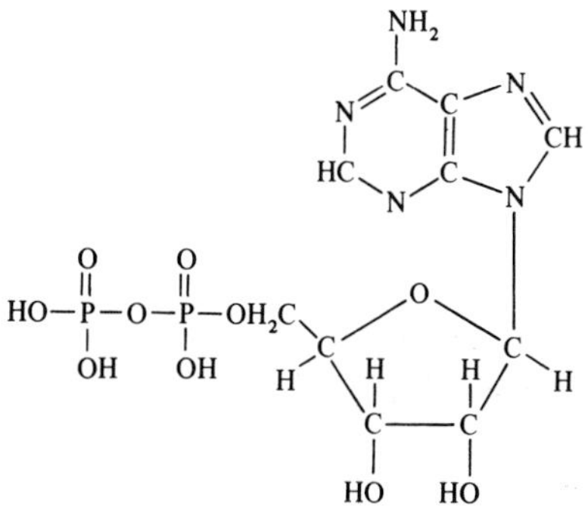

adenosine diphosphate (ADP)

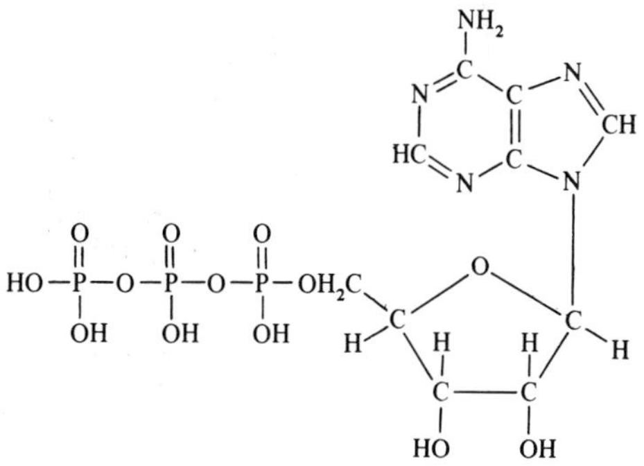

adenosine triphosphate (ATP)

These monomeric compounds are very important in biology and biophysics.

The formation of a nucleic acid occurs in principle by means of polycondensation of the nucleoside triphosphates. The incorporation of a nucleotide in the chain is accompained by the liberation of one molecule of inorganic diphosphate (pyrophosphoric acid). This process can be represented schematically by the equation

$$ATP + CTP \rightarrow AMP-CTP + H_4P_2O_7$$

The structures of the chains of RNA and DNA are shown in Figs. 2.10 and 2.11, respectively. The bond between the links of the chain consists of a phosphate group connecting the 5' position of one carbohydrate and the 3' position of another.

Like protein chains, nucleic acids are linear nonbranched chains. They are asymmetric and dextrorotatory as a result of the asymmetry of carbohydrate groups. Nitrogen bases themselves are symmetric, since purines and pyrimidines are planar systems.

The nucleotide composition of a nucleic acid can be determined by means of hydrolysis. The enzymes that split RNA and DNA are, respectively, ribonuclease (RNase) and deoxyribonuclease (DNase). The hydrolyzate, the mixture of nucleotides, is analyzed chromatographically with the help of absorption spectra in the ultraviolet region.

Determination of the primary structure of a nucleic acid is much more difficult than that of a protein. The primary structure of a nucleic acid involves both the sequence of bases and the side groups of the chain. At present not much is known about the sequences of nucleotides in DNA's, whereas

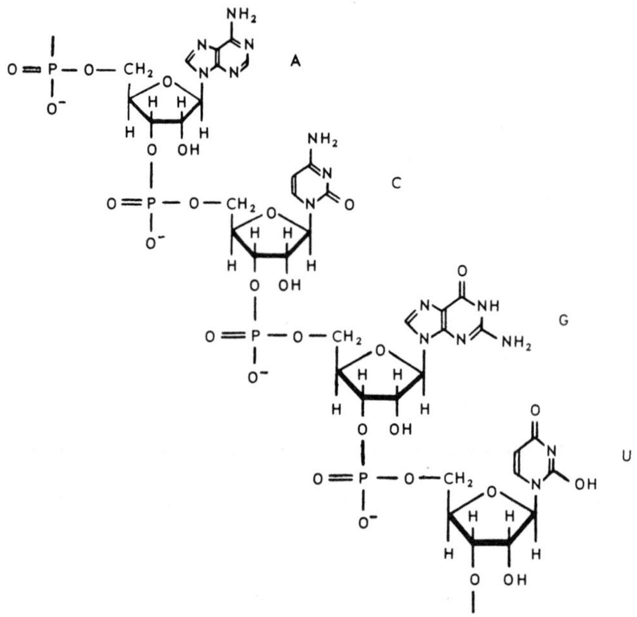


*FIG. 2.10 Structure of RNA chain.*

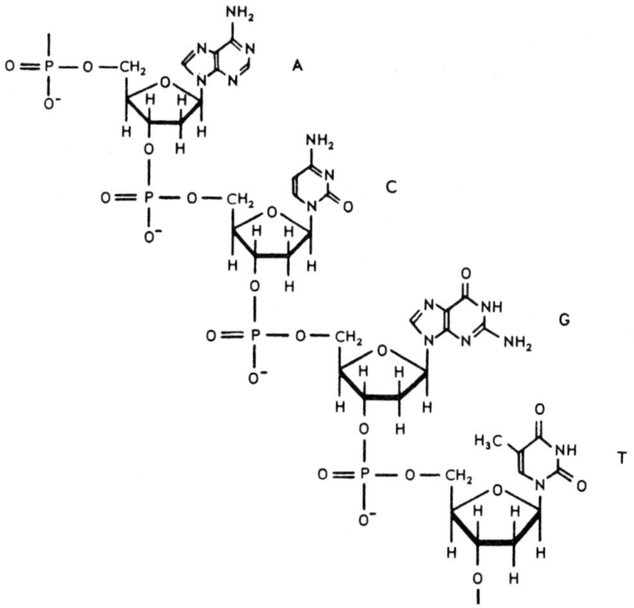


*FIG. 2.11 Structure of DNA chain.*

great progress has been made in deciphering the primary structures of the relatively short chains of the transfer RNA's (cf. Chapter 9).

Chemists are able to synthesize polynucleotide chains [25, 26]. Their work with synthetic polynucleotides has significance for molecular biology (Chapters 8 and 9). While investigating the nucleotide composition of nucleic acids, Chargaff discovered some important rules: the content of A in DNA is always equal to that of T, and the content of G is equal to that of C. Hence

$$\frac{A + G}{T + C} \cong 1$$

In other words, the amount of purine bases in DNA is equal to that of pyrimidine bases. On the other hand

$$\frac{G + T}{A + C} \cong 1$$

that is, the number of 6-$NH_2$ groups in the bases of DNA is equal to the number of 6-C=O groups. Table 2.5 shows the accuracy of these rules. Exceptions are observed in the cases of several phages.

However, the ratio

*TABLE 2.5*
*Molar Proportions (%) of Nucleotides in DNA of Different Origins*

| Source | A | G | C | T | MC | $\frac{A+G}{T+C+MC}$ | $\frac{G+T}{A+C+MC}$ | $\frac{G+C+MC}{A+T}$ |
|---|---|---|---|---|---|---|---|---|
| Ox (thymus gland) | 28.2 | 21.5 | 21.2 | 27.8 | 1.3 | 0.99 | 0.98 | 0.79 |
| Ox (spleen) | 27.9 | 22.7 | 20.8 | 27.3 | 1.3 | 1.02 | 1.00 | 0.81 |
| Ox (sperm) | 28.7 | 22.2 | 20.7 | 27.2 | 1.3 | 1.03 | 0.98 | 0.79 |
| Rat (bone marrow) | 28.6 | 21.4 | 20.4 | 28.4 | 1.1 | 1.00 | 0.99 | 0.75 |
| Herring | 27.9 | 19.5 | 21.5 | 28.2 | 2.8 | 0.91 | 0.92 | 0.81 |
| *Paraceutrotus lividus* | 32.2 | 17.7 | 17.3 | 32.1 | 1.1 | 1.00 | 0.97 | 0.57 |
| Wheat (grains) | 27.3 | 22.7 | 16.8 | 27.1 | 6.0 | 1.00 | 1.00 | 1.02 |
| Yeast | 31.3 | 18.7 | 17.1 | 32.9 | -- | 1.00 | 0.95 | 0.55 |
| *Escherichia coli* | 26.0 | 24.9 | 25.2 | 23.9 | -- | 1.03 | 0.95 | 1.00 |
| *Mycobacterium tuberculosis* | 15.1 | 34.9 | 35.4 | 14.6 | -- | 1.00 | 0.98 | 2.37 |
| *Rickettsia prowaczekii* | 35.7 | 17.1 | 15.4 | 31.8 | -- | 1.12 | 0.95 | 0.57 |

$$\frac{G + C}{A + T} \cong \frac{G}{A} \cong \frac{C}{T}$$

which can be called the specificity factor, is not constant. Belozersky and Spirin have shown that in the case of microorganisms this factor has varied over a broad range--from 0.45 to 2.8 [27] (see also [28]). The highest plants and animals are characterized by a specificity factor somewhat less than unity; G + C constitute approximately half of the overall content.

The species specificity of the nucleotide composition of DNA was studied in detail by Belozersky and his co-workers. Specificity is expressed not only by the relative content of

G + C but also by the content of minor bases: of 6-methylaminopurine (MAP) and methylcytosine (MC). MC is found in many different bacteria, in whose DNA its content varies from 0.06 to 0.65 mole %. The DNA of animals contains 1.5-2.0% MC, the invertebrate level is lower. The content of MC in the DNA of higher plants is high, occasionally 5-7% [29, 30]. The content of MAP in algae is typical of that for microorganisms (0.10-0.60%), and the MC level is 1.0-3.5%.

These data are important for the classification of species, but of course analysis of the species specificity of DNA requires determination of its primary structure. Belozersky and co-workers established many facts concerning this problem. Thus, in animal DNA chains, MC is concentrated mainly in the purine-MC-purine sequences. Bacteria do not contain such sequences. In *Escherichia coli* MAP is present in the pyrimidine-MAP-pyrimidine and pyrimidine-MAP-purine [29] triplets. The coefficient of specificity (i.e., the percentage of GC) is not representative; it can be the same for unrelated species.

RNA does not follow Chargaff's rules [27]. The ratio G + C/A + U varies widely, especially for insects (from 0.7 to 1.5). The content of GC is higher for the more ancient orders of insects, the AU content increases for the younger evolutionary forms. The same regularity was found for other groups of animals as well. In the case of fish G + C/A + U varies from 1.17 to 1.56 [29].

DNA is contained in chromosomes and mitochondria of the cells of multicellular organisms, in one-cell organisms and in bacteriophages. Molecular weights of DNA reach billions; these are the largest molecules yet known. RNA is contained mainly in the cytoplasm and nuclei of cells, as well as in plant viruses and phages. There are four types of RNA: the high-molecular-weight RNA's, or ribosomal RNA's (rRNA's), contained in ribosomes with molecular weights of the order of $2 \times 10^6$; messenger RNA's (mRNA's), with molecular weights from 30,000 to 60,000 and higher. (Since the average molecular weight of a ribonucleotide is 224, the shortest mRNA chain contains more than 150 nucleotides.) The third type of RNA comprises the transfer RNA's (tRNA's), with molecular weights of the order of 20,000 and containing approximately 80 nucleotides; the last type is the viral RNA's. More detailed information on the chemistry of nucleic acids and the corresponding monomers is presented in [25].

## 2.8 Carbohydrates and Lipids

The third type of biopolymer comprises carbohydrates, or polysaccharides. Polysaccharide chains are formed by monosaccharide links which have in the free monomeric state the empirical formula $C_6H_{12}O_6$. Glucose, the most important monosaccharide in plants and animals, contains, in contrast to ribose, a six-membered ring. The configuration of glucose is the asymmetric D form. Each of the five C atoms in the ring is

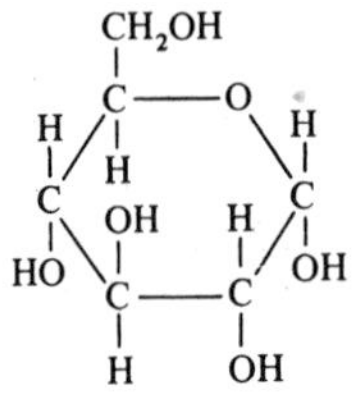


α-D-glucose

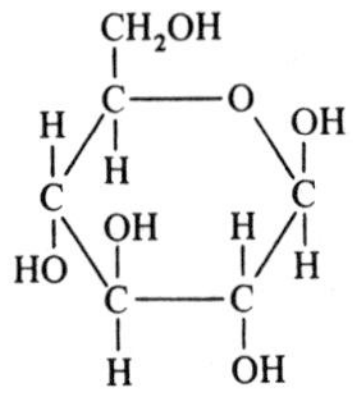

β-D-glucose

asymmetric. Along with monosaccharides, the cells contain disaccharides with the common formula $C_{12}H_{22}O_{11}$. We give as an example the structural formula of sucrose (beet sugar and cane sugar). Sucrose is formed from two monosaccharides, glucose and fructose. Mono- and disaccharides serve as energy sources in cells (discussed later).

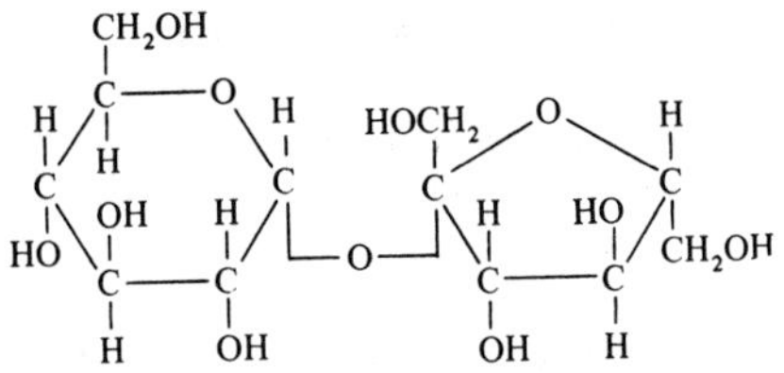

sucrose

Polysaccharides perform two very important functions. Starch (in two forms, amylose and amylopectin) and glycogen are the sources of mono- and disaccharides. Cellulose and chitin are the substances forming the skeleton or supporting structure in plants and arthropoda, respectively. In Fig. 2.12

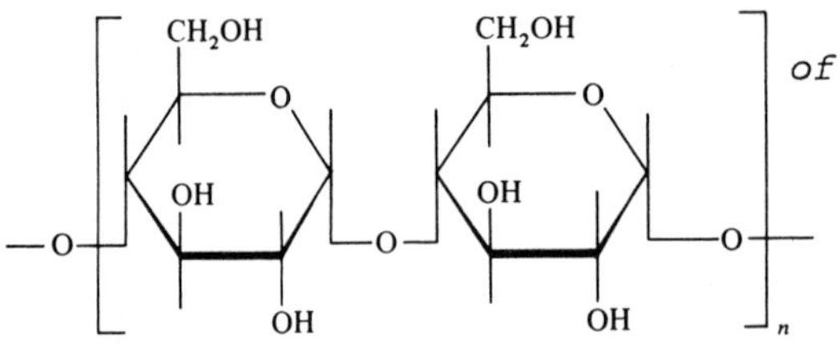

*FIG. 2.12 The structure of amylose.*

the structure of amylose, with its repeated maltose (diglucose) units, is shown. The chains of amylopectin, in contrast to amylose, are branched. The role played by starch in plants is performed by glycogen in animal organisms. Glycogen also consists of glucose units, but has a highly branched structure. Cellulose is made up of repeated units of cellobiose.

Polysaccharides contain no information in the primary structure. Their dimensions and branching vary over wide limits; the molecules have no definite length. The molecular weight of potato starch amylose is ~35,000 (200 glucose residues), that of the amylopectin from rice starch is ~500,000 with 80-90 branching points. The molecular weight of glycogen from muscles is $10^6$, that from liver is $5 \times 10^6$. Cellulose has a molecular weight of the order of 500,000 (cotton).

Polysaccharides play an important role in the external membranes of several kinds of cells, constituting the cellular shells of many bacterial species. In membranes polysaccharides form complexes with proteins and lipids (the latter are fatty substances invariably occurring in external and internal membranes). In the majority of cases membranes are made up of complexes of lipids and proteins.

Natural fats and oils are the triglycerides of the fatty acids (i.e., their glycerol esters), for instance, the triglyceride of stearic acid $H_3C(CH_2)_{16}COOH$:

$$CH_2-O-CO-(CH_2)_{16}CH_3$$
$$CH-O-CO-(CH_2)_{16}CH_3$$
$$CH_2-O-CO-(CH_2)_{16}CH_3$$

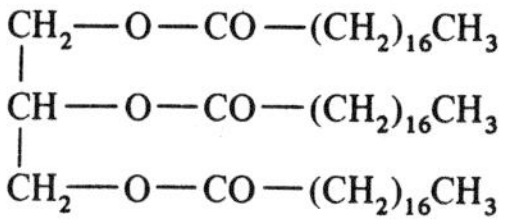

Hence, these substances contain long nonpolar hydrocarbon residues and strongly polar small groups. The functional lipids of cells (cell membranes) are more complicated substances which can contain carbohydrate, amino, and alkylamino groups. A series of important substances belongs to the phospholipids; one of them is lecithin (Fig. 2.13). Sphingolipids contain the amino alcohol sphingosine

$$H_3C-(CH_2)_{12}-CH=CH-CH-CH-CH_2OH$$
$$\qquad\qquad\qquad\qquad\;\; OH\quad NH_2$$

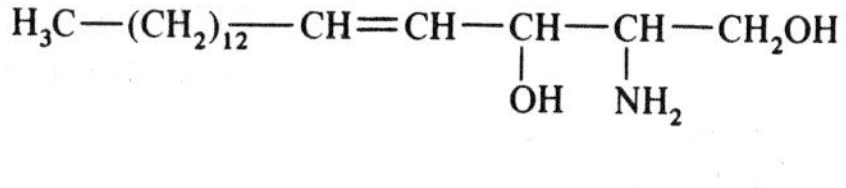

$$
\begin{array}{l}
\quad\quad\quad\quad O \\
\quad\quad\quad\quad \parallel \\
O \quad\quad CH_2-O-C-R_1 \\
\parallel \quad\quad\quad | \\
R_2-C-O-CH \quad\quad O^- \quad\quad\quad CH_3 \\
\quad\quad\quad\quad | \quad\quad\quad\quad\quad | \\
\quad\quad\quad CH_2-O-P-O-CH_2-CH_2-N^+-CH_3 \\
\quad\quad\quad\quad\quad \parallel \quad\quad\quad\quad\quad\quad | \\
\quad\quad\quad\quad\quad O \quad\quad\quad\quad\quad\quad CH_3
\end{array}
$$

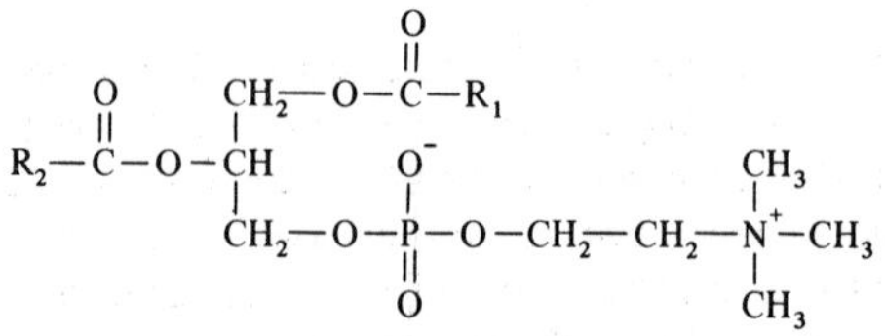

*FIG. 2.13  The structure of lecithin (phosphatidylcholine).*

An example is sphingomyelin, which also contains the residue of phosphoric acid.  Another important derivative of sphingomyelin, cerebroside, does not have phosphate group but contains a carbohydrate ring.  Gangliosides are complex sphingolipids containing sphingosine, fatty acid, one or more sugars, and neuraminic acid

$$
\begin{array}{l}
\quad\quad\quad\quad\quad\quad\quad OH \quad\quad\quad\quad OH \quad OH \\
\quad\quad\quad\quad\quad\quad\quad | \quad\quad\quad\quad\quad\quad | \quad\quad | \\
HOOC-CO-CH_2-CH-CH-CH-CH-CH-CH_2OH \\
\quad\quad\quad\quad\quad\quad\quad\quad\quad | \quad\quad | \\
\quad\quad\quad\quad\quad\quad\quad\quad\quad NH_2 \quad OH
\end{array}
$$

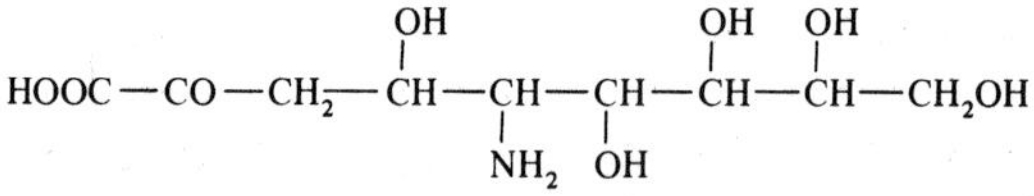

A series of important classes of complicated phospholipids is associated with phosphatidylic acid.  To them belong phosphatidycholines, lecithins, phosphatidylenolamines, and phosphatidylserines.  The formula of phosphatidylcholine is

$$
\begin{array}{l}
\quad\quad\quad H_2CO-CO-R_1 \quad O^- \quad\quad\quad\quad CH_3 \\
\quad\quad\quad\quad\quad | \quad\quad\quad\quad\quad | \quad\quad\quad\quad\quad | \\
R_2-CO-O-CH-CH_2-O-P-O-CH_2-CH_2-N^+-CH_3 \\
\quad\quad\quad\quad\quad\quad\quad\quad\quad \parallel \quad\quad\quad\quad\quad\quad\quad | \\
\quad\quad\quad\quad\quad\quad\quad\quad\quad O \quad\quad\quad\quad\quad\quad\quad CH_3
\end{array}
$$

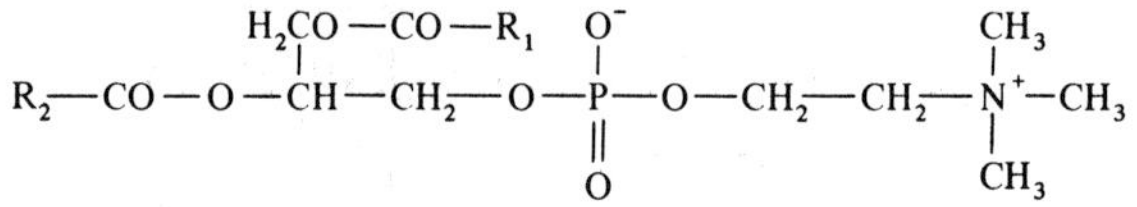

In all lipids, as in fats and oils, long nonpolar hydrocarbon chains and short strongly polar groups are present.  The general scheme of the structure of many lipids is shown in Fig. 2.14.

## 2.9  Cofactors, Vitamins, and Hormones

Proteins perform their most important enzymic function

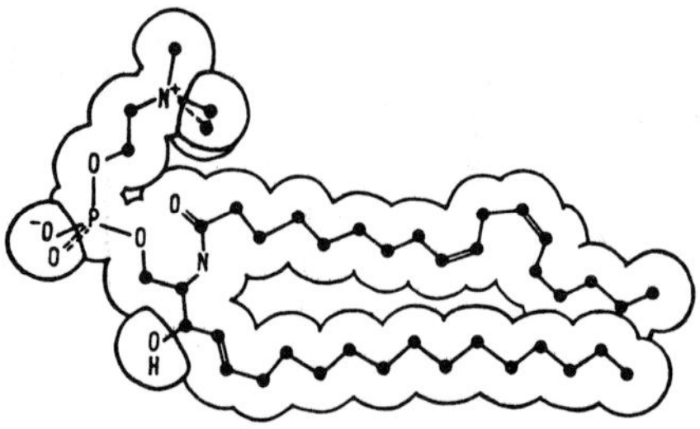

*FIG. 2.14  Scheme of the Structure of a lipid (sphingomyelin).*

mainly in complexes with low-molecular cofactors and with prosthetic groups. The latter are linked with proteins by valence bonds. Coenzymes are bound to the apoenzyme (i.e., to the protein moiety) by weaker bonds and can be transferred from one protein molecule to another. This does not always happen, however, and the distinction between coenzyme and prosthetic group is not quite definite. The enzyme as a whole, that is, the complex of apoenzyme with coenzyme is called a holoenzyme. In many cases metal ions play the role of cofactors.

The functional combination of the biopolymer (the protein) with low-molecular substances is one of the fundamentals of life processes. Cofactors are very variable but we meet here again some most important "chemical melodies" which are related to those which are "sounding" in proteins and nucleic acids. The relatively few coenzymes that are known belong to the aliphatic series. There are diphosphates: D-glucose 1,6-diphosphate, D-mannose 1,6-diphosphate, the 1,6-diphosphate of acyl-D-glucosamine, etc. participating in reactions involving the transfer of phosphate groups. Other important aliphatic compounds include lipoic acid and gluthathione, a tripeptide, γ-L-glutamyl-2-cysteinylglycine.

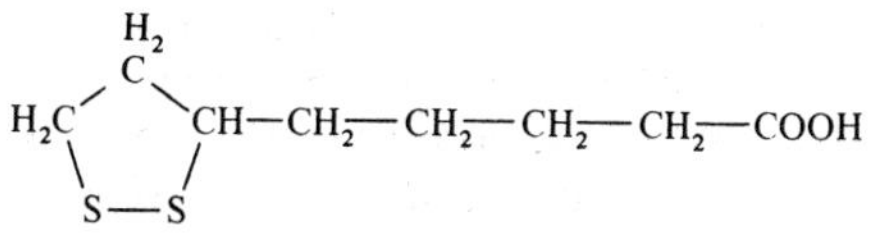


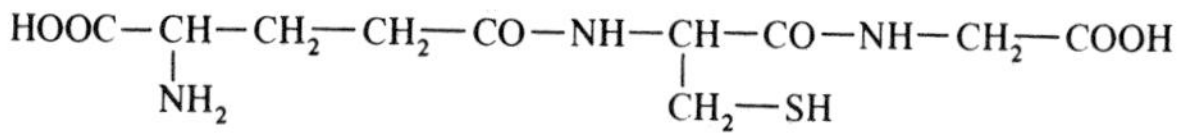


Lipoic acid participates in the enzymic carboxylation of pyruvic acid (p. 90). Glutathione serves as coenzyme in a series of reactions: the transformation of glyoxals into α-oxyacids, cis-trans isomerization, etc.

The essential participant in a series of oxidative-reductive processes is ascorbic acid, or vitamin C.

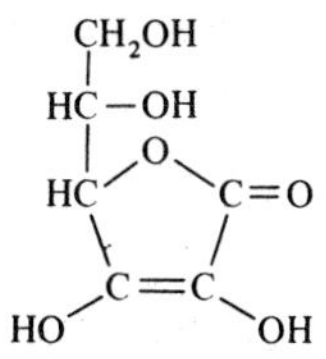


Vitamins are necessary for an organism because they serve as coenzymes or are transformed into coenzymes. They are necessary in small amounts, since their function is a catalytic one.

The majority of the most important coenzymes belong to

$\pi$-electron conjugated systems containing heterocyclic or aromatic rings. As we have already seen, nitrogen bases, nucleosides, and nucleotides belong to the same group of organic compounds. In many cases, low-molecular nucleosides and nucleotides and their derivatives act as coenzymes. ATP has to be considered one of the most important (cf. p. 71). The fundamental participants in the oxidation-reduction processes figure here: the nicotinamide coenzymes NAD and NADP and the flavin coenzymes FAD and FMN. The structural formula of the first two compounds is

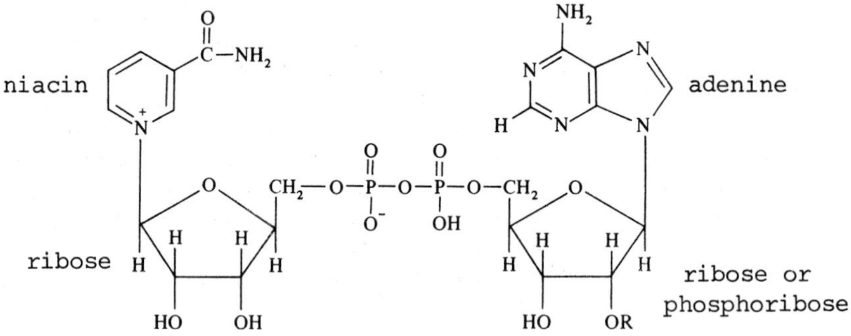


In NAD (nicotinamide adenine dinucleotide) R is H; in NADP (nicotinamide adenine dinucleotide phosphate) R is $PO_3H_2$. The functional group of niacin, contained in NAD and NADP, is an important vitamin, since an amide of nicotinic acid is an antipellagra factor

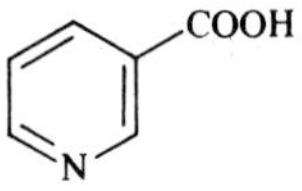

The structural formulas of the flavinic coenzymes--flavin mononucleotide (FMN) and flavin adenine dinucleotide (FAD) are

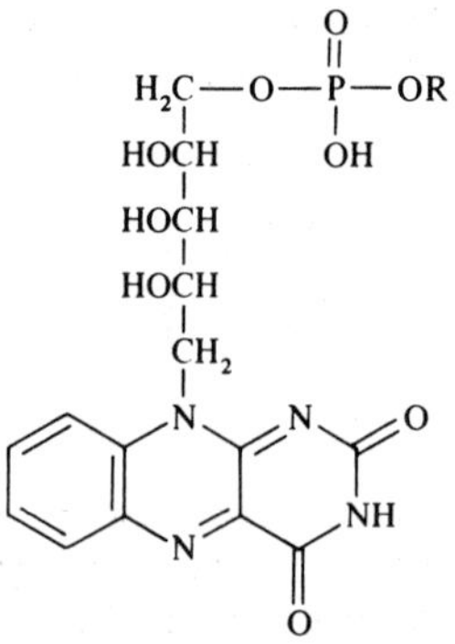

In FMN, R is hydrogen; in FAD, it is the nucleotide, adenine. Flavin coenzymes have a yellow-orange color because riboflavin, which contains a conjugated three-ring system, absorbs light in the visible region of the spectrum.

The same adenine group is contained in coenzyme A (p. 89).

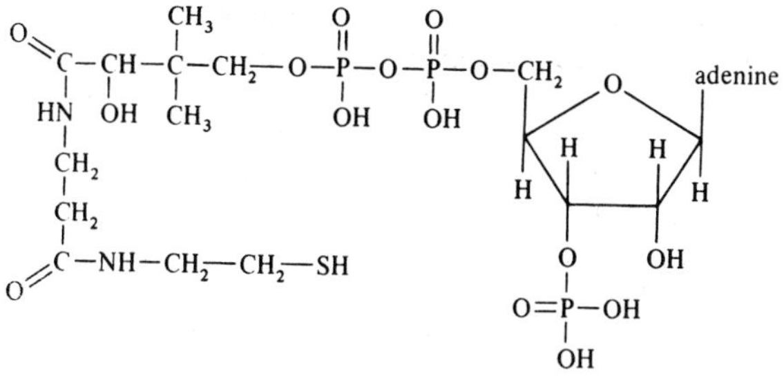


and in a series of acyladenylates

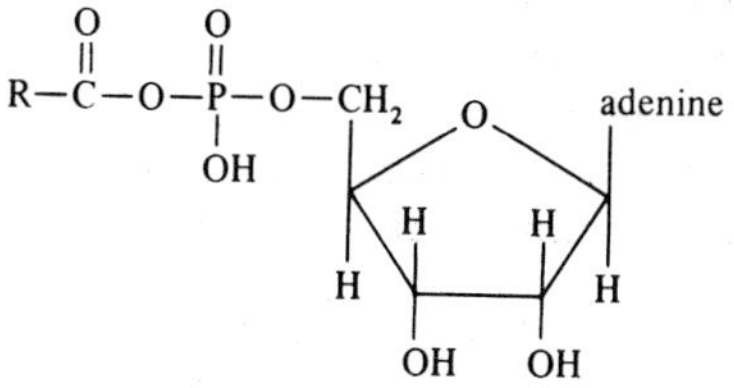


where R—CO is the residue of α-amino acid, acetic acid, β-alanine, etc. Acyladenylates participate in the activation and transfer of acylic residues. Adenine is contained in S-adenosylmethionine and in the cobamide coenzymes related to cobalamin or vitamin $B_{12}$ and containing the cobalt atom in the center of the tetrapyrrole ring.

Nucleotides are also present in the guanosine phosphate sugars; uridine phosphate sugars, amino sugars, and glucuronic acids; cytidine phosphate alcohols and amino alcohols.

Other heterocycles are parts of several important coenzymes. Pyridoxal phosphate (PALP) participates in transformations of amino acids (transamination reactions; p. 371). It is a derivative of pyridine.

The formula of thiamine pyrophosphate is

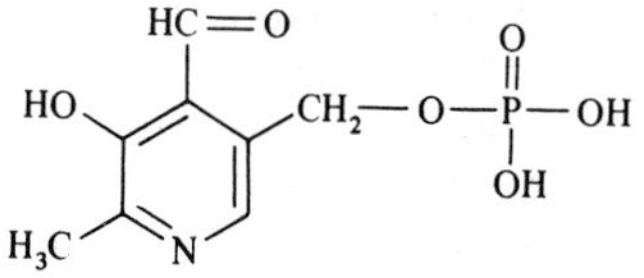

Thiamine (vitamin $B_1$) is necessary for life; its absence is responsible for the severe avitaminosis, beri-beri.

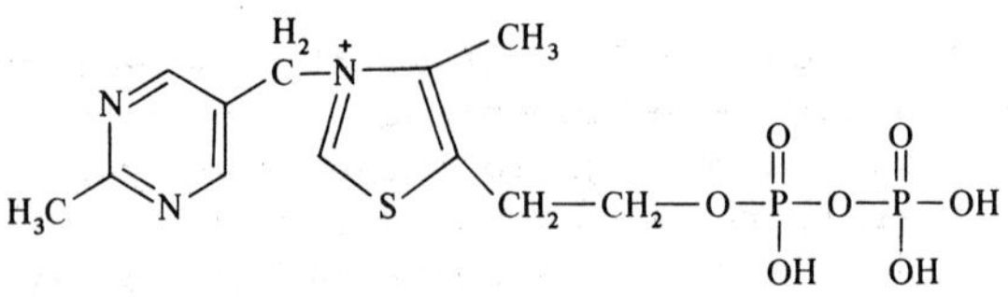

Finally, the ubiquinones (coenzyme Q) contain an aromatic quinone group and hydrocarbon chains of varying length (n = 1, ..., 10)

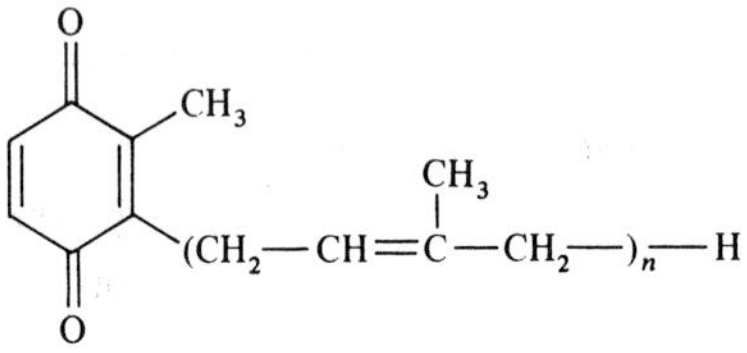

Ubiquinones take part in oxidation-reduction reactions. The structure of an ubiquinone is related to that of vitamin $K_1$, helping the coagulation of blood.

The conjugated $\pi$-electron systems of porphyrin compounds figure significantly in fundamental vital processes starting with photosynthesis. This tetrapyrrole ring occurs in the

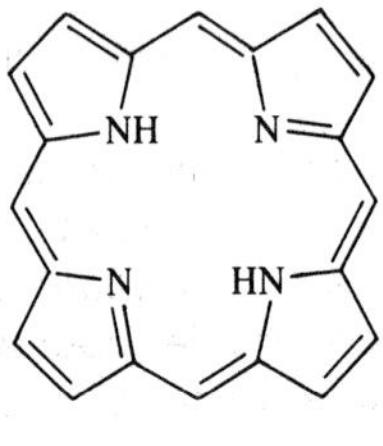

above-mentioned cobamide enzymes and in cobalamin, in the prosthetic groups of a series of the most important proteins, and chlorophyll. The structure of chlorophyll a, responsible for the initial process of photosynthesis, is shown in Fig. 2.15. Chlorophyll is a coordination compound of magnesium, whose atom is located at the center of the plane porphyrin ring.

The porphyrin complexes of iron serve as prosthetic groups for proteins that are essential participants in the process of respiration. These proteins are myoglobin, where molecular oxygen is stored (Chapter 7); hemoglobin, which transfers molecular oxygen (Chapter 7); and the cytochromes--enzymes transferring electrons in the process of oxidative phosphorylation (see the next section). The iron porphyrin group is contained also in the oxidative-reductive enzymes catalase and peroxidase.

Hemoglobin and myoglobin contain the heme group (protoheme IX), the porphyrin complex of bivalent iron (i.e.,

$$R = \begin{matrix} H_3C \\ H_3C \end{matrix} \hspace{-0.3em} \rangle CH-CH_2 \hspace{-0.3em} \left[ CH_2-\overset{\displaystyle CH_3}{\underset{\displaystyle |}{CH}}-CH_2-CH_2 \right]_2 \hspace{-0.3em} CH_2-\overset{\displaystyle CH_3}{\underset{\displaystyle |}{C}}=CH-CH_2-$$

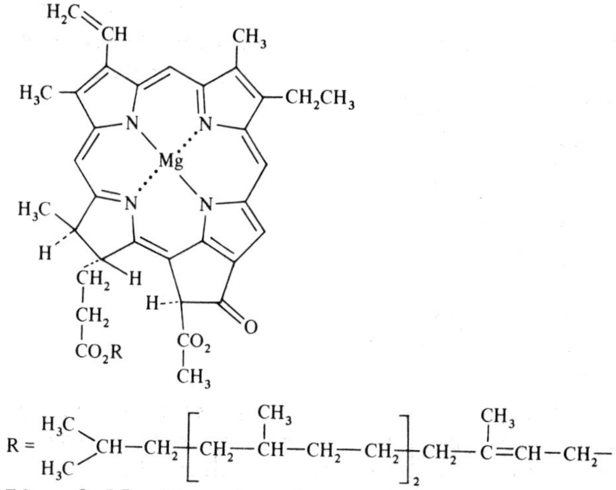

*FIG. 2.15  The structure of chlorophyll a.*

ferroporphyrin; Fig. 2.16).   In the free state, heme is oxidized rapidly and transformed into hemin, the complex of trivalent
iron (i.e., ferriporphyrin).   In an alkaline medium, ferriporphyrin links the $OH^-$ ion to iron to form hematin.   All these
compounds have the ability to interact with a series of low-
molecular substances (Chapter 7).   The heme group is also contained in some other important proteins, for instance, in lego-
globin, the protein extracted from nodular bacteria which participates in the fixation of atmospheric nitrogen, and in erythrocruorins, the respirative proteins of invertebrates.

   In iron porphyrin enzymes, heme groups with other side
groups are present.   In Fig. 2.17 the heme group of cytochrome
*c*, iron mesoporphyrin, is shown.   The chemical and physical
properties of such compounds are presented in [31, 32].   Related
compounds function in many organisms.   Thus, the blood of *Asci-
dia* contains the dipyrrole complex of vanadium.

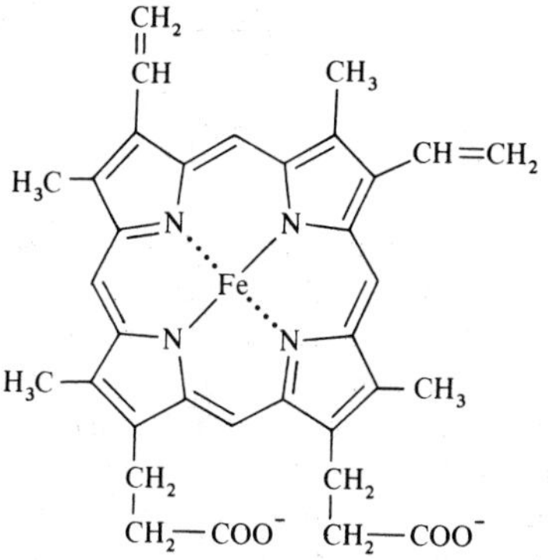

*FIG. 2.16  Heme (protopheme IX).*

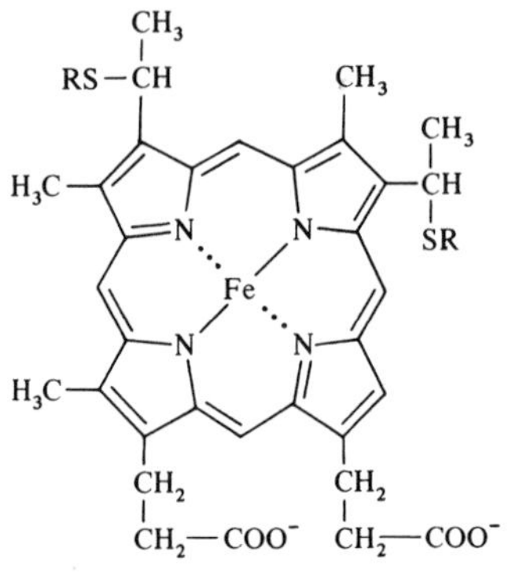


*FIG. 2.17 The heme group of cytochrome c (iron-mesoporphyrin).*

The porphyrins in biological systems are always in the form of metal complexes, which indicates the essential role of metals in biological processes. Besides the ions that are present in organisms in large amounts, such as $Na^+$, $K^+$, $Ca^{2+}$, and $Mg^{2+}$, there are metals, such as $Zn^{2+}$ and $Cu^{2+}$, that act as the cofactors of a multitude of enzymes. For some processes $Mn^{2+}$ is essential; molybdenum participates in the fixation of nitrogen. The cobamide enzymes contain $Co^{2+}$. Alkaline and alkaline earth metals are necessary for the maintenance of the ionic strength of solutions, for the biological function of proteins and nucleic acids. Phenomena such as nerve transmission and muscle contraction are determined by these ions (cf. "General Biophysics").

Returning to coenzymes and prosthetic groups, to nucleotides and porphyrins, we have to emphasize a general principle, the biological importance of the conjugated heterocyclic systems containing nitrogen, that is, the derivatives of pyridine, pyrimidine, and purine, the pyrrole compounds. The last include the bile pigments, whose basic structure is similar to that of the open porphyrin ring:

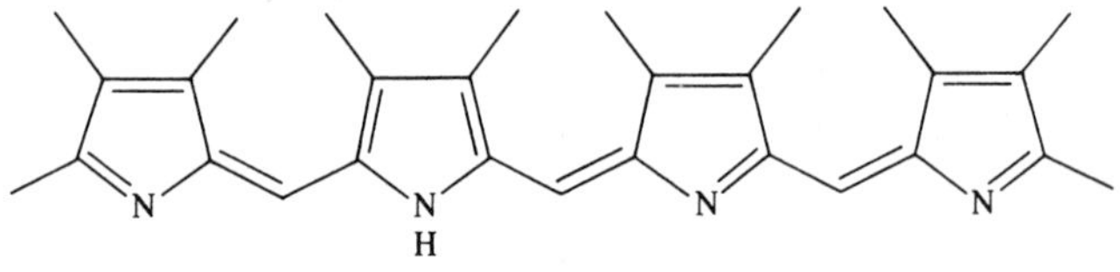


Indeed, there are facts showing that bile pigments are formed as a result of the destruction of hemoglobin. Nitrogen-containing heterocyclic rings also include such amino acids as tryptophan and histidine (p. 46) A series of natural compounds, alkaloids and others, are also heterocyclic.

Hormones, the regulatory substances in complex organisms, can be divided into two main groups of compounds. The first includes protein polypeptide hormones, the most important of which we list here.

The protein of the thyroid gland, thyroglobulin, contains iodine in the form of the amino acid tyrosine.

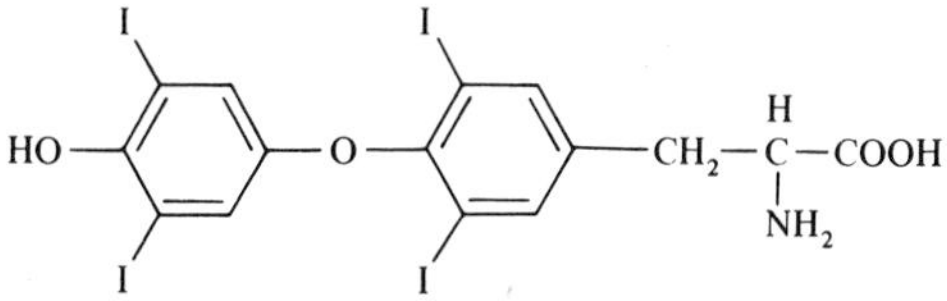

Insulin is a hormone synthesized in the Islets of Langerhans of the pancreas; it regulates the blood sugar level (Fig. 2.4).

Oxytocin and vasopressin are hormones of the pituitary. The first produces contractions of the uterus; the second elevates the blood pressure. Melanocyte-stimulating hormone (MSH), also produced by the pituitary, increases the pigmentation of frogs. It consists of two polypeptides with the structure (pig MSH):

α-MSH Acetyl-Ser-Tyr-Ser-Met-Glu-His-Phe-Arg-Trp-Gly-Lys-Pro-Val

β-MSH Asp-Ser-Gly-Pro-Try-Lys-Met-Glu-His-Phe-Arg-Trp-Gly-Ser-Pro-Pro-Lys-Asp

Adrenocorticotropin (ACTH) is the hormone of the front part of the pituitary and is important in the treatment of arthritis. The β-ACTH of pig has the structure shown in Fig. 2.4 (p. 60).

Other hormones of the front part of the pituitary--the thyrotropic hormone, prolactine; the growth hormone, gonadotropin; etc.--are also polypeptides. The pancreas produces the protein hormone glucagon, which stimulates the transformation of glycogen into glucose. There is also bradykinin, which stimulates the contraction of smooth muscles:

Arg-Pro-Pro-Gly-Phe-Ser-Pro-Phe-Arg

This list does not exhaust the protein hormones.

The second group of hormones is the steroids, compounds containing the carbon skeleton of cyclopentanephenanthrene.

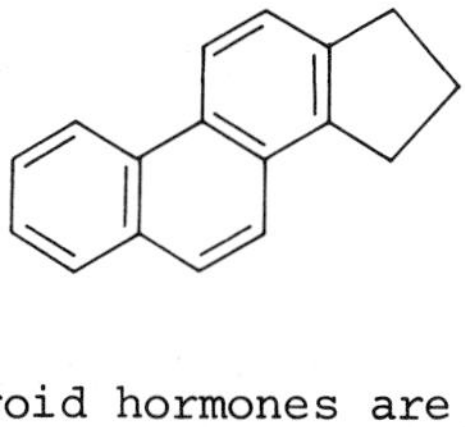

Also grouped with steroid hormones are such physiologically significant substances as cholesterol

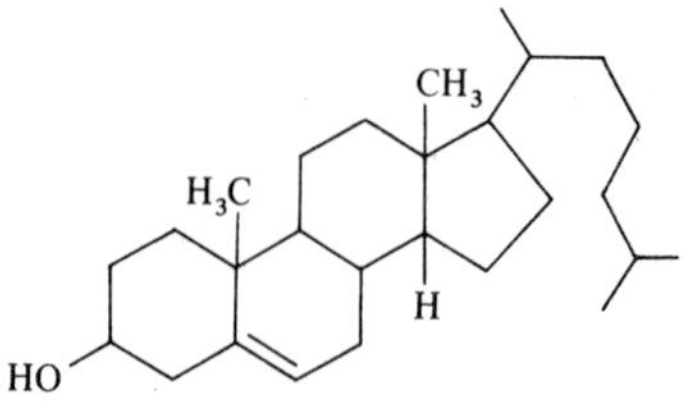


and ergosterol, which is converted by ultraviolet light into products leading to the antirickets factor vitamin $D_2$.

The most important steroid hormones are the sex hormones: estrone, progesterone, testosterone, and androsterone; and the hormone of the adrenal gland, cortisone. An important hormone of the same gland that increases arterial pressure and stimulates heart action is adrenaline, whose structure is comparatively simple:

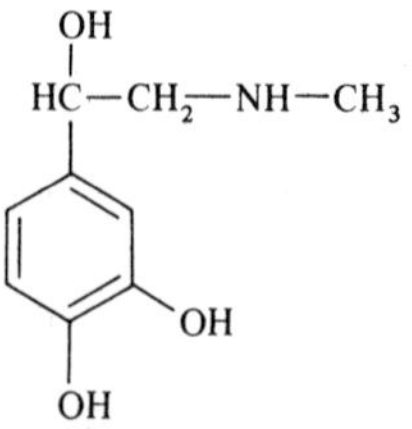


We see that hormonal activity is related to the chemical function of a limited number of chemical groups. The chemical differences between testosterol and cortisone are small, but their physiological functions are quite different. The structures and properties of hormones clearly demonstrate the biological importance of individual molecular structures.

Consider finally the structure of the noncyclical conjugated chain of β-carotene, the red pigment responsible for the orange color of carrots.

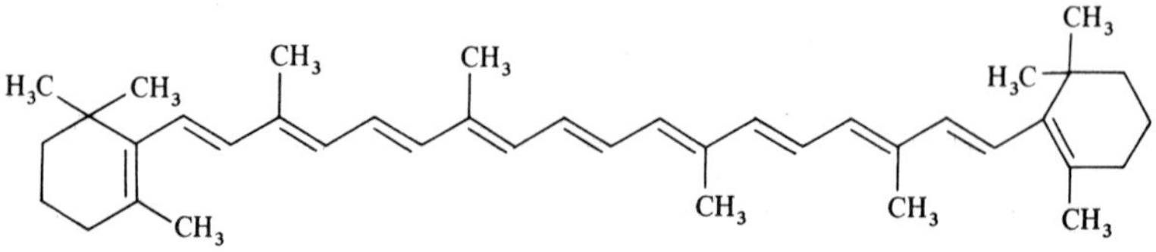


By oxidation of carotene, vitamin A, which participates in the initial act of visual reception, is formed (cf. "General Biophysics")

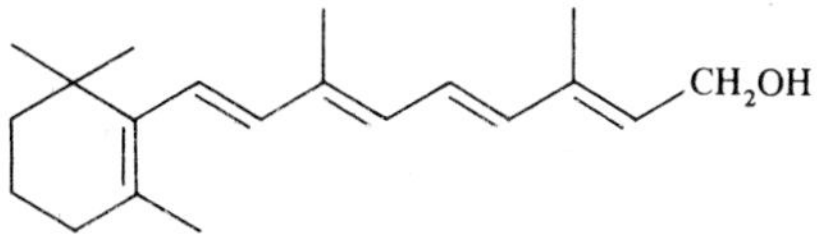

Naturally we cannot present further details concerning the structure and properties of the enormous number of natural and physiologically active compounds here. The biological roles of many of them are of great importance. Vitamins, hormones, and cofactors can be grouped under the general label bioregulators, since these substances regulate the biochemical processes in cells and organisms. Further details are contained in modern textbooks on biochemistry (e.g., [33]).

## 2.10 Some Fundamental Biochemical Processes

The fundamental biochemical processes determining the function of cells and organisms treated as open systems are, first, bioenergetic processes. The free energy necessary for the work of a cell,--that is, for the synthesis of biopolymers and functional small molecules; for cell division, for the mechanical motion of the cell and its organoids, for active transport, etc.-- is chemical energy which has as its source the light of the sun. The energy of the light quanta is transformed into chemical energy in the process of photosynthesis in green plants. Chemical energy is stored mainly in the molecules of ATP, and used for the majority of biological processes. The structural formula of ATP is shown on page 71. Two of the three phosphate bonds are high energy bonds--their hydrolytic splitting liberates an amount of free energy of the order of 10 kcal/mole. In the process of photosynthesis ATP is formed from ADP (adenosine diphosphate) and inorganic phosphate. This transformation requires electrons (from chlorophyll) and protons obtained from water. Simultaneously with the "charging of the battery or storage cell" (with ATP formation), there occurs the reduction of one of the electron carriers, NADP (p. 80), to NADPH. The chemical energy of ATP is used in plants for the synthesis of carbohydrates from $CO_2$ and $H_2O$; such synthesis requires, in addition to an energy source, a reducing agent, NADPH. If both ATP and NADPH are present, carbohydrate synthesis is also possible in the absence of light.

The total reaction for carbohydrate formation by photosynthesis can be written in the form

$$nCO_2 + nH_2O + \text{light} \rightarrow (CH_2O)_n + nO_2 + n \times 112 \text{ kcal/mole}$$

In the case of glucose formation, n = 6. This process is a very efficient one; nearly 75% of the energy of light absorbed

by the chlorophyll molecule is converted into chemical energy.
    Thus the energy of light quanta is transformed first into
ATP energy.  Then the secondary process of carbohydrate formation proceeds; these molecules store chemical energy, too.
This energy is extracted in the process of respiration and
transformed into ATP energy once more.
    Respiration is necessary for both autotrophic (mainly
photosynthesizing) and heterotrophic (i.e., obtaining nourishment from other organisms) organisms.  Respiration in the biochemical sense results in the oxidation of carbohydrates.
During the oxidation of glucose, a large amount of energy is
liberated

$$C_6H_{12}O_6 + 6O_2 \rightarrow 6CO_2 + 6H_2O + 690 \text{ kcal/mole}$$

The first phase of carbohydrate oxidation is called glycolysis.
During it, glucose is split, in a many-stage process, into two
molecules of lactic acid.  Eleven stages of the process have
been established, involving no fewer than 11 specific enzymes.
The entire reaction of the first stage can be presented in the
form

$$C_6H_{12}O_6 + 2P_n + 2ADP \longrightarrow 2H_3C \cdot CH(OH) \cdot COOH + 2ATP$$

Here $P_n$ means inorganic phosphate.  Thus the formation of two
lactate molecules from a molecule of glucose is accompanied
by phosphorylation--two molecules of ATP are formed by two
molecules of ADP and two phosphate molecules.  The reduction
of NAD to NADH and the reverse process (oxidation) take place
at the same time.  The thermodynamic balance of the process
comes about as a result of the liberation of 56 kcal/mole of
heat and the storage of 20 kcal in the 2 moles of ATP.
    After the splitting of glucose, the major part of the re-
maining energy is liberated in the second stage of the process--
the oxidation of lactate to $CO_2$ and $H_2O$ by oxygen from the air.
Glycolysis is the anaerobic transformation of glucose; the second phase is aerobic transformation, true oxidation.  Lactic
acid is oxidized into pyruvic acid.  The process is catalyzed

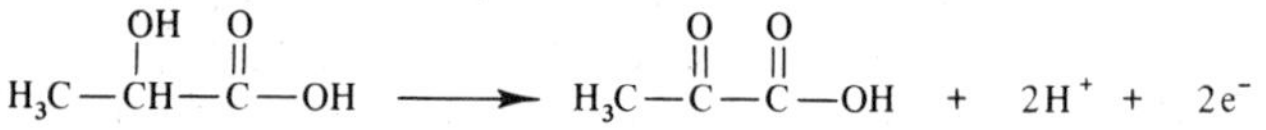

$$\underset{\text{lactate}}{H_3C-\overset{\overset{\displaystyle OH}{|}}{CH}-\overset{\overset{\displaystyle O}{\|}}{C}-OH} \longrightarrow \underset{\text{pyruvate}}{H_3C-\overset{\overset{\displaystyle O}{\|}}{C}-\overset{\overset{\displaystyle O}{\|}}{C}-OH} + 2H^+ + 2e^-$$

by the enzyme lactate dehydrogenase (LDH) acting with the coenzyme NAD.  The functional group of NAD--niacin (p. 80)--can
accept two protons and two electrons from lactate.  NAD is
reduced to NADH and one proton is liberated:

$$HC \overset{\displaystyle H}{\underset{\displaystyle C}{\overset{\displaystyle |}{\underset{\displaystyle ||}{C}}}} \overset{\displaystyle O}{\underset{\displaystyle ||}{C}} NH_2 \quad + \quad 2H^+ \quad + \quad 2e^- \quad \longrightarrow \quad \overset{\displaystyle H_2}{C} \overset{\displaystyle O}{\underset{\displaystyle ||}{C}} NH_2 \quad + \quad H^+$$

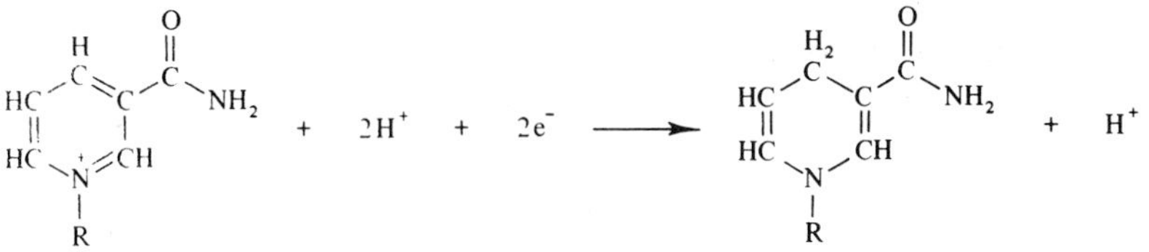

Pyruvate undergoes decarboxylation, that is, the $CO_2$ molecule is removed:

$$H_3C - \overset{O}{\overset{||}{C}} - \overset{O}{\overset{||}{C}} - OH \quad \longrightarrow \quad H_3C - \overset{O}{\overset{||}{C}} - H \quad + \quad CO_2$$

pyruvate                    acetaldehyde

Acetaldehyde interacts with a complex molecule of coenzyme A (p. 81) denoted CoASH

$$H_3C - \overset{O}{\overset{||}{C}} - H \quad + \quad CoASH \quad \longrightarrow \quad H_3C - \overset{OH}{\underset{H}{\overset{|}{C}}} - S - CoA \quad \longrightarrow$$

$$H_3C - \overset{O}{\overset{||}{C}} - S - CoA \quad + \quad 2H^+ \quad + \quad 2e^-$$

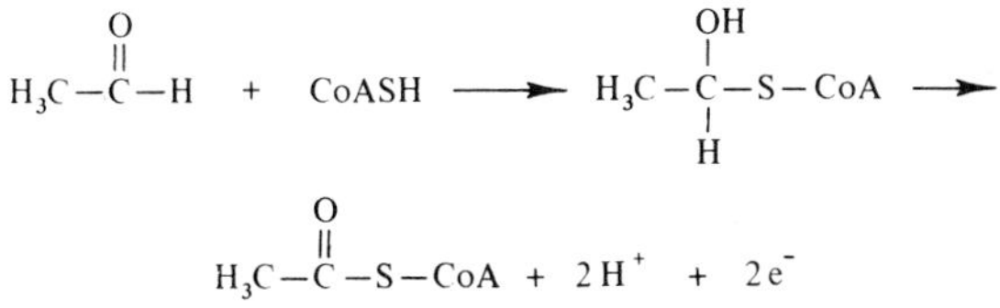

acetylcoenzyme A

and in this form enters the cycle of transformation called the citrate cycle or the Krebs cycle (Fig. 2.18). The cycle consists of a chain of reactions starting with the formation of citric acid from acetic and oxaloacetic acids. After a series of dehydrations and decarboxylations, citrate loses two carbon atoms in the form of $CO_2$, forms oxaloacetate once more, and the cycle begins again.

At the stage of acetyl-CoA formation in the metabolic process there enter, along with carbohydrates (hexoses), the products of the hydrolytic splitting of proteins and lipids, namely, amino acids, fatty acids, and glycerol. In this way, food substances are involved in the process of energy liberation. This is shown schematically in Table 2.6 [35].

What does this energy liberation mean? The electrons removed at different stages in the Krebs cycle, as during the two previous stages (the formation of lactate and pyruvate), are transferred along the "respiratory chain" of carriers. This chain involves a complex of enzymes and coenzymes, (namely, NAD), an enzyme belonging to the flavoprotein (FP) group, and a series of iron-containing enzymes--cytochromes $b$, $c$, $a$, and $a_3$. The electrons being transferred in the chain give up their

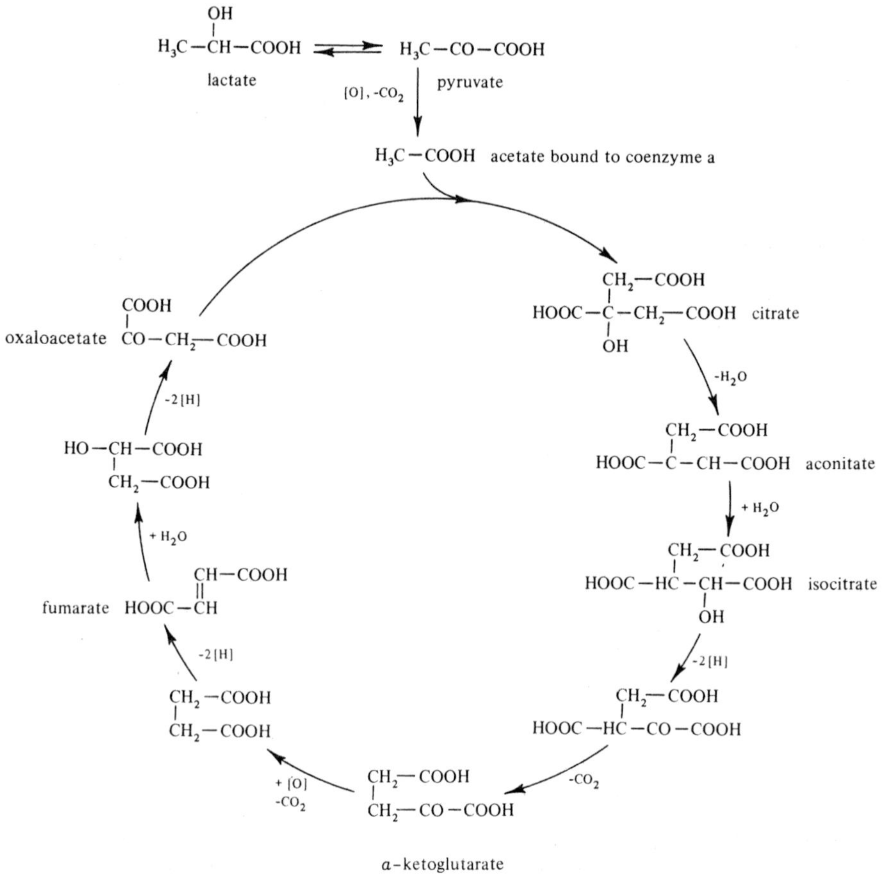


*FIG. 2.18 The citric acid cycle.*

energy to the molecules of ATP--oxidative phosphorylation occurs. Discovery of this most important phenomenon is attributed to Englehardt [36] and Belitzer [37, 38]. Finally, the electrons are transferred to oxygen (they reduce it to water).

To summarize in photosynthesis, oxygen is formed by the removal of electrons from water. Transformation of the nourishing substances in the respiration process amounts to the transfer of electrons to oxygen and its reduction to water. In this way these two processes are reversed.

Like photosynthesis, respiration is a self-consistent multistage process requiring the participation of a series of enzymes. In both cases a definite structural organization is necessary--such processes are localized in specialized cell organelles; photosynthesis, for example, occurs in chloroplasts, and respiration in mitochondria.

*TABLE 2.6*
*Three Stages of Energy Liberation from Nourishing Substances*

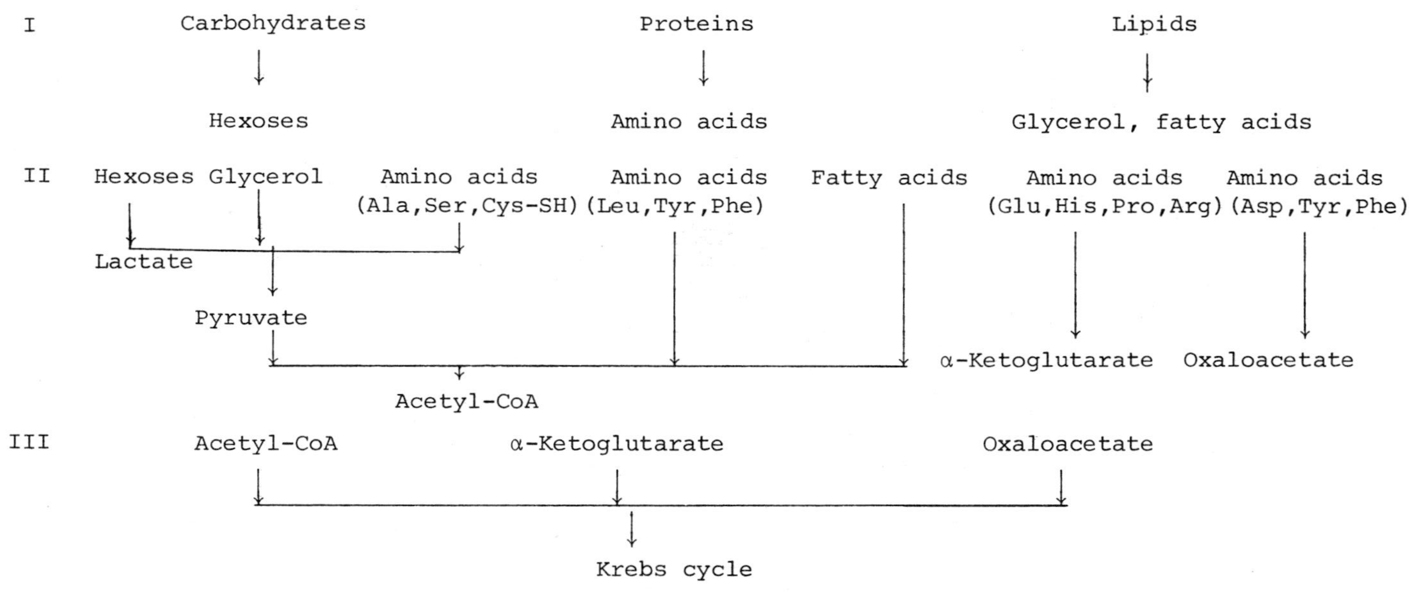

Glucose oxidation also occurs outside of the Krebs cycle, the tricarboxylic acid cycle, in the so-called pentose cycle. This process is not accompanied by phosphorylation but it, too, serves as an energy source. Glucose is phosphorylated by ATP

Glucose + ATP → Glucose 6-phosphate + ADP

and glucose 6-phosphate enters the chain of transformation involving NADP. The total process can be written in the form

$$\text{6Glucose 6-phosphate} + \text{12NADP} \rightarrow 6CO_2 + 6H_2O + \text{12NADPH} + \text{5glucose 6-phosphate}$$

With ATP, NAD and NADH are the unique primary storage cells of energy common to all living cells, and the carriers of chemical energy. The third such type of carrier is ferredoxin, which functions in the cells of plants and photosynthesizing bacteria. It is an iron-containing protein possessing high reductive ability (i.e., a high potential for oxidation and reduction). It also participates in the reduction of NADP. Cf. "General Biophysics".

Synthetic processes in cells--the synthesis of proteins, nucleic acids, purines, pyrimidines, lipids, sugars, etc.--, are as a rule endergonic processes. They require the expenditure of free energy. Biosynthesis occurs in open thermodynamic systems of the cell as a result of coupling (p. 13) with the exergonic processes of hydrolysis of ATP and the oxidation of NADH, NADPH, and ferredoxin, giving the necessary free energy. In the end, the reduced coenzymes are also formed at the expense of ATP. The fundamental biosynthetic processes proceed with the help of enzymes--kinases or synthetases.

The chemical energy in a cell is utilized for more than chemical reactions. Simultaneously, a complicated set of physicochemical processes occurs in which chemical energy is transformed into mechanical, osmotic, electrical, and light energy (bioluminescence). Investigation of the physical fundamentals of these energy transformation processes is one of the main problems of biophysics. Detailed information concerning the biochemical reaction accompanied by storage and expenditure of chemical energy is contained in [33, 39-44].

## 2.11 Quantum Biochemistry

The chemical and physical properties of atoms and molecules are determined by the structure of their electron shells interacting with atomic nuclei. Quantum mechanics is the basis of chemistry and therefore of biochemistry and biology. The general theory of the structure and properties of molecules is quantum chemistry. Correspondingly, the field of quantum-mechanical studies of the structure and properties of biologically functioning molecules is called quantum biochemistry.

Do such molecules, and primarily biopolymers, have some special electronic properties that distinguish them from other molecules and are responsible for their biological functionality? There is a school of thought in contemporary biophysics whose proponents seek the explanation of biological phenomena that occur at the molecular level in the specific electronic properties of biopolymers. Szent-Györgyi [45] and many others proceed from the suggested energy migration in proteins. It is theorized that this migration determines many biological processes, particularly muscle contraction [46-48]. In a series of papers, the "giant molecules" of biopolymers have been treated as semiconductors [49], antiferromagnets [50], and even superconductors.

A. Pullman and B. Pullman state that the majority of the substances playing dominate roles in biochemistry and representing the main active centers responsible for life processes are made up of conjugated structures rich in $\pi$-electrons [51]. Such structures, characterized by the high mobility of $\pi$-electrons, can bring about special phenomena. The same authors suggest, for instance, that the action of carcinogenic substances, such as benzo[3,4]pyrene is the result of the specific distribution of $\pi$-electrons in these molecules [51-53].

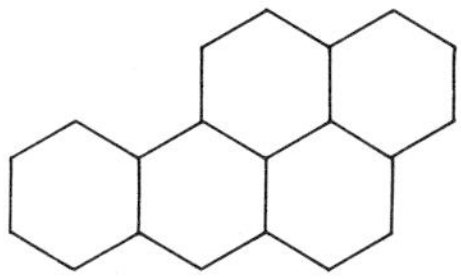

Are these suggestions well grounded? First, do biopolymers contain the kind of electrons to which the peculiar properties of a macromolecule as a total system have been ascribed by these theorists?

If the atoms are in regular positions in a chain or in a crystal, the electrons energy levels form zones occupied by electrons according to the Pauli principle. In electron conductors--metals--the free zone (the zone of conduction) directly abuts the occupied zone. In dielectrics the distance between occupied and free zones is relatively large. Electronic semiconductors are intermediate substances--in them the distance between zones is of the order ot the thermal energy kT; therefore, heating can make a semiconductor conductive. The conduction of a semiconductor is expressed by the formula

$$\sigma = \sigma_0 \exp(\Delta E/2kT) \qquad (2.35)$$

where $\Delta E$ is the interval between the zones. Hence, the problem of the semiconductivity of a biopolymer devolves upon the value of ΔE.

Evans and Gergely [54] calculated the energy zones of proteins (polypeptides) in the regular β form (cf. p. 167), treating such a system as conjugated by hydrogen bonds and using the molecular orbital method--linear combination of atomic orbitals (LCAO). They obtained ΔE values of the order of 3-4.8 eV (i.e., 70-110 kcal/mole), which is many times more than the ΔE values for genuine semiconductors. More precise calculations gave values of 5 eV and more [51]. Thus, proteins are dielectrics but not semiconductors. There is no conjugation of bonds in a polypeptide chain, no delocalization of electrons in the chain as a whole. Pi-electrons are partly delocalized only in a single peptide group —NH—CO—. Every peptide group is separated from neighboring ones by insulating —CHR— groups. Clear evidence of the absence of conjugation and therefore of semiconductivity in proteins is the absence of absorption bands in the visible and near-ultraviolet regions in their spectra.

Experimental investigations have shown the limited dependent temperature conductivity of many proteins, according to formula (2.35) with ΔE values of the order of 2-3 eV (see [51, 55]). This conductivity is dependent on the degree of hydration of the protein. There are strong arguments in favor of the hypothesis that the conductivity of proteins is determined by contaminants, particularly ionic ones, and that it has no biological implications.

The sugar-phosphate chain of a nucleic acid is also nonconjugated. Nucleic acids are dielectrics. Their ferromagnetic properties, observed by means of electron paramagnetic resonance, proved to be the result of contamination by iron-containing substances. It is very difficult to eliminate these contaminants.

Therefore, the Pullmans' hypothesis does not apply to proteins and nucleic acids, the biological functions of which, although determined by chemical structure, are not associated with the collective behavior of π-electrons. The function of proteins (primarily, the enzymic function) is based on their conformational properties. Chemical action in a protein causes a change in conformation. The major practical problem in quantum biochemistry consists in investigating electron-conformation interactions (ECI) (cf. [56, 57] and (p. 234). Bioregulators are really mainly π-electron systems whose regulating action on proteins is determined by ECI [56, 58].

The displacement of electrons in biopolymers is a result of the excitation produced by the absorption of light or short-wave radiation. The migration of energy and electrons is expressed in the optical and biological properties of biopolymers. However, in the absence of radiation, in "dark biology," these electronic effects do not exist [59, 60].

Until now. contemporary quantum biochemistry was limited to static problems and did not treat the most interesting and important (but also most complicated) problems of biochemical dynamics, particularly of ECI. With the help of the methods of quantum chemistry, the electron structures of the most important classes of biological molecules can be calculated. A. Pullman and B. Pullman [51-53] used the simplest and roughest method of quantum chemistry, molecular orbital linear combination of atomic orbitals (MO LCAO), with which it is possible to get estimations of the delocalization energy of electrons, the distribution of atomic charges, the bond orders (i.e., the measure of $\pi$-electron participation in bonds), and the indices of free valence (i.e., the characteristics of the nonsaturated atom). Figure 2.19 presents schematically the electron characteristics that can be ascertained by the MO method and the possibilities for using them.

The MO LCAO method gives semiquantitative estimates of the foregoing properties. Knowing them, we can understand many chemical and optical phenomena (cf. [61-64]). More rigorous methods of quantum chemistry (cf. [65]) have not been widely used in biology until now.

Consider some results obtained by quantum-mechanical calculations applied to biological molecules. Figure 2.20 shows the main electronic indices of the nitrogen bases of nucleic acids [51]. Dotted lines show the hydrogen bonds (see section 4.4). In Fig. 2.21 the scheme of the main chemical and physicochemical properties of nitrogen bases is given, based on quantum-mechanical calculations [51]. The calculations show that the maximal stabilization by electron resonance is more characteristic for adenine than for guanine. It is possible that purines are therefore more stable toward radiation than pyrimidines. The highest occupied molecular orbitals are also typical of purines, and this explains their properties as electron donors, their ability to form charge-transfer complexes with a series of acceptors. The most basic nitrogen atom in the ring is $N_7$ of guanine, $N_1$ of adenine. Outside the ring the $NH_2$ group of adenine is less positively charged than the same group of cytosine. The most electronegative oxygen atom belongs to cytosine. The highest bond order corresponds to the $C_5$-$C_6$ bond of thymine, then $C_5$-$C_6$ of cytosine.

Quantum chemistry explains the high energy properties of phosphates, mainly of ATP. It has been found that the large amount of free energy liberated by hydrolysis of the first and second phosphate bonds is the sum of a set of contributions. First, the sum of the resonance energies of the fragments of ATP hydrolysis is larger than the resonance energy of ATP itself. Second, ATP contains a high energy of electrostatic repulsion, exceeding that of the hydrolysis products. This is determined by the particular distribution of charges in the

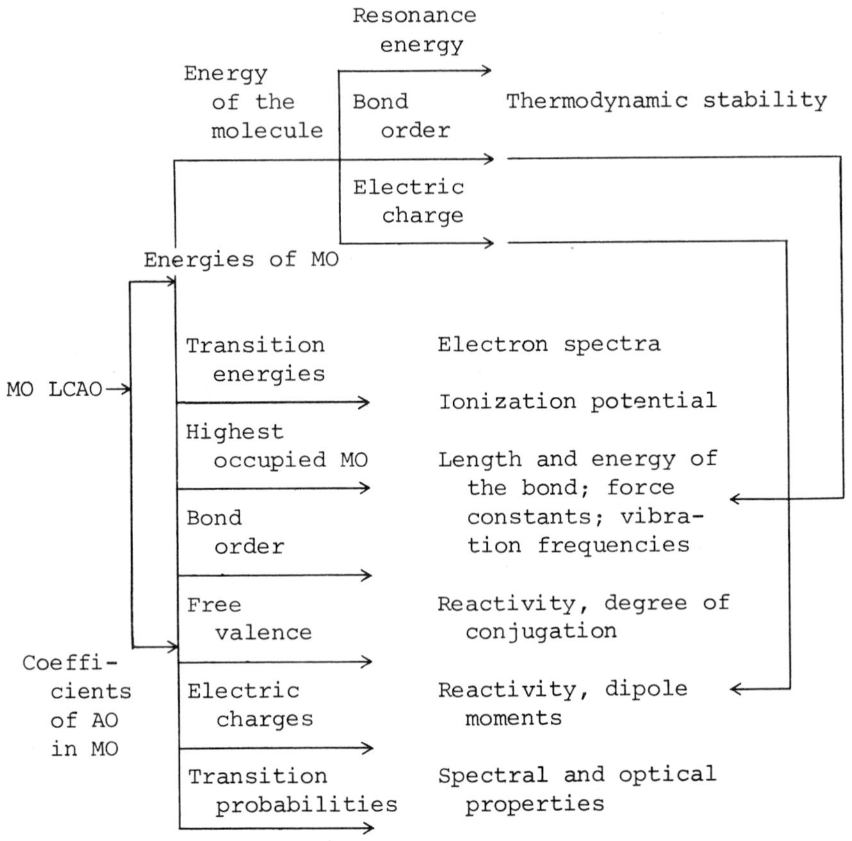


*FIG. 2.19 Electronic characteristics obtained by the molecular orbital method and their applications.*

pyrophosphate chain of ATP, shown in Fig. 2.22. For the other macroergic compounds such as phosphoenol pyruvate, the energy of the keto-enol tautometry of hydrolysis products is important in the total energy balance. Finally, in the case of carboxyl phosphates, the contribution to the free energy is given by the free energy of the ionization products. These contributions are summed up in Table 2.7 [52].

The MO LCAO method is especially suitable for calculations of the properties of π-electron systems. In proteins these are the aromatic amino acid residues Phe, Tyr, His, and Trp. Their electron structure has been studied in detail, and their properties as donors and acceptors of electrons calculated. The calculations have shown that these residues are donors rather than acceptors of electrons [51, 52]. These results are useful for the interpretation of a series of protein properties, particularly of their enzyme activity.

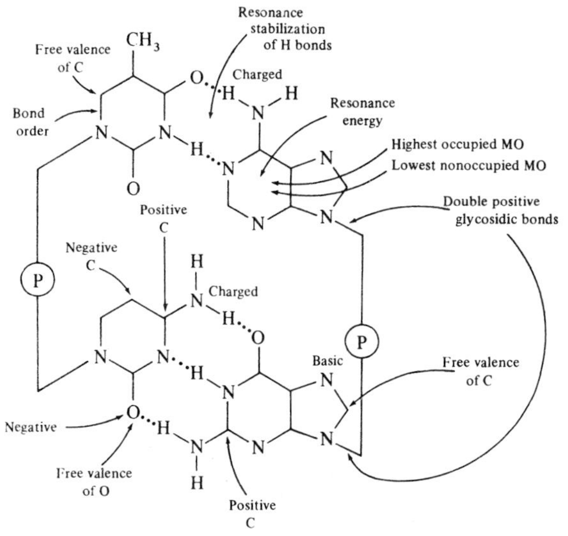


*FIG. 2.20 Electronic indices of the nitrogen bases of nucleic acids.*

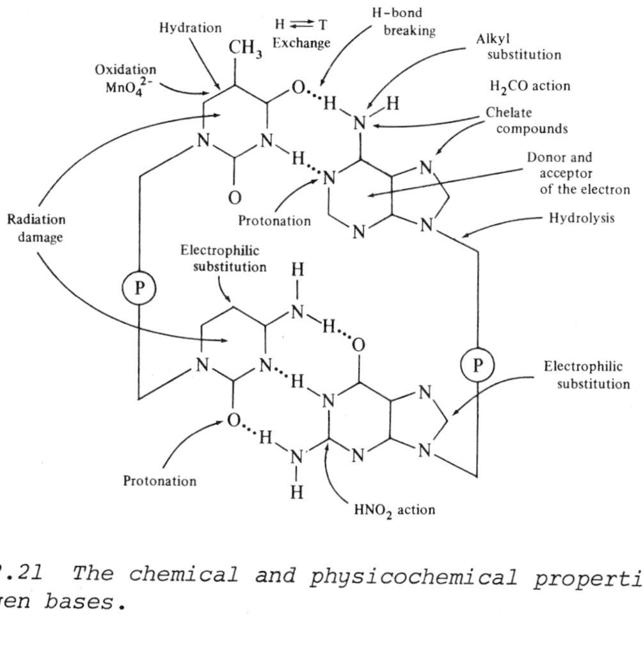


*FIG. 2.21 The chemical and physicochemical properties of nitrogen bases.*

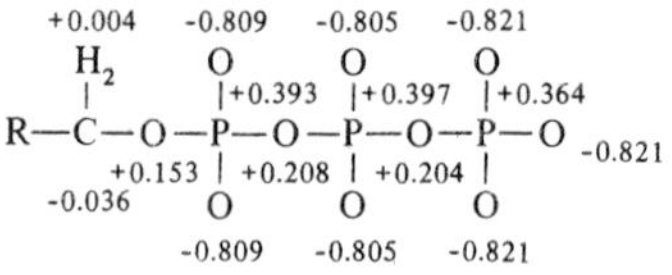


*FIG. 2.22 Charge distribution in the pyrophosphate chain of adenosine triphosphate.*

*TABLE 2.7*
*"Energy Wealth" of Phosphates*

| Substance | Experimental values | Basic hydrolyzate energy | Resonance energy | Electrostatic repulsion | Free energy of ionization products | Keto-enol tautometry | Total |
|---|---|---|---|---|---|---|---|
| ATP | 7-8 | 3 | 2.6 | 2 | | | 7.6 |
| ADP | 7-8 | 3 | 2.6 | 1.4 | | | 7 |
| Carboxyl phosphates | 10-12 | 3 | 4.6 | -0.7 | | 3.2 | 10.1 |
| Phosphoenolpyruvate | 11.5-12.5 | 3 | 1 | -0.5 | 9 | | 12.5 |
| Guanidine phosphates | 9-10 | 3 | 1.2 | -0.7 | | ? | ? |

[a]Energies in kilocalories per mole.

As was said earlier, the theory that π-electron systems play a decisive role in biology is valid for many low-molecular compounds interacting with proteins. Examples are porphyrins, quinones, carotenoids, practically all the coenzymes that participate in transfer reactions involving electrons, hydrogen, and other chemical groups. The results of calculations of the electron structures of these compounds [51, 52] are important elements in the explanation of their function. Quantum biochemistry treated as a quantum-mechanical theory of the behavior of molecules in biochemical reactions is an integral part of biochemistry. Thus, the studies of carcinogens made by A. Pullman and B. Pullman revealed the electronic properties of these compounds [51, 52]. However, the question whether such a complex phenomenon as carcinogenesis can be explained as the interaction of carcinogens with nucleic acids remains open. Unfortunately, the mechanisms of carcinogenesis are still obscure and our biological knowledge is still insufficient for the rigorous formulation of the physical problem.

Quantum chemistry in the broad sense of the word is a discipline that treats the structures of biological molecules and the changes these structures undergo during chemical reactions, in conformational changes, and as a result of mutagenesis. Direct investigation of phenomena involving electrons, their quantum-mechanical theory, is necessary in order to understand photobiological processes, particularly photosynthesis. The ideas of the modern quantum theory of solids are of great importance here. The problems of ECI that have to be solved by quantum mechanics are the most pressing [56, 57]. ECI forms the basis of enzyme action. The theory of enzymic activity, which is being developed, is a quantummechanical theory. In this connection the problems concerning the enzymes, which contain atoms of transition metals as cofactors, are of special interest. Their theoretical study must be based on the quantum chemistry of coordination compounds, i.e., on theoretical bioinorganic chemistry.

Finally, quantum mechanics serves as theoretical basis for optical and spectroscopic investigations, which have great importance for molecular biophysics. The methods of quantum chemistry are described in a series of monographs, particularly in [62-65].

## References


1. "Fundamentals of Molecular Biology. Physico-chemical Methods of Investigation, Analysis and Fractionation of Biopolymers." Nauka, Moscow, 1966 (R).
2. F. Haurowitz, "The Chemistry and Function of Proteins." Academic Press, New York, 1963.
3. R. Robinson and R. Stokes, "Electrolyte Solutions." Academic Press, New York, 1955.
4. L. Pauling and R. Corey, Proc. Nat. Acad. Sci. U.S. **37**, 235, 241, 282 (1951).
5. M. Tswett, "Selected Works." Ed. of Acad. Sci. USSR, Moscow, 1946 (R).
6. G. Samsonov, "Chromatography." Medgiz, Leningrad, 1955 (R).
7. G. Samsonov, "Sorbtion and Chromatography of Antibiotics." Ed. of Acad. Sci. USSR, Moscow, 1960 (R).
8. E. Lederer and M. Lederer, "Chromatography." American Elsevier, New York, 1957.
9. G. Ackers, Advan. Protein Chem. **24**, 343 (1970).
10. J. Bailey, "Techniques in Protein Chemistry." American Elsevier, New York, 1962.
11. C. Hirs, W. Stein, and S. Moore, J. Biol. Chem. **211**, 941 (1954).
12. F. Sanger, Biochem. J. **44**, 126 (1949).

13. M. Dayhoff and R. Eck, Atlas of Protein Sequence and Structure 1967-68. Nat. Biomed. Res. Foundation, Silver Spring, Maryland (1968); Georgetown Univ. Med. Center, Washington, D.C. (1972).
14. M. Ycas, in Symp. Informa. Theory Biol. (H. Yockey *et al.*, eds.). Pergamon, Oxford, 1958.
15. E. Margoliash and A. Scheiter, Advan. Protein. Chem. **21**, 113 (1966).
16. C. Nolan and E. Margoliash, Ann. Rev. Biochem. **37**, 727 (1968).
17. L. Pauling, H. Itano, S. Sanger, and I. Wells, Science **110**, 543 (1949).
18. V. Ingram, Nature (London) **180**, 326 (1957); "The Hemoglobins in Genetics and Evolution." Columbia Univ. Press, New York, 1963.
19. V. Ingram, "The Biosynthesis of Macromolecules." Benjamin, New York, 1965.
20. J. Fruton, "The Proteins," 2nd ed., Vol. I (H. Neurath, ed.). Academic Press, New York, 1963.
21. R. Merrifield, Sci. Amer. March, **56** (1968).
22. M. Vladimirov, Chem. and Life **N 3**, 50 (1966); **N 4**, 39 (1966) (R).
23. G. Gause, "Optical Activity and Living Matter Biodynamic." Normandy, Missouri, 1941.
24. V. Alpatov, Priroda **N 4**, 49 (1947) (R).
25. J. Davidson, "The Biochemistry of Nucleic Acids." Wiley, New York, 1960.
26. G. Khorana, in "The Nucleic Acids: Chemistry and Biology" (E. Chargaff and J. Davidson, eds.). Academic Press, New York, 1960.
27. A. Belozersky and A. Spirin, in "The Nucleic Acids: Chemistry and Biology" (E. Chargaff and J. Davidson, eds.). Academic Press, New York, 1960.
28. Ki Jong Lee, K. Wahl, and E. Barbu, Ann. Inst. Pasteur **91**, 212 (1956).
29. A. Belozersky, Nucleic Acids, Plenary Lecture All Un. Biochem. Congr., 2nd Edition FAN, Tashkent, 1969 (R).
30. B. Vaniushin, A. Belozersky, and N. Kokurina, Tr. Mosk. Obschest. Ispyt. Prir. **24**, 7 (1966) (R).
31. L. Blumenfeld and A. Purmal, in "Fundamentals of Molecular Biology Enzymes." Nauka, Moscow, 1964 (R).
32. J. Falk, "Porphyrins and Metalloporphyrins." Elsevier, Amsterdam, 1964.
33. H. Mahler and E. Cordes, "Basic Biological Chemistry." Harper, New York, and Weatherhill, Tokyo, 1968.
34. A. Lehninger, Rev. Mod. Phys. **31**, 136 (1959).
35. H. Krebs and H. Kornberg, "Energy Transformation in Living Matter." Springer, Berlin, 1957.
36. V. Engelhardt, Usp. Sovrem. Biol. **18**, 17 (1944); Izv.

Acad. Nauk SSSR Ser. Biol. **182** (1945) (R).
37. V. Belitzer, Enzymologia 6, 1 (1939).
38. V. Belitzer,and E. Tsybakowa, Biochimia **4**, 516 (1939) (R).
39. L. Ingraham, "Biochemical Mechanisms." Wiley, New York, 1962.
40. D. Green and R. Goldberger, "Molecular Insights into the Living Process." Academic Press, New York, 1967.
41. A. Lehninger, "Biochemistry." Worth Publishers, New York, 1972.
42. E. Kosower, "Molecular Biochemistry." McGraw-Hill, New York, 1962.
43. E. Racker, "Mechanisms in Bioenergetics." Academic Press, New York, 1965.
44. V. Skulachev, "Accumulation of Energy in Cell." Nauka, Moscow, 1969 (R).
45. A. Szent-Györgyi, Science **93**, 609 (1941); Nature (London) **148**, 157 (1941).
46. A. Szent-Györgyi, "Introduction to a Submolecular Biology." Academic Press, New York, 1960.
47. A. Szent-Györgyi, "Bioenergetics." Academic Press, New York, 1957.
48. A. Szent-Györgyi, "The Chemistry of Muscular Contraction." Academic Press, New York, 1947.
49. L. Brillouin, in "Horizons in Biochemistry" (M. Kasha and B. Pullman, eds.). Academic Press, New York, 1962.
50. J. Dushesne, in "Horizons in Biochemistry" (M. Kasha and B. Pullman, eds.). Academic Press, New York, 1962.
51. A. Pullman and B. Pullman, "Quantum Biochemistry." Wiley (Interscience), New York, 1963.
52. B. Pullman, "La Biochimie Electronique." Presses Univ. de France, Paris, 1963.
53. A. Pullman and B. Pullman, in "Horizons in Biochemistry" (M. Kasha and B. Pullman, eds.). Academic Press, New York, 1962.
54. M. Evans and J. Gergely, Biochim. Biophys. Acta **3**, 188 (1949).
55. D. Eley, in "Horizons in Biochemistry" (M. Kasha and B. Pullman, eds.). Academic Press, New York, 1962.
56. M. Volkenstein, Izve. Acad. Nauk SSSR Ser. Biol. **6**, 805 (1971) (R).
57. M. Volkenstein, J. Theor. Biol. **34**, 193 (1972).
58. M. Volkenstein, Biophysica **15**, 215 (1970) (R).
59. H. Longuet-Higgins, in "Electronic Aspects of Biochemistry" (B. Pullman, ed.). Academic Press, New York, 1964.
60. M. Volkenstein, "Molecules and Life." Plenum Press, New York, 1970.
61. J. Golovanov, A. Piskunov, and N. Sergeev, "Elementary Introduction to Quantum Biochemistry." Nauka, Moscow, 1969 (R).

62. C. Coulson, "Valence." Oxford Univ. Press (Clarendon), London and New York, 1953.
63. W. Kautzmann, "Quantum Chemistry." Academic Press, New York, 1957.
64. J. Murrel, S. Kettle, and J. Tedder, "Valence Theory," 2nd ed. Wiley, New York, 1970.
65. O. Sinanoglu (ed.), "Modern Quantum Chemistry." Academic Press, New York, 1965.

# Chapter 3

# The Physics of Macromolecules

## 3.1 Polymeric Chains

The special biological and physical properties of proteins and nucleic acids are determined to considerable extent by their macromolecular structure. Long chain molecules differ in many respects from small ones. Bodies made up of macromolecules have specific physical properties.

In the last decades, polymer physics and the macromolecular physics have developed considerably. Polymer physics is a broader area, since it studies not only single molecules but also the bodies that are made up of them. At present, the theoretical foundations of polymer physics are sufficiently clear; its further development depends on technology and biology. Technical polymer physics deals with problems concerning the practical application of the physicomechanical properties of polymeric materials--rubbers, plastics, and fibers. Macromolecular biophysics forms the basis of molecular biophysics. The general laws that have been established for comparatively simple synthetic macromolecules are also of fundamental importance for much more complicated biopolymers.

Synthetic macromolecules are obtained in the laboratory and in industry by means of the polymerization or polycondensation of monomers. In the first case, combining of the monomers in the chain occurs through their direct bonding, a result of breaking of the double bonds. Thus polyethylene is obtained from ethylene:

$$H_2C=CH_2 \quad + \quad H_2C=CH_2 \quad + \cdots + \quad H_2C=CH_2 \quad \longrightarrow$$

$$\diagdown_{CH_2}\diagup^{CH_2}\diagdown_{CH_2}\diagup^{CH_2}\diagdown_{CH_2}\diagup^{CH_2}\diagdown_{CH_2}\diagup$$

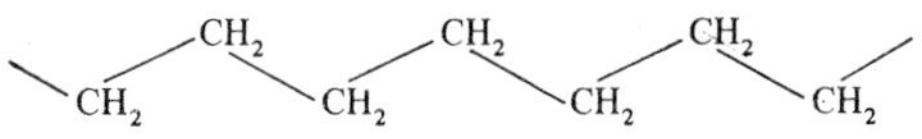

polyisoprene from isoprene:

$$H_2C=\underset{\underset{CH_3}{|}}{C}-CH=CH_2 \;+\; \cdots \;+\; H_2C=\underset{\underset{CH_3}{|}}{C}-CH=CH_2 \longrightarrow$$

$$-H_2C-CH_2 \underset{H_3C}{\overset{}{>}}C=C\underset{CH_2-CH_2}{\overset{H}{<}} \underset{}{\overset{H_3C}{>}}C=C\underset{H}{\overset{CH_2-}{<}}$$

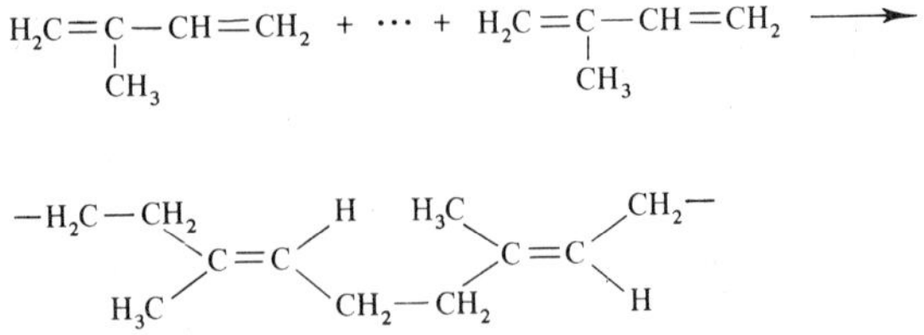

*trans*-polyisoprene (gutta-percha)

or

$$-H_2C-CH_2 \underset{H_3C}{\overset{}{>}}C=C\underset{H}{\overset{CH_2-CH_2}{<}} \underset{H_3C}{\overset{}{>}}C=C\underset{H}{\overset{CH_2-}{<}}$$

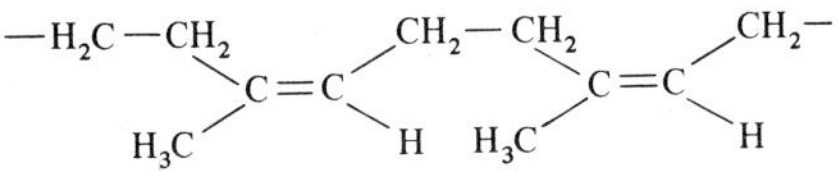

*cis*-polyisoprene (rubber)

In polycondensation (pp. 45, 72) the binding of two monomers occurs by the removal of one molecule of a substance (frequently water) per link.  An example is the polycondensation of
a diamine with a dicarboxylic acid:

$$-CO-OH \;+\; H_2N-(CH_2)_6-NH_2 \;+\; HO-CO-(CH_2)_4-CO-OH \;+\; H_2N-$$

hexamethylenediamine                         adipic acid

$$\Big\downarrow {\scriptstyle -nH_2O}$$

$$-CO-NH-(CH_2)_6-NH-CO-(CH_2)_4-CO-NH-$$

nylon 66

Linear synthetic polymers are soluble in various solvents,
forming true, not colloidal, solutions.  This property was
discovered by Staudinger [1].  The study of polymers in solutions gives information about the structure of macromolecules
that cannot be obtained from studies of the gaseous phase
because polymers are destroyed by heating.

Synthetic macromolecules are models of the main chain of
a protein or a nucleic acid.  The homopolymer macromolecule is
not an informational molecule, since it does not contain any
"text," any definite sequence of different monomeric units.
Its properties model only those properties of the biological
macromolecules which do not depend on the specificity of the
primary structure.  However, the polymeric chain can be considered a carrier of information about the electronic and conformational transformations of its structure.  These problems
will be treated in Section 3.5.

Synthetic macromolecules differ from proteins and nucleic acids not only by their lack of primary structure; synthetic polymers are heterogeneous (i.e., they contain heterogeneous molecules). The formulas of chains presented earlier are idealized formulas. Real chains often contain various groups; for instance, some of the H atoms in polyethylene are substituted by methyl groups, etc. The chains branch in a random way. Every sample of a synthetic polymer contains chains of differing lengths. Consequently, the molecular weight of a polymer is an average molecular weight of all the polymer homologs. On the other hand, all molecules of a given protein are identical in that they possess a definite molecular weight, composition, and primary structure.

A polymeric chain has to be characterized first by its configuration--the fixed arrangement of chemical bonds which can be changed only by their breaking. Thus *cis*- and *trans*-polyisoprene (p. 104) differ precisely by their configurations. Polymeric chains of the type $(-CHR-CH_2-)_n$ (e.g., polystyrene, where R is the phenyl group $C_6H_5$) can be obtained in a multitude of different configurations. Every R group can be located to the right or to the left of the main chain. If the number of monomeric links in the chain is 1000, the number of possible chains with different configurations is $2^{1000}$; hence, it is much larger than the number of macromolecules in a given sample of polymer. This means that all the macromolecules in the sample are different. Therefore, synthetic macromolecular substances do not obey a fundamental law of chemistry--the law of constancy of composition and structure. Natta discovered how to synthesize stereoregular polymers, the configurations of which are much more definite. Figure 3.1 shows the structures of the chains of isotactic and syndiotactic polystyrene. In isotactic polystyrene all the phenyl groups are on one side of a chain; in syndiotactic polystyrene their positions alternate regularly. A chain in which the positions of the R groups

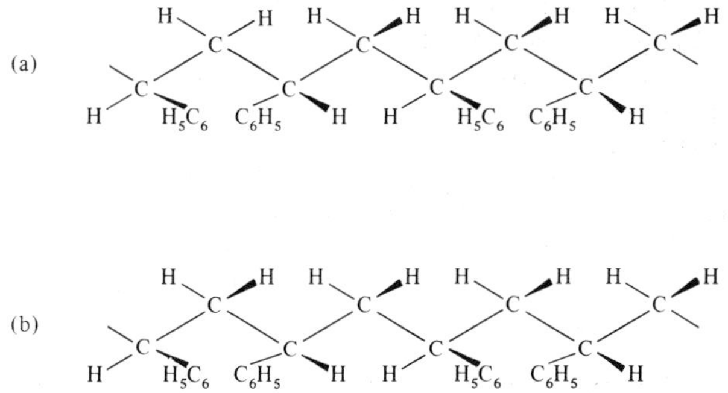

*FIG. 3.1 (a) Isotactic and (b) syndiotactic polystyrene.*

are randomly distributed is called atactic. The distinction between the optical isomers of small and large molecules is a configurational one (cf. p. 65).

Macromolecules are distinguished from small molecules by a characteristic of their physical properties that is attributable to the large number of identical links in a macromolecular linear chain. As a rule macromolecules contain single $\sigma$ bonds: C-C, C-N, and C-O. Different conformations of the chain appear as a result of rotations around these single bonds (see Sections 3.2, 3.4). The most important characteristic of a polymeric chain is its conformational lability, that is, its ability to assume a large number of different conformations. This property is reflected in peculiarities of the macroscopic behavior of polymers--in high elasticity, a property inherent only in polymers, and therefore in natural and synthetic rubbers. High elasticity is the ability of a polymer to undergo large elastic deformations (up to a few hundred percent) with a small modulus of elasticity. Like other elastic bodies, rubber that is subjected to small deformations obeys Hooke's law

$$\sigma = \varepsilon (L-L_0)/L_0 \tag{3.1}$$

Here $\sigma$ is the stress, L the length of a stretched sample, $L_0$ that of an unstretched sample, and $\varepsilon$ the modulus of elasticity. In contrast to steel, for which $\varepsilon \sim 20{,}000\ kg/mm^2$, rubber has $\varepsilon \sim 0.02 - 0.08\ kg/mm^2$ (depending on the degree of vulcanization of the rubber). Such a small modulus of elasticity is characteristic of an ideal gas. Indeed, an ideal gas is described by the equation of state

$$pV = RT \tag{3.2}$$

Let us compress the gas isothermically, increasing the pressure for dp. The decrease in volume will be dV. From (3.2) we get

$$dp = -p\ dV/V = p(L_0-L)/L_0 \tag{3.3}$$

where $L_0$ is the initial and L the final level of the piston. Equation (3.3) is identical to (3.1), with the role of the modulus $\varepsilon$ played by the pressure p. Atmospheric pressure corresponds to $\varepsilon = 1\ kg/cm^2 = 0.01\ kg/mm^2$, a value of the same order as that for rubber. An ideal gas is heated if compressed adiabatically. Similarly, rubber is heated if adiabatically stretched. This means that the entropy decreases in both these cases of deformation. The work involved in stretching rubber by a force f for dL is equal to

$$f\ dL = dF = dE - T\ dS \tag{3.4}$$

where F is the free energy, and E the internal energy. Rubber is practically incompressible. The elastic force of the isothermal stretching of rubber is

$$f = (\partial F/\partial L)_T = (\partial E/\partial L)_T - T(\partial S/\partial L)_T \tag{3.5}$$

Experiment shows that for rubber, f is really proportional to T and that the straight line f(T) crosses the origin of the coordinates. In other words

$$(\partial E/\partial L)_T \cong 0 \tag{3.6}$$

Just as the internal energy of an ideal gas does not depend on volume, the internal energy of rubber does not depend on length. The presence of the elastic force in both cases is due not to a change in the internal energy but to a change in entropy

$$f = -T(\partial S/\partial L)_T \tag{3.7}$$

This is the principal way in which the high elasticity of a polymer differs from the elasticity of a solid, say, of a steel spring, the elasticity of which is due to a change in internal energy.

The entropy elasticity of an ideal gas means that a decrease in the gas's volume is accompanied by an increase in the number of collisions of its molecules with the vessel's walls--elastic force is related to the thermal motion of particles. Compression of a gas diminishes its entropy by transforming the gas from the more probable rarefied state into the less probable compressed state. The occurrence of an apparently similar process in rubber means that rubber contains a great number of units having independent degrees of freedom, and that stretching rubber causes a transition from the more probable positions of these units to less probable ones (i.e., a decrease in entropy). Only in this respect does the analogy between rubber and an ideal gas hold.

The behavior of a polymer--specifically, of rubber-- is determined by the structure of its macromolecular chains. The independent motions of the units of a polymeric chain determine its conformational lability. A polymeric chain's flexibility is its most important property. We have to make a distinction between thermodynamic and kinetic flexibility. The first is responsible for the equilibrium properties of a polymer, particularly for the high elasticity of rubber. Thermodynamic flexibility is determined by the number of chain conformations that have equal or similar energies. Kinetic flexibility characterizes the rate of change of chain conformations, and is determined by the heights of the energy barriers that must be overcome.

## 3.2 Internal Rotation and Rotational Isomers

Classical organic chemistry held that the rotation of atomic groups around single bonds was quite free. Hence, all

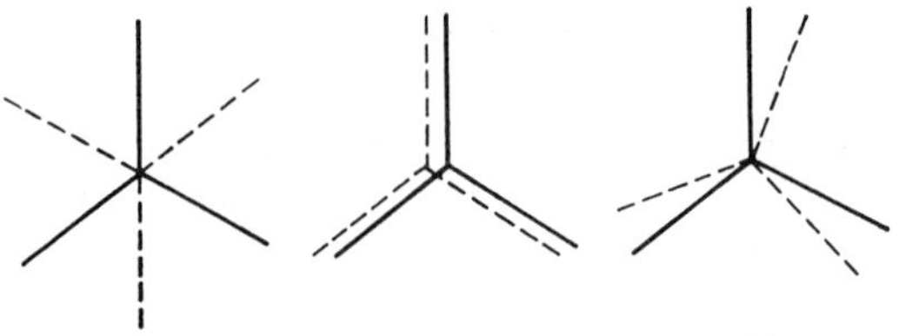


*FIG. 3.2 Positions of the C-H bonds of ethane in (a) cis, (b) trans, and (c) staggered conformations. Projections on the plane are perpendicular to the C-C bond.*

conformations of ethane produced by internal rotation possess the same free energy--a change in the angle of rotation does not require expenditure of energy. Some ethane conformations are shown in Fig. 3.2. However, investigations of the thermodynamic properties of ethane and other compounds containing single bonds, as well as structural studies performed by means of spectroscopy, nuclear magnetic resonance, etc., have shown that internal rotation is always hindered. The ethane molecule has an energy minimum at its trans conformation (Fig. 3.2a) and a maximum at its cis conformation (Fig. 3.2b). For a 120° rotation, that is, for the transition from one trans conformation to another identical one, an energy barrier of 2.90 kcal/mole must be overcome. For ethane, as for other molecules with $C_3$ symmetry, the dependence of the potential energy of the internal rotation U on the rotation angle $\phi$ can be approximately represented by the formula

$$U = \frac{1}{2} U_0 (1 - \cos 3\phi) \tag{3.8}$$

where $U_0$ is the height of the potential barrier (Fig. 3.3).

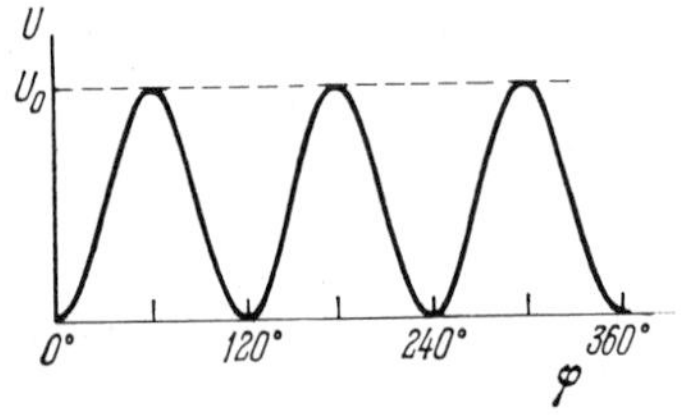


*FIG. 3.3 The potential energy of internal rotation in ethane.*

The value of $U_0$ increases if the H atoms in ethane are replaced by the bulkier atoms and groups ($CH_3$, halogens) and decreases if the rotation axis is elongated (compare $H_3C\text{-}CH_3$, $H_3C\text{-}SiH_3$, $H_3Si\text{-}SiH_3$). Some experimental values of $U_0$ are listed in Table 3.1 (see also [2]).

*TABLE 3.1*
*Heights of the Internal Rotation Barriers*[a]

| Substance | $U_0$ | Substance | $U_0$ |
|---|---|---|---|
| $H_3C-CH_3$ | 2.9 | $H_3C-OCH_3$ | 2.72 |
| $H_3C-CH_2CH_3$ | 3.4 | $H_3C-SH$ | 1.27 |
| $H_3C-CH(CH_3)_2$ | 3.9 | $H_3C-CH=CH_2$ | 1.98 |
| $H_3C-C(CH_3)_3$ | 4.4 | $H_3C-CH=O$ | 1.17 |
| $H_3C-CH_2F$ | 3.31 | $H_3C-SiH_3$ | 1.7 |
| $H_3C-CH_2Cl$ | 3.69 | $H_3Si-SiH_3$ | 1.0 |
| $H_3C-CH_2Br$ | 3.57 | $H_3C-C\equiv C-CF_3$ | 0 |
| $H_3C-OH$ | 1.07 | | |

[a]Values in kilocalories per mole.

The potential of internal rotation is due to the interactions of atoms and groups that are not bonded by valences. Rigorous quantum-mechanical calculation of $U_0$ is difficult because $U_0$ is much smaller than the total energy of the molecule and has to be calculated as the small difference between two large values--the total molecular energy in the cis conformation and the total molecular energy in the trans conformation. Approximate calculations of the hindering barrier have shown its complicated nature [2-5]. The barrier is formed by the steric van der Waals repulsion of the nonbonded atoms and by the quantum-mechanical interactions of the bonds joining the rotation axis (the effect of bond orientation). Both make the trans conformation more stable (the principle of crossed bonds). The van der Waals interaction can be estimated if the typical dependencies of the intermolecular energies on the intermolecular distances are known for specific substances. On the other hand, estimating the effect of the crossed bonds is rather difficult. A rough estimation can be made, which suggests that because of the small van der Waals radius of hydrogen, $U(\phi)$ for ethane is completely determined by the orientation effect and that this effect is the same for ethane derivatives. Then we get for the derivatives of ethane

$$U(\phi) = \frac{1}{2} U_{orient}(1 - \cos 3\phi) + \sum_{i,k} U(r_{ik}) \qquad (3.9)$$

$U(r_{ik})$ is the steric potential energy of the interaction of nonbonded atoms i and k at a distance $r_{ik}$. Obviously, $r_{ik}$ depends on $\phi$.

If a molecule contains strongly polar bonds, we have to add to the right-hand side of (3.9) terms describing the electrostatic interactions of the electron shells of the bonds joining the rotation axis. This feature is important in the case of biopolymers.

The interaction $U(r_{ik})$ of the H and C atoms of the C-H and C-C bonds can be described by empirical potentials estimated by Hill [6], Bartell [7], Kitaigorodsky [8], Kitaigorodsky and Mirskaya [9], and others on the basis of the crystallochemical and thermodynamic properties of simple hydrocarbons. The potentials can be written as (Buckingham)

$$U(r) = Ke^{-ar} - (K'/r^6) \tag{3.10}$$

The U(r) curves are shown in Fig. 3.4.

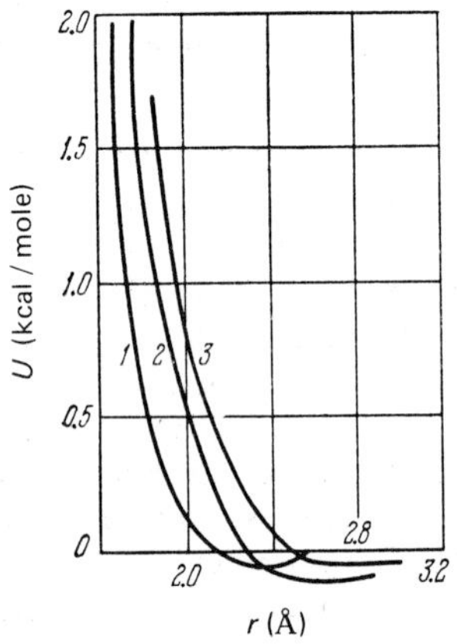


*FIG. 3.4 Interaction potentials of the nonbound H atoms of C-H bonds according to (1) Hill, (2) Bartell, and (3) Kitaigorodsky.*

Kitaigorodsky introduced a universal U(r) function, the "atom-atom potential." The interatomic potential of a pair of atoms expresses the interaction of "universal neutral atoms," and of the effective charges of nuclei, which are generally different from the atomic numbers. The Kitaigorodsky function has the form

$$U = U_{2/3}(z^{-6} - (6/\alpha)e^{\alpha}e^{-\alpha z})(11.4 - (6/\alpha)e^{\alpha/3}) \tag{3.11}$$

where $z = r/r_0$, $\alpha = ar_0$, $r_0$ is the sum of the van der Waals radii of the interacting atoms, and $U_{2/3}$ the value of U at $r = \frac{2}{3} r_0$. For C•••C, C•••H, and H•••H, Kitaigorodsky uses $U_{2/3} = 3.5$ kcal/mole and $\alpha = 13$. We get

$$U = 3.5(8600e^{-13z} - 0.004z^{-6})$$

Thus this potential contains only one parameter $r_0$. The potential gives quite satisfactory results in calculations of the

*FIG. 3.5 Potential energy of the internal rotation in n-butane*

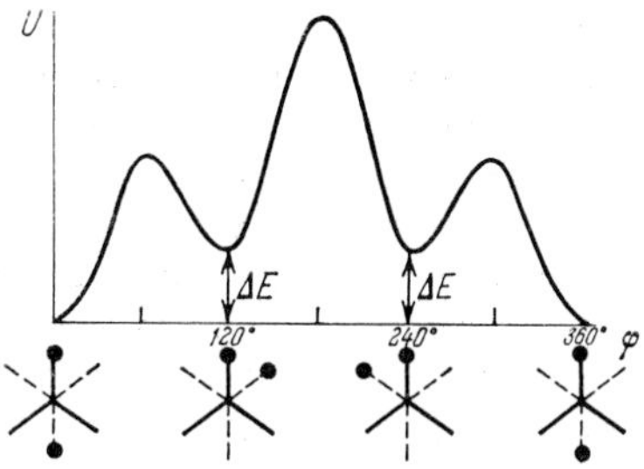


conformations of small molecules and macromolecules, particularly of those of biopolymers (cf. [10] and Section 4.2).

If, unlike ethane, the molecule lacks axial symmetry, the curve $U(\phi)$ is nonsymmetrical and cannot be described by a formula like (3.8). Thus for the molecule n-butane the curve of the dependence of the internal energy on the angle of rotation around the central C-C bond looks as shown in Fig. 3.5. In this case three minima of energy are present, the lowest one corresponding to the trans conformation. The other two, which are equal, correspond to conformations obtained from the trans position by rotating the $C_2H_5$ groups by 120° and -120°. These conformations are called gauche conformations.

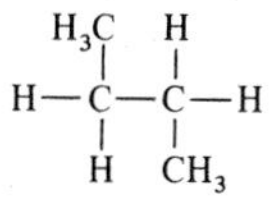

Obviously, molecules characterized by several nonequivalent energy minima exist in only these states. The rate of transition from one minimum to another is determined by the height of the barrier between them. The equilibrium amounts of n-butane molecules in the trans and gauche conformations is expressed as

$$N_t = N\frac{1}{1+2\exp(-\Delta E/RT)}, \qquad N_d = N_\ell = N\frac{\exp(-\Delta E/RT)}{1+2\exp(-\Delta E/RT)} \qquad (3.12)$$

where $N_t$ is the number of trans molecules, $N_d$ and $N_\ell$ the number of molecules in conformations rotated 120° to the right and to the left, respectively. Evidently

$$N_t+N_d+N_\ell = N$$

where N is the total number of molecules. Thus the substance appears as a dynamic mixture of conformations, which in such cases are called rotational isomers or rotamers. The content of an equilibrium mixture is determined by the difference in the rotamer energies $\Delta E$ and by temperature. If $T \to \infty$, $N_t = N_d = N_\ell = \frac{1}{3}N$. If the temperature is lowered, the substance crystalizes in the form of the single most stable rotamer and formulas (3.12) are not valid (cf. p. 118).

If the barrier heights are of the order of several kilocalories per mole, the time of rotational isomerization, (i.e., of the transition from one rotamer to another) is of the order of $10^{-10}$ sec. This estimation is obtained from the rate of expression for rotamerization based on the theory of absolute rates of reactions [11] (cf. Section 6.1).

Therefore, the rotamers cannot be separated. The presence of rotamers in an equilibrium mixture is established by studies of its chemical and physical properties. The spatial structures of rotamers differ; consequently, their vibrational spectra also differ. The spectrum of the substance is an overlapping of the rotamer spectra. In fact, during the lifetime of a rotamer, hundreds and thousands of vibrations occur (with frequencies of the order of 100-1000 $cm^{-1}$) [11]. Rotational isomerism was discovered by Kohlrausch by means of Raman spectra [2,12]. The relative intensities of the spectral lines corresponding to different rotamers depend on the content of rotamers in the mixture. This content changes with temperature; therefore, the energy differences $\Delta E$ can be determined by studying the temperature dependence of the intensities of the spectral lines. Thus, for n-butane $\Delta E \cong 600$ cal/mole. The values for a series of other molecules are given in [2].

Information about rotamers can also be obtained by means of radiospectroscopy, nuclear magnetic resonance (NMR), dipole moment measurements, etc. (cf. [2,3,13]).

The $\Delta E$ values and the energy barriers that separate the rotational isomers can be calculated with the help of the potentials of Kitaigorodsky, Hill, and others on a semiempirical base. For molecules of the n-butane type and for more complicated molecules, the rotations around several bonds have to be taken into account. In these cases the energy of internal rotation depends on several rotation angles and cannot be represented by a curve, but by a surface, generally a multidimensional one. The calculation of such a surface was first done [14] (cf. also [3]) for n-butane. It was based on formula (3.9) and on the C•••C potentials of Kitaigorodsky and H•••H potentials of Hill. Figure 3.6 shows the geodesic map expressing the dependence of the butane energy on the rotation angles

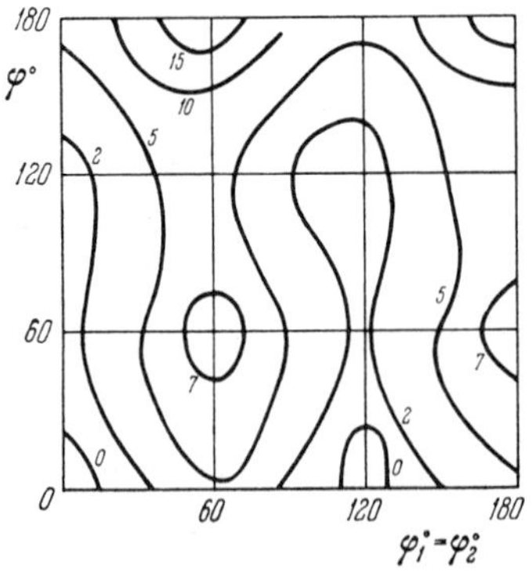


*FIG. 3.6 Dependence of the energy of internal rotation of n-butane on rotation angles. The numbers are energies in kilocalories per mole.*

$\phi$ ($-H_2C-CH_2-$) and $\phi_1 = \phi_2$ ($H_3C-CH_2-$). The values $\phi = 0, \pm 120°$, and $\phi_1 = \phi_2$ close to 0, ±120°, correspond to the energy minima. The transition from the trans to the gauche isomer requires overcoming a barrier of the order of 3 kcal/mole. $\Delta E$ was estimated to be 900 cal/mole. A similar analysis was subsequently made of the conditions of internal rotations in substances containing oxygen atoms [15], double bonds [16], and many others [17,18].

The investigation of rotational isomerism has great importance for organic and bioorganic chemistry. The chemical and physicochemical properties of molecules depend essentially on their conformations [19,20]. The main features of the physical behavior of macromolecules are due to rotational isomerism.

## 3.3 The Rotational-Isomeric Theory of Macromolecules

Internal rotations occur in every link of the macromolecular chain of polyethylene formed by single C-C bonds. This internal rotation determines the chain's flexibility and is responsible for the high elasticity of polymers.

Assume that a chain is freely jointed, that is, that its valence angles are not fixed and the rotations are free. Therefore, the set of conformations arising from rotations around the given chain atom is continuous in the angle range from 0 to $4\pi$ and the energy is not changed by the rotations. The chain can be characterized by a vector of length h directed from the first atom of the chain toward the last one (Fig. 3.7).

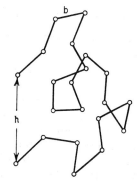

*FIG. 3.7 Scheme of a freely jointed chain.*

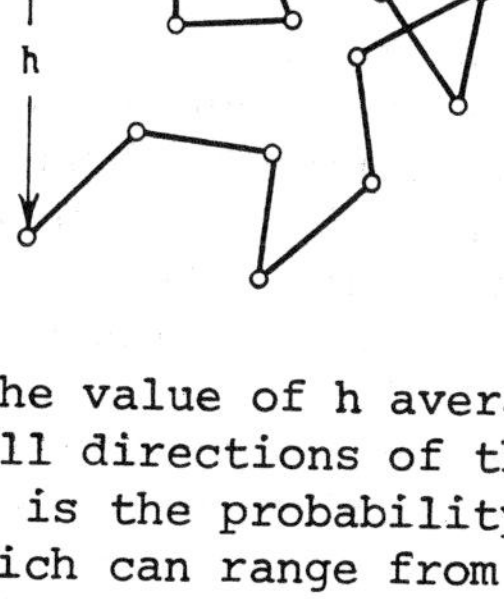


Obviously, because of thermal motion the value of h averaged over all conformations is zero since all directions of the vector are of equal probability. What is the probability distribution of the values of $h \equiv |\mathbf{h}|$, which can range from zero to the maximal length of the chain (which equals Zb (where Z is the number of links and b the length of each link)? The solution to this problem resembles that of the diffusion problem--the determination of the probability of the displacement of a Brownian particle for length h, as the result of Z paths, all of length b [21]. The distribution happens to be Gaussian--the probability of the end-to-end distance of the chain lying in the range from h to h + dh is

$$W(h)\,dh = (3/2\pi Zb^2)^{3/2} 4\pi h^2 \exp(-3h^2/2Zb^2)\,dh \tag{3.13}$$

Consequently

$$\bar{h} \cong \int_0^\infty hW(h)\,dh = 0$$

and

$$\overline{h^2} \cong \int_0^\infty h^2 W(h)\,dh = Zb^2 \tag{3.14}$$

which means that the chain is strongly coiled. Formulas (3.13) and (3.14) are valid if $Z \gg 1$.

Hence, thermal motion folds the macromolecule into a coil. Such a state is the most probable one because the elongated conformation can be formed in only one way and the coiled one in a multitude of ways. The coiled form of macromolecules in solution is confirmed by direct experimentation (discussed later).

The entropical nature of the high elasticity of rubber follows directly from this consideration. In the nonstretched state the chains are folded, forming statistical coils that correspond to maximal entropy. Stretching the coils diminishes the entropy. The entropy of the chain is

$$S = k \ln W(h) = C - (3kh^2/2Zb^2) \tag{3.15}$$

where C is a constant not depending on h. In accordance with (3.7), the elastic force produced by stretching the chain is

$$f = -T(\partial S/\partial h)_T = (3kT/Zb^2)h \tag{3.16}$$

The modulus of elasticity, like that of an ideal gas (which is equal to $p = RT/V$; p. 106), depends linearly on temperature. This is the kinetic theory of rubber, which was suggested for the first time by Kuhn.

In real macromolecules the valence angles between bonds are fixed and rotations are hindered. Let us fix the positions of the first two links of a polyethylene chain (Fig. 3.8). The third link can have different positions at the conic surface with an opening of $2\theta$ ($\pi-\theta$ is the valence angle of the C–C bonds close to the tetrahedral one 109°28'). These various positions, characterized by rotation angle $\phi$ around the second C–C bond, possess various energies $U(\phi)$. The position of the fourth bond in relation to the first two bonds is less definite, since it is located at the cone circumscribed around every position of

*FIG. 3.8 A chain with fixed valence angles.*

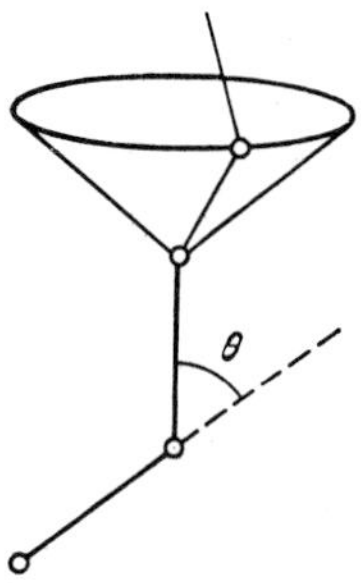


the third bond, etc. The position of a bond sufficiently remote from the first one is practically arbitrary. Therefore, a long chain folds in a coil. The macromolecule can be mentally divided into segments whose positions are not correlated. For such a system expressions (3.13)-(3.16) remain valid but Z and b designate the number and length, respectively, of the freely jointed segments, not of real bonds.

It is possible, however, to express the average square of the coil length, $h^2$, not by theoretical Z and b values but by real values: N (the number of links); $\ell$ (the length of the bond); and $\theta$ (the angle complementary to the valence angle). The vector h is the sum of the link vectors

$$h = \sum_{i=1}^{N} \ell_i \qquad (3.17)$$

Hence

$$\overline{h^2} = \sum_{i=1}^{N} \sum_{j=1}^{N} (\ell_i \ell_j) = \sum_{i=1}^{N} \overline{\ell_i^2} + 2 \sum_{i=2}^{N} \sum_{j=1}^{i=1} \overline{(\ell_i \ell_j)} \qquad (3.18)$$

Expression (3.14) follows automatically from (3.18). For the freely jointed chain $(\overline{\ell_i \ell_j}) = 0$ if $i \neq j$ and

$$\overline{h^2} = N\ell^2 \qquad (3.19)$$

In a real chain the average values of the scalar products $(\ell_i \ell_j)$ depend on $\theta$ and $\phi$. For chains with symmetrical side groups of the polyethylene type (but not that of polystyrene) rigorous calculation (cf. [2,3,5]) gives the formula (first obtained by Oka [22])

$$\overline{h^2} = N\ell^2 \frac{1 + \cos\theta}{1 - \cos\theta} \frac{1 + \eta}{1 - \eta} \qquad (3.20)$$

where $\eta$ is the average cosine of the internal rotation angle

$$\eta = \overline{\cos\phi} = \int_0^{2\pi} \exp(-U(\phi)/kT) \cos\phi \, d\phi \Big/ \int_0^{2\pi} \exp(-U(\phi)/kT) \, d\phi \qquad (3.21)$$

Formula (3.20) is valid for long chains with N >> 1. The value

$\overline{h^2}$ calculated by means of (3.20) is larger than that of a freely jointed chain (3.19)--the correlation determined by the lack of freedom of rotation and by the valence angles elongates the chain. Evidently the value of $\overline{h^2}$ can be considered a measure of the chain's thermodynamic flexibility--the smaller $\overline{h^2}$ is at given N and $\ell$, the greater the flexibility is. For a very stiff chain, characterized mainly by trans rotamers, the angle $\phi$ is near 0° and $\eta$ near to 1°. From the formula (3.20) we get

$$\overline{h^2} = N\ell^2 \frac{1 + \cos\theta}{1 - \cos\theta} \frac{2}{1 - \eta} \tag{3.22}$$

For flexible chains, calculation of $\eta$ according to (3.21) requires knowledge of the potential energy $U(\phi)$.

The theory of sizes and forms, as well as of dipole moments and anisotropic polarizabilities, of polymeric chains must be based on the physical mechanism of their flexibility. In every case flexibility is determined by rotations around single bonds For the first time the restricted motions of the links have been taken into account in the form of torsional vibrations around the trans conformations [23,24]. In this case the function $U(\phi)$ looks like

$$U(\phi) = \frac{1}{2} U_0 (1 - \cos\phi) \cong \frac{1}{4} U_0 \phi^2 \tag{3.23}$$

and calculation of $\overline{h^2}$ gives expression (3.22) since $\cos\phi$ is close to 1. However, comparison with experiment did not confirm the validity of this formula for real macromolecules.

As we have seen, the internal rotation in small molecules that lack axial symmetry produced a mixture of rotational isomers. The same is true of macromolecules. In Figs. 3.9 and 3.10 the rotamers of n-butane and polyethylene, respectively, are shown occurring as results of internal rotation around any

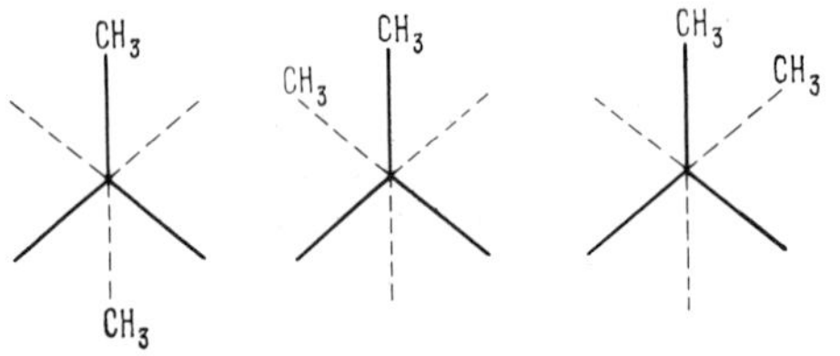


*FIG. 3.9 Rotamers of n-butane.*

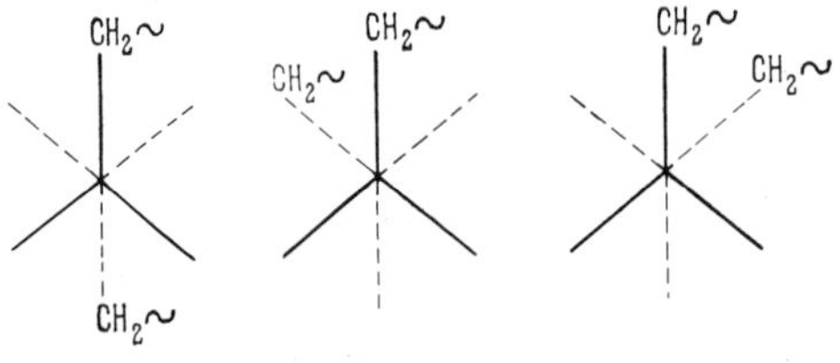


*FIG. 3.10 Rotamers of polyethylene.*

C-C bond of the chain. The distinction is that in the second case instead of the $CH_3$ group, $CH_2$ groups connected with the continuing chain are present; this is indicated in Fig. 3.10 by wavy lines.

The rotational-isomeric theory of macromolecules, which was proposed in 1951 [25], treated the macromolecule as an equilibrium mixture of rotational isomers and the internal rotation in a chain as rotational isomerization. In other words, the continuous curve $U(\phi)$ is replaced by a finite and small number of infinitely narrow wells of various depths at definite values of $\phi$.

Of course, in reality there are fluctuations, torsional vibrations around the $\phi$ positions, corresponding to rotational isomers. However, because these fluctuations are random, they compensate each other and have no influence on the average properties of a macromolecule; heance, they can be disregarded.

The rotational-isomeric theory replaces integration (3.21) by summation. Thus for polyethylene

$$\eta = \frac{\cos\phi_1 \exp(-E_1/kT) + \cos\phi_2 \exp(-E_2/kT) + \cos\phi_3 \exp(-E_3/kT)}{\exp(-E_1/kT) + \exp(-E_2/kT) + \exp(-E_3/kT)} \tag{3.24}$$

According to (3.12), $E_2 - E_1 = E_3 - E_1 = \Delta E$, $\phi_1 = 0°$, $\phi_2 = 120°$, $\phi_3 = -120°$. We get

$$\eta = \frac{1 - \exp(\Delta E/kT)}{1 + 2\exp(\Delta E/kT)} \tag{3.24a}$$

If $\Delta E$ = 600 cal/mole, T = 0°, $\eta$ = 0.37. We find, with Oka's formula (3.20), taking $\cos\theta = -\frac{1}{3}$

$$\overline{h^2} = 1.1N\ell^2 \tag{3.25}$$

The thermodynamic flexibility of macromolecules (i.e., the degree of folding of the statistical coil) is greater the smaller $\Delta E$ is (i.e., the smaller $\eta$ is). If the energy minima are distributed in a symmetrical way ($\phi_1 = 0°$, $\phi_2 = 120°$, $\phi_3 = -120°$), then at $\Delta E = 0$, $\eta = 0$ and the macromolecule behaves like a chain with free internal rotation.

Obviously the rotational-isomeric theory holds for those cases in which the energy minima are separated by barriers that are larger than kT. If this condition is not satisfied, the theory retains its validity as an approximate mathematical method, which makes it possible to replace integration by summation.

The rotational isomerism of polymers was established with the help of a series of methods, the first of which was infrared spectroscopy. For the most important polymers the barriers are in fact much larger than kT. The rotational-isomeric theory was confirmed experimentally in studies of the thermomechanical properties and the stretching of polymers (cf. p. 121). This theory, which forms the basis of the contemporary statistical physics of macromolecules, is presented in three successive monographs [2,3,5].

Concrete calculations of the flexibility of polymeric chains must be based on their chemical structure. Thus, the conformations of monomeric links in polymers of the type $(-CH_2-CHR-)_n$ (e.g., polystyrene) and $(-CH_2-CR_2-)_n$ are due mainly to interactions of the bulky side groups R. Information concerning these conformations was obtained by means of X-ray studies of crystalline polymers. Because of configurational heterogeneity and the dispersion of chain lengths, conventional polymers do not crystallize or crystallize only partially. However, stereoregular polymers crystallize well; they can even be obtained in the form of monocrystals. In bulky polymers crystallization is also incomplete.

Crystallization is hindered by kinetic factors as well as by heterogeneity. When forming a crystal, molecules have to reorient themselves. Statistically fluctuating coils cannot be crystallized--the chains would have to be stretched. Even if thermodynamic conditions favor the orientation and unfolding of the coils, these processes may require more time than the experiment permits. It is necessary to overcome the barriers of internal rotation. Whereas equilibrium thermodynamic properties of rotameric macromolecules are determined by differences in the rotamer energies, kinetic properties depend on the heights of the energy barriers. For crystallization, kinetic as well as thermodynamic flexibility is important. The heating of a polymer or its swelling in a low-molecular solvent facilitates crystallization and increases the degree of crystallinity.

The thermodynamic conditions of the crystallization and melting of polymers are directly related to rotational isomerization. The melting point is expressed as a ratio of the enthalpy difference and the entropy difference (p. 27). The increase in entropy during polymer melting is dependent on the transition between an ordered, selected chain conformation and a mixture of rotational isomers in a random coil. The entropy of mixing makes an essential contribution to $\Delta S$ [2,3].

The typical structure of polymer fibers is crystalline. Macromolecules in such fibers are regularly packed, their chains parallel to the fiber axis. Fibrous polymeric polycrystals play an important role in life and technology. Synthetic fibers (polyamides, polyesters, etc.) and natural ones (cellulose fibers, silk, and wool) all have a crystalline structure. The natural fibers are fibrillar proteins (cf. Chapter 4).

X-ray analyses of stereoregular polymers performed by Natta and Corradini [26,27] have shown that the repeating structural units of the chain in a crystal occupy geometrically

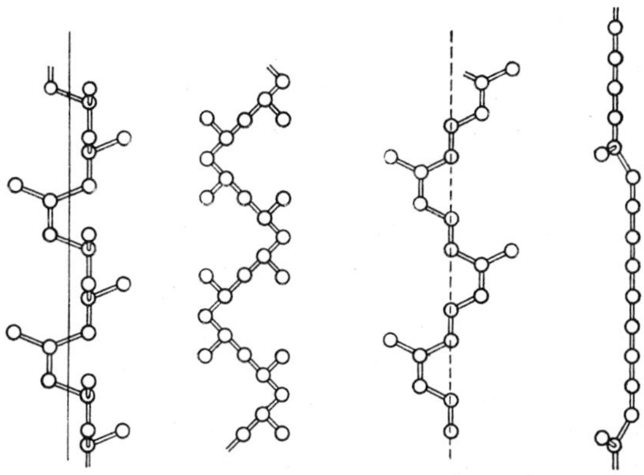


*Fig. 3.11 Conformations of crystalline chains: (a) isotactic polypropylene; (b) syndiotactic polypropylene; (c) cis-1,4-polyisoprene; (d) nylon 77.*

equivalent positions in relation to the chain's axis. The postulate of the equivalency of the geometrical positions of structural units can be formulated [3]; it makes possible the establishment of the symmetry of the chain in a crystal and the conformations of monomeric links proceeding from the chain's stereochemical structure (i.e., from its configuration). The regularity of the chain and equivalence of its links lead to the crystallization of polyamides and polypeptides in helical forms only because the structure of their monomeric links is nonsymmetrical. 1,4-Polyisoprene has to be crystallized as either a helix or a structure possessing a plane of sliding. Figure 3.11 shows the conformations of crystalline chains of some polymers. The helix as a whole can be right- or left-handed independently of the presence of chirality in the monomers. Therefore, the helix can be optically active. However, a sample of synthetic polymer obtained from symmetric monomers is optically inactive. It contains equal amounts of right and left helices, a racemic mixture.

The conformation of a stereoregular macromolecule $(-CH_2-CRR'-)_n$ is characterized by two angles of internal rotation $\phi_1$ and $\phi_2$ corresponding to rotation around the bonds $CH_2-CRR'$ and $CRR'-CH_2$. The helix can be described by the expression $(\phi_1,\phi_2)_n$ with $\phi_1 \neq \phi_2$. The equality $\phi_1 = \phi_2$ is possible only if these angles have the value zero, that is, for the planar trans conformation. If $(\phi_1,\phi_2)_n$ designates the right helix, then the left helix (mirror image) is designated by $(-\phi_2,\phi_1)_n$. These relations are described in detail in [3]. Isotactic polymers $(-CH_2-CHR-)_n$ crystallize mainly in the form of helix $3_1$, corresponding to $\phi_1 = 0°$ $\phi_2 = 120°$. It is a regular alternation of the trans and gauche rotamers. Syndiotactic polymers, on the other hand, crystallize either in the form of coplanar trans chains or in helices with the angles $\phi_1 = \phi_1' = 0°$, $\phi_2 = \phi_2' = -120°$ (polypropylene; in this case, the chain conformation has to be characterized not by two but by four rotation angles).

The melting of a crystalline helical polymer is rotational isomerization, the helix-coil transition. Similar processes are of great importance for molecular biophysics, the physics of proteins and nucleic acids (Chapters 4 and 8). Reports [15, 16,28,29] have been published of the calculations of the internal rotation energies as functions of the angles $\phi_1$ and $\phi_2$ for several polymers (cf. p. 112). Such calculations are now being done for many of the most important synthetic polymers (cf. [3, 5]) and biopolymers. The results concerning protein chains are treated in Chapter 4.

Thus the notion of the rotational isomerism of macromolecules is supported by the results of investigations of crystalline polymers. In their turn, these results are very important for the theoretical treatment of macromolecules in solution.

Fixation of definite rotational isomers for all links of the chain in a crystal ensures its long-range order. Knowing the positions of the atoms of a given monomer, we know them for any other monomer, because the arrangement of atoms is strictly periodic. At the same time there is also short-range order in the crystal--the definite arrangement of neighboring links. If the polymer is melted or dissolved, the long-range order vanishes, but there are weighty arguments insisting that the short-range order remains. It is established for the low-molecular liquids. The contemporary theory of liquids is based on the hypothesis of their quasicrystalline structure introduced by Frenkel [24]. Maintenance of the short-range order in macromolecules follows from the correspondence of the crystalline structures of polymers to a minimum in potential energy. Using these ideas, Ptitsyn and Sharonov [30] suggested that the near short-range order in a free macromolecule in a state of statistical coil is analogous to the long-range order in a one-dimensional crystal. This suggestion was confirmed by the quoted calculations of conformations and in spectroscopic and optical investigations [3]; it became the basis for rigorous calculations of the physical properties of free macromolecules using the rotational-isomeric theory.

For chains with symmetrical side groups $(-CH_2-CR_2-)_n$ the potential does not depend on the sign of rotation around single bonds, that is

$$U(\phi_1,\phi_2) = U(-\phi_1,-\phi_2)$$

where $\phi_1$ is the rotation angle around the $CH_2-CR_2$ bond and $\phi_2$ that for the rotation around $CR_2-CH_2$. Since both these bonds are identical

$$U(\phi_1,\phi_2) = U(\phi_2,\phi_1) = U(-\phi_2,-\phi_1)$$

If only one rotamer $(\phi_1,\phi_2)$ is present in the crystal, then in solution the rotamers $(\phi_2,\phi_1)$, $(-\phi_1,-\phi_2)$, and $(-\phi_2,-\phi_1)$ must also be present with the same energy. Hence, the whole set of rotamers is known and calculation of $\eta = \overline{\cos\phi}$ becomes possible.

For nonsymmetrical isotactic polymers the crystalline conformation $(\phi_1,\phi_2)$ gives $(\phi_1,\phi_2)$ and $(-\phi_2,-\phi_1)$ for the free macromolecule. For syndiotactic polymers the crystalline conformation $(\phi_1,\phi_1,\phi_2,\phi_2)$ corresponds to $(\phi_1,\phi_1)$ and $(\phi_2,\phi_2)$ in solution. Obviously the rotations in neighboring links of the chain are correlated (cooperativity is present; cf. the next section) and the average square of the chain length is expressed by

$$\overline{h^2} = e^{-\Delta U/kT}\,(\overline{h^2})_{\Delta U=0} \tag{3.26}$$

where (for the simplest case of two equivalent rotamers)

$$\Delta U = U[(\phi,\phi),(\phi,\phi)] - U[(\phi,\phi),(-\phi,-\phi)]$$

The parentheses are related to the two neighboring monomers. For chains that crystallize in a helical conformation the sequence of equal rotamers is energetically preferable to that of nonequal rotamers; in other words, $\Delta U < 0$. Correlation elongates the chain (makes it stiffer).

The $\Delta U$ values can be estimated theoretically. For polyethylene, stable conformations are (0°,0°) (trans-trans, tt); (0°,120°) and (0°,-120°) (ts); (120°,0°), (-120°,0°) (st) and (120°,120°), (-120°,-120°) (ss). Conformations (120°,-120°) and (-120°,120°), which can be designated as $(s^+s^-)$ and $(s^-s^+)$ are energetically unprofitable. In this case conformations (ts) and (st) contain nonequivalent rotamers, and calculation of the correlation is somewhat more difficult in comparison with formula (3.26). The $\Delta U$ values for (tt) and (st) are equal to zero; those for (ts) and (ss), 0.5 kcal/mole; those for $(s^+s^-)$, 2.3-2.7 kcal/mole. We get

$$\overline{h^2} = 3.2\overline{h_0^2} = 1.6N\ell^2 \tag{3.27}$$

Unlike (3.25), this value is in agreement with experiment.

For nonsymmetrical chains Oka's formula (3.20) is no more valid. Theory gives more complicated expressions containing $\overline{\sin\phi}$ as well as $\overline{\cos\phi}$. If the side groups are symmetrical, $\overline{\sin\phi} = 0$.

The rotational-isomeric theory gives a quantitative explanation of the physical properties of macromolecules in solution, of the sizes and forms of coils, of their dipole moments and their anisotropic polarizabilities. The theory, which is in agreement with experiment [2,3,5], explains the

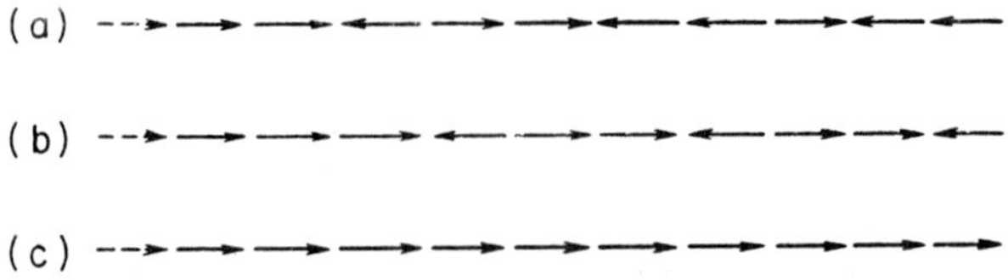


*FIG. 3.12 The stretching of a one-dimensional chain.*

physical mechanism of the stretching of elastic polymers, such as rubber. The stretching of the chain alters the set of its conformations. The mechanism of this alteration is rotational isomerization. We illustrate this statement by the one-dimensional model of a macromolecule, every link of which we represent by an arrow of length $\ell$ which can point either right or left. To one rotamer (designated t) correspond two neighboring arrows with the same directions; to another one (designated s), two neighboring arrows with opposite directions. Figure 3.12a shows a chain containing 10 links, 5 directed toward the right, 5 toward the left. The total length of the chain is a vector sum of all arrows. Hence, the length of this chain is zero. The chain contains 5 t and 5 s rotamers (in order that the total number of rotamers equal the number of links, i.e., 10, we have added the "zero" link directed toward the right; it is shown as a dashed arrow). Let us stretch the chain by a force directed to the right. Figure 3.12b shows a stretched state of the chain in which the length becomes $h = (7 - 3)\ell \equiv 4\times\ell$, but the number of rotamers remains the same. The chain is elongated as a result of the redistribution of rotamers; there is no change in their relative content. The redistribution is described by the transition

ttsststsst $\rightarrow$ tttsstssts

It is not accompanied by a change in energy, but the entropy changes, decreasing with the increase in ordering of the rotamers. However, redistribution of the t and s rotamers alone cannot effect the stretching of the chain. The completely extended chain (Fig. 3.12c) contains only t rotamers. In this case the transformation is accompanied not only by a change in entropy, but also by a change in internal energy, since $10E_t \neq 5E_t + 5E_s$.

The foregoing is the molecular approach to polymer stretching. Theory predicts that expression (3.16) for a highly elastic deformation is not rigorous. The elasticity of entropy must exist along with the elasticity of the energy because of the difference in rotamer energies $\Delta E$. In other words, the value

$$f' = (\partial E/\partial h)_T$$

has to be different from zero and depend on $\Delta E$ [2,3,31]. This theoretical prediction was confirmed experimentally. Investigation of the thermomechanical properties of rubber-like polymers by Flory and his collaborators on the basis of rotational-isomeric theory not only has shown the presence of energy elasticity, but has also yielded experimental values of $\Delta E$ that are in good agreement with the theoretical ones [3,32]. On the other hand, direct investigation of polymer stretching by means of infrared spectroscopy has shown that stretching in fact alters the relative content of rotamers [2,3,33-35].

## 3.4 The Macromolecule as a Cooperative System

As a consequence of the interdependence of the links in the chain, a macromolecule is a cooperative system. The state of a given link depends on the states of its neighbor links. Taking this interdependence into account, we get an expression for energy that contains the terms depending on the conformations of at least the two neighbor links

$$U(\Omega_1, \Omega_2, \ldots, \Omega_N) = \sum_{k=1}^{N} U(\Omega_{k-1}, \Omega_k) \tag{3.28}$$

where $\Omega_k$ denotes the conformations of the kth link.

A macromolecule is a one-dimensional cooperative system in which every link has two neighbors. Statistical treatment of such a system (the evaluation of its partition function) is much simpler than that of a two-dimensional one, and very much easier than that of a three-dimensional system. Calculations can be done on the basis of the one-dimensional Ising model (cf. p. 28). The partition function for the one-dimensional model is calculated in an analytical form with the help of a matrix method suggested by Kramers and Vannier [36] (see also [2,3,37-41].

The conformational (rotameric) partition function of a macromolecule in the absence of external forces has the form

$$Z = \sum_{\Omega_1} \sum_{\Omega_2} \cdots \sum_{\Omega_N} \exp[-U(\Omega_1, \Omega_2, \ldots, \Omega_N)/kT] \tag{3.29}$$

Summation must be done over all r rotamers of every link. If expression (3.28) is valid, then

$$Z = \sum_{\Omega_1} \sum_{\Omega_2} \cdots \sum_{\Omega_N} \prod_{j=1}^{N} g(\Omega_{j-1}, \Omega_j) \tag{3.30}$$

where

$$g(\Omega_{j-1}, \Omega_j) = \exp(-U(\Omega_{j-1}, \Omega_j)/kT \tag{3.31}$$

Assuming the polymer to be formed by repeated identical units, we can treat the values $g(\Omega_{j-1}, \Omega_j)$ as elements of the matrix

$$G = \begin{pmatrix} g(\Omega^{(1)},\Omega^{(1)}) & \cdots & g(\Omega^{(1)},\Omega^{(r)}) \\ \vdots & \vdots & \vdots \\ g(\Omega^{(r)},\Omega^{(1)}) & \cdots & g(\Omega^{(r)},\Omega^{(r)}) \end{pmatrix} \tag{3.32}$$

Let us denote

$$g(\Omega_{j-1}^{(\alpha)},\Omega_j^{(\beta)}) = G_{\Omega_{j-1}^{(\alpha)},\Omega_j^{(\beta)}} = g_{\alpha\beta}$$

Partition function (3.30) can be rewritten in the form

$$Z = \sum_{\Omega_1}\sum_{\Omega_2}\cdots\sum_{\Omega_N} G_{\Omega_0\Omega_1}G_{\Omega_1\Omega_2}\cdots G_{\Omega_{N-1}\Omega_N}$$

and according to the rule of multiplication of matrices

$$Z = \sum_{\Omega_N} (G^N)_{\Omega_0\Omega_N} \tag{3.33}$$

Introducing the cyclic condition $\Omega_0 = \Omega_N$ (if $N >> 1$, this condition plays no practical role), we get Z as the spur, that is, the sum of the diagonal terms of matrix G taken in the Nth power:

$$Z = \mathrm{Sp}(G^N) = \lambda_1^N + \lambda_2^N + \cdots + \lambda_r^N \tag{3.34}$$

where $\lambda_1,\lambda_2,\ldots,\lambda_r$ are the eigenvalues of matrix G. All elements of the matrix are positive; hence, it has the maximal eigenvalue $\lambda_1$, substantial, positive and nondegenerate (Frobenius's theorem; see [42]). If $N >> 1$

$$Z \cong \lambda_1^N \tag{3.35}$$

Thus calculation of the rotameric partition function comes down to the evaluation of the maximal root $\lambda_1$. Knowing Z, we can evaluate the equilibrium properties of a macromolecule. In this way we obtain expressions for the average square of the length of the chain in both the crystalline (helical) state and the coiled state. Indeed, the fixation of the valence angle between neighboring links determines the correlation of these links, the cooperativity of the chain. The same is true of the correlation of rotational isomers. Oka's formula (3.20) and the similar ones for chains with nonsymmetrical side groups, formula (3.26), are most easily obtained by the matrix method [3,5].

Consider here the simple example shown in Fig. 3.12, the stretching of a one-dimensional cooperative model by an external force f. Let the energy of the t rotamer (i.e., the energy of two neighboring parallel links) be $-\varepsilon$ and the energy of the s rotamer (i.e., of the two neighboring antiparallel links) be $\varepsilon$. Hence, the difference in the energies of the two rotamers is equal to

$$\Delta E = 2\varepsilon \tag{3.36}$$

We write this condition in the form

$$E_{ij} = \begin{cases} -\varepsilon\sigma_i\sigma_j & \text{if } i = j+1 \\ 0 & \text{if } i \neq j+1 \end{cases} \tag{3.37}$$

Here $\sigma_i = 1$ if the arrow is directed toward the right and $\sigma_i = -1$ if it is directed toward the left; i and j are the numbers of links. Energy $E_{ij}$ is zero if the links are not neighbors--correlation includes only neighboring arrows. If an external force directed to the right acts on the one-dimensional chain, then every arrow of the length $\ell$ receives additional energy

$$E_f = -\ell f \cos(\ell, f) = -\ell f \sigma_j \tag{3.38}$$

Since every arrow can have only two orientations, the matrix G is of the second rank (r = 2). We get, according to (3.31)

$$g(\sigma_j, \sigma_{j+1}) = \exp\left(\frac{\varepsilon}{kT}\sigma_j\sigma_{j+1} + \frac{\ell f}{kT}\sigma_j\right) \tag{3.39}$$

and the matrix G is obtained by putting into $g(\sigma_j, \sigma_{j+1})$ the values $\sigma_j$ and $\sigma_{j+1}$, equal to +1 and -1

$$G = \begin{pmatrix} g(1,1) & g(1,-1) \\ g(-1,1) & g(-1,1) \end{pmatrix} \tag{3.40}$$

Setting $a = \varepsilon/kT$, $b = \ell f/kT$, we get

$$G = \begin{pmatrix} e^{a+b} & e^{-a+b} \\ e^{-a-b} & e^{a-b} \end{pmatrix} \tag{3.41}$$

The eigenvalues obey the equation

$$\begin{vmatrix} e^{a+b}-\lambda & e^{-a+b} \\ e^{-a-b} & e^{a-b}-\lambda \end{vmatrix} = 0 \tag{3.42}$$

They are

$$\lambda_{1,2} = e^a \cosh b \pm (e^{2a}\sinh^2 b + e^{-2a})^{1/2} \tag{3.43}$$

Since $\lambda_1 > \lambda_2$, it follows from (3.35) that

$$Z \cong \lambda_1{}^N = [e^a \cosh b + (e^{2a}\sinh^2 b + e^{-2a})^{1/2}]^N \tag{3.44}$$

We have obtained a simple analytical expression for the partition function. With its help we can evaluate the mean length of the chain

$$\bar{h} = kT \frac{\partial \ln Z}{\partial f} = NkT \frac{\partial \ln \lambda_1}{\partial f} = N\ell \frac{\sinh b}{(\sinh^2 b + e^{-4a})^{1/2}} \tag{3.45}$$

The measure of cooperativity is the quantity $\varepsilon$ (i.e., a), or the difference in energies of the rotamers $\Delta E$. In the absence of cooperativity (i.e., at $a = 0$)

$$\bar{h} = N\ell \tanh b \equiv N\ell \tanh(f\ell/kT) \tag{3.46}$$

Compare this equation with formula (1.48). If the arrows are reoriented freely--in the absence of rotamers--the length of the chain is larger, the larger the force and the lower the temperature. At small forces (i.e., if $f\ell \ll kT$)

$$\bar{h} \cong N\ell^2 (f/kT) \tag{3.47}$$

or

$$f \cong (kT/N\ell^2)\bar{h} \tag{3.48}$$

that is, the length of the chain is proportional to the stretching force (Hooke's law), and the modulus of elasticity is proportional to the absolute temperature, in agreement with (3.16).

If, on the contrary, $a \gg 1$, (i.e., $\Delta E \gg kT$), then the value $\exp(-4a)$ in (3.45) can be neglected and $h = N\ell$ independently of the force. The chain is stiff and therefore elongated. We obtain the same result with large forces, if $f\ell \gg kT$. Here, too, $\sinh^2 b \gg \exp(-4a)$--the large force causes total stretching of the chain.

The mean square length of a rotameric one-dimensional chain in the absence of external force is calculated according to (3.18):

$$\overline{h^2} = N\ell^2 + 2 \sum_{i=2}^{N} \sum_{j=1}^{i-1} \overline{(\boldsymbol{\ell}_i \boldsymbol{\ell}_j)} = N\ell^2 \frac{1+\eta}{1-\eta} + 2\ell^2 \frac{\eta(\eta^N - 1)}{(1-\eta)^2} \tag{3.49}$$

where $\eta$ is the average cosine of the angle between two neighboring arrows. In our case [cf. (3.24)]

$$\eta = \frac{\cos 0° \exp(\varepsilon/kT) + \cos 180° \exp(-\varepsilon/kT)}{\exp(\varepsilon/kT) + \exp(-\varepsilon/kT)} = \tanh \frac{\varepsilon}{kT}$$

and if $N \gg 1$

$$\overline{h^2} = N\ell^2 \exp(2\varepsilon/kT) = N\ell^2 \exp(\Delta E/kT) \tag{3.50}$$

This expression is similar to (3.26). Correlation due to the energy advantage of the parallel orientations of arrows elongates the chain. The one-dimensional Ising model gives results that are very instructive for the treatment of macromolecules--

of systems with strong interactions along the chain. A one-dimensional model cannot be used for ferromagnets. Indeed, if we replace the length of the chain $\ell$ by the magnetic moment $\mu$ and the external force f by magnetic field H, we get expression (3.45) for magnetization, which does not describe the spontaneous magnetization and phase transition at the Curie point (cf. p. 29). The theory of ferromagnetism requires at least a two-dimensional model.

All this presentation shows that the rotational isomerization of polymers, both in the processes of the melting of crystals and of stretching, is a cooperative process. This thesis is very important for an understanding of the properties of biopolymers.

The statistics of one-dimensional cooperative systems has features in common with the theory of Markov's chains [21,43 44]. Markov's chains are sequences of interdependent random events. The probability of a given event in a chain depends on the previous events. In a simple Markov chain the appearance of a given event depends on the outcome of one previous event; in a complex Markov chain it depends on several previous events. The probabilities are interdependent and the Markov chain is a system with "cooperativity in time." Obviously we meet such cooperativity when studying teleological evolutionary systems.

The mathematical apparatus of Markov chains is similar to the statistical theory of a cooperative polymer chain. The chain, when strong short-range interactions are present, can be modeled by a Markov process with a memory of a finite number of steps (see [3], §10). However, physics is different here. The conformation of a macromolecule is changing continuously because of thermal motion. The conformation of every link depends on both previous and subsequent links. In this sense cooperativity in space differs from cooperativity in time. The theory of Markov chains can be applied directly to investigations of the content and sequence of links in a copolymer formed by polymerization of two or more different monomers if the probabily of binding monomers A and B to the end of a growing chain depends on the kind of monomer at the chain's end. Such copolymers can be called Markov copolymers ([5], Section IV.7).

## 3.5 The Peculiarities of the Macromolecule as a Statistical System

A macromolecular chain also differs from a Markov chain in the prohibition against self-intersection of the chain. If a Brownian particle wanders in a random way, such a prohibition does not exist--we are dealing with a sequence of steps (links) in time. A real macromolecule is made up of atoms whose interactions are repulsive at small distances. The short-range interactions of neighboring atoms in the chain are responsible for rotamerism and the correlation of rotamers. But in a randomly coiled chain there also exist long-range interactions: the mutual repulsion of atoms whose positions along the chain are remote but that are nonetheless close to one another. Repulsion can be explained by the actual volume of the atoms and by the prohibition against their mutual penetration. Consequently, the influence of mutual repulsion is called a volume effect.

Evidently volume effects perturb the distribution of the lengths of molecules, which we calculated earlier without considering these effects. If $\overline{h^2}$ is the actual mean value of $h^2$, then

$$\overline{h^2} = \alpha^2 (\overline{h^2})_0 \tag{3.51}$$

where $(\overline{h^2})_0$ is the value calculated if only short-range interactions are taken into account, and $\alpha$ is the factor expressing the change in the coil length caused by long-range interactions (i.e., volume effects).

The volume effects depend not only on the intrinsic volume of the macromolecule but also on its interaction with the solvent; $\alpha$ depends on solvent and temperature. Theoretical evaluation of $\alpha$ is rather complicated. However, it is possible to exclude the perturbation experimentally and to determine the unperturbed size of the coil.

The excluded volume increases in good solvents; in bad solvents, however, mutual attraction of the chain atoms is greater than their attraction by the solvent molecules and the volume effects decrease. If solvent and temperature are properly chosen, the final volume of a link can be exactly compensated by the mutual attraction of links. The excluded volume vanishes and $\alpha = 1$. These ideas were developed by Flory [45-47]. Evidently this phenomenon is similar to the Boyle point of a real gas. At the Boyle temperature the repulsion of molecules is compensated by their mutual attraction. Flory introduced the notion of the theta ($\theta$) point of a polymer solution. Like the Boyle point, the $\theta$ point is the temperature at which the second virial coefficient of osmotic pressure becomes zero (cf. p. 133). This means that at the $\theta$ point the osmotic pressure follows Van't Hoff's law.

Therefore theoretical calculations of the properties of macromolecules that do not take into account long-range interactions must be compared with experimental results obtained at the $\theta$ point. Another possibility consists in extrapolating the data obtained in good solvents to the $\theta$ point. These extrapolations require theoretical evaluation of $\alpha$. A practical theory was developed for this purpose (cf. [2,47,48]). Volume effects are especially important in dilute solutions in good solvents (which are most often used).

The limitations of the analogy between the macromolecular chain and a stochastic Markov process proceeding in time are also expressed in the fundamentals of macromolecular statistics. Its principal features were treated by Lifshitz [49]. A macromolecule is characterized by linear memory--the links are connected in a chain and have successive positions there. Therefore, the links, units in a statistical ensemble, are basically distinct. Every one of them has its own place in the chain and the transposition of links requires the breakage of chemical bonds. Linear memory is present both in a homogeneous homopolymeric chain and in an informational chain of biopolymer. In the second case it is expressed by the primary structure.

The fixed sequence of links implies the nontotal equilibrium of the chain. The chain structure determines the peculiarities of the fluctuative behavior. The characteristic time of the conformational relaxation of a long and flexible chain is proportional to $N^2$ (N is the number of links). The condition of chain stability is

$$N^2 \exp(-E_0/kT) \ll 1$$

where $E_0$ are the energy barriers hindering the breakage of valence bonds.

Let us denote by $\mathbf{x}_j$ ($j = 1,\ldots,N$) the coordinates of the links. Because of their connections, the positions of the neighboring links are correlated. This correlation is expressed by the functions

$$g_j = g(\mathbf{x}_j, \mathbf{x}_{j+1})$$

and the distribution function of links in the configurational space of all coordinates has the form

$$\rho(\mathbf{x}_1,\ldots,\mathbf{x}_N) = \prod_j g_j \tag{3.52}$$

Let us introduce the coordinates $y_j = \mathbf{x}_{j+1} - \mathbf{x}_j$. The functions $g_j = g(y_j)$ are normalized

$$\int g_j \, dy_{j\xi} \, dy_{j\eta} \, dy_{j\zeta} = 1$$

where $\xi$, $\eta$, and $\zeta$ are the spatial coordinates. For simplicity it can be assumed that the $g_j$ depend only on the coordinates of neighboring links, not on their orientations. The probability for the kth link to be at point x if the zeroth link is fixed at the origin of coordinates is equal to (for large k)

$$\rho_k(\mathbf{x}) \cong k^{-3/2} \exp(-x^2/4ka^2) \tag{3.53}$$

where

$$a^2 = \frac{1}{6}\overline{y^2} = \frac{1}{6}\int g(y) y^2 dy_\xi dy_\eta dy_\zeta \tag{3.54}$$

Hence, this distribution is Gaussian (cf. p. 114).

As we have seen, at $N \gg 1$ the chain folds in a statistical coil because of the fluctuations of bends. The linear size of the coil is

$$h_N \sim aN^{1/2}$$

and its volume is

$$V_N \sim a^3N^{3/2}$$

Consequently the mean density of the coil is

$$N/V_N \sim a^{-3}N^{-1/2}$$

The density of the links at point x is

$$n(\mathbf{x}) = \sum_{j=1}^{n} \delta(\mathbf{x}-\mathbf{x}_j) \tag{3.55}$$

where $\delta$ is the Dirac function, equal to $\infty$ at $x = x_j$ and 0 if $x \neq x_j$. The mean value of n is calculated with the distribution function (3.53). We get

$$n(\mathbf{x})dx_\xi dx_\eta dx_\zeta = N\rho(\mathbf{r})dr_\xi dr_\eta dr_\zeta \tag{3.56}$$

where $\mathbf{r} = \mathbf{x}/h_N$, $h_N = aN^{1/2}$, and

$$\rho(\mathbf{r}) = \frac{1}{8\pi^{3/2}} \int_0^1 \exp(-r^2/4s^2) \frac{ds}{s^{1/2}} \tag{3.57}$$

In the new r scale the density $\rho(r)$ is essentially different from zero at distances $r \sim 1$, that is, the radius of the system $r_N \sim 1$. Consider now the correlation of fluctuations. In the absence of correlations

$$\overline{n(\mathbf{x}_1)n(\mathbf{x}_2)} = \overline{n(\mathbf{x}_1)} \cdot \overline{n(\mathbf{x}_2)}$$

Hence the quantity

$$w = \frac{\overline{n(\mathbf{x}_1)n(\mathbf{x}_2)} - \overline{n(\mathbf{x}_1)} \cdot \overline{n(\mathbf{x}_2)}}{\overline{n(\mathbf{x}_1)} \cdot \overline{n(\mathbf{x}_2)}} = \frac{\rho_{12}(\mathbf{r}_1,\mathbf{r}_2)}{\rho_1(\mathbf{r}_1)\rho_2(\mathbf{r}_2)} - 1 \tag{3.58}$$

can be used as a measure of correlation.

For conventional equilibrium systems (e.g., for gas)

$$w \sim N^{-1}$$

Correlation is absent if an ensemble contains a large number of particles. On the contrary, in the case of a macromolecule the density correlation is different from zero at the same distances as the density itself. The value

$$\rho_{12}(\mathbf{r}_1,\mathbf{r}_2) = \frac{1}{4\pi^{3/2}} \int_0^1 \int_0^1 \exp\left\{-\frac{r_1^2}{4s_1} - \frac{|\mathbf{r}_1-\mathbf{r}_2|^2}{4|s_1-s_2|}\right\} \frac{ds_1\, ds_2}{s_1^{3/2}|s_1-s_2|^{3/2}} \tag{3.59}$$

decreases at distances $r \sim 1$. Therefore, the correlation radius is of the order of the radius of the system and the instantaneous density has no definite value--it is not a thermodynamic parameter, that is, a reliable quantity if $N \to \infty$. In other words, the macromolecular coil "pulsates" continuously; its spatial fluctuations are macroscopic. At the same time the coil is an ergodic system; its parameters averaged by exemplars coincide with those averaged in time. The densities $\rho(\mathbf{r})$ are just such time averages. These peculiarities are important for the understanding of the properties of globular macromolecules described in Chapter 4.

The presence of linear memory and cooperativity in the macromolecular chain determine its specific informational properties. The change in conformation of a section of the chain influences its chemical reactivity; conversely, chemical reactions at some section of the chain alter its conformational state. Events that have occurred in some section of the chain produce conformational changes in the entire chain. A macromolecular chain can serve as a channel for the transmission of information about the chemical events in some remote link. In its turn, a conformational signal generated at the beginning of the chain can reach any link of the chain and cause a variation in its chemical properties [50,51]. It is determined by electron-conformation interactions (ECI, cf. p. 94). In the language of nonequilibrium thermodynamics, this means the coupling of chemical (electronic) processes with changes in conformation. We have (cf. p. 13)

$$v_{chem} = L_{11}A_{chem} + L_{12}A_{conf}, \quad v_{conf} = L_{21}A_{chem} + L_{22}A_{conf} \tag{3.60}$$

The rate of the chemical reaction depends on the conformational affinity and *vice versa*.

## 3.6 Determination of the Molecular Weights for Macromolecules

Investigations of macromolecules of both synthetic and biological polymers require, first, determination of molecular weights (MW). In polymer solutions these determinations can be made by many methods. The methods based on lowering the freezing point and raising the boiling point (cryoscopy and ebullioscopy) are applicable only for very dilute solutions of polymers possessing small MW (100-5000). The sensitivity of

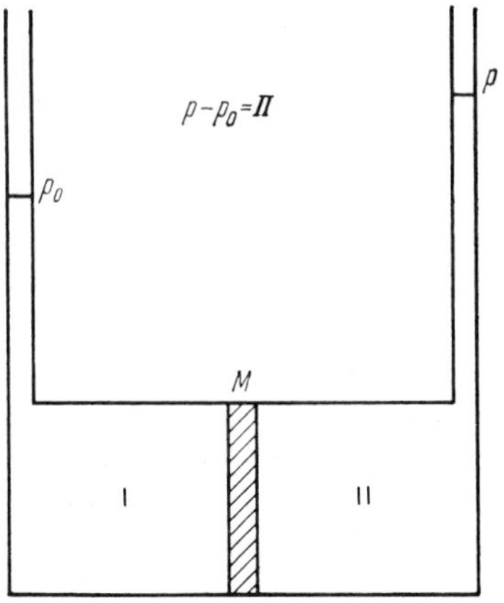


*FIG. 3.13 Scheme of an osmometer. M: membrane; I: solvent compartment, $\mu_1'(P_0)$; II: solvent and polymer, $\mu_1(P_0)$, $\mu_1''(P_0+\Pi)$.*

these methods diminishes when the MW increases, and they are not used in practice. The method of isothermal distillation, based on lowering the vapor pressure over the solution in comparison with the solvent, is satisfactory for MW of 1000-20,000 but involves many experimental difficulties [47,52]. The theoretical foundations of this method are essentially the same as those of osmometry, which is widely used in the physical chemistry of polymers [47,52,53].

The basic diagram of an osmometer is shown in Fig. 3.13. The membrane M is permeable to molecules of the solvent but not of the solute. The equilibrium condition for the solution in II and for the pure solvent in I is the equality of chemical potentials

$$\mu_1' = \mu_1'' \tag{3.61}$$

where

$$\mu_1'' = \mu_1 + \int_{P_0}^{P_0+\Pi} (\partial\mu_1/\partial p)_T \, dP \tag{3.62}$$

$\mu_1$ is the chemical potential in II at atmospheric pressure $P_0$; $\Pi$ is the osmotic pressure. But

$$(\partial\mu_1/\partial P)_T = V_1$$

$\overline{V_1}$ is the partial molar volume of the solvent which in practice is not pressure dependent. Hence

$$\mu_1'' = \mu_1 + \overline{V_1}\Pi = \mu_1' \tag{3.63}$$

In the limit when the concentration of the solute $c_2$ in II tends

*FIG. 3.14 Graph of $\Pi/c$ vs. (c) for polymethyl methacrylate in three different solvents.*

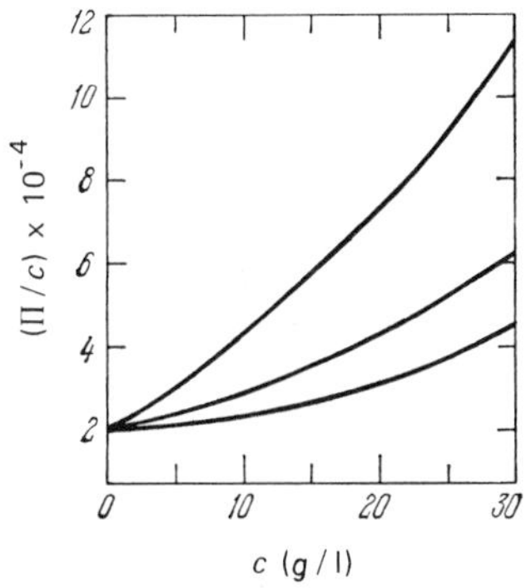


toward zero, we get

$$\Pi/c_2 = RT/M_2 \tag{3.64}$$

This is Van't Hoff's law; $c_2$ is expressed in grams per cubic centimeter; $M_2$, the MW of a polydisperse polymer, is the mean numerical MW

$$M_2 = \sum_i n_i M_i / \sum_i n_i \tag{3.65}$$

Van't Hoff's law is valid for ideal solutions at any concentration. However, solutions of polymers are never ideal (cf. [2, 47,48,52,53]); therefore the osmotic pressure has to be expressed as

$$\Pi = (RT/M)c + Bc^2 + Cc^3 + \cdots \tag{3.66}$$

We have omitted the index 2 on M and c; B, C, etc., are the second, third, etc., virial coefficients. For dilute solutions, $Cc^3 \ll Bc^2$. Drawing the graph of the dependence of $\Pi/c$ on c, we get the line crossing the ordinate at RT/M and find M. The slope of the curve at $c \to 0$ gives the value of B. Figure 3.14 shows these curves for polymethyl methacrylate in three different solvents. The sensitivity of the osmotic method decreases with increasing M but it is applicable up to MW of the order of 300,000.

Comparatively simple measurements of the viscosity of a solution also give information concerning the MW of macromolecules. Intrinsic viscosity is measured according to

$$[\eta] = \lim_{c\to 0} \frac{\eta - \eta_0}{\eta_0 c} \tag{3.67}$$

where $\eta$ is the viscosity of the solution, $\eta_0$ that of the pure solvent, and c the weight concentration of the polymer. The value of $[\eta]$ depends on M.

The viscosity of a liquid characterizes the internal

friction in a laminar flow, the rate of energy dissipation. The dissolved macromolecules perturb the field of the flow and increase the viscosity. This increase expresses the energy losses due to rotation of macromolecules in flow. The calculation is rather complicated. It can be assumed, however, that the field of flow is not perturbed but that the energy dissipates if the particle moves relative to the surrounding liquid, and it is possible to simplify the calculation. Einstein [54] formulated an expression for the viscosity of a solution containing any number of particles sufficiently removed from one another to be free of the mutual influence of the perturbations of flow produced by single particles. He got

$$\eta = \eta_0(1+\nu\phi) \tag{3.68}$$

where $\nu$ is the numerical coefficient, and $\phi$ is the ratio of the sum of the volumes of all dissolved or suspended particles to the total volume of the system. If V is the actual volume of the macromolecule and $N_1 = N_A c/M$ is the number of the particles in 1 ml, then

$$\phi = N_A cV/M$$

and the intrinsic viscosity $[\eta]$ (3.67) is expressed as

$$[\eta] = \nu(N_A V/M) \tag{3.69}$$

For hard-sphere particles Einstein got $\nu = 2.5$. Simha [55] performed the calculations for hard ellipsoids. Here $\nu > 2.5$ and depends on the ratio of the axes. A detailed review of the theory of the viscosity of polymer solutions is presented in [48] and in [53].

For hard spheres V is proportional to M; consequently, $[\eta]$ does not depend on M. This is true for any hard particles if their masses and dimensions increase in some homological series retaining similarity of form. On the other hand, for hard "sticklike" particles whose long axis increases with M but whose cross section remains constant, $\nu$ depends on M. Such a particle can be treated as an elongated ellipsoid with an axes ratio p. For $20 \leq p \leq 300$ theory gives the values $38.6 \leq \nu \leq 4278$ and

$$[\eta] = KM^{1.7}$$

where K is a constant characterizing the given polymer-solvent system.

Flexible macromolecules in solution form coils that are partially or totally permeable to the solvent. $[\eta]$ turns out to be dependent on MW

$$[\eta] = KM^a \tag{3.70}$$

where $0.5 \leq a \leq 1.0$, depending on the hydrodynamic permeability of the coil. For a totally permeable coil $a = 1.0$. On the other hand, if the coil is not permeable, $a = 0.5$. This result can be demonstrated by means of a rough calculation. The non-permeable coil can be treated as a sphere with a radius

$$r = (\overline{h^2}/6)^{1/2} = \gamma M^{0.5}$$

where $\overline{h^2}$ is the mean square distance between the ends of the chain, proportional to M (p. 115). According to Einstein's law (3.69) we get

$$[\eta] = \nu \frac{N_A 4\pi r^3}{3M} = 2.5 \frac{N_A 4\pi\gamma^3 M^{3/2}}{3M} = KM^{0.5}$$

or

$$[\eta] = \phi[(\overline{h^2})^{3/2}/M] \tag{3.71}$$

This is the Flory-Fox formula [56], which is valid for any polymer in the θ solvent (p. 128). The order of magnitude of the constant $\phi$ can be readily estimated from this calculation. If the concentration is expressed in grams per deciliter, then $\phi \sim 10^{21}$. Experiment gives, for the majority of polymers being studied, $\phi = 2.2 \times 10^{21}$. As was shown by Ptitsyn and Eisner [57], outside the θ point in a good solvent where the volume effects are importance

$$[\eta] = KM^{0.5+\varepsilon} \tag{3.72}$$

where $\varepsilon$ is usually of the order of magnitude of 0.2-0.3, but may reach as much as 0.5. At the θ point $\varepsilon = 0$.

Evidently the formula (3.70) is a universal one--it describes both stiff and flexible particles. For the hard spheres $a = 0$, and for the long stiff "pivots" $a = 1.7$, for flexible macromolecules, as we said earlier, $0.5 \leq a \leq 1.0$. The relation

$$[\eta] = KM^{1.7}$$

obtained for the long stiff particles has been found correct for solutions of tobacco mosaic virus as well as for the comparatively short polypeptide chains.

Formula (3.70) can be applied for determination of the MW of a polymer if the constants K and a are known from independent data.

The best absolute method of measuring the MW and of determining the molecular weight distribution, widely used in biophysics, biochemistry, and molecular biology, is sedimentation. It consists in the precipitation of macromolecules under the action of a centrifugal force in a centrifuge rotating at a rate of the order of $10^4$-$10^5$ turns per minute. The centrifugal acceleration rate is many times greater than the gravitational one, g. In modern ultracentrifuges it is as great as 350,000g (the number of rotations per minute is 70,000). The cell containing a polymer solution is put into the rotor of the centrifuge. The cell is a cylinder with windows made of crystalline quartz. A light beam passed through the cell and the observation of sedimentation is made by optical methods. Sedimentation in an ultracentrifuge was first applied in studies of polymers by Svedberg in 1925. A detailed account of experimental methods can be found in [48] and [58].

If the sedimentation rate is much higher than the rate of diffusion of the macromolecules, they are precipitated. In an initially homogeneous solution, two regions are formed--pure solvent and solution. A transitional zone is formed between them where the polymer concentration varies from zero to some maximal value. In the course of sedimentation this zone or boundary moves toward the bottom of the cell, that is, away from the rotation axis. The rate of movement of the boundary is measured.

The macromolecule is acted upon by a centrifugal force $V\rho_M\omega^2 x$ where V is the volume of the macromolecule, $\rho_M$ its density, $\omega$ the angular velocity, and x the distance from the rotation axis. But the macromolecule is in a solution; therefore, an Archimedean force acts on it and the effective centrifugal force becomes $V(\rho_M-\rho_0)\omega^2 x$ where $\rho_0$ is the density of the solvent. This force is equilibrated by the force of translational friction $f\dot{x}$ (f is the friction coefficient). We get, if calculated per mole,

$$N_A V\rho_M(1-\bar{V}\rho_0)\omega^2 x = f\dot{x}$$

where $\bar{V} = \rho_M^{-1}$ is the specific partial volume of the macromolecule. But

$$N_A V\rho_M = M$$

and for dilute solutions f is expressed by the coefficient of diffusion D (Einstein's formula)

$$f = RT/D \tag{3.73}$$

We obtain the Svedberg formula

$$M = RTs/(1-\bar{V}\rho_0)D \tag{3.74}$$

where s is the sedimentation coefficient

$$s = \frac{1}{\omega^2 x}\frac{dx}{dt} \equiv \frac{1}{\omega^2}\frac{d\ln x}{dt} \tag{3.75}$$

The dimension of s is time; the unit of the sedimentation coefficient is called a svedberg (1S = $10^{-13}$ sec).

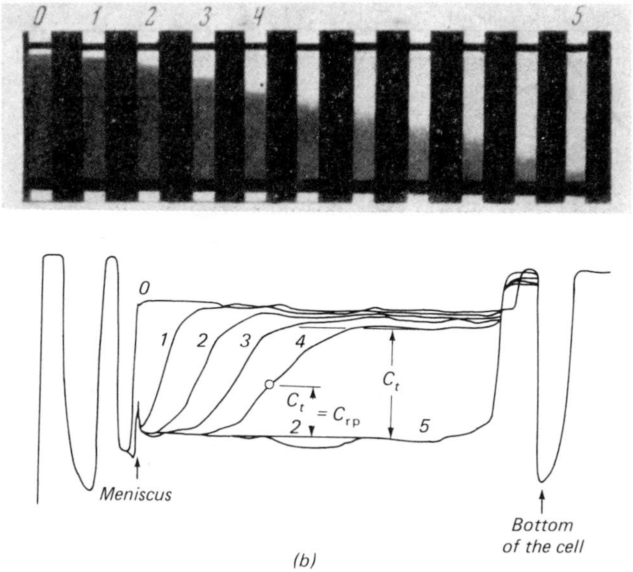


(b)

*FIG. 3.15 Sedimentation diagram for β-lactoglobulin.*

The sedimentation coefficient is determined from measurements of the shifts in the sedimentation boundary. Figure 3.15 shows the sedimentation diagram for the protein β-lactoglobulin. The successive states of the boundary are registered in ultraviolet light. By measuring s, the diffusion coefficient D, and $\rho_M$ and $\rho_0$, it is possible to determine M. The value of s depends on the concentration because of the hydrodynamic interaction of macromolecules. The value of s extrapolated toward zero concentration

$$s_0 = \lim_{c \to 0} s \tag{3.76}$$

is called the sedimentation constant. The values of M and $s_0$ vary in a parallel way but are not proportional. Table 3.2 contains some values of M and $s_0$ for proteins. An equilibrium can be established in a centrifuge because of the equality of the opposing sedimentation and diffusion flows. If the diffusion coefficient is large enough, equilibrium will be reached rather quickly and the equilibrium distribution of the substance can be observed.

If c is the concentration of the solute at a distance x from the rotation axis, the number of molecules sedimented in time dt across the unit of surface perpendicular to x is

$$c(dx/dt)\,dt$$

where dx/dt is the sedimentation rate at distance x. During the same time dt, the quantity of the dissolved substance

TABLE 3.2
Values of M and $S_0$ for Some Proteins[a]

| Protein | M | $S_0$ | Protein | M | $S_0$ |
|---|---|---|---|---|---|
| Ribonuclease | 13,700 | 1.64 | Glyceraldehyde phosphate dehydrogenase | 122,000 | 7.0 |
| Lysozyme | 14,100 | 1.9 | Human γ-globulin | 185,000 | 7.4 |
| Myoglobin | 17,600 | 2.04 | Human fibrinogen | 450,000 | 9.0 |
| Cytochrome *c* | 22,400 | 2.5 | Catalase | 240,000 | 11.0 |
| Trypsin | 23,800 | 2.5 | Phycoerythrin | 290,000 | 12.0 |
| Pepsin | 35,000 | 3.3 | Urease | 630,000 | 18.6 |
| Human serum albumin | 65,000 | 4.3 | Actomyosin of rabbit | $5\times10^6$ | 35.0 |
| Hemoglobin | 68,000 | 4.5 | Hemocyanin | $9\times10^6$ | 103.0 |

[a] Values in svedbergs.

diffused across the same cross section is

$$D\,(dc/dt)\,dt$$

At equilibrium

$$c(dx/dt) = D(dc/dx) \tag{3.77}$$

But in accordance with (3.75) and (3.74)

$$\frac{dx}{dt} = s\omega^2 x = \frac{M(1-\bar{V}\rho_0)}{RT} D\omega^2 x$$

and we obtain the second equation of Svedberg, which does not contain the diffusion coefficient

$$M = \frac{2RT}{(1-\bar{V}\rho_0)\omega^2 x}\,\frac{d\ln c}{dx} \tag{3.78}$$

Integrating this equation between two values $x_1$ and $x_2$, we obtain

$$M = \frac{2RT}{(1-\bar{V}\rho_0)\omega^2}\,\frac{\ln(c_2/c_1)}{x_2^2-x_1^2} \tag{3.79}$$

For the determination of M apply Archibald's method [48,58,59], which requires much less time and smaller amounts of the substance. With this method we measure not the equilibrium but the approach to equilibrium. The flow of the substance across the meniscus and the cell bottom is equal to zero independently of the approach to equilibrium. Hence

$$\frac{1}{x_M}\frac{1}{c_M}\left(\frac{dc}{dx}\right)_M = \frac{1}{x_B}\frac{1}{c_B}\left(\frac{dc}{dx}\right)_B = \frac{s\omega^2}{D}$$

The letter M denotes the meniscus, and B the bottom of the cell. We find

$$M = \frac{RT}{(1-V\rho_0)\omega^2}\frac{(dc/dx)_M}{x_M c_M}$$

$$M = \frac{RT}{(1-V\rho_0)\omega^2}\frac{(dc/dx)_B}{x_B c_B} \qquad (3.80)$$

For a homogeneous polymer these two values of M coincide.

In 1957, Meselsohn *et al.* [60] suggested a new method of sedimentational analysis which has become widely used in molecular biology and biophysics, particularly in studies of nucleic acids (p. 504). This method is sedimentation in a density gradient.

In a concentrated solution of a low-molecular-weight substance, particularly in a salt solution, after prolonged ultracentrifugation a concentration gradient (density gradient) will be established. If we put the macromolecules into such a solution, then because of the proportionality

$$s \sim 1-\bar{V}\rho_0$$

they will be located in that part of the cell where $s = 0$, that is, $\bar{V}\rho_0 = 1$ or

$$\rho_0 = \rho_M$$

In other words, macromolecules will be located in that part of the cell where the density of the concentrated solution coincides with the density of the macromolecules. Hence, the density $\rho_M$ is directly measured. The heterogeneous mixture of macromolecules will be separated and the spectrum is somewhat spread because of diffusion. The distribution of concentration in the band (i.e., in the zone of localization) of macromolecules is Gaussian

$$c = c_0 \exp(-x^2/2\sigma^2) \qquad (3.81)$$

where $c_0$ is the concentration at the center of the band. The dispersion of the Gaussian distribution $\sigma$ characterizes the bandwidth. Calculation shows that

$$\sigma^2 = RT\rho_0/[M(d\rho/dx)\omega^2 x] \qquad (3.82)$$

where $\rho_0$ is the density at the band center. From the dependence of ln c on $x^2$ the value $\sigma^2$ (and consequently, the molecular weight M) is found. A detailed description of this method can be found in [48,58,61].

We see that diffusion is one of the determining factors for sedimentation processes. In the method based on the measurement of the rate of sedimentation, direct determinations of the diffusion coefficient D are required. Diffusion makes it possible to determine M by the sedimentation method in a density gradient.

Moreover, the study of diffusion yields information about the mobility of macromolecules and therefore about their geometric and hydrodynamic properties. Diffusion is due to a concentration gradient. The flow of substance in direction x per second is equal to (the first equation of Fick)

$$J = -D(dc/dx) \tag{3.83}$$

But in nonstationary conditions the concentration itself changes because of diffusion. We have

$$\partial c/\partial t = -\partial J/\partial x$$

Consequently, we get Fick's second equation

$$\partial c/\partial t = (\partial/\partial x)[D(\partial c/\partial x)] \tag{3.84}$$

and if D is not dependent on x, then

$$\partial c/\partial t = D(\partial^2 c/\partial x^2) \tag{3.85}$$

The experimental problem consists in measuring D. For this purpose the concentration gradient is determined by optical methods. In a special cell, layers of the solution and of the pure solvent are formed, and the refraction or interference of light is measured. The principles and devices of modern diffusiometers are described in [48].

Einstein's formula (3.73) is correct for an ideal solution. Solutions of polymers, however, are far from ideal. Einstein's equation follows from Van't Hoff's law of osmotic pressure (3.64). From relation (3.66) we get

$$D = (1/f)(kT+2BMc+3CMc^2+\cdots) \tag{3.86}$$

Therefore D depends on concentration.

The coefficient of friction f depends on the form of the macromolecule and on its permeability by the solvent. For hard spheres the Stokes law is valid (r is the sphere radius)

$$f_0 = 6\pi\eta_0 r \tag{3.87}$$

for hard ellipsoids of rotation the coefficients of friction are different for movement of the particle in a direction parallel or perpendicular to the ellipsoidal axis. Both values

$f_\perp$ and $f_\parallel$ depend on the axial ratio p. Thus for a very elongated ellipsoid ($p > 10$)

$$f_\parallel = \frac{4\pi\eta_0 L}{2\ \ln 2p-1}, \qquad f_\perp = \frac{8\pi\eta_0 L}{2\ \ln 2p+1}$$

where L is the length of the main axis of the ellipsoid. The practically observable mean value is expressed by

$$\frac{1}{f} = \frac{1}{3}\left(\frac{1}{f_\parallel} + \frac{1}{f_\perp}\right)$$

and if $p > 10$

$$f = \frac{3\pi\eta_0 L}{\ln 2p} \tag{3.88}$$

The frictional coefficient of an ellipsoid is larger than that of a sphere of the same volume.

Formula (3.87) can be rewritten as

$$f_0 = 6\pi\eta_0 (3V/4\pi)^{1/3}$$

where V is the volume of the sphere. For an ellipsoid

$$f = 6\pi\eta_0 (3V/4\pi)^{1/3} (f/f_0) \tag{3.89}$$

The ratios $f/f_0$ for bodies of equal volume are known (cf. [43]). They depend only on p. Hence, knowing V (i.e., M and $\rho$), we can determine the value of p from measurements of f (i.e., of D). Since the macromolecules of proteins in their native state are solid particles (see Chapter 4), the study of diffusion serves as a direct method of determining their form.

The theory of the translational friction of flexible chain molecules is naturally much more complicated. It is built on the same basis as the theory of viscosity; in both cases hydrodynamic behavior is studied. The measurement of diffusion enables us to find the statistical linear dimensions of macromolecules [48].

## 3.7 Optical Methods of Investigation of Macromolecules

Important information about macromolecules, about their molecular weight, form, and size, can be obtained by means of light scattering. The scattering of light is a very important phenomenon of molecular optics; it provides valuable and varied data concerning the structure and properties of molecules, liquids, and solids [62,63].

A perfectly homogeneous medium would not scatter light; the secondary light waves radiated by the electrons of molecules are coherent and would cancel each other in all directions except those corresponding to the rules of geometrical optics. In every real medium, however, there are always fluctuations, deviations from a homogeneous distribution of the positions and orientations of molecules. Light is scattered at these fluctuations of density and orientation in gases and liquids, and concentration fluctuations in solutions.

The electric field of the light wave $\mathbf{E}_0$ induces a dipole in a volume v which is small in comparison with $\lambda^3$, where $\lambda$ is the wavelength. This dipole radiates secondary waves. The electric field of a secondary wave at a point a distance r from the dipole is

$$E_\theta = (4\pi^2/\lambda^2 r)p \sin\theta \tag{3.90}$$

where p is the induced dipole moment and $\theta$ the angle between the vectors $\mathbf{r}$ and $\mathbf{p}$ (i.e., $E_0$). The dipole moment of the volume v is

$$p = (\varepsilon - 1/4\pi)\ v\ E_0 \tag{3.91}$$

where $\varepsilon = n^2$, the square of the refraction index of the medium. In the presence of fluctuations

$$\varepsilon = \bar{\varepsilon} + \Delta\varepsilon \tag{3.92}$$

and

$$p = \bar{p} + \Delta p = (\bar{\varepsilon} - 1/4\pi)vE_0 + (\Delta\varepsilon/4\pi)vE_0 \tag{3.93}$$

Since scattering is produced only by fluctuations, the field of the light wave scattered by the volume v is equal to

$$E_\theta^s = (4\pi^2/\lambda^2 r)(\Delta\varepsilon/4\pi)E_0 v \sin\theta \tag{3.94}$$

and the observed intensity of scattered light

$$I_\theta^s = \overline{(E_\theta^s)^2} = (\pi^2/\lambda^4 r^2)(\overline{\Delta\varepsilon})^2 I_0 v^2 \sin^2\theta \tag{3.95}$$

where $I_0$ is the intensity of the incident light.

The fluctuation of $\varepsilon$ (i.e., of the refraction index n) is expressed by the fluctuation in density $\rho$ or in the number of particles N in the volume v

$$\Delta\varepsilon = (\partial\varepsilon/\partial\rho)\ \Delta\rho = (\partial\varepsilon/\partial\rho)\ \bar{\rho}\ (\Delta N/N) \tag{3.96}$$

because $\Delta\rho = \bar{\rho}\ \Delta N/N$.

The entire scattering volume V consists of a large number of microscopic volumes v and the field of the light scattered by V is the sum of the fields produced by volumes v. The total intensity will be expressed as the mean square of the entire field. The mean square of the fluctuation $\Delta N$ is [62]

$$\overleftarrow{(\Delta N)^2} = N$$

and we get

$$I_\theta = \frac{\pi^2}{\lambda^4 r^2} \sin^2\theta \left(\frac{\partial\varepsilon}{\partial\rho}\bar{\rho}\right)^2 \frac{V}{N_1} I_0 \tag{3.97}$$

where $N_1$ is the number of particles in a unit volume.

For gases

$$\varepsilon - 1 = n^2 - 1 = \text{const}\cdot\bar{\rho}$$

Hence

$$\frac{\partial\varepsilon}{\partial\rho} = \frac{\varepsilon - 1}{\bar{\rho}} = \frac{n^2-1}{\bar{\rho}} \cong \frac{2(n-1)}{\bar{\rho}} \tag{3.98}$$

For natural light we get

$$I_\theta = (2\pi^2/\lambda^4 r^2)(1+\cos^2\theta)(n-1)^2(V/N_1)I_0 \tag{3.99}$$

where $\theta$ is the angle between the incident and scattered light beams (this is Rayleigh's formula). The total intensity of light scattered in all directions by the gas volume V can be obtained by integration of expression (3.99) over the surface of a sphere of radius r

$$I^s = (32\pi^2/3\lambda^4)(n-1)^2(V/N_1)I_0 \tag{3.100}$$

Because of scattering, the intensity of the light passing through a layer of the medium of thickness $\ell$ decreases according to the rule

$$I = I_0\, e^{-h\ell} \tag{3.101}$$

where h is the turbidity coefficient, equal to the ratio of the intensity of the light scattered by a unit volume to that of the incident light

$$h = (32\pi^3/3\lambda^4)[(n-1)^2/N_1] \tag{3.102}$$

It is convenient to use the Rayleigh ratio

$$R_\theta = \frac{I_\theta r^2}{I_0 V} = \frac{2\pi^2(n-1)^2}{\lambda^4 N_1}(1+\cos^2\theta) \tag{3.103}$$

We get

$$h = \frac{16\pi}{3}\,\frac{R_\theta}{(1+\cos^2\theta)} \tag{3.104}$$

Let us consider the solution. The formula equivalent to (3.103) has the form (cf. [48,62])

$$R_\theta = \frac{2\pi^2 n_0^2 (n-n_0)^2}{\lambda^4 N_1}(1+\cos^2\theta) \qquad (3.105)$$

where n is the refraction index of the solution and n is that of the pure solvent. Expressing n and $N_1$ by the weight concentration c, we get ($N_A$ is Avogadro's number)

$$N_1 = \frac{cN_A}{M}, \qquad \frac{n-n_0}{c} = \frac{dn}{dc}$$

and

$$R_\theta = \frac{2\pi^2 r_0{}^2 c(dn/dc)^2 M}{\lambda^4 N_A}(1+\cos^2\theta) = HcM\frac{1+\cos^2\theta}{2} \qquad (3.106)$$

where

$$H = \frac{4\pi^2 n_0{}^2 (dn/dc)}{\lambda^4 N_A} \qquad (3.107)$$

The turbidity coefficient is

$$h = (8\pi/3)HcM \qquad (3.108)$$

Thus, knowing n, $n_0$, and c and measuring the turbidity coefficient, it is possible to determine the molecular weight M of a dissolved substance.

However, the analogy to gas, on which the transition from (3.103) to (3.105) is based, is valid only for a dilute ideal solution in which the fluctuations in small volumes v are independent. In real solutions of macromolecules, the swelling of the coils results in concentration-dependent interactions of the scattering centers. In this case it cannot be suggested that $(\overline{\Delta N^2}) = \overline{N}$ (p. 143), and in the calculation of $R_\theta$ or h Einstein's theory regarding the scattering of light by liquids has to be used [62,64]. Instead of (3.105) the theory gives, for a nonideal solution

$$R = \frac{2\pi^2 n_0{}^2 c(dn/dc)^2(1+\cos^2\theta)}{\lambda^4 N_A (d/dc)(\Pi/RT)} = Hc\,\frac{1+\cos^2\theta}{2(d/dc)(\Pi/RT)} \qquad (3.109)$$

where $\Pi$ is the osmotic pressure. According to (3.66)

$$R_\theta = Hc\frac{1+\cos^2\theta}{2}\,\frac{1}{(1/M)+(2B/RT)c+\cdots} \tag{3.110}$$

and

$$h = \frac{8\pi}{3}Hc\,\frac{1}{(1/M)+(2B/RT)c+\cdots}$$

or

$$\frac{8\pi}{3}\frac{Hc}{h} = \frac{1}{M} + \frac{2B}{RT}c + \cdots \tag{3.111}$$

The determination of M requires the measurement of h and H for a series of concentrations c. The extrapolation of the curve (Hc/h)(c) to $c \to 0$ gives the weight average MW for a polydisperse polymer

$$M_\omega = \frac{\Sigma_i c_i M_i}{\Sigma_i c_i} = \frac{\Sigma_i n_i M_i^2}{\Sigma_i n_i M_i} \tag{3.112}$$

[cf. (3.65)].

This method for determining M, suggested by Debye, is the more exact the greater the MW [65]. In this sense light scattering has definite advantages in comparison with osmometry. A description of a seris of nephelometers--devices for measuring the turbidity h is given in [48].

The theory just presented concerns particles whose sizes are much smaller than the wavelength of light λ. If this condition is not fulfilled, the phase differences of the secondary light waves radiated by the different points of a particle must be taken into account. Interference of waves scattered by a single particle occurs and consequently the integral intensity of scattering decreases. The angular distribution of intensity changes and it can be expressed by an additional factor P(θ), a complicated function depending asymmetrically on θ. The forward scattering is larger than the backward scattering, the Mie effect (which can be readily explained) is observed. In Fig. 3.16 the scattering particle is shown. The incident parallel rays falling at the equivalent scattering elements A and B are

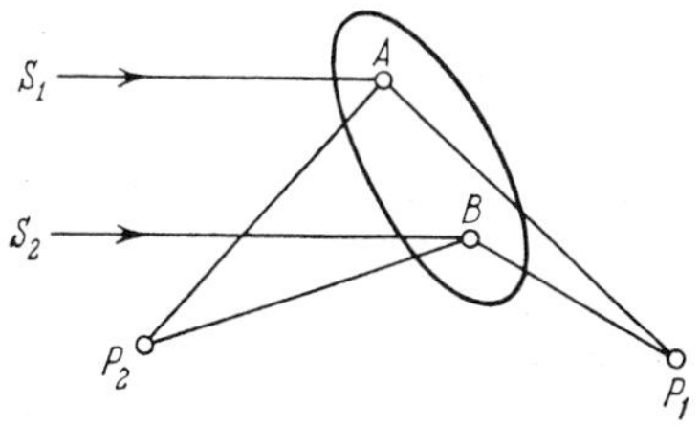


*FIG. 3.16 The Mie effect.*

denoted as $S_1A$ and $S_2B$. The distances traveled by the rays to point $P_1$ (forward) are $S_1AP_1$ and $S_2BP_1$. $S_1A < S_2B$, but $AP_1 > BP_1$. The total phase difference is therefore rather small. On the other hand, at point $P_2$ (backward), $S_2B > S_1A$ and $BP_2 > AP_2$. Hence, the phase difference is larger here and the decrease in intensity produced by the interference of scattered light is also larger. The theory of Mie is based on the consideration of diffraction phenomena [62,66]. The theory gives, for spherical particles of radius r, the expression obtained by Rayleigh

$$P(\theta) = [(3/z^3)(\sin z - z \cos z)]^2 \tag{3.113}$$

where

$$z = (4\pi r/\lambda) \sin (\theta/2)$$

This expression can be used when studying globular proteins.

For pivotlike stiff particles having a length L that is much greater than their diameter,

$$P(\theta) = (1/y)\mathrm{Si}(2y) - (\sin y/y)^2 \tag{3.114}$$

where

$$y = \frac{2\pi L}{\lambda} \sin\frac{\theta}{2}, \qquad \mathrm{Si}(2y) = \int_0^{2y} \frac{\sin t}{t}\, dt$$

The function for the statistical Gaussian coil obtained by Debye [60] has the form

$$P(\theta) = \frac{2}{x^2}(e^{-x}+x-1), \qquad x = \frac{8\pi^2}{3}\frac{\overline{h^2}}{\lambda^2}\sin^2\frac{\theta}{2} \tag{3.115}$$

All these expressions are particular cases of the general expression

$$P(\theta) = \frac{I_\theta}{I_0} = \frac{1}{N^2}\sum_{p=1}^{N}\sum_{q=1}^{N}\frac{\sin \mu r_{pq}}{\mu r_{pq}} \tag{3.116}$$

where

$$\mu = (4\pi/\lambda) \sin (\theta/2) \tag{3.117}$$

and $r_{pq}$ is the distance between two radiating points p and q in the particle. The double summation has to be performed over all points, whose number is N. For solid particles the summation can be replaced by integration; for a flexible chain the summed terms have to be averaged over all conformations. Doing so gives expression (3.115).

Since in every case the functions $P(\theta)$ contain geometrical parameters (r, L, $\overline{h^2}$), the measurement of $P(\theta)$ (i.e., the study of the asymmetry of scattering) enables us to determine the dimensions and form of a macromolecule.

As was shown by Debye in the expansion of $P(\theta)$ in series near $\mu = 0$

$$P(\theta) = 1 - a_1\mu^2 + a_2\mu^4 - \cdots \qquad (3.118)$$

the coefficient $a_1$ characterizes the mean dimensions of the particle independently of its form. The calculation gives

$$a_1 = \frac{1}{3}\,\overline{r^2} \qquad (3.119)$$

where $\overline{r^2}$ is the mean square radius of the particle. For a Gaussian coil

$$r^2 = \frac{1}{6}\,\overline{h^2} \qquad (3.120)$$

Using this property of the angular function $P(\theta)$, Zimm proposed the double-extrapolation method [67] which makes it possible to determine both the MW and the sizes of particles. The measurements of scattered intensity are made for a series of concentrations c and angles $\theta$. The diagram of the dependence of $cH/R_\theta$ (p. 144) on $\sin^2(\theta/2)+Kc$ is constructed where K is a constant, chosen in such a way that $Kc_{max}$ is of the order of magnitude of several units. The diagram contains two families of parallel straight lines [or of parallel curves, if the next expansion terms in (3.118) cannot be omitted]. The lines of one family represent the dependence of $Hc/R_\theta$ on c at various values of $\sin^2(\theta/2)$; the lines of the second family, the dependence of $Hc/R_\theta$ on $\sin^2(\theta/2)$ at various values of c. Indeed, according to (3.106) we get, for large molecules,

$$R_\theta = HcM\,\frac{1+\cos^2\theta}{2}\,P(\theta) \qquad (3.121)$$

and

$$M = \left(\frac{cH(1+\cos^2\theta)}{2R_\theta}\right)^{-1}_{c=0,\theta=0} \qquad (3.122)$$

This means that M is determined by the intersection of the straight lines $c = 0$ and $\theta = 0$ lying at the ordinate axis. The size of the particle is given by the initial slope of the line $(CH/R_\theta)_{c=0}$. Figure 3.17 shows Zimm's graph for solutions of a fraction of poly-2,5-dichlorostyrene in dioxane ($M = 16.7\times10^6$). The description of experimental methods and further details about the theory are presented in [48,53,68,87].

Valuable information about the structure of macromolecules can be obtained by means of small-angle scattering of X rays by dilute polymer solutions [58,69-71] (cf.(p. 372).

Let us consider the dynamooptical, Maxwell effect, the phenomenon of flow birefringence. Birefringence occurs if there is a velocity gradient in flow, because of the mechanical deformation of a liquid medium. In the case of laminar flow

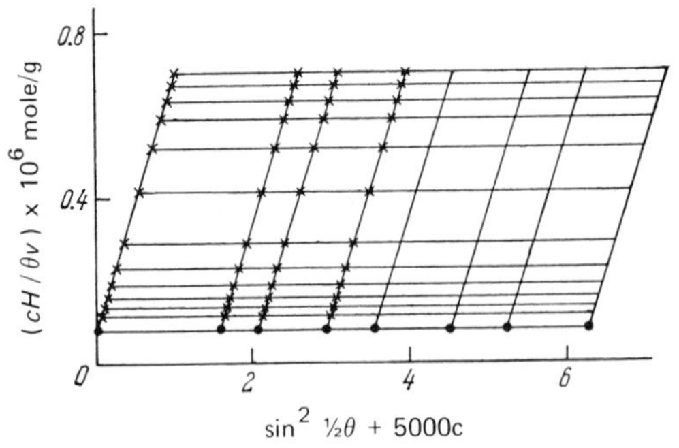


*FIG. 3.17 Zimm's graph.*

with a constant velocity there is no deformation; the liquid is displaced as a whole.

Studies of this phenomenon are done with a dynamooptimeter, a device utilizing two coaxial cylinders. As the internal cylinder (the rotor) turns, the liquid which has been poured between the two cylinders moves. A radial velocity gradient is established in the liquid, since the layer adjoining the rotor surface moves at a maximal velocity and the layer adjoining the wall of the fixed cylinder remains motionless. First, the orientations of the stiff particles in the velocity gradient field have to be considered. The theory of this phenomenon is presented in detail in [48] and [62] (cf. also [24]).

Let the particle be a "stick" of length L. The laminar flow of the liquid is directed along the x axis, the flow velocity u has a constant gradient along the y axis

$$g = du/dy$$

Consider the simple two-dimensional problem (Fig. 3.18). If the origin of the x,y coordinates coincides with the center of the stick, which itself moves in the flow, then the flow velocity relative to the center of the particle is

$$u = u_x = gy \qquad (3.123)$$

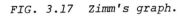

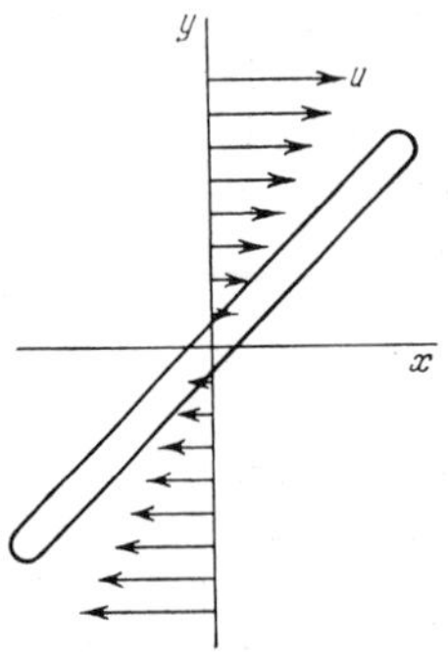


*FIG. 3.18 A sticklike particle in the gradient field of the flow velocity.*

The relative flow velocity near a point A on the particle at a distance $\ell$ from its center is equal to (Fig. 3.18)

$$u_A = g\ell \sin \phi$$

The component of this velocity along the particle axis is $u_1 = u_A \cos \phi = \frac{1}{2} g\ell \sin 2\phi$ and the component perpendicular to the stick is $u_2 = u_A \sin \phi = g\ell \sin^2 \phi$. The tangent components $u_1$ tend to stretch or to compress the particle, depending on the sign of $\phi$. The normal components $u_2$ rotate the particle, since the viscous forces form a rotatory momentum which produces rotatory movement of the particle with an angular velocity $\omega$. The value of $\omega$ can be found, if we neglect inertia, from the condition of the equality of linear velocity of the stick ends and of the normal component of the liquid velocity

$$\frac{1}{2} L \dot{\phi} = \frac{1}{2} Lg \sin^2 \phi$$

Hence

$$\omega = \dot{\phi} = g \sin^2 \phi \tag{3.124}$$

Because $\dot{\phi}$ depends on $\phi$, the rotation is not uniform. After some time interval the particle will be oriented along the flow axis x. However, Brownian thermal motion also acts on the particle. In the two-dimensional case the stationary distribution of particles over the angles $\phi$ is characterized by the function $f(\phi)$ [$f(\phi)\, d\phi$ is the relative number of particles having orientations in the range from $\phi$ to $\phi+d\phi$]. The condition of the stationary distribution has the form

$$f\dot{\phi}+D_r(\partial f/\partial \phi) = \text{const} \tag{3.125}$$

where $D_r$ is the coefficient of rotatory diffusion. If diffusion plays the principal role, f does not differ much from $f_0 = (2\pi)^{-1}$, that is, from the uniform angular distribution. We look for a solution of (3.125) in the form

$$f = f_0+f_1$$

where $f_1 \ll f_0$. We have

$$f_0\dot{\phi}+D_r(\partial f_1/\partial \phi) = C-f_1\dot{\phi} \cong C$$

and according to (3.45)

$$f_0 g \sin^2 \phi + D_r(\partial f_1/\partial \phi) = 0$$

By integration we get

$$f_1 = \left(\frac{C}{D_r} - \frac{f_0 g}{2D_r}\right)\phi + \frac{f_0 g}{4D_r} \sin 2\phi$$

Since the distribution function must not be changed by substituting $\phi \rightarrow \phi+2\pi$, the first term of this expression is zero. We get $C = f_0 g/2$ and

$$f = f_0[1+(g/4D_r) \sin 2\phi] \tag{3.126}$$

The maximum of the function f corresponds to $\phi = 45°$. The phenomenological investigation of the deformation of a liquid shows also that the principal direction of the deformation tensor forms a 45° angle with the direction of the velocity gradient.

The solution of the problem in a general three-dimensional case, and without the limitation $g \ll D_r$, is of course much more complicated (cf. [62]). The calculation shows that if $D_r \gg g$, the particle is oriented in the plane x,y containing the directions of the flow and of the gradient, and forms a 45° angle with the direction of flow, as before. If $D_r \ll g$, the particle is oriented along the flow.

If the particles are small, then in the real case $D_r \gg g$ always. For macromolecules this condition can be invalid and the orientation angle becomes dependent on the size and shape of the particles. Obviously, all these phenomena occur only in a liquid containing particles with some anisotropy of form.

A liquid containing oriented anisotropic particles is birefringent, similar to a biaxial crystal. The birefringence is measured in a direction z parallel to the axis of the dynamo-optimeter. The cause of the birefringence is the anisotropy of the polarizability of the particles. For solid ellipsoids with a ratio of large and small axes equal to p the theory gives a magnitude of birefringence equal to

$$\frac{\Delta n}{n} = g 2\pi N_1 V \frac{a_1 - a_2}{n^2} \frac{b}{15 D_r} \{1 - \frac{\sigma^2}{72} (1+ \frac{6b^2}{35}) + \cdots\} \tag{3.127}$$

where $N_1$ is the number of particles in the unit volume, V the volume of the particle, and $a_1 - a_2$ the difference between the longitudinal and transverse polarizability (i.e., the anisotropy of polarizability)

$$b = (p^2-1)/(p^2+1), \qquad \sigma = g/D_r$$

We see that $\Delta n$ becomes zero if $b = 0$ (i.e., $p = 1$, a sphere) and if $a_1 = a_2$ (isotropic polarizability). Thus, flow birefringence is expressed as a product of two factors, the optical one and the mechanical one. The second factor depends essentially on the velocity gradient g. The first is determined by the anisotropy of polarizability. In the case of macromolecules $a_1 - a_2$ is an effective anisotropy containing two contributions: the internal anisotropy $\Delta a_i$ and the anisotropy of the form $\Delta a_f$

$$(a_1-a_2)V = \Delta a = \Delta a_i+\Delta a_f \tag{3.128}$$

The internal anisotropy depends directly on the structure of the electron shell of a macromolecule. The anisotropic polarizability of a molecule can be evaluated if the anisotropic bond polarizabilities and the positions of the bonds are known. The polarizability tensor of a molecule can be expressed as the tensor sum of the polarizabilities of the bonds or of the atomic groups. This method of evaluation is called the valence-optical scheme [62,72]. The polarizability tensors for all important bonds are determined from data on molecular refraction, polarization of light scattering, and the Kerr effect [2,62,72]. In the case of a flexible macromolecule, the calculated value $\Delta a_i$ has to be averaged over all conformations [2,3,5].

The form anisotropy is different from zero if a nonspherical particle possesses a refraction index n different from that of the solvent $n_0$. As Maxwell demonstrated, the value of $\Delta a_f$ for the ellipsoid of rotation is

$$\Delta a_f = \frac{n^2-n_0^2}{4\pi+[(n^2-n_0^2)/n_0^2]L_1}V - \frac{n^2-n_0^2}{4\pi+[(n^2-n_0^2)/n_0^2]L_2}V$$

where for an elongated ellipsoid $(p > 1)$

$$L_1 = (4\pi/3)(1-e), \qquad L_2 = L_3 = (4\pi/3)(1+e)$$

$$e = \frac{1}{4(p^2-1)}\left[2p^2+4-\frac{3p}{(p^2-1)^{1/2}}\ln\frac{p+(p^2-1)^{1/2}}{p-(p^2-1)^{1/2}}\right]$$

If the particle simultaneously possesses an internal anisotropy, then its refractive indices $n_1$ and $n_2$ parallel and perpendicular to the main axis of the ellipsoid are different. Therefore

$$\Delta a = \Delta a_i+\Delta a_f$$

$$= \frac{4\pi n_0^4(n_1^2-n_2^2)+n_0^2(n_1^2-n_0^2)(n_2^2-n_0^2)(L_2-L_1)}{[4\pi n_0^2+(n_1^2-n_0^2)L_1][4\pi n_0^2+(n_2^2-n_0^2)L_2]}V \tag{3.129}$$

Having calculated $\Delta a_i$ by the valence-optical method (i.e., knowing $n_1$ and $n_2$), it is possible to determine the form of macromolecule (i.e., the p value) from measurements of the flow birefringence. If $n_0$ is chosen equal to $n_1$ or $n_2$, the effect of form can be excluded. Its value can be obtained from experiments with solvents having different $n_0$.

For small velocity gradients $\sigma \ll 1$ and formula (3.127) is simplified

$$\frac{\Delta n}{n} = \frac{2\pi N_1}{n^2}\,\frac{gb}{15D_r}\,(a_1-a_2)V \tag{3.130}$$

Intrinsic birefringence (the Maxwell constant) is

$$[n] = \left(\frac{\Delta n}{g\eta_0 c}\right)_{c\to 0,\,g\to 0} = \frac{2\pi N_A b}{15n_0 M\eta_0 D_r}\,(a_1-a_2)V \tag{3.131}$$

Dividing [n] by the intrinsic viscosity [η] (3.68), we get

$$\frac{[n]}{[\eta]} = \frac{2\pi b}{15n_0\eta_0 \nu D_r}\,(a_1-a_2)$$

The proportionality coefficient at $(a_1 - a_2)$ does not depend strongly on the nature of particle.

The measurements of birefringence give not only Δn (i.e., [n]) but also the orientation angle of the particle in flow, which determines the directions of the principal optical axes of a birefringent system. The value of the orientation angle χ is determined by the coefficients of rotatory diffusion

$$\left[\frac{\chi}{g}\right] = \lim_{g\to 0}\frac{d\chi}{dg} = \frac{1}{12D_r} \tag{3.132}$$

which means that χ characterizes the dynamics of orientation.

Flexible macromolecules are not only oriented but also deformed by flow. The tangential components of the velocity gradient stretch the coil. This stretching produces additional birefringence, namely, the photoelastic effect, whose theory is presented in [43]. The photoelastic effect can be separated from the orientational one. Consequently, flow birefringence gives valuable information about the kinetic flexibility of macromolecules. Further details concerning the Maxwell effect, along with a description of experimental methods and results, are contained in [48].

The Kerr effect (i.e., birefringence in an electric field) also makes it possible to obtain information about the internal anisotropy of macromolecules and about their form anisotropy. However, flow birefringence gives more reliable results. The Kerr effect has no real value in applications to biopolymers. Biopolymers are polyelectrolytes, that is, macromolecules with electric charges (see the next section). Therefore, they move in an electric field, and electrophoresis occurs (cf. p. 52). Stationary measurements are impossible; an impulse method has to be used, that is, measurements must be performed in synchrony with the instantaneous switching on of the field. The interpretation of results obtained in such a way is rather difficult, since these results are determined by relaxational, kinetic phenomena.

## 3.8 Polyelectrolytes

Many macromolecules, including proteins and nucleic acids, are macroions in aqueous solutions, that is, they possess many charges. These are polyelectrolytes. Biopolymers are polyampholytes, since they contain both cationic and anionic groups. The properties of a polyelectrolyte macromolecule in a solution depend on its structure, on the pH, and on the ionic strength of the solution (p. 51).

There exist both solid macroions (globular proteins, etc.) and flexible polyelectrolyte chains. Obviously, the behavior of such macromolecules depends on the interactions of their charges. The effective charges are determined by the degree of dissociation of the ionogenic groups and of the surrounding ionic atmosphere.

The conformation of a flexible polyelectrolyte chain, which was first treated in [73], is determined by minima conditions for the sum of conformational and electrostatic free energies. Charges of the same sign in a chain repel each other, and this mutual repulsion results in unfolding of the coil, that is, in an increase in its dimensions. Calculations of the electrostatic free energy of a flexible macromolecule must take into account the ionic atmosphere (p. 50). Such calculations were made by Harris and Rice [74] (see also [75]).

Consider a spatial configuration of a polyion with a given distribution of immovable charges q. The electrostatic free energy of the charge interaction is expressed by (cf. Section 2.3)

$$U_{es} = \frac{q^2}{2\varepsilon} \sum_{k=1 \neq \ell=1}^{p} \sum^{p} \sigma_k \sigma_\ell \frac{e^{-\kappa r_{k\ell}}}{r_{k\ell}} \tag{3.133}$$

where the summation is performed over all p ionogenic sites, $\varepsilon$ is the dielectric constant of the medium, $\kappa$ is the Debye-Hückel parameter (2.9), $r_{k\ell}$ is the distance between groups k and $\ell$, $\sigma_k$ and $\sigma_\ell$ are zero for noncharged groups and unity for charged ones. Using the model of a chain formed by freely jointed segments (p. 113), it is possible to express $U_{es}$ as the sum of two terms: the interaction energy between the parts of the same segment $U_{es1}$, and the interaction energy of the charges at neighboring segments $U_{es2}$ since $U_{es}$ decreases rapidly with increasing $r_{k\ell}$. The calculation gives

$$U_{es1} = \frac{zq^2}{\varepsilon b} \sum_{j=1}^{n-1} \frac{n(n-j)}{j} e^{-j\kappa b/n} \tag{3.134}$$

where z is the number of segments in the chain, b the length of the segment, and n the number of charges on one segment. We also obtain

$$U_{es2} = \frac{n^2q^2}{\varepsilon b} \sum_{s=1}^{z-1} \frac{\exp[-\kappa b \cos(\gamma_s/2)]}{\cos(\gamma_s/2)} \tag{3.135}$$

The angles $\gamma_s$ describe the conformation [$r_{k\ell}$ is replaced by $b \cos(\gamma_s/2)$]. $U_{es}$ is minimal if the polyelectrolyte chain is completely stretched. The parameter $\kappa$ increases with the ionic strength of the solution; therefore, its increase also diminishes $U_{es}$. Hence, an increase in ionic strength reduces the swelling of the coil (the stretching of the macromolecule).

Table 3.3 contains the values of the electrostatic energy of hypothetical polyelectrolyte with z = 1000 at different values of radii of inertia and different ionic strengths [53]. These values are calculated with the help of a simplified model, which suggests that the static charges in a sphere of radius R are uniformly distributed [76].

*TABLE 3.3*
*Electrostatic Free Energy*

| Ionic strength (mole/liter) | $U_{es}$ (cal/mole) | | | |
|---|---|---|---|---|
| | R = 100 Å | R = 150 Å | R = 300 Å | R = 1000 Å |
| 0.001 | $1.12\times10^7$ | $5.6\times10^6$ | $1.25\times10^6$ | $5.0\times10^4$ |
| 0.01 | $3.43\times10^6$ | $1.23\times10^6$ | $1.80\times10^5$ | 5600 |
| 0.05 | $9.4\times10^5$ | $3.0\times10^5$ | $4.05\times10^4$ | 1150 |

The Debye-Hückel theory using the linearized Poisson-Boltzmann equation for the charge density (p. 50) cannot be applied to polyelectrolytes. If the degree of ionization is not too small, the electrostatic field around the polyelectrolyte molecule is great, its energy several times larger than the thermal one. Therefore it is necessary to solve the problem with the nonlinearized Poisson-Boltzmann equation.

Flory developed a theory of the dimensions of polyelectrolyte chains that is similar to his theory of volume effects (p. 128). Electrostatic repulsion also produces swelling of the coil, depending on the ionic strength [77]. Flory suggested that the coil, together with the immobilized solvent, was electroneutral. Calculation shows that electrostatic interactions cannot transform a coiled macromolecule into a stretched one. Only swelling of the coil takes place. These results agree with experimental data (with the dependence of the intrinsic viscosity [η] on M). Ptitsyn developed a more rigorous statistical theory of charged macromolecules [78]. The initial hypothesis of this theory is that because of the screening action of counterions, only the charged groups of macromolecules whose positions along the chain are remote interact,

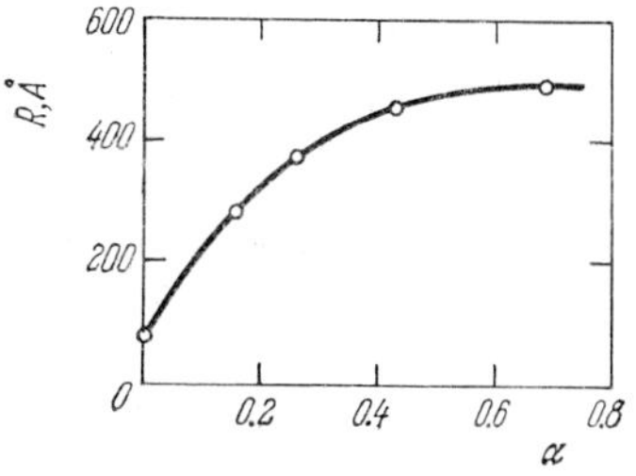

*FIG. 3.19  Dependence of the radius of inertia of polymeth-
acrylic acid macromolecules on the degree of dissociation of
carboxyl groups.*

being drawn together by chance as a result of the fluctuational
bending of the chain.  From Ptitsyn's theory it follows that
the conformational properties of charged macromolecules are
intermediate between those of the nonswollen statistical coils
and those of the rigid "sticks."  For polyelectrolytes, $h^2 \sim M^{4/3}$
and $[\eta] \sim M$, for nonswollen coils, $h^2 \sim M$ and $[\eta] \sim M^{1/2}$; and
for sticks, $h^2 \sim M^2$ and $[\eta] \sim M^2/\ln M$.  These calculated results
agree with experimental findings (particularly for denatured
DNA).  It also follows from Ptitsyn's theory that the square
of the linear swelling depends linearly on $I^{-2/3}$ (I is the ionic
strength).  This result also agrees with experiment.

    As has already been said, the maximal degree of unfolding
of a polyelectrolyte macromolecule occurs at a very low ionic
strength.  Figure 3.19 shows the influence of the charge on
the radius of inertia of macromolecules of polymethacrylic acid
These data were obtained by the method of light
scattering by salt-free solutions [79].  The
degree of dissociation of the polyacid is regulated by addition of a base.  Of course, the
ionic strength does not remain constant.

$$\left[ \begin{array}{c} COOH \\ | \\ -C-CH_2- \\ | \\ CH_3 \end{array} \right]_n$$

    The swelling of the coil increases with
dilution, since the decrease in polyelectro-
lyte concentration at a constant degree of dissociation entails
a decrease in ionic strength.  Consequently, anomalies are observed in the dependence on concentration of the reduced viscosity $\eta_{sp}/c$ (cf. p. 133) and of the inverse reduced intensity
of the light scattering (cf. p. 143).  Instead of decreasing,
the reduced viscosity increases nonlinearly with dilution.  The
$(\eta_{sp}/c)(c)$ curves can contain maxima.  Interpretation of such
data is rather difficult.  Clear information about the structure and properties of polyelectrolyte macromolecules can be
obtained by the method of isoionic dilution, in which a polyelectrolyte solution is diluted with a salt solution that has
the same ionic strength that the polyelectrolyte solution has
at maximal concentration; the aim is to maintain the total concentration of counterions.  The linear dependencies of $\eta_{sp}/c$

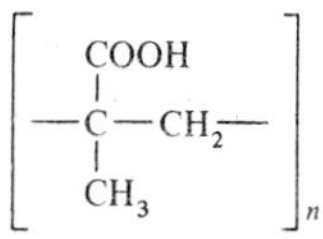

on c were obtained in these conditions (cf. [53]).

The polyelectrolyte macromolecule binds counterions. The latter are distributed in the region of the solution occupied by the macromolecules. Therefore, a polyion interacting with other polyions behaves like a neutral system and is represented by the values of the second virial coefficient, determined by means of osmometry (p. 132) or light scattering (p. 143) [80]. The counterions can be specifically bound by the ionized groups of the polyelectrolyte, and such binding depends on the chemical nature of the macroion and small ion. This specific binding, resulting in the formation of salt bonds at fixed sites in a macromolecule, has to be distinguished from the nonspecific one, the formation of the ionic atmosphere. In a salt bond the distance between the counterion and polyion is much smaller than the distance between the central ion and the counterion of the ionic atmosphere. The specific binding of counterions determines the ion-exchange properties of polyelectrolytes. These properties have important practical applications. Cross-linked insoluble polyelectrolytes swelling in water and other liquids are employed as ion-exchange resins or ionites [81]. Ionites can adsorb specific ions from solutions, and this ability is used in the purification and fractionation of different electrolytes, and to eliminate ionic contaminations from nonelectrolytes

In polyampholytes, cationic and anionic groups can serve as counterions for the formation of salt bonds.

Proteins contain acid (Glu, Asp, Tyr) and basic (Arg, His, Lys) amino acid residues (p. 45). Nucleic acids contain acid phosphate groups and basic groups N≡, =NH, $-NH_2$ of the nitrogen bases. Hence, the study of the structure and properties of biopolymers has to take into account their polyampholytic nature and therefore the pH and ionic strength of the medium. The structure of native and denatured molecules of proteins and nucleic acids depends basically on electrostatic, ionic interactions. The same factors are very important for biologically significant interactions of biopolymers with small molecules and ions.

Ion-exchange phenomena are responsible for the behavior of such supramolecular systems as biological membranes. The transport of small ions across membranes occurs by means of the sorption and desorption of these ions. The interactions of protein polyions with $K^+$, $Na^+$, $Ca^{++}$, and $Mg^{++}$ ions determine the most important biological phenomena, such as nerve excitation and muscle contraction.

These phenomena are closely associated with mechanochemical processes. Weakly cross-linked polyelectrolyte gels are able to transform chemical energy into mechanical work. Let us consider a polyanionic gel made from, for example, polymethacrylic acid swelling in water at neutral pH. The addition of alkali will increase the degree of ionization of the acid groups, their mutual electrostatic repulsion, and therefore the degree of swelling. The addition of an acid will result in contraction of the gel. The change in gel volume can be used to perform work. If a fiber is made of a weakly cross-linked polyelectrolyte, it will be shortened or elongated by the addition of an acid or alkali, and can thus do work, such as lifting and lowering a load. Conversely, the mechanical stretching of such a fiber produces variations in the ionic content of the surrounding liquid. Similar processes result from changes in ionic strength brought about by introducing a neutral salt into the swollen polyelectrolyte. Katchalsky and Oplatka applied this principle to build an engine that runs continuously; it consists of a polyelectrolyte fiber (collagen) that is submerged alternately in a salt solution and pure water (Fig. 3.20 [51]). The engine stops when, as a result of the transport of small ions by the fiber into the water, the chemical potentials of both reservoirs become equal. Such a machine can work in the opposite way by using an external source of energy, as in the transformation of a heat engine into a refrigerator. Here the engine can be used to extract salt from the solution, for instance, to desalinate seawater. Biological mechanochemical processes generally are of different character, but the swelling and contraction of a polyelectrolyte seem to be important for membrane phenomena. A more detailed account of polyelectrolyte properties is given in monographs [48,53,82].

The polyelectrolytic properties of proteins are applied for their separation and study by means of electrophoresis. If a solution containing charged particles is put in an electrostatic field, the particles will move toward the corresponding electrodes. A protein's electrophoretic motility depends on its total charge at given values of pH, ionic strength, etc. Hence, this property reveals important characteristics of a macromolecule. A series of electrophoretic methods of analyzing proteins and their mixtures has been developed. In the method of frontal electrophoresis, the rate of movement of the boundary between the solution of protein and the solvent in the electric field is measured. For preparative separation of

*FIG. 3.20 The mechanochemical engine of Katchalsky and Oplatka.*

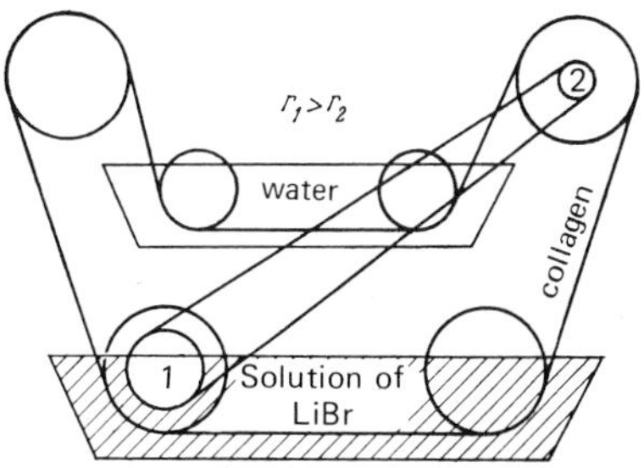

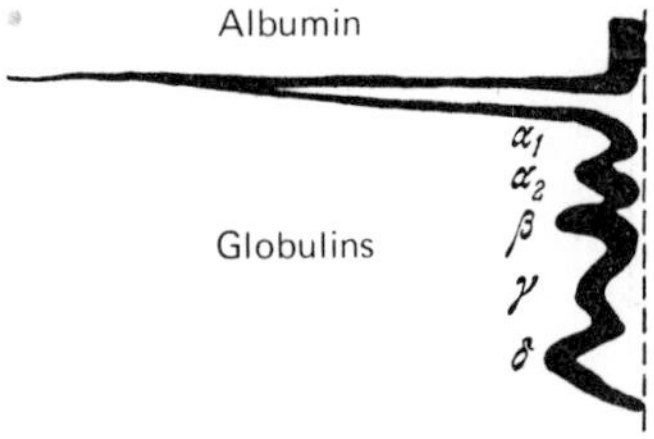


*FIG. 3.21 Electrophoregram of the proteins of human blood serum.*

mixtures the method of zonal electrophoresis is more convenient. In this method the solution under study is deposited in a thin layer between two layers of solvent. After some time, substances with different motilities will have shifted, their zones become distinct, and the products of separation can be extracted from the solution. Evidently separation is hindered by thermal convection. Convection is minimized by introducing a porous maintaining medium--gels, paper, etc. The best method for stabilizing the system seems to be the density gradient (cf. p. 139). In this case electrophoresis is performed in a vertical tube containing a buffer solution that has a density gradient.

As in the diffusion and sedimentation methods, electrophoresis is registered by optical methods. The electrophoregram obtained in this way contains a series of peaks corresponding to the different components of the mixture. Figure 3.21 shows the electrophoregram of human serum proteins. A description of the electrophoretic methods, their theory, and results of their applications are contained in several books and papers (cf. particularly [83-86]).

References

1. H. Staudinger, "From Organic Chemistry to Macromolecules." Wiley (Interscience), New York, 1970.
2. M. Volkenstein, "Configurational Statistics of Polymeric Chains." Wiley (Interscience), New York, 1963.
3. T. Birshtein and O. Ptitsyn, "Conformations of Macromolecules." Wiley (Interscience), New York, 1966.
4. L. Pedersen and K. Morokuma, J. Chem. Phys. 46, 3941 (1967).
5. P. Flory, "Statistical Mechanics of Chain Molecules," Wiley (Interscience), New York (1969).
6. T. Hill, J. Chem. Phys. 16, 399, 938 (1948).
7. L. Bartell, J. Chem. Phys. 32, 827 (1960).
8. A. Kitaigorodsky, Dokl. Acad. Sci. USSR 124, 1267 (1959) (R); 137, 116 (1961); Tetrahedron 9, 183 (1960); Acta Crystallogr. 18, 585 (1965).
9. A. Kitaigorodsky and M. Mirskaya, Krystallographia 6, 507 (1961) (R).
10. G. Ramachandran and V. Sasisekharan. Advan. Protein Chem.

23, 283 (1968).
11. M. Volkenstein, Izv. Acad. Sci. USSR Ser. Phys. **14**, 466 (1950) (R).
12. K. W. F. Kohlrausch, "Ramanspektren." Leipzig, 1943.
13. S. Mizushima, "Structure of Molecules and Internal Rotation." Academic Press, New York, 1954.
14. N. Borisova and M. Volkenstein, J. Struct. Chem. (USSR) **2**, 469 (1961) (R),
15. N. Borisova and T. Birstein, Vysokomol. Soedin. **5**, 279 (1963) (R).
16. N. Borisova, in "Vysokomol. Sojedin. Carbochain Compounds," p. 74, 1963 (R).
17. V. Dashevsky, "Conformations of Organic Molecules." Chimia, Moscow, 1974 (R).
18. E. Popov, V. Dashevsky, G. Lipkind, and S. Arkhipov, Mol. Biol. **2**, 612 (1968) (R).
19. E. Eliel, "Stereochemistry of Carbon Compounds." McGraw-Hill, New York, 1962.
20. E. Eliel, N. Allinger, S. Angiel, and G. Morrison, "Conformational Analysis." Wiley, New York, 1965.
21. S. Chandrasekhar, "Stochastic Problems in Physics and Astronomy."
22. S. Oka, Proc. Phys. Math. Soc. Japan **24**, 657 (1942).
23. S. Bresler and J. Frenkel, J. Exp. Theor. Phys. **9**, 1094 (1939) (R).
24. J. Frenkel, "Kinetic Theory of Liquids." Peter Smith, Gloucester, Massachusetts,
25. M. Volkenstein, Dokl. Acad. Sci. USSR **78**, 879 (1951); J. Phys. Chem. **26**, 1072 (1952) (R).
26. G. Natta and P. Corradini, J. Polym. Sci. **39**, 29 (1959); Nuovo Cimento, Suppl. I, **15**, 9 (1960).
27. G. Natta, P. Corradini, and P. Gauis, Makromol. Chem. **39**, 238 (1960); J. Polym. Sci. **58**, 1191 (1962).
28. N. Borisova and T. Birstein, Vysokomol. Soedin. **6**, 1234 (1964) (R).
29. N. Borisova, Vysokomol. Soedin. **6**, 135 (1964) (R).
30. O. Ptitsyn and J. Sharonov, J. Tech. Phys. **27**, 2762 (1957) (R).
31. M. Volkenstein and P. Ptitsyn, J. Tech. Phys. **25**, 649, 662 (1952) (R).
32. A. Cifferri, C. Hoeve, and P. Flory, J. Amer. Chem. Soc. **83**, 1015 (1961).
33. V. Nikitin, M. Volkenstein, and B. Volchek, J. Tech. Phys. **25**, 2486 (1955) (R).
34. V. Nikitin, B. Volchek, and M. Volkenstein, Conf. Spectrosc., **1**, 411. Ed. Lwow Univ. (1957) (R).
35. B. Volchek and V. Nikitin, J. Tech. Phys. **28**, 1953 (1958) (R).
36. H. Kramers and G. Wannier, Phys. Rev. **60**, 252 (1941).

37. E. Montroll, J. Chem. Phys. 9, 708 (1941); 10, 61 (1942); Ann. Math. Statist. 18, 18 (1947).
38. G. Newell and E. Montroll, Rev. Mod. Phys. 25, 353 (1953).
39. G. Rumer, Uspekhi Phys. Nauk. 53, 245 (1954) (R).
40. T. Hill, "Statistical Mechanics." McGraw-Hill, New York, 1956.
41. K. Huang, "Statistical Mechanics." Wiley, New York, 1963.
42. F. Gantmacher, "Theory of Matrices." Gostekhzdat, Moscow, 1953 (R).
43. A. Markov, "Selected Papers." Ed. Acad. Sci. USSR, Novosibirsk, 1951 (R).
44. B. Gnedenko, "Theory of Probabilities." Nauka, Moscow, 1969 (R).
45. P. Flory, J. Chem. Phys. 17, 303 (1949).
46. T. Fox and P. Flory, J. Phys. Chem. 53, 197 (1949); J. Amer. Chem. Soc. 73, 1904, 1909, 1915 (1951).
47. P. Flory, "Principles of Polymer Chemistry." Cornell Univ. Press, Ithaca, New York, 1953.
48. V. Tsvetkov, V. Eskin, and S. Frenkel, "Structure of Macromolecules in Solutions." Nauka, Moscow, 1964 (R).
49. I. Lifshitz, J. Exp. Theor. Phys. 55, 2408 (1968) (R).
50. M. Volkenstein, Izv. Acad. Sci. USSR Ser. Biol., N 1, 3 (1958) (R).
51. A. Katchalsky, in "Biology and Physical Sciences." Columbia Univ. Press, New York, 1969.
52. S. Bresler and V. Erusalimsky, "Physics and Chemistry of Macromolecules." Nauka, Moscow, 1963.
53. C. Tanford, "Physical Chemistry of Macromolecules." Wiley, New York, 1961.
54. A. Einstein, Ann. Phys. 19, 289 (1906); 34, 591 (1911).
55. R. Simha, J. Phys. Chem. 44, 25 (1940).
56. P. Flory and T. Fox, J. Phys. Coll. Chem. 53, 197 (1949).
57. O. Ptitsyn and J. Eisner, Vysokomel. Soedin. I, 1200 (1959) (R).
58. V. Engelhardt, ed., Fundamentals of molecular biology, "Physical Methods of Investigations of Proteins and Nucleic Acids." Nauka, Moscow, 1967 (R).
59. W. Archibald, J. Phys. Coll. Chem. 51, 1204 (1947).
60. N. Meselson, F. Stahl, and J. Vinograd, Proc. Nat. Acad. Sci. U.S. 43, 581 (1957).
61. J. Vinograd and J. Hearst, in "Fortschritte der Chemie organischer Naturstoffe." Springer-Verlag, Vienna, 1962.
62. M. Volkenstein, "Molecular Optics." Gostekhizdat, Moscow, 1951 (R).
63. I. Fabelinsky, "Molecular Scattering of Light." Nauka, Moscow, 1965 (R).
64. A. Einstein, Ann. Phys. 33, 1275 (1910).
65. P. Debye, J. Phys. Coll. Chem. 51, 18 (1947).
66. M. Born, "Optik." Springer, Berlin, 1932.

67. B. Zimm, J. Chem. Phys. 16, 1093, 1099 (1948).
68. K. Stacey, "Light-Scattering in Physical Chemistry." Academic Press, New York, 1956.
69. E. Geiduschek and A. Holtzer, Advan. Biol. Med. Phys. 6, 431 (1958).
70. A. Kitaigorodsky, "X-ray Analysis of Micro-crystalline and Amorphous Bodies." Gostekhizdat, Moscow, 1952 (R).
71. A. Guinier and G. Fournet, "Small Angle X-Ray Scattering." Wiley, New York, 1955.
72. M. Volkenstein, "Structure and Physical Properties of Molecules." Izd. Acad. Sci. USSR, Moscow, 1955 (R).
73. W. Kuhn, O. Künzle, and A. Katchalsky, Helv. Chim. Acta 31, 1994 (1948).
74. F. Harris and S. Rice, J. Phys. Chem. 58, 725, 733 (1954); J. Chem. Phys. 24, 326 (1956); 25, 955 (1956).
75. S. Rice, Biophysical science. A study program, Rev. Mod. Phys. 31, N 1 (1959).
76. J. Hermans and J. Overbeek, Rec. Trav. Chim. Pays-Bas 67, 761 (1948).
77. P. Flory, J. Chem. Phys. 21, 162 (1953).
78. O. Ptitsyn, Vysokomol. Soedin. 3, 1084, 1252, 1401 (1961) (R).
79. A. Oth and P. Doty, J. Phys. Chem. 56, 43 (1952).
80. N. Schneider and P. Doty, J. Phys. Chem. 58, 762 (1954).
81. F. Helfferich, "Ion Exchange." McGraw-Hill, New York, 1962.
82. S. Rice and M. Nagasawa, "Polyelectrolyte Solutions." Academic Press, New York, 1961.
83. G. Troitsky, "Elektrophoresis of Proteins." Ed. Kharkov Univ., 1962.
84. E. Sominsky, in "Molecular Biology. Physico-Chemical Methods of Analysis and Fractionation of Biopolymers." Nauka, Moscow, 1966 (R).
85. B. Magdof, in "Analytical Methods of Protein Chemistry." Pergamon Press, London, 1961.
86. C. Vunderly, in "Analytical Methods of Protein Chemistry." Pergamon Press, London, 1961.
87. V. Eskin, "Scattering of Light by Polymer Solutions." Nauka, Moscow, 1973 (R).

# Chapter 4

# The Physics of Proteins

## 4.1 Biological Functions of Proteins

All biochemical processes in cells and organisms require the participation of proteins. Proteins (enzymes) catalyze every chemical process in the cell. The most important protein function is this enzymic, or catalytic, one. Enzymes are essential participants in the biosynthesis of proteins, whose structure is fixed at the genetic level. Enzymes participate at all stages of protein biosynthesis. On the other hand, proteins serve as regulators of the genetic function of nucleic acids. All metabolic processes in the cell, particularly its nourishment and respiration, are catalyzed by enzymes. Enzymes perform both catalytic and regulatory functions; the so-called allosteric enzymes (Chapter 7) provide feedback in metabolic pathways.

The mechanochemical function of contractile proteins, which are responsible for cellular and intracellular motion (i.e., for cell movement, the movement of protoplasm, the movement of the cell contents during mitosis and meiosis), is closely associated with the enzymic function. Contractile proteins are enzymes; the result of their catalytic action is the transformation of chemical energy into mechanical work.

The existence of a cell as a system, the existence of functional cellular organelles, requires compartmentalization, the spatial separation of these systems by membranes possessing internally regulated permeability. Proteins (enzymes) that are contained in cell membranes in the form of lipoprotein complexes, determine the active transport of metabolites into and out of the cell, in a direction opposite to that of the concentration gradient. This function of proteins is closely related to the mechanochemical one. On the other hand, proteins catalyze the metabolic bioenergetic processes in membranes. Thus, mitochondrial enzymes localized in membranes are responsible for the biochemical processes involved in respiration, in the mechanical motion of mitochondria, and in active transport.

When we investigate multicellular organisms, we encounter other specialized functions of proteins. In such organisms proteins serve to store (myoblobin) and transport (hemoglobins, hemocyanins) oxygen. These functions are related to, but different from, that of enzymes, since they do not involve the chemical transformation of molecular oxygen. Low-molecular-weight proteins or polypeptides, such as insulin, oxytocin, and vasopressin, perform regulatory functions as hormones.

Specific proteins of higher organisms (namely, γ-globulins) protect these organisms from foreign biopolymers by acting as antibodies in immunological reactions. Special proteins contained in the skin, hair, feathers, connective tissue, etc., perform important supportive functions, providing for the nonrigid but stable interconnection of organs, thereby assuring their integrity and protection.

Proteins are indispensable participants in the storage, transmission, reception, and decoding of chemical signals. In many cases proteins themselves, or protein-containing *supra*molecular systems, serve as such signals, and as the receptors and transformers of signals. An organism's reaction to any external influence amounts to the translation of external signals into a chemical language that is made up of proteins.

The fundamental physical problem arising in the study of proteins consists in establishing the relation between their structure and properties. This is a classical problem in molecular physics. Its solution (which is easily described but difficult to effect) begins with a determination of the protein's structure, that is, of the spatial positions of the atoms and of the states of their electrons in a protein molecule. But protein molecules are the most complex of all the molecules now known. Their structure is determined not only by chemical bonds but by a series of different interactions among amino acid residues. Moreover, proteins function in aqueous surroundings, and the medium profoundly influences the structure and properties of protein molecules.

A protein macromolecule is similar to a solid because a considerable number of its atoms have fixed positions. In this sense a macromolecule of protein is an aperiodic crystal. Approaches to the treatment of such a structure based on the concepts of the physics of solids are quite natural and sound. But the protein macromolecule is also a dynamic system characterized by conformational lability. It is a kind of machine, whose behavior depends on the positions and properties of every individual element. Treatment of the dynamic properties of proteins requires the theoretical and experimental methods of macromolecular physics.

As we said earlier (p. 129), the "linear memory" in a polymer chain results in its specific physical properties [1]. The behavior of a protein molecule is determined by the properties of the chain as an integral system, by electron-conformation interactions (p. 131), and by concrete features of the primary structure--by informational characteristics coded at the genetic level. An important physical problem consists in the determination of the causal relation between the spatial structure of the protein molecule and the primary structure of the chain or chains that form this molecule.

The formation of the primary structure of the protein chain itself in a template synthesis, in which other informational macromolecules (namely, nucleic acids), participate is a complicated physical problem that entails the problem of the genetic code. The concept of the genetic code originated in the theory that there exists a specific molecular mechanism for transforming genetic information into the structural functionality of protein molecules. This theory is a physical one.

Along with enzymic catalysis, biophysics studies other dynamic properties of the protein molecules responsible for mechanochemical processes, immunological protection, oxygen storage and transport, etc. Consequently, we can speak about the physics of contractile proteins as the foundation of mechanochemistry, about the physics of immunological processes, and even about the physics of single proteins, such as myoglobin and hemoglobin.

Thus, the problems of protein physics include:

(1) theoretical and experimental investigation of the structure of protein molecules and of the supramolecular structures formed by proteins in conjunction with other substances;

(2) determination of the relation between the primary structure of protein chains and the spatial structure of protein molecules;

(3) investigation of the physical mechanisms of the biosynthesis of proteins;

(4) investigation of the physical mechanisms responsible for the biological functions of proteins.

## 4.2 Conformations of the Polypeptide Chain

An understanding of protein structure requires knowledge of the conformational structure of polypeptide chains. As we stated earlier, the structure of the peptide unit of the chain —CO—NH— is planar (p. 55). Structural parameters of this unit determined by X-ray studies of peptides and of a series of related compounds are presented in Table 4.1 [2-4], in which $C^{\alpha}$

*TABLE 4.1*
*Structural Parameters of Peptide Units*[a]
*(Bond lengths in angstroms Å)*

| | | | |
|---|---|---|---|
| $C^{\alpha}$—C | 1.53 | $<C^{\alpha}CH$ | 114° |
| C—N | 1.32 | $<OCN$ | 125° |
| N—$C^{\alpha}$ | 1.47 | $<C^{\alpha}CO$ | 121° |
| C=O | 1.24 | $<CNC^{\alpha}$ | 123° |
| N—H | 1.00 | $<CNH$ | 123° |
| $C^{\alpha}$—$C^{\beta}$ | 1.54 | $<HNC^{\alpha}$ | 114° |
| $C^{\alpha}$—H | 1.07 | $<XC^{\alpha}Y$ | 109.5° |

[a] From [2].

is the carbon atom of the amino acid residue that enters the chain; $C^{\beta}$, the first carbon atom of the radical R, which can be represented as $-C^{\beta}H_2-R'$; and X and Y are atoms bound by $C^{\alpha}$ in both the main chain and the side chain. The completely stretched chain (without deformations of the valence angles and elongations of bond lengths) possesses a trans conformation, characterized by zero values of the angle of rotatation $\phi$ around the bond N—$C^{\alpha}$ and of the angle of rotation $\psi$ around the bond $C^{\alpha}$—C. The rotation angle around the C—N peptide bond connecting neighboring peptide units is denoted by $\omega$. It also equals zero in the trans conformation. However, this conformation is not the most stable one. Hydrogen bonds are formed between the oxygen atoms of the C=O carbonyl groups and hydrogen atoms of the N—H imino groups with energies of the order of 5 kcal/mole (cf. Section 4.4).

Determination of the most stable conformation of the polypeptide chain requires minimization of the total energy, including the energy of intramolecular hydrogen bonds.

Pauling and Corey determined the stablest conformations of the polypeptide chain by using the results of X-ray studies of the peptide groups and the theoretical treatment of close packing of the chains with the maximal number of hydrogen bonds [5, 6]. There are three such conformations.

The most important conformation is the $\alpha$-helix, shown in Fig. 4.1. It is a helix characterized by rotation around the axis by 100° and translation along the axis by 1.5 Å for every peptide unit. One complete rotation around the helix axis corresponds to 3.6 peptide units and to translation along the axis equal to 5.4 Å. Hydrogen bonds link the carbonyl group of a given unit with the imino group of the fourth unit preceding it. All C=O and N—H groups, excluding the terminal ones, are linked by hydrogen bonds. The structure is very compact and in principle it can be formed by all amino acid residues except prolyl, which does not contain the NH group and therefore does not form a hydrogen bond.

*FIG. 4.1 The α-helix.*

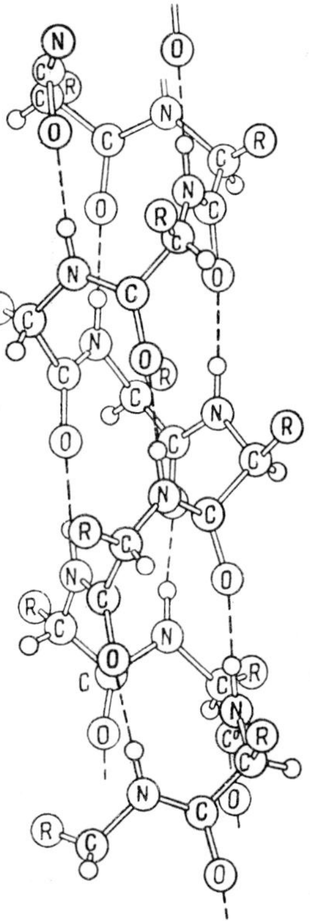


The α-helix can be either right- or left-handed. In the first case $\phi$ = 132°, $\psi$ = 123°; in the second case $\phi$ = 228°, $\psi'$ = 237° (the angles are referred to the coplanar trans position. In the α-helix, the —CO—NH— peptide groups maintain their planar structure. This structure is one that determines the α-helical conformation.

Other conformation with maximal saturation of hydrogen bonds are the parallel and antiparallel β-forms shown in Figs. 4.2 and 4.3. These are regular conformations not of a single chain but of an aggregate of chains forming the pleated sheet structure. The chains in the β-form do not have planar trans structure (Fig. 4.4). In the parallel conformation, $\phi$ = 61°, $\psi$ = 293°; in the antiparallel one, $\phi$ = 38°, $\psi$ = 325°.

The β-form can also exist in a single polypeptide chain because of its systematic bonds. A schematic picture of such a form, called the cross-β-form, is shown in Fig. 4.5. The internal rotation angles at the bending sites obviously have values different from those characteristic of the ordered parts of the chain.

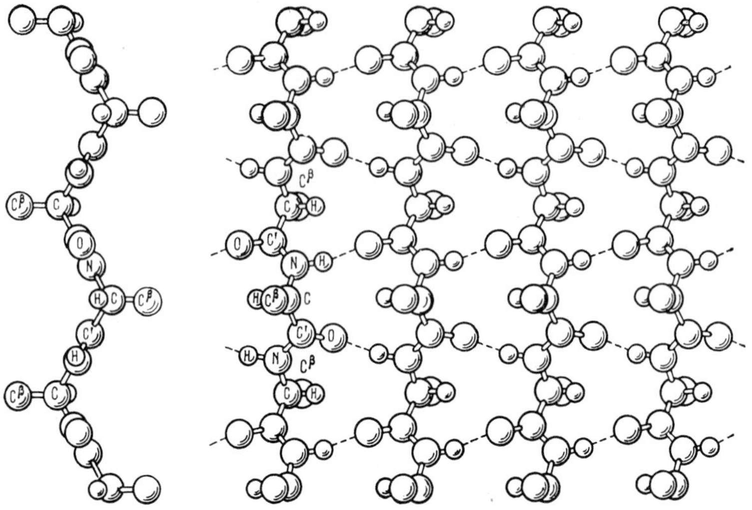


FIG. 4.2 The parallel β-form.

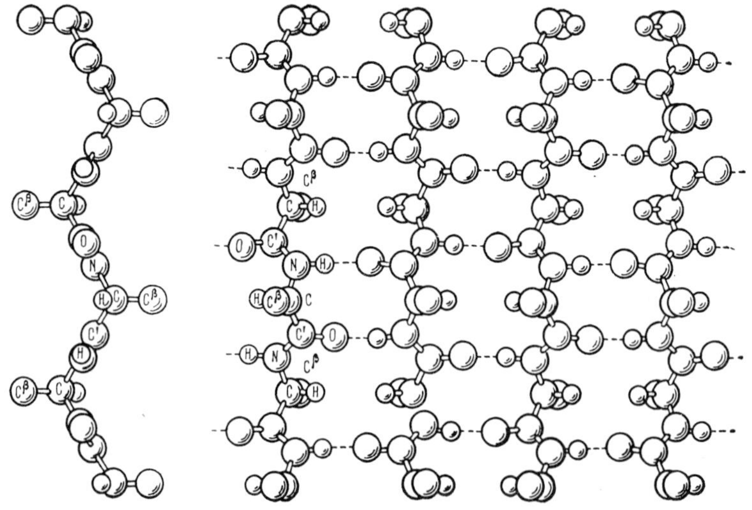


FIG. 4.3 The antiparallel β-form.

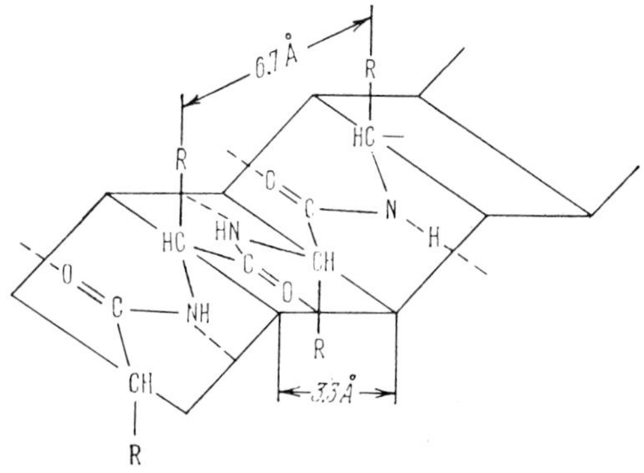

*FIG. 4.4 The general structure of the polypeptide chain in the β-form.*

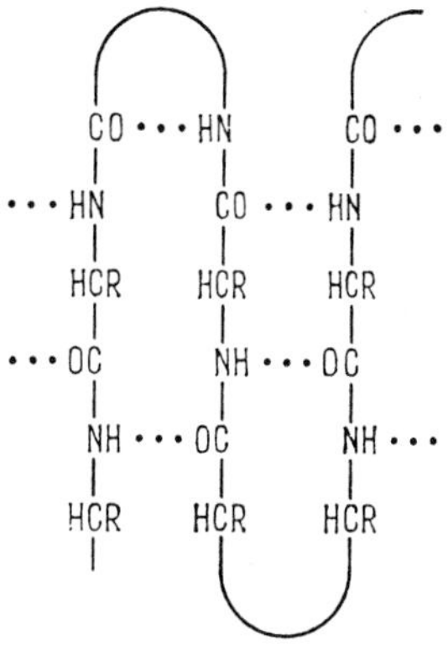

*FIG. 4.5 The cross-β-form.*

The essential feature of the polypeptide chain is the stabilization of the selected conformation in solution by hydrogen bonds. As was said earlier, the selection of one rotamer of a conventional macromolecule occurs in the crystalline polymer containing many macromolecules (p. 111). In this sense the α-helix and cross-β-form are similar to crystals, corresponding to one- and two-dimensional ones.

The α-helix and β-forms are the most important, but not the only, conformations of polypeptide chains. The polyamino acid polyglycine assumes either the β-form or form II, in which

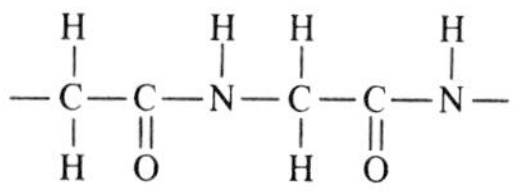

all the polyglycine chains are parallel and each possesses a screw axis of the third order. Transition from one peptide unit to the neighboring one is performed by longitudinal translation by 3.1 Å and by a 120° rotation around the helix axis. The chains are packed in a hexagonal structure, with every chain connected to six neighbors by hydrogen bonds. These features of the structure were established by Crick and Rich by means of X-ray studies [7]. The structure of crystalline polyglycine II is shown in Fig. 4.6 [7, 8]. This structure is characterized by the angles $\phi = 100°$ and $\psi = 330°$. The explanation of polyglycine's unusual structure lies in the peculiarity of glycyl, which, unlike all the other amino acid residues, does not contain a bulky R radical.

Other specific conformations of polypeptide chains are found in studies of fibrillar proteins (Section 4.11). The conformation of a protein chain as a complete system is called its secondary structure.

Let us consider the dependence of polypeptide chain energy on the angles of internal rotation $\phi$ and $\psi$, the corresponding steric maps [2, 4, 9]. The conformational energy of the chain is determined by interactions of the chemically nonbonded atoms. Because of the peptide group's planar structure, the $\phi_i$, $\psi_i$ values for the given ith unit do not depend on the values $\phi_{i-1}$, $\psi_{i-1}$ of the neighboring unit [2, 10]. If angles

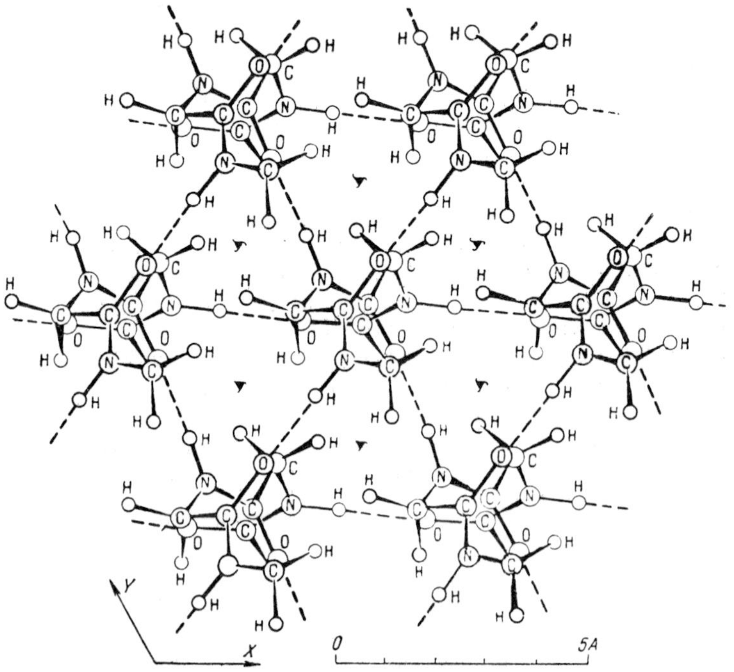


*FIG. 4.6 The structure of crystalline polyglycine II.*

$\phi_i$, $\psi_i$ vary in the range of values not forbidden by steric overlapping of the peptide group atoms, and if angles $\phi_{i-1}$, $\psi_{i-1}$ vary simultaneously, there is no combination of these four angles that would constitute steric interaction of group i with group i-2. In this sense the cooperativity of the polypeptide chain is restricted by nearest neighbors. This makes is possible to treat separately the conformational energies for single amino acid residues [2]. Obviously, the steric map for a given residue depends on the nature of the radical R involved. Thus it is possible to consider the interactions in a given pair of peptide groups as typical for the amino acid residue connecting these groups. Such calculations were first made by Ramachandran and his co-workers [11] in a study of the dipeptide glycyl-L-alanine. In other words, the Indian scientists evaluated the conformational (steric) map for alanyl. Their calculation was based on a simple theory in which the atoms were considered hard spheres characterized by the van der Waals radii obtained from data on the interatomic distances in molecular crystals. In Table 4.2 the normal, generally observed distances between atoms in crystals are shown, along with minimal distances observed in single cases.

*TABLE 4.2*
*The Contact Interatomic Distances in Polypeptides*[a]

| Pair of atoms | Normal distance (Å) | Minimal distance (Å) |
|---|---|---|
| C•••C | 3.2 | 3.0 |
| C•••O | 2.8 | 2.7 |
| C•••N | 2.9 | 2.8 |
| C•••H | 2.4 | 2.2 |
| O•••O | 2.8 | 2.7 |
| O•••N | 2.7 | 2.6 |
| O•••H | 2.4 | 2.2 |
| N•••N | 2.7 | 2.6 |
| N•••H | 2.4 | 2.2 |
| H•••H | 2.0 | 1.9 |

[a] From [2].

Using the potential for hard spheres, both the allowed and forbidden regions of the values of $\phi$ and $\psi$ can be found. Figure 4.7 shows the results of computations made in [11]. The solid lines outline regions allowed at normal interatomic distances; the broken lines, those allowed at the minimal interatomic distances listed in the third column of Table 4.2. Shown on the "map" are the conformations of the right and left $\alpha$-helix ($\alpha^R$ and $\alpha^L$), the parallel and antiparallel $\beta$-forms ($\beta^P$ and $\beta^a$), polyglycine II (II), and collagen (c). Similar maps were obtained for a series of other amino acid residues

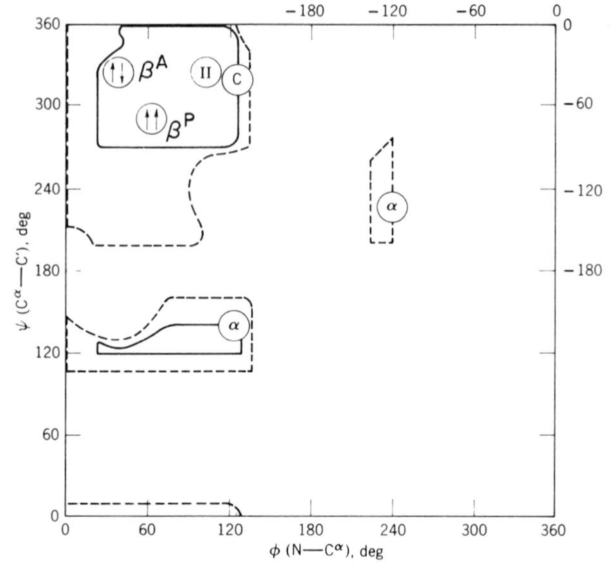


*FIG. 4.7 Steric map for L-alanine. Shown are right ($\alpha^R$) and left ($\alpha^L$) α-helix conformations; parallel ($\beta^P$) and anti-parallel ($\beta^A$) β forms; polyglycine II (II); collagen (C).*

[4, 12, 13]. Naturally, an increase in the size of radical R reduces the area of the allowed φ and ψ values.

Calculations of the conformational energies require more exact potentials. The potential is the sum of the effect of the crossed bonds and the effect of orientation (p. 109). The general form is

$$u(\phi,\psi) = \frac{1}{2}u_\phi^\circ(1-3\cos\phi) + \frac{1}{2}u_\psi^\circ(1-3\cos\psi) + \sum_{i,k} u(r_{ik}) + u_c \quad (4.1)$$

The potential $u(r_{ik})$ can be expressed either in the form of (3.10) (p. 110) or, according to Lennard-Jones, as

$$u(r_{ik}) = \frac{d_{ik}}{r_{ik}^{12}} - \frac{e_{ik}}{r_{ik}^{6}} \quad (4.2)$$

$u_c$ characterizes the Coulombic electrostatic interaction between neighboring pairs of peptide groups, determined by their large dipole moments, which can reach 3.7 D.

The barriers of internal rotation $u_\phi^\circ$ and $u_\psi^\circ$ can be obtained from the results of investigations of low-molecular-weight model compounds. Apparently, $0.48 < u_\phi^\circ < 1.3$ kcal/mole (0.48 is the value for $CH_3$—CO—OH, 1.3 for $CH_3$—CO—Cl).

According to the estimation of Scott and Sheraga [14], $u_\psi^0$ has an even smaller value (<0.2 kcal/mole), but Brant and Flory [10] give 1.5 kcal/mole. In every case the barriers $u_\phi^0$ and $u_\psi^0$ are considerably lower than for saturated hydrocarbons.

The attraction of nonbonded atoms, which gives a negative contribution to $u(r_{ik})$, is determined by van der Waals dispersion forces (cf. the following sections). The coefficients $e_{ik}$ in (4.2) or the similar coefficients $K'$ in expression (3.10) depend on the polarizabilities of atoms and on the number of their external electrons [4]. The constants for $u(r_{ik})$ in the Lennard-Jones form (4.2) are presented in Table 4.3.

*TABLE 4.3*
*Constants of the Lennard-Jones Potential*[a]

| Atoms | $P_{ik}$ (kcal/mole Å) | $r_{ik}$ (Å) Minimum | $10^{-3} \times d_{ik}$ (kcal/mole = Å$^{12}$) |
|---|---|---|---|
| H-H | 47 | 2.40 | 4.5 |
| C-C | 370 | 3.40 | 286 |
| N-N | 363 | 3.10 | 161 |
| O-O | 367 | 3.04 | 145 |
| C-O | 367 | 3.22 | 205 |
| C-N | 366 | 3.25 | 216 |
| C-H | 128 | 2.90 | 38 |
| N-O | 365 | 3.07 | 153 |
| H-O | 124 | 2.72 | 25 |
| H-N | 125 | 2.75 | 27 |

[a] From [4].

Calculation of the electrostatic energy $u_c$ is difficult, since the value of the effective dielectric constant $\varepsilon$ is not known.

$$u_c = \sum_{i,k} \frac{q_i q_k}{\varepsilon r_{ik}} \tag{4.3}$$

where $q_i$, $q_k$ are the charges on atoms i and k. If there are no other chain atoms or solvent molecules between the interacting atoms, $\varepsilon$ is determined by the polarizabilities of the interacting atoms and the reactive field of the surrounding medium. At such distances $\varepsilon \sim 3.5$ [10]. Calculations of $u_c$ have been made by means of the partial charges of atoms chosen in such a way as to give correct values of the dipole moments of the bonds and of the entire peptide group [15]. Poland and Sheraga [16] used the $\sigma$ and $\pi$ charges of atoms evaluated by the MO method (cf. p. 95) for this purpose.

The semiempirical function $u_{HB} = u_{ik} + u_c$ can be written for O and N atoms linked by hydrogen bonds. This function is determined by the values of hydrogen bond energy and by the equilibrium distances O···H-N [4, 5, 16].

Expression (4.1) does not take into account distortions of the bond lengths and valence angles; the values of these distortions are determined by the force constants, which can be obtained from the frequencies in the vibrational spectra. Apparently these distortions, which have not yet been calculated, are not large.

Expression (4.1) describes an isolated polypeptide chain. In an aqueous solution it is surrounded by the solvate layer. Treatment of the interaction of nonbonded atoms has to take into account the change in the free energy of the water molecules removed from this layer when interatomic contact occurs. This change is considerable for the polar, and especially for the charged, groups and atoms [4, 17]. In reality, minimization of the sum of the intramolecular potential energy and of the free energy of the solvent is required for estimation of stable conformations [18].

Calculations for a series of polypeptide chains and for such polypeptides as gramicidin S, oxytocin, and vasopressin were made by Sheraga *et al.* and Flory *et al.* by means of computers. Figures 4.8-4.11 show the conformational maps obtained by Flory and his co-workers [2, 19] with $U_C$ and without $U_C$

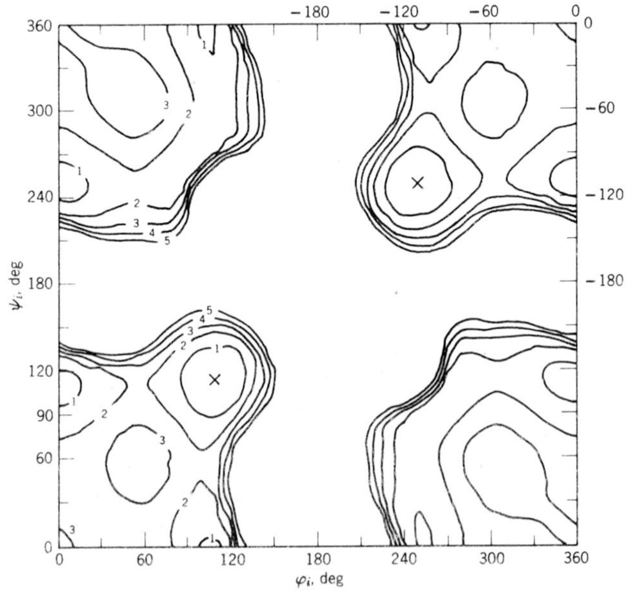


*FIG. 4.8 Conformational energy of Gly calculated without* $U_C$.

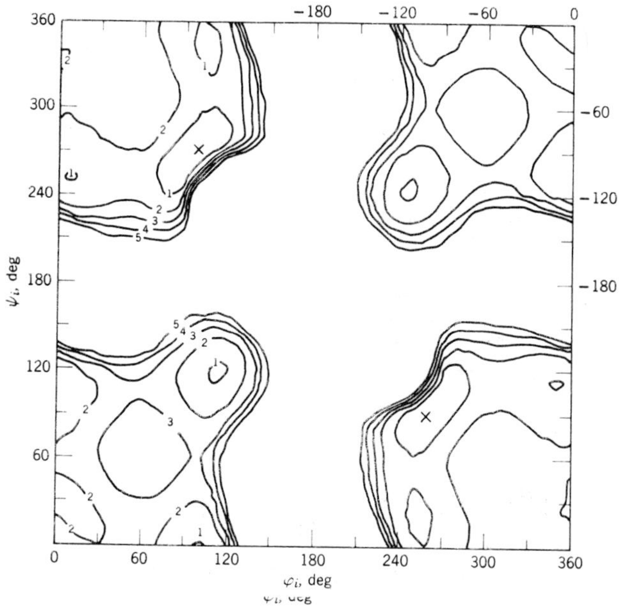


FIG. 4.9 Conformational energy of Gly calculated with $U_C$.

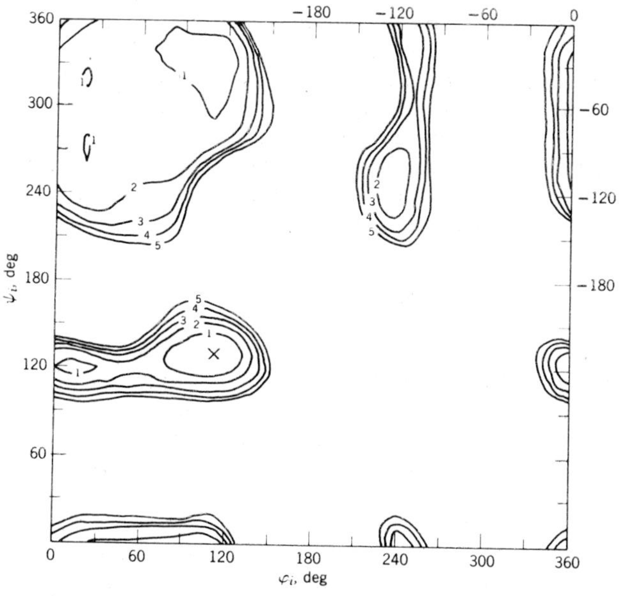


FIG. 4.10 Conformational energy of Ala calculated without $U_C$.

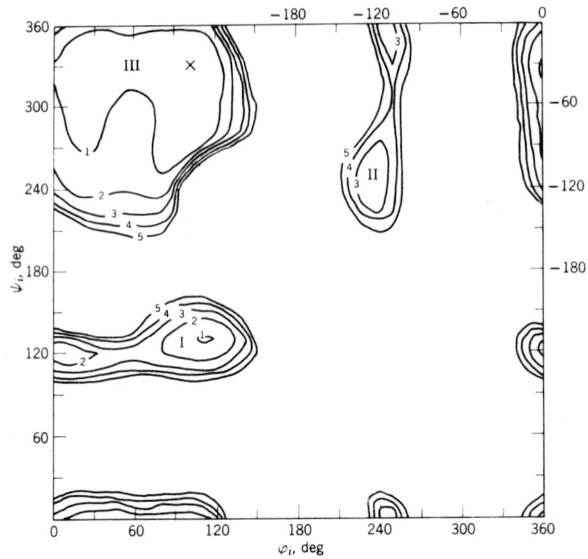


*FIG. 4.11 Conformational energy of Ala calculated with $U_C$.*

for glycine and alanine. The maps include the "geodesic lines" corresponding to energies from 1 to 5 kcal/mole; the lines for higher energies are not shown. The lowest energy values are indicated by crosses. Taking $U_C$ into account changes the positions of these minima but does not influence the general pattern much. Comparison of Figs. 4.10 and 4.11 with the map of allowed regions $\phi,\psi$ for alanyl is rather instructive. More rigorous calculations using potentials that do not differ in principle from the atom-atom potentials of Kitaigorodsky (p. 110) gave results that do not differ essentially from those presented [20].

## 4.3 Van der Waals Forces

The functionality of a substance in conventional chemistry depends primarily on the strong interactions on the chemical valence bonds. The interactions responsible for the transmission and reception of chemical signals in biological systems, on the other hand, are mainly weak nonvalent interactions. The cell (and organism) exists in soft conditions of low temperature and normal pressure. In many cases, biochemical processes bring about subtle rearrangements of chemical bonds that are not accompanied by considerable changes in the free energy. However, the total contribution of the weak interactions in these changes can be compared with the "chemical" one.

A biological system is a condensed system whose very

existence is determined by weak, nonchemical bonds. In rough terms it can be said that the cell is a molecular, but not an ionic or atomic, aperiodic crystal. It can also be said that the links of a biopolymer are in a condensed state in a macromolecular or in a supramolecular structure. Being connected by chemical bonds, the links of a biopolymeric chain form a secondary structure stabilized by weak nonvalent interactions. The functional structure of a biopolymer and also of a biologically active low-molecular-weight compound is the conformational structure created by weak interactions.

Nonchemical forces are responsible for the existence of the condensed liquid or solid state; they determine the properties of the phase surface (surface tension), the phenomenon of adsorption, etc. The heat of evaporation of liquids can serve as a rough measure of these interactions, being many times larger than the heat of melting of a crystal.

The van der Waals forces, nonspecific interactions between atoms and molecules, are electric forces. They are created by interactions of the outer electron shells of atoms and molecules. Two main physical quantities characterize these electron shells: the dipole moment (a vector) and the polarizability (a tensor). Of course, along with the dipole moment, the higher multipoles--the quadrupole and octupole moments, etc.--must also be taken into account. Their role, however, is usually very small.

The dipole moment of a molecule or of an atomic group expresses the asymmetry of the distribution of electric charges in an electrically neutral system. The order of magnitude of the dipole moment of a small molecule corresponds to the product of the electron charge ($4.8\times10^{-10}$ esu) and the bond length ($\sim 10^{-8}$ cm). We get $10^{-18}$ esu-cm, a unit called a debye (D). Polarizability expresses the deformability of an electron shell under the action of an external electric field, static field, or alternant field of a light wave. The field E induces an electric dipole moment p in an electron system

$$p = aE$$

The coefficient a, which has the dimensions of volume is called the polarizability. The order of magnitude of the polarizability of atoms and small molecules is the same as that of their volumes, namely $10^{-24}$ $cm^3$.

The usual van der Waals interactions of polar molecules determining their mutual attraction can be expressed as sums of three contributions--the orientation effect (Keesom), the induction effect (Debye), and the dispersion effect (London). If the molecules lack a permanent dipole moment, only dispersion forces act between them.

The energy of interaction of two dipoles is inversely proportional to the cube of the distance r between them

$$U_{or} = \frac{1}{r^3}\left\{p_1 p_2 - \frac{3(p_1 r)(p_2 r)}{r^2}\right\} \tag{4.4}$$

If two dipoles are oriented "head to tail," that is, if all three vectors $p_1$, $p_2$, and r are collinear, then

$$U_{or} = -2p_1p_2/r^3 \tag{4.5}$$

If the dipolar molecules are in a state of thermal motion in the gas or liquid, the expression (4.4) has to be averaged over all mutual orientations of the dipoles, taking into account the Boltzmann factor

$$\overline{U(r,T)} = \frac{\iint U(r,\Omega_1,\Omega_2)e^{-U(r,\Omega_1,\Omega_2)/kT}\, d\Omega_1\, d\Omega_2}{\iint e^{-U(r,\Omega_1,\Omega_2)/kT}\, d\Omega_1\, d\Omega_2} \tag{4.6}$$

where $\Omega_1$ and $\Omega_2$ characterize the orientations of both dipoles. If $U_{or} << kT$, we get

$$U_{or} = -2p_1{}^2p_2{}^2/3kTr^6 \tag{4.7}$$

The energy of the orientation interaction is inversely proportional to the sixth power of the intermolecular distance.

The induction effect is the interaction of the permanent dipole of one molecule with the dipole of another one, induced by the field of the first dipole. These dipoles are parallel. The field of dipole p in a molecule at a distance r from p is (if p and r are parallel)

$$E = 2p/r^3$$

and the energy of the inductive interaction is

$$U_{ind} = -aE^2/2 = -2ap^2/r^6 \tag{4.8}$$

Like $U_{or}$, the energy of this attraction is also inversely proportional to the sixth power of the intermolecular distance, but it does not depend on temperature.

Formulas (4.4)-(4.8) are valid only for point dipoles; that is, dipoles whose size is small in comparison with the distance between them. This condition can be approximately written as

$$p << re$$

where e is the charge of the electron. If p = 1 D, then p/e is 0.2 Å, a value that is actually much less than interatomic distances, which are of the order of 1 Å. Such an estimation is, however, approximate; in many cases in which the value of p itself is small enough, the charges forming the dipole are localized at rather remote atoms. If the point dipole condition is not obeyed, it is necessary to calculate monopole interactions.

This classical theory is adequate for estimating the corresponding energies. Rigorous quantum-mechanical theory requires application of perturbation theory in the first and second approximations. Orientations are characterized by quantum numbers. Thus, averaging over all dipole orientations is expressed in quantum mechanics by averaging over magnetic quantum states. The general character of dependence on r and p remains the same, and the order of magnitude of the effect is maintained but this does not mean that the classical theory and quantum theory are completely equivalent. In quantum mechanics specific resonance forces arise due to the removal of degeneration of the wave functions (i.e., to their hybridization).

Rigorous quantum-mechanical calculations of the orientational and inductive forces become necessary in studies of some fine effects in polar gases, such as the pressure broadening of lines in the microwave spectrum or transfer phenomena.

In contrast to the electrostatic orientation and inductive forces, dispersion forces have no classical analog. These forces are the most important and universal kind of nonvalence interaction of atoms and molecules.

Hellman [21] and Feynman [22] formulated an important theorem. The forces acting on the nuclei of atoms and molecules determined by the quantum-mechanically calculated potential energy surfaces coincide with those calculated on the basis of classical electrostatics if the distribution of the electron density is known. In other words, if this distribution is found from the solution of the Schrödinger equation, then the forces acting on the nuclei can be found by means of the classical expression

$$F_{ab} = \iint \frac{\rho_a(r_i)\rho_b(r_k)}{r_{ik}^3} r_{ik} \, dr_i \, dr_k \qquad (4.9)$$

where a and b denote two molecules and $\rho$ is the electron density. In fact, London's theory of dispersion forces is based on this theorem.

The elementary quantum-mechanical calculation of the energy of dispersion forces presented in many textbooks and monographs (cf. e.g., [23]) is based on regarding electrons as harmonic oscillators. An electrostatic dipole-dipole interaction of two oscillators occurs. Hence, instead of the original vibration with frequency $\omega_0$, there arise two normal vibrations, the frequencies of which differ from $\omega_0$ in direct relation to the intensity of the interaction. Consequently, the zero-point energy $E_0^0$ is changed (the quantum-mechanical effect!).

Instead of

$$E_0^0 = 6(\hbar\omega_0/2) \tag{4.10}$$

(the factor 6 enters because every electron oscillator has three degrees of freedom), the instantaneous electric interaction makes the zero-point energy equal to

$$E_0 = \frac{\hbar\omega_0}{2}\left\{2\left[\left(1 + \frac{e^2}{kr^3}\right)^{\frac{1}{2}} + \left(1 - \frac{e^2}{kr^3}\right)^{\frac{1}{2}}\right] + \left(1 + \frac{2e^2}{kr^3}\right)^{\frac{1}{2}} + \left(1 - \frac{2e^2}{kr^3}\right)^{\frac{1}{2}}\right\} \tag{4.11}$$

where e is the electron charge, k the elasticity coefficient of the oscillator, and r the distance between oscillators. If $e^2/kr^3 \ll 1$,

$$E_0 \cong 3\hbar\omega_0 - \frac{3}{4}\hbar\omega_0(e^4/k^2r^6) \tag{4.12}$$

Hence the interaction energy is

$$U_{dis} = -\frac{3}{4}\hbar\omega_0(e^4/k^2r^6) \tag{4.13}$$

Let us put an electron oscillator in the electric field **E**. The force acting at the electron e**E** is balanced by the elastic force k**r**. Hence, the induced dipole moment it

$$p = er = (e^2/k)\ E = aE$$

where a is the polarizability of the oscillator, equal to

$$a = e^2/k = e^2/m_0^2 \tag{4.14}$$

Putting a into $U_{dis}$, we get

$$U_{dis} = -\frac{3}{4}\hbar\omega_0(a^2/r^6) \tag{4.15}$$

The energy of the dispersion interaction is also inversely proportional to the sixth power of the distance between the interacting systems.

Dispersion forces are the forces of interaction of "instantaneous" dipoles arising because of the quantum-mechanical motion of electrons in systems lacking permanent dipole moments. More rigorous calculation of the dispersion interaction, performed by either the perturbation method or the variational method [21, 23-25], gives the expression

$$U_{dis} = -\frac{3}{2}\frac{I_1I_2}{I_1+I_2}\frac{a_1a_2}{r^6} \tag{4.16}$$

where $a_1$, $a_2$ are the polarizabilities of both molecules and $I_1$, $I_2$ are their ionization potentials (i.e., the energies necessary to remove an electron). If both molecules are identical, then

$$U_{dis} = -\frac{3}{4}I(a^2/r^6) \tag{4.17}$$

In the simplified theory the quantum of the vibrational energy of an electron oscillator, $h\omega_0$, is substituted for I. We also present the approximate expression for $U_{dis}$ obtained by Slater and Kirkwood [26].

$$U_{dis} = -\frac{3}{2r^6}\frac{e}{m^{\frac{1}{2}}}h\frac{a_1a_2}{(a_1/n_1)^{\frac{1}{2}} + (a_2/n_2)^{\frac{1}{2}}} \tag{4.18}$$

where $n_1$ and $n_2$ are the actual numbers of external electrons of the interacting systems.

These expressions are valid at distances that are much less than the velocity of light divided by the frequency $\omega_0$. This condition is obeyed at conventional intermolecular distances in condensed systems. At larger distances (of the order of hundreds of angstroms) we have to take into account the retardation effects caused by the finite velocity of the electromagnetic field. These effects reduce $U_{dis}$ [23].

The total van der Waals interaction of gas molecules with permanent dipoles p and polarizabilities a calculated per pair of molecules is expressed as the sum of expressions (4.7), (4.8), and (4.17).

$$U = U_{or} + U_{ind} + U_{dis} = -\frac{2p^4}{3kTr^6} - \frac{2ap^2}{r^6} - \frac{3}{4}I\frac{a^2}{r^6} \tag{4.19}$$

This is precisely the sum which is contained in the potentials (3.10) and (4.2). The largest contribution to U is made by dispersion forces except in cases with very large dipole moments. Table 4.4 presents the energies of the van der Waals forces for simple molecules [i.e., the coefficients $e_{ik}$ in expression (4.2)].

The calculations of total energies agree well with experimental values of the heats of evaporation. The van der Waals energy of the densely packed molecules in a liquid or solid is a quantity of the order of several kilocalories per mole.

In the foregoing presentation, polarizability has been treated as a scalar quantity. In reality, the polarizability of a nonspherical electron system--a molecule or atomic group--is a tensor quantity possessing different values in different directions in the molecule. This fact has to be taken into account when treating interactions at small distances, particularly in cases of densely packed molecules (i.e., in crystals and liquids). The very packing of molecules in molecular crystals is due to anisotropic dispersion forces [27]. Thus, the strongest interaction of planar π-electron systems--of aromatic compounds, nitrogen bases (p. 69), etc.--corresponds to their parallel positions. These interactions are of great importance for the secondary structure of nucleic acids (Chapter 8).

TABLE 4.4

*Van der Waals Interaction of Simple Molecules*

| Atoms and molecules | $a \times 10^{24}$ ($cm^3$) | $p \times 10^{18}$ | I (eV) | $p_{ik} \times 10^{12}$ (erg) | | |
|---|---|---|---|---|---|---|
| | | | | $U_{or}$ | $U_{ind}$ | $U_{dis}$ |
| He | 0.205 | 0 | 24.5 | 0 | 0 | 1.49 |
| Ne | 0.39 | 0 | 25.7 | 0 | 0 | 7.97 |
| Ar | 1.63 | 0 | 17.5 | 0 | 0 | 69.5 |
| Kr | 2.46 | 0 | 14.7 | 0 | 0 | 129 |
| Xe | 4.0 | 0 | 12.2 | 0 | 0 | 273 |
| $H_2$ | 0.81 | 0 | 14.5 | 0 | 0 | 11.4 |
| CO | 1.99 | 0.1 | 14.3 | 0.003 | 0.057 | 67.5 |
| HCl | 2.63 | 1.03 | 13.4 | 18.6 | 5.4 | 111 |
| HBr | 3.58 | 0.78 | 12.1 | 6.2 | 4.05 | 185 |
| HI | 5.4 | 0.38 | 10.5 | 0.35 | 1.68 | 370 |
| $NH_3$ | 2.24 | 1.5 | 11.7 | 84 | 10 | 70 |
| $H_2O$ | 1.48 | 1.84 | 18 | 190 | 10 | 47 |

The theory of van der Waals forces presented here is the dipole approximation theory. In classical expressions and in the potentials used in calculating dispersion forces, multipoles contribute to the energy of attraction terms that decrease more rapidly with increasing distance, that is, as $r^{-8}$, $r^{-10}$, etc.

If a molecule is in an excited energy state, interaction with an identical but nonexcited molecule can produce resonance--the transfer of excitation energy. This phenomenon results in resonance interaction forces which decrease with distance as $r^3$. These interactions play an essential role in the optical properties of molecules, particularly of biopolymers (cf. Chapter 5).

We have considered the van der Waals, nonchemical attractive forces of electron systems. Repulsion (i.e., the positive term in the potential of intermolecular forces) appears only at small distances [e.g., the term proportional to $r^{-12}$ in the potential (4.2)]. The origin of this repulsion, which can be interpreted as the impossibility of the spatial overlapping of two electron systems, is explained by quantum mechanics. The repulsion of systems with saturated valences characterizes the saturability of chemical bonds. Let us consider the interaction of the molecule $H_2$ with an atom H. The spins of electrons in $H_2$ are antiparallel. We have

$$H\uparrow(1)\!-\!H\downarrow(2) \quad H\uparrow(3)$$

Exchanging spin 3 with spin 2, we get

$$H\uparrow \; H\uparrow \quad H\downarrow$$

But the state with parallel spins is a state of repulsion; such an exchange increases the energy. The exchange $1 \rightleftarrows 3$ does not destroy the molecule $H_2$ as a stable system but does result in an increase in energy, since spins 1 and 3 are parallel and the coordinate wave function is asymmetric (i.e., corresponds to the repulsion state). In both cases exchange results in repulsion between the molecule $H_2$ and atom H. The repulsion forces are exchange forces. Their theoretical calculation is possible in simple cases [23], but for practical purposes, it is convenient to use an empirical potential. For instance the model of hard spheres with van der Waals radii can be used.

## 4.4 The Hydrogen Bond and the Structure of Water

Just as important, biologically and biophysically, as the van der Waals forces are hydrogen bonds. We have already seen that hydrogen bonds stabilize the secondary structure of polypeptide chains. The same is true with regard to the conformational structure of nucleic acids and carbohydrates. Biopolymers function in aqueous surroundings. Water is a liquid possessing a number of anomalies, all due to hydrogen bonds.

The notion of the hydrogen bond was introduced by Latimer and Rodebush in 1920 [28], with the aim of explaining the properties of associated liquids, the most important of which is water. The hydrogen atoms contained in the groups O—H, N—H, H—F, H—Cl, and sometimes those in S—H and C—H, form specific bonds with the O, N, F, or Cl atoms of the same (or another) molecule

$$>\mathrm{N{-}H}\cdots\mathrm{O{=}C}<$$

The existence of such bonds is deduced from a series of physical and physicochemical properties of the substance. Thus, intermolecular hydrogen bonds are responsible for the special properties of associated liquids. Associated substances are characterized by large heats of evaporation, high melting and boiling points, and broad intervals between them. Table 4.5 compares four substances formed by isoelectronic molecules. Methane is a nonassociated liquid which does not contain hydrogen bonds; the other three substances are associated.

Table 4.6 compares the properties of two isomeric substances--ethanol and dimethyl ether. Ethanol, which contains the OH group, forms hydrogen bonds; dimethyl ether does not have this ability. Hydrogen bonds are the reason for the large values of dielectric constants and deviations of the dipole moments from the vector sums of bond moments. The following are the dielectric constants of typical associated liquids at 20°C: water, 80; HCN, 95; formamide, 84; methanol, 33; ammonia, 15.5.

*TABLE 4.5*

*Properties of Isoelectronic Liquids*

| Substance | $T_m$ (°K) | $T_b$ (°K) | Heat of evaporation (kcal/mole) | Molar volume ($cm^3$/mole) |
|---|---|---|---|---|
| Hydrogen fluoride (HF) | 181 | 292 | 7.20 | 20.2 |
| Water ($H_2O$) | 273 | 373 | 9.72 | 18.0 |
| Ammonia ($H_3N$) | 195 | 240 | 5.57 | 20.8 |
| Methane ($H_4C$) | 89 | 112 | 2.21 | 34.0 |

*TABLE 4.6*

*Properties of Ethanol and Dimethyl Ether*

| Substance | $T_m$ (°K) | $T_b$ (°K) | Heat of evaporation (kcal/mole) |
|---|---|---|---|
| Ethanol ($C_2H_5OH$) | 161 | 351 | 10.19 |
| Dimethyl ether ($(CH_3)_2O$) | 135 | 249 | 4.45 |

Structural investigations of crystals that contain hydrogen bonds show that a hydrogen bond linking two electronegative atoms A and B usually reduces the distance between them compared with the sum of the van der Waals radii. Corresponding data are presented in Table 4.7 [29]. The sum of van der Waals radii concerns only atoms A and B, for example, atoms O and N in the O—H···N bond. The values in the last column were estimated by Pimentel and McClellan from spectroscopic data [29].

Hydrogen bonding is well expressed in spectra--vibrational (infrared and Raman), electronic, and nuclear magnetic resonance (NMR) spectra. The characteristic frequencies of hydrogen-containing groups, such as the O—H group, are lowered considerably if this hydrogen forms a hydrogen bond. Thus, in monomeric formic acid this frequency is 3682 $cm^{-1}$; in the dimer formed by two hydrogen bonds

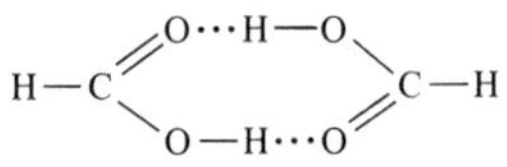


the frequency is lowered to 3080 $cm^{-1}$. Such dimers of carbonic acids are stable even in the vapor phase. Other frequencies of the monomer are also changed. The infrared absorption bands of O—H groups are significantly broadened if

TABLE 4.7

*Interatomic Distances in Crystals*

| A—H···B | Type of Compound | Sum of van der Waals radii of A and B (Å) | $R_{AB}$ (Å) | Estimate of compression produced by hydrogen bond (Å) |
|---|---|---|---|---|
| O—H···O | Inorganic acids | 2.80 | 2.55 | 0.65 |
| | Carbonid acids | | 2.63 | 0.57 |
| | Phenols | | 2.67 | 0.53 |
| | Alcohols | | 2.74 | 0.46 |
| | Ice | | 2.76 | 0.40 |
| O—H···N | | 2.90 (3.05) | 2.80 | 0.5 |
| N—H···O | Ammonium salts | 2.90 (3.05) | 2.88 | 0.5 |
| | Amides | | 2.93 | 0.4 |
| | Amines | | 3.04 | 0.3 |
| N—H···N | | 3.00 (3.30) | 3.10 | 0.4 |
| O—H···Cl | | 3.20 | 3.08 | -- |
| N—H···F | | 2.85 (3.00) | 2.78 | -- |
| N—H···Cl | | 3.30 (3.45) | 3.21 | -- |
| F—H···F | | 2.70 | 2.44 | -- |

hydrogen bonds are formed, and their intensities are greatly increased. The causes of these phenomena were investigated in a series of theoretical papers (cf. [30-32]). The spectroscopic features of hydrogen bonds are described in detail in the monograph [29]. Hydrogen bonding markedly influences some electronic transitions, as in the case of pyridine interacting with water [33]. It shifts to the shorter wave lengths the so-called $n\ \pi^*$ transitions, which enables us to distinguish them from $\pi\ \pi^*$ transitions (p. 275). In proton magnetic resonance spectra, specific chemical shifts produced by hydrogen bonds are readily observable. These spectra, like the infrared ones, provide outstanding methods for studying hydrogen bonds. The foregoing discussion pertains to both intermolecular and intramolecular hydrogen bonds. Figure 4.12 shows bonds of both types, in water and in salicylic acid.

The hydrogen bond energy can be determined from the thermodynamic properties of the substances involved, from spectroscopic data, etc. Thermodynamic functions can be expressed by the equilibrium constant

$$K = \frac{\text{activity of compound}}{\text{activities of reagents}} \cong \frac{[A\text{—}H\cdots B]}{[AH][B]} \tag{4.20}$$

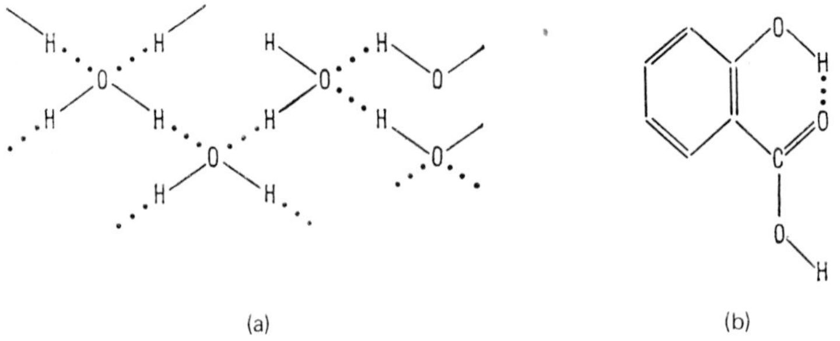


*FIG. 4.12 (a) Intermolecular hydrogen bonds in water and (b) intramolecular hydrogen bond in salicylic acid.*

$$\Delta G = \Delta H - T\,\Delta S = -\,RT \ln K \tag{4.21}$$

$$\Delta H = RT^2 \left(\frac{\partial \ln K}{\partial T}\right)_P \tag{4.22}$$

Characteristic values of $\Delta H$ for hydrogen bonds are presented in Table 4.8, and we also present some estimates [34]: here F—H···F, 7; O—H···O, 3-6; N—H···N, 3-5; N—H···F, 5; O—H···N, 4.7; C—H···O, 2.6; N—H···O, 2.3 kcal/mole. The value of $\Delta S$ for O—H···O is 14-18 cal/mole deg.

A hydrogen bond is formed by hydrogen linked with carbon in only a few cases [HCN, chloroform ($CHCl_3$) with pyridine and triethylamine, etc,].

The nature of the hydrogen bond is unusual. Atomic H is not equally bound by atoms A and B. X-ray studies do not enable us to determine the position of the hydrogen atom directly (p. 262). It can be directly determined, however, by means of neutronography, as was done in the case of ice in [35]. It was found that the notation O—H···O has real meaning--the O—H bond is shorter, and therefore stronger, than the H···O bond, as is shown in Fig. 4.12. If the temperature increases, the positions of the O—H and H···O bonds can be interchanged. Quantum-mechanical calculations as well as general considerations indicate that the curve of the dependence of the energy on the position of the atom H in the formic acid dimer has two symmetrical minima separated by a barrier. Hence, if the temperature increases, the hydrogen atom can migrate from one oxygen atom to the other (the tunnel effect).

Hydrogen bonding always occurs between two electronegative atoms. An early hypothesis held that the origin of the hydrogen bond was purely electrostatic [36, 37]. Theoretical estimations of the electrostatic interaction did, in fact, give energy values for the hydrogen bond that agreed with experimental determinations. However, a purely electrostatic origin

*TABLE 4.8*

*ΔH Values for Different Types of Hydrogen Bonds*

| Type | Substance | $-\Delta H$ (kcal/mole) |
|---|---|---|
| O—H···O | | |
| | Water $H_2O$ | Gas 4.4-5.0, liquid 2.8 |
| | Methanol $CH_3OH$ | Gas 3.2-7.3, liquid 4.7 |
| | Ethanol $C_2H_5OH$ | Gas and liquid 4.0 |
| N—H···N | | |
| | Ammonia $NH_3$ | Gas 3.7-4.4 |
| | Amines ($CH_3NH_2$, $C_2H_5NH_2$) | Gas 3.1-3.6 |
| F—H···F | | |
| | Hydrogen fluoride HF | Gas 6.7-7.0 |
| C—H···N | | |
| | Hydrogen cyanide HCN | Gas 3.3, liquid 4.6 |

of the hydrogen bond contradicts the spectroscopic data--infrared intensities as well as dipole moments and NMR spectra [33]. Quantum-mechanical calculation shows that along the electrostatic contribution to the hydrogen bond energy, we must include that resulting from the delocalization of electrons. The wave function of the O—H···O system treated as a four-electron system (two electrons of the O—H bond and two lone pair electrons at the second O atom) can be expressed as a linear combination of the functions corresponding to the following five valence structures [34]

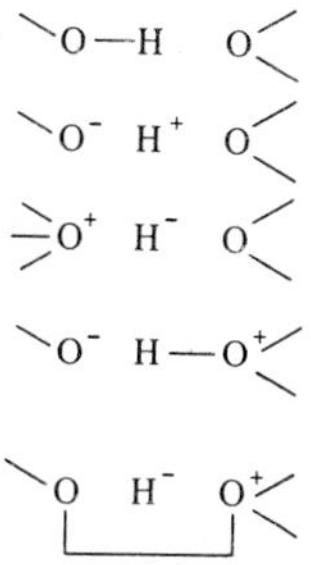

covalent

ionic

ionic

charge transfer, long H—O$^+$ bond

charge transfer, O—O bond

that is

$$\Psi = c_1\psi_1 + c_2\psi_2 + c_3\psi_3 + c_4\psi_4 + c_5\psi_5 \qquad (4.23)$$

The electrostatic interaction is described by structures 1,2, and 3. The delocalization energy [i.e., the decrease in energy as compared with structure (1)] produced by hybridization with four other structures has been estimated as 8 kcal/mole [38, 39].

Both electrostatic interaction and delocalization reduce

the length of the hydrogen bond. This decrease is hindered by repulsion, whose energy estimated semiempirically is approximately 8.4 kcal/mole. The dispersion interaction ultimately gives a contribution of the order of 3 kcal/mole [34, 40]. Thus, the total hydrogen bond energy is the sum of the electrostatic energy (6 kcal/mole), delocalization energy (8 kcal/mole), and dispersion energy (3 kcal/mole) minus the energy of repulsion (8.4 kcal/mole). For O—H···O this gives 8.6 kcal/mole instead of the experimental value (for ice) of 6.1 kcal/mole.

Sokolov developed a quantum-mechanical theory of hydrogen bonding based on the concept of a donor-acceptor bond resulting from the collectivization of lone pair electrons. In the system A—H···B the $A^-$—H bond occupies an "intermediate position" between the ionic bond $A^-H^+$ and the covalent bond. Along with electrostatic interactions, two new factors arise. First, because of a decrease in electron density near an H atom the repulsion decreases, which is characteristic of the system covalent bond A—H and atom B. Second, there occurs an additional attraction between H and B produced by a rearrangement of the electron density of atom B in the field of atom H, similar to donor-acceptor interactions. Using these ideas, Sokolov obtained a quantitative explanation of the spectroscopic features of the hydrogen bond [32]. In this work it is shown that the processes of inter or intramolecular proton transfer, particularly tautomeric transformations (p. 70) always begin with the formation of a hydrogen bond.

Hydrogen bonds are responsible for the structure of water and for its properties. Consider first the structure of ice, that is, of its usual modification, called "ice I" (in contrast to the other polymorphic forms--no fewer than eight such forms exist at increased pressure [41]). Ice I is hexagonal; every oxygen atom of its lattice lies at the center of a tetrahedron at whose apices lie the neighboring oxygen atoms. The O—O distance is 2.76 Å (cf. p. 185). Every $H_2O$ molecule is linked to four neighbor molecules by hydrogen bonds. The elementary cell contains four molecules. The ice lattice is loose; its density is low because the coordination number (i.e., the number of nearest neighbors) is small, equal to 4. Therefore, ice is lighter than its melt. This property of ice is not unique; the crystals of diamond, silicon, and germanium have the same property, since they have similar structures.

The ice lattice is a molecular lattice. Studying its structure, Pauling came to the conclusion that the order of the crystal is nonperfect even at 0°K [42]. All positions of the $H_2O$ molecules in the lattice are equally probably if the following conditions are obeyed: (1) the $H_2O$ molecules retain their integrity; (2) every $H_2O$ molecule is oriented in such a way that two of its H—O bonds are directed toward two of the four nearest oxygen atoms; (3) the orientations of the neighboring $H_2O$ molecules are such that only one H atom lies on the straight line connecting two neighbor oxygen atoms. Evidently every H atom can have two positions: near the given O atom, and near the nearest neighbor O atom. These hypotheses were confirmed by studies of the structure of ice by means of neutronography (cf. Section 5.1) [43]. One mole of ice contains 2 $N_A$ hydrogen atoms. They can have $2^{2N_A}$ different positions. Of $2^4$ modes of placing four H atoms near the given O atom, 6 modes correspond to the $H_2O$ molecule, but not to $H_3O^+$, etc. (Fig. 4.13). The total number of configurations of the crystal is

$$W = 2^{2N_A}(6/16)^{N_A} = (3/2)^{N_A} \tag{4.24}$$

Hence the residual entropy is equal to

$$S_0 = k \ln W = R \ln (3/2) = 0.805 \text{ eu} \tag{4.25}$$

This calculation gave excellent agreement with the experimental value $S_0 = 0.82 \pm 0.15$ eu. More rigorous calculations gave results that do not differ essentially from those obtained by Pauling [44, 45].

The icelike structure is maintained in liquid water but with considerable distortion [36]. Following Eisenberg and Kantzmann [41], let us consider the states of solid and liquid water. The molecules in the crystal perform vibrations, rotations, and comparatively rare translational transitions (diffusion). A photograph made during time $\tau$, which is much less than the period of vibrations $\tau_V$ (in ice $\tau_V \sim 2 \times 10^{-13}$ sec), gives the picture shown in Fig. 4.14a--the instantaneous, or I, structure of ice. In the photograph obtained during time $\tau$, much larger than $\tau_V$ but much less than the time of rotational diffusion $\tau_D \sim 10^{-5}$ sec, vibrations will be averaged and we shall see regular positions of nonregularly oriented

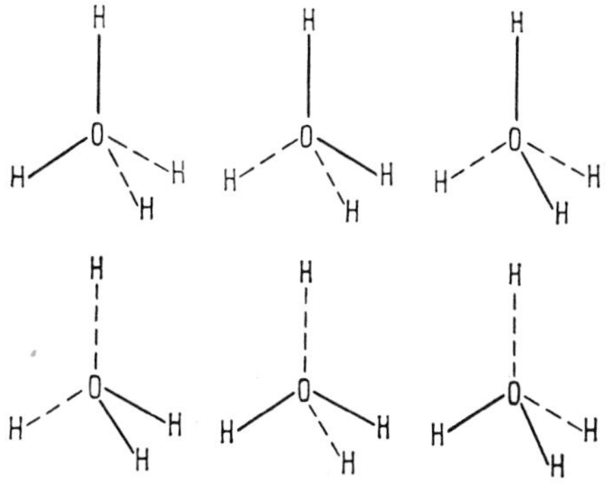

FIG. 4.13 Disposition of hydrogen atoms.

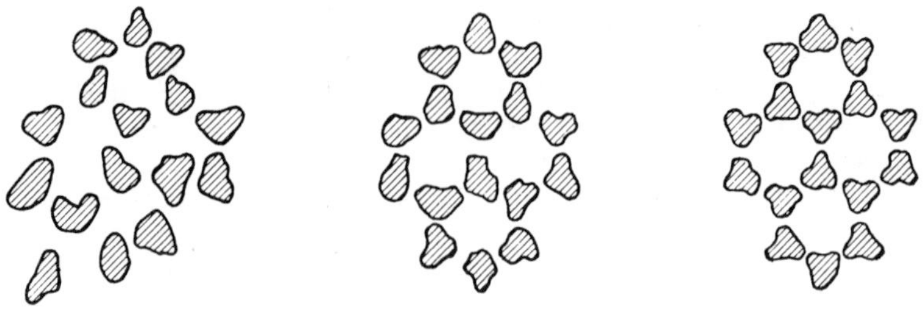


*FIG. 4.14 (a) Instantaneous, (b) vibrationally averaged, and (c) diffusionally averaged water structure.*

molecules. This is the vibrationally averaged V structure (Fig. 4.14b). Finally, if $\tau >> \tau_D$, we shall see the diffusionally averaged D structure of ice (Fig. 4.14c).

In the liquid, the I and V structures are like those of the crystal, but the D structure is totally destroyed by molecular motion. A more informative picture could be obtained by putting the camera inside a given $H_2O$ molecule and recording the surrounding molecules during its motion. The picture obtained in this mental experiment can be called the D structure of liquid, which will be ordered to some extent. Such a D structure is the average of the V structures.

Using various methods of investigation, characterized by different times $\tau$, we shall obtain information about the structure--in a different sense of this word--about the I, V, or D structure (Fig. 4.15). Naturally, thermodynamic properties are related to the D structure of a liquid. The same can

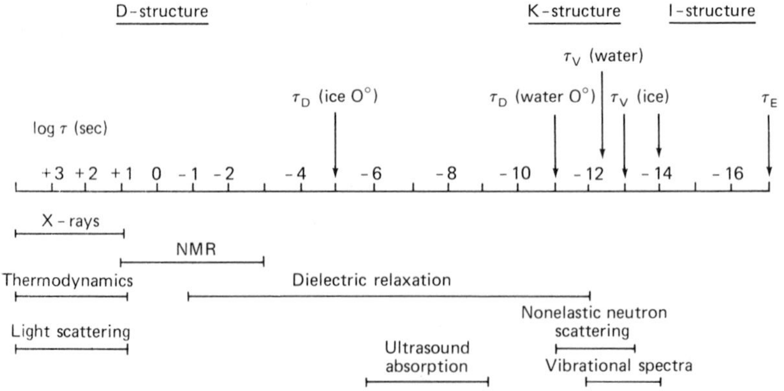


*FIG. 4.15 Time scale for molecular processes in ice and liquid water. $\tau_D$, time of displacement of a molecule; $\tau_V$, time of vibration; $\tau_E$, time of electron motion.*

be said about roentgenography and light scattering. This approach can be applied to any fluctuating condensed system, particularly to the macromolecules of proteins.

X-ray studies of liquid water give the radial distribution function (c.f. Section 5.1), that is, the relative numbers of molecules at the given distances from one another. If the temperature varies from 4°C to 100°C, the main maximum of the radial function is shifted continuously from 2.82 Å to 2.94 Å [46]. The coordination number over the entire range is ~4.4 (in liquid Ne it is 8.6, in liquid Ar, 10.5). Hence, liquid water is a quasicrystalline system; every molecule posesses an average of 4 neighbors.

The D structure of water is the spatial average of its V structures described by various models. Nemethy and Sheraga [47] interpreted the radial function, postulating that water is a mixture of monomeric molecules and hydrogen-bonded clusters. The clusters dissociate continuously and are formed again; thus every molecule possesses the same average surrounding. Samoilov [48] suggested that the $H_2O$ monomers are located in the cavities of a quasicrystalline lattice formed by hydrogen bonds (cf. also [46]). Pople [49] proposed a model with "bent", but not broken, hydrogen bonds. The same idea was employed in the model of Bernal [50], where linked molecules with quaternary coordination formed in water (but not in ice) a nonregular network containing rings. Many rings contain five molecules of water (the angle H-O-H in $H_2O$ is close to 108°, the angle in a regular pentagon), others contain four or more molecules. All these models agree more or less with the observed radial function [41]. However, it is difficult to choose a definite model at present.

The estimated energies of hydrogen bonds in liquid water obtained from thermodynamic and spectroscopic data vary from 1.3 to 4.5 kcal/mole [41]. The most reliable values are probably 2.5-2.8 kcal/mole [51].

All these models explain the peculiar properties of water, and primarily the dependence of specific volume on temperature, in nearly the same way. The minimum of specific volume at 4°C is due to the competition of two processes. The first one is the loosening or destruction of the ordered icelike structure with a small coordination number 4 accompanied by a decrease in volume. It is the continuation of melting. The second process is the thermal expansion of the liquid due to an increase in the mixture of monomers and clusters model, the clusters are thought to have a larger volume. In the model of Samoilov, the decrease in volume is explained by the filling of cavities by monomer molecules. In Pople's model, the bending of hydrogen bonds results in the rapprochement of the neighbor molecules (i.e., in a decrease in volume).

Nemethy and Sheraga [47, 52] suggested that every $H_2O$ molecule can occupy one of five energy levels, depending on the number of hydrogen bonds formed--0, 1, 2, 3, or 4. They calculated the partition function of such a system, expressing the numbers of the 4-, 3-, and 2-bonded molecules in clusters through the average number of molecules in the cluster and the fraction of 1-bonded molecules. The clusters have a molecular volume and expansion coefficient equal to those of ice. For nonbonded molecules these quantities are determined from experimental values of the molecular volume of the liquid at 0°, 4°, and 25°C. The molecular volume calculated in such a way has a minimum at 4°C, and satisfactory agreement with experiment was also obtained for $D_2O$. However, models that suggest the mixture of distinct kinds of water do not agree with a series of spectroscopic facts. These facts negate the possibility of treating clusters as real icelike structures [41]. It seems that there exists a broad, practically continuous distribution of water structures. The corresponding quantitative theory has not yet been derived.

Syrnikov used the method of graph theory to calculate the partition function of liquid water [172]. This topological method makes it possible to take into account the states of both the free and the hydrogen-bonded molecules. The method can be applied to the study of the influence of solutes. Quantitative calculations require knowledge of the probabilities of hydrogen bond formation which are yet unknown. However, this method seems to be promising.

Rahman and Stillinger [173] have presented a theoretical study of the structures formed by rigid molecules with a postulated potential which describes the formation of the hydrogen bonds. A model that differs from the icelike structure is obtained which explains satisfactorily a series of properties of water.

The thermodynamic properties of water have been calculated [174] with the help of the Monte Carlo method using the atom-atom potentials of Kitaigorodsky. The hydrogen bond has been described by a Morse potential and the energy of the hydrogen bond has been taken as 5.5 kcal $mole^{-1}$. The calculations of thermodynamic functions for the temperatures 300, 320, and 350°K were in agreement with experiment for the internal energy, heat capacity, and free energy. This method makes it possible to determine the location and orientation of the $H_2O$ molecules in liquid.

The method introduced in [174] has been applied effectively in the calculation of the thermodynamic functions of the water-plus-methane and water-plus-hard-spheres systems [173-175]. The statistical mechanics of hydrophobic interactions has been developed in these papers; in particular, the dependence of the energy of hydrophobic interactions on the distances between the particles has been obtained. The further development of such investigations can be important for portein physics.

Let us consider briefly the electric properties of water. Its large dielectric constant is explained not only by the polarity of single molecules but also by the correlation of their mutual orientations [53], that is, by the presence of hydrogen bonds. Its low electric conduction is evidently due to the small content ($2 \times 10^{-9}$) of strongly hydrated $H^+$ and $OH^-$ ions ($H_9O_4^+$, etc.). Investigations of the motilities of $H^+$ and $OH^-$ ions has shown that they are essentially larger than those of the other univalent ions in water. For $H^+$, $u_+ = 3.62 \times 10^{-3}$ $cm^2\ v^{-1}\ sec^{-1}$; for $OH^-$, $u_- = 1.98 \times 10^{-3}$; at the same time, for $Na^+$, $u_+ = 0.53 \times 10^{-3}$ and for $Cl^-$, $u_- = 0.79 \times 10^{-3}$. This can be explained by the ability of hydrogen bonds to provide rapid proton transfer:

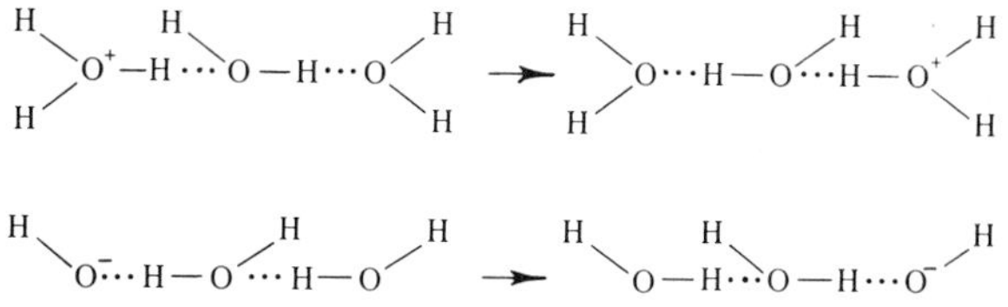

The motilities of the $H^+$ and $OH^-$ ions in ice are higher because of the better ordering of hydrogen bonds [41, 44, 54].

## 4.5 Helix-Coil Transitions

The regular conformations of the polypeptide chains, stabilized by hydrogen bonds, such as α and β forms, are stable only at definite conditions. The variation in temperature, solvent, and pH of the medium result in order-disorder transitions, the transformation of the regular chain conformation into a random coil. These processes can be most conveniently studied in model systems, using synthetic polyamino acids.

Many polyamino acids, particularly polyglutamic acid (PGA) and its derivative poly-γ-benzyl glutamate (PBG)

$$-[\mathrm{CO{-}CH{-}NH}]_n- \quad \text{with side chain on CH: } \mathrm{CH_2{-}CH_2{-}CO{-}O{-}CH_2{-}C_6H_5}$$

form α-helices in solutions, and this is proved by all of their hydrodynamic and optical properties. Doty has shown that the helix-coil transition is very abrupt [55, 56]. Figure 4.16 shows the dependence of the degree of ionization, intrinsic

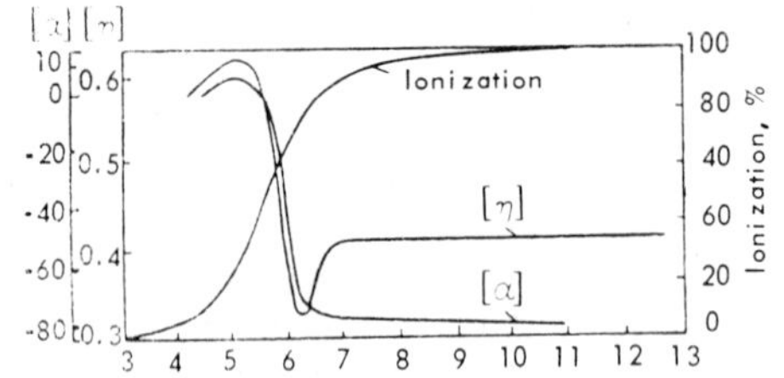


*FIG. 4.16 Helix-coil transition in polyglutamic acid.*

viscosity [η], and specific optical rotation [α] of PGA on pH. A sharp decrease in viscosity and specific rotation occurs in the region near pH 6. PGA cannot be dissolved in organic solvents; PBG dissolves and forms α-helices in dichloroethane, chloroform, and formamide. This is shown by means of light scattering and viscosimetry and directly by electron microscopy. On the other hand, in solvents that form hydrogen bonds with PBG, for example, in trifluoro- and trichloroacetic acids, the PBG macromolecules are in the form of random coils. Such solutions have low viscosity and flow birefringence. Gradual change of the solvent content (binary mixture) results in the sharp helix-coil transition in some narrow range (Fig. 4.17).

The sharpness of the transition shows its cooperative nature. In this sense the helix-coil transition is like the melting of a one-dimensional crystal. The cause of cooperativity is obvious if we consider the structure of the α-helix. The conformations of the peptide units are interdependent because the hydrogen bond linking the C=O group of the ith unit with the N—H group of the (i - 4)th unit constrains the conformations of units i - 1, i - 2, and i - 3. The liberation of a given peptide unit, meaning a gain in entropy, requires the breakage of no less than three successive hydrogen bonds. Therefore, peptide units can be liberated only in a cooperative way.

The theory of helix-coil transitions is based on Ising's model (pp. 28, 123). The problem is to evaluate the partition function of the α-helix. This problem has been solved in

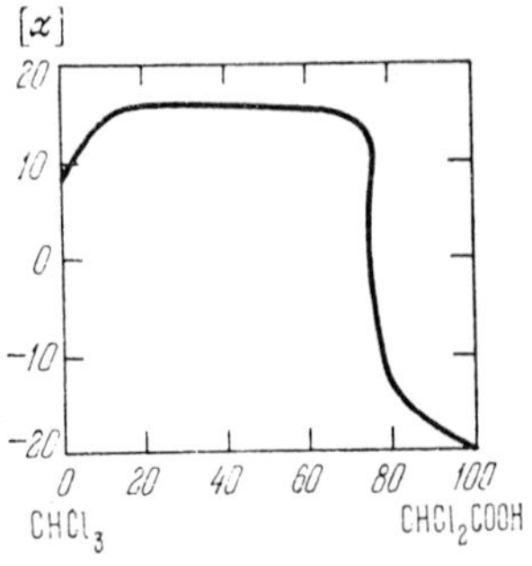


*FIG. 4.17 Optical activity of poly-γ-benzyl glutamate as a function of the solvent composition.*

many papers [57-61]. The simplest and clearest presentation of the theory was given by Zimm and Bragg [57] (see also [62, 63]).

Every peptide unit can exist in the free or in the hydrogen bonded state. Let us denote the first state of the ith unit by $\mu_i=0$, the second state by $\mu_i=1$. The free energy of the chain depends on the $\mu_i$ values; the conformations of the four sequential units are interdependent. Therefore

$$F(\mu_1, \mu_2, \cdots, \mu_N) = \sum_{i=1}^{N} F(\mu_{i-3}, \mu_{i-2}, \mu_{i-1}, \mu_i) \tag{4.26}$$

As always (p. 124), we assume that the chain is long enough, $N \gg 1$, and we do not consider the end effects. Evidently the free energy of the free unit does not depend on the states of the preceding units, that is

$$F_{free} \equiv F(\mu_{i-3}, \mu_{i-2}, \mu_{i-1}, 0) \tag{4.27}$$

The free energy of a bound unit depends on the states of the preceding units. The free energy of a bound unit following the bound one is

$$F_{bound} \equiv F(\mu_{i-3}, \mu_{i-2}, 1, 1) \tag{4.28}$$

$F_{bound}$ is independent of the states of the (i-3)th and (i-2)th units, since the constraints of these units determined by the (i-1)th one are taken into account by $\mu_{i-1} = 1$.

Hence, the free energy change if the number of bound units is increased by one at the expense of the neighboring nonbound unit is equal to

$$\Delta F = F_{bound} - F_{free} = F(\mu_{i-3}, \mu_{i-2}, 1, 1) - F(\mu_{i-3}, \mu_{i-2}, \mu_{i-1}, 0) \tag{4.29}$$

If a bound unit appears following three successive free units, it requires free energy in addition to $\Delta F$. The binding of a unit means the initiation of a helix; it constrains four units simultaneously. We get

$$F(0, 0, 0, 1) = F_{bound} + F_{init} \tag{4.30}$$

where $F_{init}$ is due to a decrease in entropy of the bound peptide groups. Assuming that the rotation of a free unit around every bond corresponds to three rotamers, we get

$$F_{init} = 2 \times 300 \times 4 \times \ln 3 = 2.5 \text{ kcal/mole}$$

a value that is much larger than RT.

Evidently the free energy required for the liberation of one or two units between bound units has to be large, because in reality such liberation does not occur. The units remain in the helix and the enthalpy used for the breakage of hydrogen bonds is not compensated by an increase in entropy. Therefore, it can be assumed that

$$F(\mu_{i-3}, 1, 0, 1) \to \infty \qquad F(1, 0, 0, 1) \to \infty \tag{4.31}$$

Using these values of F it is possible to obtain the expression for the partition function Z. The contribution of a given state of the chain to Z contains the following multipliers:

(1) Factor 1 for every unit in the free state ($\mu_i = 0$). In other words, its free energy is considered zero

$$\exp(-F_{free}/kT) = 1$$

(2) Factor

$$s = \exp(-\Delta F/kT) \tag{4.32}$$

for a unit in the bound state ($\mu_i = 1$). The term s has the sense of the equilibrium constant in the reaction of hydrogen bond formation by the unit following the bound one.

(3) Factor

$$\sigma = \exp(-F_{init}/kT) \tag{4.33}$$

for every bound unit following three or more free units. The term $\sigma$ is the equilibrium constant for the reaction of formation of one breakage in a sequence of hydrogen bonds. If $F_{init} = 2.5$ kcal/mole, then $\sigma \simeq 10^{-2}$. Evidently $\sigma$ plays the role of a cooperativity parameter.

(4) According to (4.31), factor 0 for every bound unit following the free ones if their number is less than three

Hence

$$Z = \sum_{\{\mu_i\}} \exp(-F\{\mu_i\}/kT)$$

$$= \sum_{\{\mu_i\}} \prod_{i=1}^{N} s^{\mu_i} \sigma^{\mu_i(1-\mu_{i-1})} \left[1 - \delta_{\mu_i,1}\delta_{\mu_{i-1},0} \times (1 - \delta_{\mu_{i-2},0}\delta_{\mu_{i-3},0})\right] \tag{4.34}$$

where $\delta$ is the Kronecker delta. The summation has to be performed over all sets $\{\mu_i\}$.

As we have seen (p. 124), Z expressed as a sum of products can be represented as the spur of the Nth power of some matrix. In this case the matrix has rank $2^3=8$ because the states of four successive links, every one of which is in one of two states $\mu_i$=0 or 1, have to be taken into account. Zimm and

Bragg have analyzed this problem and shown that practically the same results can be obtained by means of considerable simplification. If the states of not four, but only two, successive links are considered, the rank of the matrix decreases from eight to two. The connection between the states of the ith and (i - 2)th and (i - 3)th units is given only by conditions (4.34). If we omit them, we get

$$F(\mu_i, \mu_2, \cdots, \mu_N) = \sum_{i=1}^{N} F(\mu_{i-1}, \mu_i) \tag{4.35}$$

and

$$F_{free} = F(\mu_{i-1}, 0) \tag{4.36}$$

$$F_{bound} = F(1, 1) \tag{4.37}$$

$$F_{bound} + F_{init} = F(0,1) \tag{4.38}$$

Assuming the free energy of the nonbound unit to be zero, we obtain the partition function where every bound unit contributes s, the free unit contributes 1, and the first bound unit following one or more free units contributes $\sigma$. We obtain

$$Z = \sum_{\{\mu\}} \prod_{i=1}^{N} s^{\mu_i} \sigma^{\mu_i(1 - \mu_{i-1})} = \mathrm{Sp}(\mathbf{P}^N) \tag{4.39}$$

Matrix **P** has the form

$$\mathbf{P} = \begin{array}{c|cc} \mu_{i-1} \backslash \mu_i & 0 & 1 \\ \hline 0 & 1 & \sigma s \\ 1 & 1 & s \end{array} \tag{4.40}$$

Its characteristic equation is

$$(\lambda-1)(\lambda-s) = \sigma s \tag{4.41}$$

Let us examine two extreme cases--total cooperativity and total lack of cooperativity. Cooperativity is maximal if $F_{init} \to \infty$ or $\sigma = 0$. We have two roots $\lambda = 1$ and $\lambda = s$ in this case, and the partition function is

$$Z = 1 + s^N \tag{4.42}$$

If $N >> 1$

$$Z \cong \begin{cases} s^N & s>1 \\ 1 & s<1 \end{cases} \tag{4.43}$$

The value of Z changes sharply if s = 1, and the cooperative helix-coil transition occurs. The fraction of bound peptide groups is

$$\theta = \frac{1}{N}\frac{\partial \ln Z}{\partial \ln s} = \frac{s^N}{1 + s^N} \tag{4.44}$$

and if $N >> 1$

$$\theta \cong \begin{cases} 1 & s > 1 \\ 0 & s < 1 \end{cases} \tag{4.45}$$

Cooperative transition occurs according to the all-or-none principle at a point $s = 1$. If $\sigma$ is small but not zero, the solution of (4.41) also gives the transition, not at a point, but in some range of s values, which is narrower the smaller $\sigma$ is. Therefore, there is some region of coexistence of the helical and nonordered parts of the chain; the all-or-none principle is no longer valid.

In the absence of cooperativity, $F_{init} = 0$ and $\sigma = 1$. The roots of Eq. (4.41) are equal to $1 + s$ and 0. The partition function is

$$Z = (1 + s)^N \tag{4.46}$$

that is, the simple product of partition functions of independent units--the breakage and formation of hydrogen bonds by every unit occur independently. The fraction of bound peptide groups is

$$\theta = \frac{1}{N}\frac{\partial \ln Z}{\partial \ln s} = \frac{s}{1 + s} \tag{4.47}$$

The dependence of $\theta$ on s is a smooth one--there is no sharp transition. Figure 4.18 shows the $\theta(s)$ curves for $\sigma = 10^{-4}$, $\sigma = 10^{-2}$, and $\sigma = 1$. Evidently the dependence of $\theta$ on s means, according to (4.32), dependence on temperature. The temperature of the helix-coil transition can be called the melting point of the helix. The melting corresponds to $s = 1$, or

$$\Delta H - T_m \Delta S = 0 \tag{4.48}$$

where $\Delta H$ is the difference in enthalpies of the coil and helix,

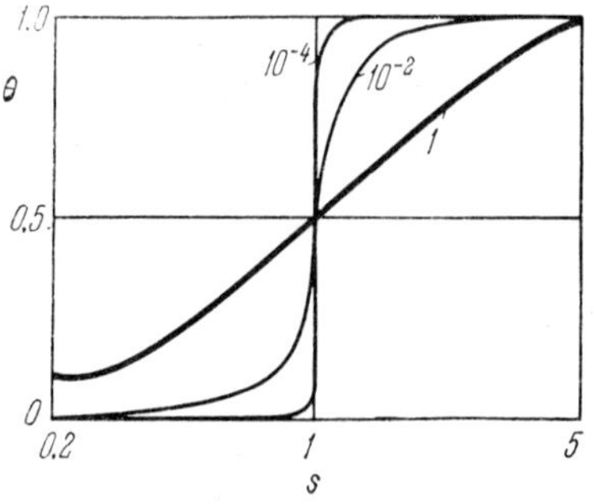


*FIG. 4.18 Theoretical $\theta(s)$ curves.*

that is, the heat of transition. Calculation shows that the region of transition at a small but finite $\sigma$ value is given by the condition [62]

$$\Delta T/T_m = 2\sigma^{\frac{1}{2}}(kT_m/|\Delta H|) \qquad (4.49)$$

In the case of total cooperativity, expression (4.47), corresponding to a noncooperative transition (i.e., to the independent change of state of single peptide units), is replaced by expression (4.44), describing the cooperative transformation of all N units. In intermediate cases, when $0 < \sigma < 1$, $\theta(s)$ can be written approximately as

$$\theta = s^n/(1 + s^n) \qquad (4.50)$$

where $1 < n < N$. Let $s^n = \tilde{s}$. The value $\tilde{s}$ is the equilibrium constant for a monomolecular reaction in which n units participate, that is

$$\tilde{s} = \exp[-(\Delta H_{eff} - T\,\Delta S_{eff})/kT] \qquad (4.51)$$

This value depends on n. Calculation shows that in the transition range $\Delta H_{eff}$ actually does not depend on temperature, and hence $\tilde{s}$ depends on $1/kT$ exponentially. If $\tilde{s} = s^n$, then it can be assumed

$$\Delta H_{eff} = n\,\Delta h \qquad (4.52)$$

where $\Delta h$ is the enthalpy change required for the liberation of one link, and n is of course related to the cooperativity parameter $\sigma$: if $\sigma = 1$, then $n = 1$; if $\sigma = 0$, then $n = N$.

Let us examine this relation. The general solution of the characteristic equation (4.41) is

$$\lambda_{1,2} = \frac{1}{2}(1 + s) \pm \left[\frac{1}{4}(1 - s)^2 + s\sigma\right]^{\frac{1}{2}} \qquad (4.53)$$

If $N \gg 1$, the partition function is expressed only by the larger root $\lambda_1$ (with a plus sign)

$$Z \cong \lambda_1^N \qquad (4.54)$$

The fraction of bound units is

$$\theta = \frac{1}{N}\frac{\partial \ln Z}{\partial \ln s} \cong \frac{\partial \ln \lambda_1}{\partial \ln s} = \frac{\lambda_1 - 1}{2\lambda_1 - 1 - s} \qquad (4.55)$$

But according to (4.47)

$$\theta = \tilde{s}/(1 + \tilde{s})$$

Comparing these two expressions, we get

$$\tilde{s} = (\lambda_1 - 1)/(\lambda_1 - s) \qquad (4.56)$$

The difference in enthalpies because of (4.51) is equal to

$$\Delta H_{eff} = -\frac{d \ln \tilde{s}}{d(1/kT)} = -\frac{d \ln s}{d(1/kT)} \frac{d \ln \tilde{s}}{d \ln s} = \Delta h \frac{d \ln \tilde{s}}{d \ln s} \qquad (4.57)$$

The calculation gives

$$\frac{d \ln \tilde{s}}{d \ln s} = \frac{s}{(\lambda_1 - 1)(\lambda_1 - s)} \frac{2(\lambda_1 - 1)(\lambda_1 - s) + \sigma(1-s)}{2\lambda_1 - 1 - s} \qquad (4.58)$$

and in the transition range where $s \cong 1$

$$\frac{d \ln \tilde{s}}{d \ln s} \cong \frac{2s}{2\lambda_1 - 1 - s} \cong \frac{1}{\sigma^{1/2}} \qquad (4.59)$$

Hence

$$\Delta H_{eff} \cong \frac{\Delta h}{\sigma^{1/2}} \qquad (4.60)$$

and the number of units that perform the transition together is inversely proportional to the square root of the cooperativity parameter

$$n \cong 1/\sigma^{1/2} \qquad (4.61)$$

the value of $\Delta H_{eff}$ can be found from the slope of the $\theta(1/kT)$ curve in the transition range. Since in this region $\tilde{s} \cong 1$, we have

$$\frac{d\theta}{d(1/kT)} = -\frac{d\theta}{d \ln \tilde{s}} \Delta H_{eff} = -\frac{\tilde{s}}{(1+\tilde{s})^2} \Delta H_{eff}$$

$$\cong -\Delta H_{eff} = -\frac{\Delta h}{4\sigma^{1/2}} \qquad (4.62)$$

Optical methods (Chapter 5) make possible the direct determination of $\theta$ as a function of temperature and therefore of $\Delta h/\sigma^{1/2}$. For determination of the separate values of $\Delta h$ and $\sigma$ other data are required. They can be obtained by the study of the dependence of $T_m$ on the degree of polymerization N. The theory of transitions in polypeptide chains with finite N values is developed in [64, 65]. A decrease in N shifts the transitions toward larger s values (i.e., toward coil formation). Simultaneously, the range of the values of s (or T) corresponding to transition is broadened. The concrete dependencies of $\theta$ on s at various N give the parameters required.

The melting curves for PBG at N = 1500, 46, and 26 are given [66]. It was found that $\Delta h = 890 \pm 130$ cal/mole and $\sigma = 2 \times 10^{-4}$.

Table 4.9 contains the transition characteristics of three synthetic polypeptides. We see that $\Delta h$ can be both positive and negative. In the first case the helical state is less advantageous than the coiled one and the helix-coil transition occurs if the temperature is lowered but not if it is

*TABLE 4.9*
*Characteristics of Helix-Coil Transitions of Polypeptides*[a]

| | Poly-γ-benzyl-L-glutamate | Poly-L-glutamic acid (65% ionized) | Poly-L-lysine (20% ionized) |
|---|---|---|---|
| Solvent | $CHCl_2COOH$ + 1, 2 $C_2H_4Cl_2$ (4:1) | 0.2 M in $H_2O$ NaCl + dioxane 2:1, pH 5.90, 25°C | $H_2O$, pH 10.10, 22°C |
| N | 1600 | 260 | 1500 |
| $\Delta h/\sigma^{\frac{1}{2}}$(kcal/mole | +70 ± 14 | -50 ± 0.5 | -5.3 ± 0.5 |
| Δh(cal/mole) | +1000 ± 200 | -70 ± 7 | -75 ± 7 |
| Δs eu | +3.2 ± 0.6 | -0.23 ± 0.02 | -0.25 ± 0.02 |
| σ | $2\times10^{-4}$ | $<2\times10^{-4}>$ | $<2\times10^{-4}>$ |

[a]From [62].

raised. The simultaneous positive value of ΔS, that is, the increase of entropy in the helix in comparison with that in the coil, can be explained only by a decrease in solvent entropy. The negative values of h and S in two other cases correspond to the helix-coil transition if the temperature is raised and the entropy of the helix is lower than that of the coil. These cases have to be considered as the melting of helix.

The theory of a helix-coil transition produced not by a rise in temperature but by a change in the solvent or pH has the same foundation [62]. In the first case it is necessary to consider three states of the peptide unit instead of two. Every unit can lack hydrogen bonds, can have intramolecular hydrogen bonds, and can form intermolecular hydrogen bonds with the solvent. The intermolecular bonding may be assumed to be noncooperative, in which case every nonbound unit contributes to Z the multiplier $1 + e^{\Delta\mu/kT}$, where Δμ is the difference in chemical potentials of the link that has an intermolecular hydrogen bond and of the link that is in the free state. (It is the free energy change due to the formation of a hydrogen bond [58, 61]). Consequently, the condition of transition depends on Δμ. The sharpness of the transition depends on σ, as before. If the concentration of specifically bound solvent molecules increases, $T_m$ is lowered. If in the absence of intermolecular interaction the melting point is $T_m^0$, then in the presence of an active solvent [62]

$$T_m = \frac{T_m^0}{1 + (kT_m^0/|\Delta H|)\ \ln(1 + e^{\Delta\mu/kT})} \tag{4.63}$$

If $\Delta\mu \to -\infty$, that is, when the state with intermolecular hydrogen bonds is not realizable, $T_m^0 = T_m$. The larger $\Delta\mu$ is, the smaller $T_m$ is.

Polypeptides formed by ionizable amino acid residues (e.g., polyglutamic acid or polylysine) undergo helix-coil transitions if the pH is changed. The transition can be observed and investigated by means of the above-mentioned optic and hydrodynamic methods and by potentiometric titration, which gives the ionization degree $\alpha$ (Fig. 4.19). The theory of transitions of polypeptides with ionizable groups was developed in [61, 67]. The latter work is also based on the Ising model.

In a chain with ionizable links, every link has to be characterized not only by the parameter $\mu_i = 0,1$, corresponding to the free and the bound link, respectively, but also by a second parameter $\eta_i = 0,1$, corresponding to the noncharged and the charged group, respectively. The partition function has the form

$$Z = \sum_{\{\mu_i\}} \exp(-F\{\mu_i\}/kT) \sum_{\{\eta_i\}} \prod_{i=1}^{N} a^{\eta_i} \exp[-F^{(e)}_{\{\mu_i\}}(\{\eta_i\})/kT] \tag{4.64}$$

where $F\{\mu_i\}$ is the free energy of the noncharged chain at the given set of $\mu_i$ values, and a, the ratio of activities of the charged and noncharged links, is related to pH of the medium

$$\log a = \pm(\text{pH} - \text{pK}) \tag{4.65}$$

(cf. p. 49). The plus sign is used for an acid, and the minus sign for a basic group. $F^{(e)}_{\{\mu_i\}}(\{\eta_i\})$ is the free energy of the electrostatic interaction of charged groups at given sets of $\mu_i$ and $\eta_i$. The expression for Z takes into account the mutual repulsion of charged groups; the amount of this repulsion depends naturally on the conformation, that is, on the set $\{\mu_i\}$.

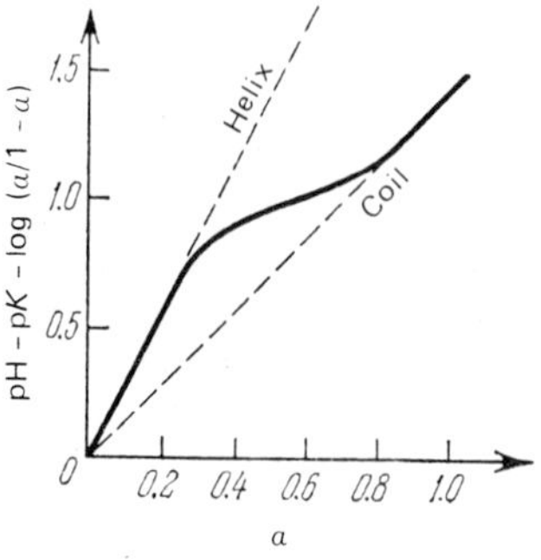


*FIG. 4.19 Polypeptide titration curve.*

Calculation shows that the repulsion in the helical conformation is larger than that in the coiled one. Hence, charging the chain fosters the helix-coil transition. The evaluation of a partition function by the Ising method allows us to determine the decrease in $T_m$ if the degree of chain ionization $\alpha$ increases and to find the titration curve (Fig. 4.19) clearly showing the helix-coil transition.

If the polypeptide chain is not homogeneous, but contains both acid and basic groups, unwinding of the chain can occur in the acid and alkaline pH regions. Consider the simplest model of a copolymer containing one ionizable group for every three nonionizable ones with a regular alternation of the acid and basic amino acid residues (A and C). Such a copolymer is a better model of a protein than the homogeneous polyamino acid. Calculation of the partition function for this system gives a bell-shaped dependence of $\theta$ on pH [68]. In the case of total cooperativity, if $\sigma = 0$, the bell transforms to a rectangle whose axis of symmetry crosses the abscissa at the point where the pH $= \frac{1}{2}(pK_A + pK_C)$. The width of the rectangle or of the bell depends strongly on s and therefore on temperature.

The form of the $\theta$(pH) curve is different for various sequences of acid and basic links. The results of investigations of various chain models are reported in [68]. If a chain contains some blocks formed by residues of one kind, the curve can be concave instead of convex. The appearance of two maxima located near the pK's of corresponding groups is also possible.

The theory developed in [68] explains the dependence of the degree of helicity of muscle proteins on the pH of the medium; the dependence was also studied by Lowey [69]. A correlation has been established between the form of the curves and the relative content of anionic and cationic residues in tropomyosin and other muscle proteins.

Experimental study of the dependence of helicity on pH for poly-L-lysine, poly-L-glutamic acid, and their copolymer gave the results shown in Fig. 4.20 [70, 71]. Theoretical calculations agree with these results. The electrostatic energy of the copolymer was evaluated and it was shown that its dependence on pH is determined by the difference in electrostatic interactions of the monomer, the charges arising because of the difference in the sizes of the side chains.

It is important to know whether helix-coil transitions can actually be treated like phase transitions of the first kind, that is, like melting. Can a real phase equilibrium exist in this case? Landau and Lifshitz [72] showed the impossibility of such an equilibrium in a one-dimensional system. The free energy of a two-phase one-dimensional system is

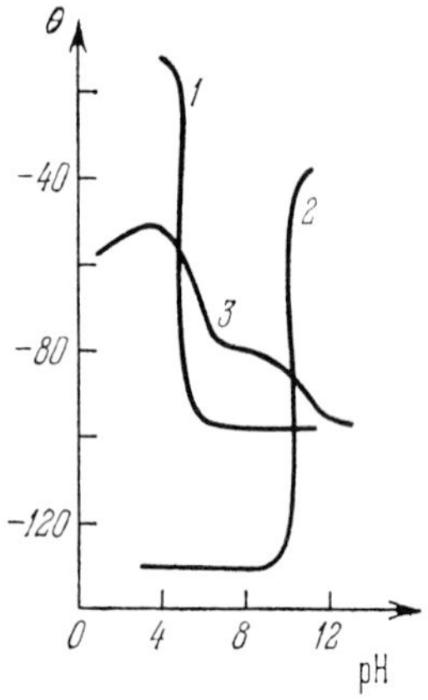


*FIG. 4.20 Dependence of specific rotation on pH: (1) poly-L-glutamic acid; (2) poly-L-lysine; (3) their copolymers.*

$$F = N\theta F_1 + N(1-\theta)F_2 + mkT \ln(m/eN) + m\psi \tag{4.66}$$

where $\theta$ is the fraction of phase 1, and $F_1$ and $F_2$ are the respective free energies of links in the first and second phases. The logarithmic term expresses the mixing entropy of m points of contact between two phases with N links of the chain, and $\psi$ is the surface tension energy at these points. We obtain

$$\partial F/\partial m = kT \ln(m/N) + \psi \tag{4.67}$$

If $m < N$, F decreases when m increases, attaining a minimum at $\partial F/\partial m = 0$, that is

$$m = Ne^{-\psi/kT} \tag{4.68}$$

Hence, both phases will mix until they are separated into small finite regions obeying condition (4.68). Consequently, phase equilibrium in a one-dimensional system is impossible; a phase transition is impossible, too.

The helix-coil transition is similar to melting but is not a genuine phase transition. Calculation of the average fraction of links at junctions of the free and bound regions gives

$$m/N = \sigma^{\frac{1}{2}} = e^{-F_{init}/2kT} \tag{4.69}$$

Comparing (4.69) and (4.68), we see that $F_{init}/2$ plays the role of surface energy. If $F_{init}$ is finite (i.e., if $0 < \sigma < 1$), the transition is not absolutely sharp. We get a phase transition only if $\sigma = 0$ or $F_{init} \to \infty$, which is impossible. The transition from β form to coil has a different character because the β form is two-dimensional. The theory of such transitions has been developed by Birshtein *et al.* [73] for the cross-β form of the polypeptide chain. Poly-S-carbobenzoxymethyl-L-cysteine has such a form [74]. The partition function of a planar system formed by structured and nonstructured regions is calculated in [73]. The structured regions are stiff regions with nonsaturated hydrogen bonds ("sticks") and the regions involved in the β form. An ensemble of chains of different lengths N is considered. The partition function of this generalized ensemble is

$$Z(\mu) = \sum_{N=1}^{\infty} Z(N) \exp -(\mu N/kT) \tag{4.70}$$

where μ is the chemical potential of the link. For large N

$$Z(N) = \exp(-F/kT) \equiv \exp(\mu_0 N/kT) \tag{4.71}$$

The transition from β form to coil corresponds to some critical value of μ. The theory gives the boundaries between two phases--of the coil with an admixture of β form and of the regular β form. Thus the phase diagram is obtained. Depending on the contact energy of two units (i.e., on the hydrogen bond energy) the transition is of either the first or the second kind. The first case seems more realistic.

The theory makes possible the derivation of analytical expressions for the monomer fractions in structured and nonstructured forms, depending on the system parameters and on the temperature. The principal physical conclusion is that in the case of stiff chains, weak attraction is sufficient for a sharp change in the characteristics of the system (transitions of the first kind). This is in agreement with Flory's theory of the crystallization of stiff polymeric chains [75]. It was shown recently that the formation of compact structures by stiff chains in a cubic lattice occurs according to the all-or-none principle [76].

Consequently, the weak interactions in polypeptide systems (which are the models of proteins) are responsible for their strongly cooperative behavior--for cooperative transitions similar to phase transitions. Further details concerning the theory and experimental studies of these transitions are contained in the monographs [62] and [63].

## 4.6 Protein Globules and Hydrophobic Interactions

Secondary structures--α-helices and β forms--exist in pure form in polyamino acids which are not informational macromolecules. These structures themselves are monotonic and periodic.

α-Helices and β forms are the elements of spatial structure in protein molecules that can exist in parts of a protein chain. The spatial structure of protein as we indicated earlier, is biologically functional. In many cases this structure is globular, and in such cases it is called tertiary structure. The hierarchy of protein structures comprises primary, secondary, tertiary, and quaternary structure. The last is an aggregate of a number of globules in a molecular or supramolecular system (examples include hemoglobin, whose molecule contains four globules, and tobacco mosaic virus protein, which contains 2000 identical globular particles).

However, dividing secondary and tertiary protein structure is misleading, since in each case we have a unique spatial structure containing regular regions and nonordered links. This does not mean the negation of the hierarchical principle in biology. Hierarchy arises in any system consisting of many identical elements if these elements interact. This problem has been studied by Bernal [77].

Two basic principles can be formulated. First, a multitude of identical molecules is synthesized in the cell. Second, the probability of formation of a more complex structure from its elements increases and the number of possible ways this formation can occur decreases if the structures can be divided into a finite series of substructures that are subsequently incorporated into one another. Hierarchy is manifest in, for example, the Cosmos: stars form galaxies, galaxies constitute a metagalaxy, etc. In biophysics, we deal with atoms, which form peptides; peptides, which form protein chains; protein chains, which form quaternary structures; etc.

The features of a structure at every level of organization are determined by the geometrical properties of the structures at the preceding level and by the forces action between these structures. The formation of a more complex structure occurs automatically, as a result of "self-assemblage," a phenomenon whose principles are not yet clear. Solution of the self-assemblage problem should make it possible, for instance, to predict the macroscopic structure of muscle from the chemical structure of muscle proteins. The explanation of the structure of crystals formed by small molecules, on the basis of a knowledge of their structure is the solution of the self-assemblage problem in a much simpler case (cf. [27]).

However, the "tertiary structure" of the protein globule cannot be considered as formed by secondary structure--the very formation of ordered regions of the protein chain is related to the formation of the entire spatial structure.

In contrast to the random coil, a protein globule is not a loose fluctuating system, but a compact, densely packed, regular system, an "aperiodic crystal." The dense globular structure of a protein molecule is manifest in the low viscosity of proteins in solution. The intrinsic viscosity $[\eta]$ of protein has the order of magnitude of several hundredths deciliter/gram. (cf., e.g., [78]). The corresponding specific volume is much less than the specific volumes of conventional polymers, which form loose coils in solutions, and is close to the specific volume of dry protein. This is confirmed by studies of proteins by means of sedimentation, diffusion, light scattering, roentgenography, low-angle X-ray scattering, and electron microscopy.

Since amino acid residues are polyfunctional, various kinds of forces are involved in the formation of globules. The features of a globule can be only partly represented by the model of a homogeneous chain because in this case the diversity of links and interactions plays a major role.

The only strong interactions in globules are chemical disulfide bonds between Cys residues. The presence of several disulfide cross links between the monomers of one chain or between several chains (as in insulin, p. 60) puts considerable constrain on the possible chain conformations. However, it would be impossible to speak about globules if the only interactions involved were disulfide bridges. If such were the case, the polypeptide chain would be similar to that of vulcanized rubber, which maintains the properties of random coil despite the presence of di- and polysulfide bridges.

Globule is formed by weak van der Waals forces (Section 4.3), hydrogen bonds (Section 4.4), and the electrostatic attraction of charged inogenic groups (salt bonds). The so-called hydrophobic interactions are especially important and are considered in detail later. A complicated interplay among all these forces results in the formation of a dense globule having a stable regular structure in aqueous solution at physiological values of pH and ionic strength. Both energy and entropy factors are important here. If the regions of the polypeptide chain are stiff enough (as the α-helical regions are), the formation of some elements of a compact structure becomes possible in the absence of energy interactions. This situation has been studied by Flory [75] in the cases of concentrated polymer solutions and the crystallization of polymers.

In a concentrated solution, polymeric chains cannot exist in various conformations independently. Only those conformations are possible in which every link of the chain has those positions in space that are not occupied by other links or by the links of other molecules.

If $\omega$ is the number of possible conformations of the link, the total number of conformations of a single chain containing N links is equal to $\omega^N$. For n independent chains, the number of conformations is $\omega^{Nn}$. In a solution with the volume fraction of polymer v the number of conformations of every macromolecule decreases $e^{-vN}$ times and the total number of conformations is [75]

$$Z = (\omega e^{-v})^{Nn} \tag{4.72}$$

In the nondiluted polymer v = 1. If the chain flexibility

expressed by $\omega$ is small (if $\omega \to e$), then $Z \to 1$. The nonordered state is characterized by $Z > 1$. The value $Z = 1$ means ordering (i.e., crystallization at $v = 1$ or formation of the liquid-crystalline phase at $v < 1$). Evidently we consider here only the entropy factor--the ordering of chains can occur in the absence of any forces (excluding the forces that hinder internal rotation) if the chains are stiff enough.

An important problem in protein physics consists in determining the optimal structure of a globule formed by the chain or chains of the known primary structures in aqueous surroundings (Section 4.9). Examination of the globule in a vacuum makes no physical sense. Proteins function in water, and water influences the dissolved molecules. The structure of a protein macromolecule in aqueous solution is distinctive and differs from that in other surroundings. The solution of the above mentioned problem has to take into account the properties of both the protein and the medium. Such an approach corresponds to the ideology of modern physics, which disapproves of the isolation of an element from the general system of interactions. A systemic approach is necessary.

In the first place, water influences the hydrogen bonds in globules. Stabilization of the elements of secondary structure and of the spatial structure as a whole requires the gain of free energy of intramolecular hydrogen bonds in comparison with that of hydrogen bonds with solvent molecules. However, these effects are not large. Klotz and Franklin found that the formation of hydrogen-bonded dimers of methylacetamide $H_3C{-}NH{-}COCH_3$ was energetically profitable in organic solvents but not in water [79, 80]. The heat of formation of the dimer in $CCl_4$ is 4.0 kcal/mole, in $CHCl_3$ it is 1.6 kcal/mole, in water nearly zero. Ptitsyn and Skvortsov estimated the energy of hydrogen bond formation in polyglutamic acid and polylysine by analyzing the conditions of the helix-coil transition in aqueous solution [81]. In these cases, too, the effective bond energy is small, approximately 0.2 kcal/mole. It may be thought thought that the role of water in the formation of a globule is quite different.

The great importance of hydrophobic interactions for the globularization of proteins had been recognized in 1944 [82, 83]. The hydrocarbon, nonpolar radicals of amino acid residues are mainly in contact one with another but not with water. The polar radicals, on the other hand, interact with water. This interaction results in the folding of a protein macromolecule. Nonpolar, hydrophobic residues have to be located inside the globule, and the polar hydrophilic residues must be at its surface in contact with the water.

This idea is based on the low solubility of nonpolar substances, such as hydrocarbons, in water. On the other hand, many facts show that molecules containing both polar and nonpolar groups that are in aqueous surroundings are arranged in such a way that the polar groups make contact with the water and the nonpolar ones are remote from the water. The classical experiments of Langmuir, who studied monomolecular layers of fatty acids on the surface of water, showed that the polar carboxyl groups of molecules are immersed in the water and the nonpolar hydrocarbon radicals are directed out of the water [84]. The same situation is characteristic of the structure of soap micelles in colloidal solutions of soaps--the hydrophobic groups are located inside the micelle and the hydrophilic ones on its surface.

Globular proteins are denatured (Section 4.8) by low polarity organic solvents, which are not as strongly hydrogen bonded as water. The action of these solvents is determined by contact with hydrophobic groups, destroying hydrophobic interactions. The denaturing action of alcohols increases with the increase in size of an aliphatic radical [85, 86]. The strong denaturing action of urea is also explained by the weakening of hydrophobic interactions. The concept of hydrophobic interactions as the most important factor determining the spatial protein structure was developed by Kautzmann [87].

The physical nature of hydrophobic interactions is unusual. The low solubility of hydrocarbons in water is explained not by an increase in energy but by a decrease in entropy. Consequently, the solubility of hydrocarbons in water decreases when the water is heated. Experiment shows that the energy (enthalpy) of dissolved aliphatic hydrocarbons is lowered; the process of dissolving is exothermic. However, this decrease in energy is paralleled by an increase in free energy because of the entropy lowering. Table 4.10 presents the values of $\Delta S$, $\Delta H$, and $\Delta G$ for the transfer of hydrocarbons from a nonpolar solvent into water [87]. $\Delta S_u$ is the total entropy change if a liquid hydrocarbon is dissolved in water up to the molar concentration x

$$\Delta S_u = \Delta S + R \ln x \qquad (4.73)$$

where R ln x is the entropy of mixing. Hence

$$\Delta G = - RT(\ln K + \ln x) \qquad (4.74)$$

where K is the equilibrium constant of the solution. The values of $\Delta H$ and $\Delta S_u$ are determined from the temperature dependence of K. The formation of soap micelles in water is also accompanied by a considerable decrease in entropy [88].

Butler has shown that these effects characterize a change in the state of water but not of a hydrocarbon [89]. They can be treated as the intrusion of nonpolar molecules into structured (icelike) water regions and into denser nonstructured regions. In the first case molecules in the cavities of the loose structure form clathrates; additional contacts decrease

*TABLE 4.10*

*Changes in Thermodynamic Quantities of Hydrocarbons Transferred from Nonpolar Solvent into Water*

| Original system | T (°K) | $\Delta S_u$ (eu) | $\Delta H$ (cal/mole) | $\Delta G$ (cal/mole) |
|---|---|---|---|---|
| Methane ($CH_4$) in benzene | 298 | -18 | -2800 | +2600 |
| Methane in ether | 298 | -19 | -2400 | +3300 |
| Methane in $CCl_4$ | 298 | -18 | -2500 | +2900 |
| Ethane ($C_2H_6$) in benzene | 298 | -20 | -2200 | +3800 |
| Ethane in $CCl_4$ | 298 | -18 | -1700 | +3700 |
| Ethylene ($C_2H_4$) in benzene | 298 | -15 | -1610 | +2920 |
| Acetylene ($C_2H_2$) in benzene | 298 | - 7 | - 190 | +1870 |
| Liquid propane $C_3H_8$ | 298 | -23 | -1800 | +5050 |
| Liquid n-butane $C_4H_{10}$ | 298 | -23 | -1000 | +5850 |
| Liquid benzene $C_6H_6$ | 291 | -14 | 0 | +4070 |
| Liquid toluene $C_7H_8$ | 291 | -16 | 0 | +4650 |
| Liquid ethylbenzene $C_8H_{10}$ | 291 | -19 | 0 | +5500 |
| Liquid m- or p-xylene $C_8H_{10}$ | 291 | -20 | 0 | +5800 |

the energy. In the second case the increase in the number of hydrocarbon-water contacts is accompanied by a decrease in water-water contacts corresponding to lower energy. This results in an increase in energy. Consequently, hydrocarbons are more soluble in the structured regions and the equilibrium has to be shifted toward greater structuring, which entails a lowering of entropy [87, 90]. This purely qualitative reasoning does not show that the decrease in entropy has to be larger than the decrease in energy. It must be emphasized that the concept of hydrophobic interactions contradicts the hypothesis that the protein molecule is surrounded by ordered layers of "structured" water which stabilizes the native, globular conformation [91-94]. This is a questionable hypothesis, since the formation of such ordered layers would result not in a decrease but in an increase of the free energy.

A series of attempts have been made to construct a quantitative theory of hydrophobic interactions based on the models of water structure. As has been said (p. 192), Sheraga *et al.* proposed a model taking into account a set of possible states of $H_2O$ molecules: without hydrogen bonds, and with 1, 2, 3, and 4 hydrogen bonds. The last state corresponds to the icelike structure. The coordination numbers of such molecules change in reverse sequence--the largest corresponds to the nonbound state where the molecules can be closely packed, the smallest one--equal to 4--to the icelike structure (p. 188) [47, 52, 95-100]. The intrusion of a hydrophobic molecule or group increases the coordination number of molecules in the icelike structure and consequently decreases the energy of molecules with four hydrogen bonds by some value $\Delta E_1$. Conversely, the energies of the other four states increase, presumably by a value $\Delta E_2$ (Fig. 4.21). Nemethy and Sheraga estimated two parameters $\Delta E_1$ and $\Delta E_2$ from experimental data and found that for aliphatic hydrocarbons $\Delta E_1 = -0.03$, $\Delta E_2 = 0.31$ kcal/mole; for aromatic ones $\Delta E_1 = -0.16$, $\Delta E_2$ 0.18 kcal/mole. By evaluating the corresponding expression for the partition function, it is possible to calculate the values of $\Delta G$, $\Delta H$, $\Delta S$, $\Delta C_p$ for the contacts in any pairs of nonpolar side chains of the protein molecule. The estimated values vary, of course, for various groups. At 25°C $\Delta G$ ranges from -0.2 to -1.5 kcal/mole, $\Delta H$ from 0.3 to 1.8 kcal/mole, $\Delta S$ from 1.7 to 11 eu, and $\Delta C_p$ from -10 to -50 cal/mole-deg. As Ptitsyn emphasized [101], if we take into account that the number of hydrophobic residues is $\frac{1}{3}$-$\frac{1}{2}$ of all residues in the majority of proteins [102] and that the nonpolar parts of polar side chains can also participate in hydrophobic interactions, the free energy gain in the formation of a hydrophobic nucleus can be quite sufficient for stabilization of the globular structure. These quantitative estimations agree with qualitative considerations. However, as we said earlier (p. 192), this simplified theory is not corroborated by spectroscopy.

*FIG. 4.21 Scheme of the energy levels in water and in an aqueous solution of a hydrocarbon.*

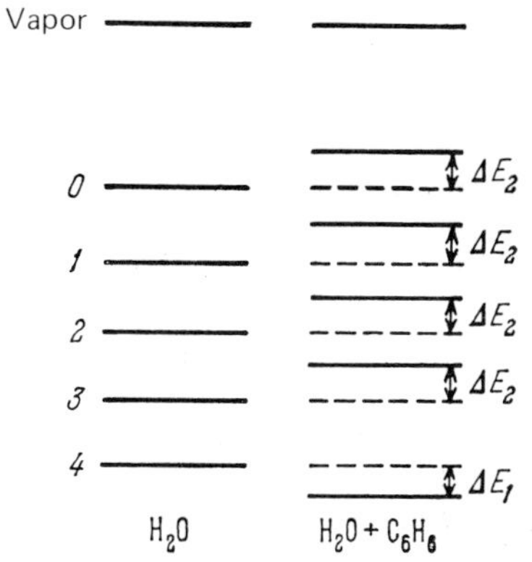

Obviously there is no sense in speaking about "hydrophobic forces." No specific forces are involved here; we are dealing with conventional van der Waals forces and hydrogen bonds, and with the structure of water as a whole. In the next section we shall discuss the influence of aqueous surrounding, that is, first, of hydrophobic interactions, on the various forces acting in globules, on the fine structure of globules. Let us consider now a rough estimation of the hydrophobic interaction's effect on the form a globule. Such an estimation was proposed by Fisher [103], who divided all amino acid residues into two groups: polar or hydrophilic (Arg, Asp, Glu, His, Lys, Ser, Thr, Tyr), and nonpolar or hydrophobic (the remaining 12).

For simplicity we assume that all residues have approximately the same volume. Hydrophobic interactions force the hydrophobic residues to be located inside the globule, whereas hydrophilic residues are positioned at its surface. The approximate form of the globule can be found by means of elementary calculation. We assume that the external layer of a globule formed by hydrophilic residues is a monomolecular one, having a constant thickness d. If the globule is spherical, then the volume of this layer is

$$V_e = (4\pi/3)[r^3 - (r-d)^3] \tag{4.75}$$

where r is the radius of the globule. The internal volume of the globule, that is, the volume of the hydrophobic residues, is

$$V_i = (4\pi/3)(r-d)^3 \tag{4.76}$$

The ratio of hydrophilic and hydrophobic residues is $p = V_e/V_i$. Hence, for a spherical globule we get

$$p_s = \frac{r^3}{(r-d)^3} - 1 \tag{4.77}$$

and

$$V_e = V\frac{p}{p+1} \tag{4.78}$$

where $V = V_e + V_i$ is the total volume of the globule. On the other hand, $V_e \cong Ad$ where A is the surface of the hydrophobic nucleus of a globule. Fisher takes $d \cong 4$ Å. Hence

$$p = \frac{A}{(V/4)-A} \quad (A \text{ in } \text{Å}^2) \tag{4.79}$$

The smaller V (i.e., the MW of the protein) is, the larger its relative hydrophility (polarity) must be. If the V are small, situations are possible in which the hydrophobic residues are not covered by the hydrophilic ones. For a sphere with r = d, $p_s \to \infty$. Figure 4.22 presents a comparison of theoretical

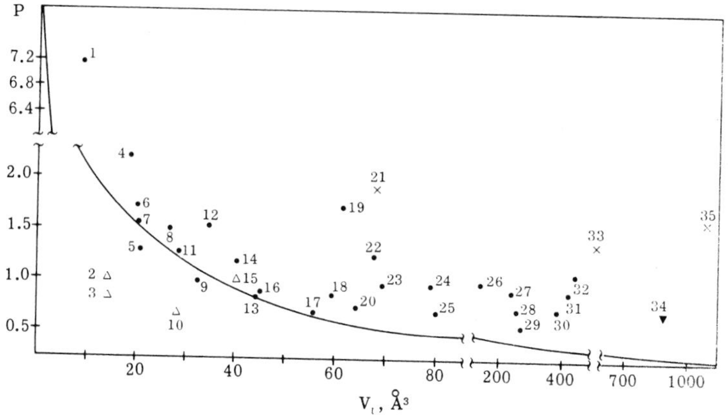


*FIG. 4.22 The value of p (calculated from data on the composition of 35 proteins) as a function of V.*

curve (4.79), calculated for the sphere, with experimental values of V and p for a series of proteins. Experimental points lie near the theoretical curve but as a rule are higher. A globule can be spherical only if $p = p_s$. Usually $p>p_s$ and the globule's form is ellipsoidal. The condition of minimal surface-to-volume ratio is not fulfilled. If $p>p_s$, the number of hydrophilic residues is larger than the minimal one required to protect the hydrophobic nucleus from its aqueous surroundings. If $p>>p_s$, the structure is fibrillar as in 21-tropomyosin, 33-fibrinogen, 35-myosin rather than globular.

On the other hand, if $p<p_s$, the hydrophilic residues do not screen the hydrophobic ones. Some hydrophobic residues remain on the surface, unprotected from the solvent. Hydrophobic interactions of these residues can lead to aggregation of the globules, which results in the formation of quaternary structure. In Fig. 4.22 these cases are denoted by triangles. Aggregation shifts these points to the right.

Later Fisher examined the conditions of hydration of protein molecules [104]. He suggested that the surface of a protein is covered by a monomolecular aqueous layer in which the $H_2O$ molecules are homogeneously distributed. A molecule of $H_2O$ occupies a volume of 33.3 $Å^3$, equivalent to the protein surface of 10.3 $Å^2$. Hence the number of grams of water per square angstrom of protein surface is

$$g = \frac{18 \quad g\ mole^{-1}}{10.3 \quad Å^2 \times 6.02 \times 10^{23} \quad mole^{-1}} = 2.9 \times 10^{-24} \quad g/Å^2$$

The hydrated protein surface is A. Since $V_e = Ad$

$A = V_e/4 \quad Å^2$

Hence, the number of grams of water per gram of protein is

$$gA = \frac{1}{4}\, gV_e = 0.725 \times 10^{-24} V_e = 0.725 \times 10^{-24} V[p/(p+1)] \quad (4.80)$$

Calculations performed for 34 proteins have shown that the value of gA varies from 0.22 to 0.33 for proteins with MW from 12,500 to 320,000. The average value is gA = 0.28. This means that the surface of the protein macromolecule increases approximately proportionally to its volume, and the ratio A/V is approximately constant, independent of p (in the 34 proteins examined p varies from 0.7 to 1.7). Comparison of these data with the calculations presented earlier shows that $p/p_s$ increases approximately linearly with V (i.e., the MW). The larger the volume of a macromolecule, the larger is its deviation from the spherical form. However, the volume V in expression (4.80) is calculated from the MW of protein using the rough notion of the equality of the specific volumes (per unit molecular weight) of all residues. Actually, the linear dependence of gA on p/(p+1) is not always valid. Some very large proteins are known to be spherical.

The results of these works agree well with the hypothesis regarding the fundamental role of hydrophobic interactions. An attempt is made in the works of Fisher to avoid the difficulties of the rigorous treatment of a protein molecule as an ensemble containing 20 different kinds of units. Instead of 20 types, only 2 types are considered. Reality is of course much more complicated, and such simplification is valid only for rough estimations, since it ignores not only the variety of polar and nonpolar residues but also their given sequence along the chain. Evidently, nonpolar, hydrophobic residues can be located on the surface of a globule if they are adjacent to the hydrophilic residues in the chain. On the other hand, it can be thought that the nucleus of a globule must be mainly hydrophobic because it stabilizes the structure. In the next section we will show that this notion is confirmed by experiments.

The separation of amino acid residues into two classes is conditional. Actually we have to introduce the degree of hydrophobicity of a residue expressed quantitatively. Tanford suggested taking as a measure of hydrophobicity the free energy change ΔG per side chain of a free amino acid in the transfer of an amino acid from $C_2H_5OH$ in water [105]. Table 4.11 contains the ΔG values determined experimentally by Tanford and estimations of various contributions to these quantities [106]. For Gly, ΔG is taken as zero because Gly does not contain a side chain. The first 10 amino acids can be assumed to be hydrophobic, the second 10 hydrophilic. Evidently, this classification does not agree with that based on polarity (i.e., on dipole moments). Strongly polar Arg has the same hydrophobicity as nonpolar Ala because of the bulky hydrocarbon side chain.

TABLE 4.11

*Hydrophobicity of Amino Acid Residues*

| | | Contribution | | |
|---|---|---|---|---|
| Residue | ΔG(cal/mole) | Hydrophobic | Polar | Aromatic group |
| Trp | 3000 | 5850 | -600 | -3200 |
| Ile | 2970 | 2600 | | |
| Tyr | 2870 | 4500 | -600 | -1600 |
| Phe | 2650 | 4500 | | -1600 |
| Pro | 2600 | 1950 | | |
| Leu | 2420 | 2600 | | |
| Val | 1690 | 1950 | | |
| Lys | 1500 | 2600 | -600 | |
| His | 1400[a] | 2600 | -1200 | |
| Met | 1300 | 1950 | | |
| Ala | 730 | 650 | | |
| Arg | 730 | 1950 | -1800 | |
| Cys | 650 | | | |
| Glu | 550 | 1300 | -1200 | |
| Asp | 540 | 650 | -1200 | |
| Thr | 440 | 1300 | -600 | |
| Ser | 40 | 650 | -600 | |
| Gly | 0 | | | |
| Asn | -10 | 650 | -1200 | |
| Gln | -100 | 1300 | -1200 | |

[a]Estimation.

The degree of hydrophobicity of residues can give information concerning the stabilization of the globular protein molecule by hydrophobic interactions. However, such estimations are not sufficient. The actual structure of the globule must be considered and the total balance of existing interactions must be taken into account. Fisher's simplified theory is of course not satisfactory. The distribution of residues between the external layer and the nucleus of the globule must be determined by the free energies of their interactions with the neighboring residues and with water. Brandts has developed Fisher's ideas in a more rigorous form [107], distinguishing three kinds of residues: hydrophobic, located mainly inside the globule (Ala, Val, Leu, Phe, etc.); hydrophilic, located mainly on its surface (including particularly all charged residues); and neutral (Gly, Ser, Cys, and probably Asn and Gln), which can be positioned both on the surface of and inside the globule. The numbers occupying the internal and external sites for every kind of residue are weighted according to the free energy of transfer of a residue from outside into the interior of the globule with statistical degeneration. This makes it possible to write the partition function of the globule and to develop the statistical-thermodynamic theory of denaturation (Sections 4.7 and 4.8).

## 4.7 Structure and Stability of Globules

The following are sources of information about the structure of globules:

(1) Data from X-ray studies, which in the ideal case give the position of every atom in the protein molecule (Section 5.1);

(2) Data from optical and spectroscopic measurements, which enable us to estimate the fractions of residues in the α-helical, cross-β, and nonordered forms (Section 5.4 etc.);

(3) Data obtained by means of deuterium-exchange studies, giving the fraction of hydrogen-bonded residues;

(4) Data obtained by means of canonical methods of macromolecular studies, giving the size and form of globules and the charges on the globule surface (Chapter 3);

(5) Theoretical and semiempirical calculations.

Let us examine the structures of myoglobin and hemoglobin. These important proteins were the first studied by X-ray diffraction. Their properties will also be discussed in Chapter 7. Figure 4.23, the structure of myoglobin [108], shows the helical and nonhelical regions. (The molecule contains the prosthetic heme group, indicated by an arrow.) The number of helical regions is eight. Nearly 75% of the total of 153 residues belong to these helical regions which are indicated by the letters A, B, ..., H (beginning from the N-end of the chain). The residues in every α-helical region are denoted as A1, A2, ..., An, etc. Nonordered regions situated between helices are labeled AB, CD, etc.; the end regions are NA and HC [109]. The hemoglobin molecule has a quaternary structure containing four subunits, two denoted as α and two as β. The structure of every subunit is similar to that of myoglobin.

Detailed analysis of the structure of these proteins yields important results [110-113]. The interior of the myoglobin molecule contains closely packed nonpolar side chains of amino acid residues. The same can be said about every hemoglobin subunit [114]. The number of internal residues is 36, including two His linked with the heme group. Counting Gly but not Pro, the total number of nonpolar residues in chains of horse hemoglobin is 72 (in each of the two α chains) and 78 (in the β chains). Many Gly and Ala residues, being weakly hydrophobic, are located on the surface of the molecule. Large nonpolar side chains, which are not inside the globule, are buried in a groove near the surface and therefore have minimal

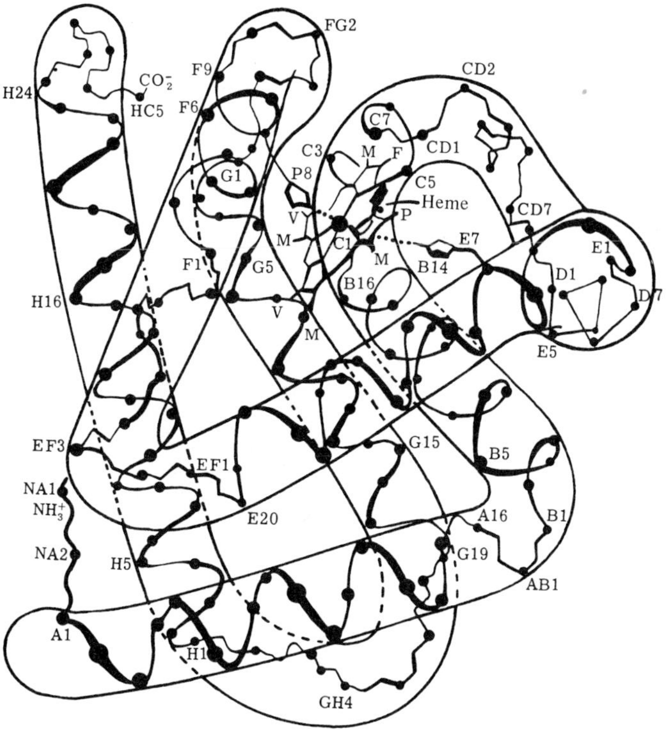


*FIG. 4.23 Myoglobin structure, according to Perutz. A, B, C, D, E, F, G, and H are α-helical regions.*

contact with water. All side chains ionized at neutral pH are located on the surface of the globule. The same is true of other polar side chains, excluding His; the latter is linked with heme and Thr C4, which are hydrogen bonded [110]. Only five or six of the 77 polar groups of myoglobin (including Trp) are situated inside the globule; all the others are on its surface [114]. The study of hemoglobin from different species of vertebrates (primates, horse, pig, rabbit, llama, carp, lamprey) and of myoglobins of the sperm whale and man has shown that the substitutions of the 33 internal residues maintain their nonpolar character in most cases. This is shown in Table 4.12 [111]. These residues do not come into contact with water. On the contrary, only 10 residues, invariantly nonpolar, are on the globule's surface.

Investigations of myoglobin and hemoglobin confirm the fundamental principle presented in the foregoing section: the extreme importance of the hydrophobic interactions that form the nonpolar nucleus of the globule. On the other hand, these results present important features of the helical regions of the protein chain in globules. Analysis of the A, B, E, G,

*TABLE 4.12*

*Substitutions of 33 Internal Residues*

| Index of residue | Observed residues | Index of residue | Observed residues |
|---|---|---|---|
| A 8 | Val, Ile, Leu | E 15 | Val, Leu, Phe |
| 11 | Ala, Val, Leu | 18 | Gly, Ala, Ile |
| 12 | Trp, Phe | 19 | Val, Leu, Ile |
| 15 | Val, Leu, Ile | F 1 | Leu, Ile, Phe, Tyr |
| B 6 | Gly | FG 5 | Val, Ile |
| 9 | Ala, Ile, Ser, Thr | G 5 | Phe, Leu |
| 10 | Leu, Ile | 8 | Val, Leu, Ile |
| 13 | Met, Leu, Phe | 11 | Ala, Val, CysH |
| 14 | Leu, Phe | 12 | Leu, Ile |
| C 4 | Thr | 16 | Leu, Val, Ser |
| CD 1 | Phe | H 8 | Leu, Phe, Met, Trp, Tyr |
| 4 | Phe, Trp | 11 | Ala, Val, Phe |
| D 5 | Val, Leu, Ile, Met | 12 | Val, Leu, Phe |
| E 4 | Val, Leu, Phe | 15 | Val, Phe |
| 8 | Gly, Ala | 19 | Leu, Ile, Met |
| 11 | Val, Ile | 23 | Tyr |
| 12 | Ala, Leu, Ile | | |

and H α-helices shows the periodic positions of nonpolar residues [111]. The helical sequences are oriented in such a way that these residues appear to be located in the nucleus of the globule. Spiralization of the polypeptide chain is thermodynamically advantageous for a series of amino acid residues, since it provides saturation of the hydrogen bonds. But such an α-spiralization (and the formation of β forms) is simultaneously determined by hydrophobic interactions. The secondary structure is stabilized by the spatial (tertiary) structure of protein.

Ptitsyn examined the influence of hydrophobic interactions on the helicity of the polypeptide chain [101]. Many data show such an influence on the structure of synthetic polyamino acids. Fasman analyzed the stability of helical polymers under the action of dichloro- and difluoroacetic acids and showed that the stability of poly-L-methionine ($-CH_2-CH_2-S-CH_3$), poly-L-alanine ($-CH_3$), and polyleucine ($-CH_2-CH(CH_3)_2$) was much higher than that of poly-δ-carbobenzoxy-L-lysine ($-(CH_2)_4-NH-COO-CH_2-C_6H_5$) and poly-γ-benzyl-L-glutamate ($-(CH_2)_2-COO-CH_2-C_6H_5$) [115]. Incorporation of the nonpolar side chains into water-soluble polypeptides increases the stability of their helical conformations. This agrees with results obtained in studies of other synthetic polypeptides [116-120].

The salt bonds in proteins are also stabilized by

the aqueous medium, since their formation liberates the oriented water molecules surrounding the charged groups. Therefore, the formation of a salt bond is accompanied by an increase in the entropy of water. This gain in free energy is greater than that due to the Coulombic interaction of charges [87]. However, the influence of water on the salt bond is different from the hydrophobic interaction--salt bonds are strengthened, but the hydrophobic ones become weaker, when aqueous solvents are added [101].

X-ray investigations of other proteins also confirm the ideas presented concerning the effect of aqueous surroundings on globular structure. As a rule, a globule contains a hydrophobic mucleus.

Thus the physical nature of the interactions responsible for the formation of a globule is more or less clear. However, the problem of theoretical calculation of the globular structure is very far from solution; only a few serious attempts at such calculations have been made.

Scheraga examined the theoretical aspects of the globule problem [4, 121], suggesting that the native conformation of the polypeptide chain in solution corresponds to the minimum of the sum of the potential energies of all intramolecular interactions and of the free energy of the protein-solvent interaction. It can be assumed that the native conformation of the globule corresponds to minimal free energy (cf. [122]). However, the native state contains a set of similar but nonidentical conformations because of the lability of nonordered regions. This makes the situation rather difficult.

Scheraga calculated the partition function of the globule-solvent system, taking into account both the intrachain interactions and interactions with the solvent. The function is integrated over all conformations of the system. The potential energy of the intrapeptide interactions can be estimated from steric maps. Electrostatic interactions are calculated according to Coulomb's law, the main difficulty being the estimation of the dielectric constant. At small distances between charges in the protein globule we can take $\varepsilon \cong 3.0$ [123]. It is reasonable to use a Morse-type potential for hydrogen bonds. The energy required for the variation in valence angles can be found from spectroscopic values of the elasticity constants for bending vibrations. A detailed description of these potentials is presented in Scheraga's review article [4].

The solvent (water) influences the polypeptide energy in two ways. First, there is an additional set of terms describing the interactions of the polypeptide atoms with the solvent. Their approximate estimation has been given by Gibson and Scheraga [123]. The quantity of water removed from the surroundings of every atom is calculated. Corresponding data are presented in Table 4.13. Second, the presence of the solvent

*TABLE 4.13*

*Free Energy of Removal of Water Molecules*

| Atom or group | Solvation number[a] | Free energy of removal of solvent (kcal/mole) |
|---|---|---|
| H | 2 | 0.31 |
| O (carbonyl) | 4 | 0.94 |
| O (hydroxyl) | 6 | 0.84 |
| $O^-$ (carboxyl) | 5 | 4.80 |
| N (amide) | 2 | 0.63 |
| $NH_3^+$ (amine) | 5 | 15.40 |
| $N^+$ (imidazole) | 3 | 3.30 |
| $N^+$ (guanidine) | 6 | 1.20 |
| CH (aliphatic) | 2 | -0.13 |
| $CH_2$ (aliphatic) | 3 | -0.13 |
| $CH_3$ (aliphatic) | 3 | -0.13 |
| C (aromatic | 2 | 0.11 |
| CH (aromatic) | 3 | 0.11 |
| S | 6 | -0.17 |

[a] Number of $H_2O$ molecules.

modifies the energy of interaction of every pair of polypeptide atoms. Such calculations are not performed for proteins yet, since they obviously involve enormous difficulty. For investigation of the denaturation process (Section 4.1) the simplified partition function can be used. According to Schellmann, the partition function of a one-dimensional polypeptide containing both helical and nonordered regions can be expressed in the form

$$Z_{(1)} = 1 + \exp\left(-\frac{3\ \Delta S}{R}\right) \sum_{n=4}^{N} (N-n+1)\left\{\exp \frac{(\Delta h - T\ \Delta S)}{RT}\right\}^{n-3} \quad (4.81)$$

where N is the number of links in the chain and n the number of residues in a helical region, $\Delta h \cong 1000$ cal/mole, $\Delta s \cong 3.2$ eu--the differences in the enthalpies and entropies of the helical and nonhelical links. There are N-n+1 ways to form the helix containing n links from the chain with N residues. Three hydrogen bonds are missing from the ends of the chain. For a three-dimensional system this factor has to be introduced into Z, taking into account all effects arising if the side chains are put into a spherically folded system. Therefore [107]

$$Z_{(3)} = Z_{(1)} q(n) \quad (4.82)$$

As we said earlier, three kinds of residues can be considered: hydrophobic (h); hydrophilic (p); and neutral (n).

We get

$$n = n^h + n^p + n^n \text{ and } n^h = n_i^h + n_e^h, \quad \text{etc.}$$

where i and e denote the residues located inside and on the surface of the sphere. The number of ways that three kinds of residues can be distributed between internal and external "cells" is

$$\Omega(n) = \frac{(n_i^h + n_e^h)!}{n_i^h!\, n_e^h!} \frac{(n_i^p + n_e^p)!}{n_i^p!\, n_e^p!} \frac{(n_i^n + n_e^n)!}{n_i^n!\, n_e^n!} \tag{4.83}$$

The numbers of internal and external "cells" can be estimated according to Fisher (p. 212):

$$N_i = \frac{4\pi}{3} \frac{(r-d)^3}{V_{aa}}, \qquad N_e = \frac{4\pi}{3} \frac{[r^3-(r-d)^3]}{V_{aa}}$$

where $V_{aa}$ is the volume of one residue $\cong 125$ Å$^3$, and d = 4 Å. We get [107]

$$q(n) = \Omega(n)[\exp(-\Delta f^h/RT)]^{n_i^h} [\exp(-\Delta f^p/RT)]^{n_i^p} [\exp(-\Delta f^n/RT)]^{n_i^n} \tag{4.84}$$

where $\Delta f^h$, $\Delta f^p$, $\Delta f^n$ are the changes in free energy corresponding to the transfer of one residue from the surface to the interior of the globule. Brandts takes $\Delta f^h = 25.746T - 0.20265T^2 + 0.00077576T^3$ (data for Val), $\Delta f^p = \infty$, and $\Delta f^n = 0$. Consequently, all p residues are fixed at the surface. Calculation shows that the fraction of submerged hydrophobic residues depends on N more smoothly than in Fisher's model.

These statistical-mechanical approaches are useful mainly for examining the processes of total destruction of globules, that is, of denaturation (cf. the next section). Again we emphasize that the protein globule is not a statistical but a dynamic system, like an engine, whose work depends on the exact positions and interactions of all its elements.

On the other hand, the globular structure of a polymeric chain is characterized by some peculiar features. Lifshitz [1] treated the homopolymeric globule as a system with linear memory (p. 129). The energy of the chain can be expressed as

$$E_M = E_{\alpha_1} + u_{\alpha_1\alpha_2} + E_{\alpha_2} + u_{\alpha_2\alpha_3} + \cdots + u_{\alpha_{N-1}\alpha_N} + E_{\alpha_N} \tag{4.85}$$

where $\alpha_i$ denotes the state of the ith link of the chain containing N links and M denotes the totality of all $\alpha_i$ numbers. The partition function of this chain has the form

$$Z_N = \sum_M \exp(-E_M/kT) = \sum_{\alpha_i} \exp[(-E_{\alpha_1}+E_{\alpha_N})/kT] \prod_i G_{\alpha_i\alpha_{i-1}}$$

$$= \sum_{\alpha,\beta=1}^{p} \exp[-(E_\alpha + E_\beta)/2KT\ (G^{N-1})_{\alpha\beta} \tag{4.86}$$

where G is the matrix of elements $G_{\alpha\alpha'}$ (i.e., the transition matrix) and p the number of various states (rotamers) of every link. This expression is like (3.30)(p. 123). The eigenvalues of the matrix are found from the secular equation

$$||G_{\alpha\beta} - \lambda\delta_{\alpha\beta}|| = 0 \tag{4.87}$$

and for N >> 1

$$Z_N = \lambda_m^N \tag{4.88}$$

where $\lambda_m$ is the maximal eigenvalue.

A polypeptide chain can form a globule under the action of a field, depending on the coordinates of the links. In this case we must write, instead of $G_{\alpha_i\alpha_{i-1}}$,

$$G(x,x') = Cg(x,x')\exp(-E(x)/kT) \tag{4.89}$$

where E(x) is the energy of a link in absence of the field, C is the normalizing factor, $\mathbf{x}' = \mathbf{x} + \ell\xi$, where $\ell$ is the length of the link, and g(x,x') is the correlation function. In the absence of an external field (index 0) we have

$$g_0(x,x') = g_0(x-x')$$

If a field is present $g_0(x-x')$ transforms into $g_0(x-x')\exp(-U(x)/kT)$, where U(x) is the link energy in the external field. The partition function then has the form

$$Z_N(x) = C^N \int\cdots\int \exp(-E/kT) \prod_{j=1}^{N} g_0(x_{j+1}-x_j)$$

$$\times \exp(-U(x_j)/kT)\ d^3x_2\cdots d^3x_{N-1} \tag{4.90}$$

Let us write

$$C^{-N}Z_N(x) = \psi_N(x) \tag{4.91}$$

Then

$$\psi_{N+1}(x) = \exp(-U(x)/kT)\hat{g}_0\psi_N(x)$$

where

$$\hat{g}_0\psi_N(x) = \int g_0(x-x')\psi_N(x')\ d^3x'$$

Hence

$$\psi_{N+1}(x) = \exp(-U(x)/kT)\int g(y)\psi_N(x+y)\; d^3y$$

and using the condition

$$\int g(y)\; d^3y = 1$$

we obtain

$$\int g(y)\psi(x+y)\; d^3y = \psi(x) + \nabla\psi\int g(y)y\; d^3y + \frac{1}{2}\nabla_i\nabla_j\psi\int y_i y_j g(y)\; d^3y + \cdots \qquad (4.92)$$

Averaging this expression, we get

$$\psi_{N+1}(x) \cong \exp(-U(x)/kT)\,[\psi_N(x) + a^2\nabla^2\psi_N(x)] \qquad (4.93)$$

where $a^2 = \frac{1}{6}\langle y^2\rangle$. In the absence of an external field, $U(x) = 0$ and

$$\psi_{N+1} - \psi_N \cong a^2\nabla^2\psi_N \qquad (4.94)$$

or if $N \gg 1$

$$\partial\psi_N/\partial N = a^2\nabla^2\psi_N \qquad (4.95)$$

If $\psi_0(x)$ is $\delta(x)$, then $\psi_N \sim N^{-\frac{3}{2}}\exp(-x^2/4a^2N)$ and we get a Gaussian distribution.

In the presence of a field, if $N \gg 1$ we get instead of Eq. (4.93),

$$\partial\psi_N/\partial N = \psi_N[\exp(-U/kT)-1] + a^2\exp(-U/kT)\nabla^2\psi_N \qquad (4.96)$$

We look for the solution of Eq. (4.96) in the form

$$\psi_N(x) = \sum_q \exp(-\lambda_q N)C_q\psi^{(q)}(x_0)\psi^{(q)}(x) \qquad (4.97)$$

We have

$$\exp(-\lambda_q)\psi^{(q)}(x) = \exp(-U(x)/kT)\hat{g}_0\psi^{(q)}(x)$$

Hence

$$\exp(-\lambda_q)\psi^{(q)} = \exp(-U/kT)(\psi^{(q)} + a^2\nabla^2\psi^{(q)})$$

and

$$\nabla^2\psi^{(q)} + (1/a^2)[1 - \exp(-U/kT-\lambda_q)]\psi^{(q)} = 0 \qquad (4.98)$$

Let us rewrite the expression (4.97) in the form

$$\psi_N = \exp(-\lambda_0 N)\psi^{(0)}(x_0)\psi^{(0)}(x)$$

$$+ \sum_{q \neq 0} \exp[-(\lambda_q - \lambda_0)N]\psi^{(q)}(x_0)\psi^{(q)}(x) \quad (4.99)$$

If $\lambda_1, \lambda_2, \cdots > \lambda_0$ and $N >> 1$, we have

$$\psi_N \cong \exp(-\lambda_0 N)\psi^{(0)}(x_0)\psi^{(0)}(x) \quad (4.100)$$

Therefore

$$Z_N = Z_N^{(0)} \exp(-\lambda_0 N) \quad (4.101)$$

and the free energy is

$$F = F_0 - kT \ln Z_N/Z_N^{(0)} = kT\, N \ln \lambda_0 \quad (4.102)$$

where $F_0$ is the free energy in the absence of an external field at $U = 0$.

The functions $\psi^{(q)}(x)$ describe the state of the links. Let us now examine Eq. (4.98). We denote $\phi = U/kT$ and assume that $\phi - \lambda << 1$ in all cases when $\psi$ differs considerably from zero. Then

$$\nabla^2\psi + \frac{\lambda - \phi}{a^2}\psi = 0 \quad (4.103)$$

where the index q is omitted. This equation is similar to that of Schrödinger. However, here the parameter $\lambda$ characterizes the free energy of the system which results in the "quantized" values for the corresponding values of the potential energy $\phi$.

In the case of a "spherical potential well"

$$\phi = \begin{cases} 0 & |x| > R \\ -U_0/kT & |x| < R \end{cases} \quad (4.104)$$

Then if $|x| < R$,

$$\nabla^2\psi + \kappa^2\psi = 0 \quad (4.105)$$

If $U_0 >> kT$, we get the eigenfunctions

$$\psi = \frac{\sin \kappa R}{\kappa R}, \quad \kappa R = m\pi, \quad m = 1,2,\ldots \quad (4.106)$$

and the eigenvalues

$$\kappa^2 = \frac{\lambda + (U_0/kT)}{a^2} = \left(\frac{m\pi}{R}\right)^2 \quad (4.107)$$

For the ground level, m = 1 and

$$\lambda = -U_0/kT + (\pi a/R)^2 \tag{4.108}$$

Hence

$$F - F_0 = -NU_0 + NkT(\pi a/R)^2 \tag{4.109}$$

Let us find the equation of state of a globule. The pressure is

$$p = -\frac{\partial F}{\partial V} = -\frac{\partial F}{4\pi R^2\,\partial R} = \frac{2}{3}\pi^2 NkT\left(\frac{a}{R}\right)^2\frac{1}{V} \tag{4.110}$$

or

$$pV = \frac{2}{3}\pi^2(a/R)^2 NkT \tag{4.111}$$

The origin of the elastic force p opposing the deformation of the globule is a purely entropic one, as in the case of a random coil (p. 107).

If the potential well has an arbitrary form, the potentials for the lowest levels can be taken as parabolic and we obtain the equation for the harmonic oscillator

$$\nabla^2\psi + (1/a^2)(\lambda - \phi_0 - \tfrac{1}{2}\beta^2x^2)\psi = 0 \tag{4.112}$$

with the eigenvalues

$$E_m = (\lambda_m - \phi_0)/a^2 = \omega(m + \tfrac{1}{2}), \qquad m = 0,1,2,\ldots \tag{4.113}$$

where

$$\omega = \beta/a$$

For the ground level, m = 0 and

$$\lambda_0 = \phi_0 + \frac{1}{2}a\beta \tag{4.114}$$

The expression $\phi_0 + \frac{1}{2}\beta^2x^2$ can be rewritten in the form

$$\frac{U_0}{kT} + \frac{1}{kT}\left|\frac{\partial^2 U}{\partial x^2}\right|_{x=0}\frac{x^2}{2}$$

Thus $\beta^2 = (1/kT)\left|\partial^2U/\partial x\right|_{x=0}$ and consequently

$$\lambda_0 = \frac{U_0}{kT} + \frac{1}{2}a\left(\frac{1}{kT}\left|\frac{\partial^2 U}{\partial x^2}\right|_{x=0}\right)^{\frac{1}{2}} \tag{4.115}$$

We see that a sufficiently strong field U promotes the coil-globule transition, which is a kind of phase transition. If the temperature is high enough, there is no level of the free energy in the potential well; the first level arises at some critical temperature $T_c$.

The statistical analysis performed in [178] shows that the globule has a specific structure. If $T > T_c$, or in the absence of interactions and an external field, the chain maintains the state of a loose statistical coil and the fluctuations of its density are of the same order as the density itself. The coil performs continuous macroscopic pulsations and the density cannot be considered as a thermodynamic quantity (p. 131). The globule is formed under the action of a compressing field of sufficient strength or as the result of the volume interaction. The structure of the globule depends on the character of the field or of the interactions.

Since the correlation function is isotropic, a globule formed in the absence of an external field contains a dense spherical nucleus. Around this nucleus the density decreases continuously toward zero, therefore the globule has an "edge of the forest". If the temperature increases, the density at the edge decreases like the vapor density over a drop of liquid. Since the links are interconnected in the chain, at a sufficiently low temperature the short chain can form a globule without an edge directly from the coil. It is shown in [178] that the transitions between three possible states of a homopolymeric chain (coil, globule with an edge, globule without an edge) are phase transitions of the first kind, and the corresponding phase diagram is calculated.

The nucleus of a globule is similar to a crystal but not to a liquid; it has zero configurational entropy. The self-organization of the globular structure means the selection of one optimal microstructure. The qualitative treatment of a heteropolymer (cf. [178]) shows that only a small fraction of all possible sequences of monomers can form the particular tertiary structure whose ordering is similar to that of an aperiodic crystal. Only such a structure can be of biological importance.

Let us find the conditions of formation for an ordered tertiary globular structure. The chain can exist in many different conformations with the same energy E (the value of E includes the chain energy and the entropy of any deformation changing the tertiary structure). The value $E_0$ corresponds to the optimal ordering of the chain conformation. If the changes of conformations are large, $E - E_0$ is of the order of the volume energy $N\varepsilon$. The conformational entropy S, which is a function of E, is expressed by the logarithm of the number of conformations corresponding to the energy E. If $E - E_0$ is small, then

$$S(E) = (1/\gamma)(E-E_0) = N\varepsilon/\gamma$$

where $\gamma$ is determined by the energy of side interactions per link. We obtain the total free energy

$$F \cong N\varepsilon(1-T/\gamma)$$

Hence if $T>\gamma$ the system will contain many conformational states; i.e., the tertiary structure is like a liquid. The formation of a crystal-like structure requires $T<\gamma$. In this case also local deviations from order can arise in the system, changing the entropy by k ln c, where c is the concentration of local defects. In the globule the defects are related to topological changes, to changes of the total conformation. The characteristic defect energy $\tilde{E}$ can be different for a defect in the depths of the globule ($\tilde{E}_1$) or on its surface ($\tilde{E}_2$). Inside the globule the minimal concentration of defects is proportional to $N^{-1}$ and on the surface, to $N^{-2/3}$. Therefore the "volume" and "surface" topological deviations change the free energy by $\tilde{E}_1 - kT \ln N$ and $\tilde{E}_2 - \frac{2}{3} kT \ln N$, respectively. Hence the stability requirement has the form

$$N < N_{crit} \simeq \exp(\tilde{E}/kT) \tag{4.116}$$

and a very long chain cannot form one globule; it must form several connected globules. Perhaps this is the reason for the relatively small dimensions of the monoglobular protein molecules that usually contain about 150 amino acid residues and have a molecular weight of the order of 20,000.

The compactness of the globule and the sufficiently high surface tension necessary for the self-organization of a crystal-like structure determine the rigid limitations of the primary protein structure.

References [1,178] are of primary importance in protein physics. The statistical-thermodynamic analysis based on consideration of the linear memory in the chain explains the general properties of protein globules. Development of these ideas concerning heteropolymer chains is necessary. On the other hand, the solution of the problem of self-organization of a globule requires study of the kinetic factors. Obviously the self-organization of a long chain must involve a combination of the kinetic and thermodynamic requirements. The energetically optimal conformation of a chain must have the simplest topology and has to be kinetically achievable.

## 4.8 Denaturation of Proteins

As we said earlier, even comparatively soft reactions, which do not break peptide bonds, can result in the loss of biological activity by proteins. This denaturation of protein can be provoked by heating, mechanical action (ultrasound), changes in pH, and the action of such agents as urea. Denaturation consists in the destruction of the spatial structure of the protein molecule, with retention of the primary structure of the chains. The denatured molecule appears in the state of a random coil, either with constraints imposed by disulfide cross links or without them. For globular proteins, the denaturation process amounts to a globule-coil transition.

The study of denaturation provides information concerning the nature and stability of native structure. However, interpretation of the experimental data is rather difficult because of great variety of interactions inside the globule and at its surface; transition parameters give only some integral values. A globule has a fixed compact structure, forming an "aperiodic crystal." The globule-coil transition differs from helix-coil transitions and transitions from β form to coil because it occurs in a three-dimensional system. Cooperativity of transition is determined not only by short-range but also by long-range interactions and by geometric packing factors [2].

Figure 4.24 shows the results of an investigation of the thermal denaturation of chymotrypsinogen conducted by Brandts and Hunt [125,126]. The transition is indicated by the change in the extinction coefficient at 293 nm. The process is reversible and the curves are equilibrium curves. This transition is typical for globular proteins. The transition region is approximately 10°C. However, the similarity to the helix-coil transition is deceptive. Careful analysis of the dependence of the degree of denaturation on temperature shows that the finite breadth of the transition cannot be explained by the presence of partly denatured molecules that are in equilibrium with native and entirely denatured ones. Transition occurs in one step and in this sense is similar to a phase transition of the first kind [127].

Calorimetric studies give direct information about the thermodynamic parameters of denaturation. Privalov has shown that when some globular proteins (albumins, myoglobin,

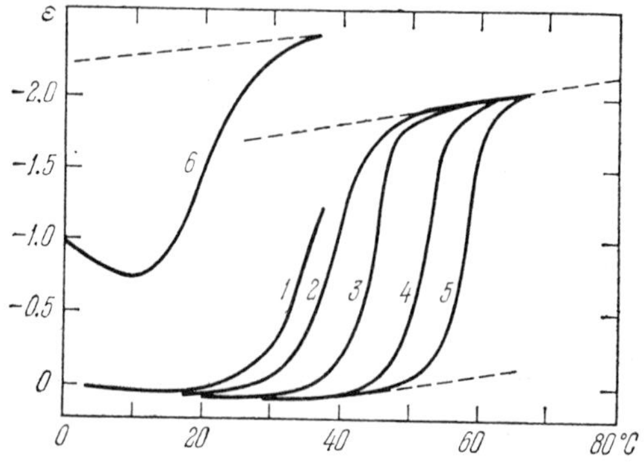


*FIG. 4.24 Dependence on T of the extinction coefficient at 293 nm of chymotrypsinogen solution. Numbers are the pH values. The lower dashed curve is the native form, and upper the denatured form.*

chymotrypsinogen, ribonuclease) are heated, heat absorption occurs in two stages. The first stage--predenaturation--is characterized by an increase in heat capacity without a jumplike change in enthalpy, and the second one is conventional denaturation, a phase transition [128]. The denaturation enthalpy depends strongly on pH, as is shown in Table 4.14. The spatial structures of these proteins are known and the enthalpy of denaturation per mole of hydrogen bonds can be calculated. For all three proteins this value is 1.4-1.5 kcal/mole. Privalov concludes that hydrogen bonds play a determining role in the denaturation process. His conclusion cannot be considered a final one--it is based on a limited number of data and contradicts the general ideas concerning the structure of globules.

*TABLE 4.14*

*Temperatures and Enthalpies of Denaturation*

| Chymotrypsinogen | | | Ribonuclease | | | Myoglobin | | |
|---|---|---|---|---|---|---|---|---|
| pH | T(°C) | ΔH (kcal/mole) | pH | T(°C) | ΔH (kcal/mole) | pH | T(°C) | ΔH (kcal/mole) |
| 2.3 | 43 | 78 | 2.4 | 36 | 52 | 12.2 | 50 | 73 |
| 2.6 | 49 | 102 | 3.3 | 47 | 66 | 12.0 | 54 | 80 |
| 2.8 | 51 | 110 | 3.7 | 50 | 73 | 11.5 | 63 | 100 |
| 3.4 | 58 | 130 | 4.4 | 54 | 77 | 11.3 | 67 | 117 |
| 4.0 | 61 | 140 | 6.0 | 59 | 89 | 11.0 | 72 | 134 |
| 5.0 | 62 | 148 | | | | 10.7 | 78 | 170 |

Multistage denaturation has been observed in some cases (e.g., the denaturation of paramyosin by guanidine chloride [129]). Identification of intermediate states and the treatment of the multistage process involves considerable difficulty.

Typical values for the thermodynamic parameters of denaturation are presented in Table 4.15. T* is the temperature of maximal stability when the equilibrium constant $K_D$ of the process native protein ⇄ denatured protein is the minimal one. If $T<T^*$, $K_D$ increases with decreasing T; if $T>T^*$, $K_D$ increases with increasing T. The methods for determining of $K_D$ have been described by Tanford [130], who proposed a method for determining the differences in free energy of denatured and native protein ΔG based on a study of denaturation in urea solution [131]. In this process we have

$$-\Delta G_u = RT \ln K_u = RT \ln ([D]/[N]) \qquad (4.117)$$

where [D] and [N] are the concentrations of denatured and native protein. We are interested in the value $\Delta G_{H_2O}$ for the

*TABLE 4.15*

*Typical Values of Thermodynamic Parameters*

| Protein | $\Delta G^{a}$ | $\Delta H^{a}$ | $\Delta S^{b}$ | $\Delta C_p{}^{b}$ | T* (°C) |
|---|---|---|---|---|---|
| Ribonuclease, thermal transition, pH 25, 30°C | 0.9 | 57 | 185 | 2000 | -9 |
| Chymotrypsinogen, thermal transition, pH 3.0, 0.01 M Cl, 25°C | 7.3 | 39 | 105 | 2600 | 10 |
| Myoglobin, thermal transition, pH 9.0, 25°C | 13.6 | 42 | 95 | 1400 | 0 |
| Lactoglobulin, 5 M urea, pH 3.0, 25°C | 0.6 | -21 | -72 | 2150 | 35 |

[a] $\Delta G$ and $\Delta H$ in kcal/mole.
[b] $\Delta S$ and $\Delta C_p$ in cal/mole-deg.

aqueous solution. Employing the cycle

$$
\begin{array}{ccc}
\text{Native protein in water} & \xrightarrow{\Delta G_{H_2O}} & \text{Denatured protein in water} \\
\Delta G_N \downarrow & & \downarrow \Delta G_D \\
\text{Native protein in urea solution} & \xrightarrow[\Delta G_u]{} & \text{Denatured protein in urea solution}
\end{array}
$$

we get

$$\delta\ \Delta G = \Delta G_u - \Delta G_{H_2O} = \Delta G_D - \Delta G_N \tag{4.118}$$

The value of $\delta\ \Delta G$ can be expressed as

$$\delta\ \Delta G = \sum_i \alpha_i n_i\ \Delta g_i \tag{4.119}$$

where $n_i$ is the number of groups of type i in protein, $\Delta g_i$ their contribution to the free energy of transition, and $\alpha_i$ a numerical factor which depends on the degree of accessibility of the group i to the solvent in the native conformation.

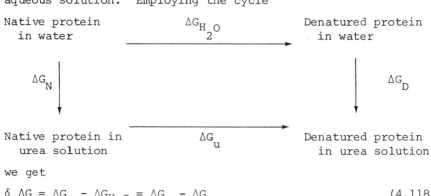

According to Tanford, theory agrees with experiment if we take, for polar groups, $\alpha_i = 0.25$; for hydrophobic, $\alpha_i = 0.75$; and for peptide bonds, $\alpha_i = 0.50$. The values $\Delta g_i$ are found from the solubilities of amino acids in water and in the urea solution of given concentration.

Let $C_m^*$ be the concentration of urea corresponding to [D] = [N], that is, the semidenaturation point. Then according to (4.117) $\Delta G_u = 0$ and we get

$$\Delta G_{H_2O} = -(\delta\ \Delta G)_{C_m^*} = -\sum_i \alpha_i n_i (\Delta g_i)_{C_m^*} \qquad (4.120)$$

This is a convenient and simple method for determining the conformational stability of the protein.

The study of denaturation kinetics (particularly by means of relaxational methods, p. 470) makes it possible to estimate the activation parameters of transitions. The values $H^{\ddagger}$ are usually of the order of 40-80 kcal/mole, $S^{\ddagger}$ 40-200 eu, $G^{\ddagger}$ 20-25 kcal/mole [132].

The free energy of denaturation is the sum of many contributions. Scheraga writes [133]

$$\Delta G = \Delta G_\alpha + \Delta G_H + \Delta G_{Np} + \Delta G_e + \Delta G_m + \Delta G_x + \cdots \qquad (4.121)$$

where $\Delta G_\alpha$ is determined by the unfolding of $\alpha$-helices, $\Delta G_H$ by breakage of hydrogen bonds linking the protein chains of neighboring macromolecules, $\Delta G_{Np}$ by the change in hydrophobic interactions, $\Delta G_e$ by the change in electrostatic interactions, $\Delta G_m$ by the swelling of the formed coil, $\Delta G_x$ by the breakage of cross links between the helices existing in the crystalline phase.

$\Delta G_\alpha$ and $\Delta G_x$ do not depend on the pH; other terms connected with the presence of ionogenic groups depend on the pH. We have

$$\Delta G_\alpha = (N-4)\Delta H_0 - (N-1)T\ \Delta S_0 \qquad (4.122)$$

where N is the number of links in the $\alpha$-helix and $\Delta H_0$ and $\Delta S_0$ are the changes in enthalpy and entropy in the helix-coil transition per link of an infinite chain. The $\alpha$-helix containing N links has N - 4 hydrogen bonds which constrain the motility of N - links. It follows from (4.122) that the $\alpha$-helix is stable (i.e., $\Delta G_\alpha > 0$) only if N is rather large.

$$\Delta G_H = -kT \sum_{i,j} \ln (1-x_{ij}) \qquad (4.123)$$

where $x_{ij}$ is the fraction of molecules having side hydrogen bonds between groups i and j, equal to

$$x_{ij} = K_{ij}\{1 + K_{ij} + (K_1/[H^+]) + ([H^+]/K_2)\}^{-1} \qquad (4.124)$$

where $K_{ij}$ is the equilibrium constant of ij bond formation.

$K_1$ and $K_2$ are the ionization constants of the donor and acceptor groups not forming hydrogen bonds.

$$\Delta G_m = -2pkT \ln (1 + Kc) \tag{4.125}$$

where K is the equilibrium constant for the interaction of NH or CO groups of protein with small molecules in solution. Thus, urea forms hydrogen bonds with protein and favors denaturation. In this case c is the molar urea concentration and p the number of residues of "crystalline" chain which can interact with urea after the destruction of the helix. Of course, the water molecules interact with protein too, but this interaction has already been taken into account in $\Delta G_\alpha$, $\Delta G_H$, and $\Delta G_{Np}$. If c is small

$$\Delta G_m = -2pkTKc \tag{4.126}$$

The term $\Delta G_x$ is purely entropic:

$$\Delta G_x = -T\,\Delta S_x = kT\,\frac{3}{2}\,\nu(\ln n + 3) \tag{4.127}$$

where $\nu$ is the number of cross-linked helices and n the number of residues between cross links.

Determination of $\Delta G_{Np}$ and $\Delta G_e$ is more difficult. It can be suggested that these terms are rather small (we investigate $\Delta G$ but not $\Delta H$ and $\Delta S$). According to the estimation of Shellmann [134], $\Delta H_0 = 1.5$ kcal/mole, $\Delta S_0 = 4.2$ eu. The value of $\Delta H_0$ is considerably smaller than the conventional energies of hydrogen bonds. $\Delta H_0$ is the difference of energies of the groups linked by intramolecular and intermolecular hydrogen bonds. Privalov's estimation is similar [128].

At large N

$$\Delta G_\alpha \cong N(\Delta H_0 - T\,\Delta S_0) \tag{4.128}$$

and if $T = 300°K$, $\Delta G_\alpha = N(1500 - 1260) = 240N$ cal/mole, which is less than RT. In the presence of urea $\Delta G_m$ is negative, that is $\Delta G_m$ lowers the stability of native protein. On the other hand, $\Delta G_{Np}$ increases the stability. If the protein does not have side hydrogen or hydrophobic bonds, the temperature of transition $T_{tr}$ can be estimated from the condition

$$\Delta G = (N-4)\Delta H_0 - (N-1)T_{tr}\,\Delta S_0 - T_{tr}\,\Delta S_x = 0 \tag{4.129}$$

and

$$T_{tr} = \frac{(N-4)\,\Delta H_0}{(N-1)\Delta S_0 + \Delta S_x} \tag{4.130}$$

The value of $\Delta S_x$ is also proportional to N--the longer the chain, the more cross links can be formed. Hence, $\nu = \beta N$. If $N \gg 1$, $T_{tr}$ does not depend on N. If $\Delta H_0$ and $\Delta S_0$ are known, $\Delta S_x$ can be determined from $T_{tr}$. $\Delta G_{Np}$ is estimated as 0.5 kcal/mole. The $\Delta G_H$ and $\Delta G_e$ values are of the same order of magnitude.

The fraction of denatured substance is

$$x = \frac{\exp(-\Delta G/kT}{1 + \exp(-\Delta G/kT)} \qquad (4.131)$$

At the transition point $\Delta G = 0$ and $x = 0.5$. A detailed examination of hydrogen bonds and hydrophobic interactions in the protein-water system is presented in the paper [102]. The quantitative estimates of $\Delta H$, $\Delta S$, and $\Delta G$ agree with the ones presented here.

The sharpness of a thermal transition is characterized by the derivative

$$(dx/dt)_{T_{tr}} = \Delta H/4RT_{tr}^2 \qquad (4.132)$$

where

$$\Delta H = (N-4)\Delta H_0 + \Delta H_H \qquad (4.133)$$

and

$$\Delta H_H = \sum x_{ij}\left\{-\Delta H_{ij} + \frac{(K\ /[H^+])\Delta H_1 - ([H^+]/K_2)\Delta H_2}{1 + (K_1/[H^+]) + ([H^+]/K)}\right\} \qquad (4.134)$$

In (4.134) $x_{ij}$ is defined by Eq. (4.124), $H_{ij}$ is the heat of formation of hydrogen bond ij, $\Delta H_1$ and $\Delta H_2$ are the ionization heats corresponding to the equilibrium constants $K_1$ and $K_2$. $T_{tr}$ does not depend on the cooperation of helical regions. On the contrary, the sharpness of transition depends on this cooperation. If the transition is performed simultaneously by r participants, then

$$(dx/dT)_{T_{tr}} = r\Delta H/4RT_{tr}^2 \qquad (4.135)$$

The theoretical consideration presented is in general agreement with experiment (cf., e.g., [133]). Many experimental data concerning denaturation are contained in the monograph of Joly [132] and the reviews of Tanford [130] and Brandts [107], which also contain the theoretical analysis.

We see that denaturation can be treated as a cooperative transition process between two states, the native and the denatured one. Brandts provided the foundation for this concept [107]. Both observed macroscopic states must be considered as the results of averaging over a totality of microscopic states. Every microscopic state belongs to either the native or the denatured state, or to both of them. Using partition function (4.82), where q is given by (4.84)(p. 221), Brandts estimated the change in the number n of hydrogen-bonded links with temperature. He considered the "protein" with N = 200, containing 40% hydrophobic, 20% hydrophilic, and 40% neutral residues. The distribution function of microstates characterized by the number of hydrogen bonds is clearly bimodal. If the temperature changes from 20°C to 50°C, the probabilities of the states with $n \sim 200$ decrease sharply, and simultaneously the probabilities of the states with $n \sim 0$ increase strongly.

Investigations of denaturation transitions give information about the degree of stability of native protein (the change in free energy), and about the cooperativity of interactions in protein (the sharpness of transition). These characteristics are not always correlated.

Of course, studies of denaturation do not reveal the structure of the "machine," the protein globule. Denaturation means destruction of the machine. However, this destruction provides some information about the stability of internal interactions, about electron-conformation interactions (ECI) in protein (cf. p. 400).

The equilibrium treatment can be rigorously applied only in cases of reversible denaturation. In the majority of cases the denaturation observed is partly or totally irreversible. However, a series of reversible denaturation phenomena is known. If denaturation proceeds under soft conditions (e.g., if it is produced by a slow and moderate increase in temperature), then renaturation of a series of proteins can be observed (i.e., regeneration of their native structure and properties upon reversal of the denaturing action).

It is obvious that total reversibility of denaturation can be observed for proteins which do not contain groups participating in the denatured state in irreversible chemical reactions (e.g., oxidation of SH groups) [101]. Renaturation was established for ribonuclease [135], taka-amylase A [136], and α-amylase [137]. The works of Anfinsen *et al.* have shown that renaturation of proteins with broken disulfide bonds is also possible [138-140]. It can be concluded that denaturation can actually be treated as a thermodynamic conformational transition and that the native structure of protein corresponds to some relative minimum of free energy, if not to an absolute one.

## 4.9 Primary Structure of the Polypeptide Chain and Spatial Structure of the Globule

The problem of the relation between the primary structure of the polypeptide chain and three-dimensional globular structure is one of the most important problems in protein physics. Biological functionality inheres in the native spatial structure of the molecule, but only the primary structure is coded genetically. Should there not be a definite connection between these two kinds of structures, the fundamental ideas of molecular biology would have to be revised. The facts quoted on p. 234 concerning the renaturation of proteins provide the experimental foundation for such a connection, which follows from general theoretical considerations (cf. p. 590).

A series of papers reports a comparative analysis of spatial protein structures, determined by X-ray diffraction, and their primary structures; the analysis was meant to reveal the amino acid residues forming α-helical regions of the globule and those hindering spiralization. Guzzo [141] divided amino acids into "helical" and "nonhelical" ones, using data on hemoglobin and myoglobin (cf. also [142]). Prothero [143] expanded this classification, having examined a larger number of proteins, and formulated criteria for the formation of helical segments, depending on the definite kinds of residues contained in segment lengths. Shiffer and Edmundson [144] proposed a model describing spiralization at various positions of residues. They suggested that the periodic accumulation of hydrophobic and hydrophilic residues observed in hemoglobin and myoglobin [111], p. 218)and in lysozyme [145] is responsible for stabilizing the helical regions of globules. Peritti *et al*. [146] developed a statistical method for predicting secondary structure. They determined the frequencies of appearance of a pair of residues that are 0,1,... residues apart in the helical and nonhelical regions in hemoglobin and myoglobin and predicted the helical regions in lysozyme.

On the basis of a statistical analysis of the amino acid content and sequence of the polypeptide chain regions having different secondary structures, Ptitsyn [147,148] suggested the classification of amino acids and the method for predicting the secondary structure of proteins. An individual tendency to enter the helical region ("helical potential") was ascribed to every amino acid. This tendency does not depend on the neighboring residues in the first approximation.

This hypothesis was challenged earlier by Kotelchuk and Scheraga [149] on the basis of a study of conformational energies. It was shown that the interaction of the amino acid side chain with the amide group at its C-terminal was the most important one. As a rule this interaction is not dependent on the side chains of the neighboring residues. The conformational energies of every residue in the three conformations with lowest energy (right helix, left helix, and antiparallel β form) were compared. According to [149], the residues for which the right helices correspond to the lowest energy are "helical," in other cases the residues are nonhelical. Pane and Robson [150] calculated the helical potentials of amino acids, theorizing that local interactions between the side chains and the main chain are the essential ones.

Ptitsyn and Finkelstein [148] also estimated the "β-structure potential" of amino acids. Positive helical potential is ascribed to Ala, Leu, Met, and negative to Asp, Cys,

Ser, Tyr, Thr, and Gly; zero potential is attributed to the remaining residues. Such amino acid behavior can be explained as follows. Amino acids with bulky hydrocarbon side chains (Ile, Leu, Phe) have a strong positive β-structural potential; amino acids with charged side groups (Asp, Glu, Arg, His, Lys) and Pro have a strong negative β-structural potential. Other hydrophobic residues have a small positive β-potential, whereas polar residues and Gly have a small negative β-potential. Using the data obtained for myoglobin, lysozyme, papain, and α-chymotrypsin, Ptitsyn and Finkelstein predicted the structures of the α and β chains of hemoglobin, ribonuclease A, subtilysine BPN', and the carboxypeptidase fragment. Agreement with experiment was obtained for 74% of the residues.

A more substantive approach was suggested in the work of Lim [151]. As we stated earlier, the separate examination of the secondary and tertiary structures of protein has no exact physical meaning, since the secondary structure is an element of spatial structure determined by spatial structure (p. 206). Lim treats the globule as consisting of a hydrophobic nucleus and polar shell. Hydrophobic residues themselves cannot form more than one twist of the helix because coiling is hindered by interactions with water. Only those hydrophilic regions that are adjacent to the helix "tied" to the nucleus can twist. Totally hydrophobic regions are helical only if they are located in the nucleus. That mixed regions are helical if the hydrophilic residues are on the surface of the globule and the hydrophobic ones inside it is confirmed particularly by studies of hemoglobin (p. 218). For helices, "the staples," which contain hydrophobic residues and are located in the positions i, i + 4, are typical.

It turns out to be reasonable to classify residues according to their properties as just described (Lim [151]).

| | |
|---|---|
| Small residues | Gly, Ala; |
| Middle hydrophobic | Cys, Val, Met, Leu, Ile, Pro; |
| Big hydrophobic | Phe, Tyr, Trp; |
| Small hydrophilic | Asp, Asn, Glu, Gln, Ser, Thr; |
| Big hydrophilic | Lys, Arg, His. |

A series of empirical rules has been formulated, describing the participation or nonparticipation of these classes of residues in helices, and reasonable agreement with structural data has been obtained for 12 proteins.

Lim's work [151] differs significantly from other studies in that it examines long-range interactions, which seem to be the determining ones.

Robson and Pane [152] attempted to apply information theory to this problem. They treated the sequence of residues in the chain and the sequence of their conformations in a globule as two messages connected by some translational code. By means of computers, the "spiralizing information" was determined for 11 proteins. Using only information regarding single residues, these researchers were able to predict helical regions. Some discrepancies were eliminated by introducing corrections taking into account information in pairs of residues.

According to [152], the "spiralizing information" contained in a single residue is expressed as the difference in its energies in the helical and nonhelical structures depending on the interaction of the side group with the main chain.

Examination of steric energy maps shows that the information in the pair of residues is expressed by the loss of a hydrogen bond at the COOH end of an induced $\alpha_{\pi}$ conformation. This loss forces the subsequent residues to turn into nonhelical conformations. This approach seems somewhat limited, however, since it does not take into account the physics of the globule.

Esipova and Tumanian [153] studied not helicity but the relation between the primary and spatial structures of protein.

Tertiary structure is characterized by the sites of turns in the main chain. In other words, the chain is approximated by a line, the regions of its maximal curvature are determined, and the causes of local turns in the chain are investigated. The bending of the chain is caused by the residues which occur at the turn. The change in curvature can be determined by hydrogen bonds. A turn in the chain means a sharp change in the values of the angles $\phi$ and $\psi$ (p. 166). For most residues the $\phi$ and $\psi$ values that correspond to the right $\alpha$-helix are the most advantageous. However, Ser, Asn, Trp, Try, and Lys are characterized by energy minima in the conformations of the left $\alpha$-helix and $\beta$ form.

Gly produces the maximal flexibility. It can be considered a "gimbal" in the protein molecule. It can also be assumed that nonhydrophobic residues, particularly Ser, have to occur at the turns.

In [153] the structures of lysozyme, chymotrypsin, myoglobin, elastase, ribonuclease, papain, and carboxypeptidase were analyzed. The hypotheses just presented were confirmed. Thus, in lysozyme, 17 turns in the chain were determined and practically all Gly residues were found to participate in the turns. It is established that the presence of Gly is a sufficient but not necessary condition for a sharp turn. In 7 of 17 turns in the regions of maximal curvature Ser, Asp, Arg, and Trp are present. These residues are the neighbors of "turning" Gly. Similar regularities were found for chymotrypsin, elastase, ribonuclease, and papain. Of the 19 Gly in $\alpha$-chymotrypsin, 17 participate in the turns; in elastase 23 Gly of 25 are located in turns. In addition to these residues, Thr, Asn, Lys, and Gln are also found in turns. Inclusion of one of these residues together with Gly appears to be important for turning. An essential role is played by hydrogen bonds linking the side radical with main chain (Ser).

Reliable statistical estimates show the participation in turns of Gly, Ser, Asn, and Arg. The participation of Tyr, Thr, and Asp is less well established. Leu, Val, Ala, avoid the turns, as do Phe, Pro, and His.

In proteins with low α-helicity, Gly, which does not form the turn, is usually the neighbor of Val and Ile. In myoglobin, however, Gly is usually the neighbor of Glu and Ala, stabilizing the α-helix. Therefore, these Gly cannot destroy the helix and form turns. On the other hand, Gly in myoglobin and in similar proteins provides the contacts between α-helical segments. Therefore, tertiary structure stabilizes α-helices that include Gly.

Consequently, the tertiary structure of protein can be modeled by a set of approximately rectilinear segments connected by joints. The so-called nonordered segments are the most important elements of spatial structure.

We quote some other works treating the same problem [179, 180]. Ptitsyn approached the problem of the relation of the primary and spatial structure of a protein globule by using a physical hypothesis about its formation [181]. He suggested that the globule self-organization is a result of some directed process. The study of the protein renaturation shows that the program of self-organization is coded by the primary structure. The self-organization occurs stepwise; the more complex and stable structures are formed in sequential steps. A fluctuating "nucleus" of the next stage is formed at the preceding stage. The native structure of the protein molecule is built of discrete structured regions like blocks. Their formation at a preceding stage is not changed but stabilized by the next stage.

According to Ptitsyn, at the first stage fluctuating regions of the secondary structure are formed in the unfolded polypeptide chain, and their formation is determined by local interactions such as hydrogen bonds. At the second step these fluctuating regions collapse, forming a compact globule. This process is determined by the long-range, nonspecific interactions of the side groups with the surrounding medium. At the third step this intermediate compact structure is rearranged into the unique native structure of the globular protein. This unique structure is determined by specific long-range interactions of the spatially close amino acid residues.

Quantitative investigation based on this hypothesis made it possible to analyze the pathways of self-organization of proteins, particularly of myoglobin [182]. The theory explained the positions of the five long and two short helical regions of the chain. The calculations of conformational energies of the system at the formation stages of the "crystallization centers" have been made. The spatial structure of myoglobin thus established is in satisfactory agreement with experimental data.

This theory must be developed quantitatively taking into account the kinetics of the self-organization. The experimental approach to the problem includes the study of renaturation kinetics. Information about the stages of renaturation can be obtained by means of nuclear magnetic resonance.

## 4.10 Fibrillar Proteins

This chapter has been devoted mainly to globular proteins, which possess a great variety of structures and functions. Fibrillar proteins are not so diversified; they have specific structures and perform special functions. They comprise structural and contractile proteins. The first play the role of supporting and protecting the components of such tissues as tendons, cartilage, bones, etc. (collagens) and epidermis, hair, wool, horns, etc. (keratins). The second are contained in working substances of mechanochemical systems, particularly muscles (myosin).

In contrast to the majority of globular proteins, fibrillar (i.e., fiber-forming) proteins do not function in solution (the cytoplasm of cells) but form supramolecular tissue systems. The performance of a structural function requires precision of structure--structural regularity at all levels, starting with the primary one.

Animal skin contains collagen (in the dermis) and keratin (in the epidermis). Extraction by a cold solution of salts, or by dilute acetic acid (pH 3.9), as well as by alkaline solutions, transfers part of the collagen into solution. These soluble molecules are practically identical in all kinds of extraction experiments. The part of collagen that is soluble in acids is called procollagen.

Collagenic fibers (fibrils) are formed by aggregation of procollagen chains. In the process of bone growth (osteogenesis), these fibrils play the role of condensation centers for the growth of needlelike crystals of hydroxyapatite, $Ca_{10}(PO_4)_6(OH)_2$. Collagen fibers are not soluble in water, but prolonged heating with water transforms collagen into soluble gelatin as a result of the hydrolysis of some of the peptide bonds and therefore of the destruction of the long protein chains.

The fundamental problem in the physics of collagen and of other fibrillar proteins is to find out how the features of primary structure determine specific fibrillar properties.

The amino acid composition of collagen is unusual: 33% of all the residues are Gly; 12% Pro; and 10% the noncanonical residue oxyprolyl (Opro); oxylysyl (Olys), contained in collagen in amounts of 0.3-1.2%. Collagen also contains up to 10%

Ala and much smaller amounts of other amino acids. The content of Tyr, His, Cys, Met, Val, and Phe is especially low, less than 1%. Thus nearly two thirds of all residues in collagen are Gly, Pro, Opro, and Ala, which means that less information is stored in collagen than in globular proteins; collagen molecules are not actually informational ones.

The sequence of residues in collagen can be expressed in the form $(Gly\text{-}x\text{-}x)_n$, where x is any residue. The sequences Gly-Pro-Opro, Gly-Pro-Ala, and Gly-Ala-Opro are the most frequent ones [154].

The structure of collagen was established by means of X-ray studies by Ramachandran *et al.* [155,156] and by Rich and Crick [7,8,157]. Collagen is similar to synthetic polyglycine and polyproline. The structure of polyglycine is described on p. 170.

The collagen molecule is formed by three polyglycine-type chains coiled together in a helix (Fig. 4.25). The Gly residues are located near the central axis of the triple helix and the other residues face outward and can bind neighbor chains. The molecular weight of such a triple helix (procollagen) is about 360,000; each of the three chains contains nearly 1000 residues. The length of the molecule is 290 nm, its diameter is 1.5 nm.

It has actually been shown that the structure of a synthetic polypeptide $(Gly\text{-}Pro\text{-}Opro)_n$ with a molecular weight of

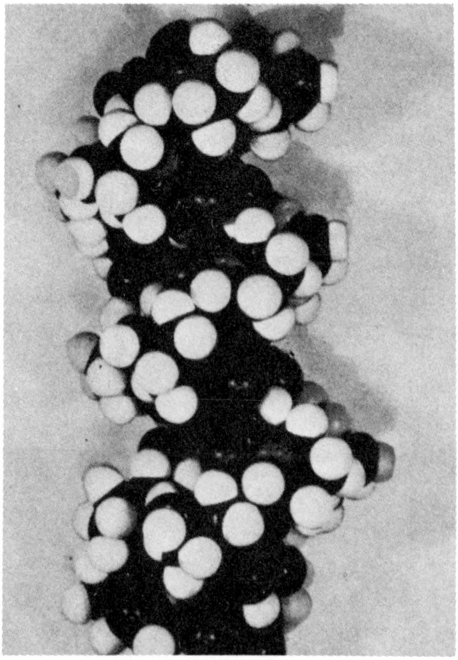

*FIG. 4.25 The structure of collagen.*

several thousand is similar to that of collagen. The similarity was ascertained by means of X-ray diffraction, infrared spectroscopy, and spectropolarimetry [158,159].

The scheme of aggregation of procollagen molecules in supramolecular fibrils is shown in Fig. 4.26. Electron micrographs of native collagen fibers show the transverse lines corresponding to this scheme (Fig. 4.27). The period of these lines is 640 Å.

Miller and Ray suggested another model of collagen fibrils in which the tropocollagen molecules are grouped in braids twisted from five threads [160]. This model also agrees with X-ray data; therefore, the problem requires further investigation.

A theoretical calculation of collagen structure based on the minimization of conformational energy was made by Tumanian [161]. This theory takes into account the stabilizing role of water (cf. [162]). If collagen is carefully heated in water, the fibers irreversibly contract at 62-63°C to one third their original size. Apparently this transition, a conformational one, is due to cooperative destruction of hydrogen bonds and hydrophobic interactions. Procollagen undergoes thermal transition in solution. Privalov and Tiktopulo investigated this transition by means of calorimetry [163].

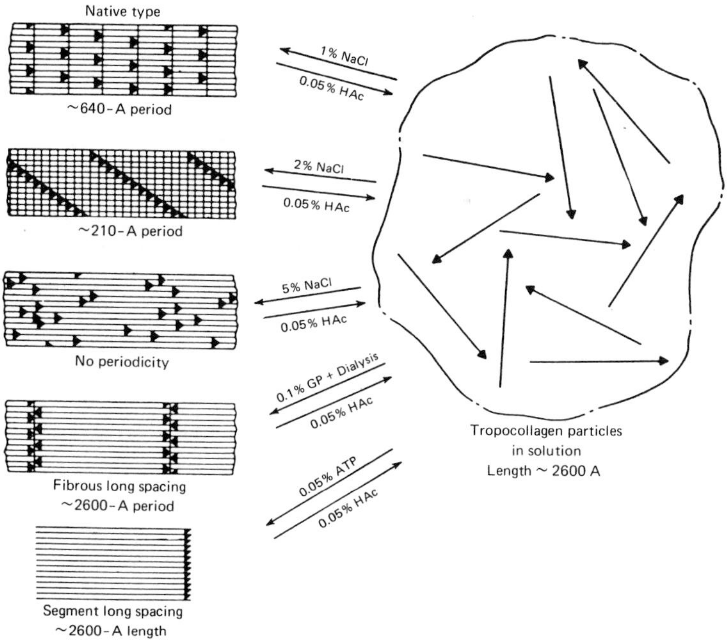


*FIG. 4.26 Aggregation of procollagen molecules.*

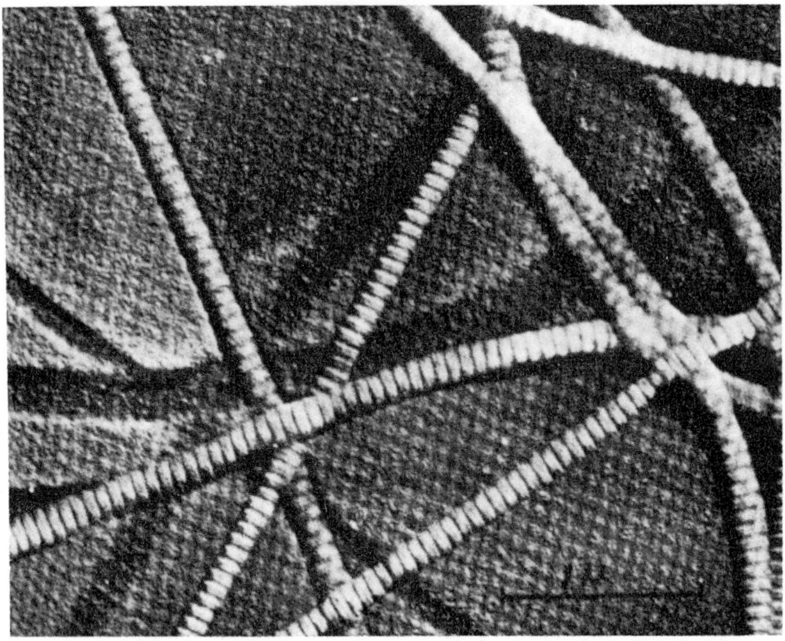

*FIG. 4.27 Electron micrograph of a collagen fiber.*

Table 4.16 contains the denaturation parameters of tropocollagen at pH 3.5 depending on the content of imino acid residues (Pro, Opro). A strong increase in Q, ΔH, and ΔS with an increase in the imino acid residue content cannot be explained by suggesting that the native tropocollagen structure is stabilized only by intramolecular hydrogen bonds, since imino acid residues do not form such bonds. Now can we save the situation by attributing the determining role to hydrophobic interactions. The breaking of hydrogen bonds is an exothermic, not an endothermic, process. Electrostatic interactions cannot play any significant role, since ΔH is practically independent of pH. Privalov and Tiktopulo [163], following Tumanian's theory, suggest that tropocollagen is stabilized by the adjoining aqueous structure. It is actually known that the removal of water results in the destruction of collagen structure and that addition of water restores this structure [162]. Physical methods, particularly nuclear magnetic resonance [164] attest to the existence of ordered water structure bound with collagen. Recently a study of the structure of collagen disclosed the formation by the water molecule [165] of an additional hydrogen bond per triplet.

The dependence of denaturation enthalpy on the imino acid residue content shows the linear heterogeneity of tropocollagen, whose molecules involve regions with various degrees of

*TABLE 4.16*

*Thermodynamic Characteristics of the Denaturation of Tropocollagen*

| Tropocollagen from | Imino acid content per 1000 residues | $T_{trans}$ (°C) | $Q_{trans}$ (cal/g) | ΔH (cal/mole) | ΔS (eu) |
|---|---|---|---|---|---|
| Rat | 226 | 40.8 | 16.8 | 1530 | 4.9 |
| Pike | 199 | 30.6 | 13.6 | 1240 | 4.1 |
| Merlang | -- | 21.5 | 9.7 | 880 | 3.0 |
| Cod | 155 | 20.0 | 8.2 | 750 | 2.6 |

ordering. It follows also that the complicated character of heat absorption by a salt solution of tropocollagen is related to some predenaturational conformational transitions in conditions similar to physiological ones (cf. also [166]).

Many results concerning the structure and properties of collagen are contained in the review article [167]. The methods of X-ray analysis of fibrillar systems are described in Section 5.2.

The problem stated above cannot be considered solved; despite the large number of works devoted to collagen, we do not yet have a complete theory of its structure and properties. Arriving at such a theory is important in two respects. First, collagen is one of the most important proteins in animal organisms. Second, the relative simplicity of its structure and properties (in comparison with those of globular proteins) makes collagen a very valuable model for general studies of the structure and properties of proteins.

We now consider another important fibrillar protein--keratin. The macroscopic properties of keratin-containing biological structures--hair, wool, feathers, horn, nails, and hoofs--demonstrate its high stability and nonsolubility. Studies of keratin show that these features are attributable, first, to the large number of disulfide cross links connecting polypeptide chains. The keratin of human hair and of wool contains 11-12% cysteine (i.e., 3% sulfur).

The length of a keratin fiber depends strongly on its water content (this is the basis for the hair hydrometer). These fibers are elastic and can be stretched. In the thirties Astbury performed his classical investigation of the structure of wool keratin by X-ray diffraction [168]. The structure depends strongly on stretching. For the keratin of a non-stretched fiber a periodicity of 5.1 Å is typical (this is so-called α-keratin). Stretching produces the β-keratin structure; the periodicity becomes 3.33 Å and transverse periods of 4.65 and 9.7 Å appear. Astbury theorized that α-keratin contains regular bends of polypeptide chains that are straightened by stretching. Actually, it was shown later that β-keratin has the canonical β form, whereas α-keratin contains a great deal of α-helical substance (cf. [169]).

Keratin is a complicated protein. If disulfide bonds are broken as the result of oxidation or reduction, a soluble substance is formed. Two fractions--one sulfur rich, the other sulfur poor--can be separated. The sulfur-poor fraction contains fibrillar molecules (i.e., is able to form fibers); the sulfur-rich fraction contains globular molecules. It can be suggested that these globular molecules serve as cross links in keratin fibers, whose protofibrils consist of protein containing little sulfur. Investigations by X-ray diffraction and electron microscopy have shown that the basic unit of the fiber is the cylindrical microfibril with a diameter of 75 Å composed of proteins with a low sulfur content. Regularly packed regions of these proteins are α-helices coiled by pairs. Figure 4.28 shows the hypothetical model of the molecular organization of α-keratin. Protein is heterogeneous and contains two main components in the ratio 2:1. Single molecules can be located either in series (Fig. 4.28a,b) or parallel to each other (Fig. 4.28c). Helical regions of molecules are coiled either by pairs or in the triple superhelix. Experiments indicate large periods along the fiber equal to 200 Å in agreement with the models in Fig. 4.28 b,c, but not with (Fig. 4.28a). Nonhelical regions regulate the common packing in an unknown way.

The microfibril as a whole contains 11 such protofibrils--double or triple helices. Two of them are situated at the center of the microfibril and nine at its periphery (Fig. 4.28 a). This 9:2 ratio is characteristic of a series of biological supramolecular structures consisting of fibrillar biopolymers. The microfibril has a homogeneous structure. The lengths of the helical regions forming a protofibril do not vary much in different species, but the composition of nonhelical regions varies markedly. The principal variations in mammalian keratins are due to the kind of microfibril packing and to the quantity and content of the sulfur-rich component, that is, of the matrix joining the microfibrils together. The properties of this matrix are considerably changed by swelling in water.

β-keratin has been studied much less. It cannot be separated into two fractions. Approximately half of β-keratin possesses the β structure; the other half is not ordered. Microfibrils with a diameter of approximately 35 Å are made up of two interlaced molecular threads. Every thread contains four molecules per period, each having a length of 25 Å. Further details concerning the structure of keratins can be found in [16] and [170,171].

Keratins are of special interest for biophysics because

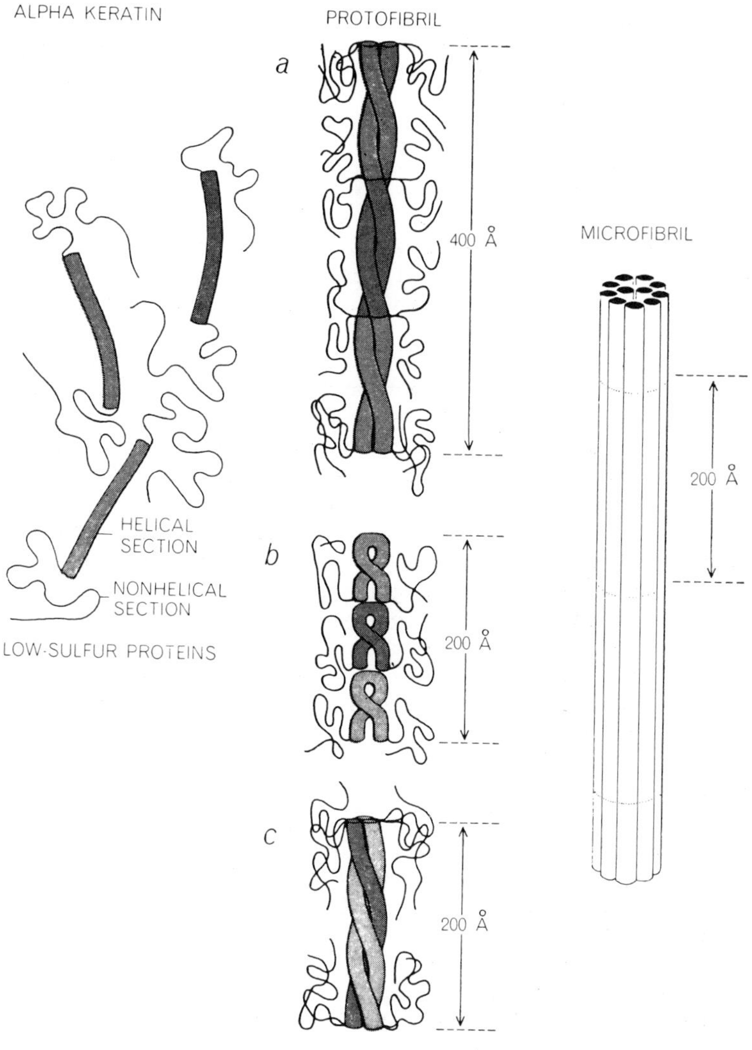


*FIG. 4.28 Model of α-keratin.*
From "Keratins" by R.D.B. Fraser. 

they are good models of complicated supramolecular oriented structures. The problem of the relation of this structure and its properties to the primary structures of corresponding proteins is far from being solved.

We have to mention another fibrillar protein--silk fibroin. Fibroin has a relatively simple amino acid composition and in this respect is more similar to collagen than to keratin. It contains 42.8 weight % Gly, 33.5 Ala, 16.3 Ser, 11.9 Tyr. Other amino acids are present only in small amounts or are absent (Cys, Met). X-ray diffraction patterns of fibroin are similar to that of β-keratin; the main conformation of fibroin is the β form.

The structure of fibroin is similar to a polypeptide $(\text{Gly-Ala})_n$, (silk of *Bombyx mori*) and polyalanine (silk of *Lyssax*). The sequences Gly-Ala-Ser-Tyr are typical for silk fibroin; large regions contain the turning pairs Gly-Ala.

Thus, the specialized structural functions of fibrillar proteins are attributable to their specific oriented structure. The biological role of these proteins is not only a protective one, as in the cases of keratin and fibroin; collagen is also necessary for osteogenesis, and myosin is enzymically active--it catalyzes the hydrolysis of ATP.

A discussion of the properties of myosin, as well as of another fibrillar protein of muscles--F-actin, will be given in the general examination of mechanochemical processes in "General Biophysics."

## References

1. I. Lifshitz, J. Exp. Theor. Phys. 55, 2408 (1968) (R).
2. P. Flory, "Statistical Mechanics of Chain Molecules." Wiley (Interscience), New York, 1969.
3. V. Sasisekharan, in "Collagen" (N. Ramanathan, ed.). Wiley (Interscience), New York, 1962.
4. H. Scheraga, Advan. Phys. Org. Chem. 6, 103 (1968).
5. L. Pauling, R. Corey, and H. Branson, Proc. Nat. Acad. Sci. U.S. 37, 205 (1951).
6. L. Pauling and R. Corey, Proc. Nat. Acad. Sci. U.S. 37, 241, 729 (1951).
7. F. Crick and A. Rich, Nature (London) 176, 780 (1955).
8. A. Rich, Biophysical science. A study program, Rev. Mod. Phys. 31, N 1 (1959).
9. G. Ramachandran (ed.), "Conformations of Biopolymers," Vol. I, Academic Press, New York, 1967.
10. D. Brant and P. Flory, J. Amer. Chem. Soc. 87, 663, 2791 (1965).
11. G. Ramachandran, C. Ramakrishnan, and V. Sasisekharan, J. Mol. Biol. 7, 95 (1963).

12. C. Ramakrishnan and G. Ramachandran, Biophys. J. 5, 909 (1965).
13. S. Leach, G. Nemethy, and H. Scheraga, Biopolymers 3, 591 (1965).
14. R. Scott and H. Scheraga, J. Chem. Phys. 45, 2091 (1966).
15. T. Ooi, R. Scott, G. Vanderkooi, and H. Scheraga, J. Chem. Phys. 46, 4410 (1967).
16. D. Poland and H. Scheraga, Biochemistry 6, 3791 (1967).
17. K. Gibson and H. Scheraga, Proc. Nat. Acad. Sci. U.S. 58, 420 (1967).
18. S. Lifson and I. Oppenheimer, J. Chem. Phys. 33, 109 (1960).
19. D. Brant, W. Miller, and P. Flory, J. Mol. Biol. 23, 47 (1967).
20. G. Ramachandran and V. Sasisekharan, Advan. Protein Chem. 23, 283 (1968).
21. H. Hellmann, "Quantum Chemistry." Gostekhizdat, Moscow, 1937 (R).
22. R. Feynman, Phys. Rev. 56, 340 (1939).
23. J. Hirshfelder, C. Curtiss, and R. Bird, "Molecular Theory of Gases and Liquids." Wiley, New York and Chapman and Hall, London, 1954.
24. F. London, Trans. Faraday Soc. 33, 8 (1937).
25. H. Margenau, Rev. Mod. Phys. II, I (1939).
26. J. Slater and J. Kirkwood, Phys. Rev. 37, 682 (1931).
27. A. Kitaigorodsky, "Organic Crystallochemistry." Ed. Acad. Sci. USSR, Moscow, 1955 (R).
28. W. Latimer and W. Rodebush, J. Amer. Chem. Soc. 42, 1419 (1920).
29. G. Pimentel and A. Mc Clellan, "Hydrogen Bond." Freeman, San Francisco, Calfiornia, 1960.
30. B. Stepanov, J. Phys. Chem. 19, 507 (1945); 20, 407 (1946) (R).
31. M. Volkenstein, M. Eliashevitch, and B. Stepanov, J. Phys. Chem. 24, 1158 (1950) (R).
32. N. Sokolov, Usp. Phys. Nauk 57, 205 (1955) (R).
33. G. Bradley and M. Kasha, J. Amer. Chem. Soc. 77, 4462 (1955).
34. C. Coulson, Research 10, 149 (1957).
35. E. Wollen, W. Davidson, and C. Shull, Phys. Rev. 75, 1348 (1949).
36. J. Bernal and R. Fowler, J. Chem. Phys. I, 515 (1933).
37. M. Magat, Ann. Phys. Paris 6, 108 (1936).
38. C. Coulson and U. Danielsson, Ark. Phys. 8, 239, 245 (1954).
39. H. Tsubomura, Bull. Chem. Soc. Japan 27, 445 (1954).
40. E. Verwey, Rec. Trav. Chim. Pays-Bas 60, 887 (1941).

41. D. Eisenberg and W. Kautzmann, "The Structure and Properties of Water." Oxford Univ. Press (Clarendon), London and New York, 1969.
42. L. Pauling, J. Amer. Chem. Soc. 57, 2680 (1935).
43. S. Peterson and H. Levy, Acta Crystallogr. 10, 70 (1957).
44. L. Onsager and M. Dupuis, in "Rendiconti della Sculoa Internazionale di Fisica Enrico Fermi," Corso X. Termodinamika dei Processi Irreversible, Bologna, 1960.
45. J. Nagle, J. Math. Phys. 7, 1484 (1966).
46. A. Narten, M. Danford, and H. Levy, Discuss. Faraday Soc. **43**, 97 (1967).
47. G. Nemethy and H. Scheraga, J. Chem. Phys. 36, 3382, 3401 (1962).
48. O. Samoilov, "Structure of Aqueous Solutions of Electrolytes and Hydratation of Ions." Ed. Acad. Sci. USSR, Moscow, 1957 (R).
49. J. People, Proc. Roy. Soc. **A205**, 163 (1951).
50. J. Bernal, Proc. Roy. Soc. **A280**, 299 (1964).
51. G. Walrafen, J. Chem. Phys. **44**, 1546 (1966); **48**, 244 (1968).
52. G. Nemethy and H. Scheraga, J. Chem. Phys. **41**, 680 (1964).
53. G. Oster and J. Kirkwood, J. Chem. Phys. II, 175 (1943).
54. M. Eigen, Angew. Chem. (Int. Ed.) 3, I (1964).
55. P. Doty, Biophysical science. A study program, Rev. Mod. Phys. **31**, N 1, (1959).
56. P. Doty, Collect. Czech. Commun. Spec. Issue 22, 5 (1957).
57. B. Zimm and J. Bragg, J. Chem. Phys. 28, 1246 (1958); **31**, 546 (1959).
58. J. Gibbs and E. Di Marzio, J. Chem. Phys. 28, I (1958); 30, 271 (1959).
59. S. Rice, A. Wada, and E. Geiduschek, Discuss. Faraday Soc. 25, 130 (1958).
60. T. Hill, J. Chem. Phys. 30, 383 (1959).
61. L. Peller, J. Phys. Chem. 63, 1194 (1959).
62. T. Birshtein and O. Ptitsyn, "Conformations of Macromolecules." Wiley (Interscience), New York, 1966.
63. D. Poland and H. Scheraga, "Theory of Helix-Coil Transition in Biopolymers." Academic Press, New York, 1970.
64. B. Zimm and J. Bragg, J. Chem. Phys. **31**, 526 (1959).
65. S. Lifson and A. Roig, J. Chem. Phys. **34**, 1963 (1961).
66. B. Zimm, P. Doty, and K. Iso, Proc. Nat. Acad. Sci. U.S. **45**, 1601 (1966).
67. B. Zimm and S. Rice, Mol. Phys. 3, 391 (1960).
68. M. Volkenstein and S. Fishman, Biophysica II, 956 (1966); 12, 14 (1967) (R).
69. S. Lowey, J. Biol. Chem. 240, 2421 (1965).

70. P. Doty, K. Imahori, and E. Klemperer, Proc. Nat. Acad. Sci. U.S. **44**, 424 (1958).
71. E. Blout and M. Idelson, J. Amer. Chem. Soc. 80, 4909 (1958).
72. L. Landau and E. Lifshitz, "Statistical Physics," 2nd ed. Addison-Wesley, Reading, Massachusetts, 1969.
73. T. Birshtein, A. Eliashevitch, and A. Skvortsov, Mol. Biol. 5, 204 (1971) (R).
74. E. Anufrieva *et al.*, Biophysica 10, 918 (1965) (R); J. Polymer Sci. N 16, 3533 (1968).
75. P. Flory, Proc. Roy. Soc. **A234**, 60 (1956).
76. A. Eliashevitch and A. Skvortsov, Mol. Biol. 5, 204 (1971) (R).
77. J. Bernal, "Origin of Life," Universe, New York, 1967.
78. J. Yang, Advan. Protein Chem. **16**, 323 (1961).
79. J. Klotz, Brookhaven Symp. Theoret. Biol. 13, 25 (1960).
80. J. Klotz and J. Franklin, J. Amer. Chem. Soc. **84**, 3461 (1962).
81. O. Ptitsyn and A. Skvortsov, Biophysica **10**, 909 (1965) (R).
82. S. Bresler and D. Talmud, Dokl. Acad. Sci. USSR **43**, 326, 367 (1944) (R).
83. S. Bresler, Biochimia **14**, 180 (1949) (R).
84. P. Rehbinder, "Surface Active Substances." Ed. Acad. Sci. USSR, Moscow, 1961 (R).
85. V. Belitzer and E. Khodorova, Ukr. Biochem. J. **21**, 34 (1949) (R).
86. A. Kurono and K. Hamaguchi, J. Biochem. **56**, 432 (1962).
87. W. Kauzmann, in "The Mechanism of Enzyme Action" (W. McElroy and B. Glass eds.) Greenwood, Westport, Connecticut, 1954; Advan. Protein Chem. **14**, I (1959).
88. E. Goddard, C. Hoeve, and G. Benson, J. Phys. Chem. **61**, 593 (1957).
89. J. Butler, Trans. Faraday Soc. **33**, 235 (1937).
90. H. Frank and M. Evans, J. Chem. Phys. **13**, 507 (1945).
91. F. Johnson, H. Eyring, and M. Polissar, "The Kinetic Basis of Molecular Biology." Wiley, New York, 1954.
92. B. Jacobson, W. Anderson, and J. Arnold, Nature (London) **173**, 772 (1954).
93. A. Szent-Györgyi, "Bioenergetics." Academic Press, New York, 1957.
94. M. Klotz, Science **128**, 815 (1958).
95. H. Scheraga, J. Phys. Chem. **65**, 1071 (1961); Ber. Bunsenges. Phys. Chem. **68**, 838 (1964).
96. G. Nemethy and H. Scheraga, J. Phys. Chem. **66**, 1773 (1962).
97. H. Scheraga, G. Nemethy, and I. Steinberg, J. Biol. Chem. **237**, 2506 (1962).

98. I. Steinberg and H. Scheraga, J. Biol. Chem. 238, 172 (1963).
99. D. Poland and H. Scheraga, Biopolymers 3, 283, 305 (1965).
100. M. Bixon, H. Scheraga, and S. Lifson, Biopolymers 1, 419 (1963).
101. O. Ptitsyn, Usp. Sovrem. Biol. 63, 3 (1967) (R).
102. G. Nemethy, I. Steinberg, and H. Scheraga, Biopolymers 1, 43 (1963).
103. H. Fisher, Proc. Nat. Acad. Sci. U.S. 51, 1285 (1964).
104. H. Fisher, Bichim. Biophys. Acta 109, 544 (1965).
105. C. Tanford, J. Amer. Chem. Soc. 84, 4240 (1962); 86, 2050 (1964).
106. P. Dunnill, Biophys. J. 8, 865 (1968).
107. J. Brandts, in "Structure and Stability of Biological Macromolecules" (S. Timasheff and G. Fasman, eds.) M. Dekker, New York, 1969.
108. R. Dickerson, in "The Proteins" (H. Neurath, ed.), Vol. 2, 2nd ed. Academic Press, New York, 1964.
109. J. Kendrew, H. Watson, B. Strandberg, R. Dickerson, D. Phillips, and V. Shore, Nature (London) 190, 663 (1961).
110. M. Perutz, J. Mol. Biol. 13, 646 (1965).
111. M. Perutz, J. Kendrew, and H. Watson, J. Mol. Biol. 13, 669 (1965).
112. H. Muirhead, J. Cox, L. Mazzarella, and M. Perutz, J. Mol. Biol. 28, 117 (1967).
113. W. Bolton, J. Cox, and M. Perutz, J. Mol. Biol. 33, 283 (1968).
114. J. Kendrew, Brookhaven Symp. Theor. Biol. 15, 216 (1962); Ber. Bunsenges. Phys. Chem. 68, 721 (1964).
115. G. Fasman, in "Polyamino Acids, Polypeptides and Proteins" (M. Stahmann, ed.) Univ. Wisconsin Press, Madison, Wisconsin, 1962.
116. G. Fasman, E. Bodenheimer, and C. Lidblow, Biochemistry 3, 155, 1665 (1964).
117. H. Sage and G. Fasman, Biochemistry 5, 286 (1966).
118. R. Kulkarni and E. Blout, J. Amer. Chem. Soc. 84, 3871 (1962).
119. N. Lotan, A. Yaron, and A. Berger, Biopolymers 4, 365 (1966).
120. W. Gratzer and P. Doty, J. Amer. Chem. Soc. 85, 1193 (1963).
121. K. Gibson and H. Scheraga, Physiol. Chem. and Phys. I, 109 (1969).
122. J. Schellman and C. Schellman, in "The Proteins" (H. Neurath, ed.), Vol. 2, Academic Press, New York, 1964.
123. K. Gibson and H. Scheraga, Proc. Nat. Açad. Sci. U.S. 58, 420, 1317 (1967).

124. N. Go, M. Go, and H. Scheraga, Proc. Nat. Acad. Sci. U.S. 59, 1030 (1968).
125. J. Brandts, J. Amer. Chem. Soc. 86, 4291 (1964); 87, 2759 (1965).
126. J. Brandts and L. Hunt, J. Amer. Chem. Soc. 89, 4826 (1967).
127. J. Schellmann, C.R. Trav. Lab. Carlsberg. Ser. Chim. 29, 223, 230 (1955); J. Phys. Chem. 62, 1485 (1958).
128. P. Privalov, Biophysica 8, 3 (1965); 15, 206 (1970) (R).
129. L. Riddiford, J. Biol. Chem. 241, 2792 (1966).
130. G. Tanford, Advan. Protein Chem. 23, 121 (1968); 24, 1 (1970).
131. G. Tanford, J. Amer. Chem. Soc. 86, 2050 (1964).
132. M. Joly, "A Physico-Chemical Approach to the Denaturation of Proteins." Academic Press, New York, 1965.
133. H. Scheraga, in "Polyamino Acids, Polypeptides and Proteins" (M. Stahmann, ed.) Univ. Wisconsin Press, Madison, Wisconsin, 1962.
134. J. Schellmann, C. R. Trav. Lab. Carlsberg Ser. Chim. 30, 450 (1958).
135. T. Takagi and T. Isemura, J. Biochem. 52, 314 (1962).
136. A. Imanishi, K. Kekiuchi, and T. Isemura, J. Biochem. 54, 89 (1963).
137. C. Anfinsen, Symp. 4, Int. Congr. Biochem., 5th.
138. E. Haber and Ch. Anfinsen, J. Biol. Chem. 237, 2175 (1961).
139. Ch. Epstein and Ch. Anfinsen, J. Biol. Chem. 237, 2175 (1962).
140. K. Imai, T. Takagi, and T. Isemura, J. Biochem. 53, I (1963).
141. A. Guzzo, Biophys. J. 5, 809 (1965).
142. M. Volkenstein, "Enzyme Physics." Plenum Press, New York, 1969.
143. J. Prothero, Biophys. J. 3, 367 (1966).
144. M. Schiffer and A. Edmunson, Biophysics 7, 121 (1967).
145. D. Phillips, Proc. Nat. Acad. Sci. U.S. 57, 484 (1967).
146. P. Periti, G. Quagliarotti, and A. Liquori, J. Mol. Biol. 24, 313 (1967).
147. O. Ptitsyn, Mol. Biol. 3, 627 (1969) (R); J. Mol. Biol. 42, 501 (1969); Usp. Sovrem. Biol. 69, 26 (1970) (R).
148. O. Ptitsyn and A. Finkelstein, Dokl. Akad. Sci. USSR 195, (1970); Biophysika 15, 757 (1970) (R).
149. D. Kotelchuk and H. Scheraga, Proc. Nat. Acad. Sci. U.S. 61, 1163 (1968); 62, 14 (1969); 63, 615 (1969); 65, 810 (1970).
150. R. Pain and B. Robson, Nature (London) 227, 62 (1970).
151. V. Lim, Dokl. Akad. Sci. USSR 203, 480 (1972) (R).
152. B. Robson and R. Pain, J. Mol. Biol. 58, 237 (1971).
153. N. Esipova and V. Tumanian, Mol. Biol. 6, 840 (1972)(R).

154. W. Grassmann *et al.*, Z. Physiol. Chem. **323**, 48 (1961).
155. G. Ramachandran and G. Kartha, Proc. Indian Acad. Sci. **42**, 215 (1955); Nature (London) **174**, 269 (1954); **176**, 593 (1955); **177**, 710 (1956).
156. G. Ramachandran and V. Sasisekharan, Biochim. Biophys. Acta **109**, 314 (1965).
157. A. Rich and F. Crick, J. Mol. Biol. 3, 483 (1961).
158. M. Millionova and N. Andreeeva, Biophysica. 3, 259 (1958); **4**, 374 (1959) (R).
159. N. Andreeva, V. Debabov, M. Millionova, V. Shibnev, and J. Chirgadze, Biophysica. 6, 244 (1961) (R).
160. A. Miller and J. Wray, Nature (London) 230, 437 (1971).
161. V. Tumanian, Mol. Biol. 5, 499 (1971) (R).
162. N. Esipova and J. Chirgadze, in "State and Role of Water in Biological Systems." Nauka, Moscow, 1967 (R).
163. P. Privalov and E. Tiktopulo, Biopolymers 9, 127 (1970).
164. H. Berendsen and C. Michelsen, Fed. Proc. 25, 998 (1966); Ann. N.Y. Acad. Sci. **125**, 365 (1965).
165. G. Ramachandran and R. Chandrasekharan, Biopolymers 6, 1649 (1968).
166. R. Rigby, in Symp. Fibrous Proteins, Australia, 219 (1967).
167. W. Harrington and P. von Hippel, Advan. Protein Chem. **16**, 1 (1961).
168. W. Astbury, Advan. Enzymol. 3, 63 (1943).
169. R. Fraser, Sci. Amer. 221(2), 87 (1969).
170. H. Lundgren and W. Ward, in "Ultrastructure of Protein Fibers" (R. Borasky, ed.) Academic Press, New York, 1963.
171. W. Crewther, R. Fraser, F. Lennok, and H. Lindley, Advan. Protein Chem. 20, 191 (1965).
172. Y. Syrnikov, J. Struct. Chem. 7, 15, 665 (1966); in "Structure and Role of Water in Living Organism," Vol. 3, Ed. by Leningrad Univ., 1970 (R).
173. A. Rahman and F. Stillinger, J. Chem. Phys. 55, 3336 (1971).
174. G. Sarkisov and V. Dashevsky, J. Struct. Chem. 13, 199 (1972) (R).
175. G. Sarkisov, M. Dakhis, G. Malenkov, and V. Dashevsky, Dokl. Acad. Sci. USSR 205, 638 (1972) (R).
176. G. Sarkisov, G. Malenkov, and V. Dashevsky, J. Struct. Chem. **14**, 3 (1973) (R).
177. V. Dashevsky, "Conformations of Organic Molecules." Chimia, Moscow, 1974 (R).
178. J. Lifshitz and A. Grossberg, J. Exp. Theor. Phys. 65, 2403 (1973) (R).
179. D. Denisov, L. Drosdov-Tikhomirov, and D. Grigorieva, J. Theor. Biol. **41**, 431 (1973).
180. S. Galaktionov, The Study of the Spatial Molecular Structure of Proteins and Peptides. Thesis, Minsk, 1973 (R).

181. O. Ptitsyn, Dokl. Acad. Sci. USSR **210**, 1213 (1973); Vestn. Acad. Sci. USSR N 5, 57 (1973) (R).

182. O. Ptitsyn and A. Rashin, A Model of Self-Organization of a Myoglobin Molecule, Preprint, Acad. Sci. USSR, Inst. of Protein Res., Poustchino-on-Oka (1973).

# Chapter 5

# X-Ray Analysis, Optics, and Spectroscopy of Biopolymers

## 5.1 X-Ray Structural Analysis

Contemporary natural science utilizes two important techniques to investigate the structure of matter. These techniques are chemistry and optics in the broad sense of the word, i.e., studies of interaction of matter with light in the entire accessible range of wavelengths--from X rays to radio waves. Chemistry deciphers the primary structure of protein chains and also the structure of the functional sites of protein globules, particularly the active sites of enzymes (Chapter 6). However, chemistry (biochemistry) cannot determine the spatial structure of a protein or a nucleic acid. X-ray analysis, on the other hand, gives direct information about the structure of matter, i.e., about the positions of atoms in molecules.

X rays, electromagnetic waves with lengths of the order of 1 Å, are scattered by the electron shells of atoms. The interference of waves scattered by matter results in a definite diffraction pattern. Scattering by a crystal can be treated as a "reflection" of X rays at the planes of the crystalline lattice. Reflection occurs if the scattered waves are in phase, that is, if the path difference is equal to an integral number n of waves. If the distance between the planes is d, then the diffraction (reflection) is given by the well-known Bragg expression

$$n\lambda = 2d \sin \theta \qquad (5.1)$$

where $\theta$ is the angle between the incident ray and reflecting crystal plane.

Diffraction of X rays by crystals can be observed because the lattice's interatomic distances d have the same order of magnitude (1-4 Å) as the wavelengths $\lambda$ (frequently Cu $K_\alpha$ radiation which has $\lambda$ = 1.54 Å is used).

Assume that the scattering system consists of two centers

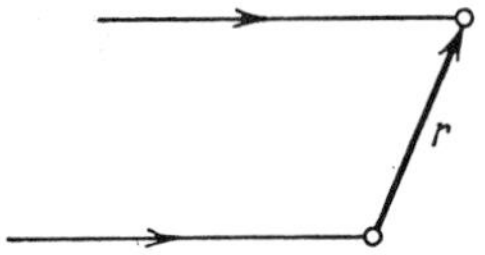


*FIG. 5.1 A two-center scattering system.*

(atoms) a distance r from each other (Fig. 5.1). The incident plane wave excites the electrons of the centers and they become the sources of secondary spherical waves (cf. p. 141). The electrical field of the incident wave at point r is

$$E = A \exp[i(\mathbf{k}\mathbf{r}+\phi)] \tag{5.2}$$

where A is the wave amplitude, $\phi$ is the phase, and $\mathbf{k}$ is the wave vector, which has the direction of the ray

$$|\mathbf{k}| = 2\pi/\lambda$$

Let us find the path difference of waves scattered by two centers as shown in Fig. 5.1. Placing the origin of the coordinate system at one of the centers, we get

$$\mathbf{k}\mathbf{r}-\mathbf{k}_0\mathbf{r} = \mathbf{s}\mathbf{r} \tag{5.3}$$

The wavelength $\lambda$ is not changed and vector $\mathbf{s} = \mathbf{k}-\mathbf{k}_0$ is not zero because the directions of vectors $\mathbf{k}$ and $\mathbf{k}_0$ are different. If the incident wave has amplitude A, then the scattering center emits the wave

$$E' = Af \exp[i(\mathbf{s}\mathbf{r})] \tag{5.4}$$

where f is the "scattering strength" of the center.

Vector $\mathbf{s}$ is normal to the reflecting plane. It follows from Fig. 5.1 that

$$s = (4\pi \sin \theta)/\lambda \tag{5.5}$$

Diffraction by a system of N scattering centers is expressed as a sum of expressions like (5.4). The quantity

$$F(\mathbf{s}) = \sum_{j=1}^{N} f_j \exp[i(\mathbf{s}\mathbf{r}_j)] \tag{5.6}$$

is called the scattering amplitude. If the scattering centers are continuously distributed, we have to substitute for sum (5.6) an integral of the Fourier type

$$F(\mathbf{s}) = \int \rho(\mathbf{r}) \exp[i(\mathbf{s}\mathbf{r})]\, dv_r \tag{5.7}$$

where $\rho(\mathbf{r})$ is the time-averaged electron density of the system, and $dv_r$ is the element of scattering volume.

If the system is a single spherically symmetrical atom, integral (5.7) assumes the form

$$f(s) = \int_0^\infty 4\pi r^2 \rho(r) \frac{\sin sr}{sr} dr \tag{5.8}$$

This expression for the scattering strength of an atom depends directly on the electron density distribution $\rho(r)$. The quantity $f(s)$ is called the atomic factor. Its values for all atoms at various s have been tabulated. If $s \to 0$, that is, at zero scattering angle

$$f(0) = \int_0^\infty 4\pi r^2 \rho(r)\, dr = N \tag{5.9}$$

where N is the number of electrons in the atom. The higher the atomic number, the stronger is the scattering of X rays; hydrogen has the lowest number.

The richest diffraction pattern, and consequently the most detailed information about interatomic distances, is obtained in the study of crystals. The positions of the atoms in a crystal are periodic; the electron density $\rho(x,y,z)$ is a periodic function of the coordinates x,y,z. The Fourier integral, which resembles (5.7), has the form

$$F_{hk\ell} = \frac{1}{abc} \int_0^a \int_0^b \int_0^c \rho(x,y,z) \exp[2\pi i(\frac{hx}{a} + \frac{ky}{b} + \frac{\ell z}{c})]\, dx\, dy\, dz$$
$$= \frac{1}{V_0} \int \rho(\mathbf{r}) \exp[2\pi i(\mathbf{Hr})]\, dv_r \tag{5.10}$$

Here a, b, c are the periods of the crystalline lattice, that is, the edge lengths of the elementary cells, and $V_0$ is the cell volume. The vector $H = s/2\pi$ has the components h/a, k/b, $\ell$/c, where h, k, $\ell$ are integers. If scattered by a periodic structure, the amplitude is nonzero only at these values of components of the H vector. For the component of the path difference vector s in the x direction, we get, according to (5.5),

$$2\pi(h/a) = (4\pi \sin\theta)/\lambda$$

and from condition (5.1),

$$h\lambda = 2a \sin\theta$$

In general, if vector **r** is equal to one of three axial vectors **a**, **b**, **c**, we obtain three Laue conditions:

$$H_{hk\ell}(\mathbf{a}/h) = 1, \quad H_{hk\ell}(\mathbf{b}/k) = 1, \quad H_{hk\ell}(\mathbf{c}/\ell) = 1 \tag{5.11}$$

Vectors **a**, **b**, **c** define the crystalline lattice. Integers h, k, $\ell$, the Miller indices, define all possible crystal planes "reflecting" X rays. The corresponding interplane distances

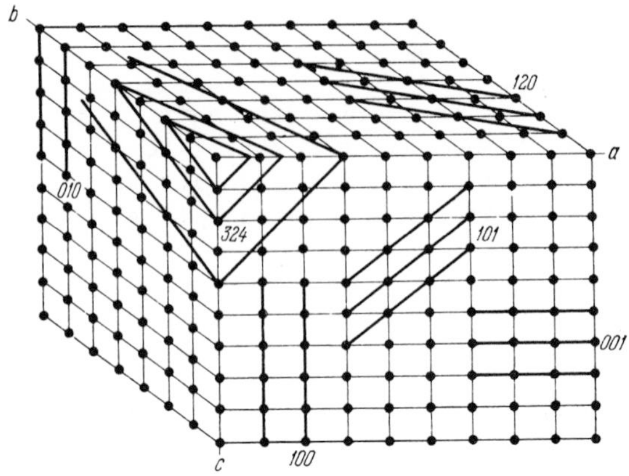


*FIG. 5.2 Cross sections of reflecting planes in an orthogonal crystalline lattice.*

$d_{hk\ell}$ depend on these indices. Figure 5.2 shows the cross sections of various systems of reflecting planes in an orthogonal crystalline lattice.

The dimension of the components of vectors $H_{hk\ell}$ is length$^{-1}$. These vectors are given in reciprocal space. The distribution of points where the scattering amplitude is nonzero and equal to $F_{hk\ell}$ is periodic in this reciprocal space and forms the reciprocal lattice, which can be used for calculations. Every point of the reciprocal lattice $hk\ell$ is characterized by a vector

$$H_{hk\ell} = ha^*+kb^*+\ell c^* \tag{5.12}$$

Vectors **a***, **b***, **c*** obey the conditions

$$\mathbf{a^*b} = \mathbf{a^*c} = \mathbf{b^*a} = \mathbf{b^*c} = \mathbf{c^*a} = \mathbf{c^*b} = 0 \tag{5.13}$$

and

$$\mathbf{a^*a} = \mathbf{b^*b} = \mathbf{c^*c} = 1 \tag{5.14}$$

Laue conditions (5.11) follow directly from (5.12)-(5.14). For an orthogonal lattice we get

$$a^* = a^{-1}, \quad b^* = b^{-1}, \quad c^* = c^{-1}$$

From these expressions we get

$$|H_{hk\ell}| = 1/d_{hk\ell}$$

and the direction of vector $H_{hk\ell}$ is perpendicular to the crystal planes described by Miller indices h, k, $\ell$. A geometric interpretation of reflection conditions (5.11) is obtained by forming a reciprocal lattice and drawing the sphere of radius $\lambda^{-1}$

*FIG. 5.3 Ewald's sphere.*

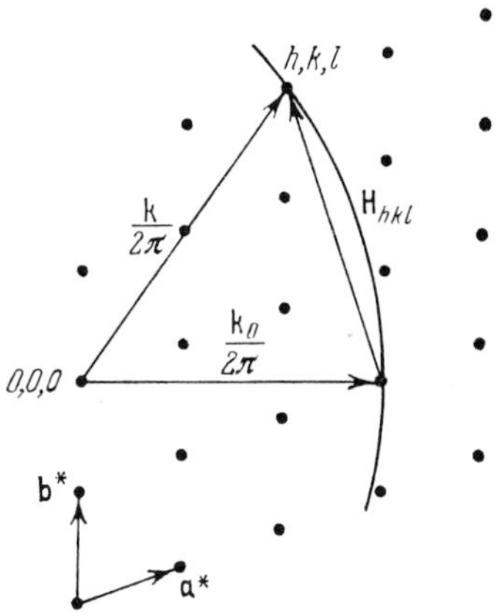


in such a way that this sphere (called the reflection sphere or Ewald's sphere) intersects the origin of coordinates h, k, ℓ = 0, 0, 0, and the radius directed from the center of the sphere to the coordinate origin coincides with the incident ray direction $\mathbf{k}_0$ (Fig. 5.3). If one of the points h, k, ℓ of the reciprocal lattice appears on the sphere surface, then the vector connecting the points 0, 0, 0 and h, k, ℓ will be vector $H_{hk\ell}$. According to (5.3) the vector directed from the sphere center to the point h, k, ℓ will be the vector of the "reflected" ray **k**. Coordinates of the point h, k, ℓ give the indices of the reflecting plane. Hence only points on the sphere surface obey the diffraction conditions.

The points of the reciprocal lattice of conventional low molecular-weight crystals are far apart because the periods of lattice a, b, c are small, and consequently the periods a*, b*, c* are large. The increase in the number of intersections of Ewald's sphere with the lattice points requires the use of various wavelengths λ, or the crystal must be rotated or vibrated (methods of rotating or vibrating crystal). Periods in protein monocrystals are large; therefore, the probability of intersections of Ewald's sphere by the reciprocal lattice points is great. The fixed protein monocrystal gives a rich diffraction pattern in monochromatic X rays. Figure 5.4 shows an X-ray diffraction pattern of sperm whale myoglobin; this picture was obtained photographically, but ionization methods give more precise results. The diffraction pattern is a reproduction of the reciprocal crystal lattice. Reflected rays are directed along the generator of a cone that has its apex at the center of the Ewald sphere. These rays intersect the photographic film by second-order curves (Fig. 5.5). To obtain undistorted pictures, the crystal and the film are moved synchronously in such a way that the film plane remains parallel to the corresponding plane of the reciprocal lattice.

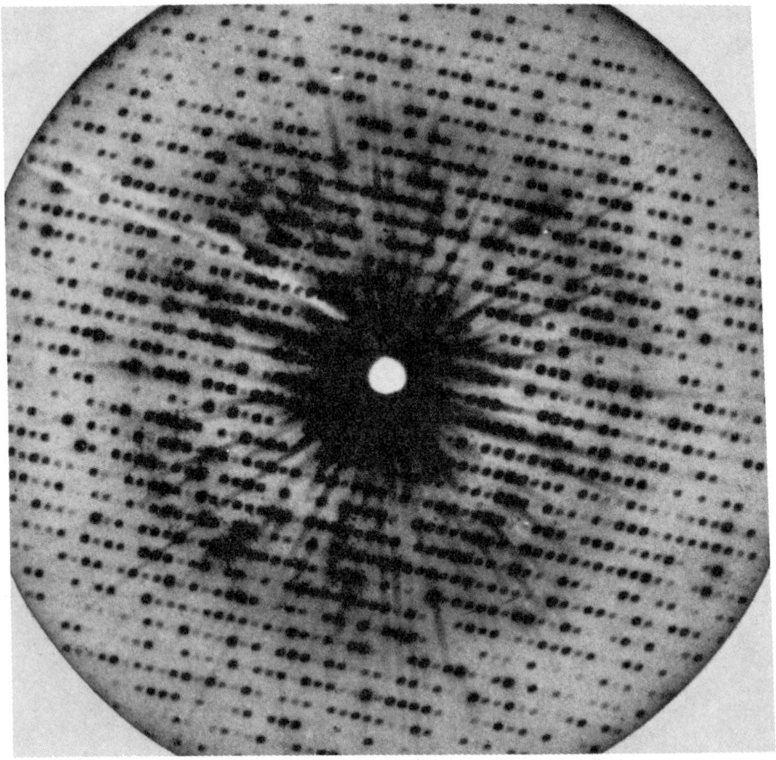

FIG. 5.4 X-ray diagram of sperm whale myoglobin.

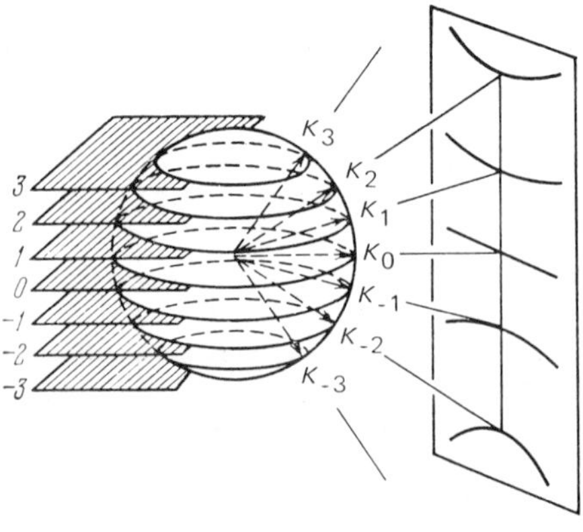


FIG. 5.5 Scheme of the formation of an X-ray diagram.

Analysis of its diffraction pattern establishes the electron density distribution in a crystal. Integral (5.7), expressing the scattering amplitude, can be rewritten in the form

$$F(s_x s_y s_z) = \iiint_{-\infty}^{+\infty} \rho(xyz)\exp[i(s_x x+s_y y+s_z z)]dx\ dy\ dz \tag{5.15}$$

Knowing the function $F(s_x s_y s_z)$, we can calculate values of $\rho(xyz)$ by means of the inverse Fourier transform

$$\rho(xyz) = \iiint_{-\infty}^{+\infty} F(s_x s_y s_z)\exp[-i(s_x x+s_y y+s_z z)]ds_x\ ds_y\ ds_z \tag{5.16}$$

or

$$\rho(r) = \int F(s)\exp[-i(s/r)]dv_s \tag{5.17}$$

If $\rho$ is a periodic function, representing a crystal, the integral (5.10) or the corresponding sum can be transformed in a similar way

$$\rho(xyz) = \frac{1}{V}\sum_h \sum_k \sum_\ell F_{hk\ell} \exp[-2\pi i(\frac{hx}{a} + \frac{ky}{b} + \frac{\ell z}{c})] \tag{5.18}$$

Experiment gives the moduli of the structural amplitudes F but not their phases. Actually F(s) is a complex quantity

$$F = |F|e^{i\phi}$$

Without knowledge of the phases, the structure of the object cannot be determined. As Perutz wrote [1], the X-ray pattern of a crystal is a hieroglyphic to which we lack the key. The method for determining phases developed by Perutz in his investigations of proteins consists in the binding of heavy atoms, such as mercury atoms, by the crystal-forming substance. For example, a heavy atom, possessing large scattering strength, produces considerable alteration in the intensity of diffraction spots. Phase can be determined from the difference between the amplitude noted in the presence of the heavy atom and that noted in its absence. The phase problem can be solved using protein derivatives containing several heavy atoms. A necessary condition is the conservation of protein structure after the incorporation of heavy atoms. In other words, this is the method of isomorphous substitution: mercurial derivatives of proteins give crystals isomorphous to those of unsubstituted proteins (cf. [2]).

Experiment gives the scattering intensity

$$I(s) = F(s)F^*(s^*)$$

Therefore, the function that relates the diffraction pattern directly to the crystal structure is not (5.18) but

$$Q(r) = \frac{1}{V}\sum_h \sum_k \sum_\ell |F_{hk\ell}|^2 \cos 2\pi\left(\frac{hx}{a} + \frac{ky}{b} + \frac{\ell z}{c}\right) \quad (5.19)$$

This function (Patterson's function) is the Fourier transform of the scattered intensity. The vector connecting any two atoms in $\rho(r)$ (5.18) is represented in function $Q(r)$ by the vector of the same orientation, beginning at the coordinate origin. Patterson's function is centrosymmetrical, which means that every vector $\mathbf{r}_{ij}$ corresponds to a vector $\mathbf{r}_{ji}$ of the same value but with opposite direction. Patterson's function has maxima if

$$x = x_i - x_j, \quad y = y_i - y_j, \quad z = z_i - z_j$$

where $x_i$, $y_i$, $z_i$ and $x_j$, $y_j$, $z_j$ are the coordinates of any two atoms in an elementary cell. The heights of the maxima are proportional to the product of atomic numbers $N_iN_j$ and, if the distances $r_{ij}$ are repeated in the structure, to the number of these repetitions. If atom i is a heavy one, then $N_i$ is large and the corresponding maxima are high. Therefore, Patterson's function is especially suitable for determinations of coordinates of heavy atoms in a crystal.

X-ray analysis ultimately allows the determination of the electron density distribution from observed intensities of diffraction maxima with the help of the Fourier series. Density functions $\rho(xyz)$ are represented in the form of "geodesic maps" on which lines are drawn through the points with equal $\rho$ values. These maps are projections of the three-dimensional density distribution on a plane and represent microscopic sections of molecules. Because of the small structural factor, hydrogen atoms are not directly visible on maps of electron density. Their positions can be determined by bends of isolines or with the help of special differential methods.

Before the 1940s X-ray analysis of comparatively simple substances confirmed their structures as determined by chemical methods and gave quantitative information concerning interatomic distances. In 1944 Dorothy Hodgkin deciphered for the first time the structure of penicillin, which had not been established by chemists. The penicillin molecule contains 23 atoms besides those of hydrogen. Later Hodgkin established the structure of vitamin $B_{12}$. In this case the coordinates of 93 atoms were determined. Thus X-ray analysis became firmly established as an independent research method, and was an important tool for the study of the most complicated molecules--proteins. The initiator of this field of molecular biophysics was Bernal and the greatest achievements in protein studies are those of the Cambridge scientific school. Bragg, Kendrew, and Perutz are outstanding innovators in this field. In 1957 Kendrew determined the spatial structure of the first protein, myoglobin (Section 4.7). The myoglobin molecule contains more than 2500 atoms.

Protein crystals contain large amounts of water and are investigated in the mother solution. Bernal and Hodgkin applied this method for the first time, and they obtained tens of thousands of clear diffraction reflections in X-ray patterns. The number of reflections can reach hundreds of thousands. Deciphering such complicated diffraction patterns is long and difficult work that can be performed only by means of computers. For an exact determination of the phase of every reflection it is necessary to measure its intensity several times for diffraction from both pure protein and its heavy-atom derivatives. The calculations involve tens of millions of figures.

Spatial density distribution can be visualized by superimposing a series of contour maps drawn on sheets of transparent plastic. Such a picture for myoglobin is shown in Fig. 5.6. The final result of this procedure is a spatial model of the protein

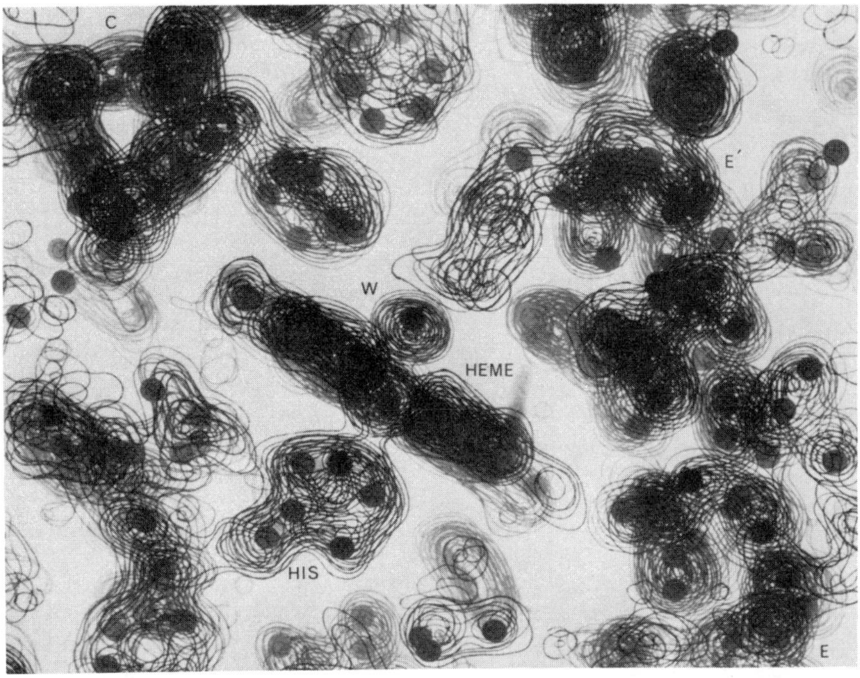


*FIG. 5.6 Ths spatial distribution of electron density in myoglobin.*

molecule in which the positions of all the atoms are established. Such a model is shown on page 217.

The principal question in the X-ray analysis of proteins is whether the molecular structure in a crystal coincides with that in the solution where the protein performs its functions. Evidently a negative answer to this question would invalidate the results of X-ray analysis and deprive them of biophysical meaning. If the protein globule is not an "aperiodic crystal" but a strongly fluctuating system, crystallization would mean the selection of one or of several conformations from among the large number of conformations in solution. But a globule has fixed structure. Crystalline proteins contain large amounts of water and are studied in the mother solution. The results of X-ray analysis of protein crystals agree with results of optical examination of the same protein in solution. In particular, the degrees of helicity determined by both methods coincide (see Section 5.7). What is more, the biological functionality--enzymic activity--of proteins is preserved in crystals because they are strongly hydrated. Up to this point there are no grounds to doubt a positive answer to the question. Proteins subjected to lyophilic drying, however, possess altered structure and their X-ray diffraction patterns are very poor.

Only a short account of the physical principles of the X-ray analysis of globular proteins is given here; a detailed presentation of theory and methods can be found in a series of monographs and papers [3-12]. Some dozens of globular proteins have already been investigated by X-ray analysis. A series of enzymes has been studied and the structures of complexes formed by enzymes with inhibitors and substrate analogs have been established (Chapter 6). Because it is a direct method of investigation, X-ray analysis yields rich and valuable information about protein structure, and leads to general conclusions concerning the structure and properties of these molecules. Table 5.1 presents a list of some of the proteins that have been investigated by means of X-ray analysis.

A number of other proteins and polypeptides have been investigated with lower resolution ($\gamma$- and $\delta$-chymotrypsin, carboanhydrase, pepsin, trypsin, rubredoxin, glucagon, oxytocin, $\gamma$-globulin, etc.). Beautiful complex supramolecular structures composed of catalase have also been studied [30]. Doubtless the number of protein molecules investigated will increase greatly in the near future.

Let us consider briefly two methods of structural analysis having much in common with the X-ray method. They are electronography and neutronography. According to the de Broglie expression

$$\lambda = h/mv$$

*TABLE 5.1*
*X-Ray Analysis of Proteins*

| Protein | Origin | Molecular weight | Maximal resolution (A) | Reference |
|---|---|---|---|---|
| Myoglobin | Sperm whale | 17800 | 1.4 | [13] |
| Oxyhemoglobin | Horse | 67000 | 2.8 | [14] |
| Deoxyhemoglobin | Man | 67000 | 2.8 | [15] |
| Erythrocruorin | | | 2.8 | [16] |
| Cytochrome c | Horse | 12400 | 2.7 | [17] |
| Lysozyme | Hen's egg | 14400 | 2.0 | [18] |
| Ribonuclease A | | | 2.0 | [19] |
| Ribonuclease | | 14000 | 2.0 | [20] |
| Carboxypeptidase A | Ox | 34300 | 2.0 | [21] |
| Chymotrypsinogen | | 25000 | 2.5 | [22] |
| α-Chymotrypsin | | 23000 | 2.0 | [23] |
| Papain | | 22000 | 2.8 | [24] |
| Nuclease | Staphylococcus aureus | 16000 | 2.0 | [25] |
| Rubredoxin | | 6000 | 2.5 | [26] |
| Lactate dehydrogenase | | 140000 | 2.8 | [27] |
| Insulin | Pig | 5733 | 2.8 | [28] |
| Subtilisin | | | 2.5 | [29] |

where $\lambda$ is the wavelength, h is Planck's constant, m is the particle mass, and v its velocity; the scattering of any particles by atoms can result in diffraction phenomena if the particles move with suitable velocity. Electrons with $\lambda$ of the order of 0.05 Å and neutrons with $\lambda$ of the order of 1-1.5 Å (i.e., of the same order as that of X rays) are used. The theory of diffraction of crystals in these cases is similar in principle to the one presented, but there are considerable methodological differences.

If the electron wavelength is small, the radius of Ewald's sphere is large (p. 259) and it degenerates into a plane. This phenomenon simplifies the theory of electron diffraction patterns since the pattern becomes a direct picture of the planar cross section of the reciprocal crystal lattice. Atomic factors for electron scattering are also proportional to atomic numbers but their absolute values are much greater than those for X rays. In other words, the interaction of electrons with matter is much stronger than that of X-ray quanta. Therefore, electrons are strongly absorbed by matter, and for structural studies we have to use very thin films, whose thickness is of the order of $10^{-5}$-$10^{-6}$ cm. The sizes of crystals studied by X rays are of the order of $10^{-1}$ cm. Investigations must be performed in a high vacuum, which makes it impossible to use electronography in the study of globular proteins in their native state (because the vacuum would dry the protein). However, electronography yields valuable results in the study of fibrillar protein structures, polymers, and other amorphous bodies. The essential advantage of electronography is that with this technique it is possible to locate hydrogen atoms. A detailed account of electronography is given in [31,32].

For neutronographic investigations, atomic reactors that give powerful beams of neutrons monochromatized by reflection from a crystal plate (e.g., $CaF_2$) have to be used. Diffracting neutrons, which are registered by counters, are scattered not by the electron shell of an atom but by its nucleus, and the atomic scattering factor is determined by the specific proton-neutron nuclear structure and not by the atomic number. Therefore, the atomic factors of isotopes differ considerably. The atomic scattering factor for hydrogen (proton) is not the smallest one; it is higher than that of a number of heavy elements. Hence, neutronography makes it possible to locate hydrogen atoms; with this technique the structure of ice was ascertained (p. 188). It would appear that neutronography will play an important role in future investigations of biopolymers. For further details see [33].

## 5.2 X-Ray Analysis of Fibrillar Structures

In many cases biopolymers form structures that are ordered and periodic in one dimension. Fibrillar proteins (e.g., collagen, (p. 240) and DNA are in this group. X-ray patterns of such structures are characterized by specific features.

If the periodicity of the chain polymer is determined by a vector c, then, similar to (5.15), the structural amplitude F(s) can be written as

$$F(s) = F[X,Y,(\ell/c)] \tag{5.20}$$

where X,Y are coordinates of reciprocal space. In other words, the structural amplitude is nonzero only at the "layer planes" of reciprocal space, where $Z = \ell/c$. We are concerned with the one-dimensional reciprocal lattice. Instead of (5.18) we get

$$F(s) = F[X,Y,(\ell/c)] = \sum_j f_j \exp i[x_jX+y_jY+z_j(\ell/c)] \tag{5.21}$$

The use of an Ewald sphere like that in Fig. 5.7 shows that the diffraction pattern represents layer lines. Their positions obey Bragg's condition for a one-dimensional structure (Fig. 5.8):

$$\ell\lambda = c \sin \phi \tag{5.22}$$

Figure 5.8 shows the X-ray diffraction pattern of the lithium salt of DNA in the so-called B form (p. 489). The layer lines

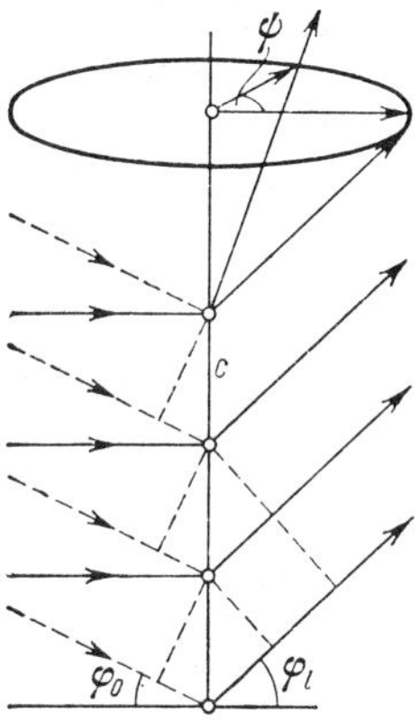


*FIG. 5.7 Formation of an X-ray diagram from a fibrillar structure.*

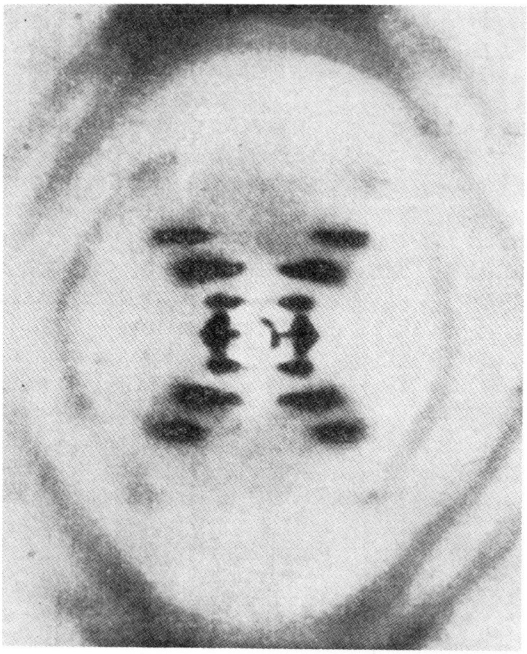

*FIG. 5.8 X-ray diagram of the Li salt of DNA.*

are well pronounced. Of course, it is not the pattern of a single Li-DNA molecule, but of its axial texture. Axial texture exists in a system with one-dimensional order where the axes of the molecules are approximately parallel to some definite direction, while the other axes are oriented randomly. Texture has one symmetry axis of the order of infinity.

One-dimensional periodic systems can be treated as a special case of ordered three-dimensional systems (crystals). The determination of the electron density distribution from the distribution of intensities in the diffraction pattern is made, as earlier, by means of Fourier synthesis. A detailed account of the theory is presented in the monograph of Vainstein [9]; here we limit ourselves to some fundamentals.

Instead of the Cartesian coordinates, it is convenient to use the cylindrical coordinates for one-dimensional systems, that is, radius r, polar angle $\psi$, and height z. The relationship is

$$x = r\cos\psi, \quad y = r\sin\psi, \quad z = z$$

and in the reciprocal space

$$X = R\cos\Psi, \quad Y = R\sin\Psi, \quad Z = Z$$

The Fourier integral (5.10) has the form, if these coordinates are used,

$$F(R,\psi,Z) = \int_0^\infty \int_0^{2\pi} \int_{-\infty}^{+\infty} \rho(r,\psi,z)\exp\{2\pi i[Rr\cos(\psi-\Psi)+zZ]\}r\,dr\,d\psi\,dz \tag{5.23}$$

and its Fourier transform is

$$\rho(r,\psi,z) = \int_0^\infty \int_0^{2\pi} \int_{-\infty}^{+\infty} F(R,\Psi,Z)\exp\{-2\pi i[Rr\cos(\psi-\Psi)+zZ]\}R\,dR\,d\Psi\,dZ \tag{5.24}$$

For one-dimensional periodic structures (chain molecules), the dependence of $\rho$ on r and $\psi$ can be separated. The two-dimensional density distribution in the plane perpendicular to Z can be calculated, as shown by theory, with the help of the Bessel functions. The most important models (continuous cylinder, hollow cylinder with walls of finite and infinitely small thickness, pillar structures) have been studied in detail [9].

If there is a periodicity along the z axis, the integral (5.23) becomes

$$F_\ell[R,\Psi,(\ell/c)] = \int_0^\infty \int_0^{2\pi} \int_0^c \rho_M(r,\psi,z)\{\exp 2\pi i[Rr\cos(\psi-\Psi)+z(\ell/c)]\}$$
$$\times\, r\,dr\,d\psi\,dz \tag{5.25}$$

Here $\rho_M$ is the density distribution in the elementary repeating grouping. Integration by z has to be performed from 0 to c, that is, in the limits of one period. $F_\ell$ is nonzero only at layer lines satisfying the condition $Z = \ell/c$.

Helical structures are especially important in biopolymer physics. For a continuous helix of radius $r_0$ and the repeat distance C, the following conditions are obeyed (in cylindrical coordinates).

$$r = r_0, \quad \psi = 2\pi z/C \tag{5.26}$$

In the simplest model the density along the helix can be taken as unity, that is,

$$\rho(r,\psi,z) = \delta(r-r_0)\ \delta[\psi-(2\pi z/C)]$$

where $\delta$ is the sign of the $\delta$ function. The integral (5.25) becomes

$$F_\ell[R,\Psi(\ell/c)] = \int_0^\infty \int_0^{2\pi} \int_0^c \exp\{2\pi i[Rr_0 \cos((2\pi z/C)-\psi)+(\ell z/C)]\} \times r\,dr\,d\psi\,dz \tag{5.27}$$

Evaluation of this integral gives

$$F_{\ell=n}[R,\Psi,(\ell/c)] = 2\pi r_0 I_n(2\pi R r_0)\exp\{in[\Psi+(\pi/2)]\} \tag{5.28}$$

where $I_n$ is the nth-order Bessel function. The modulus of this expression possesses cylindrical symmetry. The intensity distribution $|F|^2$ at the layer line of number $|\ell| = n$ is proportional to the square of the nth-order Bessel function. Since the radius $R_1$ of the first maximum of $I_n$ increases with increasing n, the intensity distribution has a crosslike form (Fig. 5.8). The origin of such an oblique cross can be easily understood if we examine the positions of the most densely populated planes of atoms in the helix. The largest values of scattered intensities correspond to perpendiculars to these planes in reciprocal space (Fig. 5.9). The presence of an oblique cross in the diffraction pattern testifies directly to the helical structure of the polymer under investigation. Just such photos led to the discovery of the double-helical structure of DNA, which played such an important role in the development of modern biology and biophysics [34].

However, it would be erroneous to think that mere observation of the röntgenogram solves the problem. The system under study is an aggregate of chain molecules possessing as a whole much less ordering than a three-dimensional crystal. Consequently, the diffraction pattern does not have many reflections. A series of methods for deciphering such pictures has been developed, including the "trial-and-error" method, in which, on the basis of structural-physical considerations and the analysis of atomic models, a trial model of the system is constructed and the intensity distribution is calculated. The agreement of the calculated distribution with the observed one proves the correctness of the model. If the agreement is good, the phases can be calculated and Fourier synthesis can be performed. For fibrillar structures it is convenient to use cylindrically symmetrical Patterson functions (cf. [9]).

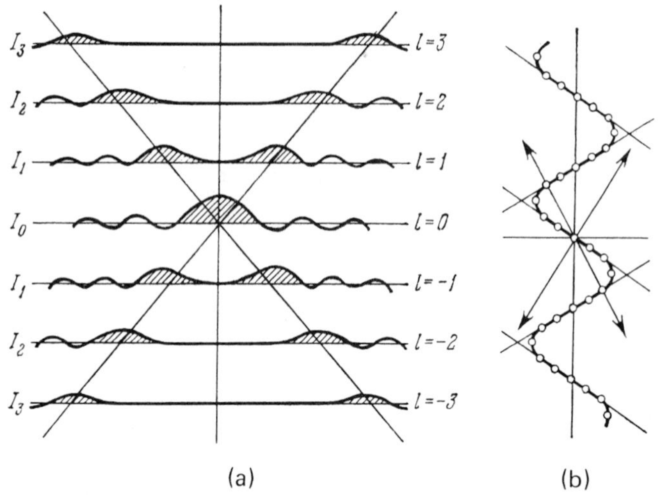


*FIG. 5.9 (a) Distribution scheme of Bessel functions determining the intensity values at the layer lines. (b) Scheme of the correspondence of the crosslike positions of the main intensity maxima at the layer lines to the cross of perpendiculars to the dense atom series.*

In aggregates of chain molecules various degrees of order are present, from genuine crystals with axial texture, to amorphous polymers with disordered molecules. Theoretical studies sometimes allow the observer to judge when the diffraction patterns show violations of order produced by shears, bends, and deviations from the parallel packing of macromolecules. Polymer systems form paracrystals in many cases; these lack real three-dimensional order, but consist of chain molecules displaced and rotated parallel to each other. Paracrystals can be ideal and nonideal. In the latter case, the coordination number is preserved, but there are deviations from the parallel packing of the chains. A series of fibrous proteins (α- and β-keratin, collagen), cellulose, and some other fibrous substances manifest conventional distortion of order, in which approximately parallel orientations of the molecular axes are preserved. Cellulose has been studied in detail. The studies suggesting its amorphous structure on the basis of qualitative treatment not related to rigorous calculations proved to be erroneous. Kitaigorodsky and Tsvankin conducted a rigorous investigation of cellulose X-ray diffraction patterns and showed that this natural polymer has paracrystalline structure with limited dimensions of ordered regions (of the order of 100 Å) [35]. At the same time, some derivatives of cellulose form genuine monocrystals. The multiformity of molecular and supramolecular biopolymeric structures

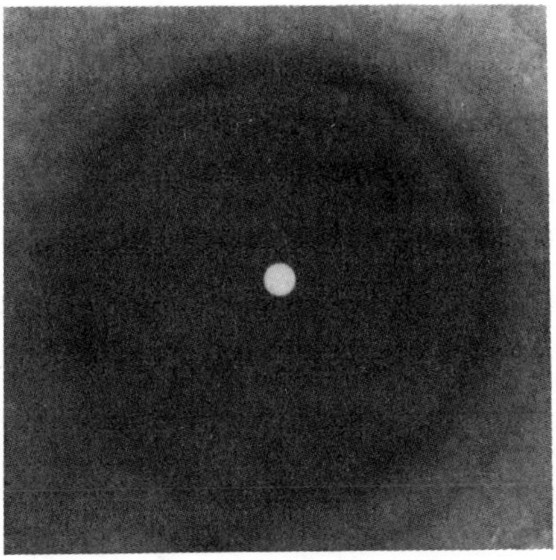

*FIG. 5.10 Debye-Scherrer diffraction pattern.*

is great. Their investigation is very important for biophysics.

Synthetic amorphous polymers, such as rubbers, produce diffraction patterns in the form of several concentric rings (Debye-Scherrer diagrams, Fig. 5.10). Such a picture, being much poorer than the Laue diagram of a crystal, contains a diffuse ring (amorphous halo), the diameter of which is determined by preferential distances between scattering centers. If an amorphous polymer is stretched, texture arises and instead of homogeneous rings some more or less extensive arcs are observed near the meridian or equator of the ring. Similar pictures are given by fibrillar proteins, such as keratin, myosin, epidermin, and fibrinogen [see Astbury [36]] (cf. p. 243). Further perspectives of X-ray analysis of biologically functioning substances are revealed in two fields of investigations. First is the ultimate method of direct phase determination using strong reflections [37]. This avoids isomorphous substitution. Second, the possibilities of isomorphous substitution are far from being exhausted; this method has to be developed further. The use of partially isomorphous derivatives seems particularly important.

Along with the study of biological macromolecules, structural investigation of the supramolecular biological systems in their native state (membranes, muscular fibers, etc.) are needed. The possibilities of these investigations are dependent on the development of high-speed X-ray analysis, that is, the creation of powerful sources of X radiation with polarized and nondivergent beams. Apparently the synchrotron radiation, which arises if electrons are centripetally accelerated in a magnetic field, can be utilized effectively in this context. Unlike X rays and thermal radiation, the synchrotron radiation beam is characterized by great power, small divergence, and high polarization.

## 5.3 Scattering of X Rays by Macromolecules in Solution

Discrete low-angle scattering of X rays is a particular case of diffraction by crystals. Small angles corresponding to observed interference represent lattice periods much larger than the wavelength.

Diffuse low-angle scattering makes it possible to investigate macromolecules in solutions. Summary scattering by single, randomly oriented macromolecules is observed (i.e., the scattered intensity is averaged over all possible orientations of the molecules). The information that can be obtained is limited. Scattering phases are lost and mixed; therefore, only a function similar to the Patterson function can be formed. However, this function enables us to characterize the form and size of the scattering system. The situation is similar to light scattering by solutions of macromolecules (p. 141). This limited information can be obtained much more easily than by X-ray analysis of crystals.

Only the electronic radius of inertia and (if the intensity is measured in an absolute scale) the molecular weight of a molecule are measured directly. The radius of inertia does not, however, represent the form of a molecule, which has to be determined by means of model calculations and the comparison of theoretical curves with experimental ones. The fundamentals of the theory are presented in [38,39].

Let a particle, consisting of centers (atoms) with scattering strengths $f_i$, scatter X rays. Then the scattered intensity, being the square of the amplitude, is

$$I(h) = A_e^2(h) \sum_k \sum_i f_k f_i \cos(hr_{ki}) \qquad (5.29)$$

where $h = (2\pi/\lambda)(s-s_0)$ is the vector proportional to the difference between the unit vectors of the scattered and incident waves.

$$I_e(h) = A_e^2(h) = 7.9 \times 10^{-26} I_0 (1+ \cos^2 2\theta)/2r^2 \qquad (5.30)$$

is the scattering intensity of one electron, where $I_0$ is the intensity of the incident beam, r is the distance between particle and observer, and $\theta$ is the scattering angle.

In the case of randomly oriented particles, using $\overline{\cos(hr)} = \sin(hr)/hr$, we obtain the Debye formula (p. 146)

$$I(h) = I_e(h)\left(\sum_k f_k [\sin(hr_k)]/hr_k\right)^2 \qquad (5.31)$$

For small angles the second factor on the right-hand side can be expanded in a series of powers of h

$$F^2(h) = \sum_k \sum_j [1-(h^2/3!)r_{kj}{}^2+(h^4/5!)r_{kj}{}^4+\cdots] \qquad (5.32)$$

Since $\Sigma_k f_k = \Sigma_j f_j = N$, the number of electrons in the particle, the first term in the expansion is $N^2$ and, having chosen the center of inertia as the reference point, we get the approximate expression for the average intensity

$$\overline{F^2(h)} = N^2 \exp(-z^2 R_0{}^2/3) \qquad (5.33)$$

where the electronic radius of inertia is

$$R_0 = (\Sigma f_k r_k{}^2)/\Sigma f_k \qquad (5.34)$$

In order to calculate the scattering by n particles, we multiply the intensity by n. However, at zero scattering angle all centers scatter in phase and the intensity is proportional to $n^2$.

At increased values of $hR_0$, higher-order terms, which give information about the form of the macromolecule, contribute. However, it is impossible to arrive at definite conclusions about the form using only the scattering indicatrix; some specific hypothesizing about the form must be done. Thus, molecular asymmetry can be determined by first hypothesizing that the molecule in question can be approximated by an ellipsoid or a cylinder, and then comparing the experimental curve with the family of theoretical curves calculated for these bodies at different ratios of axes.

For greatly elongated structures, analysis is rather difficult because all the theoretical curves corresponding to extreme asymmetry are practically identical. On the other hand, it is possible to determine the cross section, particularly in those cases in which the cylindrical structure of the molecule is already known (DNA in solution). Multiplying the scattering intensity by the corresponding scattering angle yields the function hI(h), which characterizes the cross section of the molecule and determines (as in the determination of $R_0$) the radius of inertia of the cross section [40]. Multiplying the intensity by $h^2$ also permits us to separate the vectors corresponding to the molecule's cross dimensions. This method was successfully used by Ptitsyn and Fedorov for interpretation of the scattering indicatrix [41].

The choice of a model is based on the data of other methods, on the results of physicochemical investigations. Various methods for treating the curve of low-angle scattering may be combined in such a way that the parameters obtained by various calculations coincide. The method of Ritland *et al.* [42] uses the value $R_0M^{-1/3}$, which is a function only of the ratio of the axes of particles of different molecular weights. By constructing a nomogram expressing the dependence of $R_0M^{-1/3}$ on the axial ratio, it is possible to determine this ratio if $R_0$ and M are known. The agreement of experimental and theoretical values proves the correctness of the model.

The use of low-angle diffuse X-ray scattering is especially convenient for proteins of moderate molecular weights. The morphology of many proteins has been studied, particularly pepsin [43], trypsin [44], and aspartate aminotransferase [45]. The size and form of transfer RNA have been determined [46]. The low-angle scattering method is helpful in many areas of molecular biophysics, one more instance being the study of soluble antigen-antibody complexes [47].

Methods have also been developed that make it possible to estimate the volume [48] and surface-to-volume ratio directly from the low-angle scattering curve.

The scattering of X rays at comparatively large angles yields information about conformational short-range order in synthetic polymers and globular proteins [49,50]. It has been found that the parameter $(4\pi/\lambda) \sin \theta$ for intensity maxima has different values for α-helices and β forms of proteins. Nonordered regions do not give maxima at all; consequently, this method is suitable for denaturation studies.

## 5.4 Electronic Spectra of Biopolymers

All amino acid residues and peptide (amide) groups —CO—NH— absorb light in the ultraviolet region of the spectrum. Aromatic residues (Trp, Phe, Tyr) have characteristic absorption bands near 280 nm.

The most interesting region for spectroscopy of synthetic polypeptides and proteins is the short-wave region 240-185 nm, in which the absorption bands of the peptide bond lie. The monomer of the polypeptide chain is similar to the amide molecule $R_1$—CO—NH—$R_2$. Studies of the electronic spectra of simple amides (formamide, methylacetamide, myristamide) have revealed the general scheme of electronic transitions in the amide group (Fig. 5.11) [51-54]. The scheme is also based on theoretical calculations using both the simple method of Hückel [55] and the self-consistent-field method [56-58]. Figure 5.12 shows the wave functions of an amide group for the most labile electrons. The $\pi_2$ level corresponds to the bonding, and the $\pi^*$ level to the nonbonding, orbital of CO; the $\pi_1$ level corresponds to the nonbonding orbital of nitrogen, level n to the state of a lone pair of oxygen electrons. The n $\sigma^*$ is the Rydberg atomic transition in oxygen.

According to the Bouguer-Lambert-Beer law, the intensity of light passing through an absorbing layer of thickness $\ell$ is equal to

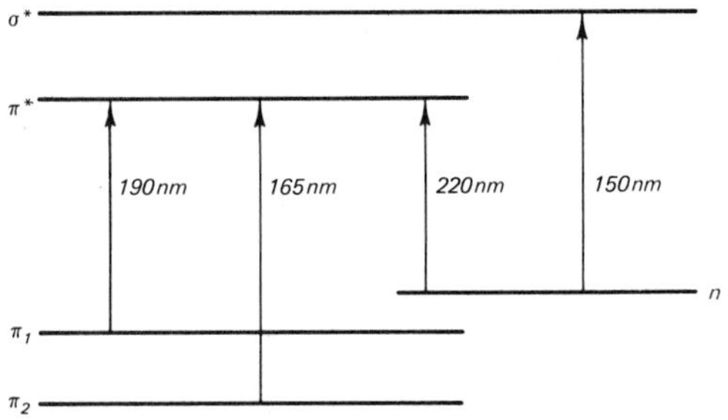


*FIG. 5.11 Electronic transitions in the amide group.*

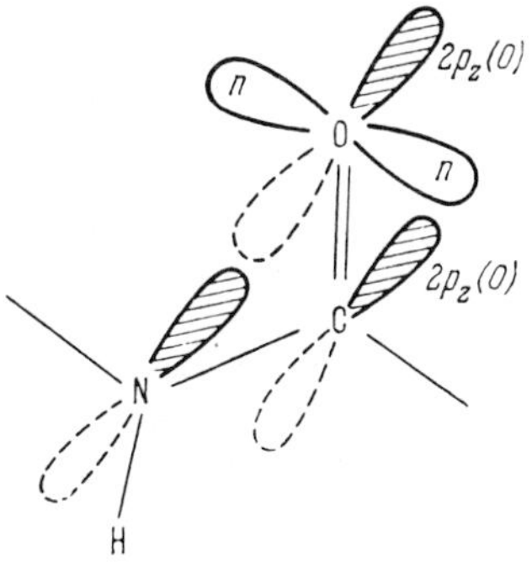


*FIG. 5.12 Wave functions of an amide group.*

$$I = I_0 e^{-\varepsilon C \ell} \tag{5.35}$$

where $I_0$ is the intensity of the incident light, C is the concentration of the absorbing substance expressed in moles per liter, and $\varepsilon$ is the molar absorption coefficient. Absorption in a given band corresponding to the transition $0 \to j$ is expressed by the so-called oscillator strength $f_{0j}$ determined by the absorption intensity

$$f_{0j} = 4.32 \times 10^{-9} \int \varepsilon_{0j} \, d\nu' \tag{5.36}$$

$\nu'$ is the wave number expressed in reciprocal centimeters; integration is performed over the whole absorption band. Theory gives

$$f_{0j} = (2m/3\hbar e^2) \, \omega_{0j} \, p_{0j}^2 \tag{5.37}$$

where e and m are the charge and mass, respectively, of the electron, $\omega_{0j}$ is the circular frequency of transition, and $p_{0j}$ is the matrix element of the electric dipole moment corresponding to the transition $0 \to j$

$$p_{0j} = (0|p|j) \equiv \int \psi_0 p \psi_j^* \, d\tau \tag{5.38}$$

The value $p_{0j}^2$ is called the transition dipole strength.

The n $\pi^*$ transition of the amide group has low dipole strength, since the electron clouds of the n and $\pi^*$ states are nearly perpendicular and do not overlap to any considerable extent (Fig. 5.12).

Theoretical calculations give the following transition wavelengths: n $\pi^*$, 2340; $\pi_1$ $\pi^*$, 1605; $\pi_2$ $\pi^*$, 1347 Å. In the myristamide spectrum, 2200, 1850, and 1600 Å bands are observed [54]; in the formamide spectrum 1717 and 1345 Å [53].

The transition dipole moment of the $\pi_1$ $\pi^*$ transition lies in the HNCO plane and forms a 9° angle with the line connecting the O and N atoms. The n $\pi$ transition is perpendicular to this plane [51,59].

In polypeptide chains, the spectra of amide groups are changed because of the electronic interaction of these groups. In particular, the levels of end groups of the chain differ from those of internal groups [51]. Of greatest interest for biophysics are the effects--Davydov's splitting and hypochromism--which arise because of the exciton resonance interaction.

In formulating his theory of the spectra of molecular crystals, Davydov showed that in a regular sequence of chromophore groups, the transfer of resonance excitation energy can occur between their excited energy levels. Therefore, in a regular system the motion of an excitation wave (an exciton) is possible. Interaction results in splitting of the energy levels, which forms a broad zone. Selection rules allow transitions in the absorption and emission of light not at any levels in the zone but at rigorously defined ones. The polarization of bands is determined by the symmetry of the regular system as a whole [60,61].

Resonance interaction is possible even in a dimer. In this case the original level splits into two levels. Figure 5.13 shows the scheme of this splitting for the cases of

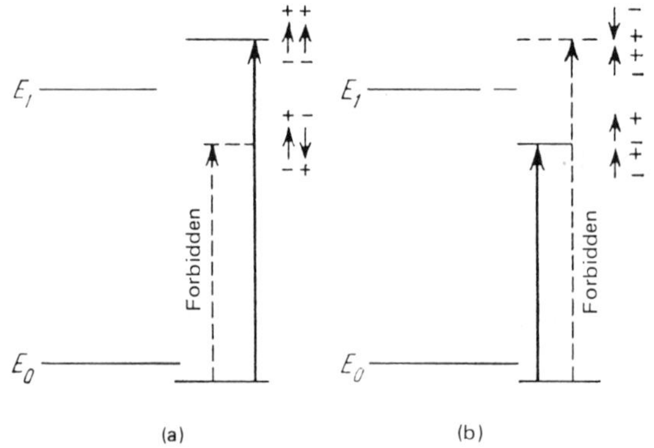


*FIG. 5.13 Energy levels for a system of (a) parallel and (b) transition dipoles. Solid arrows indicate allowed transitions; broken arrows, forbidden ones.*

parallel and collinear positions of the transition dipoles.

Moffitt made a theoretical calculation of the exciton splitting for the $\pi_1$ $\pi^*$ 190 nm transition of the amide group in an $\alpha$-helix [62]. In agreement with Davydov's theory, this band is split into two components, one polarized along the helix axis and the other perpendicular to it. The frequency of the first band is 2700 $cm^{-1}$ lower than that of the second one, in good agreement with experiment.

Along with exciton splitting, we have to consider that the peptide bonds in proteins are surrounded by charged atomic groups and by groups having permanent dipole moments and polarizabilities--the peptide bonds are in a condensed medium. Interactions of chromophores with the medium alter the transition frequencies. Neporent and Bakhshiev performed detailed experimental and theoretical investigations of the influence of intermolecular interactions on the electronic spectra of complicated molecules (cf., e.g., [63,64]). They found that an increase in the dipole moment and polarizability decreases the transition frequencies. The change in the medium produced by denaturation results in the transfer of peptide groups from the surrounding with the larger polarizability into that with the smaller one, a transfer that must diminish the transition frequency. Such effects were actually observed [65].

Especially interesting and indicative are the intensity changes resulting from exciton effects. Resonance interaction results in a redistribution of transition intensities. In the case of two collinear transition dipoles, the transition with the smaller frequency (longer wavelength) increases its intensity at the expense of the intensity of the short-wave transition. Hyperchromism of the long-wave band ensues. In the case of parallel transition dipoles, on the other hand, the intensity of the long-wave band becomes lower and that of the short-wave band higher. Hypochromism of the long-wave band occurs. Just this effect is observed in spectra of $\alpha$-helical polypeptides and proteins and in spectra of native nucleic acids. If the dipoles are perpendicular, there is no redistribution of intensities.

The theory of hypochromism is based on electrostatic dipole-dipole and polarization interactions of the transition dipoles in a regular biopolymer system. These calculations can be improved if the charge (monopole) interactions instead of the dipole ones are considered. Calculation gives, as the ratio of oscillator strengths in the stiff (native) and disordered (denatured) systems,

$$\frac{f_{oa}}{f_{oa}^{(0)}} = 1 - \frac{3\lambda_a^2 e^2}{2\pi^2 mc^2} \sum_{i=1}^{N} \sum_{j\neq i} \sum_{b\neq a} \frac{f_{ob}^{(0)} \lambda_b^2 G_{ij} \mathbf{e}_i \mathbf{e}_j}{\lambda_a^2 - \lambda_b^2} \tag{5.39}$$

where $f_{oa}$ is the oscillator strength for the group with stiff conformation; $f_{oa}^{(0)}$, $f_{ob}^{(0)}$ those for single groups; $e_i$, $e_j$ the unit vectors describing the orientations of the interacting groups, whose total number is N. The value $G_{ij}$ expresses monopole interaction

$$G_{ij} = \sum_s \sum_t \frac{\rho_{is}\rho_{jt}}{R_{is,jt}} \tag{5.40}$$

where $\rho_{is}$, $\rho_{jt}$ are the transition monopoles of the groups i and j.

A review of the various forms of hypochromism theory is given by Weissbluth [66] (cf. also [67,71]). Classical or semiclassical models (cf. p. 289) lead to results equivalent to those obtained with the help of the quantum-mechanical exciton models. Classical models treat oscillators bound by dipole-dipole interactions; exciton theory applies the theory of time-independent perturbations. The theory of time-dependent perturbations can also be employed.

Figure 5.14 shows experimental results obtained for polyglutamic acid [59]. At pH 4.9 the polymer is α-helical; at pH 8 it has the coiled form. Two bands, the result of Davydov's splitting and pronounced hypochromism, are readily observable. The $f_{oa}/f_{oa}^{(0)}$ ratio is 0.7, in agreement with theory (cf. [67]).

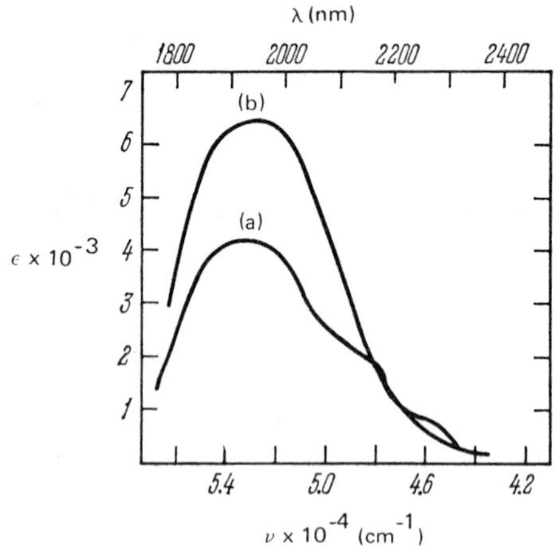


*FIG. 5.14 Absorption spectra of polyglutamic acid; (a) α-helix, (b) coil.*

The dependence of the hypochromic effect on chain length was studied by Goodman *et al.* [72]; its dependence on ionic strength by McDiarmid and Doty [73]. Table 5.2 contains some data on the hypochromism of polyamino acids [51].

Obviously the vanishing of hypochromism in the helix-coil transition at denaturation can give a quantitative measure of the α-helicity of a protein. Because of the difficulties of

*TABLE 5.2*
*Molar Absorption of Polyamino Acids in α-Helical and Disordered Forms*[a]

| | Molar absorption | | | | Partial oscil- lator strength |
|---|---|---|---|---|---|
| | At maximum | 190 nm | 197 nm | 205 nm | |
| Coil | | | | | |
| Poly-L-glutamic acid | | | | | |
| without salt | 7100 | 7000 | 6550 | 3500 | 0.115 |
| ionic strength 0.0004 | | (7200) | (6600) | (4850) | |
| ionic strength 2.0 | | (6050) | (5350) | (3700) | |
| Poly-L-lysine | 7100(6600) | 6900 (6500) | | 3300 | 0.106 |
| Mean values | | 6950 | | 3400 | 0.110 |
| α-Helix | | | | | |
| Poly-L-glutamic acid | 4200 | 4200 | 3300 | 2150 | 0.0778 |
| Poly-L-lysine | 4400 | 4400 | 3500 | 2300 | 0.0720 |
| Mean values | 4300 | 4300 | 3400 | 2200 | 0.0749 |

[a]Data from [74] (except those in parentheses, which are from [73]).

spectrophotometric measurements in the far-ultraviolet region around 200 nm, this method is not widely used. On the other hand, the technique can be rather simply and effectively used in the case of nucleic acids, whose long-wavelength absorption bands lie near 260 nm. These bands, due to $\pi\ \pi^*$ transitions, are characterized by transition dipoles located in the planes of the nitrogen bases. Table 5.3 contains the absorption band parameters of nitrogen bases [71]. The angle in the last column is measured anticlockwise from the line connecting positions

*TABLE 5.3*
*Absorption Bands of Nitrogen Bases*

| Base | Maxima (nm) | Half width (nm) | Transition dipole moment (D) | Direction of moment |
|---|---|---|---|---|
| Adenine | 260.0 | 26.0 | 3.90 | −2° |
| | 240.0 | 20.0 | 1.68 | −156° |
| Cytosine | 271.0 | 30.5 | 3.04 | 82° |
| | 240.0 | 28.0 | 2.63 | 180° |
| Guanine | 277.5 | 23.5 | 2.46 | −83° |
| | 251.0 | 29.0 | 3.87 | 165° |
| Uracil | 262.5 | 31.5 | 3.29 | 173° |
| | 240.0 | 19.0 | 1.16 | 160° |

3 and 6 in purines or from the line connecting positions 1 and 4 in pyrimidines.

Parallel transitions in the 260-nm region are $\pi\ \pi^*$ transitions with molar absorption coefficients of 8,000-10,000. In the 190-nm region lie more intensive, also parallel, $\pi\ \pi^*$ transitions with absorption coefficients of 20,000-25,000. Long-wavelength transitions near 280 nm are perpendicular $n\ \pi^*$ transitions with small absorption coefficients (<5,000) [75].

The planes of nitrogen bases are parallel in the double helix of DNA (p. 490) and nearly parallel (7° angle) in the double-helical regions of RNA (p. 500). The hypochromism of the 260-nm band of DNA may reach 40%. Quantum-mechanical calculations of the first $\pi$-electron levels of nitrogen base pairs in DNA were made in a series of papers, particularly [75] and [76].

## 5.5 The Theory of Optical Activity

Spectropolarimetry is one of the most important and accessible methods for studying biopolymers. Optical activity--rotation of the plane of polarization of light passing through an asymmetric medium--was first observed by Arago in 1811. Fresnel (1820) demonstrated that optical activity is equivalent to circular birefringence: it is due to the difference in velocities of light with right and left circular polarization. In the right wave, the electric intensity vector **E** in the ray arriving at the observer's eye rotates clockwise; in the left one the vector rotates in the opposite direction. Consequently, for the right (D) and left (L) wave passing along axis z

$$E_y = iE_x(D), \quad E_y = -iE_x(L) \tag{5.41}$$

The phase of $E_y$ of the right wave differs from that of $E_x$ by $\pi/2$ ($i \equiv e^{i\pi/2}$); in the case of the left wave, by $-\pi/2$. The vector of electric induction in the medium is

$$D = n^2E \tag{5.42}$$

where n is the refraction index. Let us represent the right and left waves by the expressions $D_D = D_x - iD_y$, $D_L = D_x + iD_y$. If $D_D = 0$, we get the L-wave condition $D_y = -iD_x$; and if $D_L = 0$, we have that of the D wave, $D_y = iD_x$. Let the wave entering a substance possess amplitude 1 and linear polarization parallel to the x axis, that is, $D_x = 1$, $D_y = 0$, $D_D = D_L = 1$. The velocities of the right and left waves in the substance are different (i.e., $n_D \equiv n_+$ is different from $n_L \equiv n_-$. Then the waves leaving a layer of the substance of thickness $\ell$ are described as

$$D_D = \exp[-(2\pi i/\lambda)n_-\ell], \quad D_L = \exp[-(2\pi i/\lambda)n_+\ell] \tag{5.43}$$

where $\lambda$ is the wavelength.

Let $n = \frac{1}{2}(n_+ + n_-)$, $\Delta n = n_- - n_+$. We have

$$D_x = (1/2)\exp[-(2\pi i/\lambda)n\ell]\{\exp[(2\pi i/\lambda)(n-n_+)\ell]$$

$$+\exp[(2\pi i/\lambda)(n-n_-)\ell]\}$$

$$= \exp[-(2\pi i/\lambda)n\ell]\cos(\pi\,\Delta n/\lambda)\ell \quad (5.44)$$

and

$$D_y = (1/2i)\exp[-(2\pi i/\lambda)n\ell]\{\exp[(2\pi i/\lambda)(n-n_+)\ell]$$

$$-\exp[(2\pi i/\lambda)(n-n_-)\ell]\}$$

$$= \exp[-(2\pi i/\lambda)n\ell]\sin(\pi\,\Delta n/\lambda)\ell \quad (5.45)$$

These expressions give

$$D_y/D_x = \tan\phi = \tan(\pi\,\Delta n/\lambda)\ell \quad (5.46)$$

which means that the plane of polarization is rotated by the angle

$$\phi = (\pi/\lambda)(n_L - n_D)\ell \quad (5.47)$$

A linearly polarized light wave can always be decomposed into two waves, one having right and the other left circular polarization. If in a medium one of these waves outstrips the other, the plane of polarization of the incident wave--the vector D being the vector sum of $D_D$ and $D_L$--is rotated by the angle $\phi$.

Specific rotation of a solution containing c g/cm$^3$ of the optically active substance is

$$[\alpha] = \phi/\ell c \quad (5.48)$$

If $\ell$ is measured in decimeters and $\phi$ in radians (i.e., $\phi$ = 180 $\chi/\pi$), then

$$[\alpha] = \frac{180}{\pi}\,\frac{10}{c}\,\frac{\chi}{\ell} \quad \text{rad-cm}^3\ \text{dm}^{-1}\ \text{g}^{-1} \quad (5.49)$$

For a pure substance we have to replace c by the density $\rho$.

The quantity [M] is called the molecular rotation

$$[M] = \frac{M}{100}[\alpha] = \chi\,\frac{18\ M}{\pi\ c} \quad (5.50)$$

where M is the molecular weight of the rotating substance.

In the absorption range, the index of refraction is a complex quantity

$$\tilde{n} = n - i\kappa \quad (5.51)$$

where $\kappa$ is the absorption index. The intensity of the light that passed through a layer of thickness $\ell$ is equal to [cf. (5.35)]

$$I = I_0 \exp[-(4\pi\kappa/\lambda)\ell] \tag{5.52}$$

where $I_0$ is the initial intensity. Circular birefringence is closely related to circular dichroism--the absorption of light with right and left circular polarization is different, $\kappa_L \neq \kappa_D$. The angle of rotation is a complex quantity, too:

$$\tilde{\phi} = (\pi/\lambda)(\tilde{n}_L-\tilde{n}_D)\ell = (\pi/\lambda)(n_L-n_D)\ell-(i\pi/\lambda)(\kappa_L-\kappa_D)\ell \tag{5.53}$$

Because of the nonequal absorption of right and left waves, an optically active substance in the absorption region not only rotates the plane of polarization of light but also transforms linearly polarized light into elliptically polarized light. The measure of ellipticity is the quantity

$$\theta = (\pi/\lambda)(\kappa_L-\kappa_D) \tag{5.54}$$

Molecular ellipticity is

$$[\theta] = \theta\,\frac{18}{\pi}\frac{M}{C} = 2.303\,\frac{4500}{\pi}(\varepsilon_L-\varepsilon_D) \equiv 3300(\varepsilon_L-\varepsilon_D) \tag{5.55}$$

where $\varepsilon_L$, $\varepsilon_D$ are the molar absorption coefficients, determined by a condition equivalent to (5.52) (cf. Eq. (5.35))

$$I = I_0\, e^{-\varepsilon C\ell} \tag{5.56}$$

where C is the concentration in moles per liter.

Theory has to explain the origin of the quantities $n_L-n_D$ and $\kappa_L-\kappa_D$, and determine the relation of these quantities to the structure of matter. In conventional optics (i.e., the theory of the refraction and dispersion of light), the sizes of molecules are considered to be infinitely small in comparison to the wavelength $\lambda$ of light. In other words, the phase differences of a light wave at different points in a molecule are not taken into account. Quantities of the order of $r/\lambda$, where r (the length) is of the order of molecular size, are considered infinitely small. For small molecules and visible light $r/\lambda \sim 10^{-3}$. In the treatment of optical activity the phase differences of a light wave at different points in a molecule have to be taken into account [77-79].

The extremely high sensitivity of optical activity to transformations in molecular structure can be explained if we understand that spectropolarimetry is a kind of molecular interferometry. The classical theory of this phenomenon, based on the electron harmonic oscillator model developed by Born and Oseen (1915), shows that three conditions have to be fulfilled for optical activity to occur. First, as was said earlier, the distances between different electrons in a molecule interacting with light waves have to be nonvanishing in comparison to $\lambda$. Second, these electrons have to interact. Third, the molecule

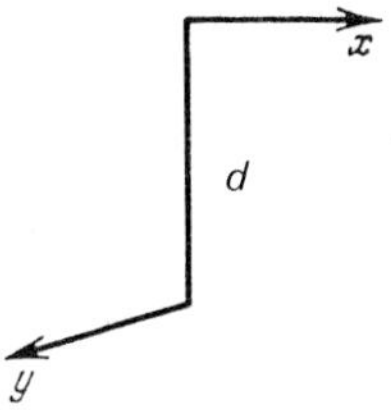


*FIG. 5.15 Kuhn's two-electron model.*

has to be chiral (p. 65), that is, it must not have a plane or center of symmetry. Consequently, the simplest classical model of an optically active molecule, devised by Kuhn, contains two interacting electrons, some distance apart, moving along nonparallel directions, for instance, along directions x and y fixed in the molecule (Fig. 5.15). A detailed presentation of the classical theory is given in the monographs [78,79].

A rigorous quantum-mechanical theory of optical activity was formulated by Rosenfeld [80]. This theory, which applies the perturbation theory of quantum mechanics, is detailed in [79,81-85]; only some fundamentals are presented here.

Let us treat a molecule as an electronic system characterized by nonperturbed wave functions, and an electromagnetic field as a perturbation. Since the electric field of light waves is alternating, we have to use the theory of time-dependent perturbations. The theory expresses the molecular constants--polarizability and magnetic susceptibility--as matrix elements of electric and magnetic dipole moments (i.e., as transition moments). In the usual nonmagnetic medium, equations of the electromagnetic field have the form

$$D = \varepsilon E, \qquad B = \mu H \cong H \tag{5.57}$$

where H and B are, respectively, the intensity and induction of the magnetic field; the dielectric permeability $\varepsilon = n^2$; and the magnetic permeability $\mu = 1$. As is shown by the theory, in an isotropic opticaly active medium

$$D = \varepsilon E - g\dot{H}, \qquad B = H + g\dot{E} \tag{5.58}$$

and (c is the velocity of light)

$$n_D = \varepsilon^{1/2} - (2\pi gc/\lambda), \qquad n_L = \varepsilon^{1/2} + (2\pi gc/\lambda) \tag{5.59}$$

Rotation of the polarization plane is expressed by the constant g (gyration)

$$\phi = (\pi/\lambda)(n_L - n_D)\ell = (4\pi^2 c/\lambda^2) g\ell \tag{5.60}$$

On the other hand,

$$D = E + 4\pi P, \qquad B = H + 4\pi M \tag{5.61}$$

where P is the electric polarization vector, and M is the

magnetization vector. These vectors are expressed as the summary electric and magnetic dipoles of the unit of volume, $\mathbf{P} = N_1\mathbf{p}$, $\mathbf{M} = N_1\mathbf{m}$. These dipoles can be expressed as

$$\mathbf{p} \cong a\,\mathbf{E}' - \frac{\beta}{c}\dot{\mathbf{H}}, \qquad \mathbf{m} = \frac{\beta}{c}\dot{\mathbf{E}}' \tag{5.62}$$

where

$$\mathbf{E}' = \mathbf{E} + \frac{4\pi}{3}\mathbf{P} \tag{5.63}$$

is the effective electric field acting on a molecule in an isotropic medium, $a$ is the polarizability of the molecule, and $\beta$ is the molecular parameter of optical activity. Comparing formulas (5.58), (5.62), and (5.63), we get

$$\frac{\varepsilon-1}{\varepsilon+2} = \frac{n^2-1}{n^2+2} = \frac{4\pi}{3}N_1 a \tag{5.64}$$

the Lorentz-Lorenz equation, and

$$g = \frac{4\pi}{3}N_1\frac{\beta}{c}\frac{n^2+2}{3} \tag{5.65}$$

Hence

$$\phi = \frac{16\pi^3 N_1\beta}{\lambda^2}\,\frac{n^2+2}{3}\,\ell \tag{5.66}$$

Quantum-mechanical expressions of $a$ and $\beta$ for a molecule in its ground electron state with quantum number 0 have the form

$$a = \frac{2}{3h}\sum_j \frac{\nu_j|(0|\mathbf{P}|j)|^2}{\nu_j^2-\nu^2} \tag{5.67}$$

$$\beta = \frac{c}{3\pi h}\sum_j \frac{\mathrm{Im}\{(0|\mathbf{P}|j)(j|\mathbf{m}|0)\}}{\nu_j^2-\nu^2} \tag{5.68}$$

Here $h$ is Planck's constant; $(0|\mathbf{P}|j)$ and $(j|\mathbf{m}|0)$ are the matrix elements of the electric and magnetic moment for the transition between states 0 and $j$

$$(0|\mathbf{P}|j) \equiv \int \psi_0\mathbf{p}\psi_j^*\,d\tau, \qquad (j|\mathbf{m}|0) \equiv \int \psi_j\mathbf{m}\psi_0^*\,d\tau$$

where $\nu_j$ is the transition frequency and $\nu_0$ is the frequency of the incident light. Summation is done over all states $j$. The symbol Im shows that $\beta$ contains only imaginary terms of the complex products $(0|\mathbf{P}|j)(j|\mathbf{m}|0)$.

The quantity $D_j = |(0|\mathbf{P}|j)|^2$ is the dipole strength, which is directly related to the intensity of absorption (cf. p. 275). The quantity

$$R_j = \mathrm{Im}\{(0|\mathbf{p}|j)(j|\mathbf{m}|0)\} \tag{5.69}$$

is called the rotatory strength.

The sum of the dipole strengths over all $0 \to j$ transitions is a positive constant proportional to the number of electrons in the molecule. The sum of the rotatory strengths, by contrast, is equal to zero

$$\sum_j R_j = \sum_j \mathrm{Im}\{(0|\mathbf{p}|j)(j|\mathbf{m}|0)\}$$

$$= \mathrm{Im}\{(0|\mathbf{pm}|0)-(0|\mathbf{p}|0)(0|\mathbf{m}|0)\} = 0 \tag{5.70}$$

because the diagonal matrix elements $(0|\mathbf{pm}|0)$, $(0|\mathbf{p}|0)$, and $(0|\mathbf{m}|0)$ are substantial (i.e., their imaginary parts are zero). The rotational strength becomes zero if the molecule contains a center or plane of symmetry. In these cases all states of the molecule are divided into even and odd, depending on the preservation or change of the sign of the wave function when reflected in a center or in a plane. Operators $\mathbf{p}$ and $\mathbf{m}$ have the form, in Cartesian coordinates,

$$\mathbf{p} = e(\mathbf{i}x+\mathbf{j}y+\mathbf{k}z)$$

$$\mathbf{m} = \frac{eh}{4\pi mci}\left\{\mathbf{i}\left(y\frac{\partial}{\partial z}-z\frac{\partial}{\partial y}\right)+\mathbf{j}\left(z\frac{\partial}{\partial x}-x\frac{\partial}{\partial z}\right)+\mathbf{k}\left(x\frac{\partial}{\partial y}-y\frac{\partial}{\partial x}\right)\right\} \tag{5.71}$$

where e and m are, respectively, the charge and mass of the electron and $\mathbf{i}$, $\mathbf{j}$, $\mathbf{k}$ are unit vectors. Let us consider a centrosymmetrical molecule. When reflected at the center, $\mathbf{p}$ changes its sign and $\mathbf{m}$ preserves its sign. Therefore, $(0|\mathbf{p}|j)$ is nonzero only at transitions from odd to even states and vice versa. By contrast, $(j|\mathrm{m}|0)$ is nonzero only for transitions between states of equal symmetry. Hence, the scalar products of these matrix elements are zero for any transitions. The equality $R_j = 0$ for molecules with the plane of symmetry is proved in an analogous way. Electric and magnetic comultipliers cannot be simultaneously different from zero in the absence of chirality [81]. The signs of $R_j$ of right and left enantiomers are opposite. The order of magnitude of $R_j$ is that of the product of the electric and magnetic moments of the electronic structure. The order of magnitude of $(0|\mathbf{p}|j)$ is 1 D = $10^{-18}$ esu-cm, the magnetic moment of the electron, $9.3 \times 10^{-20}$ esu. Hence, the order of quantity $R_j$ is $10^{-38}$ esu. It is convenient to use the reduced rotatory strength, expressed by numbers of the order of unity

$$[R] = \frac{100}{\mu_B p_D} R$$

where $\mu_B$ is Bohr's magneton and $p_D$ = 1 D.

The Rosenfeld formula (5.68) is derived for regions far from intrinsic absorption and it describes the normal dispersion of optical activity. In the absorption region, damping has to be taken into account, and we obtain a complex expression for β near the absorption band $\nu_j$ (cf., e.g., [83,85,86]).

$$\beta_j = \frac{c}{3\pi h} \frac{R_j}{\nu_j^2 - \nu^2 + i\nu\Gamma_j}$$

$$\equiv \frac{c}{3\pi h} \frac{R_j(\nu_j^2-\nu^2)}{(\nu_j^2-\nu^2)^2+\nu^2\Gamma_j^2} - \frac{ic}{3\pi h} \frac{R_j\nu\Gamma_j}{(\nu_j^2-\nu^2)^2+\nu^2\Gamma_j^2} \tag{5.72}$$

where $\Gamma_j$ characterizes the half width of the absorption band in the scale of frequencies. The first term in (5.72) describes rotation of the polarization plane; the second one, circular dichroism (p. 282). In a similar way, we obtain the expression for polarizability near the absorption band

$$a_j = \frac{2}{3h} \frac{\nu_j(\nu_j^2-\nu^2)D_j}{(\nu_j^2-\nu^2)^2+\nu^2\Gamma_j^2} - i\,\frac{2}{3h} \frac{D_j\nu\Gamma_j}{(\nu_j^2-\nu^2)^2+\nu^2\Gamma_j^2} \tag{5.73}$$

where the first term describes the dispersion of the refractive index, and the second, the absorption of light.

Near the absorption band, anomalous dispersion of optical activity (ADOA) and circular dichroism (CD) are observed simultaneously. Model curves for a right-rotating and a left-rotating substance are shown in Fig. 5.16. Corresponding

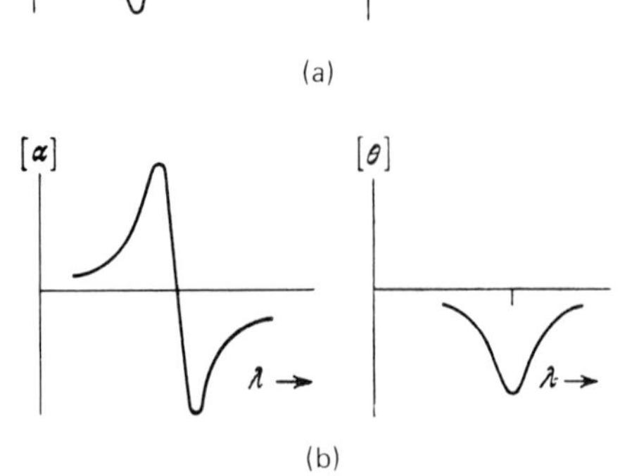

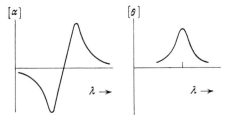

*FIG. 5.16 Anomalous dispersion of optical activity (left) and circular dichroism (right) for (a) a right-rotating substance and (b) a left-rotating substance.*

curves for polarizability and absorption look like the ADOA and CD curves for the right-rotating substance. ADOA is usually called the Cotton effect.

Obviously the dipole strength $D_j$ can be determined experimentally from both absorption and dispersion, and rotatory strength can be ascertained from CD and ADOA. Theory gives

$$D_j = \frac{3hc}{8\pi^3 N_1} \int_0^\infty \frac{k_j(\nu)}{\nu} d\nu \tag{5.74}$$

$$R_j = \frac{3hc}{8\pi^3 N_1} \int_0^\infty \frac{\theta_j(\nu)}{\nu} d\nu \tag{5.75}$$

where $k_j$ is the absorption coefficient of the $0 \rightarrow j$ band

$$k_j = \frac{1}{\ell} \ln \frac{I_0}{I_j}$$

and $\theta_j$ is the ellipticity (p. 282). The parallelism between the phenomena of dispersion and absorption and ADOA and CD is presented in Table 5.4 [86]. The Kramers-Kronig relations mentioned in Table 5.4 are of fundamental importance for optics and for physics in general [85-89].

*TABLE 5.4*
*Fundamental Characteristics of Optical Phenomena*

| Usual phenomena | Rotatory phenomena |
|---|---|
| 1. Partial absorption band due to $0 \rightarrow j$ transition is characacterized by three parameters: | 1. Partial band of circular dichromism due to $0 \rightarrow j$ transition is characterized by three parameters: |
| a. Wavelength (or frequency) of absorption maximum $\lambda_j$ (or $\nu_j$); | a. Wavelength (or frequency) of maximal ellipticity $\lambda_j$ (or $\nu_j$); |
| b. Half width of the band $\Gamma_j$; | b. Half width of the band $\Gamma_j$; |
| c. Measure of intensity, dipole strength $D_j$ [integral (5.74)]; | c. Measure of intensity, rotatory strength $R_j$ [integral (5.75)]; |
| 2. $D_j$ is expressed as $\|(0\|\mathbf{p}\|j)\|^2$ | 2. $R_j$ is expressed by formula (5.69) |
| 3. Substantial and imaginary parts of the refractive index (refraction and absorption) are related by Kramers-Kronig relations. | 3. Substantial and imaginary parts of optical activity (circular birefringence and circular dichroism) are related by Kramers-Kronig relations. |

Let us assume that a linear system is subjected to some time-dependent action which can be expressed as a Fourier integral [85]

$$A(t) = \int_{-\infty}^{+\infty} A_0(\nu)\, e^{2\pi i\nu t}\, d\nu \tag{5.76}$$

Reaction of the system is expressed by the function B(t) containing the complex function $\chi(\nu)$

$$\begin{aligned} B(t) &= \int_{-\infty}^{+\infty} \chi(\nu) A_0(\nu)\, e^{2\pi i\nu t}\, d\nu \\ &= \int_{-\infty}^{+\infty} [\chi_1(\nu) - i\chi_2(\nu)] A_0(\nu)\, e^{2\pi i\nu t}\, d\nu \end{aligned} \tag{5.77}$$

Let us take $A(t) = B(t) = 0$ at $t < 0$ (causality). Since the system is linear in the limit as $\nu \to \infty$

$$\chi_1(\nu) \to \chi_\infty, \qquad \chi_2(\nu) \to 0$$

Here $\chi_1$ is an even and $\chi_2$ an odd function of $\nu$. We get

$$\chi_1(\nu) - \chi(\infty) = \frac{2}{\pi} \int_0^\infty \frac{\nu'\chi_2(\nu')}{\nu'^2 - \nu^2}\, d\nu' \tag{5.78}$$

$$\chi_2(\nu) = -\frac{2}{\pi} \int_0^\infty \frac{\chi_1(\nu') - \chi(\infty)}{\nu'^2 - \nu^2}\, d\nu' \tag{5.79}$$

These are Kronig-Kramers relations which establish the relation between $\chi_1$ and $\chi_2$ and the possibility of their mutual recalculation. Thus, if an electric circuit is considered, A is the voltage, B the current, $\chi_1$ the resistance, and $\chi_2$ the reactance. In our case **E'** and **H** in expressions (5.62) are periodic causes, and **p** and **m** are periodic effects. If expressions (5.73) and (5.72) are rewritten in the form

$$a = a_1 - ia_2, \qquad \beta = \beta_1 - i\beta_2 \tag{5.80}$$

then

$$\begin{aligned} a_1(\nu) &= \frac{2}{\pi} \int_0^\infty \frac{\nu' a_2(\nu')}{\nu'^2 - \nu^2}\, d\nu', \qquad & a_2(\nu) &= -\frac{2\nu}{\pi} \int_0^\infty \frac{a_1(\nu')}{\nu'^2 - \nu^2}\, d\nu' \\ \beta_1(\nu) &= \frac{2}{\pi} \int_0^\infty \frac{\nu' \beta_2(\nu')}{\nu'^2 - \nu^2}\, d\nu', \qquad & \beta_2(\nu) &= = \frac{2\nu}{\pi} \int_0^\infty \frac{\beta_1(\nu')}{\nu'^2 - \nu^2}\, d\nu' \end{aligned} \tag{5.81}$$

Calculation of ellipticity from rotation and vice versa (and analogous calculations for refraction and absorption) is made using the formulas

$$\phi(\nu) = \frac{2\nu^2}{\pi}\int_0^\infty \frac{\theta(\nu')}{\nu'(\nu'^2-\nu^2)}\,d\nu' \tag{5.82}$$

$$\theta(\nu) = -\frac{2\nu^3}{\pi}\int_0^\infty \frac{\phi(\nu')}{\nu'^2(\nu'^2-\nu^2)}\,d\nu' \tag{5.83}$$

We see that the rotatory strength can be determined from both CD and ADOA (cf. also [90]).

Normal dispersion of rotation (or refraction) in regions far from that of absorption is an expression of CD (or absorption) in absorption bands. The formula for normal dispersion (5.68) [correspondingly, (5.67)] contains the rotatory strengths (dipole strengths) characterizing CD (absorption).

Rosenfeld's formula does not enable us to calculate directly the optical activity if the molecular structure is known, because to do this we have to know the entire set of energy levels of the molecule and the corresponding wave functions. It is necessary to use approximate methods to calculate $\beta$.

A molecule can be mentally divided into atomic groups that do not exchange their electrons. For instance, in the case of a protein these groups are single peptide groups and amino acid radicals. Then $\beta$ is expressed as a sum [91]

$$\beta = \beta_1+\beta_2+\beta_3+\beta_4 \tag{5.84}$$

where $\beta_1$ contains the contributions of single groups, equal to zero if these groups are not chiral (e.g., the $CH_3$ group); $\beta_2$ contains the "one-electron" terms, arising because of the "mixing" of electric and magnetic dipole transitions of one group, that is in an asymmetric field of other groups; $\beta_3$ is the result of the "mixing" of the magnetic dipole transition of one group and the electric dipole transition of another one; and finally, $\beta_4$ is due to simultaneous electric dipole excitation of different groups. The most important contributions are $\beta_2$ and $\beta_4$.

The expression for $\beta_4$ was first derived in the quantum-mechanical theory of Kirkwood [92]. Later it was shown that it can also be derived in a purely classical way [79,93]. Consider the electric dipole-dipole interaction of anisotropically polarizable groups in a molecule. We get

$$\beta_4 = \frac{1}{6}\sum_{k\neq\ell} (\alpha_{k1}-\alpha_{k2})(\alpha_{\ell 1}-\alpha_{\ell 2})(\mathbf{r}_{k\ell}[\mathbf{k}\boldsymbol{\ell}])(\mathbf{k}\mathbf{T}_{k\ell}\boldsymbol{\ell}) \tag{5.85}$$

Here k and $\ell$ are the numbers of groups assumed to be axially symmetrical; **k** and **$\ell$**, unit vectors directed along the group axes; $\alpha_{k1}$, the polarizability of group k along its axis; $\alpha_{k2}$, that in the perpendicular direction; $r_{k\ell}$ is the distance between groups k and ℓ. The last factor describes dipole-dipole interaction (cf. p. 178)

$$(\mathbf{k}T_{k\ell}\boldsymbol{\ell}) = \frac{1}{r_{k\ell}^{3}}\{(\mathbf{k}\boldsymbol{\ell}) - \frac{3(\mathbf{k}\mathbf{r}_{k\ell})(\boldsymbol{\ell}\mathbf{r}_{k\ell})}{r_{k\ell}^{2}}\} \quad (5.86)$$

It is easy to see that the presence of a center or plane of symmetry makes (5.85) equal to zero. The contribution of $\beta_4$ is due to polarizabilities of atomic groups and is directly dependent on their geometrical positions (i.e., on the configuration and conformation of the molecule). Term $\beta_4$ shows the high susceptibility of the optical activity for conformational changes.

The sense and derivation of formula (5.85) can be demonstrated by classical calculation using a simple model [94]. Let the molecule be a dimer, similar to the two-oscillator model of Kuhn (Fig. 5.15), containing two similar linear particles k and ℓ a distance r apart and polarizable only along their axes. The particles (rods) lie in the planes parallel to x,y under the angle 2γ. The energy of electrical interactions of dipoles $\mathbf{p}_k$ and $\mathbf{p}_\ell$ induced in the particles is

$$V_{k\ell} = (\mathbf{p}_k T_{k\ell} \mathbf{p}_\ell) \quad (5.87)$$

The effective field acting on particle k is equal to

$$E_k' = E - T_{k\ell} p_\ell \quad (5.88)$$

and that acting on particle ℓ is equal to

$$E_\ell' = E - T_{k\ell} p_k \quad (5.89)$$

where

$$p_k = \alpha_k E_k', \qquad p_\ell = \alpha_\ell E_\ell'$$

($\alpha_k$, $\alpha_\ell$ are the polarizabilities of the particles). For our model, $(\mathbf{r}_{k\ell}[\mathbf{k}\boldsymbol{\ell}]) = r \sin 2\gamma$, $(\mathbf{k}T_{k\ell}\boldsymbol{\ell}) = (\cos 2\gamma)/r^3 = T$, and according to (5.85)

$$\beta_4 = \frac{1}{6}\alpha_k\alpha_\ell r \sin 2\gamma \frac{\cos 2\gamma}{r^3} \quad (5.90)$$

Let us obtain now the same expression in a direct way. The electric field of a wave directed along the x axis makes the electrons of both particles oscillate in phase. In every particle there is induced the dipole

$$p_x = \alpha(E \cos\gamma - Tp_x) \quad (5.91)$$

The polarizability of an oscillating electron can be found

from its equation of motion

$$\ddot{x}+\omega_0{}^2x = (e/m)E = (e/m)E_0e^{i\omega t} \tag{5.92}$$

We have

$$x = \frac{e}{m}\frac{E}{\omega_0{}^2-\omega^2}$$

and

$$p_x = ex = \frac{e^2}{m}\frac{E}{\omega_0{}^2-\omega^2} = \alpha E$$

The polarizability is

$$\alpha = \frac{e^2}{m}\frac{1}{\omega_0{}^2-\omega^2} \tag{5.93}$$

Inserting the expressions for T and $\omega$ into (5.91), we find

$$p_x = \frac{e^2}{m}\frac{\cos\gamma}{\omega_0{}^2+(e^2/m)(\cos 2\gamma/r^3)-\omega^2}E = \alpha_x E\cos\gamma \tag{5.94}$$

The resulting dipole moment of both particles is equal to

$$P_x = 2p_x\cos\gamma \tag{5.95}$$

Comparison of $\alpha_x$ with $\alpha$ shows that the dipole-dipole interaction changes the frequency

$$\omega_x{}^2 = \omega_0{}^2 + \frac{e^2}{m}\frac{\cos 2\gamma}{r^3} \tag{5.96}$$

The field directed along the y axis provokes oscillation of particles with opposite phases. We get

$$p_y = \alpha(E\sin\gamma + Tp_y) \tag{5.97}$$

and

$$p_y = \frac{e^2}{m}\frac{\sin\gamma}{\omega_0{}^2-(e^2m)(\cos 2\gamma/r^3)-\omega^2}E = \alpha_y E\sin\gamma \tag{5.98}$$

$$P_y = 2p_y\sin\gamma \tag{5.99}$$

$$\omega_y{}^2 = \omega_0{}^2 - \frac{e^2}{m}\frac{\cos 2\gamma}{r^3} \tag{5.100}$$

Electric polarizational interaction of particles produces splitting of the frequency of their oscillations, equal to

$$\omega_x - \omega_y \cong \frac{\omega_x^2 - \omega_y^2}{2\omega_0} = \frac{e^2}{m} \frac{\cos 2\gamma}{\omega_0 r^3} \tag{5.101}$$

Instead of one vibration with frequency $\omega_0$, two vibrations with frequencies $\omega_x$ and $\omega_y$ arise, polarized in perpendicular directions. The summary oscillator strength remains 1, since $\cos^2 \gamma + \sin^2 \gamma = 1$.

Our model is chiral, and we obtain a magnetic moment under the action of a field

$$m = \frac{1}{2c} [r\dot{P}] = \frac{1}{c} \beta \dot{E} \tag{5.102}$$

The magnetic moment component due to vibration with frequency $\omega_x$ directed along the y axis is

$$m_{(x)} = -\frac{1}{2c} r \sin(\gamma)\dot{p}_x = -\frac{\alpha_x}{4c} r \sin(2\gamma)\dot{E} = \frac{1}{c} \beta_x \dot{E} \tag{5.103}$$

and that due to vibration $\omega_y$

$$m_{(y)} = \frac{1}{2c} r \cos(\gamma)\dot{P}_y = \frac{\alpha_y}{4c} r \sin(2\gamma)\dot{E} = \frac{1}{c} \beta_y \dot{E} \tag{5.104}$$

Hence

$$\beta_x = -\frac{\alpha_x}{4} r \sin 2\gamma, \quad \beta_y = \frac{\alpha_y}{4} r \sin 2\gamma, \quad \beta_z = 0 \tag{5.105}$$

The quantity $\beta_4$ is the average value of $\beta_x$, $\beta_y$, and $\beta_z$. We have

$$\begin{aligned} \beta_4 &= \frac{1}{3} (\beta_x + \beta_y) = \frac{1}{12} r \sin 2\gamma (\alpha_y - \alpha_x) \\ &= \frac{1}{12} r \sin 2\gamma \frac{e^2}{m} \left( \frac{1}{\omega_x^2 - \omega^2} - \frac{1}{\omega_y^2 - \omega^2} \right) \end{aligned} \tag{5.106}$$

The rotatory strengths of vibrations $\omega_x$ and $\omega_y$ are of equal absolute value but their signs are opposite. In agreement with (5.67) their sum is zero. Transforming (5.106), we obtain

$$\begin{aligned} \beta_4 &= (1/12) r \sin 2\gamma (e^2/m)(\omega_x^2 - \omega_y^2)[(\omega_y^2 - \omega^2)(\omega_x^2 - \omega^2)]^{-1} \\ &= (1/6) r \sin 2\gamma (\cos 2\gamma / r^3)(e^4/m^2)[(\omega_y^2 - \omega^2)(\omega_x^2 - \omega^2)]^{-1} \\ &= (1/6)\alpha^2 r \sin 2\gamma (\cos 2\gamma / r^3) \end{aligned} \tag{5.107}$$

*quod erat demonstrandum*--our result is formula (5.90).

Thus, the polarization contribution $\beta_4$ to the optical activity is expressed by the product of the polarizabilities. Frequencies $\omega_x$ and $\omega_y$ in the dispersion formula correspond to electric dipole transitions responsible for the polarizability.

Therefore, these are transitions with large dipole strengths, corresponding to strong absorption bands. However, rotatory strength is determined by products $(0|\mathbf{p}|j)(j|\mathbf{m}|0)$. Large values of $(0|\mathbf{p}|j)$ and hence of $D_j$ usually correspond to small $(j|\mathbf{m}|0)$, and vice versa. Weak absorption bands with small $(0|\mathbf{p}|j)$ but large $(j|\mathbf{m}|0)$ can influence the optical activity. Such bands do not contribute considerably to absorption and polarizability. The term $\beta_4$ does not take into account weak bands.

Assumptions have been made in the calculation of $\beta_4$ that are not always valid. The dipole-dipole interaction expression (5.86) assumes that the distance between dipoles is much larger than their sizes (p. 178). This assumption may not be valid inside the molecule. On the other hand, polarization theory is applicable in studies of the influence of intermolecular interaction on optical activity [95].

Consider the term $\beta_2$, which can be calculated by means of the so-called one-electron model suggested in the work of Condon *et al.* [96] (see also [83,84,97]). This theory is based on the quantum-mechanical Rosenfeld formula but it is limited by consideration of the electrons of chromophoric groups responsible for long-wavelength absorption bands. A chromophore (e.g., C=O, peptide bond —NH—CO—, etc.) is itself symmetrical and optically nonactive. But in the asymmetrical milieu of neighboring atoms it makes a contribution to $\beta$. The problem is solved by means of perturbation theory. The perturbation potential of atoms can be made up of the potentials of the central, dipole, and ionic forces. It is necessary to dispose of nonperturbed wave functions of the chromophore. The one-electron model actually reproduces the dispersion of optical activity and especially ADOA (Cotton effect) and CD in the absorption band of the chromophore. Using this model, it becomes possible to calculate the rotatory strengths of chromophore groups in a satisfactory way.

A change in conformation alters the perturbation potential. Therefore, the one-electron theory may be effectively applied in conformational analysis. In a series of works of Djerassi and others, data obtained for ketones, terpenes, steroids, etc., were interpreted on the basis of the one-electron theory. Conformations of many substances were determined and important general rules were formulated [98-100]. Detailed analysis of the relations between the four terms of $\beta$ has been carried out in references [101,102].

Modern intensive development of spectropolarimetry is directly related to studies of natural compounds, which are becoming the most important topics in organic chemistry and molecular biology. Applications of the dispersion of optical activity, expecially of ADOA and CD, have fostered this development. Earlier, only specific rotation at one wavelength was used. We have to emphasize that the first systematic studies of the dispersion of optical activity were made by Chugajev, who found important regularities (cf. e.g., [103]). Only later were these ideas developed by Kuhn [79,104].

The modern devices for the measurement of rotation and CD, spectropolarimeters and dichrographs, must meet rather high requirements. High sensitivity (rotation measurements up to $10^{-4}$ deg) is necessary, and the ability to work in a broad range of wavelengths. Thus for proteins measurements in the 180- to 230-nm region are required. These requirements are realized in the best modern devices, which are described in [100].

## 5.6 The Theory of Optical Activity of Biopolymers

In the case of polymers it is reasonable to express optical activity and CD quantities per monomer. Instead of [M], the mean monomer rotation is used

$$[m] = \frac{M_0}{100}[\alpha] \tag{5.108}$$

where $M_0$ is the MW of the monomer. The influence of the solvent can be accounted for in a rough way by introducing the Lorentz correction. Therefore, instead of [m], the reduced mean monomer rotation is used

$$[m'] = \frac{3}{n^2+2}[m] = \frac{3}{n^2+2}\frac{M_0}{100}[\alpha] \tag{5.109}$$

where n is the refraction index of the medium. For water at 20°C, $3/(n^2+2)$ varies from 0.7945 (at 600 nm) to 0.7306 (at 185 nm).

The double chirality of α-helices of proteins and polypeptides, due to the asymmetry of L-amino acid residues and the asymmetry of the helix itself, determines the DOA and CD of such molecules. Conformational changes of proteins produce sharp changes in the DOA and CD.

Far from the absorption region, the rotation of any molecular system is expressed as a sum of one-term Drude formulas [cf. (5.68)]

$$[m'] = \sum_i \frac{a_i\lambda_i^2}{\lambda^2-\lambda_i^2} \tag{5.110}$$

Let us expand this expression in a series of $(\lambda^2-\lambda_0^2)^{-1}$ where $\lambda_0$ has to be defined [105]:

$$[m'] = \sum_i \left\{ \frac{a_i\lambda_i^2}{\lambda^2-\lambda_0^2} + \frac{a_i\lambda_i^2(\lambda_i^2-\lambda_0^2)}{(\lambda^2-\lambda_0^2)} + \frac{a_i\lambda_i^2(\lambda_i^2-\lambda_0^2)^2}{(\lambda^2-\lambda_0^2)^3} + \cdots \right\} \tag{5.111}$$

The series converges rapidly if $\lambda_i << \lambda$. Let us rewrite (5.111) in the form

$$[m'] = \frac{a_0\lambda_0^2}{\lambda^2-\lambda_0^2} + \frac{b_0\lambda_0^4}{(\lambda^2-\lambda_0^2)^2} + \frac{c_0\lambda_0^6}{(\lambda^2-\lambda_0^2)^3} + \cdots \tag{5.112}$$

where

$$a_0\lambda_0^2 = \sum_i a_i\lambda_i^2, \qquad b_0\lambda_0^4 = \sum_i a_i\lambda_i^2(\lambda_i^2-\lambda_0^2)$$

$$c_0\lambda_0^6 = \sum_i a_i\lambda_i^2(\lambda_i^2-\lambda_0^2)^2$$

If $b_0 = c_0 = \cdots = 0$, then Eq. (5.110) becomes the one-term Drude formula

$$[m'] = \frac{a_0\lambda_0^2}{\lambda^2-\lambda_0^2} = \frac{K}{\lambda^2-\lambda_0^2} \tag{5.113}$$

If $b_0 \neq 0$, but $c_0 = \cdots = 0$, then

$$[m'] = \frac{a_0\lambda_0^2}{\lambda^2-\lambda_0^2} + \frac{b_0\lambda_0^4}{(\lambda^2-\lambda_0^2)^2} \tag{5.114}$$

In the coil state the DOA of polyamino acids is satisfactorily described by (5.113) with $\lambda_0 = 268$ nm. For the α-helix Moffitt derived a theoretical formula, similar to (5.114) [106], which agrees well with experiment. Let us rewrite it in the form

$$[m'](\lambda^2/\lambda_0^2-1) = a_0 + \frac{b_0}{(\lambda^2/\lambda_0^2-1)} \tag{5.115}$$

The function $[m'](\lambda^2/\lambda_0^2-1)$ must depend linearly on $(\lambda^2/\lambda_0^2-1)^{-1}$ if the correct parameter $\lambda_0$ is chosen. Agreement with experiment occurs at $\lambda_0 = 212$ nm and $b_0 = -630$, whereas $a_0 = 630$-$650$. Equation (5.114) can be treated as empirical. Moffitt has applied the theory of the exciton spectrum of regular polymers in his calculations (cf. p. 276). As has already been said, Moffitt obtained the splitting of the amide bands of an α-helix into components parallel and perpendicular to its axis. For one such pair of components we get the two-term Drude equation

$$[m'] = \frac{a_1\lambda_1^2}{\lambda^2-\lambda_1^2} + \frac{a_2\lambda_2^2}{\lambda^2-\lambda_2^2} \quad (5.116)$$

which coincides with (5.114) if $a_1$ and $a_2$ have opposite signs and

$$\lambda_0^2 = \frac{\lambda_1^2|a_1\lambda_1|^{1/2}+\lambda_2^2|a_2\lambda_2|^{1/2}}{|a_1\lambda_1|^{1/2}-|a_2\lambda_2|^{1/2}}$$

Hence, in general $\lambda_0 \neq \frac{1}{2}(\lambda_1+\lambda_2)$. However, if $|a_1\lambda_1^2| = |a_2\lambda_2^2|$, then $\lambda_0 \cong \frac{1}{2}(\lambda_1+\lambda_2)$ and if $|a_1| = |a_2|$, then $\lambda_0^2 = \lambda_1^2+\lambda_2^2-\lambda_1\lambda_2$. In both these cases we can write $\lambda_{1,2} = \lambda_0[1\mp(\Delta/2)]$ [107] and $\lambda_0 \cong \frac{1}{2}(\lambda_1+\lambda_2)$ if $\Delta \ll 1$.

Let us examine the theoretical calculation in more detail. If there are no interactions between monomers in a regular homopolymer (N-mer), the wave function which expresses its electronic ground state has the form

$$\psi_0 = \prod_{i=1}^{N} \phi_{i0} \quad (5.117)$$

where $\psi_{i0}$ is the wave function of the ith monomer. At excitation of the ith monomer $0 \to a$, we get

$$\psi_{ia} = (\psi_0/\phi_{i0})\phi_{ia} \quad (5.118)$$

The wave functions $\psi_{ia}$ are degenerate. The functions of an excited state of the polymer are expressed as linear combinations

$$\psi_{ak} = \sum_{i=1}^{N} c_{iak}\psi_{ia} \qquad (k = 1,2,\cdots,N) \quad (5.119)$$

Because of resonance interaction the excited level a is split into N levels, forming a zone. The energies of the zone levels $\varepsilon_{ak}$ and the coefficients $c_{iak}$ are found by solving the system of N equations

$$\sum_{i=1}^{N} c_{iak}V_{i0a,j0a} - \varepsilon_{ak}c_{jak} = 0 \qquad (j,k = 1,2,\cdots,N) \quad (5.120)$$

where the matrix element

$$V_{i0a,j0a} = (\phi_{ia}{}^*\phi_{i0}|V_{ij}|\phi_{ja}\phi_{j0}{}^*) \quad (5.121)$$

represents the exchange of excitation between the ith and jth monomers. $V_{ij}$ is the dipole-dipole interaction potential of the excited monomers. The energies of zone levels are

$$\varepsilon_{ak} = \varepsilon_a + \varepsilon'_{ak} + G \qquad (5.122)$$

where $\varepsilon_a$ is the energy of the $0 \to a$ transition in a monomer, $\varepsilon'_{ak}$ is the energy of the kth sublevel of the zone, and the quantity

$$G = \sum_{j \neq i} \{(|\phi_{ia}|^2 |V_{ij}| |\phi_{j0}|^2) - (|\phi_{i0}|^2 |V_{ij}| |\phi_{j0}|^2)\} \qquad (5.123)$$

represents the difference in the interaction energy of the excited and nonexcited ith monomer with the others.

The transition dipole strength is

$$D_{ak} = \frac{1}{N} \left| (\psi_0 | \sum_{i=1}^{N} p_i | \psi_{ak}) \right|^2 \qquad (5.124)$$

and the rotatory strength is

$$R_{ak} = \frac{1}{N} \operatorname{Im}\{ (\psi_{ak} | \sum_{i=1}^{N} m_i | \psi_0)(\psi_0 | \sum_{i=1}^{N} p_i | \psi_{ak}) \} \qquad (5.125)$$

This exciton theory is equivalent to applying the polarization approximation (p. 289) to degenerate states. In the case of degeneration, optical activity already arises in the zero approximation of perturbation theory.

The presence of long-range order in a linear chain makes it possible to simplify the system of equations (5.120) by taking into account translational symmetry and using cyclic boundary conditions. This is precisely the method used by Moffitt. However, it has since been shown that cyclic boundary conditions cannot be used in calculations of the rotatory strength (though they are necessary in calculating dipole strength) [108] and a system of equations of order N has to be solved. The solution has the form

$$[M_a] = \frac{b_a \omega_a^2 \omega^2}{(\omega_a^2 - \omega^2)^2} = \frac{b_a \lambda_a^2 \lambda^2}{(\lambda^2 - \lambda_a^2)^2} \qquad (5.126)$$

where

$$b_a = b_a^{(1)} + b_a^{(2)}$$

$$b_a^{(1)} = -\frac{192 N_1}{h^2 c^2} |p_{0a}|^2 \rho e_{at} e_{av} \sum_{j \neq 1} \sum_i V_{i0a,j0a} \sin^2 \frac{\pi (j-i)}{Q}$$

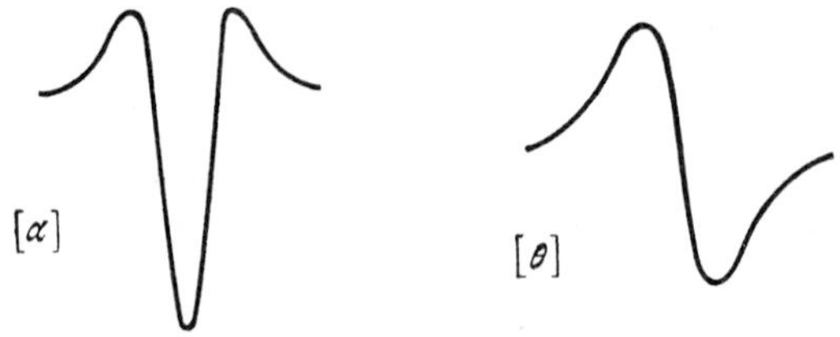


*FIG. 5.17 ADOA and CD calculated for an α-helix.*

$$b_a^{(2)} = \frac{48N_1}{h^2c^2}|p_{0a}|^2 z(e_{at}^2+e_{ar}^2) \sum_{j\neq i} \sum_i V_{i0a,j0a}(j-i) \frac{\sin 2\pi(j-i)}{Q}$$

and $\rho$ is the diameter of the α-helix; z its step; $e_{av}$, $e_{at}$, $e_{ar}$ cosines of the angles formed by the transition moment $p_{0a}$ in relation to the axis of the helix; its tangent; and its radius. Q is the number of monomers per spire.

Expression (5.126) corresponds to symmetrical ADOA and asymmetrical CD curves (Fig. 5.17). The use of cyclic boundary conditions results in the loss of term $b_a^{(2)}$.

Developing these ideas, Tinoco and other authors calculated the rotatory strength and CD of α-helices in a more rigorous way [109-111]. Not only the resonance interaction of equal electron transitions has been examined, but also the nonexciton interaction of various transitions. As has already been said, exciton interaction is taken into account by the zero-order wave functions of perturbation theory. For estimation of a nonexciton contribution it is necessary to use first-order wave functions of the perturbation theory.

Actual evaluation of rotatory strengths $R_{ak}$ depends on the kind of $0 \to a$ transition. If it is an optically strong transition, then $p_{0a}$ is large; if a weak one, $p_{0a}$ is small but $m_{a0}$ may be large. For strong transitions the exciton contribution $R_{ak}^e$ can be singled out due to interactions of $0 \to a$ transitions in different monomers

$$R_{ak}^e = \frac{\omega_a}{2c} \sum_{j=i} \sum_i c_{iak}c_{jak}(r_i-r_j)[m_{i0a}m_{j0a}] \quad (5.127)$$

$r_i - r_j$ is the distance between links i and j; k = 1,2,...,N is the number of the level produced by exciton splitting. The transition frequencies $\omega_a$ and coefficients $c_{iak}$ are determined from eigenvalues and eigenvectors of the Nth-order matrix, whose elements are determined by the interaction energy of strong transitions of different monomers. The sum of the exciton rotatory strengths over all values of k from 1 to N is zero--these contributions are "conservative."

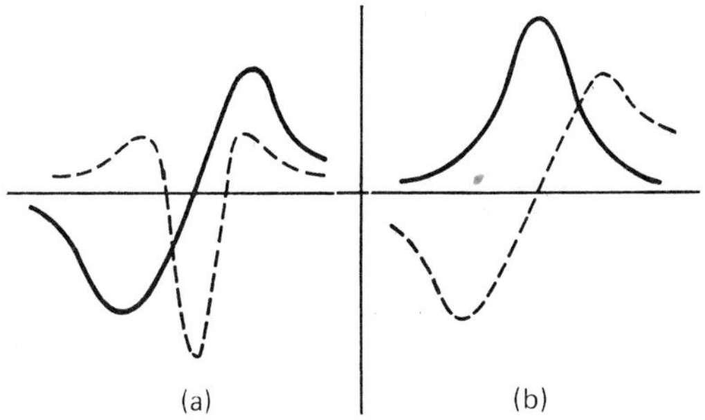


*FIG. 5.18 ADOA and CD curves for the (a) exciton transition (conservative contribution) and (b) nonexciton transition (nonconservative contribution).*

The nonexciton interaction of the 0 → a transition with other allowed 0 → b transitions gives contributions of the type

$$R_{ak}^{\ b} = -\frac{1}{c\hbar}\sum_i \sum_j \sum_{\ell \neq j} c_{iak} c_{jak} \frac{\omega_a \omega_b V_{j0a,\ell 0b} (r_\ell - r_i)[m_{\ell 0b} m_{i0a}]}{\omega_b^2 - \omega_a^2} \tag{5.128}$$

Transitions in the far-ultraviolet region of the spectrum can influence considerably the rotatory strength of the 0 → a transition. Estimation of corresponding $R_{ak}{}^b$ contributions can be done by means of polarization theory. Nonexciton contributions are not conservative [112,113]. Figure 5.18 shows typical DOA and CD curves for the exciton (conservative) and nonexciton (nonconservative) contributions [113].

Contributions of weak transitions to rotation can be evaluated without considering the splitting (i.e., by summation over k). These contributions are

$$R_{b,a}^{(1)} = -\frac{1}{\hbar}\sum_{j \neq i}\sum_i \frac{\mathrm{Im}(V_{iab,j00} m_{ia0} p_{i0b})}{\omega_b - \omega_a} \tag{5.129}$$

$$R_{b,a}^{(2)} = -2\sum_{j \neq 1}\sum_i \omega_b \frac{\mathrm{Im}(V_{i0a,j0b} m_{ia0} p_{j0b})}{\omega_b^2 - \omega_a^2} \tag{5.130}$$

$R_{ba}^{(1)}$ is the fundamental term in the one-electron theory of optical activity (p. 293).

Exciton theory has been applied in calculations of the ADOA and CD of the α-helix [111], β form [114], and poly-L-proline [115]. See also [116,117].

In [118] detailed calculations have been made of the contributions to α-helical CD and ADOA due to π π* transitions of peptide groups. Both exciton and nonexciton interactions have been taken into account. Table 5.5 shows the results obtained.

*TABLE 5.5*
*Calculated Values of Exciton and Nonexciton Rotatory Strengths R*[a]

| N | Δλ (nm) | $R_{exciton}$ | $R_{nonexciton}$ |
|---|---|---|---|
| 20 | -5.2 | -221 | 3.8 |
| | -0.5 | 321 | - 8.6 |
| | +7.2 | -100 | -17.1 |
| 40 | -4.5 | -393 | |
| | -1.6 | 496 | |
| | +7.7 | -103 | |

[a] Values are given in $10^{-40}$ erg $cm^3$; calculations are for exciton interactions in π π* transitions and for nonexciton interactions in an α-helix.

Δλ is the difference between the wavelength of the exciton components of the π π* transition and the wavelength $\lambda^{\alpha}_{\pi\pi*}$ of the given group in the absence of transition interactions. The sums of $R_{exciton}$ are zero. Good agreement with the experimental CD value for the π π* transition (190 nm) is obtained if the half width Γ of the band is 3500 $cm^{-1}$ (∿13 nm). These results agree with those obtained in the work [114], though only the interactions of the 10 nearest neighbors were considered there, whereas in [118] all interactions in the chain were taken into account. Nonexciton contributions due to interaction of the π π* transition with other transitions of the peptide group and with anisotropically polarized groups of the α-helix are presented in the last column of the table. We see that nonexciton contributions are comparatively small in this case

Figure 5.19 shows the calculated and experimental CD curves of an α-helix in the region from 190 to 230 nm. Agreement is quite satisfactory in the 190 to 210 nm region. Two observed bands of CD, the positive at 190 nm and the negative at 208 nm, are due to splitting of the exciton contribution of the π π* transition. The negative CD band at 220 nm is due to the weak n π* transition. Calculation gives $R_{n\pi*} = -3\times10^{-40}$ erg $cm^3$ [111,114] and experiment $-15\times10^{-40}$ erg cm . The discrepancy is due to the nonexactness of the wave functions used in [111] and [114].

Far from the absorption region, the DOA of an α-helix is satisfactorily represented by the Moffitt expression (5.114) with $b_0 = -630$. For a random coil $b_0$ is small and the DOA is described by the simple Drude formula. Another regular form of the polypeptide chain, the β form, is in principle also

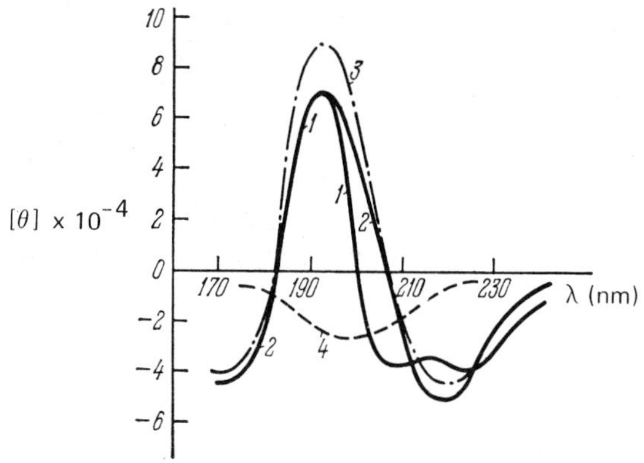


*FIG. 5.19 Calculated (1) and experimental (2) CD curves for an α-helix. Curve 1 is the sum of an exciton (3) and a nonexciton (4) contribution.*

represented by Moffitt's formula but with a small value of $b_0$. The main distinction from the coil is the increased right rotation of the plane of polarization [119-127]. Theoretical calculation of the DOA of the parallel and antiparallel β forms is done in [128], and that of the CD in [118]. The calculation gives the value -30 for $b_0$, which is in good agreement with experiment. Detailed analysis of the CD of the π π* transition gives the results listed in Table 5.6.

*TABLE 5.6*
*Rotatory Strengths*[a]

| Contribution | Δλ (nm) | Exciton contribution | Nonexciton contribution | Sum |
|---|---|---|---|---|
| Parallel β form | -2 | +4.4 | +17.0 | +21.4 |
| | +9.5 | -4.4 | + 1.9 | - 2.5 |
| Sum | | 0 | +18.9 | +18.9 |
| Antiparallel β form | -13.5 | -1.6 | + 9.6 | + 8.0 |
| | + 1.5 | +1.6 | +14.0 | +15.6 |
| Sum | | 0 | +23.6 | +23.6 |

[a]Values in $10^{-40}$ erg $cm^3$.

In the case of the β form, the nonexciton contribution dominates. In experimental CD spectra of the antiparallel β forms of poly-L-lysine and silk fibroin [125], a positive CD band is observed in the 190 to 200 nm π π* transition region with $R = 15.25\times10^{-40}$ erg $cm^3$, and a weaker negative band with $R = -6\times10^{-40}$ erg $cm^3$ in the 215 to 220 nm region. The latter band is due to the n $\pi^*$ transition [129].

The general theory of the optical activity of coiled macromolecules, based on the statistical averaging over all conformations of the tensor which determines the optical activity, is developed in [130]. The theory of the CD of random coils of polypeptides shows the possibility of using the exciton theory. In this case comparatively small ensembles of short nonordered chains can be considered. Both exciton and nonexciton contributions are essential [131].

To explain the optical activity of nucleic acids, the phenomenon of induced optical activity (IOA) has to be examined. Symmetrical (i.e., nonchiral) molecules of dyes bound by $\alpha$-helical polypeptides show ADOA and CD in their absorption bands. This effect vanishes upon denaturation of the $\alpha$-helix complex with dye [132,133]. Several ways of binding the dye (a symmetrical chromophore) by the $\alpha$-helix have been suggested: (1) interaction of a nonaggregated molecule of the dye with a peptide residue near the asymmetrical center; (2) formation of a superhelix by the dye molecules on the $\alpha$-helix; (3) aggregation of the dye molecules at some site of the $\alpha$-helix to form a helical polymer [132,133]. Experiments made with acridine orange, in concentrations very small by comparison with the protein concentration, confirm the first model [134], as does the IOA of prosthetic groups and coenzymes. The ADOA and CD in the absorption bands of pyridoxal phosphate (the coenzyme of aspartate aminotransferase) served as a valuable information source concerning the structure of the active site of this enzyme (p. 373) [135]. Figure 5.20 shows the ADOA curves of deoxy- and oxyhemoglobin in the absorption bands of the prosthetic heme group [136]. Under the influence of the structural asymmetry of a biopolymer, asymmetry of the electron shell of the chromophore arises. Similar results have been obtained in studies of the ADOA of complexes formed by DNA with acridine dyes [137]. In these cases, asymmetrical dimers of the dye

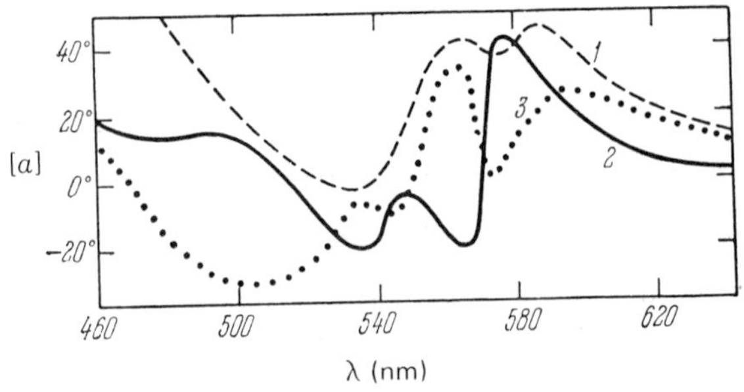


*FIG. 5.20 The ADOA of (1) deoxyhemoglobin, (2) oxyhemoglobin, and (3) carboxyhemoglobin.*

molecules were observed in the complex of DNA, possessing symmetrical ADOA curves. Quantum-chemical calculation, which applies the dye molecule model as a potential well undergoing asymmetrical perturbation, corroborates these ideas [138].

Actually, IOA is considered in the one-electron model (p. 293). The IOA theory was perfected recently [139] when it was shown that in the rigorous theory of IOA the electronic-vibrational problem in adiabatic approximation has to be solved. Changes in the equilibrium positions of atomic nuclei due to asymmetrical perturbations have to be taken into account. The rotatory strength is expressed by a formula of the type

$$R_{v_1,\cdots,v_s} = \{A + \sum_a B_a Q_a{}^{(0)} + \sum_a C_a y_a[1-(v_a/y_a{}^2)]\}W_{v_1,\cdots,v_s} \tag{5.131}$$

where $Q_a{}^{(0)}$ is the zero amplitude of the ath normal vibration; $v_a$, the vibrational quantum number; $y_a = \Delta Q_a/Q_a{}^{(0)}$, the relative change in amplitude; W, a quantity proportional to the intensity of absorption. In the one-electron model, the two-atom chromophore group was considered (p. 293) and the new correction terms became zero. Theory shows that the CD spectrum in the case of IOA can differ from the absorption band if $v_a \neq 0$. This is actually observed.

It has to be emphasized that consideration of the vibronic structure of electron absorption bands in the theory of optical activity is an important, as yet unsolved, problem (cf. [85]).

The optical activity of nucleosides, nucleotides, DNA and RNA is characterized, first, by the Cotton effects (i.e., the ADOA) and CD in the absorption bands, due to nitrogen bases, near 260 nm. Planar nitrogen bases are not optically active. ADOA and CD arise because these bases link up with asymmetrical sugars. Hence, these effects are due to IOA. Actually, the substitution for β-ribose or β-deoxyribose of the corresponding α compound changes the sign of the effect [140]. The ADOA of 5'-deoxymononucleotides is shown in Fig. 5.21 [141]. (This effect depends strongly on the pH of the medium.)

Johnson and Tinoco developed a simplified theory of the CD of polynucleotides and, hence, of DNA and RNA [142]. Both exciton (conservative) and nonexciton (nonconservative) contributions are taken into account. It is suggested that the CD for every monomer can be expressed as $\nu f(\nu-\nu_k)$, where f describes the form of the band and $\nu_k$ is the frequency at its maximum. Then the CD of a polymer is

$$\varepsilon_L-\varepsilon_D = \nu \sum_{k=1}^{N} R_k f(\nu-\nu_k) \tag{5.132}$$

The quantity $(\varepsilon_L-\varepsilon_D)/\nu$ is expanded in a series near the mean

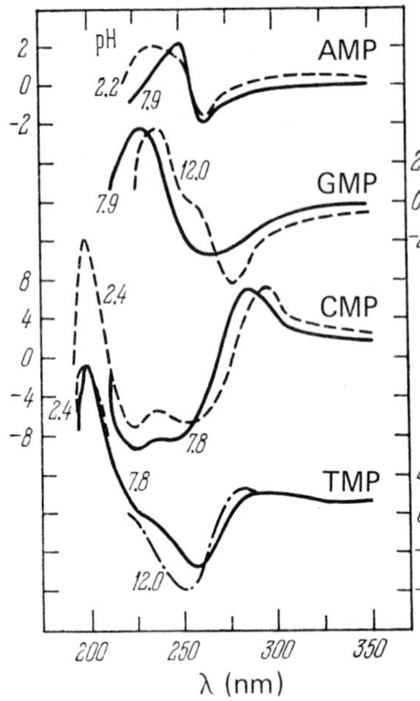


*FIG. 5.21 The ADOA of 5'-deoxymononucleotides. Numerals adjacent to curves designate the pH of the medium.*

frequency ν of the actual spectral region

$$\frac{\varepsilon_L-\varepsilon_D}{\nu} = \sum_{K=1}^{N} R_K f(\nu-\bar{\nu}) - \sum_{K=1}^{N} R_K \frac{\partial f(\nu-\bar{\nu})}{\partial \nu} (\nu_K-\bar{\nu})+\cdots \tag{5.133}$$

Such a presentation enables us to avoid solving the eigenvalue problem. In calculating rotatory strengths, the potential of monopole interactions is employed

$$V_{i0a,j0b} = \sum_{s,t} \frac{\rho_s^{i0a}\rho_t^{j0b}}{r_{st}}$$

where $\rho_s^{i0a}$ is the monopole charge of group s of monomer i corresponding to the $0 \rightarrow a$ transition, and $r_{st}$ is the distance between group s of monomer i and group t of monomer j.

The theory gives the CD value

$$\varepsilon_L-\varepsilon_D = \frac{\pi\nu}{c} f(\nu-\bar{\nu}) \sum_{ij}{}' \sum_{a} \nu_{i0a}\Big(\sum_{s,t} \frac{\rho_s^{i0a}}{r_{st}^3} r_{st}\Big)\alpha_t^j \times P_{i0a}\cdot R_{ij}$$

$$+ \frac{\pi\nu}{2c} \frac{\partial f(\nu-\bar{\nu})}{\partial \nu} \sum_{ij} \sum_{a,b} \nu_{i0a}\Big(\sum_{s,t} \frac{\rho_s^{i0a}\rho_t^{j0b}}{r_{st}}\Big) R_{ij}\cdot P_{i0a}\times P_{j0b} \tag{5.134}$$

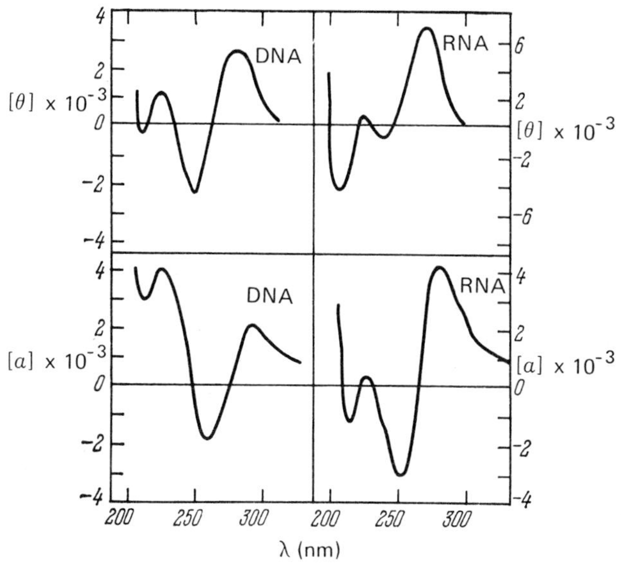


*FIG. 5.22 CD and ADOA curves of DNA and RNA.*

where $\alpha_t^j$ is the polarizability of the tth group of the jth monomer. Calculation requires only knowledge of the monomer properties and of their mutual orientation. The function $f(\nu-\bar{\nu})$ is taken as the average over the spectra of the bases. The satisfactory agreement of theory and experiment shows that the CD of polynucleotides is due mainly to $\pi\ \pi^*$ and not to $n\ \pi^*$ transitions. Figure 5.22 shows the CD and ADOA curves for DNA and RNA [143]. These effects are rather small due to mutual compensation of rotatory strengths of a series of bands.

## 5.7 Spectropolarimetry of Biopolymers

We see that various conformations of polypeptides possess different DOA, ADOA, and CD curves (Fig. 5.23). Table 5.7 [144] contains characteristic values of maxima and minima of ADOA curves.

Thus measurements of the ADOA give information about the conformation of a system. Evidently the CD gives clearer results, since CD is due solely to the rotatory strength of the given-band. A pure α-helix is characterized by ellipticities $[\theta] = 3{,}300\ (\varepsilon_L-\varepsilon_D)$ equal for 206 nm, -36,000; for 222 nm, -38,000; and for 190 nm, +71,000 [145]. These three CD bands are typical for an α-helix. A good measure of α-helicity is $[\theta]_{222}$, the residual ellipticity of protein in a disordered state. In the work [145] a series of examples of such

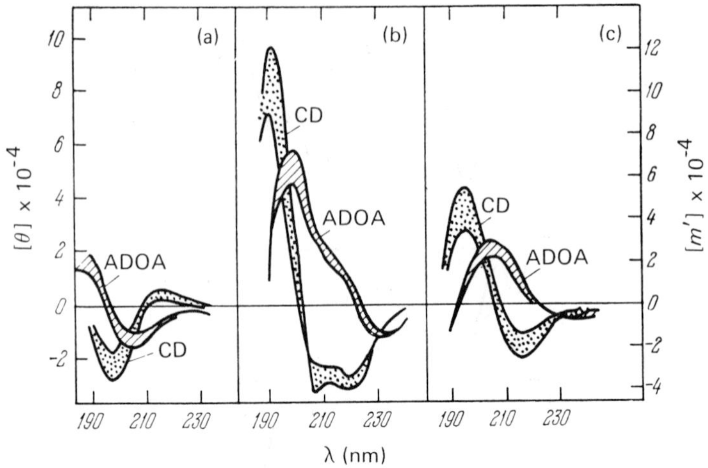


*FIG. 5.23 ADOA curves of L-polypeptides in (a) coil, (b) α-helix; (c) β form.*

*TABLE 5.7*
*ADOA Curves of Polypeptides*

| Conformation | Minimum (nm) | Intersection point $\lvert [M] = 0 \rvert$ (nm) | Maximum (nm) |
|---|---|---|---|
| α-Helix | 232–233 | 224 | 198–199 |
| | 182–184 | 190 | |
| β Form | 229–230 | 220 | 205 |
| | 190 | 196 | |
| Nonordered | 238 (small) | -- | 228 (small) |
| | 204–205 | 198 | 189 |
| Poly-L-proline II | 210 | 203 | 194 |

estimations is presented (Table 5.8; see also [146]).

The CD makes it possible to determine not only α-helices, but also β forms and nonordered links in protein. An empirical method has been suggested for this purpose [147]. The CD curves for three proteins (ribonuclease, lysozyme, and myoglobin), for which the content of α, β, and nonordered forms (x, y, z) are known from X-ray data, are measured. We obtain a set of equations for the three proteins

$$[\theta]_1 = x_1[\theta]_\alpha + y_1[\theta]_\beta + (1-x_1-y_1)[\theta]_r$$

$$[\theta]_2 = x_2[\theta]_\alpha + y_2[\theta]_\beta + (1-x_2-y_2)[\theta]_r$$

$$[\theta]_3 = x_3[\theta]_\alpha + y_3[\theta]_\beta + (1-x_3-y_3)[\theta]_r$$

From experimental values of $[\theta]_1$, $[\theta]_2$, and $[\theta]_3$ and from the

*TABLE 5.8*
*CD of Proteins in Different Conditions*

| Protein | Solvent | $[\theta]_{222}\times10^4$ | Degree of $\alpha$-helicity $x_\alpha$ |
|---|---|---|---|
| Lysozyme | pH 7.0, phosphate buffer | -1.30 | (1.30-0.29)/3.8=0.27 |
| | 8 M urea | -1.26 | (1.26-0.29)/3.8=0.26 |
| | 5 M guanidine-HCl | -0.29 | 0 |
| | 2-Chloroethanol | -3.0 | (3.0 -0.29)/3.8=0.7 |
| Insulin | pH 7.0, phosphate buffer | -1.46 | (1.46-0.64)/3.8=0.21 |
| | 8 M urea | -0.64 | 0 |
| Hemoglobin | pH 7.0, phosphate buffer | -3.1 | (3.1 -0.34)/3.8=0.72 |
| | 8 M urea | -0.34 | 0 |

values of the x and y parameters, $[\theta]_\alpha$, $[\theta]_\beta$, $[\theta]_r$ are estimated for any $\lambda$. With this method, values of x and y that agree well with the results of X-ray analysis have been found for carboxypeptidase A and $\alpha$-chymotrypsin.

Conformational determinations are more accessible if DOA curves far from the absorption region are used. Applying the Moffitt equation, we can estimate the $\alpha$-helicity $x_\alpha$ as the ratio of the measured $b_0$ value to the canonical one

$$x_\alpha = b_0/b_0^\alpha = -b_0/630$$

On the other hand, the constant $a_0$ for a partly helical protein can be expressed as

$$a_0 = a_0^{aa}+x_\alpha a_0^\alpha$$

where $a_0^{aa}$ is due to single amino acid residues. In other terms

$$[M] = a_0^{aa}\frac{\lambda_1^2}{\lambda^2-\lambda_1^2} + x_\alpha\left(a_0^\alpha\frac{\lambda_0^2}{\lambda^2-\lambda_0^2} + b_0^\alpha\frac{\lambda_0^4}{(\lambda^2-\lambda_0^2)^2}\right) \qquad (5.135)$$

and, expanding the first term into a series in $(\lambda^2-\lambda_0^2)^{-1}$, we obtain

$$[M] = a^{aa}\frac{\lambda_0^2}{\lambda^2-\lambda_1^2} +b_0^{aa}\frac{\lambda_0^4}{(\lambda^2-\lambda_0^2)^2} +x_\alpha\left(a_0^2\frac{\lambda_0^2}{\lambda^2-\lambda_0^2} +b_0^\alpha\frac{\lambda_0^4}{(\lambda^2-\lambda_0^2)^2}\right)$$

(5.136)

Consequently

$$x_\alpha = \frac{b_0 - b_0^{aa}}{b_0^{\alpha} - b_0^{aa}} = -\frac{b_0 - b_0^{aa}}{630}$$

Usually $b_0^{aa}$ is rather small, of the order of ±50. In the work [148] another expression has been applied which gives practically the same results. The value of $a_0^{\alpha}$ is 630-650 deg cm$^2$ decimole$^{-1}$. The value of $a^{aa}$ can be found from the DOA of denatured protein.

The dependence of $[M](\lambda^2/\lambda_0^2-1)$ on $(\lambda^2/\lambda_0^2-1)^{-1}$ is rectilinear. The slope of the straight line is $x_\alpha b_0^{\alpha}$ and it intersects the ordinate at $a_0^{aa}+x_\alpha a^{\alpha}$. Hence, $x_\alpha$ can be determined in two independent ways. It has to be emphasized that aromatic amino acid residues greatly influence the value of Moffitt constants.

Another empirical method for determining $x_\alpha$ applies the one-term Drude formula

$$[M] = K/(\lambda^2-\lambda_0^2)$$

suggesting linear dependence of $\lambda_0$ on $x_\alpha$. For $x_\alpha = 0.4$, $\lambda_0 = 268$ nm; for $x_\alpha = 0$ (coil), $\lambda_0 = 212$ nm.

Blout and Shechter suggested the application of a kind of two-term Drude formula [130,149,150]. Instead of an effective $\lambda_0$ value, this expression contains experimentally observable absorption maxima $\lambda_1 = 193$ nm and $\lambda_2 = 225$ nm

$$[M] = \frac{a_{193}\lambda_1^2}{\lambda^2-\lambda_1^2} + \frac{a_{225}\lambda_2^2}{\lambda^2-\lambda_2^2} \qquad (5.137)$$

where $a_{193}$ is positive and $a_{225}$ negative. These constants appear to be linearly interdependent. In aqueous solutions of polyamino acids (dielectric constant $\varepsilon > 30$)

$$a_{225} = -0.55a_{193}-430 \qquad (5.138)$$

and in organic solvents ($\varepsilon < 30$)

$$a_{225} = -0.55a_{193}-280 \qquad (5.139)$$

However, Eq. (5.139) is less reliable [144]. Using the data for poly-L-glutamic acid in water at pH 4 (helix) and at pH 7 (coil), Shechter and Blout determine $x_\alpha$ as

$$x_\alpha = (a_{193}+750)/3650 \qquad (5.140)$$

or

$$x_\alpha = -(a_{225}+60)/1990 \qquad (5.140a)$$

It is suggested that these values coincide and, hence, experimental values of $a_{193}$ and $a_{225}$ must obey the linear expression (5.138) if the protein contains only α-helical and nonordered conformations. The deviation from linearity indicates the presence of other structures (e.g., the β form). Actually this method (like the method using $a_0$ if the Moffitt equation is applied) is rather rough for such conclusions. Further details are given in [133] and [144].

In Table 5.9, the values of $x_\alpha$ obtained by various methods for a series of proteins are compared. As indicated in the table, these data are generally in reasonable agreement with the results of X-ray analysis of proteins. Thus, DOA, ADOA, and CD are effective in determinations of the α-helicity of proteins and polypeptides. On the other hand, important results can be obtained by investigating the IOA of complexes of proteins with dyes and metal ions, and the IOA of coenzymes and prosthetic groups. Such investigations give information about conformations, and when combined with chemistry they make it possible to decipher events occurring at the active sites of enzymes (cf. Chapter 6 and [135]).

*TABLE 5.9*
*Degree of α-Helicity of Proteins in Aqueous Solution*

| | Determined from | | | |
|---|---|---|---|---|
| Protein | $b_0$ | $a_0$ | $a_{193}$ | $a_{225}$ |
| Tropomyosin | 0.90 | 0.90 | 0.85 | 0.85 |
| Hemoglobin[a] | 0.70-0.80 | -- | -- | -- |
| Myoglobin[a] | 0.70-0.80 | -- | -- | -- |
| Serum albumin | 0.45 | 0.60 | 0.55 | 0.55 |
| Insulin | 0.40 | 0.60 | -- | -- |
| Fibrinogen | 0.35 | -- | 0.35 | 0.30 |
| Ovalbumin | 0.30 | 0.50 | -- | -- |
| Lysozyme[a] | 0.30 | 0.40 | -- | -- |
| Histone | 0.20 | 0.30 | -- | -- |
| Ribonuclease[a] | 0.15 | 0.15 | -- | -- |
| Chymotrypsinogen | 0.10 | 0.10 | -- | -- |
| β-Lactoglobulin | 0.0-0.1 | 0.30 | 0.30 | 0.20 |

[a]Proteins studied by X-ray analysis.

Recently studies of polypeptide optical activity in the near-infrared region of the spectrum, in aqueous solutions up to 1.2 μm and in organic ones up to 2.2 μm have been undertaken [151]. Since optical activity is determined by the ratio of molecular dimensions to wavelength (p. 282), these effects are small in the infrared region. Rotatory strengths corresponding to vibrational rather than electronic transitions are also small. However, the results obtained seem promising; the new method may be useful for ascertaining the β-form content of proteins.

ADOA and CD are different for DNA and RNA (p. 305) and make it possible to study their conformations. Theoretical interpretation of CD of DNA and RNA in aqueous solutions is in agreement with conformations which have been determined by X-ray analysis.

The study of oligonucleotides, which has revealed the differences in the CD and DOA of 16 dinucleotides [152,153], justifies the hope that spectropolarimetry can yield information about the primary structure of nucleic acid fragments. Apparently the optical activity of one-chain polymers can be expressed as the sum of dimer contributions [154].

Spectropolarimetry has yielded valuable information about the state of nucleic acids in ribosomes [155,157], viruses [157], nucleohistones [158], etc. Further details are presented in Chapter 8.

Along with natural optical activity, rotation of the plane of polarization of light and CD in a magnetic field appear to be promising methods with which to investigate biopolymers, primarily heme-containing proteins. These phenomena are described in Sections 7.3 and 7.4.

## 5.8 Luminescence of Biopolymers

Complex molecules are characterized by a large number of vibrational degrees of freedom, by strong interaction of vibrations, and therefore by the continuous transfer of vibrational energy from one degree of freedom to another. Consequently, the vibrational levels of complex molecules usually form a continuum. Electronic and vibrational motions interact. Separation of these kinds of motion, achieved with a high degree of approximation in the case of simple molecules, becomes impossible for complicated molecules. This is due to the large supply of vibrational energy, which can be close to the value sufficient for excitation of the electron shell (cf. also p. 393) [159,160].

Consequently, fluorescence spectra, like absorption spectra, are diffuse and lack fine details (this is not true of the linear fluorescence spectra obtained by Shpolsky under special conditions [161]). It is not the wavelengths of band maxima, but the intensities, polarizations, etc., which are informative.

Intensity of fluorescence can be expressed by the value of the quantum yield $\gamma$. Assume that there are $n(0)$ excited molecules at a time $t = 0$. If the probability of transition from an excited to a nonexcited state, with emission of a light quantum, is f and the probability of nonradiational transition is g, then

$$dn = -fn\ dt - gn\ dt \tag{5.141}$$

and

$$n(t) = n(0)e^{-(f+g)t} \tag{5.142}$$

The mean lifetime of an excited state

$$\tau = \int_{t=0}^{\infty} t\ dn \Big/ \int_{t=0}^{\infty} dn = 1/(f+g) \tag{5.143}$$

Fluorescence intensity decreases exponentially. In the majority of cases $\tau$ is of the order of $10^{-8}$-$10^{-9}$ sec, but it can be much larger, too.

Quantum yield is expressed as the ratio of the number of radiational transitions to the total number of transitions

$$\gamma = f/(f+g) \tag{5.144}$$

Because of interactions with the solvent, an equilibrium distribution of molecules over their vibrational energy supplies will be established. This distribution does not depend on the surplus of vibrational energy obtained at excitation and therefore on the wavelength $\lambda_{ex}$ of exciting light. Hence, neither $\tau$ nor $\gamma$ depends on $\lambda_{ex}$ (Vavilov's law).

Fluorescence of complicated molecules (in particular, of dyes) is polarized even if the incident light is natural. A detailed theory of polarized luminescence was developed by Vavilov ([162], see also [163]). Incident light is absorbed by the molecules, which possess definite orientation in relation to the electric vector **E** of the light wave. Later, absorption energy is emitted as the result of another electronic transition, which generally has another orientation in the molecule (i.e., another transition dipole direction). Luminescence is polarized if the lifetime of the excited state (i.e., the time of energy transfer from the absorbing to the emitting dipole) is small in comparison with the reorientation time of the molecule.

The degree of polarization is expressed as

$$P = \frac{I_z - I_x}{I_z + I_x} \tag{5.145}$$

where $I_z$ and $I_x$ are radiation intensities which correspond to the components of the electric vector along the z and x axes; the incident ray is directed along the y axis. Simple calculation shows that the limiting value of P for fixed molecules is $\frac{1}{3}$ if the incident light is natural. The rotational motion of molecules depolarizes the radiation. Theory leads to the following expression [162,163] (for natural incident light):

$$\frac{1}{P} = \frac{1}{P_0} + \left( \frac{1}{P_0} + \frac{1}{3} \right) \frac{kT}{V\eta} \tau \qquad (5.146)$$

Here P is the observed polarization, $P_0$ is its limiting value in the absence of depolarization (i.e., at $T \to 0$ or at infinite viscosity, $\eta \to \infty$), V is the molecular volume, and $\tau$ is the lifetime of the excited state. Thus, polarization decreases with increasing motility of the radiating molecule and can serve as a measure of motility.

Nonradiational transitions, which make $\gamma$ smaller than unity, amount to the interconversion and degradation of light energy [164]. A molecule excited by photon $h\nu_a$ to some level E* can expend this energy in fluorescence with probability f of photon $h\nu_f$; can effect the internal conversion of energy E* into vibrations, with accompanying degradation into heat; or can undergo transition to a metastable triplet level $E_T$ without emission, spending part of the energy of excitation and vibrations. Later, energy $E_T$ can be emitted in the form of a phosphorescence photon $h\nu_p$ or be transformed into vibrational energy and degraded into heat. All these processes are represented schematically in Fig. 5.24.

Another essential cause of a decrease in quantum yield is the quenching of luminescence, which can be brought about by foreign substances, particularly oxygen. Of special interest for biophysics are the processes manifested in concentration quenching and concentration depolarization of luminescence.

An increase in the concentration of fluorescent molecules up to $10^{-4}$ mole/liter results in a sharp decrease in the quantum

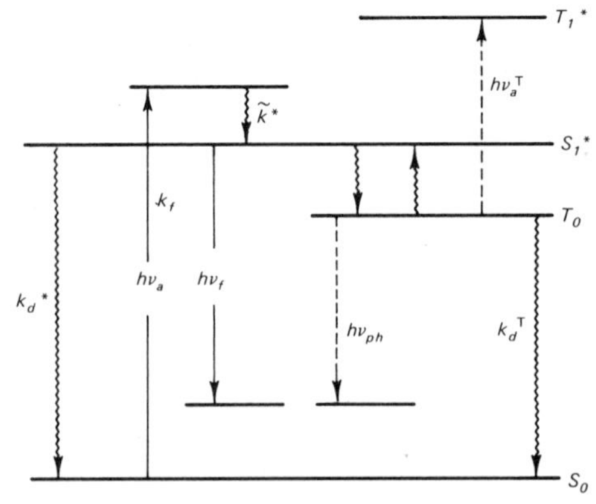


*FIG. 5.24 The scheme of energy transitions in fluorescence and phosphorescence. $S_0$, $S_1$* are singlet levels; $T_0$, $T_1$*, triplet levels; $h\nu$, quanta; k, rate constants. Wavy lines show transitions without radiation.*

yield and polarization of radiation. Both are explained by the migration of excitation energy between molecules. We do not refer here to radiation with subsequent reabsorption of light, but to the direct transfer of excitation energy because of resonance interaction of molecules. A requisite for the condition of such interaction is the strong overlapping of absorption and emission bands of molecules. Interaction decreases naturally with increasing intermolecular distance. These phenomena have been treated phenomenologically by Vavilov [165], and a rigorous quantum-mechanical theory of resonance has been proposed by Förster [166,167]. See also [168].

Resonance interaction that determines energy migration is similar to the exciton interaction in molecular crystals studied by Davydov (p. 276). The distinction between these phenomena is a quantitative one. Migration of energy occurs at a high interaction energy, one that is much larger than the breadth of the electronic-vibrational band. In this case of strong interaction, energy migration occurs rapidly and it is possible to ignore the influence of vibrations. Consequently, in contrast to the exciton spectra of molecular crystals, which occur at weak interactions (interactions in which the energy is much less than the band breadth, but much larger than the breadth of a single vibrational sublevel), vibronic structure is not observed in spectra of resonance-interacting molecules.

Energy migration produces depolarization of luminescence. The molecule which absorbs the quantum is oriented in space in some definite way. Should radiation occur in the same molecule, it would be characterized by a definite P value. However, during the time between absorption and radiation, multiple acts of energy transfer are possible to molecules of somewhat different orientations. After some time, all excited molecules will lose their original orientations (those that existed at the moment of absorption), and the radiation will become totally depolarized.

It must be emphasized that energy migration is possible between both identical and nonidentical molecules if the absorption bands of molecules of one kind overlap those of fluorescence of molecules of another kind. The notion of energy migration arising in the luminescence studies has been widely used for the explanation of a series of biological processes. As it has been said, many biological processes proceed at a high rate, according to the "all-or-none" principle (p. 33). Obviously, the true explanation of these processes must make use of the notion of cooperativity. Such explanations of many problems have actually been offered. However, hypotheses about energy migration in a biopolymer or in the surrounding "structured water" are numerous in the literature, starting with Szent-Györgyi's well-known book [169]. These migrational ideas lack a real physical foundation but can be important in studies of electron-conformation interactions. On the other hand, energy migration is evidently important for biology and the biophysics of photosynthesis.

Investigation of the features of luminescence related to energy migration is of great interest for biophysics. Energy transfer between regions of a biopolymer makes it possible to obtain information about the relative positions of these regions. The luminescence of biopolymers has been described in a series of monographs (cf., e.g., [170-172]).

Cyclic $\pi$-electron systems are frequently luminescent. The aromatic amino acid residues in proteins (Phe, Tyr, and Trp) are different in this respect. Their corresponding amino acids are characterized by the spectral properties presented in Table 5.10. The fluorescence of Phe is practically nonobservable. The radiation of Tyr and Trp depends essentially on the pH and polarity of the medium--on the protonization of neighboring groups. In native proteins $\gamma_{Tyr}$ is strongly lowered, whereas $\gamma_{Trp}$ can be increased up to 0.32. Denaturation by urea equalizes these values, increasing $\gamma_{Tyr}$ and decreasing $\gamma_{Trp}$ Hence, quantum yields of Tyr and Trp are conformationally susceptible parameters. This is explained by the dependence of the luminescence quenching of Tyr and Trp on the medium, particularly on its hydrophilic or hydrophobic nature. Residues exposed by conformational transitions are also more easily quenched by foreign substances. Studies of the intrinsic luminescence of proteins (aromatic residues) have yielded important information about their conformational properties as well as about their interactions with small molecules and ions [173-180]. Aromatic residues serve therefore as susceptible labels.

*TABLE 5.10*

*Spectral Parameters of Aromatic Amino Acids*

| Amino acid | Absorption $\lambda_{max}$ (nm) | Molecular absorption coefficient | Fluorescence $\lambda_{max}$ (nm) | Quantum yield |
|---|---|---|---|---|
| Phenylalanine | 257 | 200 | 282 | 0.04 |
| Tyrosine | 275 | 1300 | 303-304 | 0.21 |
| Tryptophan | 280 | 5600 | 353 | 0.205 |

As has been said, the polarization of luminescence depends on the motility of the luminophore. By determining P, we can estimate the relaxational characteristics of a macromolecule. The theory of polarized luminescence of polymers is developed in the works [181,182]. The mean square time of rotational relaxation in a macromolecule $\tau_r$ can be determined from the dependence of the P of the polymer solution on the solvent viscosity with the help of the expression

$$\frac{1}{P} = \frac{1}{P_0} + \left( \frac{1}{P_0} + \frac{1}{3} \right) \frac{3\tau}{\tau_r} \tag{5.147}$$

[cf. (5.146)]. $\tau$ is the time of radiation. For the simplest case of stiff particles, $\tau_r \sim \eta/T$. This method has been used for studies of the motility and conformational transitions of a series of synthetic polymers (polymethacrylic and polyacrylic acids, polymethyl methacrylate, etc.) with luminescence labels both at the side chains and the main chain [183-185]. The activation energies of conformational changes have been established; they reveal that hydrophobic interactions strongly influence intramolecular motility.

The polarized luminescence method has been employed in an effective way in studies of complexes of nucleic acids and small molecules, particularly molecules of acridine dyes. These studies, too, have yielded valuable information about the conformational structure of nucleic acids (cf. Chapter 8).

Conformational transitions and their dynamics are manifested also in other characteristics of luminescence. Thus the affinity of various conformations of $\alpha$-chymotrypsin for the luminescent substance proflavin influences the intensity of luminescence [186]. The positions of maxima in the luminescence spectra of polar molecules in condensed polar media depend on the relaxational properties of the microenvironment. Theory gives this expression relating the position of the center of gravity of the fluorescence band $\nu_c$ with the quantities $\tau$ and $\tau_r$ [187].

$$\nu_c = \nu_\infty + (\nu_0 - \nu_\infty) \frac{\tau_r}{\tau_r + \tau} \tag{5.148}$$

where $\nu_\infty$ is the value of $\nu_c$ at $\tau_r << \tau$, $\nu_0$ is that at $\tau_r >> \tau$. The values of $\nu_c$ were used for studies of the relaxational properties of a series of polymers with luminescent labels [188].

## 5.9 Vibrational Spectra of Biopolymers

Frequencies, intensities, and polarizations of bands in vibrational spectra provide information about molecular structure. Vibrational spectra are observed either as infrared (IR) absorption (and emission or reflection) spectra, or as Raman spectra (RS). In applications to macromolecular substances, IR spectroscopy has proved more important than RS. Only quite recently, due to the possibilities of laser techniques, have clear and expressive RS of polymers and biopolymers been obtained (p. 323).

The theory of vibrational spectra of small polyatomic molecules is well developed [189-194]. It is based on the possibility of separate treatment of slow nuclear vibrations and fast electronic transitions in accordance with the Born-Oppenheimer theorem [195]. The frequencies of harmonic nuclear vibrations are calculated by solving the classical problem with direct use of molecular symmetry. Electrooptical parameters determining intensities and polarizations in vibrational spectra are evaluated by means of a theory first proposed in the 1940s [196] and discussed in detail in [189] and [190]. This theory employs the so-called valence-optical scheme, which assumes that every chemical bond can be characterized by its own dipole moment and polarizability (cf. also [197]). The dipole moment of a molecule is the vector sum of bond moments; the polarizability of a molecule, the tensor sum of anisotropic bond polarizabilities. The intensity of a given band in the IR spectrum is determined by the change in the molecular dipole moment in a given normal vibration; in the Raman spectrum, intensity depends on the change in polarizability.

A molecule consisting of N atoms has 3N - 6 normal vibrations (3N - 5 in the case of a linear molecule). Because all the atoms of a molecule are linked, they all participate in every normal vibration. Situations may occur in which the degree of participation of one or several atoms in a given vibration is much greater than that of the rest of the atoms. Such a vibration is characteristic of the given atomic group; its frequency (like the intensity of the spectral band) is only weakly dependent on the rest of the atoms in the molecule. The presence of a characteristic spectral band or line attests to the presence of the group to which it corresponds.

In the IR spectra of proteins and polypeptides, the bands of the amide group (peptide bond) —CO—NH— are observed. Their study is based on the spectra of model low-molecular-weight compounds (amides). Experimental study of acetamide ($H_3C$—CONH—$CH_3$) and related compounds has disclosed the amide group spectrum presented in Table 5.11 ([198-201]). Bands A and B are due to splitting of the stretching vibration of the N—H bond, whose frequency coincides with the overtone of vibration II--the effect of the so-called Fermi resonance (cf., e.g., [182]). The forms of these vibrations are shown in Fig. 5.25. Bands A, B, I, and II are very susceptible to hydrogen bonds, whose presence produces considerable decrease in the frequency and broadening of bands.

In examining polymer spectra, the vibrational interactions of the single monomer links have to be taken into account. An ordered polymer is similar in this respect to a molecular crystal. Only those vibrations can be observed in spectra which occur in phase in all elementary cells. A general theory that applies this idea (the "factor-group approximation") has been

TABLE 5.11
*Amide Group Vibrations*

| Band | Frequency $cm^{-1}$ | Character of vibration[a] | Degree of participation of groups |
|---|---|---|---|
| Amide A | 3300 | N—H (v), Fermi resonance with 2 × amide II band | |
| Amide B | 3100 | | |
| Amide I | 1597-1672 | C=O (v) | 1650: C=O (v)<br>80%, C—N (v)<br>10%, —N—H (d)<br>10% |
| Amide II | 1480-1575 | —N—H (d) in plane; C—N (v) | 1560: —N—H (d)<br>60%, C—N (v)<br>40% |
| Amide III | 1229-1301 | C—N (v); —N—H (d) in plane | 1300: C=O (v)<br>10%, C—N (v)<br>30%, N—H (v)<br>30%, O=C—N (d)<br>10% |
| Amide IV | 625-767 | O=C—N (d) in plane | 625: O=C—N (d)<br>20% |
| Amide V | 640-800 | =N—H (d), out of plane | |
| Amide VI | 537-606 | —C=O (d), out of plane | |
| Amide VII | 200 | Torsional vibration around C—N | |

[a]Key: (v), valent vibration (i.e., stretching of the bond); (d), deformational (bending) vibration which changes the valence angle.

proposed in the works [202] and [203] (see also [204-208]). The polymer chain treated as "one-dimensional space group" is characterized by the number of normal vibrations 3pg − 4 where p is the number of atoms in the chemical repeating group and g the number of these groups in the one-dimensional crystallographic unit cell. Thus, for the trans chain of polyglycine I (β form)

$—CH_2—CO—NH—CH_2—CO—NH—$

the chemical group contains five particles (if $CH_2$ is treated as one particle) and the crystallographic cell contains two such groups; hence, $3pg - 4 = (3 \cdot 5 \cdot 2) - 4 = 26$. The observable vibrations are $9A_1$, $9B_1$, $4A_2$, $4B_2$ ($A_1$, $B_1$ are, respectively, symmetrical and nonsymmetrical planar vibrations; $A_2$, $B_2$, nonplanar vibrations). All are active in IR spectra except $A_2$ (cf. [209,210]).

Calculation of the vibrational frequencies, and therefore interpretation of the vibrational spectra, of polypeptides and

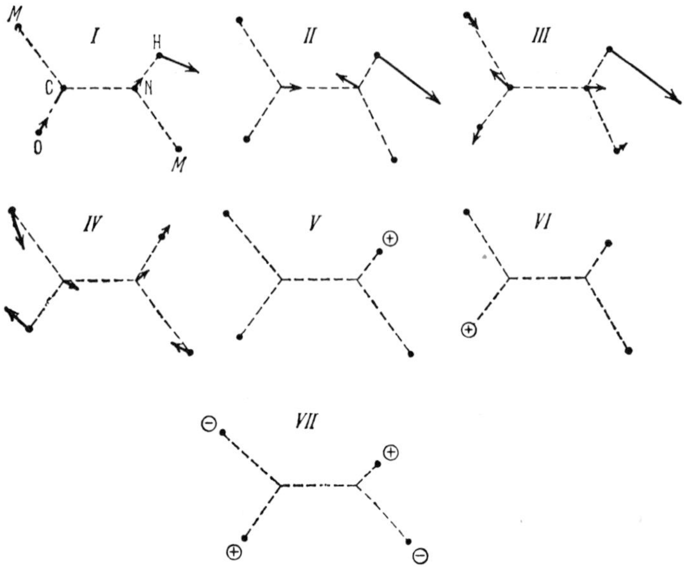


*FIG. 5.25 Calculated normal vibrations of the amide group in N-methylacetamide.*

proteins takes weakly bound oscillators into consideration on the grounds of perturbation theory [198,201,211]. The vibrational frequency is expressed by the formula

$$\nu(\delta,\delta') = \nu_0 + \sum_s D_s \cos(s\delta) + \sum_{s'} D_{s'} \cos(s'\delta') \qquad (5.149)$$

where $\nu_0$ is the unperturbed vibration frequency of an isolated group, and $\delta$ the phase difference between the vibrational motions of identical groups in the chain. $D_s$ is determined by the term describing the interaction with the sth group of the chain in expressions of potential and kinetic energy. $D_{s'}$ and $\delta'$ concern the interchain vibrations and have a sense analogous to that of $D_s$ and $\delta$. Summation can be performed over the neighboring groups [212,213].

Let us examine the results of such calculations for the α-helix and parallel and antiparallel β forms. For a nonordered chain, the interactions compensate each other on average and $\nu = \nu_0$. For an infinitely long α-helix, IR absorption arises in vibrations with phase difference $\delta = 0$ ("parallel band") or in vibrations with $\delta = \theta$ ("perpendicular band"), where $\theta$ is the angle of rotation around the helix axis for a transition from one peptide group to the next (i.e., $\theta = 2\pi/3.6$). Taking three neighboring groups into account, we obtain

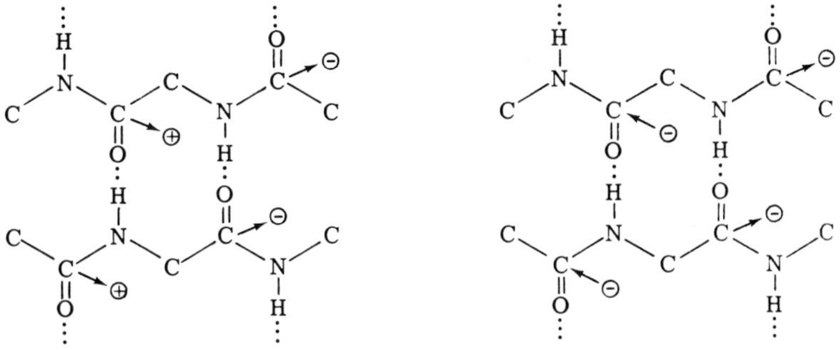


*FIG. 5.26 Vibrations of the parallel β form.*

$$\nu(0) = \nu_0+D_1+D_2+D_3 \qquad (\parallel)$$

$$\nu(\theta) = \nu_0+D_1 \cos\theta + D_2 \cos 2\theta + D_3 \cos 3\theta \qquad (\perp) \tag{5.150}$$

Thus, every vibration of the α-helix link leads to the appearance of two frequencies.

In the parallel β form, $\delta = 0$ and $\pi$ for vibrations in phase and out of phase, respectively. Both intra- and interchain interactions have to be taken into account. We obtain two active vibrations with frequencies

$$\nu(0,0)_P = \nu_0+D_1+D_1' \qquad (\parallel)$$

$$\nu(\pi,0)_P = \nu_0-D_1+D_1' \qquad (\perp) \tag{5.151}$$

These vibrations are shown in Fig. 5.26.

Similarly, for the antiparallel β form we get four vibrations with frequencies

$$\nu(0,\pi)_A = \nu_0+D_1-D_1' \qquad (\parallel)$$

$$\nu(\pi,0)_A = \nu_0-D_1+D_1' \qquad (\perp\text{, in plane}) \tag{5.152}$$

$$\nu(\pi,\pi)_A = \nu_0-D_1-D_1' \qquad (\perp\text{, out of plane})$$

$$\nu(0,0)_A \quad \text{nonactive}$$

These vibrations are shown in Fig. 5.27. The last vibration is nonactive, since it is centrosymmetrical and is not accompanied by a change in the dipole moment.

Table 5.12 presents an interpretation of corresponding amide I and amide II spectral bands based on calculations of Krimm. It indicates that the values of observed frequencies

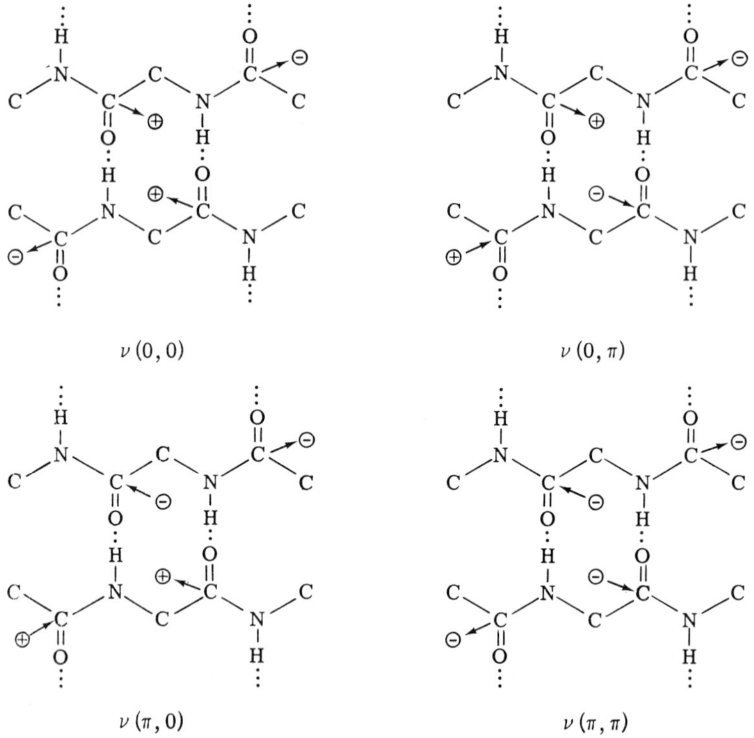


*FIG. 5.27 Vibrations of the antiparallel β form.*

*TABLE 5.12*
*Amide I and Amide II Bands*[a]

| Conformation | Vibration notation | Frequency ($cm^{-1}$) Amide I | Amide II |
|---|---|---|---|
| Disordered | $\nu_0$ | 1658 | 1520 |
| Nylon 66 | $\nu_N$ | 1640 | 1540 |
| Antiparallel β form | $\nu\|\|(0,\pi)_A$ | 1685 (w) | 1530 (s) |
| | $\nu\perp(\pi,0)_A$ | 1632 (s) | 1510 (w) |
| | $\nu\perp(\pi,\pi)_A$ | 1668 (w) | 1550 (w) |
| Parallel β form | $\nu\|\|(0,0)_P$ | 1648 (w) | 1530 (s) |
| | $\nu\perp(\pi,0)_P$ | 1632 (vw) | 1530 (w) |
| α-Helix | $\nu\|\|(0)_\alpha$ | 1650 (s) | 1516 (w) |
| | $\nu\perp(\theta)_\alpha$ | 1646 (w) | 1546 (s) |

[a]Key: (s), strong; (w), weak; (vw), very weak spectral band.

in the IR spectrum give some information about chain conformation. Some data obtained for aqueous protein solutions [214] are shown in Table 5.13.

*TABLE 5.13*
*Amide I and Amide II Bands in Aqueous Protein Solutions*

| Protein | Frequency ($cm^{-1}$) Amide I | Amide II | Main conformation |
|---|---|---|---|
| Myoglobin | 1652 | 1545 | α-Helix |
| β-Lactoglobulin | 1632 | 1530 | Antiparallel β form |
| Denatured β-lactoglobulin | 1656[a] | | Nonordered |
| Denatured α-casein | 1656 | | Nonordered |

[a]At high pH, 1570.

Parallel (||) and perpendicular (⊥) bands correspond to dipole moment vibrations parallel and perpendicular to the axis of the polymer chain (Fig. 5.28). Obviously, spectra of solutions or of nonoriented films do not yield information about the polarization of vibrations. On the other hand, if the spectra of proteins and polypeptides are measured in anisotropic media (oriented films and fibers) with the help of

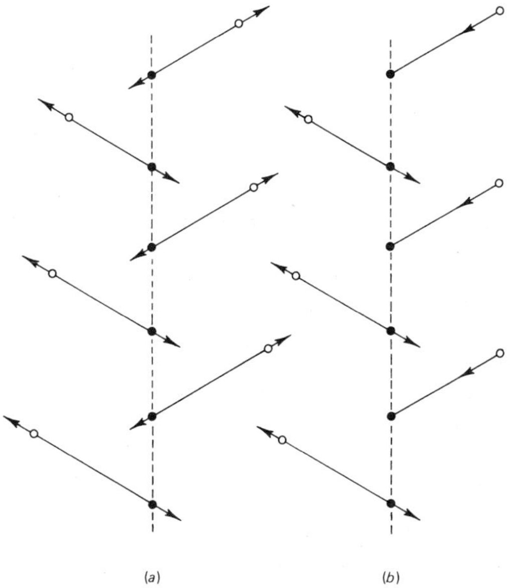


*FIG. 5.28 (a) In-phase vibrations polarized along the polymer chain axis; (b) antiphase vibrations polarized perpendicular to the chain axis.*

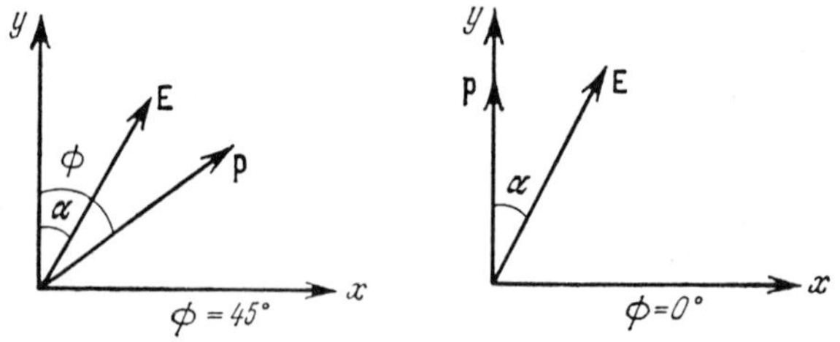


*FIG. 5.29 Positions of vectors E and P.*

polarized IR radiation, the polarization of vibrations can be determined in a direct way. Infrared dichroism is characterized by the dichroic ratio, that is, the ratio of the absorption coefficients of IR light with electric vector directions perpendicular and parallel to some chosen direction. Since the amount of light energy absorbed is proportional to the square of the scalar product of the vector of the vibrating dipole moment P and the vector of the electric field intensity of light wave E

$$A \sim |PE|^2 \cos^2 \alpha$$

the dichroic ratio is

$$R = (k_y/k_x) = (n_y/n_x) \cot^2 \phi \qquad (5.153)$$

where $n_y$ and $n_x$ are refraction indexes for waves vibrating along the y and x axes, and $\phi$ is the angle between the y axis and the dipole moment. If $\phi = 45°$, $k_y = k_x$, $R = 1$, and polarization is absent; if $k_x = 0$, $\phi = 0°$ and $R \to \infty$. We get total polarization (Fig. 5.29). The measurement of dichroism enables us to estimate the directions of vibrations and therefore to obtain information about the conformations (secondary structure) of protein chains. The theory of vibrational spectra of polymer chains of finite length is developed in [208].

Chirgadze has developed a method of quantitative analysis of the secondary structure of globular proteins in aqueous solutions based on measurements of the intensities of the amide I band [270-273]. The protein is studied in solution in heavy water in the spectral region 1500-1800 $cm^{-1}$. The frequencies of the amide I band for the deutero form of the peptide group are: for the $\alpha$-helix, 1636 $cm^{-1}$; for the antiparallel $\beta$ form, 1620 and 1690 $cm^{-1}$; and for a disordered conformation, 1650 and 1676 $cm^{-1}$. This method has been successfully applied to several proteins and has given results which agree well with those obtained by X-ray diffraction.

Apart from determinations of secondary structure, IR spectra are useful in studies of the ionization of amino acid residues, the kinetics of deuterium exchange, etc. Further details about the Infrared spectra of proteins are contained in [198,201,215].

Infrared spectroscopy of nucleic acids is less informative, because of the greater complexity of their chemical structure, the lack of sufficient characteristicity in their bands, etc. We list some typical vibrations of nucleic acids [201] in Table 5.14.

*TABLE 5.14*
*Infrared Spectra of Nucleic Acids*

| Frequencies $cm^{-1}$ | Vibration notations[a] | Vibrating groups |
|---|---|---|
| 2800-3500 | ν(OH) | $H_2O$, sugar |
| | ν(NH) | Nitrogen bases |
| | ν(CH) | Sugar and bases |
| 1500-1800 | ν(C=O) | Bases, mixed vibrations |
| | ν(C=N) | |
| | ν(C=C) | |
| | δ –NH | |
| | δ HOH | $H_2O$ |
| 1200 | $\nu(PO_2^-)$ | Antisymmetric, stretching $PO_2$ vibration |
| 1000-1100 | $\nu(PO_2^-)$ | Symmetric, stretching $PO_2$ vibration |
| | ν(CO) | Sugar |
| 700-1000 | ν(PO) | Phosphate |
| | ν(CO) | Sugar |
| | τ(NH) | Nonplanar bending vibration of base |
| 300-600 | | Skeletal deformations of the chain |

[a]Key: ν, stretching; δ and τ, bending vibrations.

The development of laser techniques has made it possible to obtain perfect Raman spectra of biopolymers. Beautiful spectra of lysozyme and chymotrypsin are presented in [216]. The measurements of low-frequency Raman bands (with frequencies up to 50 $cm^{-1}$) seem to be rather important. These bands are sensitive to conformational changes and perhaps characterize conformational vibrations (cf., e.g., [274]).

## 5.10 Spectra of Nuclear Magnetic and Electron Paramagnetic Resonance

Paramagnetic resonance spectra are successfully applied in studies of biopolymers, particularly of enzymes. Let us describe the nature of paramagnetic resonance--of nuclear (NMR) and electron (EPR) resonance.

Introduce a particle with magnetic moment $\mu$ into the constant magnetic field $H_0$. The moment will precess around the field direction with Larmor frequency

$$\omega_H = (\mu/p)H_0 = \gamma H_0 \tag{5.154}$$

where p is the mechanical momentum of the particle (e.g., electron spin). Let the particle be simultaneously subjected to the action of a weak alternating field $H_1$ directed perpendicular to $H_0$ (we assume that $H_0$ coincides with the z axis. The linear polarized field $H_1$ can be decomposed into components with right and left circular polarization (p. 281). One of these components coincides with the precession direction. If the frequency $\omega$ of the field $H_1$ becomes equal to precession frequency $\omega_H$, the action of field $H_1$ on the magnetic moment becomes especially great and resonance occurs. The energy of the precessing magnet is $-\mu H_0$. The increase in the angle between $\mu$ and $H_0$ produced by field $H_1$ increases the energy. The maximal effect corresponds to $\omega = \omega_H = \gamma H_0$. This phenomenon was discovered by Zavoisky [217].

This is an elementary classical description. Quantum mechanics shows that the energy levels of a particle having a magnetic moment are split in a magnetic field according to the rule of spatial quantization.

Let us begin with NMR. Atomic nuclei possessing odd value of one or both of the quantities--mass and atomic number--have momentum, and therefore nuclear spin. $C^{12}$, $O^{16}$, etc., lack nuclear spin, whereas it is present in H, D, $C^{13}$, $F^{19}$, etc.; hence, the latter nuclei possess magnetic moments.

If the nuclear spin number is J, the momentum is equal to $[J(J+1)]^{1/2}\hbar$, and the magnetic moment $[J(J+1)]^{1/2}\gamma\hbar$. $\gamma$ is the gyromagnetic ratio, equal to $g\mu_0$, where g is the splitting factor and $\mu_0$ the nuclear magneton. Because of Zeeman splitting, the magnetic field $H_0$ produces 2J + 1 energy levels

$$E_m = -\gamma\hbar H_0 m \qquad (m = J, J-1, \ldots, -J) \tag{5.155}$$

with distances $\gamma\hbar H_0$ between them. For a proton, $J = \frac{1}{2}$ and two levels arise, corresponding to spin directions parallel and antiparallel to the field $H_0$. If the field $H_1$ oscillates with frequency $\omega$, resonance occurs upon absorption of a quantum $\hbar\omega$ equal to the interlevel distance

$$\hbar\omega = \gamma\hbar H_0 = 2\mu H_0 \tag{5.156}$$

Here $\mu = g\mu_0 J$ is the maximal determining component of the magnetic moment. Condition (5.156) is equivalent to (5.154).

If $J > \frac{1}{2}$, the number of levels is larger but the resonance condition remains the same, since only transitions between neighboring levels are allowed.

A nuclear magneton has the value

$$\mu_0 = e\hbar/2M_pc = 5.0493 \times 10^{-24} \text{ erg/G} \tag{5.157}$$

where e is the charge of proton, $M_p$ its mass, and c the velocity of light. The factor g is dimensionless. For a proton, g = 5.58. If the field intensity $H_0$ is 10,000 G, the resonance frequency for protons is 42.6 MHz.

At a given temperature T, the number of protons in the lower level is somewhat larger than that in the higher one. The ratio of populations of the lower and higher levels is expressed by the Boltzmann factor

$$\exp(-\hbar\omega_0/kT) = \exp(-2\mu H_0/kT) \cong 1-(2\mu H_0/kT) \qquad (5.158)$$

because $2\mu H_0 \ll kT$ at room temperature and actual $H_0$ values. In fact, if $H_0 = 10^4$ G, the Boltzmann factor is equal to $1 - 14 \times 10^{-6}$. If the $H_1$ field is switched on, transitions occur from the lower level to the higher one (absorption) and vice versa (spontaneous emission) (Fig. 5.30). If the probabilities of both processes are equal, rapid saturation--equalization of the populations of both levels--must occur and absorption will stop. However, this is not observed in a real substance. Obviously, a process must exist in the system of spins that allows the spins to give up their energy without radiation. It is a relaxational process, continuously returning the spin system to the equilibrium state, corresponding to the Boltzmann distribution. This occurs because of the interaction of the nuclear spins with the lattice--with other nuclei surrounding the given one, and which are in a state of thermal motion. If field $H_1$ is switched off, the energy is transformed into thermal energy of the lattice, spin-lattice relaxation occurs. The change in the population of the levels at a time t after the switching off of field $H_1$ is expressed by the equation

$$\Delta n(t) = \Delta n(0)\exp(-t/T_1) \qquad (5.159)$$

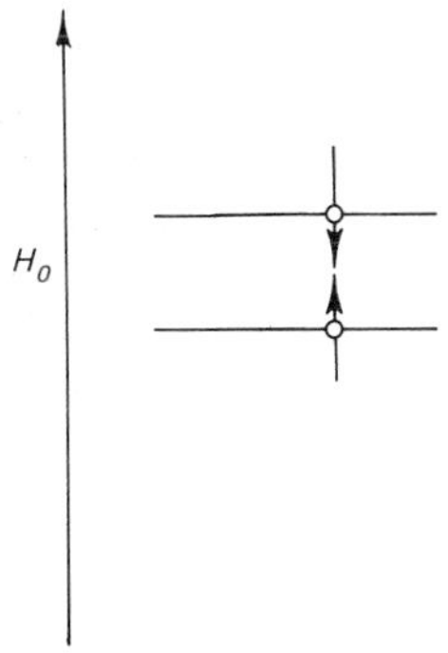


*FIG. 5.30 Splitting of the proton spin level in a magnetic field.*

where $\Delta n(t)$ is the surplus of protons at the higher level in comparison with the equilibrium population, and $\Delta n(0)$ is the same at $t = 0$. $T_1$, the spin-lattice relaxation time, depends on the concentration of magnetic nuclei, the motility of the molecules, and the temperature. In crystals, $T_1$ is of the order of minutes; in gases and liquids, of the order of seconds and less. The presence of paramagnetic impurities can reduce $T_1$ to $10^{-4}$sec.

According to the uncertainty principle of quantum mechanics, the width of an NMR spectral line is determined by the lifetime of the nucleus at the given energy level. The uncertainty of the resonance frequency is

$$\Delta\omega \sim \Delta E/\hbar \sim 1/\tau \tag{5.160}$$

Due to spin-lattice relaxation, the line width becomes of the order of $T_1^{-1}$.

Side by side with the spin-lattice relaxation there exists a second process, namely, the direct interaction of fixed atomic nuclei, or spin-spin interaction. In addition to $H_0$, every spin is acted upon by the local field $H_\ell$, produced by the neighboring nuclei.

$$H_\ell = (\mu/r^3)(3\cos^2\theta - 1) \tag{5.161}$$

where r is the distance from the nucleus to the given point, and $\theta$ is the angle formed by vectors $\mu$ and **r**. Consequently, the resonance condition assumes the form

$$\omega = \gamma(H_0 + H_\ell) \tag{5.162}$$

different from expression (5.156), which is valid for an isolated spin. The order of magnitude of $H_\ell$ is several gauss. The contribution of the spin-spin interaction to the line width is of the order of $H_\ell\gamma$. However, spin-spin line broadening comes about not only because the field that occurs is inhomogeneous; it also reflects the energy exchange between protons that occurs when the phases of their precession coincide; hence, spin-spin interaction shortens the protons' lifetime in a given state. A second characteristic time $T_2$ has to be introduced: the spin-spin relaxation time, corresponding to both contributions to the line width that are attributable to spin-spin interaction. For solids, $T_1 \gg T_2$ and the line width is determined by $T_2$ only

$$\Delta\omega \sim 1/T_2$$

In nonviscous liquids $T_2$ is of the same order as $T_1$ and has values of the order of 0.1 Hz

$$\Delta\omega \sim (1/T_1) + (1/T_2) + \Delta'\omega$$

where $\Delta'\omega$ is the effect due to the inhomogeneity of field $H_0$.

NMR spectra of liquids and solutions contain a series of comparatively narrow lines corresponding to structurally non-equivalent protons. If the protons are equivalent (e.g., the two protons in $H_2O$), only one line is observed. However, the protons of $CH_3$, $CH_2$, and OH in $C_2H_5OH$ differ in frequency from one another. This difference occurs because of the different action on the spin produced by the electron shell of each nucleus. Electrons precess in a direction opposite to that of the nuclei, thereby forming a secondary magnetic field H' proportional to $H_0$. The value of this field near the nucleus is

$$H' = -\sigma H_0 \tag{5.163}$$

and the nucleus is acted on by an effective field $H_{eff}$

$$H_{eff} = H_0 + H' = H_0(1-\sigma) \tag{5.164}$$

The constant $\sigma$ is called the screening constant. The values of H' (i.e., of $\sigma$) are nonequal along various directions in the molecule; $\sigma$ is anisotropic. In liquids, molecules change their orientations rapidly; therefore, we deal with an average value of $\sigma$.

The position of an NMR spectral line referred to some standard line is called the chemical shift. As a reference substance for proton resonance in organic compounds, tetramethylsilane (TMS), $(CH_3)_4Si$, is used. Chemical shifts are expressed as dimensionless quantities

$$\delta = (\Delta\omega/\omega_0)10^6 = (\Delta H/H_0)10^6 \tag{5.165}$$

For the protons of TMS, $\delta = 0$, and the shifts in the PMR (proton magnetic resonance) lines lying at lower $H_{eff}$ are considered as positive. The values of $\delta$ vary from +18 to -5; they are characteristic for various atomic groups.

If the PMR of biopolymers is studied, the water-soluble standard 2,2-dimethyl-2-silanepentane-5-sulfonic acid (DDS) is used. In this case, aliphatic protons have shifts from -0.5 to -2.0, aromatic ones from -0.6 to -8.5.

Figure 5.31 shows the PMR spectrum of 1,1,2-trichloroethane obtained at low resolution. Two peaks are observed, corresponding to the protons of the $CH_2Cl$ and $CHCl_2$ groups which have different shifts. The ratio of intensities is 2:1. At high resolution, the first line is split into two, and the second one into three, components (Fig. 5.32). Hyperfine (multiplet) structure, which arises because of the magnetic interaction of nuclei transferred by bond electrons (i.e., indirect spin-spin interaction, is observed. The distances between the components do not depend on $H_0$. Protons of the $CHCl_2$ group can exist in two states, with spins $\frac{1}{2}$ and $-\frac{1}{2}$. Hence, either one or the other

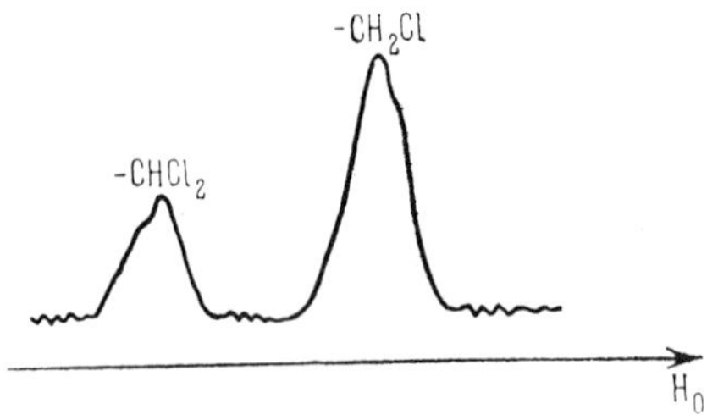


*FIG. 5.31 Proton magnetic resonance spectrum of 1,1,2-trichloroethane at low resolution.*

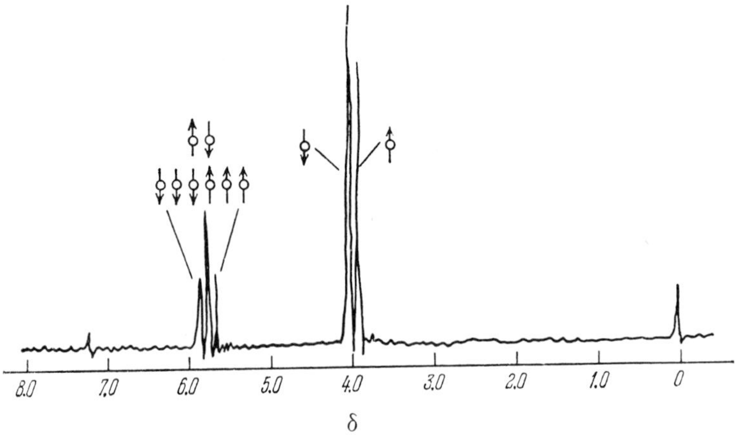


*FIG. 5.32 Proton magnetic resonance spectrum of 1,1,2-trichloroethane at high resolution.*

local field acts on the $CH_2Cl$ group protons and the PMR line of $CH_2Cl$ is subject to doublet spin-spin splitting. In the methylene group three nonequivalent states of the proton pair are possible: $\frac{1}{2}$, $\frac{1}{2}$; $\frac{1}{2}$, $-\frac{1}{2}$ (or $-\frac{1}{2}$, $\frac{1}{2}$); and $-\frac{1}{2}$, $-\frac{1}{2}$. Correspondingly, the PMR line of the methene proton is split into a triplet.

The theory of chemical shifts and hyperfine structure makes it possible to obtain quantitative data concerning the positions of nuclei and their interactions, whereas the study of NMR spectra at low resolution gives information about the dynamic, relaxational properties of molecules (i.e., about the motility of the atomic groups and of the entire molecule). NMR theory, experimental methods (which are not considered here), and applications in chemistry and physics are described in a series of monographs [218-226].

The PMR spectrum of a protein consists of many overlapping lines that can be separated if high resolution devices (with frequencies up to 330 MHz) and selective deuteration of the protein are used [227]). Jardetzky and other investigators have successfully applied these methods to a series of problems [228-230]. The observation of chemical shifts has yielded important details about the interaction of pharmacological substances (sulfamines and antibiotics) with proteins [231]. The helix-coil transitions in polyamino acids are clearly manifested by proton resonance chemical shifts [232]. NMR also provides information about hapten-antibody interactions [233].

Important results have been obtained in enzyme studies, particularly of ribonuclease, staphylococcus nuclease, and lysozyme. Figure 5.33 shows the proton resonances of nuclease, in the region that corresponds to aromatic amino acids, at four pH values [234]. Peaks H2 correspond to lower fields, peaks H4 to higher ones. Five peaks are resolved, which are sensitive to pH variation. Two of them, H2a and H2b, belong to the same His residue, as is confirmed by a series of facts. The relative intensities of these peaks and the distances between them vary with the pH. This variation can be explained by slow conformational change, which affects the surroundings of His H2. This change is local, since other peaks are not changed. The distance between H2a and H2b defines the lower limit of the lifetime τ of the His residue in every conformation

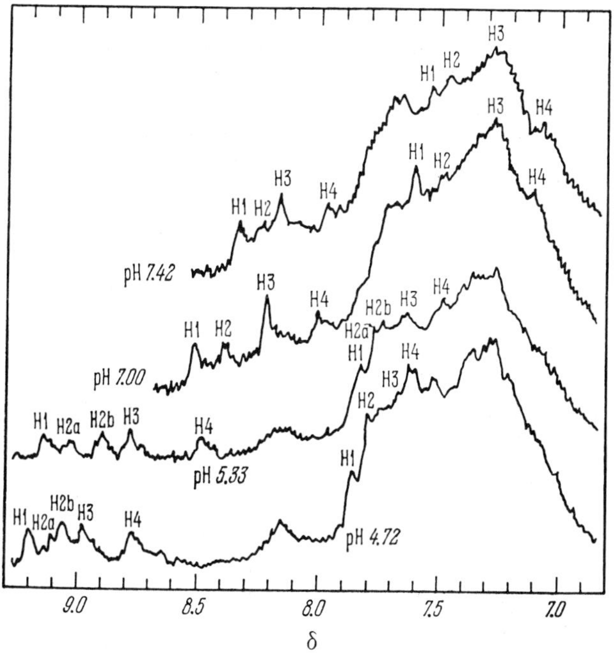


*FIG. 5.33 NMR spectra of the aromatic region of staphylococcus nuclease at various pH.*

$\tau \gg 1/\Delta\omega \sim 40$ msec

The broadening of the peaks and the features of their titration curves at pH 6.3-7.5 are explained by an increase in the rate of mutual transformation of two conformations. Their equilibrium constant K is determined as the ratio of the areas of peaks H2b and H2a. At pH 4.7, K = 0.5; at pH 5.0, K = 0.7, at pH 7.5, K = 3-5 [234].

Thus NMR enables us to investigate conformational transitions. The study of lysozyme has revealed interesting features about its denaturation [235]. In studies of ribonuclease, the resonances of four separate His residues and their exact location have been determined [236]. The action of enzyme inhibitors on these residues has been investigated and a model of an enzyme-inhibitor complex has been constructed on the basis of data obtained by means of NMR spectra [237].

Investigations of hemoglobin, myoglobin, and their complexes with various ligands have provided important confirmation of X-ray data. This finding establishes that the structures of these proteins in crystal and in solution are identical (cf. p. 264). The absence of direct interaction of the heme-groups in hemoglobin has been established (cf. p. 420). Fine interactions between amino acid residues have been investigated (cf. [230]). These few examples by no means exhaust the biopolymer studies already performed by means of NMR; they merely indicate the great potential of NMR spectroscopy (cf. also [238])

Let us now consider EPR. An electron has spin $s = \frac{1}{2}$; its energy level is split by the field $H_0$ into two levels at a distance, corresponding to the resonance condition,

$$\hbar\omega = g\mu_B H_0 \tag{5.166}$$

where $\mu_B$ is the Bohr magneton, which is 1836 times larger than the nuclear one

$$\mu_B = e\hbar/2m_e c = 0.93 \times 10^{-20} \text{ erg/G} \tag{5.167}$$

where $m_e$ is the mass of the electron. The g factor for the free electron is 2.0023. Using these values, we find the resonance frequency to be

$$\nu \text{ (MHz)} = 2.8 H_0 \text{ (G)}$$

that is, if $H_0 = 10^4$ G, $\nu_{e\ell} = 2.8 \times 10^{10}$ sec$^{-1}$ ($\nu$ of proton = $4.26 \times 10^7$ sec$^{-1}$).

In the majority of organic compounds, the electron spin moments cancel each other, their total spin is zero, and EPR is not observable. On the other hand, EPR is quite evident in the case of free radicals and of molecules having an odd number of electrons. EPR spectra serve as our main source of informatic about the structure of free radicals and their interactions with surrounding particles.

If only one unpaired electron is present in a substance, then

$$\hbar\omega = g\mu_B(H_0+H_1) \tag{5.168}$$

Local field $H_1$ is formed by the magnetic moments of neighboring nuclei, particularly of $N^{14}$. This nucleus has J = 1; hence, projections of the nuclear spin in the direction of field $H_0$ correspond to the values m = 1, 0, -1. The local field acting on the electron has three values and the EPR peak is split into a triplet. EPR is usually observed with a klystron as the source of microwave radiation (field $H_1$) with frequencies in the 9000-MHz region. In this case resonance is observed at 3200 G. The local fields producing hyperfine splitting are of the order of tens of gauss.

The interaction of electron spin with permanent and local fields that we have just described can be expressed by the Hamiltonian

$$\mathcal{H} = \mu_B SgH_0+\hbar STJ-g_N\mu_0 JH_0 \tag{5.169}$$

Here S and J are the momentum operators of the electron and nuclear spins, respectively; g is the g factor tensor of the electron (the g factor is anisotropic); T is the dipole interaction tensor of the electron and nuclear spins; $g_N$ is the g factor of the $N^{14}$ nucleus. The first term in (5.169) represents the interaction of the electron spin momentum with the external field; the second, the "hyperfine" interaction between electron and nucleus; and the third, the interaction of the nitrogen nucleus with the external field. The observable spectral lines correspond to allowed transitions between the eigenstates of this Hamiltonian.

In general, the nuclear spin J produces 2J + 1 equidistant lines of hyperfine structure in the EPR spectrum. Consider as an example the stable free radical diphenylpicrylhydrazyl. The spectrum contains five lines with relative intensities 1:2:3:2:1. This finding can be explained by assuming that the unpaired electron is delocalized between two nitrogen atoms. The summary spin of two nuclei is 2; projections in the field direction are +2, +1, 0, -1, -2. The values 2 and -2 can be obtained in only one way (2 = 1 + 1, -2 = -1 -1), the values 1 and -1 in two ways (1 = 1 + 0 = 0 + 1, -1 = -1 + 0 = 0 - 1), and the zero value in three ways (0 = 0 + 0 = +1 - 1 = -1 + 1).

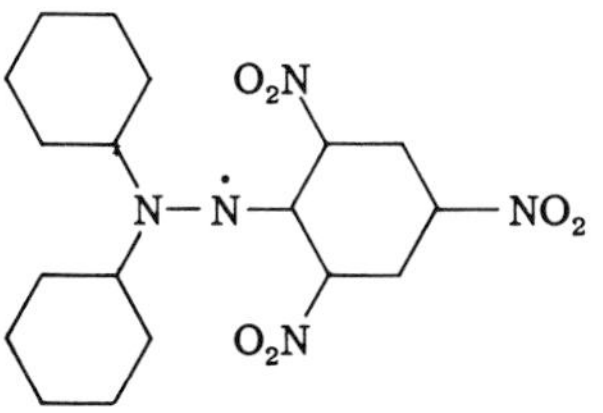

This discussion concerns hyperfine structure in the spectrum. Fine structure arises because of the anisotropy of the g factor in crystals. If the summary spin is larger than $\frac{1}{2}$, the Zeeman levels are no longer equidistant, and the spectrum is altered due to spin-orbital interaction. Instead of one line, a group of lines is observed whose positions and intensities depend on the orientation of the field $H_0$ in relation to the crystal axes. Fine structure is not resolved in liquids and solutions; there is only some broadening of the lines.

Line width in the EPR spectrum is a consequence of spin-spin and spin-lattice relaxation, as in NMR. The spin-spin relaxation time $T_2$ characterizes the equilibration rate of the magnetic moments of all paramagnetic particles; the spin-lattice relaxation time $T_1$, the rate of reestablishment of equilibrium between the spin system and lattice vibrations. $T_2$ is practically independent of temperature and is determined by spin concentration; $T_1$ increases rapidly with decreasing temperature. Evidently, both $T_1$ and $T_2$ are determined by the motility of spin-containing particles and surrounding molecules. A detailed presentation of EPR theory and experimental methods is contained in a series of monographs (cf., e.g., [231-242]).

Electron paramagnetic resonance has long been used to solve biological and biophysical problems. Free radicals are formed in a series of enzymic oxidation-reduction reactions and under the action of radiation. Free radicals are formed from coenzymes (e.g., semiquinones from flavins and flavoproteins) and substrates, in particular by oxidation of ascorbic acid, hydroquinone, etc., by peroxidase. Short-wave irradiation of organic compounds and biological systems also produces free radicals that can be registered by the EPR method. Therefore, this method is very useful in radiobiology and in studies of photobiological processes. The formation of free radicals in biological systems and in model biochemical reactions is discussed in [240,243-245].

In recent years EPR techniques for studying biopolymers have been developed that are based on the spin-label method, in which a stable free radical containing an unpaired electron is linked to a protein. Nitroxyl radicals are widely used for this purpose [246]. Spin-labeled proteins were first obtained in 1965 [247]. Nitroxyl radicals can be bound, depending on the group R, to various groups of proteins, particularly SH groups. In [248] and [249] the label is reported to have been bound by Cys-β93 of hemoglobin (Hb), alkylating the SH group. Hb has been crystallized and the dependence of EPR spectra on crystal orientation in a magnetic field has been studied. Figure 5.34 shows the EPR spectra of labeled $Hb(CO)_4$

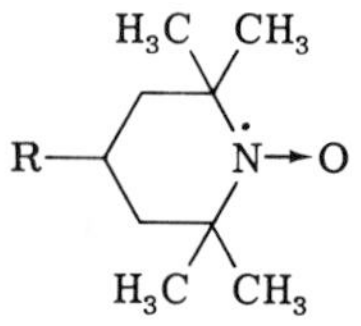

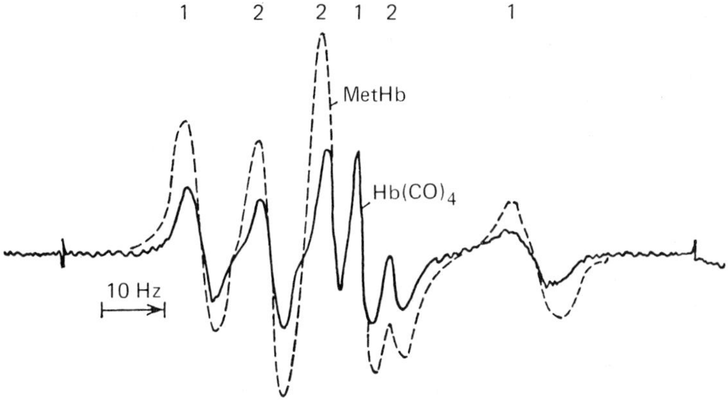


*FIG. 5.34 EPR spectra of (1) Hb(CO)₄ and (2) methemoglobin.*

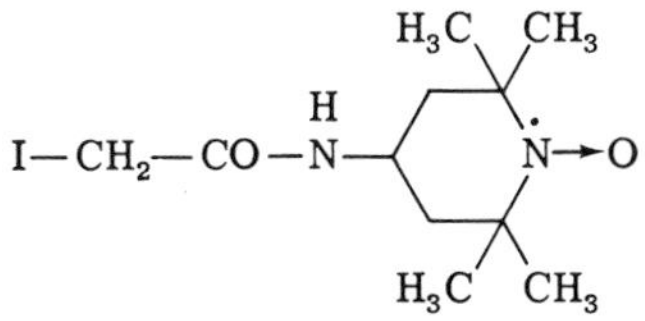

and methemoglobin at pH 7.0. Signals 1 and 2 correspond to different crystal orientations; their intensities depend on the ionic content of the crystallizing solution. The coincidence of these signals for $Hb(CO)_4$ and MetHb indicates the conformational similarity of both proteins near Cys-β93.

A label's motility is indicated by the EPR spectrum. Molecular motion makes the Hamiltonian (5.169) time dependent

$$\mathcal{H}(t) \equiv \langle\mathcal{H}(t)\rangle + \{\mathcal{H}(t) - \langle\mathcal{H}(t)\rangle\} \tag{5.170}$$

where $\langle\mathcal{H}\rangle$ is the mean value and { }, the fluctuation of $\mathcal{H}(t)$. If the basic features of the spectrum are due to $\langle\mathcal{H}\rangle$, a high speed of motion is indicated. Thus, a nitroxyl radical giving three sharp lines with hyperfine splitting equal to $\frac{1}{3}$ $(T_x + T_y + T_z)$ is subjected to fast torsional motion with reciprocal correlation time $\tau^{-1}$ that is large in comparison with the maximal anisotropic member in $\mathcal{H}$, namely, $|T_z - T_x|$. The influence of { } on EPR lines is examined in [250-252]. The time dependence of $\mathcal{H}$ makes it possible to study conformational changes in macromolecules. Hb in solution has been studied in detail in [253,254]; it has been found that the label "feels" two conformations of protein at its oxygenation. Varying the labels could reveal the presence of a local region in the Hb molecule, whose structure depends on the oxygenation state of more than one heme group. Hence, interactions of Hb subunits have been established (cf. Section 7.1).

Lichtenstein and co-workers used labels of different length and flexibility in their studies of the structure and conformational motility of a series of proteins--lysozyme, myoglobin, albumins, etc. [255-258]. The work reported in [259] established changes in the EPR spectra of aspartate-aminotransferase labeled by nitroxyl radicals, forming complexes with the substrate and similar compounds. These alterations show an increase in radical motility in the complex--the correlation time of the original protein is $8.0 \pm 0.4 \times 10^{-10}$ sec; of the protein ligand, $5.1 \pm 0.4 \times 10^{-10}$ sec.

A model developed to facilitate theoretical examination of the motion of a label bound to a protein by a flexible "leg" (the "rotator on rotator" model [260]) has made it possible to calculate the correlation time as a function of internal rotations. Use of a double label enables us to determine the distances between interacting radicals and hence the geometry of conformational changes (cf. [255,261,262]). The spin-label method is also very effective in studies of the structure of biological membranes and of conformational phenomena in membranes [263,264].

Valuable information seems to be obtainable by means of nuclear relaxation studies of biopolymers which contain a paramagnetic label. Relaxation time depends on the interaction of nuclear and electron spins and hence on the distance between them (T is proportional to $r^6$). Therefore, information about molecular geometry and molecular motions can be obtained [265]. In [266] the EPR and NMR of alcohol dehydrogenase labeled with the analog of nicotinamide adenine dinucleotide are investigated. It was found that the label competes with NADH at the enzyme binding site, is strongly immobilized by the protein, and changes the relaxation time of the protons of water, T being dependent on the presence of alcohol. The binding site of alcohols by this enzyme has been established, and the kinetic and geometrical characteristics of the system have been estimated

The possibilities of such EPR methods in the study of different relaxations of different regions of the spectrum seem to be important [263]. Double electron resonance can be useful for determining interlabel distances [267]. For further details, see [263,268,269].

Investigation of the structure, conformational transitions, and internal dynamics of complex biopolymer molecules requires methods with application of sensitive labels. In this sense, spin labels are similar to luminescent ones.

## References


1. M. Perutz, Sci. Amer. 211(5), 69 (1964).
2. C. Blake, Advan. Protein Chem. 23, 59 (1968).
3. M. Buerger, "X-ray Crystallography." Wiley, New York, 1942.
4. A. Guinier, "X-ray Analysis of Crystals."

5. A. Kitaigorodsky, "Roentgenostructural Analysis." Gostekhzdat, Moscow, 1950 (R).
6. A. Kitaigorodsky, "Roentgenostructural Analysis of Small-crystalline and Amorphous Bodies." Gostekhzdat, Moscow, 1952 (R).
7. A. Kitaigorodsky, "Theory of Structural Analysis." Ed. Acad. Sci. USSR, Moscow, 1957 (R).
8. M. Buerger, "Vector Space." Wiley, 1959.
9. B. Vainstein, "Diffraction of Roentgen Rays at Chain Molecules." Ed. Acad. Sci. USSR, Moscow, 1963 (R).
10. B. Vainstein, Usp. Phys. Nauk **88**, 527 (1966) (R).
11. N. Andreeva, in "Fundamentals of Molecular Biology. Physical Methods of Investigations of Proteins and Nucleic Acids." Nauka, Moscow, 1967 (R).
12. N. Andreeva, J. All Un. Mendeleev Chem. Soc. **16**, 369 (1971) (R).
13. J. Kendrew, Biophysica **8**, 273 (1963) (R).
14. M. Perutz, H. Muirhead, J. Cox, L. Goman, F. Methews, McGandy, and L. Webb, Nature (London) **219**, 29 (1968).
15. W. Bolton and M. Perutz, Nature (London) **228**, 551 (1970).
16. R. Huber, O. Epp, and H. Formanek, Naturwissenschaften **56**, 362 (1969).
17. R. Dickerson et al., J. Biol. Chem. **246**, 1515 (1971).
18. C. Blake, D. Koenig, G. Mair, A. North, D. Phillips, and V. Sarma, Nature (London) **206**, 757 (1965).
19. G. Kartha, J. Bello, and D. Harker, Nature (London) **213**, 862 (1967).
20. H. Wykoff, D. Tsernoglou, A. Hanson, P. Knox, B. Lee, and F. Richardson, J. Biol. Chem. **245**, 305 (1970).
21. W. Lypscomb, G. Reeke, J. Hartsuck, P. Quiocho, and B. Bethge, Phil. Trans. Roy. Soc. **B257**, 177 (1970).
22. S. Freer, J. Kraut, J. Robertus, H. Wright, and N. Xuong, Biochemistry **9**, 1997 (1970).
23. T. Steitz, R. Henderson, and D. Blow, J. Mol. Biol. **96**, 337 (1969).
24. J. Drenth, J. Jansonius, R. Koekock, H. Swen, and B. Welters, Nature (London) **218**, 929 (1968).
25. A. Arnon, G. Bier, F. Cotton, E. Hazen, D. Richardson, and J. Richardson, Proc. Nat. Acad. Sci. U.S. **64**, 420 (1969).
26. J. Heriott, L. Sicker, L. Jensen, and M. Lovenberg, J. Mol. Biol. **50**, 391 (1970).
27. M. Adams et al., Nature (London) **227**, 1098 (1970).
28. M. Adams et al., Nature (London) **224**, 491 (1968).
29. C. Schubert, H. Wright, R. Alden, and J. Kraut, Nature (London) **221**, 235 (1969).
30. B. Vainstein, Usp. Phys. Nauk 109, 545 (1973) (R).

31. S. Pinsker, "Diffraction of Electrons." Ed. Acad. Sci. USSR, Moscow, 1947 (R).
32. B. Vainstein, "Structural Electronography." Ed. Acad. Sci. USSR, Moscow, 1956 (R).
33. G. Bacon, "Neutron Diffraction." 2nd ed., Oxford Univ. Press, London and New York, 1962.
34. J. Watson, "The Double Helix." Atheneum, New York, 1968.
35. A. Kitaigorodsky and D. Tsvankin, Vysokomol. Sojedin. I, 269, 279 (1958); Crystallographia **4**, 625 (1959) (R).
36. W. Astbury, Proc. Roy. Soc. **B134**, 303 (1947); **141**, 1 (1953).
37. G. Rosenbaum, K. Holmes, and J. Witz, Nature (London) **230**, 434 (1971).
38. O. Kratky, Progr. Biophys. Mol. Biol. **13**, 194 (1965).
39. H. Brumberger (ed.), "Small-Angle X-Ray Scattering." Gordon and Breach, New York, 1967.
40. O. Kratky, B. Paletta, G. Porod, and K. Strohmaier, Z. Naturforsch. **126**, 287 (1957).
41. O. Pritsyn and B. Fedorov, Dokl. Acad. Sci. USSR **153**, 882 (1963) (R).
42. H. Ritland, P. Keasberg, and W. Beeman, J. Chem. Phys. **18**, 1237 (1950).
43. A. Vasina, G. Frank, and V. Lemajichin, J. Mol. Biol. **14**, 373 (1965).
44. R. Kajushina, N. Soifenov, I. Kuranova, and N. Konareva, Biophysica **12**, 611 (1967) (R).
45. N. Esipova, A. Dembo, V. Tumanian, and O. Polianovsky, Mol. Biol. **2**, 527 (1968) (R).
46. V. Tumanian, N. Esipova, and L. Kiselev, Dokl. Acad. Sci. USSR **168**, 211 (1966) (R).
47. P. Laggner, O. Kratky, W. Palm, and A. Holasek, FEBS Lett. **15**, 220 (1971).
48. G. Porod, Kolloid Z. **124**, 83; **125**, 51 (1951).
49. W. Wunderlich and R. Kirste, Ber. Bunsenges. Phys. Chem. **68**, 645 (1964).
50. A. Grigoriev, L. Volkova, and O. Ptitsyn, Vysokomol. Soedin. **B11**, 232 (1969); **A12**, 1363 (1970) (R); FEBS Lett. **14**, 189 (1971); **15**, 217 (1971).
51. W. Gratzer, in "Poly-α-Amino Acids" (G. Fasman, ed.). Dekker, New York, 1967.
52. J. Ham and J. Platt, J. Chem. Phys. **20**, 335 (1952).
53. H. Hunt and W. Simpson, J. Amer. Chem. Soc. **75**, 4540 (1953).
54. D. Peterson and W. Simpson, J. Amer. Chem. Soc. **79**, 2375 (1957).
55. J. Ladik, Nature (London) **202**, 1208 (1964).
56. M. Suard, G. Berthier, and B. Pullman, Biochim. Biophys. Acta **52**, 254 (1961).

57. M. Suard, J. Chim. Phys. 62, 79 (1965).
58. S. Yomosa, Biopolym. Symp. 1, 1 (1964).
59. I. Tinoco, A. Halpern, and W. Simpson, in "Polyamino Acids, Polypeptides and Proteins" (M. Stahmann, ed.). Univ. of Wisconsin Press, Madison, Wisconsin, 1962.
60. A. Davydov, "Theory of Light Absorption in Molecular Crystals." Ed. Acad. Sci. Ukr. SSR, Kiev, 1951; "Theory of Molecular Excitons." Nauka, Moscow, 1968 (R); Plenum Press, New York, 1971.
61. V. Agranovitch, "Theory of Excitons." Nauka, Moscow, 1968 (R).
62. W. Moffitt, Proc. Nat. Acad. Sci. U.S. 42, 736 (1956).
63. B. Neporent and N. Bakhshiev, Opt. Spectrosc. 5, 64 (1958); 8, 777 (1960) (R).
64. N. Bakhshiev, Opt. Spektrosk. 10, 717 (1961); 16, 821 (1964); 19, 535 (1964) (R).
65. S. Yanari and F. Bovey, J. Biol. Chem. 235, 2818 (1960).
66. M. Weissbluth, Quart. Rev. Biophys. 4, 1 (1971).
67. H. DeVoe, Biopolym. Symp. 1, 251 (1964).
68. R. Bullough, J. Chem. Phys. 43, 1927 (1965); 48, 3712 (1968).
69. H. DeVoe and I. Tinoco, J. Mol. Biol. 4, 518 (1962).
70. R. Nesbet, Mol. Phys. 7, 211 (1964); Biopolym. Symp. 1, 129 (1964).
71. W. Johnson and I. Tinoco, Biopolymers 7, 727 (1969).
72. M. Goodman, I. Listowsky, Y. Masuda, and F. Boardman, Biopolymers 1, 33 (1963).
73. R. McDiarmid and P. Doty, J. Phys. Chem. 70, 2620 (1966).
74. K. Rosenheck and P. Doty, Proc. Nat. Acad. Sci. U.S. 47, 1785 (1961).
75. V. Danilov, N. Jeltovsky, V. Ogloblin, and V. Pechenaja, J. Theor. Biol. 30, 559 (1971).
76. V. Danilov and N. Jeltovsky, Theor. Chim. Acta 19, 384 (1970).
77. M. Born, "Optik." Springer, Berlin, 1932.
78. M. Volkenstein, "Molecular Optics." Gostekhizdat, Moscow, 1951. (R).
79. D. Caldwell and H. Eyring, "The Theory of Optical Activity." Wiley (Interscience), New York, 1971.
80. L. Rosenfeld, Z. Phys. 52, 161 (1928).
81. H. Eyring, J. Walter, and G. Kimball, "Quantum Chemistry." Wiley, New York, 1964.
82. W. Kautzmann, "Quantum Chemistry. An Introduction." Academic Press, New York, 1957.
83. E. Condon, Usp. Physicheskich Nauk. 19, 380 (1938) (R).
84. M. Volkenstein, Usp. Chim. 9, 1090, 1252 (1940) (R).
85. A. Moscowitz, Advan. Chem. Phys. 4, 67 (1962).

86. A Moscowitz, in "Optical Rotatory Dispersion and Circular Dichroism in Organic Chemistry" (G. Snatzke, ed.). Heyden, London, 1967.
87. R. de L. Kronig, J. Opt. Soc. Amer. 12, 547 (1926).
88. H. Kramers, Atti congr. int. fis. Cimo 2, 545 (1927).
89. J. Macdonald and M. Brachman, Rev. Mod. Phys. 28, 393 (1956).
90. A. Moscowitz, Rev. Mod. Phys. 32, 440 (1960).
91. D. Caldwell and H. Eyring, Ann. Rev. Phys. Chem. 15, 281 (1964).
92. J. Kirkwood, J. Chem. Phys. 5, 479 (1937).
93. M. Volkenstein, Dokl. Acad. Sci. USSR 71, 447, 643 (1950) (R).
94. A. McLachlan and M. Ball, Mol. Phys. 8, 581 (1965).
95. V. Aslanian and M. Volkenstein, Opt. Spektrosk. 7, 208 (1959) (R).
96. E. Condon, J. Walter, and H. Eyring, J. Chem. Phys. 5, 753 (1937).
97. W. Kauzmann, J. Walter, and H. Eyring, Chem. Rev. 26, 339 (1940).
98. C. Djerassi, "Optical Rotatory Dispersion." McGraw-Hill, New York, 1960.
99. L. Veluz, M. Legrand, and M. Grosjean. "Optical Circular Dichroism." Verlag Chemie, Weinheim, 1965.
100. G. Snatzke (ed.), "Optical Rotatory Dispersion and Circular Dichroism in Organic Chemistry." Heyden, London, 1967.
101. M. Kruchek, in "Optika i Spektroskopia," Vol. 2. Ed. Acad. Sci. USSR, Moscow, 1963; Opt. Spektrosk. 17, 545, 794 (1964) (R).
102. M. Volkenstein and K. Konstantinavichus, Opt. Spektrosk. 23, 77 (1967 (R).
103. M. Volkenstein and A. Efros, Usp. Chim. 19, 602 (1950) (R).
104. W. Kuhn and H. Freudeuberg, Hand. Jahrbuch Chem. Phys. 8, Teil 3 (1930).
105. J. Yang, Proc. Nat. Acad. Sci. U.S. 53, 438 (1965).
106. W. Moffitt, J. Chem. Phys. 25, 467 (1956); Proc. Nat. Acad. Sci. U.S. 42, 736 (1956).
107. W. Moffitt and J. Yang, Proc. Nat. Acad. Sci. U.S. 92, 596 (1956).
108. W. Moffitt, D. Fitts, and J. Kirkwood, Proc. Nat. Acad. Sci. U.S. 43, 723 (1957).
109. I. Tinoco, Advan. Chem. Phys. 4, 113 (1962).
110. I. Tinoco, R. Woody, and D. Bradley, J. Chem. Phys. 38, 1317 (1963).
111. J. Schellmann and P. Oriel, J. Chem. Phys. 37, 2114 (1962).
112. I. Tinoco, J. Chem. Phys. 65, 91 (1968).
113. W. Gratzer and D. Cowburn, Nature (London) 222, 426 (1969).

114. R. Woody and I. Tinoco, J. Chem. Phys. **46**, 4927 (1967).
115. E. Pysh, Proc. Nat. Acad. Sci. U.S. **56**, 825 (1966); J. Mol. Biol. 23, 587 (1967).
116. N. Go, Proc. Phys. Soc. Japan 21, 1579 (1966).
117. R. Harris, J. Chem. Phys. **43**, 959 (1965).
118. V. Zubkov and M. Volkenstein, Mol. Biol. **4**, 598 (1970) (R).
119. E. Anufrieva et al., Biophysica 10, 918 (1965) (R).
120. G. Fasman and E. Blout, J. Amer. Chem. Soc. 82, 2262 (1960).
121. J. Bradbury, A. Elliott, and W. Hanby, J. Mol. Bio. 5, 487 (1962).
122. S. Ikeda, H. Maeda, and T. Isemura, J. Mol. Biol. **10**, 223 (1964).
123. K. Imahori and I. Yahari, Biopolym. Symp. **1**, 421 (1964).
124. P. Sarkar and P. Doty, Proc. Nat. Acad. Sci. U.S. 55, 981 (1966)
125. E. Iizuka and J. Yang, Proc. Nat. Acad. Sci. U.S. **55**, 1175 (1966).
126. R. Townsend, T. Kumasinsky, S. Timasheff, G. Fasman, and B. Davidson, Biochem. Biophys. Res. Commun. **23**, 163 (1966).
127. B. Davidson, N. Tooney, and G. Fasman, Biochem. Biophys. Res. Commun. **23**, 156 (1966).
128. M. Volkenstein and V. Zubkov, J. Struk. Chim. **8**, 284 (1967) (R).
129. V. Zubkov and M. Volkenstein, Dokl. Acad. Sci. USSR **175**, 942 (1967) (R).
130. T. Birstein, V. Zubkov, and M. Volkenstein, J. Polym. Sci. A-2 **8**, 177 (1970).
131. V. Zubkov, T. Birstein, J. Milevskaja, and M. Volkenstein, Mol. Biol. **4**, 715 (1970) (R).
132. E. Blout and L. Stryer, Proc. Nat. Acad. Sci. U.S. **45**, 159 (1959); J. Amer. Chem. Soc. **83**, 1411 (1961).
133. E. Blout, J. Carver, and E. Shechter, in "Optical Rotatory Dispersion and Circular Dichroism in Organic Chemistry" (G. Snatzke, ed.). Heyden, London, 1967.
134. I. Bolotina and M. Volkenstein, in "Molecular Biophysics." Nauka, Moscow, 1965 (R).
135. Y. Torchinsky, in "Uspekhi Biologitcheskoi Chimiï," Vol. 8. Nauka, Moscow, 1967.
136. T. Li and B. Johnson, Biochemistry **8**, 3638 (1969).
137. V. Permogorov, J. Lazurkin, and S. Shmurak, Dokl. Acad. Sci. USSR **155**, 1440 (1964) (R).
138. K. Konstantinavichus and M. Volkenstein, Opt. Spektrosk. **23**, 80 (1967) (R).
139. M. Frank-Kamenetzky and A. Lukashin, Opt. Spektrosk. **30**, 1092 (1971) (R).
140. J. Yang and T. Samejima, Progr. Nucl. Acids Res. Mol. Biol. **9**, 224 (1969).

141. J. Yang, T. Samejima, and P. Sarker, Biopolymers **4**, 623 (1966).
142. B. Johnson and I. Tinoco, Biopolymers **7**, 727 (1969).
143. P. Sarkar, B. Wells, and J. Yang, J. Mol. Biol. **25**, 563 (1967).
144. J. Yang, in "Poly-α-Amino Acids" (G. Fasman, ed.). Dekker, New York, 1967.
145. H. Hashizuma, M. Shiraki, and K. Imahori, J. Biochem. Japan **62**, 543 (1967).
146. S. Beychok, in "Poly-α-Amino Acids" (G. Fasman, ed.). Dekker, New York, 1967.
147. V. Saxena and D. Wetlaufer, Proc. Nat. Acad. Sci. U.S. **68**, 969 (1971).
148. P. Urnes and P. Doty, Advan. Protein Chem. **16**, 401 (1961).
149. E. Shechter and E. Blout, Proc. Nat. Acad. Sci. U.S. **51**, 695, 794 (1964).
150. E. Shechter, J. Carver, and E. Blout, Proc. Nat. Acad. Sci. U.S. **51**, 1029 (1964).
151. S. Veniaminov and J. Chirgadze, Opt. Spektrosk. **23**, 813 (1967); Dokl. Acad. Sci. USSR **195**, 722 (1970) (R).
152. M. Warshaw and I. Tinoco, J. Mol. Biol. **20**, 29 (1966).
153. G. Zavilgelsky, Mol. Biol **1**, 703 (1967) (R).
154. C. Cantor and I. Tinoco, Biopolymers **5**, 84 (1967).
155. R. Cotton, P. McPhie, and W. Gratzer, Nature (London) **216**, 864 (1967).
156. C. Bush and H. Scheraga, Biochemistry **6**, 3036 (1967).
157. P. Oriel and J. Koenig, Arch. Biochem. **127**, 274 (1968).
158. P. Oriel, Arch. Biochem. **115**, 577 (1966).
159. B. Neporent, Dokl. Acad. Sci. USSR **119**, 682 (1958) (R).
160. B. Stepanov, "Luminescence of Complex Molecules." Ed. Acad. Sci. Belorussian SSR, Minsk, 1955 (R).
161. E. Shpolsky, Usp. Phys. Nauk. **71**, 215 (1960); **80**, 255 (1963) (R).
162. S. Vavilov, "Collected Papers," Vol. 1. Ed. Acad. Sci. USSR, Moscow, 1952 (R).
163. P. Feofilov, "Polarized Luminescence of Atoms, Molecules and Crystals." Fizmatgiz, Moscow, 1959 (R).
164. A. Terenin, "Photonics of Molecules of Dyes and Related Compounds." Nauk, Moscow, 1967 (R).
165. S. Vavilov, "Microstructure of Light." Ed. Acad. Sci. USSR, Moscow, 1950 (R).
166. Th. Förster, in "Comparative Effects of Radiation" (M. Burton, J. Kirby-Smith, and J. Magee, eds.). Wiley, New York, 1960.
167. Th. Förster, Discuss. Faraday Soc. **27**, 7 (1959); Radiat. Res. Suppl. **2**, 326 (1960).
168. E. McRae and W. Siebrand, J. Chem. Phys. **41**, 905 (1964).
169. A. Szent-Györgyi, "Bioenergetics." Academic Press, New York, 1957.

170. S. Konev, "Electronic-Excited States of Biopolymers." Nauka i Technika, Minsk, 1965 (R).
171. G. Barenboim, A. Domanskii, and K. Turoverov, "Luminescence of Biopolymers and Cells." Plenum Press, New York, 1969.
172. Yu. Vladimirov, "Photochemistry and Luminescence of Proteins." Nauka, Moscow, 1965 (R).
173. G. Weber, Biochem. J. **75**, 335, 345 (1960); **79**, 29 (1961).
174. E. Burstein, Biophysica **9**, 184 (1964); **13**, 433, 718 (1968) (R).
175. Y. Vladimirov and Li-Chin-Go, Biophysika **7**, 270 (1962) (R).
176. N. Vedenkina and E. Burstein, Biophysika **15**, 965 (1970); Mol. Biol. **4**, 743 (1970) (R).
177. E. Busel and E. Burstein, Biophysika **15**, 993 (1970) (R).
178. E. Busel, T. Bushueva and E. Burstein, Opt. Spektrosk. **29**, 501 (1970) (R).
179. R. Chen, H. Edelhoch, and R. Steiner, in "Physical Principles and Techniques of Protein Chemistry" Part A (S. J. Leach, ed.). Academic Press, New York, 1969.
180. I. Ostashevsky, Biophysika **16**, N 4 (1971); Mol. Biol. **6**, 3 (1972) (R).
181. G. Weber, Biochem. J. **52**, 145 (1952).
182. P. Wahl and G. Weber, J. Mol. Biol. **30**, 371 (1967).
183. E. Anufrieva, M. Volkenstein, M. Krakoviak, and T. Sheveleva, Dokl. Acad. Sci. USSR **182**, 361 (1968); **186**, 854 (1969) (R).
184. E. Anufrieva, M. Volkenstein, J. Gottlieb, M. Krakoviak, I. Torchinsky, and T. Sheveleva, Izv. Acad. Sci. USSR Ser. Phys. **34**, 518 (1970) (R).
185. E. Anufrieva, M. Volkenstein, J. Gottlieb, M. Krakoviak, S. Skorokhodov, and T. Sheveleva, Dokl. Acad. Sci. USSR **194**, 1108 (1970) (R).
186. V. Samokish, E. Anufrieva, and M. Volkenstein, Mol. Biol. **2**, 374 (1968) (R).
187. N. Bakhshiev, J. Mazurenko, and I. Piterskaja, Izv. Acad. Sci. USSR Ser. Phys. **32**, 1360 (1969) (R).
188. E. Anufrieva, M. Volkenstein, and V. Samokish, Dokl. Acad. Sci. USSR **195**, 1215 (1970).
189. M. Volkenstein, M. Eliashevitch, and B. Stepanov, "Vibrations of Molecules," Vol. 1,2. Gostekhizdat, Moscow, 1949 (R).
190. M. Volkenstein, L. Gribov, M. Eliashevitch, and B. Stepanov, "Vibrations of Molecules." Nauka, Moscow, 1972 (R).
191. G. Herzberg, "Infrared and Raman Spectra of Polyatomic Molecules." Van Nostrand-Reinhold, Princeton, New Jersey, 1945.

192. E. Wilson, J. Decius, and P. Cross, "Molecular Vibrations." McGraw-Hill, New York, 1957.
193. L. Sverdlov, M. Kovner, and E. Krainov, "Vibrational Spectra of Polyatomic Molecules." Nauka, Moscow, 1970 (R).
194. M. Sushtshinsky, "Raman Spectra of Molecules and Crystals." Nauka, Moscow, 1969 (R).
195. M. Born and R. Oppenheimer, Ann. Phys. 84, 457 (1927).
196. M. Volkenstein, Dokl. Acad. Sci. USSR 30, 784 (1941); J. Exp. Theor. Phys. 11, 642 (1941) (R).
197. M. Volkenstein, "Molecular Optics." Gostekhizdat, Moscow, 1951 (R).
198. T. Mijazawa, in "Polyamino Acids, Polypeptides and Proteins" (M. Stahmann, ed.). Univ. Wisconsin Press, Madison, Wisconsin, 1962.
199. T. Mijazawa, T. Shimanouchi, and S. Mizushima, J. Chem. Phys. 24, 408 (1956); 29, 611 (1958).
200. T. Shimanouchi, I. Nakagawa, J. Hiraishi, and N. Ishii, J. Mol. Spectrosc. 19, 78 (1968).
201. H. Suzi, in "Structure and Stability of Biological Macromolecules" (S. Timasheff and G. Fasman, eds.). Dekker, New York, 1969.
202. C. Liang, S. Krimm, and G. Sutherland, J. Chem. Phys. 25, 543, 549 (1956).
203. C. Liang and S. Krimm, J. Chem. Phys. 25, 563 (1956).
204. Y. Gottlieb, Opt. Spectrosc. 7, 294 (1959); 9, 319 (1960); Vysokomole. Sojedin. 1, 474 (1959) (R).
205. Y. Gottlieb and L. Kudinskaja, Opt. Spectrosc. 10, 335 (1961); 13, 59 (1962) (R).
206. V. Boitsov and Y. Gottleib, Opt. Spectrosc. 2, 128, 135 (1963) (R).
207. L. Gribov and T. Abilova, Opt. Spectrosc. 23, 374, 535 (1967); 26, 915 (1969) (R).
208. O. Zubkova, L. Gribov, and A. Shabadash, J. Appl. Spectrosc. 16, 2, 306 (1972) (R).
209. K. Fukushima, Y. Ideguchi, and T. Mijazawa, Bull. Chem. Soc. Japan 36, 1301 (1963).
210. S. Suzuki, Y. Iwashita, T. Shimanouchi, and M. Tsuboi, Biopolymers 4, 337 (1966).
211. T. Mijazawa, J. Chem. Phys. 32, 1647 (1960).
212. T. Mijazawa and E. Blout, J. Amer. Chem. Soc. 83, 712 (1961).
213. S. Krimm, J. Mol. Biol. 4, 528 (1962).
214. H. Susi, S. Timasheff, and L. Stevens, J. Biol. Chem. 242, 5460 (1967).
215. J. Chirgadze, "Infrared Spectra and Structure of Polypeptides and Proteins." Nauka, Moscow, 1965 (R).
216. R. Lord and N. Yu, J. Mol. Biol. 50, 509 (1970); 51, 203 (1970).

217. E. Zavoisky, J. Phys. USSR 9, 245, 447 (1945); 10, 170, 197 (1946).
218. E. Andrew, "Nuclear Magnetic Resonance." Cambridge Univ. Press, London and New York, 1955.
219. J. Pople, W. Schneider, and H. Bernstein, "High-Resolution Nuclear Magnetic Resonance." McGraw-Hill, New York, 1959.
220. J. Roberts, "Nuclear Magnetic Resonance." McGraw-Hill, New York, 1959.
221. J. Roberts, "An Introduction to the Analysis of Spin-Spin Splitting in High-Resolution NMR Spectra." Benjamin, New York, 1961.
222. I. Bhakka and D. Williams, "Application of NMR in Organic Chemistry." Pergamon, Oxford, 1960.
223. A. Lösche, "Kerninduktion." Deutscher Verlag der Wissensch., Berlin, 1957.
224. I. Slonim and A. Ljubimov, "Nuclear Magnetic Resonance in Polymers." Chimia, Moscow, 1966 (R).
225. J. Axley, J. Finey, and L. Satkliff, "Spectroscopy of High Resolution Nuclear Magnetic Resonance."
226. A. Koltsov and V. Ershov, "Nuclear Magnetic Resonance in Organic Chemistry." Leningrad Univ. Press, Leningrad, 1968 (R).
227. J. Markley, I. Putter, and O. Jardetzky, Science 151, 1249 (1968).
228. G. Roberts and O. Jardetzky, Advan. Protein Chem. 24, 448 (1970).
229. C. McDonald and W. Phillips, in "Biological Macromolecules" (G. Fasman and S. Timasheff, eds.), Vol. 3. Dekker, New York, 1970.
230. K. Wüthrich and R. Shulman, Phys. Today 23(4), 43 (1970).
231. O. Jardetzky, Naturwissenschaften, Heft 7, 149 (1967).
232. J. Markley, D. Meadows, and O. Jardetzky, J. Mol. Biol. 27, 25 (1967).
233. A. Burgen, O. Jardetzky, J. Metcalfe, and N. Wade-Jardetzky, Proc. Nat. Acad. U.S. 58, 447 (1967).
234. J. Markley, M. Williams, and O. Jardetzky, Proc. Nat. Acad. Sci. U.S. 65, 645 (1970).
235. J. Cohen and O. Jardetzky, Proc. Nat. Acad. Sci. U.S. 60, 92 (1968).
236. D. Meadows, G. Roberts, and O. Jardetzky, J. Mol. Biol. 45, 491 (1969).
237. G. Roberts, E. Dennis, D. Meadows, J. Cohen, and O. Jardetzky, Proc. Nat. Acad. Sci. U.S. 62, 1151 (1969).
238. V. Bystrov and V. Sakharovsky, J. All-Un. Mendeleev Chem. Soc. 16, N 4, 380 (1971) (R).
239. S. Altshuler and B. Kozyrev, "Electronic Paramagnetic Resonance." Fizmatgiz, Moscow, 1961 (R).
240. D. Ingram, "Free Radicals as Studied by Electron Spin Resonance." Butterworths, London and Washington, D.C., 1958.

241. G. Pake, "Paramagnetic Resonance." Benjamin, New York, 1962.
242. D. Ingram, "Biological and Biochemical Applications of Electron Spin Resonance." Adam Hilger, London, 1969.
243. M. S. Blois, Jr. et al. (eds.), Free radicals in biological systems, Proc. Symp., Stanford, 1960, Academic Press, New York, 1961.
244. B. Commoner, "Light and Life." Johns Hopkins Press, Baltimore, Maryland, 1961.
245. L. Kajushin, K. Lvov, and M. Pulatova, "Investigations of Paramagnetic Centers of Irradiated Proteins." Nauka, Moscow, 1970 (R).
246. E. Rozantsev, "Free Aminoxyl Radicals." Chimia, Moscow, 1970 (R).
247. T. Stone, P. Buchman, P. Nordio, and H. McConnell, Proc. Nat. Acad. Sci. U.S. **54**, 1010 (1965).
248. H. McConnell and C. Hamilton, Proc. Nat. Acad. Sci. U.S. **60**, 776 (1968).
249. H. McConnell, W. Deal, and R. Ogata, Biochemistry **8**, 2580 (1969).
250. H. McConnell, J. Chem. Phys. **25**, 709 (1956).
251. D. Kivelson, J. Chem. Phys. **33**, 1094 (1960).
252. A. Hudson and G. Luckhurst, Chem. Rev. **69**, 191 (1969).
253. S. Ogawa and H. McConnell, Proc. Nat. Acad. Sci. U.S. **58**, 19 (1967).
254. S. Ogawa, H. McConnell, and A. Horwitz, Proc. Nat. Acad. Sci. U.S. **61**, 401 (1968); Nature (London) **220**, 787 (1968).
255. G. Lichtenstein, in "Uspekhi Biochimii." Nauka, Moscow, 1971 (R).
256. G. Lichtenstein, T. Troshkina, J. Akhmedov, and V. Shuvalov, Mol. Biol. **3**, 413 (1969) (R).
257. G. Lichtenstein, J. Grebenshtshikov, P. Bobojanov, and J. Kokhanov, Mol. Biol. **4**, 682 (1970) (R).
258. Y. Kokhanov, Y. Akhmedov, G. Lichtenstein, and L. Ivanov, Dokl. Acad. Sci. USSR **205**, 372 (1972) (R).
259. V. Timofeev, O. Polianovsky, M. Volkenstein, and G. Lichtenstein, Biochim. Biophys. Acta **220**, 357 (1970).
260. D. Wallach, J. Chem. Phys. **41**, 5228 (1968).
261. G. Lichtenstein, Mol. Biol. **2**, 234 (1968) (R).
262. G. Lichtenstein, A. Pivovarov, P. Bobojanov, E. Rozantsev, and N. Smolina, Biophysika **13**, 396 (1968) (R).
263. H. McConnell and B. McFarland, Quart. Rev. Biophys. **3**, 91 (1970).
264. W. Hubbell and H. McConnell, Proc. Nat. Acad. Sci. U.S. **61**, 12 (1968); **63**, 16 (1969).
265. A. Mildvan and M. Cohn, Advan. Enzymol. **33**, 1 (1970).
266. A. Mildvan and H. Weiner, Biochemistry **8**, 552 (1969); J. Biol. Chem. **244**, 2465 (1969).

267. J. Hyde, J. Chien, and J. Freed, J. Chem. Phys. **48**, 4211 (1968).
268. M. Cohn, Quart. Rev. Biophys. **3**, 61 (1970).
269. V. Jakovlev, J. All-Un. Mendeleev Chem. Soc. **16**, N 4, 391 (1971) (R).
270. Y. Chirgadze and N. Nevskaja, Dokl. Acad. Sci. USSR **208**, 447 (1973) (R).
271. Y. Chirgadze, Biophysika **14**, 792 (1969) (R).
272. Y. Chirgadze and A. Ovsepian, Mol. Biol. **6**, 721 (1972) (R); Biopolymers **11**, 2179 (1972); **12**, 637 (1973).
273. Y. Chirgadze and S. Veniaminov, Biopolymers **12**, 1337 (1973).
274. W. Peticolas et al., Proc. Nat. Acad. Sci. U.S. **69**, 1467 (1972).

# Chapter 6

# The Physics of Enzymes

## 6.1 Chemical Kinetics and Catalysis

Enzymes are the catalysts of biochemical reactions. They participate in and accelerate the reaction but are not consumed in the reaction.

As a rule, chemical reaction rates depend strongly on temperature. Empirically this dependence can be expressed by the Arrhenius formula for the rate constant

$$k = A\, e^{-E^*/RT} \tag{6.1}$$

where $E^*$ is the activation energy and A is a preexponential factor. Hence

$$\ln k = \ln A - (E^*/RT) \tag{6.2}$$

that is, ln k depends linearly on $T^{-1}$ and $E^*$ can be found from the slope of this straight line. A more rigorous equation is

$$\ln k = \ln A + (C/R) \ln T - (E^*/RT) \tag{6.3}$$

For a reversible reaction

$$A \underset{k_{-1}}{\overset{k_1}{\rightleftharpoons}} B$$

the equilibrium constant is equal to the ratio of the rate constants

$$K = [B]_e/[A]_e = k_1/k_{-1} \tag{6.4}$$

A thermodynamic argument is

$$\ln K = -\Delta G/RT = -(\Delta H/RT) + (\Delta S/R) \tag{6.5}$$

where $\Delta G$, $\Delta H$, and $\Delta S$ are the changes in standard free energy, enthalpy, and entropy, respectively. For a limited temperature range

$$\Delta H_T = \Delta H_{T_0} + \Delta C_p (T-T_0)$$

$$\Delta S_T = \Delta S_{T_0} + \Delta C_p \ln (T/T_0) \qquad (6.6)$$

where $\Delta C_p$ is the mean difference in the heat capacities of the reactants and products. Hence

$$\ln K = \frac{\Delta S_{T_0} - \Delta C_p}{R} + \frac{\Delta C_p}{R} \ln \frac{T}{T_0} - \frac{\Delta H_{T_0} - T_0 \Delta C_p}{RT} \qquad (6.7)$$

Comparing (6.7) and (6.3) we obtain

$$\ln A_1 - \ln A_{-1} = (\Delta S_{T_0} - \Delta C_p)/R$$

$$C_1 - C_{-1} = \Delta C_p \qquad (6.8)$$

$$E_1^* - E_{-1}^* = \Delta H_{T_0} - T_0 \Delta C_p$$

The difference in activation energies of forward and reverse reactions is dependent on the enthalpy difference $\Delta H$, and the ratio of the preexponential factors is dependent on the entropy difference $\Delta S$. If $\Delta C_p = 0$, Arrhenius's equation (6.1) is valid. We see that it can be rewritten in the form

$$k = \nu e^{-G^*/RT} = \nu e^{S^*/R} e^{-H^*/RT} \qquad (6.9)$$

where $G^*$, $H^*$, and $S^*$ are, respectively, the free energy, enthalpy, and entropy of activation. Identifying $H^*$ with $E^*$, we get the preexponential factor in the form

$$A = \nu e^{S^*/R} \qquad (6.10)$$

Actually

$$E^* = H^* + RT^2 (\partial \ln \nu)/\partial T \qquad (6.11)$$

and the identification is valid only if $\nu$ does not depend on T.

The physical sense of the exponential expression (6.9) becomes clear in the theory of absolute reaction rates (transition state or the activated complex theory) developed by Eyring. It is suggested that the reaction does not violate the Boltzmann distribution of molecules. The dependence of the potential energy of a reacting system on the positions of the atoms can be represented by a multidimensional surface. Figure 6.1 shows such a surface for the reaction of three atoms

$$X + Y{-}Z \rightarrow X{-}Y + Z$$

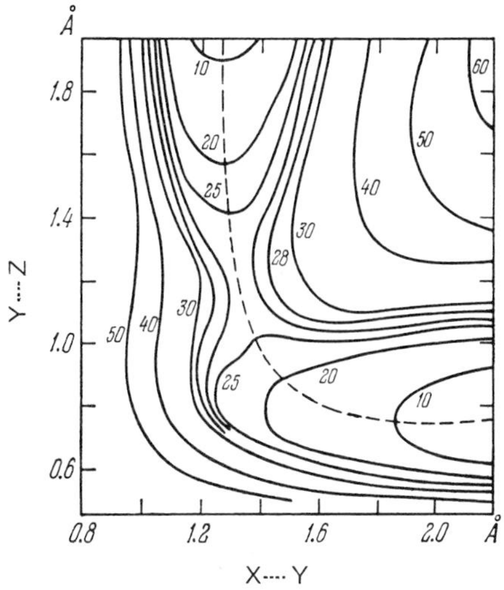


*FIG. 6.1 Potential energy surface for the reaction of three atoms.*

This is a map of the energy as a function of the distances $r_1$(X--Y) and $r_2$(Y--Z). At the top and on the right-hand side of the figure the potential energy curves of molecules XY and YZ are shown. The reaction proceeds by crossing over an energy barrier. The energy state of the system is changed in the most advantageous way, along the dotted curve in Fig. 6.1. To calculate the reaction rate we must determine the number of systems crossing the saddle point of the potential energy surface (in Fig. 6.1 the energy at the saddle point in conventional units equals 26). The state of the system at this point is called the transition state or activated complex. For the simplest reaction

$H + H_2 \rightarrow H_2 + H$

this state corresponds to a symmetrical linear system of three hydrogen atoms

H•••H•••H

According to the principles of statistical mechanics, the number of systems in some interval $\delta$ at the top of the barrier, lying on the route of reaction, must be determined (Fig. 6.2). The reaction rate is expressed as the number of systems crossing the barrier per unit time. The mean velocity of the forward motion is

$$\bar{v} = \int_0^\infty e^{-mv^2/2kT} v\, dv \Big/ \int_0^\infty e^{-mv^2/2kT} dv$$

$$= (kT/2\pi m)^{1/2} \qquad (6.12)$$

(where m is the mass of the system) and the mean time of crossing

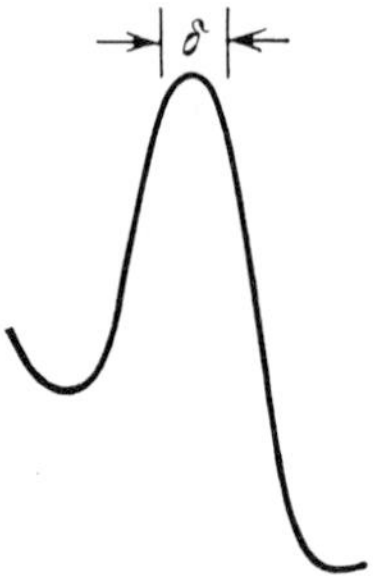


*FIG. 6.2 Activation barrier.*

the barrier is

$$\tau = \delta/\bar{v} = \delta(2\pi m/kT)^{1/2} \tag{6.13}$$

The reaction rate is equal to the number c' of systems per unit volume which are at the top of barrier (i.e., in the interval δ), divided by τ

$$\nu = c'/\tau = (c'/\delta)(kT/2\pi m)^{1/2} \tag{6.14}$$

On the other hand,

$$\nu = k\, c_A\, c_B \cdots \tag{6.15}$$

where k is the rate constant and $c_A$, $c_B$, ⋯ are the concentrations of the reagents. From (6.14) and (6.15) we obtain

$$k = \frac{c'}{c_A c_B \cdots} \left(\frac{kT}{2\pi m}\right)^{1/2} \frac{1}{\delta} = K' \left(\frac{kT}{2\pi m}\right)^{1/2} \frac{1}{\delta} \tag{6.16}$$

where K' is the equilibrium constant for the transition state or the activated complex of A, B, ⋯. The equilibrium constant is expressed by means of partition functions (see, e.g., [3]):

$$K' = Z_{ac}/Z_A Z_B \cdots \tag{6.17}$$

where $Z_{ac}$ is the partition function of the activated complex. For the translational motion of a particle, the partition function is equal to

$$Z_{trans} = (V/N_A)(2\pi mkT/h^2)^{3/2} \tag{6.18}$$

where h is Planck's constant, m is the mass of the particle, and $V/N_A$ is the occupied volume. Let us put $V/N_A = \delta^3$. $Z_{trans}$ is the partition function for 3 degrees of freedom of translational motion. But our system possesses only 1 degree of freedom, that along the reaction route. Therefore,

$$Z_{ac} = Z_{ac}' \; \delta(2\pi mkT/h^2)^{1/2} \tag{6.19}$$

where $Z_{ac}'$ is the partition function of the activated complex for all degrees of freedom except that of translational motion along the reaction route. From (6.17) and (6.19) we obtain

$$k = (kT/h)(Z_{ac}'/Z_A Z_B \cdots) \tag{6.20}$$

The partition function is related to the free energy by the expression

$$Z = e^{-G/RT}$$

Hence

$$k = \frac{kT}{h} \frac{e^{-G'/RT}}{e^{-G_A/RT} e^{-G_B/RT} \cdots} = \frac{kT}{h} e^{-F'-G_A-G_B\cdots/RT}$$

$$= \frac{kT}{h} e^{-G^*/RT} \tag{6.21}$$

where $G^*$ is the surplus of free energy in the activated complex in comparison with the sum of the free energies of the reagents. In other words, $G^*$ is the free energy of activation.

The derivation of expression (6.21) suggests that having reached the activation barrier, the system is sure to pass over it. However, this is not always so. An additional factor, the transmission coefficient $\kappa \leq 1$, has to be introduced. The final expression of the rate constant of a homogeneous gaseous reaction has the form

$$k = \kappa(kT/h) e^{S^*/R} e^{-H^*/RT} \tag{6.22}$$

which is equivalent to Eq. (6.3).

The preexponential (frequency) factor in Eq. (6.22), $\kappa(kT/h)$, has an order of magnitude of $10^{13}$–$10^{14}$ $sec^{-1}$ at room temperature if $\kappa \cong 1$. The value $\kappa = 1$ corresponds to an adiabatic reaction in the sense suggested by Ehrenfest. The process is called adiabatic if the parameters of the mechanical system are changed so slowly that the quantum numbers characterizing its motion remain unchanged. In the case of a chemical reaction, atomic nuclei move much more slowly than electrons. Every nuclear configuration can be considered fixed relative to electron motion. If the process is nonadiabatic, then $\kappa < 1$ and can be as small as $10^{-5}$. The theory of such reactions was developed in the works of Landau [4], Landau and Lifshitz [5], and Zener [6]. Observed $\kappa$ values of gaseous reactions in the majority of cases are close to 1. This means that these reactions are adiabatic.

Theoretical calculation of k (i.e., from $Z_{ac}'$, $Z_A$, $Z_B$, $\cdots$)

can in principle be done by quantum-chemical methods. However, such calculations have been performed only for the simplest cases.

It must be emphasized that the multiplier kT/h cannot be treated as the lifetime of an activated complex. Such treatment results in a paradox--in a time interval of $10^{-13}$-$10^{-14}$ sec a statistical equilibrium cannot be established. Actually, the time of existence of the system in a state corresponding to a value of the reaction coordinate from x to x + $\delta$ depends upon the choice of $\delta$ and is proportional to this quantity. In this sense, the lifetime of an activated complex has to be treated in the same way as the time of existence of states considered in the Maxwell-Boltzmann theory. The smallness of h/kT does not imply the impossibility of the establishment of an equilibrium between the activated complex and reagents [7]. Obviously, the theory of the activated complex concerns a definite state of an atomic system in the process of transformation and does not mean that there exists some metastable complex which can be investigated by the usual methods of physics and chemistry.

The significance of the theory presented here consists in its physical treatment of kinetic processes, giving a foundation for the Arrhenius formula and a rough estimate of the quantities contained in it. This theory can be directly applied to gaseous chemical reactions and to the classical problems of physical kinetics (viscosity, diffusion). It also permits us to examine some heterogeneous processes, such as adsorption and reactions at phase surfaces (cf. [2,8]).

The strong influence of a catalyst on the reaction rate is due to a change in the activation energy. A positive catalyst lowers the activation barrier.

Heterogeneous and homogeneous catalysis must be distinguished. In the former, the catalyst forms a separate phase and reaction occurs at the phase interface, that is, on the surface of the catalyst. Homogeneous catalysis occurs when catalyst and reagents are in the same phase (e.g., in solution). Catalysis proceeds by way of some intermediate compound.

The process of heterogeneous catalysis consists of the adsorbtion of the reacting molecules at the surface of the catalyst, their subsequent reaction, and the desorption of the reaction products. Adsorption brings the reacting molecules into close contact, alters the structure of their electron shells, and thereby lowers the activation energy. As has been shown by Balandin in his "multiplet theory of catalysis" [9], the most important role in this process is played by the geometrical, structural fit of the catalyst surface and the molecules being adsorbed. A metal catalyst is crystalline. If the symmetry and interatomic distances of the crystal lattice correspond to the molecular geometry of the reagents, these molecules can be effectively adsorbed and transformed to the reactive state because of electronic interaction with the atoms of the metal. Thus the reaction of benzene hydrogenation

$$C_6H_6 + 3H_2 \rightarrow C_6H_{12}$$

is catalyzed by platinum, nickel, and certain other metals but not by iron, silver, etc. The benzene molecule is a regular hexagon with C—C bond lengths of 1.4 Å. Atoms at the surface of nickel and of other effective catalysts are also positioned in the form of hexagons and the distances between them are close to the bond lengths of benzene [10]. On the other hand, noncatalyzing metals either are not hexagonal, or the dimensions of their atoms are not suitable for the benzene reaction.

By the process of adsorption on the metal, the bonds of benzene are changed. The scheme represents the breaking of the π bonds in benzene and the formation of new bonds with the metal atoms by the π electrons. In this state benzene reacts readily with hydrogen. After the reaction, the products are desorbed and the sites on the catalyst surface are ready for use again.

H—H

$$
\underset{\cdots Ni\cdot\cdot Ni\cdots}{\overset{H\quad H}{\underline{\cdots}C\overset{\cdots}{\cdots}C\underline{\cdots}}} \longrightarrow
\underset{Ni\cdot\cdot Ni}{\overset{H\quad H}{-\underset{:}{C}-\underset{:}{C}-}}
\qquad
\underset{\cdots Ni\cdot\cdot Ni\cdots}{\overset{H\quad H}{-\underset{:}{C}-\underset{:}{C}-}} \longrightarrow
\underset{\cdots Ni\cdot\cdot Ni\cdots}{\overset{H_2\quad H_2}{-C-C-}}
$$

Enzymes perform their catalytic function either as molecules in solution or in the supramolecular structures of cells. The sorption of reagents (called substrates in this case) and reaction occur at the surface of a protein molecule. In this sense, enzymic catalysis is similar to heterogeneous catalysis and the supramolecular enzymic systems actually form the second phase. However, the active surface of the protein globule is convoluted--as shown by X-ray analysis, the substrate molecules enter a cavity in the body of the globule (pp. 371, 372). There is rigorous stoichiometry of interaction; as a rule, one protein globule interacts with one molecule of substrate or other ligand. An actual intermediate compound is formed: an enzyme-substrate complex (ESC) whose structure and properties can be studied by physical methods. Consequently, enzymic catalysis must be considered homogeneous catalysis effected by large globular molecules which possess specific properties.

The enzyme can be treated as a "black box," which transforms an input signal--substrate molecule--into an output signal--product molecule. There are two ways of investigating the construction and working mechanism of the black box: the study of the molecular structure of the enzyme and the ESC by physical and chemical means, and the study of the kinetics of enzymic reactions.

## 6.2 Kinetics of Simple Enzymic Reactions

Let us examine the simplest enzymic reaction, the conversion of one substrate S into one product P. The formation of one or several enzyme-substrate complexes according to the scheme

$$F_0 + S \underset{k_{-1}}{\overset{k_1}{\rightleftharpoons}} F_1 \underset{k_{-2}}{\overset{k_2}{\rightleftharpoons}} F_2 \cdots \underset{k_{-n}}{\overset{k_n}{\rightleftharpoons}} F_n \underset{k_{-(n+1)}}{\overset{k_{n+1}}{\rightleftharpoons}} F_0 + P$$

can be postulated. $F_0$ denotes free enzyme and $F_i$ $(i = 1,2,\cdots,n)$ the enzyme-substrate complexes. The kinetic equations for this system have the form

$$\begin{aligned}
\dot{S} &= k_{-1}F_1 - k_1 S F_0 \\
\dot{F}_0 &= k_{-1}F_1 - (k_1 S + k_{-(n+1)}P)F_0 + k_{n+1}F_n \\
\dot{F}_1 &= k_1 S F_0 + k_{-2}F - (k_{-1}+k_2)F_1 \\
&\vdots \\
\dot{F}_i &= k_i F_{i-1} + k_{-(i+1)}F_{i+1} - (k_{-i}+k_{i+1})F_i \\
&\vdots \\
\dot{P} &= k_{n+1}F_n - k_{-(n+1)}PF_0
\end{aligned} \tag{6.23}$$

Here S, P, $F_i$ denote the concentrations of the corresponding substances and the overdot indicates the time derivative of a quantity. In a closed system the total amount of enzyme remains constant

$$E = \sum_{i=0}^{n} F_i = \text{const} \tag{6.24}$$

Hence

$$\dot{E} = \sum_{i=0}^{n} \dot{F}_i = 0 \tag{6.25}$$

In addition, the total amount of substrate and product remains constant

$$S + \sum_{i=1}^{n} F_i + P = \text{const} \tag{6.26}$$

It follows from (6.26) and (6.25) that

$$\dot{P} = \dot{F}_0 - \dot{S} \tag{6.27}$$

The solution of Eqs. (6.23) (i.e., evaluation of the concentrations S, P, and $F_i$ as functions of time) is rather difficult in general. However, the system can reach the steady state, corresponding to

$$F_i = \text{const} \qquad (i = 0,1,\cdots,n) \tag{6.28}$$

and

$$\dot{F}_i = 0 \qquad (i = 0,1,\cdots,n) \tag{6.29}$$

The conditions necessary for the occurrence of the stationary state require analysis. Walter treats this state as having simultaneous extrema of the functions $F_i(t)$, $(i = 0,1,\cdots,n)$ and shows that at $n > 1$ these conditions cannot be rigorously fulfilled [11]. However, this is not the point. The stationary state is approximated at practically every n value if only one condition is obeyed; namely, that the concentration of the substrate is large in comparison with that of the enzyme, $S >> E$. Let us show that such a condition is actually sufficient for the steady state. Examining the simplest case, $n = 1$, and assuming for simplicity that $k_{-2} = 0$, we have

$$F_0 + S \underset{k_{-1}}{\overset{k_1}{\rightleftharpoons}} F_1 \xrightarrow{k_2} F_0 + P$$

The kinetic equations are

$$\begin{aligned} \dot{S} &= -k_1F_0S + k_{-1}F_1 \\ \dot{F}_1 &= k_1F_0S - (k_{-1}+k_2)F_1 \\ \dot{P} &= k_2F_1 \end{aligned} \tag{6.30}$$

Let us eliminate $F_0$ with the help of (6.25) (i.e., $F_0 = E-F_1$). We obtain

$$\dot{F}_1 = k_1ES - (k_{-1}+k_2+k_1S)F_1 \tag{6.31a}$$

$$\dot{S} = -\dot{F}_1 - k_2F_1 \tag{6.31b}$$

Integrate (6.31a) considering S as practically constant, that is, treating S as a slowly changing parameter. At $t = 0$ when enzyme is added, $F_1(0) = 0$--all of the enzyme is in the state $F_0$. We obtain

$$F_1 = \frac{k_1ES}{k_{-1}+k_2+k_1S}\left(1 - e^{-t/\tau}\right) \tag{6.32}$$

where

$$\tau^{-1} = k_{-1} + k_2 + k_1 S$$

$F_1$ approaches its stationary-state value, corresponding to $\dot{F}_1 = 0$, with the time constant $\tau$ (which is smaller as the value of S increases)

$$\dot{F}_1 = k_1 ES\, e^{-t/\tau} \rightarrow 0 \tag{6.33}$$

putting the values of $F_1$ from Eq. (6.32) and $\dot{F}_1$ from Eq. (6.33) into Eq. (6.31b), we find

$$\dot{S} = -k_1 ES\, e^{-t/\tau} - \frac{k_1 k_2 ES}{k_{-1}+k_2+k_1 S}(1 - e^{-t/\tau}) \tag{6.34}$$

Hence $\dot{S}$ approaches

$$\dot{S}_{st} = -\frac{k_1 k_2 ES}{k_{-1}+k_2+k_1 S} \tag{6.35}$$

Let us show that the relative rate of change of S is actually small if S >> E. The value of $\dot{S}$ depends monotonically on time and is a maximum at t = 0

$$\dot{S}(0) = -k_1 ES$$

Therefore, the maximum change in S is

$$\Delta S = \dot{S}(0)\tau = -\frac{k_1 ES}{k_{-1}+k_2+k_1 S}$$

It is easy to see that at $E \ll S$

$$|\Delta S| \ll S$$

On the other hand

$$\Delta F_1 = \dot{F}_1(0)\tau = \frac{k_1 ES}{k_{-1}+k_2+k_1 S}$$

and

$$\Delta F_1/E \gg \Delta S/S$$

The conditions for the stationary state require only that S >> E and do not depend on rate constants.

It can be shown that in the case of any number of intermediate complexes the stationary state will be approached if S >> E. All $F_i$ values corresponding to solutions of the system of nonhomogeneous differential equations tend exponentially toward stationary values. The stationary state can be approached because there are two time scales corresponding to fast ($F_i$) and slow (S) variables. A parameter characterizing the ratio of these two scales is $ES^{-1}$ (cf. [12,13]).

According to (6.35), the stationary-state rate for the case n = 1 is

$$v \equiv \dot{P} = -\dot{S} = \frac{k_1k_2SE}{k_{-1}+k_2+k_1S} \tag{6.36}$$

The same expression can be obtained directly if we put

$$v = k_2F_1$$

and evaluate $F_1$ from Eq. (6.31a) with $\dot{F}_1 = 0$. If the reverse reaction cannot be omitted, then

$$v = \frac{k_1k_2SE-k_{-1}k_{-2}PE}{k_{-1}+k_2+k_1S+k_{-2}P} \tag{6.37}$$

For many reactions $k_{-2}$ is actually small. In these cases Eq. (6.36) is valid and can be rewritten in the form

$$v = k_2SE/(K_M+S) \tag{6.38}$$

where

$$K_M = (k_{-1}+k_2)/k_1 \tag{6.39}$$

Equation (6.38) is called the Michaelis-Menten equation, and the constant $K_M$ the Michaelis constant. The function v(S) described by (6.38) is similar to the Langmuir isotherm (Fig. 6.3). This is a smooth curve, tending asymptotically toward a maximum value of the rate as $S \to \infty$

$$v_{max} = k_2E \tag{6.40}$$

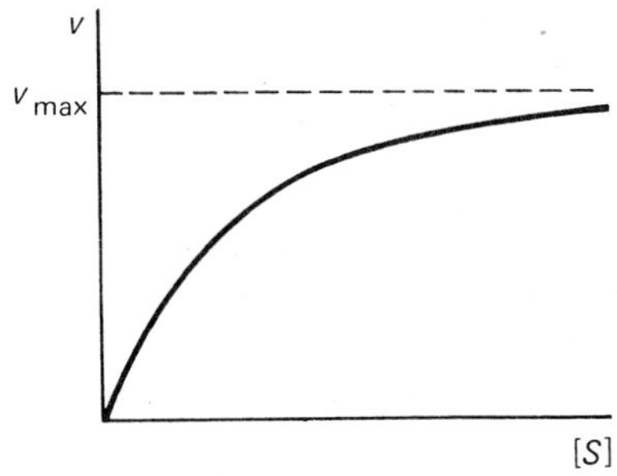


*FIG. 6.3 Michaelis-Menten curve.*

Hence

$$v = v_{max} S/(K_M + S) \tag{6.41}$$

At $v = 0.5v_{max}$, $S = K_M$. The Michaelis constant is numerically equal to the substrate concentration, which corresponds to a reaction rate of half the maximum rate. For further details see [14-16].

It is convenient to transform Eq. (6.38) according to Lineweaver and Burk [17]

$$\frac{1}{v} = \frac{K_M}{v_{max}} \frac{1}{S} + \frac{1}{v_{max}} \tag{6.42}$$

Drawing the graph of $1/v$ versus $1/S$, we find $1/v_{max}$ from the intersection point with the ordinate and $K_M$ from the slope of the straight line $v^{-1}(S^{-1})$.

The Michaelis constant can be expressed in the form

$$K_M = (k_2/k_1) + K_S$$

where $K_S = k_{-1}/k_1$. In many cases $k_2/k_1$ is relatively small. Table 6.1 presents the values of $K_M$ and $K_S$ as calculated by Jakovlev [14] from the data of Gutfreund.

*TABLE 6.1*

*The Values of $K_M$ and $K_S$ for Hydrolysis Reactions*

| Enzyme | Substrate | $K_M$ (M) | $K_S$ (M) |
|---|---|---|---|
| Trypsin | Benzoyl-2-arginine | $1.0\times10^{-5}$ | $0.6\times10^{-5}$ |
| Chymotrypsin | Ethyl ester of acetyl-2-phenylalanine | $1.0\times10^{-4}$ | $0.9\times10^{-4}$ |
| Phycin | Ethyl ester of benzoyl-2-arginine | $1.5\ 10^{-2}$ | $1.2\times10^{-2}$ |

Equation (6.41) can be rewritten in the form

$$\frac{v_{max}}{v} - 1 = \frac{K_S}{S} + \frac{v_{max}}{k_1 ES}$$

At a given value of S the dependence of $(v_{max}/v) - 1$ on $v_{max}$ is represented by the straight line intersecting the ordinate at $K_S/S$ and the abscissa at $-k_1EK_S$. The tangent of the slope angle is $1/k_1ES$. Since $v_{max} = k_2E$, all three rate constants $k_1$, $k_{-1}$, and $k_2$ can be derived from the graphs obtained at different values of S and E. Thus, for the reaction of dehydrogenation of succinic acid with the use of dehydrogenase,

$$
\begin{array}{ccc}
\underset{\text{CH}_2\text{COOH}}{\overset{\text{CH}_2\text{COOH}}{\underset{\big|}{\big|}}} & \xrightarrow{\ -2\,\text{H}\ } & \underset{\text{HCCOOH}}{\overset{\text{HOOCCH}}{\underset{\|}{\|}}}
\end{array}
$$

Succinic acid                    Fumaric acid

$k_1E = 1.15\times10^{-2}$ sec$^{-1}$, $k_{-1}E = 3.6\times10^{-7}$ mole sec$^{-1}$, and $k_2E = 5.15\times10^{-6}$ mole sec$^{-1}$, $K_S = 3.13\times10^{-5}$ mole [18].

Consider now the monosubstrate reaction with participation of a modifier--of a substance interacting with the enzyme and its complexes and changing considerably the formation rate of the product. If the modifier decreases the reaction rate, it is called an inhibitor. Let us denote its concentration by I. In Fig. 6.4 the schemes of two simple processes are shown. In (a), free enzyme forms both a reactive complex $F_1$ and a nonreactive complex $F_2$, binding the inhibitor. The inhibitor competes with the substrate at the sorption region (active site) of the enzyme.

*FIG. 6.4  Schemes of processes with (a) competitive and (b) noncompetitive inhibition.*

(a)

(b)

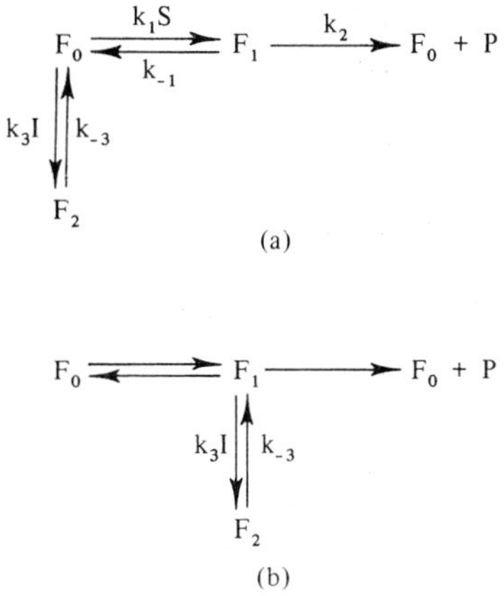

In scheme (b) the nonreactive complex $F_2$ is formed as a result of the reaction of an already formed enzyme-substrate complex $F_1$ with the inhibitor. This is noncompetitive inhibition.

Steady-state kinetic equations for the first case are

$$\dot{F}_0 = -(k_1S+k_3I)F_0 + (k_{-1}+k_2)F_1 + k_{-3}F_2 = 0$$

$$\dot{F}_1 = k_1SF_0 - (k_{-1}+k_2)F_1 = 0$$

$$\dot{F}_2 = k_3IF_0 - k_{-3}F_2 = 0 \qquad\qquad (6.43)$$

$$E = F_0 + F_1 + F_2$$

$$v = k_2F_1$$

Their solutions are

$$F_1 = \frac{k_1 S}{k_{-1}+k_2} F_0, \quad F_2 = \frac{k_3 I}{k_{-3}} F_0, \quad E = \left(1 + \frac{k_1 S}{k_{-1}+k_2} + \frac{k_3 I}{k_{-3}}\right) F_0$$

Hence

$$v = \frac{k_2 k_1 S}{k_{-1}+k_2} F_0 = \frac{k_2 k_1 SE}{(k_{-1}+k_2)[1+(k_1 S/(k_{-1}+k_2))+(k_3 I/k_{-3})]}$$

$$= \frac{k_2 SE}{K_M+S+K_M K_I I} = \frac{v_{max} S}{K_M+S+K_M K_I I} \tag{6.44}$$

where $K_I = k_3/k_{-3}$. The maximum rate is not changed in the case of competitive inhibition, but the initial rate (at small S) decreases.

In the second case, the equations have the form

$$\dot{F}_0 = -k_1 SF_0 + (k_{-1}+k_2)F_1 = 0$$

$$\dot{F}_1 = k_1 SF_0 - (k_{-1}+k_2+k_3 I)F_1 + k_{-3}F_2 = 0$$

$$\dot{F}_2 = k_3 IF_1 - k_{-3}F_2 = 0 \tag{6.45}$$

$$E = F_0 + F_1 + F_2$$

$$v = k_2 F_1$$

The solution is

$$v = k_2 SE/(K_M+S+K_I SI) \tag{6.46}$$

Noncompetitive inhibition decreases the maximum rate:

$$v_{max} = k_2 E/(1+k_I I) \tag{6.47}$$

We have considered only two limiting cases of inhibition. Frequently more complicated cases of inhibited reactions (or activated ones if the modifier accelerates the reaction) appear with a mixture of competitive and noncompetitive inhibition, etc. These processes, even in stationary conditions, require the solution of more complicated equations. Mathematical methods for the stationary kinetics of complicated enzymic reactions are described in Section 7.6.

At nonstationary conditions, the solution of even simple kinetic equations is rather difficult. They can be solved in numerical form with the aid of computers. However, calculations concerning the early stages of enzymic reactions (the prestationary regime) are simpler [11]. It can be assumed here that almost all of the enzyme and substrate are in the free state. Let us rewrite Eqs. (6.30) in the form

$$F_0 = \frac{\dot{F}_1}{k_1 S} + \frac{k_{-1}+k_2}{k_1 S} F_1, \qquad F_1 = \frac{\dot{P}}{k_2}$$

It follows from the right-hand equation that

$$\dot{F}_1 = \ddot{P}/k_2$$

where the double overdot indicates the second derivative. Eliminating $F_0$ from the left-hand equation with the help of $E = F_0+F_1$, we obtain

$$\ddot{P} + k_1(K_M+S)P = k_1 k_2 ES \tag{6.48}$$

At the initial moment

$$k_1 k_2 ES(0) = \ddot{P}(0)$$

Actually, $\ddot{P}(0) = k_2\dot{F}_1(0)$, $F_1(0) \cong 0$, and $E \cong F_0(0)$. Hence, $\dot{F}_1(0) = k_1 S(0) F_0(0) \cong k_1 S(0) E$ and $\ddot{P}(0) = k_1 k_2 ES(0)$. Double integration at $S = S(0)$ gives

$$P(t) = \frac{k_2 ES(0)}{K_M+S(0)} \left[t + \frac{e^{-k_1[K_M+S(0)]t} - 1}{k_1[K_M+S(0)]}\right] \tag{6.49}$$

At increasing t, P(t) tends asymptotically toward the straight line

$$P(t) = \frac{k_2 ES(0)}{K_M+S(0)} \left[t - \frac{1}{k_1[K_M+S(0)]}\right]$$

which intersects the t axis at $k_1^{-1}[K_M+S(0)]^{-1}$ and has the tangent of the slope angle $k_2 ES(0)/[K_M+S(0)]$.

By investigating the prestationary kinetics it is possible to evaluate the constants $k_1$ and $k_2$ if $K_M$ is known from the stationary measurements. Thus, all three rate constants can be determined [11,19].

## 6.3 Thermodynamics of Enzymic Reactions

The thermodynamic characteristics of enzymic reactions can be determined with the help of Eq. (6.22). Rate constants found in kinetic experiments are used if the adiabaticity of the process can be assumed. Free energy, enthalpy, and entropy of

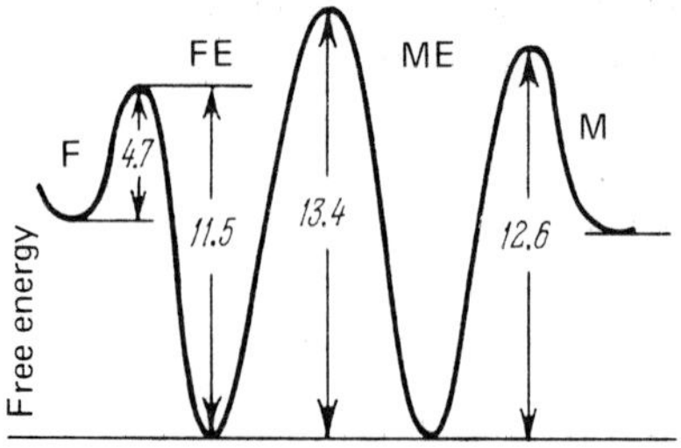


*FIG. 6.5 Free energy curve for the fumaric acid ⇄ malic acid reaction.*

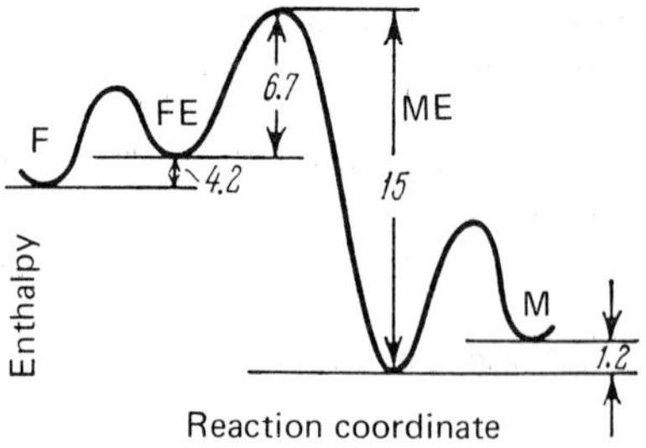


*FIG. 6.6 Enthalpy curve for the fumaric acid ⇄ malic acid reaction.*

activation are determined from the temperature dependence of k. According to (6.22)

$$\frac{d \ln k}{dT} = \frac{1}{T} + \frac{H^*}{RT^2} = \frac{H^*+RT}{RT^2} \tag{6.50}$$

This expression coincides with the Arrhenius equation if

$$E^* = H^* + RT$$

On the other hand,

$$G^* = H^* - TS^* = -RT \ln k + RT \ln(kT/h)$$

A study of the dependence of ln k on T makes it possible to determine $G^*$, $S^*$, and $H^*$. Figures 6.5 and 6.6 show the free energy and enthalpy curves derived in this way for the fumarase-catalyzed transformation of fumaric acid into malic acid [20].

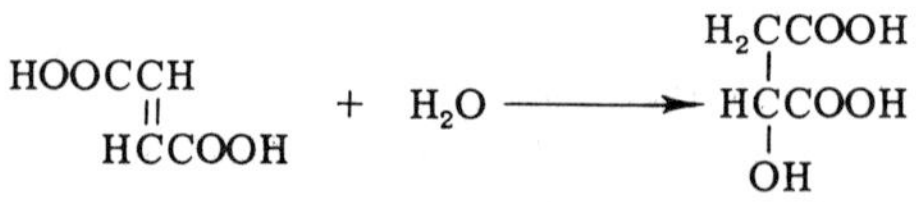


The pronounced difference in these two curves testifies to the important role of the entropy factor. This factor can be so essential in enzymic reactions that the reaction path cannot go through the saddle point corresponding to higher energy, but must go through the point corresponding to higher energy and to higher entropy [21].

In Table 6.2 the values of the activation parameters of several enzymic reactions, described for stationary-state conditions by a formula of the type of (6.38), are listed, and also

*TABLE 6.2 Thermodynamic Parameters of Simple Enzymic Reactions*

| Enzyme | Substrate | $G_1$* | $H_1$* | $S_1$* | $G_2$* | $H_2$* | $S_2$* | ΔG | ΔH | ΔS |
|---|---|---|---|---|---|---|---|---|---|---|
| Chymotrypsin | Methyl-hydrocinnamate | 17.8 | 10.9 | -23.2 | 19.7 | 16.2 | -11.8 | -1.9 | -5.3 | -11.4 |
| | Methyl-D-β-phenyl lactate | 16.6 | 2.5 | -47.2 | 18.7 | 14.5 | -14.2 | -2.2 | -12.0 | -33.0 |
| | Methyl-L-β-phenyl lactate | 14.7 | 3.2 | -38.5 | 17.5 | 10.5 | -23.4 | -2.8 | -7.3 | -15.1 |
| | Benzyl-L-tyrosine ethyl ester | 10.7 | 0.2 | -38.5 | 15.0 | 8.6 | -21.4 | -3.3 | -8.4 | -17.1 |
| | Benzyl-L-tyrosine amide | 15.9 | 3.1 | -43.0 | 17.9 | 14.0 | -13.0 | 9.9 | -9.9 | -30.0 |
| Acetylcholine exterase | Acetylcholine | 2.5 | 14-19 | 34-52 | 8 | 14-19 | 16-34 | -5.5 | 0 | 18.5 |
| | Dimethylaminoethylacetate | 5.5 | 6.7-8 | 4-8 | 9.9 | 6.7-8 | -(16.5-10.5) | -4.4 | 0 | 14.6 |
| | Aminoethylacetate | 9.7 | 9.5 | -0.6 | 12.2 | 9.5 | -9 | -2.5 | 0 | 8.4 |
| Carboxypeptidase | Carbobenzoxy-L-tryptophan | 16 | 21 | 17 | 19.3 | 16 | -11 | -3.4 | 5 | -28 |
| | Carbobenzoxy-L phenylalanine | 11.8 | 8.5 | -11 | 14.4 | 8.8 | -18 | -2.5 | -0.4 | -7 |
| | Carbobenzoxyglycyl-L-tryptophan | 11.9 | 9.3 | -8.5 | 15.3 | 9.3 | -20 | -3.4 | 0 | -11.5 |
| Pepsin | Carbobenzoxy-L-glutamyl-L-tyrosine ethyl ester | 18.3 | 22.5 | 14.1 | 22.1 | 20.1 | -6.5 | -4.7 | 1.4 | 20.6 |
| | Carbobenzoxy-L-glutamyl-L-tyrosine | 18.8 | 19.6 | 2.6 | 23.1 | 16.6 | -21.8 | -4.3 | 3.0 | 24.4 |
| Urease | Urea | 8.2 | 6.2 | -6.8 | 11.3 | 9.1 | -7.2 | -3.2 | -2.9 | 0.9 |
| ATPase | ATP | 7.3 | 20.4 | 44.0 | 14.4 | 12.4 | -8.0 | -7.5 | 8 | 52 |

the differences ΔG, ΔH, and ΔS corresponding to the "equilibrium constant" $k_2/k_1$ [21]. Free energies and enthalpies are given in kcal $mole^{-1}$, entropies in cal $mole^{-1}$ $deg^{-1}$. Let us remember that at 300°K, 33.3 entropy units (e.u.) are equivalent to a free energy of 10 kcal $mole^{-1}$. The table shows that entropy factors are actually very important (see also [22]).

Blumenfeld has examined the reliability of thermodynamic data obtained from kinetic measurements [23].

The temperature dependence of the rate constants of enzymic reactions is characterized by a series of peculiarities. In many cases maximum values of the constants are observed at temperatures much lower than the denaturation temperature. The nature of these maxima is not clear. The temperature range for kinetic measurements is rather narrow. The activation parameters depend strongly on the ionic strength of the solution; they are considerably altered by the substitution of the correct enzyme derived from a different source. In these cases, the rate constants themselves may remain practically unchanged, but both the pre-exponential factor and activation energies can be considerably altered. Finally, a linear relationship of $H^*$ and $S^*$ is observed:

$$H^* = \alpha^* + T_k S^* \tag{6.51}$$

This effect will be discussed later.

Let us assume that the true activation barrier for a given stage of an enzymic process depends on temperature because of the cooperative behavior of an organized enzyme globule. In other words, the structure of the globule itself depends on temperature. Assume also that this dependence is not accompanied by a corresponding dependence of the true value of $S^*$ on temperature. In a narrow temperature range the dependence $H^*(T)$ can be assumed to be linear. We have

$$H^* = H_0{}^* - bT, \qquad b > 0$$

Then the Arrhenius equation has the form

$$k = Ae^{b/R}\, e^{-H_0{}^*/RT}$$

and experiment gives not the real $H^*$ value but $H_0{}^*$, and not the real value of $S^*$ but $S_0{}^* = S^* + b$. This can be written

$$H_0{}^* = H^* + T(S_0{}^* - S^*) = F^* + TS_0{}^*$$

$H_0{}^*$ is obtained by means of extrapolation of the weak temperature dependence of $H^*$ to $T \to 0°K$. If, for example, $H^*$ increases slowly in the range from 20°C to 30°C, $H_0{}^*$ can increase considerably. We see that if the changes in the real activation parameters are small and the measurements are performed within a narrow temperature range, the apparent $H_0^*$ and $S_0^*$ values are linked linearly, the value of $\alpha^*$ in Eq. (6.51) is the free energy of activation, and $T_k$ is the temperature of the experiment.

These peculiarities in the temperature behavior of k express the structural transformations of the globule. We see that the measured activation parameters seem to be effective values. Determination of the real values of these parameters requires detailed study of an enzyme's construction and of the mechanism of its action. Kinetic measurements do not suffice for this purpose.

Several authors have discovered and studied the so-called compensation effect [24-26]. Enthalpy changes in processes occurring in aqueous solution are frequently proportional to the entropy changes. This applies also to enthalpy and entropy of activation. The relationship is similar to (6.51)

$$\Delta H = \alpha + T_k \, \Delta S \tag{6.52}$$

and applies to processes involving the participation of small molecules, for example, the ionization of weak electrolytes, solvation of ions and nonelectrolytes, hydrolysis and oxidation-reduction reactions, and quenching of indole fluorescence. In all cases the "compensation temperature" $T_k$ lies between 250° and 315°K. The change in free energy is

$$\Delta G = \Delta H - T\,\Delta S = \alpha + (T_k - T)\,\Delta S \tag{6.53}$$

and the free energy of activation is

$$G^* = H^* - TS^* = \alpha^* + (T_k{}^* - T)S^* \tag{6.54}$$

The values of $\alpha$ and $\alpha^*$ and also of $T_k - T$ and $T_k{}^* - T$ are comparatively small. The effect consists of the mutual compensation of considerable changes in enthalpy and entropy, resulting in comparatively small changes in free energy.

The changes in H and S on the dissolution of nonpolar substances in water can be represented by the formulas

$$\Delta H(T,x_2) \cong \Delta H(T_0,0) + \int_0^{x_2} \frac{\partial\, \Delta H(T_0,x_2)}{(\partial x_2)_{T_0}}\, dx_2 + \int_{T_0}^{T} \Delta C_p(T,x_2)\, dT$$

$$\Delta S(T,x_2) \cong \Delta S(T_0,0) + \int_0^{x_2} \frac{\partial\, \Delta S(T_0,x_2)}{(\partial x_2)_{T_0}}\, dx_2 + \int_{T_0}^{T} \Delta C_p(T,x_2)\, \frac{dT}{T} \tag{6.55}$$

where $T_0$ is the reference temperature and $x_2$ the concentration. If (6.52) is valid, then

$$\alpha = \Delta H(T_0,0) - T_k(T_0,0) + \int_0^{x_2} \frac{\partial\, \Delta H}{(\partial x_2)_{T_0}}\, dx_2 - T_k \int_0^{x_2} \frac{\partial\, \Delta S}{(\partial x_2)_{T_0}}\, dx_2$$

$$+ \int_{T_0}^{T} \Delta C_p(T,x_2)\, dT = T_k \int_{T_0}^{T} \frac{\Delta C_p(T,x_2)}{T}\, dT \qquad (6.56)$$

$\alpha$ is a small, approximately constant value if $\Delta C_p$ does not depend strongly on temperature.

Lumry and Rajender suggest that the compensation effect is attributable to properties of water. Assume that the reaction $A \rightarrow B$ is accompanied by a change in the state of n water molecules

$$A + nW_1 \rightarrow B + nW_2$$

Then

$$\Delta H = \Delta H_{A\rightarrow B} + n\, \Delta H_{W_1\rightarrow W_2}$$

$$\Delta S = \Delta S_{A\rightarrow B} + n\, \Delta S_{W_1\rightarrow W_2}$$

If $H_{A\rightarrow B}$ and $S_{A\rightarrow B}$ are small and the $W_1 \rightarrow W_2$ transition is like a phase transition, then

$$T_k = \Delta H_{W_1\rightarrow W_2}/\Delta S_{W_1\rightarrow W_2}$$

and

$$\Delta G_{W_1\rightarrow W_2}(T_k) = 0$$

The compensation effect is observed in enzymic processes. Thus, in the hydrolytic splitting of N-acetate-L-tryptophan ethyl ester by chymotrypsin, $\Delta G$ is very small and $\Delta H$ and $\Delta S$ are rather large. Indeed, many data presented in the last three columns of Table 6.2 attest to compensation. The binding of a series of inhibitors by acetylcholinesterase is accompanied by compensation--in these processes $\Delta H$ varies from -7 to +2 kcal mole$^{-1}$, and $\Delta S$ from -10 to +20 cal mole$^{-1}$ deg$^{-1}$ [26]. If the assumption of the determining role of water is correct, then we have to establish the relation between the behavior of the protein molecules and the surrounding aqueous structure. Lumry and Rajender claim that this relation manifests itself via the change in the volume of the protein molecule during the reaction. As is shown in Sections 6.5 and 6.7, enzymic activity is related to conformational transformations in proteins, and consequently the globules can change their volumes. The change in the energy of the water-protein system can be presented in the form

$$dE = T\,dS - p\,dV + W_W\,dV_W + W_p\,dV_p + \sigma\,dA \tag{6.57}$$

where $W_W$ is the work involved in changing the free volume of water by unity, $W_p$ is the analogous quantity for the protein, $\sigma$ is the work of changing the surface that separates the protein molecules and the water by unity. The change in free energy is

$$dG = dH - T\,dS - S\,dT = dE + p\,dV + V\,dp - T\,dS - S\,dT$$

and according to (6.57)

$$dG = -S\,dT + V\,dp + W_W\,dV_W + W_p\,dV_p + \sigma\,dA$$

and at constant T and p

$$dG = W_W\,dV_W + W_p\,dV_p + \sigma\,dA \tag{6.58}$$

In the enzymic reactions studied by Vaslow and Doherty [27], the maximum measured change in H compensated by $\Delta S$ is of the order of $-27$ kcal mole$^{-1}$; correspondingly, $\Delta S \sim -100$ cal mole$^{-1}$ deg$^{-1}$. The total maximum change in the water volume estimated from structural relaxation is $-83$ ml mole$^{-1}$, which is less than 0.5% of the protein volume. Direct observation of this effect is rather difficult. However, these suggestions seem reasonable. The change in protein conformation, the displacement of water from an internal cavity of the globule by the substrate molecule, can result in the rearrangement of the aqueous medium.

The "aqueous" compensation effect can be physiologically important for warm-blooded organisms, and therefore for biological evolution. The smallness of the free energy changes because of compensation indicates that the corresponding enzymic processes are not very sensitive to changes in the temperature of the surrounding medium.

However, the existence of the compensation effect and its explanation cannot be considered rigorous. This effect can be explained simply by the incorrect use of the Arrhenius equation, as has been shown by Blumenfeld (p. 364; [23,134]). The question will remain open until reliable determinations of real thermodynamic parameters are performed. Nevertheless, a consideration of the compensation effect indicates some important features of enzymic reactions.

As has already been emphasized, an understanding of the structure and properties of protein molecules is impossible if their aqueous milieu is not taken into account (Chapter 4). The treatment of enzymes in aqueous solution as a whole system is also necessary for the study of the thermodynamics of enzymic processes. Such a systemic approach typifies contemporary methodology in physics. Isolation of protein molecules from their aqueous medium is not permissible in rigorous theory. This is true of kinetics, too. The theory of absolute reaction rates, which employs the assumptions that are valid for gaseous reactions, can be used only for rough estimates. The theory of enzymic reactions requires examination of the physical properties of the medium.

## 6.4 Chemical Aspects of Enzymic Activity

The essential features of enzymic catalysis are high activity and specificity in relation to a substrate or a group of substrates. The molecular activity of enzymes is expressed as the number of substrate molecules transformed by one molecule of enzyme in one minute at the optimal substrate concentration. Molecular activity can also be expressed as the turnover rate (Warburg), which is the number of moles of substrate transformed per mole of enzyme per minute.

Methods for extracting and purifying enzymes are well advanced, and many enzymes have been obtained in the pure crystalline form [15,30,31]. Thus it is possible to study reactions and structures of enzymes *in vitro*. The situation *in vitro* differs from that in the cell because the cell is an open system and a great number of enzymic reactions, including coupled reactions, proceed there simultaneously. However, the study of enzymes *in vitro* provides a strong foundation for understanding the corresponding biological processes.

When an ESC (enzyme-substrate complex) is formed, the small substrate molecule (or molecules) is (are) bound stoichiometrically to a large enzyme molecule. Evidently the substrate binds directly to some specific small region of the enzyme molecule, called the active site. The nature of the active site, that is, the nature of the amino acid residues and their positions, as well as of the cofactors, participating at the active site, can be established by chemical and physical methods. The alterations in activity produced by chemical modifications of a protein make it possible to determine the functional groups of the active site. Information about its structure can be obtained by means of spectrophotometry, spectropolarimetry, NMR and EPR spectroscopy (in the latter a paramagnetic label is introduced), etc. X-ray structural analysis reveals the actual geometric pattern of enzyme-substrate interaction.

The diversity of amino acid residues and atomic groups of cofactors determines the polyfunctional properties of the active site, its ability to bind molecules of the substrate or modifier, and the catalytic activity [31]. Thus in the cases of esterases and esterolytically active proteinases, the presence at the active site of the functional Ser residue, acylated in an intermediate stage of the process, has been determined. Active seryl is present in pseudocholinesterase, phosphoglucomutase, chymotrypsin, trypsin, and a series of other enzymes. Figure 6.7a shows the binding of the substrate acetylcholine (AcCh)

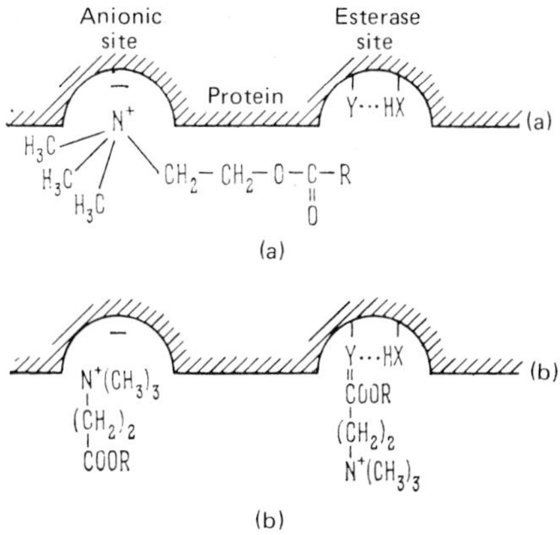


*FIG. 6.7 Scheme of the interaction of acetylcholinesterase with acetylcholine. (a) Binding of acetylcholine by acetylcholinesterase, (b) inhibition of acetylcholinesterase by a high concentration of acetylcholine.*

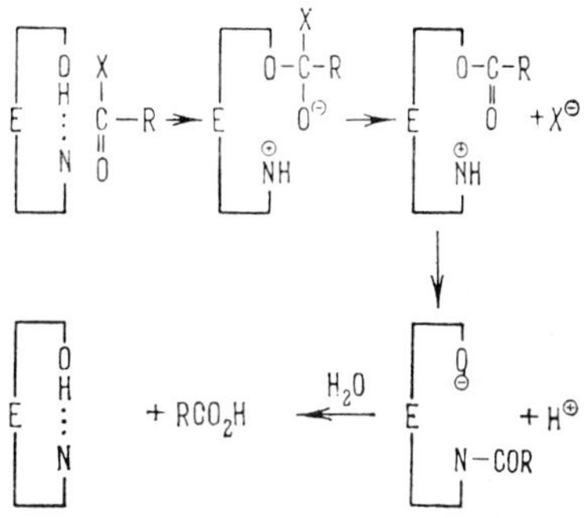

*FIG. 6.8 Simplified scheme of esterase action.*

by acetylcholinesterase (AcChE), and Fig. 6.7b inhibition of AcChE by high substrate concentration [32]. The "esterase" site of AcChE contains the nucleophilic group Y and dissociating acid group HX. The acylic residue is bound by the Y group in an intermediate reaction and is chipped off by the action of water.

A simplified scheme of the action of esterase is shown in Fig. 6.8 [33]. According to this scheme, esterases perform bifunctional acid-base catalysis. Oriented and coupled action of nucleophilic and electrophilic functional groups on the substrate molecule occurs [34].

The correlated action of different enzyme groups on the substrate can also be demonstrated by the example of creatinekinase [33,34], which catalyzes the reaction

$$ATP^{4-} + \text{creatine} \rightleftharpoons ADP^{3-} + \text{phosphocreatine}^{2-} + H^+$$

Figure 6.9 is a model of the transition state of this reaction. The $Mg^{2+}$ ion serves as cofactor. The SH group plays an especially important role at the active site. The enzyme is inactivated by iodoacetamide and iodoacetate. The rate of SH group alkylation does not depend on the pH in the range 6.0-10.0. The initial rate of reaction with iodoacetate is strongly dependent on ionic strength. In the presence of $Mg^{2+}$ and of both substrates, the SH groups are protected from the action of

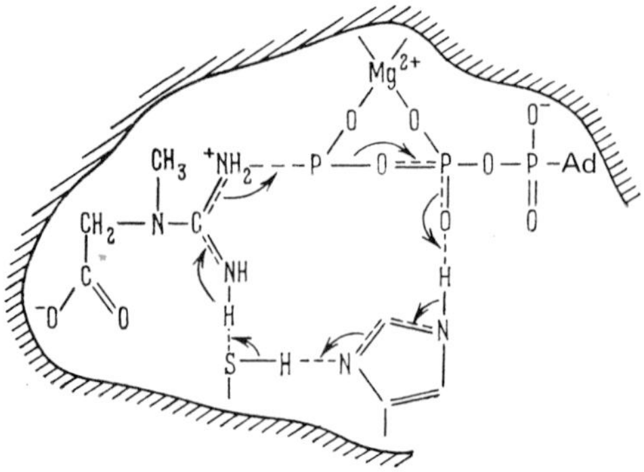

FIG. 6.9  Scheme of the
action of creatinekinase.

blocking reagents.  These facts show that the SH group is probably hydrogen bound by one of the basic groups of the enzyme.
Because of the easy proton transfer in acid-base systems linked
by a hydrogen bond, such systems have high catalytic activity.
      These few examples show that an exact mutual orientation
of the functional groups of enzyme and substrate or modifier is
important for ESC formation.  Since it is formed by L-amino acid
residues, the active site is stereospecific.  With the help of
labeled atoms, it has been shown that reactions of molecules of
the type CAABB proceed asymmetrically on the enzyme surface.
This is particularly true in the transformation of aminomalonic
acid (containing labeled carbon in one of two carboxyls) into
glycine.  That the reaction occurs in only one of the chemically

$$\begin{array}{ccc}
^*\text{COOH} & & ^*\text{COOH}\\
| & & |\\
\text{H}_2\text{N}-\text{C}-\text{H} & \longrightarrow & \text{H}_2\text{N}-\text{C}-\text{H}\\
| & & |\\
\text{COOH} & & \text{H}
\end{array}$$

and geometrically identical COOH groups can be explained by the
asymmetry of the active site.  Since the Y and Z groups binding
with carboxyls are different, the reactivity of the two carboxyls
is also different [34,35].  Enzymes distinguish stereoisomers and
in general the optical isomers of a given substrate do not act as
substrates.
      Direct X-ray structural investigations of the complexes of
enzymes with substrate analogs which are competitive inhibitors
give results that are in good agreement with chemical theories.
At present few such investigations have been carried out, but
those that exist are sources of especially valid information.
      Phillips and co-workers, having ascertained the structure
of lysozyme, also investigated the structure of lysozyme complexes with inhibiting substrate analogs (oligosaccharides) [36].
They showed that the ligand enters the cavity that exists in the
lysozyme molecule, and they established the nature of the contacts

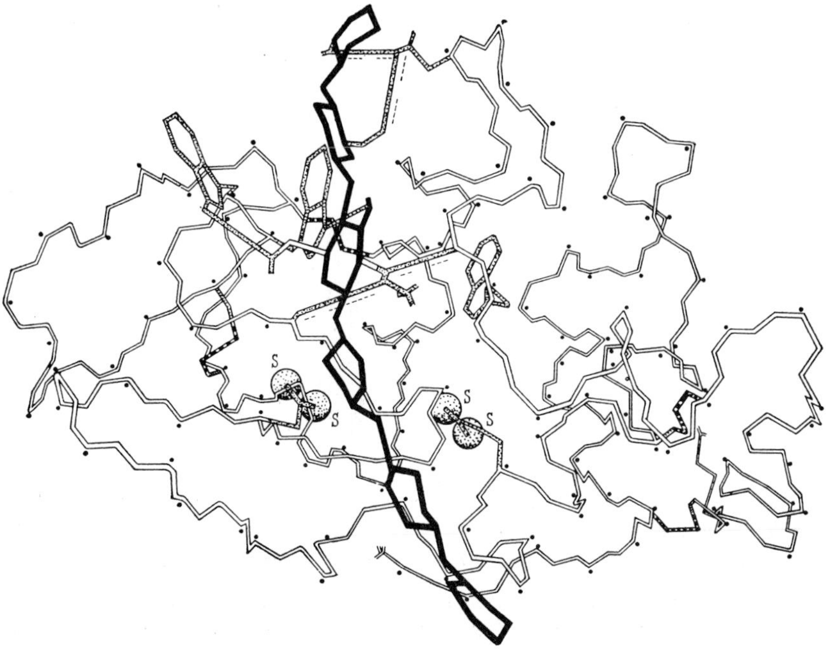


*FIG. 6.10 Structure of the complex of lysozyme with β-N-acetylglucosamine.*

between the functional groups of the enzyme and ligand. Figure 10 shows the structure of the lysozyme complex with β-N-acetyl-glucosamine. The detailed deciphering of enzyme substrate interactions is reported in [37,38].

By means of X rays, the complexes of carboxypeptidase with the substrate glycyl-L-tyrosine and with such inhibitors as β-(p-iodophenol)-propionate have been investigated (Fig. 6.11) [39]. An atom of Zn acts as cofactor of this enzyme. Ligands enter the protein molecule cavity and are bound in a region containing Zn--this atom participates directly in the work of the active site (cf. p. 387).

Consider one more example--the structure and function of aspartate aminotransferase (AAT), which has been studied in detail by Braunstein and his co-workers. Aminotransferases contain a coenzyme, pyridoxal phosphate (PLP). The general theory of the action of such enzymes proposed by Braunstein and Shemiakin [40,41] holds that PLP linked with a protein reacts with the amino acid to form the so-called Schiff base, or aldimine

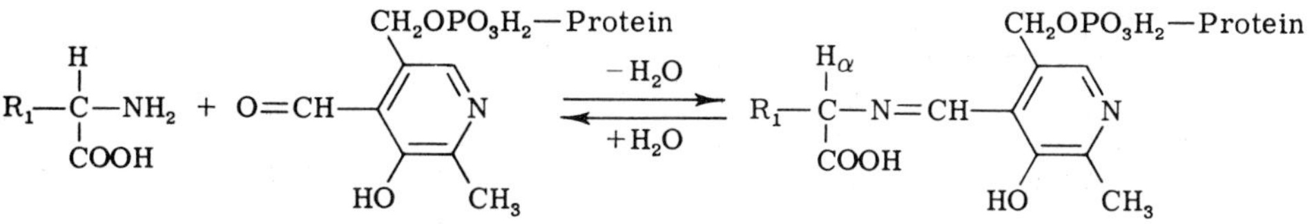

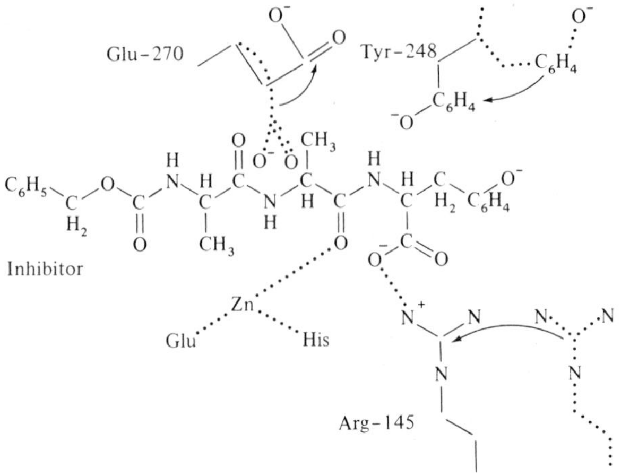


*FIG. 6.11 Structure of the complex of carboxypeptidase with glycyltyrosine. Arg-145, Tyr-248, and Glu-270 are shown before substrate bonding (broken lines) and after it.*

In the Schiff base the distribution of electrons in the amino acid residue is changed essentially and the $H_\alpha$ atom dissociates readily. In transamination reactions an aldimine undergoes tautomeric transition into a ketimine

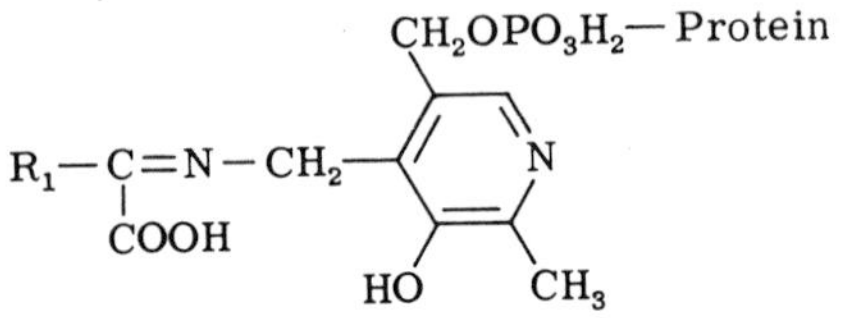

which is hydrolyzed by water, forming a ketoacid

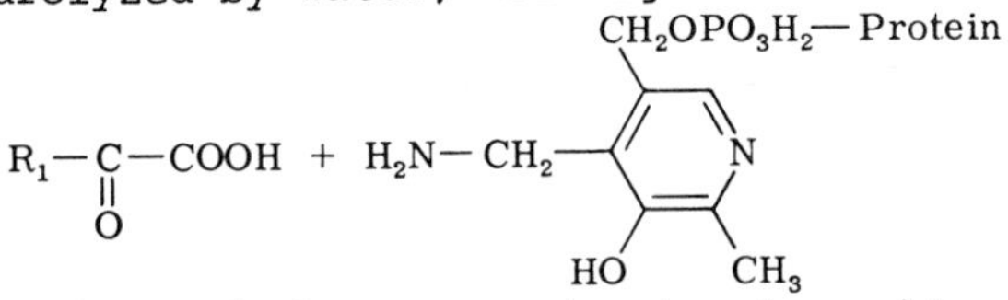

The resulting form of the enzyme is phosphopyridoxamine. It reacts with another ketoacid, containing $R_2$; the initial enzyme is regenerated and a new amino acid is formed. A similar scheme was devised independently by Snell *et al.* [42,43].

Practically every reaction catalyzed by enzymes can also occur in the absence of enzymes but, of course, at a much slower rate. In many cases the stages of enzymic reactions can have corresponding low-molecular-weight models, "congruent model systems" [44].

Transamination reactions have been studied in a system containing PLP, heavy metal ions, and substrates. The addition of a weak base to a system containing pyridoxal and amino acid completely suppresses all reactions except the breaking of the C-H bond, and such a model produces only transamination [45,46]. In the work [45] individual rate constants were determined for the stage of aldimine formation. The values reported for the reaction of amino acid (glutamate) with the anionic, dipolar, and cationic forms of the model compound 3-hydroxypyridine-4-aldehyde are: $k_{an} = 80.2$ liter mole$^{-1}$ min$^{-1}$, $k_{dip} = 1.12\times10^4$ liter mole$^{-1}$ min$^{-1}$, $k_{ca} = 2.3\times10^6$ liter mole$^{-1}$ min$^{-1}$. The enzymic reaction rate constant is much larger: $k = 10^8$ liter mole$^{-1}$ min$^{-1}$. Theoretical calculation shows that the rate of nucleophilic binding to the carbonyl group is increased $10^3$-$10^4$ times if the bimolecular reaction is transformed into the monomolecular one with appropriate spatial positions of the interacting groups [48]. It can be thought that the enzyme provides a similar vector orientation of these groups in all successive stages of the process and stabilizes the most active ionic forms of the substrates, coenzymes, and functional groups of the active site at the corresponding stages [49].

On the basis of this hypothesis, Ivanov and Karpeisky proposed a dynamic model of an AAT catalyzed reaction [49]. The 362-nm absorption band of this enzyme is replaced by a 426-nm band if the pH is lowered. This effect is due to ionization of the phenol group of the coenzyme. The $pK_a$ of the transition is equal to 6.2.

These bands are optically active. The optical activity of PLP is induced by an apoenzyme [50] (cf. p. 302). An enzyme with pyridoxamine phosphate instead of PLP has a 330-nm band. The coenzyme serves as an informative spectral label. Since adequate substrates of AAT contain two carboxyl groups, it can be suggested that the specificity of the enzyme is due to interaction of the substrate $COO^-$ with complementary cationic enzymic groups. The proof is that dicarbonic acids are competitive inhibitors of AAT. ε-N-pyridoxal lysyl has been isolated from an enzyme treated with sodium borohydride and subjected to acid hydrolysis. The phosphate group of PLP is also linked to an apoenzyme (Fig. 6.12). Study of the absorption spectra and circular dichroism of AAT in various forms (Fig. 6.13) provided the method for identifying the proton-donor group of Apo-AAT that binds the N atom of

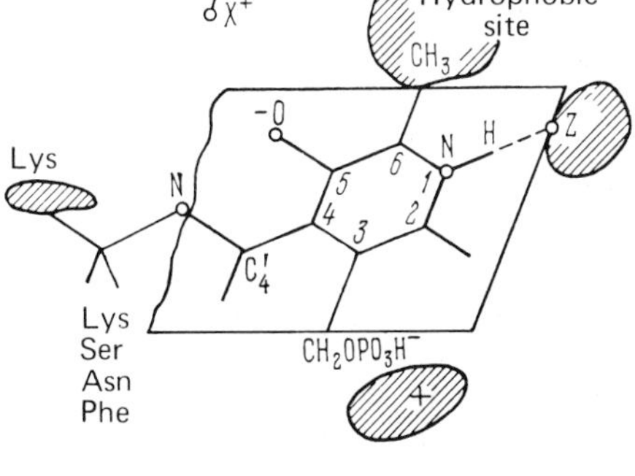


*FIG. 6.12 Binding of a coenzyme by the active site of aspartate aminotransferase (AAT).*

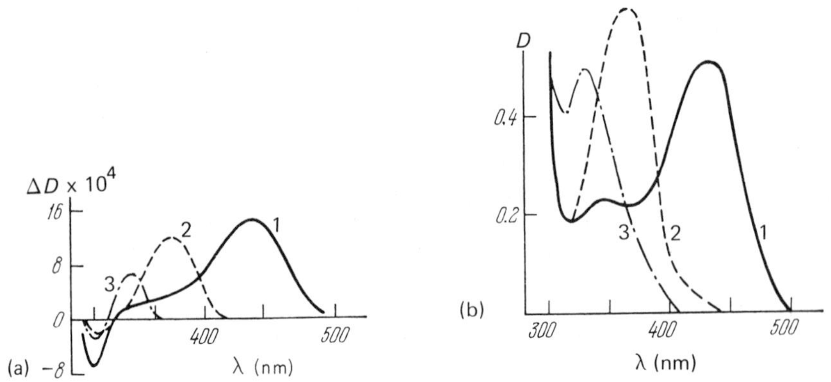


*FIG. 6.13 (a) Circular dichroism spectra and (b) absorption spectra of various forms of the free enzyme: (1) aldimine form at pH 5.2; (2) aldimine form at pH 8.3; (3) amino form. D-optical density; ΔD, difference in D for left and right circular polarized light.*

pyridine [49,51,52]. This group is Tyr. Study of the kinetics of association of Apo-AAT with a series of alkyl analogs of PLP showed that the methyl group of PLP had been in close contact with the apoenzyme [52]. Finally, the electrostatic interactions of the phenolic group of PLP with the apoenzyme have been established. Ivanov and Karpeisky [49] proposed a model of the binding of all functional groups of coenzyme with protein (cf. Fig. 6.14). The conjugated coenzyme system possesses high energy and high reactivity at the active site.

The model of the binding of PLP by an apoenzyme makes possible the detailed examination of enzymic transamination [49]. The first stage of the process is the noncovalent binding of amino acid by the active site. The second stage is the nucleophilic binding of the amino group of the substrate by the >C=N− aldimine bond. Obviously the negatively charged phenolic group of the coenzyme serves as a proton acceptor. Thus, the nonionized nucleophilic amino group becomes the neighbor of the highly reactive cationic form of the coenzyme. The spatial position of aspartic acid fixed by the active site must correspond to the cis conformation of carboxyls, since maleic acid (cis compound) is the competitive inhibitor of AAT, in contrast to fumaric acid (trans compound). This position is shown in Fig. 6.14. It follows from this model that the reaction coordinate for nucleophili binding to a carboxyl is directed along the $p_z$ orbital of carboxy lic carbon. With the help of spatial molecular models, the enzyme-substrate-aldimine structure has been devised (Fig. 6.15). The formation of the substrate aldimine occurs as a result of rotation of the coenzyme ring by ∿40° around the axis, connecting the 2-$CH_3$ and 5-$CH_2$ groups (cf. Fig. 6.15).

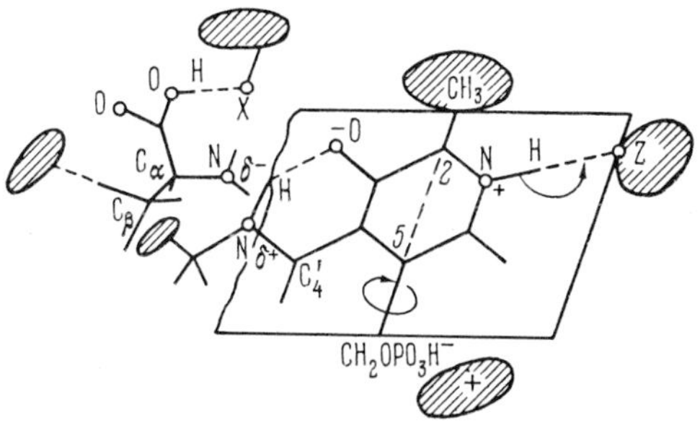

*FIG. 6.14 Suggested structure of ESC for AAT.*

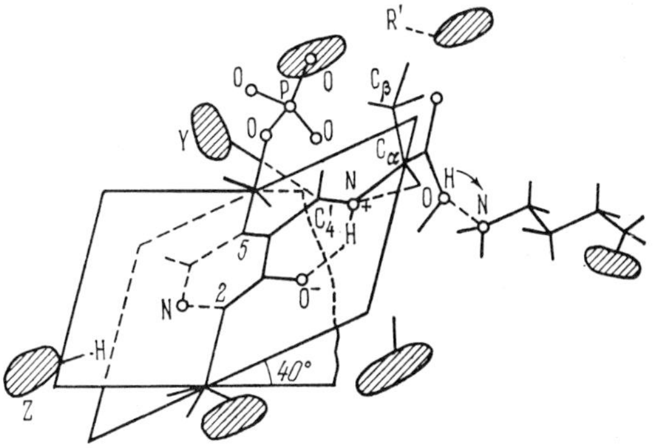

*FIG. 6.15 Suggested structure of aldimine-enzyme-substrate complex.*

The general scheme of the process is shown in Fig. 6.16. The stages already described and the subsequent ones are confirmed by a series of experimental facts [47]. The proposed theory explains the most important chemical properties of AAT.

Ivanov and Karpeisky consider multipoint binding to be the cause of stabilization of an active electronic configuration that is thermodynamically disadvantageous in a homogeneous solution, and the cause of proper orientation of reacting groups. The existence of at least two spatially different structures at the active site (in the case of AAT, of two different orientations of the coenzyme ring) fulfills the necessary conditions in a multistage enzymic reaction. Optimal conditions for each successive stage are provided by the structural transformation that occurs at the preceding stage. This means that equalizations of the free energy levels of intermediate compounds occurs at the active site. In the case of AAT this effect is manifested by the ESC absorption spectrum (Fig. 6.17), which contains the

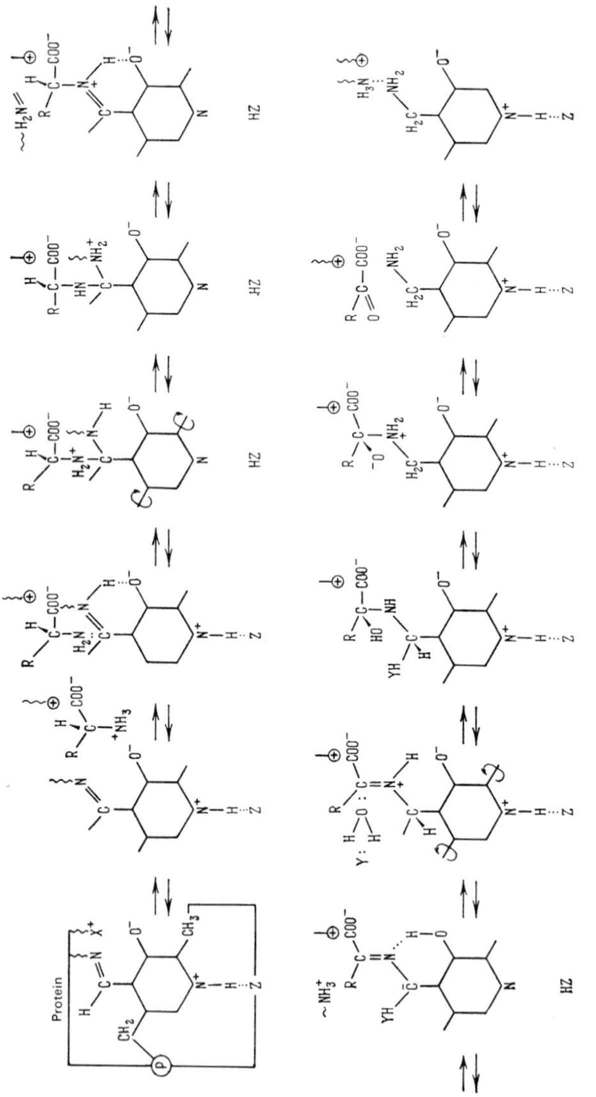

*FIG. 6.16 Scheme of events at the active site of AAT.*

*FIG. 6.17 Absorption spectrum of AAT in the presence of L-aspartate.*

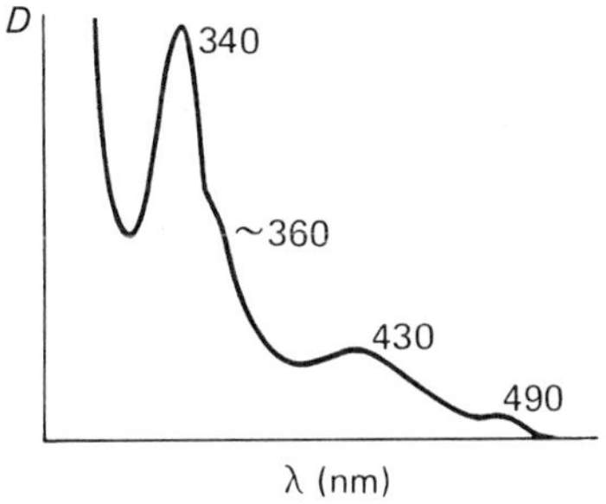


bands of practically all the important intermediate compounds. Only the 330-nm band is observed in the corresponding congruent system.

This example demonstrates the kind of information about enzyme-substrate complexes that can be obtained by means of spectrophotometry and spectropolarimetry. The methods of NMR, EPR, and luminescence, as well as chemical relaxational methods (Section 7.7) are also very valuable here.

A stereochemical picture representing the situation at the active site is obtained on the basis of many different experimental data with the aid of atomic models. Recently, new mathematical methods using computers have been used in calculations of this picture [53].

We see that there is a real possibility of deciphering the interactions between the active site and ligands. The chemical aspect of the problem concerns the qualitative interaction pattern; the behavior of the functional groups of enzyme and cofactors can be investigated. Examination of interactions at the active site shows that the activation energy must be lower than that of the congruent reaction, but the quantitative value of this reduction remains unknown. Are these interactions sufficient for a quantitative explanation of enzymic activity?

The existence of an active site having a definite structure that is formed in an organized protein globule in such a way that the active site contains amino acid residues that are very remote from one another along the protein chain is chemically significant. Analysis of events at the active site does not elucidate the role of the globule as a whole system nor explain the influence of the medium.

Undoubtedly the whole globule participates in catalysis. It was claimed that a considerable part of the protein chain can be cut off without altering enzymic activity (cf., e.g., [54]). These data have since been disproved [55,56].

Braunstein summarized the qualitative factors responsible for catalytic enzyme action [57]:

(1) High affinity of enzyme and substrate (i.e., high probability of ESC formation, equivalent to a sharp increase in

reagent concentration in conventional conditions--rapprochement effect.

(2) Exact mutual orientation of reagents, cofactors, and active site (orientation effect). In a conventional homogeneous reaction the probability of exact mutual orientation of three or more interacting molecules is very small.

(3) Action of nucleophilic and electrophilic groups at the contact site of ESC (effect of synchronous intramolecular acid base catalysis).

The cooperation of several catalyst groups can be modeled by some phenomena of homogeneous acid base catalysis. Thus, 2-hydroxypyridine catalyzes the mutarotation of glucose, breaking its six-membered ring and inducing the tautomeric transition

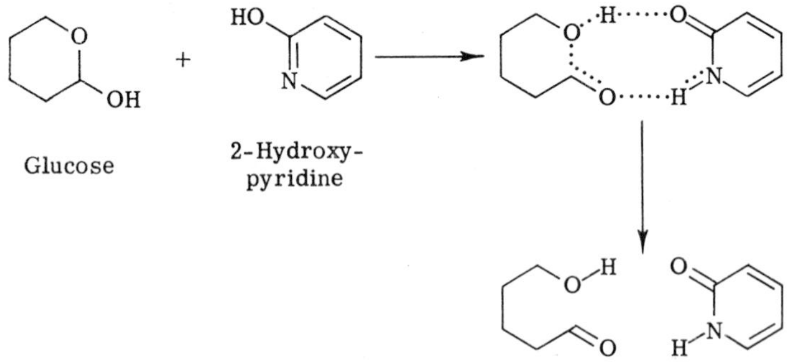


The same reaction with a mixture of pyridine and phenol

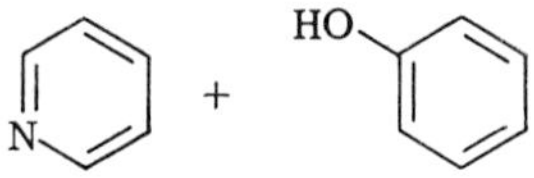

proceeds 7000 times more slowly [58].

These three effects greatly enhance the absolute reaction rate, since higher-order reactions that have low probability, such as the close meeting of three or more molecules, are replaced by efficient first-order reactions (reactions of polyfunctional intramolecular catalysis).

(4) Substrate activation due to a redistribution of the electron density under the action of electroactive enzyme groups (polarization effect).

(5) A change in protein conformation produced by interaction with the substrate (induced contact effect).

The role of the first four factors is demonstrated by the examples already described. The fifth factor will be examined in the next section.

The data of contemporary physical organic chemistry enable us to estimate the orders of magnitude of these and other catalytic factors (i.e., the acceleration of a reaction due to these factors). These estimations are presented in Table 6.3 [59].

*TABLE 6.3*
*Estimation of Catalytic Factors for Two Hypothetical Reactions*

| | Reaction 1 | Reaction 2 |
|---|---|---|
| Characteristics of reaction | | |
| Substrate concentration in moles | $S_1=10^{-3}$ $S_2=10^{-4}$ | $S_1=10^{-3}$ $S_2=55$ (water) |
| Charge of substrate | Noncharged | Noncharged |
| Charge in transition state | Small charge | Total positive charge |
| Catalytic residues | One acidic, one basic | One acidic, one negative charged group |
| Covalent intermediate compound | Yes | No |
| Factors estimated from the known chemical data | | |
| Factor of substrate rapprochement | $5.5\times10^3$ | $10^{-2}$ |
| General factor of basic catalysis | 4 | -- |
| General factor of acidic catalysis | 10 | 20 |
| Factor of rapprochement of catalytic groups | $3\times10^3$ | $3\times10^3$ |
| Factor of intermediate covalent compound | 10 | -- |
| Solvent effect | $10^{-2}$ | $10^{-5}$ |
| Effect of ion pairs formation | $10^2$ | $10^5$ |
| Product of these factors | $7\times10^9$ | $6\times10^2$ |
| The turnover number of enzyme divided by reaction rate of subststrates in absence of catalyst | $7\times10^{18}$ | $6\times10^{14}$ |
| The multiplier which has to be explained | $10^9$ | $10^{12}$ |

In Table 6.4 the rates of real enzymic and nonenzymic reactions are compared. The nonenzymic rates have been multiplied by a factor corresponding to the rapprochement effect. For two

*TABLE 6.4*
*Comparison of Enzymic Reactions with Their Nonenzymic Analogs*

| Enzyme | Nonenzymic analog | Enzymic rate $v(sec^{-1})$ | Non-enzymic rate $v_o(sec^{-1})$ | $v/v_o$ |
|---|---|---|---|---|
| Lysozyme | Acetal hydrolysis, basic catalysis | $5\cdot10^{-1}$ | $3\cdot10^{-9}$ | $2\cdot10^{8}$ |
| Chymotrypsin | Amide hydrolysis, basic catalysis | $4\cdot10^{-2}$ | $1\cdot10^{-5}$ | $4\cdot10^{2}$ |
| β-Amylase | Acetal hydrolysis, basic catalysis | $1\cdot10^{3}$ | $3\cdot10^{-9}$ | $3\cdot10^{11}$ |
| Fumarase | Alkene hydrogenation, acidic and basic catalysis | $5\cdot10^{2}$ | $3\cdot10^{-9}$ | $2\cdot10^{11}$ |

molecules of the size of $H_2O$ molecules, this factor is equal to 55 (i.e., to the molar concentration of water). The molar concentration of pairs of interacting molecules is equal to the product of the molar concentrations of the reagents divided by 55 and multiplied by the coordination number in aqueous solution [60,61]. Cf. also [62].

The problem is to fill the "gap" (i.e., to explain enzymic catalysis quantitatively). Is there any fundamental factor responsible for the high rate of enzymic reactions or is the concerted action of all factors necessary? At present we have no answers to these questions, but some ways of investigating them can be outlined.

## 6.5 Conformational Properties of Enzymes

Protein macromolecules have a conformational variability that makes possible specific interactions between enzymes and substrates or other ligands. Some protein conformations can be assumed to bind to substrate more effectively than others. The binding can select the substrate conformation. Karush attributed the ability of serum albumin to bind various substances to the "conformational adaptability" of this protein [63].

It can be suggested that those conformations of protein and substrate are stabilized in ESC which correspond to mutual structural fit, providing optimal value of interaction energy [64,65]. Structural fit in ESC formation can be defined as dynamic, induced fit. Therefore, when an ESC is being formed, changes in the actual conformation of the protein and substrate (or of one of them) can occur. Vaslow and Doherty observed conformational changes in chymotrypsin during the binding of molecules of substrates and competitive inhibitors [66]. Structural fit in an ESC is somewhat similar to the fit in heterogeneous catalysis (p. 353). On the basis of his multiplet theory, Balandin proposed a qualitative scheme of the structural fit of enzyme, coenzyme, and substrate [67,68].

It can also be suggested that a large change in free energy that occurs at ESC formation is due to conformational effects. The decrease in entropy that occurs during the process of selection of a definite conformation can be compensated by a decrease in enthalpy as the result of interaction and by an increase in the entropy of other regions of the globule. As we have seen, physical analysis of such events occurring in a protein requires consideration of phenomena occurring in the aqueous medium.

The theory of induced structural fit in enzymic catalysis was developed by Koshland [69]. The first problem of the theory was the explanation of the specificity of enzymes catalyzing bond transfer reactions of the type

$$B{-}X + Y \rightleftharpoons B{-}Y + X$$

The old notion of a static system with a structural fit, the "key-lock" idea of Fisher, explained enzyme specificity not by flexibility, but by a rigidity determining the attraction of a definite substrate molecule and the steric repulsion of slightly different analogs. A series of facts contradicts this model. Thus, water and other small hydroxyl-containing molecules are nonreactive in hydroxyl-transfer reactions catalyzed by phosphorylases and kinases. Larger hydroxyl-containing molecules, on the other hand, are substrates in these cases. Frequently, ligands that are adsorbed well by an active site lack reactivity despite the high reactivity of similar molecules. There are certain small molecules that are not adsorbed by the same active site that strongly adsorbs their larger analogs. Phosphotransacetylase acts on acetate, propionate, and butyrate, but not on formate; and β-glucosidase acts on glucosides but not on 2-deoxyglucosides. Other facts are presented in [64,69].

Koshland suggested that [70]:

(1) The substrate induces changes in the enzyme geometry as it penetrates the active site.

(2) An exact mutual orientation of catalytic groups is required for enzymic action.

(3) The substrate induces this orientation by changes in the enzyme geometry.

This situation is presented schematically in Fig. 6.18. Figure 6.18c,d demonstrates why molecules similar to substrate molecules but differing in size are nonreactive. According to Koshland's model, the specificity and catalytic activity of an enzyme are related but have different mechanisms, and a reaction can occur only when the sites of adsorption and catalysis are

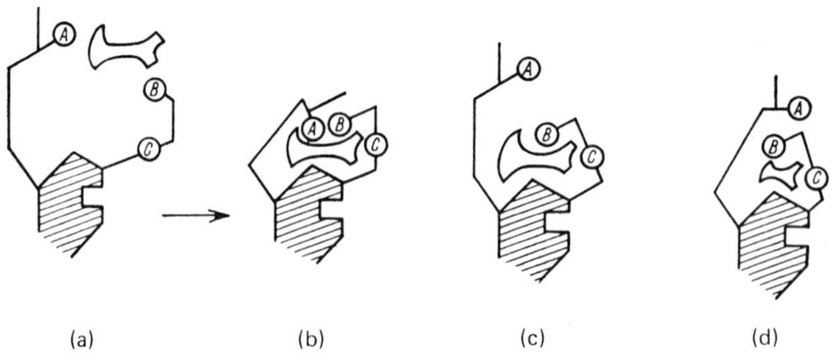


*FIG. 6.18 Scheme of induced fit. (a) Enzyme; (b) enzyme-substrate complex; (c,d) complexes with substrate analogs.*

properly positioned in relation to the substrate molecule. Koshland illustrated this situation by the example of β-amylase [71], an enzyme that acts on the end groups of amylose but not on other glucoside bonds of the polysaccharide. Cycloamyloses are competitive inhibitors of this enzyme, as is illustrated in Fig. 6.19. A reaction is seen to proceed only when the adsorbing and catalytic groups A, X, B of the enzyme are at definite spatial positions. Cf. also [72,73].

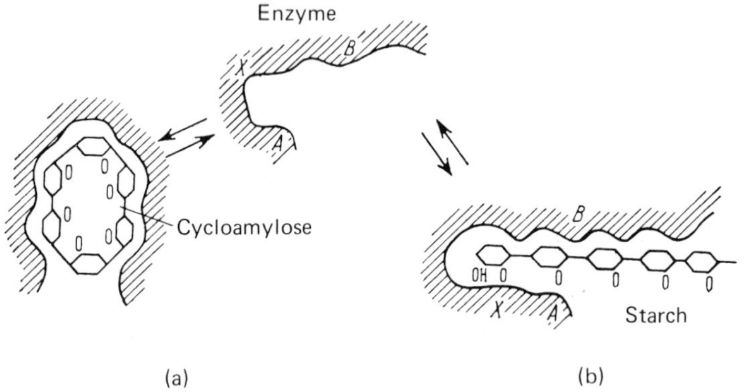


*FIG. 6.19 Scheme of the action of β-amylase. (a) Nonreactive complex; (b) reactive complex.*

The modifier action model is shown in Fig. 6.20 [72]. The inhibitor is competitive if group B is essential for substrate binding and noncompetitive if this group is a catalytic one.

A scheme showing the selection of protein conformation by a substrate appears in Fig. 6.21 [59] where H denotes hydrophobic groups, + and - the charged groups.

Koshland proposed models of the orientation effect in ESC. This effect was calculated by applying the hypothesis that a

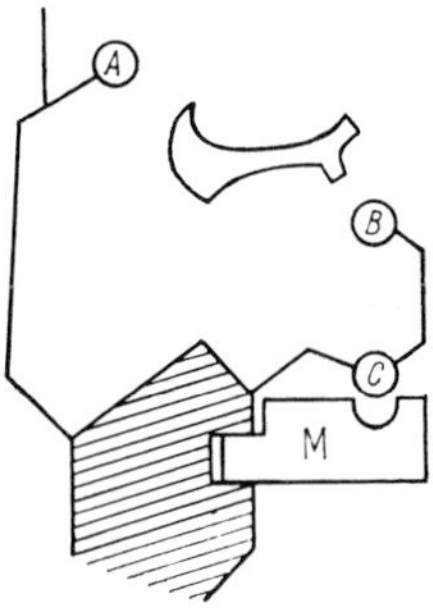


*FIG. 6.20 Scheme of modifier action.*

system's energy depends strongly on orientation angle [62]. It has since been shown that such strong dependence cannot exist [74]. For a $10^5$-fold increase in the reaction rate, the activation barrier has to be lowered by 7 kcal $mole^{-1}$. Such a change in energy is possible if there are 10°-15° rotations around the double bonds or 5°-10° changes in the valence angles. But the estimates in [62] suggest angle changes of only 0.1°.

Confirmation of the qualitative ideas of Koshland's theory requires proof of the actual conformational changes produced by protein-ligand interaction and demonstration that these changes are actually the causes of the observed increase in the reaction rate.

A series of facts attests to the occurrence of conformational changes of the enzyme (cf. [68,71]) in the presence of

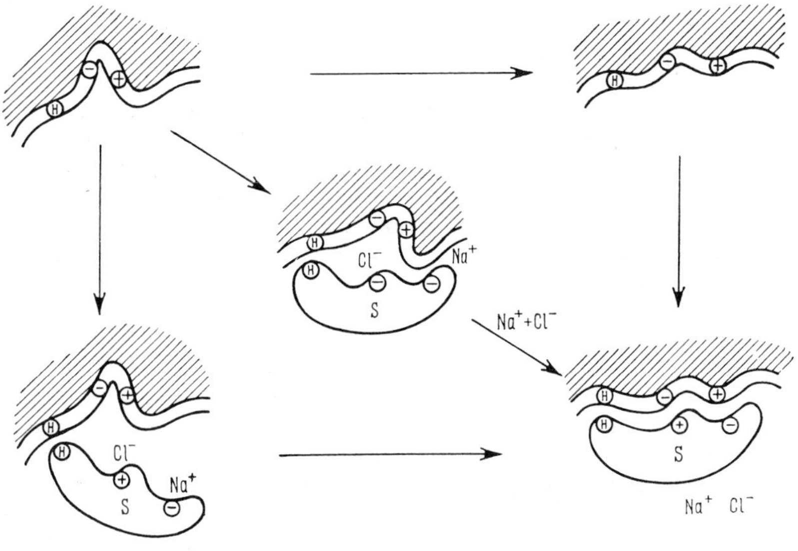


*FIG. 6.21 Scheme of the selection of protein conformation by a substrate: H, hydrophobic groups; + and −, charged groups.*

substrates some enzymes become stiffer, others more labile--their thermal denaturation becomes easier [75]. Substrates induce dissociation of glutamate dehydrogenase [76] and hexakinase [75]. The reactivity of amino acid residues changes under the action of a substrate; thus, the iodination of penicillinase is enhanced by substrate [77]. All these results can be explained by Koshland's model [79]. See also the reviews [80,81].

The absorption spectrum of chymotrypsin is altered by interaction with a substrate and these changes can be interpreted as due to conformational changes [82]. Similar phenomena have been observed in a series of other cases (cf., e.g., [83]). Conformational changes are manifested in the luminescence spectra of aromatic amino acid residues and of molecules of dyes adsorbed by proteins [cf. [84]). Conformational changes in phosphoglucomutase produced by interaction with a substrate (glucose 6-phosphate) are quite evident in both absorption and luminescence spectra [68,85].

Spectropolarimetry reveals structural transformations occuring at the active site and in the globule as a whole. The studies of aspartate-aminotransferase have already been discussed (p. 373; cf. also [68,86]). In a series of enzymes, changes in α-helicity have been observed to arise upon interaction of the enzyme with a substrate, a coenzyme, and other ligands (cf. [68]). Detailed studies of the DOA of lactate dehydrogenase and glyceraldehyde phosphate dehydrogenase have shown that binding of a coenzyme (NAD, p. 80) considerably changes the degree of α-helicity of these proteins [68,87].

Information about conformational changes is also provided by the EPR spectra of enzymes that contain paramagnetic labels, and by NMR spectra [Chapter 5).

We see that Koshland's basic assumption is confirmed in many cases. However, these results do not yet show that a conformational change in a protein leads to the attainment of a structural fit. Optical methods alone are not sufficient for the establishment of structural fit; they have to be combined with chemical ones and with thorough theoretical analysis (see the model of AAT action, pp. 371-377). Direct information concerning changes in enzyme and substrate structure during formation of the ESC can be obtained by means of X-ray structural analysis [88].

It has to be emphasized that Koshland's theory does not posit a large change in the protein conformation. If the structure of the globule predetermines the fit with the substrate, these changes may be small. On the other hand, structural fit can also be provided by a conformational change in the substrate.

This is just the situation observed in the case of lysozyme. As X-ray analysis shows [36,88], the substrate enters the cavity already existing in the enzyme molecule (p. 371); the cavity narrows, "squeezing" the substrate molecule. The amino acid residue displacements in the protein are rather small but observable--the Trp-62 residue is displaced by 0.75 Å. A simultaneous change in conformation occurs in the substrate--the carbohydrate rings rotate slightly around the glycosidic bond.

The presence of a cavity appropriate for a substrate has also been discovered in a series of other enzymes by means of X-ray analysis. The active site of ribonuclease located in the cavity contains the basic residues Lys and Arg adjacent to phosphate groups of RNA (ribonuclease catalyzes hydrolytic RNA splitting). The catalytic site in the same cavity is made up of two His residues acting as basic catalysts [89,90]. The residues Cys-25 and His-159 are active in the cavity found to exist in papain [91].

X-ray structural analysis reveals the events that occur when chymotrypsin is formed by the activation of the chymotrypsinogen. Activation consists of breakage of the peptide bond Arg-15-Ile-16 (cf., for instance, [15,64]). It has been established that the breaking of the bond is followed by the binding of the amino group of Ile with γ-COOH of Asp-194 and formation of a deep cavity. The surroundings have a small dielectric constant and the ionic bond $COO^{-}$ $^{+}NH_3$ is very strong. The active site of chymotrypsin contains Ser-195, His-57 close to the surface of the cavity; the position of Asp-102 is deeper. The carboxyl of Asp-102 forms a hydrogen bond with the imidazole of His-57, which becomes strongly polarized and attracts the hydroxyl proton of Ser-195. The region determining the chymotrypsin specificity in relation to the aromatic side chains is located in a deep cavity near the active site. The scheme of the transformation of chymotrypsinogen into chymotrypsin is shown in Fig. 6.22 (Cf. [88,92,93]).

Direct confirmation of Koshland's theory was obtained by carboxypeptidase studies [39,94,95] (p. 371). The introduction of glycyltyrosine produces considerable displacement of the amino acid residues of this enzyme. The active site looks like a deep cavity similar to a "spider mouth" [88] with "mandibles" ready to direct the substrate toward the Zn atom. One "mandible," containing Tyr-248, is directed toward the NH group of the substrate; the Arg-145 of the second interacts with the substrate carboxyl; the Glu-270 of the third with the amino end group of the substrate. This is shown schematically by Fig. 6.23.

Obviously Koshland's theory is not to be understood as a theory of flexible enzyme structure. It would be better to compare an enzyme with an engine whose functional parts are subjected to discrete displacements (changes in locations) necessary for structural fit with ligands. The construction of such an engine is characterized not by plasticity but by elasticity. The model of a protein globule consisting of "hinged elastic pivots" (p. 237) agrees with the facts obtained in enzyme studies.

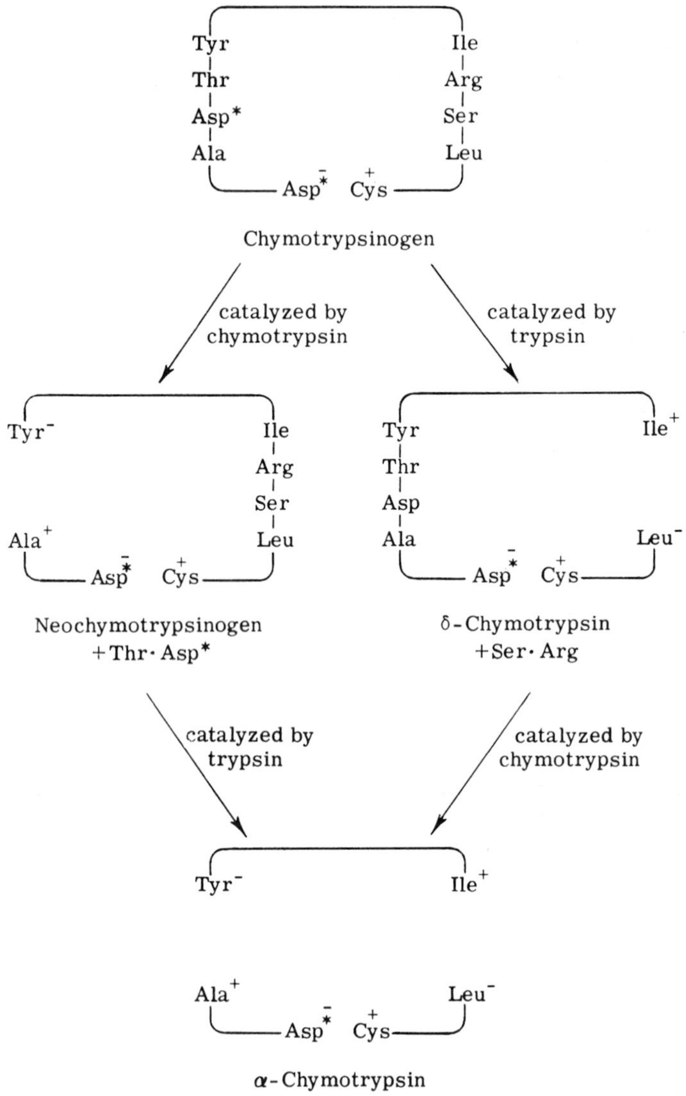


*FIG. 6.22 Scheme of the transformation of chymotrypsinogen into chymotrypsin.*

Structural fit actually seems to be achieved in ESC in the cavity of the enzyme molecule. Comparison of all structures investigated by means of X rays shows that their only common feature consists in the existence of this cavity. The internal surface of the cavity is made up mainly of nonpolar residues. Because of hydrophobic interactions, polar residues are located outside (p. 209). The nonpolar interior of a protein molecule

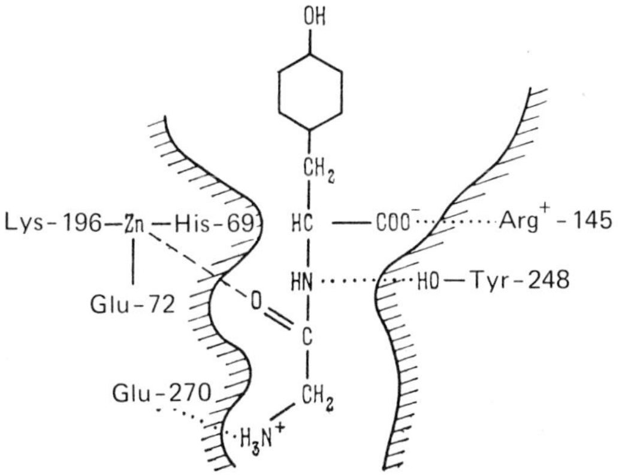


*FIG. 6.23 Scheme of the active site of carboxypeptidase.*

has a small dielectric constant and therefore enhances electric interactions [88]. The meaning of this fact is discussed later (p. 396).

## 6.6 Action of the pH of the Medium on Enzymes

Obviously enzymic activity must depend strongly on the pH of the medium. Protein macromolecules contain ionogenic groups. The same is true of the majority of substrates, modifiers, and coenzymes. The possible effects of pH were classified in detail by Webb [16], cf. [68]. A change in the pH can change the state of ionogenic groups at the active site or of neighboring groups; it can change the state of the nonenzymic components of the system and influence the globular structure locally or as a whole.

An enzyme can exist in various states of ionization that are in equilibrium. Let us consider the simplest model, that of an acid system with two ionization states [96]. We have

$$AH_2 \underset{K_1}{\rightleftharpoons} AH^- \underset{K_2}{\rightleftharpoons} A^{2-}$$

where $K_1$ and $K_2$ are the first and second ionization constants, respectively. The sum of the concentrations of all three forms is

$$A_t = [AH_2] + [AH^-] + [A^{2-}] \tag{6.59}$$

Constants $K_1$ and $K_2$ are

$$K_1 = \frac{[AH^-][H^+]}{[AH_2]}, \qquad K_2 = \frac{[A^{2-}][H^+]}{[AH^-]} = \frac{[A^{2-}][H^+]^2}{[AH_2]K_1} \tag{6.60}$$

Hence

$$A_t = [AH_2]f = [AH^-]f^- = [A^{2-}]f \tag{6.61}$$

where f, $f^-$, and $f^{2-}$ are the so-called Michaelis functions:

$$\begin{aligned} f &= 1 + (K_1/[H^+]) + (K_1K_2/[H^+]^2) \\ f^- &= 1 + ([H^+]/K_1) + (K_2/[H^+]) \\ f^{2-} &= 1 + [[H^+]/K_2) + ([H^+]^2/K_1K_2) \end{aligned} \tag{6.62}$$

Let us apply this model to enzymes, assuming that the active site can exist in these three forms but that only one of them, namely $AH^-$, is reactive. Then the reaction rate is proportional to the concentration of this form

$$v = k[AH^-] \tag{6.63}$$

and we get

$$v = \frac{kA_t}{1 + ([H^+]/K_1) + (K_2/[H^+])} \tag{6.64}$$

The dependence of v on $H^+$ (i.e., on the pH) goes through a maximum corresponding to

$$[H^+] = (K_1K_2)^{1/2}$$

or

$$pH = \frac{1}{2}(pK_1+pK_2) \tag{6.65}$$

Actually in many cases the dependence of enzymic activity on the pH has a bell-like shape. This is explained by the phenomenological theory presented, which includes convenient formulas [15,16]. However, the dependence of the reaction rate on the pH can be treated in another way. Kirkwood and Shumaker examined the fluctuations of the electric charges in an enzyme molecule [97,98]. If the free energies of various ionization states do not differ greatly, the charges can be displaced (can fluctuate). These fluctuations can lead to additional electrostatic enzyme-substrate interaction.

Let U be the potential energy of an ESC with the given charge distribution. Let us average this energy over all distributions

$$e^{-W/kT} = \left\langle e^{-U/kT} \right\rangle \tag{6.66}$$

We have

$$W = \langle U \rangle - (2kT)^{-1}[\langle U^2 \rangle - \langle U \rangle^2] + \tag{6.67}$$

In the absence of fluctuations $\langle U^2 \rangle = \langle U \rangle^2$ and $W = \langle U \rangle$. Assume that the protein contains n basic groups with charges $z_i e$. If we denote the dipole moment of the substrate molecule by p, the distance between the ith group of the active site and substrate by $R_i$, the angle between p and $R_i$ by $\gamma$, and the dielectric constant by $\varepsilon$, we get

$$U = \sum_{i=1}^{n} \frac{(z_i+x_i)ep \cos \gamma_i}{\varepsilon R_i^2} \tag{6.68}$$

where $x_i = 1$ if the proton is bound by the ith group, and $x_i = 0$ if it is not bound. Let us evaluate W. We have

$$\langle U \rangle = \sum_{i=1}^{n} \frac{z_i+\bar{x}_i}{\varepsilon R_i^2} ep \cos \gamma_i \tag{6.69}$$

and

$$\bar{x}_i = \frac{[A_iH]}{[A_iH]+[A_i]} = \frac{[H^+]}{[H^+]+K_i} \tag{6.70}$$

where

$$K_i = [A_i][H^+]/[A_iH] \tag{6.71}$$

In a similar way we find

$$\langle U^2 \rangle = \sum_i \overline{(z_i^2+2z_i\bar{x}_i+x_i^2)} \frac{e^2p^2 \cos^2 \gamma_i}{\varepsilon^2 R_i^4}$$

$$+ \sum_{i,j}{}' \overline{(z_i+x_i)(z_j+x_j)} \frac{e^2p^2 \cos \gamma_i \cos \gamma_j}{\varepsilon^2 R_i^2 R_j^2} \tag{6.72}$$

where

$$\overline{x_i^2} = \overline{x_j^2}, \quad \overline{x_i x_j} = \bar{x}_i \bar{x}_j = [H^+]^2/([H^+]+K_i)^2 \tag{6.73}$$

Hence

$$\langle U^2 \rangle - \langle U \rangle^2 = \sum_i (\overline{x_i^2}-\bar{x}_i^2) \frac{e^2p^2 \cos^2 \gamma_i}{\varepsilon^2 R_i^4} \tag{6.74}$$

where

$$\overline{x_i^2} - \bar{x}_i^2 = \frac{[H^+]}{[H^+]+K_i} \left(1 - \frac{[H^+]}{[H^+]+K_i}\right) = \frac{[H^+]K_i}{([H^+]+K_i)^2} \tag{6.75}$$

Thus

$$W = \sum_{i=1}^{n} \left(z_i + \frac{[H^+]}{[H^+]+K_i}\right) \frac{ep \cos \gamma_i}{\varepsilon R_i^2}$$

$$- \frac{1}{2kT} \sum_{i=1}^{n} \frac{K_i [H^+]}{([H^+]+K_i)^2} \frac{e^2 p^2 \cos^2 \gamma_i}{\varepsilon^2 R_i^4} + \cdots \qquad (6.76)$$

The energy due to charge fluctuations can be written in the form

$$\Delta W = - \frac{n_\alpha e^2 p^2}{4\varepsilon^2 R_\alpha^4 kT} \frac{K_\alpha [H^+]}{([H^+]+K_\alpha)^2} \qquad (6.77)$$

Here $n_\alpha$, $R_\alpha$, and $K_\alpha$ are some effective values.

Because of the charge fluctuations, both the Michaelis constant (p. 357) and the rate constant $k_2$ of the final reaction of product formation are altered. It can be assumed that

$$\ln(k_2/k_2^0) = - (W^*-W_0^*)/kT = - \Delta W^*/kT \qquad (6.78)$$

where $W^*$ and $W_0^*$ are the activation energies in the presence and absence of charge fluctuations, respectively. If $\Delta W^*$ is expressed by formula (6.77), then

$$\ln \frac{k_2}{k_2^0} = \frac{n_\alpha e^2 \Delta p^2}{4\varepsilon^2 R_\alpha^4 k^2 T^2} \frac{K_\alpha [H^+]}{([H^+]+K_\alpha)^2} \qquad (6.79)$$

where $\Delta p^2$ is the difference of squares of the substrate dipole moment in the activated and in the initial state. Expression (6.79) gives the bell-shaped dependence of $k_2$ on the pH. The maximum of $k_2$ corresponds to pH = $pK_\alpha$.

The curve $k_2$(pH) is explained here by causes quite different from those in the Michaelis theory. Direct proofs of the existence of charge fluctuations are not known. Kirkwood and Shumaker showed that fluctuations must give additional components in the spectrum of relaxation times. It has since been shown that this is not correct [99] and it is difficult at present to indicate a method for the experimental demonstration and study of charge fluctuations.

Kirkwood considered the mechanism of fluctuational enzyme-substrate interaction as a universal one, determining the general phenomenon of enzymic activity [97,98,100]. We know now that enzymic activity is due to a series of different factors and that conformational phenomena are of great importance here. Kirkwood's theory does not solve the problem. However, the mechanisms suggested by Michaelis and that proposed by Kirkwood both treat actual polyelectrolyte properties. The significance of these mechanisms remains uncertain and it is still difficult to obtain quantitative estimates.

One more aspect of pH influence has to be considered--the direct influence of the pH on conformational protein structure. If it is assumed that an induced enzyme-substrate contact takes place, then the bond rotations are accompanied by charge displacements. Of course, electrostatic interactions (including ion-pair formation) of the side chains can be as important as the states of the hydrogen bonds in α-helices and β forms. Evidently the hydrogen ion concentration must influence the conformational properties of a protein. It must also act on the hydrophobic interactions which form the globule. The dependence of hydrophobic interactions on the pH has not been adequately studied.

As already said (p. 201-205), a rigorous analysis of the pH on the α-helicity of polypeptides is possible. Corresponding model calculations [101] have shown that the shape of the dependence of α-helicity on pH depends on the sequence of anionic and cationic groups in the chain. If these groups alternate regularly and are separated by some constant number of neutral groups, the curve has the shape of a symmetric bell with a maximum at pH $= \frac{1}{2}(pK_A + pK_C)$ (p. 203). In other cases a minimum instead of a maximum and the more complicated shapes is possible.

Study of the DOA of lactate dehydrogenase at different pH values has shown a bell-shaped dependence on the pH of the constants in the Drude and Moffitt formulas. Figure 6.24 shows the corresponding curves and curves of the dependence on pH of

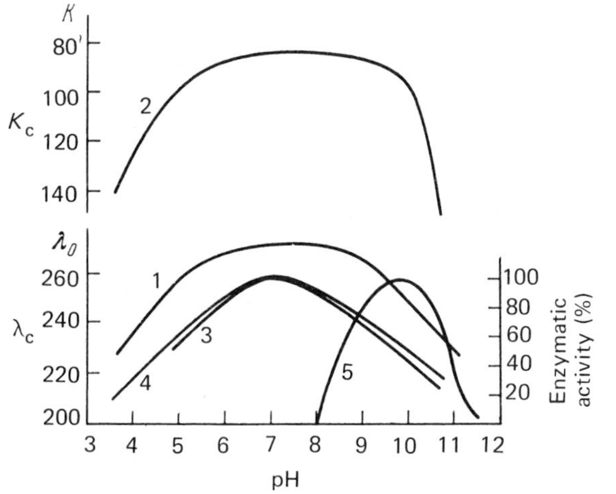


*FIG. 6.24 Action of the pH on lactate dehydrogenase: 1, pH dependence of $\lambda_0$ (Drude's formula); 2, pH dependence of $K_C$ (Drude's formula); 3, pH dependence of $\lambda_0$ for LDH + NADH; 4, pH dependence of the enzymic activity of LDH for the reaction pyruvate → lactate; 5, the same for the reverse reaction.*

the rates of the forward and reverse reactions (pyruvate $\rightleftarrows$ lactate). For the forward reaction the correlation between enzymic activity and α-helicity of the enzyme has been established. The maximum rate corresponds to maximum helicity. The correlation becomes considerably better in the presence of the coenzyme NAD, which changes the secondary enzyme structure; these changes depend on the pH. Analysis of the DOA according to Blout and Shechter (p. 308) showed that experimental points lie on the line of dependence of $A_{225}$ on $A_{193}$ corresponding to $\varepsilon < 30$ (i.e., to an organic solvent). This means that the α-helical regions are in hydrophobic surroundings. If the pH is changed, the experimental points are displaced along the same line. Hence, changes in the α-helicity due to a change in the pH in the studied range of pH values are not accompanied by unfolding of the globule.

Similar results were obtained for glyceraldehyde phosphate dehydrogenase, but the correlation in this case is not so clear as in the case of the first enzyme. For further details see [68,87]. A series of data concerning the influence of pH on secondary protein structure were presented by Joly [102].

These results confirm the general ideas concerning the determining role of conformational phenomena in enzymic activity. In this respect they agree with Koshland's theory. The correlation of the α-helicity and reaction rate of lactate dehydrogenase is very demonstrative. Of course, there are no grounds for considering such a correlation necessary in all cases.

However, at present we have no method, theoretical or experimental, with which to treat the relative role of all three mechanisms: that suggested by the phenomenological theory of Michaelis and Davidsohn; that proposed by Kirkwood and Shumaker; and the conformational one. It must be noted that along with the charge fluctuations, studied by Kirkwood, there must exist spatial fluctuations due to the motility of polypeptide chains.

## 6.7 Physical Aspects of Enzymic Activity

The main unsolved problem in enzyme physics and physical chemistry is that of a quantitative explanation of the high catalytic activity of enzymes. Undoubtedly all the effects discussed earlier contribute to enzymic activity. However, the quantitative estimations presented are not reliable enough and do not solve the problem.

Concerned here is the general mechanism of enzymic activity determined by the physical properties of the globule as a complete system. We do not yet know whether such a mechanism exists, or whether enzymic activity is due only to the chemical and physicochemical factors considered earlier. Evidently either a positive or a negative answer to the questions of the existence of a general physical mechanism would be significant.

It has to be emphasized that the properties of the globule as a whole, except for the existence of a cavity containing nonpolar residues, are not taken into account in Koshland's theory nor in the chemical theories of interactions at the active sites in enzymes.

It seems unreasonable to look for such a mechanism in the semiconductive or other electronic properties of globules. Arguments disproving the hypotheses about the special electronic properties of proteins have already been presented (p. 94), and these arguments are certainly applicable to the overwhelming majority of enzymes. However, we have to agree with Perutz's assertion that special mechanisms can be essential in enzymes that participate in electron transfer or in the transformation of various forms of energy, such as the energy of photons into chemical energy [88].

A protein globule is a dynamic system, a kind of engine, whose work is determined by concrete details of construction. Eigen wrote that when studying proteins we encounter molecules that appear to have a degree of "intelligence," unlike their inorganic counterparts, which simply say "yes" when reacting or "no" when they do not react [103]. However, it is not only appropriate but necessary to examine this engine as a whole physical system.

It is natural to suppose the ability of the globule to serve as an energy reservoir. The energy of thermal motion or the energy supplied to a globule by sorption of substrate is converted into ESC energy and therefore the activation energy is lowered. Nontotal ordering of the globule and small differences in the free energies of the ordered and disordered states (of the order of 1 kcal $mole^{-1}$) indicate the presence of conformational fluctuations [104,105]. Indirect proof of fluctuations consists in the considerable exchange of deuterium with the hydrogen of peptide bonds CO—NH at temperatures much lower than the denaturation temperature at which hydrogen bonds are broken [104]. The presence of fluctuations is also manifested by the greater stiffness of the ESC in comparison with the free enzyme; the ESC is more stable to splitting by trypsin [105-107]. Apparently the substrate binding diminishes the conformational motility of the globule. The presence of globular fluctuations follows from the general phenomenological theory of polymer globules developed by Lifshitz (p. 221).

The storage of fluctuation energy in ESC and the lowering of the activation energy of the process due to this storage seem understandable on the basis of the general theory of monomolecular decay. Consider a chemical bond incorporated in a complicated system of other bonds. The system as a whole is subjected to thermal vibrations. There is a finite probability of energy. storage in the given bond sufficient for its breakage. An isolated bond does not possess such a possibility. We wonder whether the incorporation of a substrate in an ESC can determine the possibility of the storage of energy in the substrate bonds.

This tempting idea is, however, erroneous. As was shown by the theory of monomolecular decay [1,108], the accumulation of vibrational energy does not influence the activation energy necessary for bond breakage; it acts only on the preexponential factor which represents the average value of the frequency of a vibrational system [cf. (6.10)]. The average frequency of an ESC cannot be very different from that of the substrate vibrations [68].

Another suggestion is that the energy liberated when a substrate is adsorbed can be transformed into elastic vibrational energy of the globule treated as a drop of liquid. The frequencies of such vibrations belong to the hypersound region (maximum Debye frequency of the order of $10^{13}$ $sec^{-1}$). Standing waves in the drop can form an antinode in the region of the active site and the energy of acoustical vibrations can activate the substrate molecule [109]. Quantitative estimations based on this idea have shown that the energy of elastic oscillations of the globule can actually be as large as 5-10 kcal $mole^{-1}$ and can considerably lower the effective activation barrier [108]. However, this hypothesis has not been proved; in particular, it does not explain why the vibrational energy does not dissipate into the surrounding medium.

The nonreliability of the "liquid-drop" hypothesis does not eliminate the possibility that the substrate adsorption energy is not converted into lowering of the effective activation energy [110]. The formation of enzyme-substrate structural fit brings molecules of both protein and substrate into a state of tension. It can be said that the substrate molecule becomes "stretched on the rack" [21].

Let us assume that the size of an unstretched substrate molecule is $\ell_0$, and the size of the enzyme cavity contacting the substrate is $\ell$. The deformation of the substrate molecule is x, that of the enzyme molecule y. We have

$$x + y = \ell - \ell_0$$

The condition of the equality of elastic forces is

$$k_s x = k_e y$$

where $k_s$ and $k_e$ are elastic coefficients of the substrate and enzyme. We get

$$x = \frac{\ell - \ell_0}{1 + k_s/k_e}, \qquad y = \frac{k_s}{k_e} \frac{\ell - \ell_0}{1 + k_s/k_e}$$

The elastic energy of the substrate responsible for the decrease in the activation energy is equal to

$$\Delta E = \frac{k_s x^2}{2} = \frac{k_s}{2}\left(\frac{\ell - \ell_0}{1 + k_s/k_\ell}\right)^2 \tag{6.80}$$

The order of magnitude of $k_e$ corresponds to the product of the linear dimensions of the globule and its elasticity modulus, $L\varepsilon$. For a protein $L \sim 50$ Å, $\varepsilon \sim 10^{10}$ ergs $cm^{-3}$. Hence, $k_e \sim 5\times10^3$ dynes $cm^{-1}$. The maximum energy of elastic deformation has to accumulate at the weakest part of the substrate molecule. Deformation of valence angles occurs much more easily than that of valence bonds [111]. The energy accumulated at such degrees of freedom in the molecule can be transmitted to valence bonds and decrease the energy necessary for their breakage. The elasticity coefficient corresponding to low-frequency bending vibrations ($\nu \sim 10^{13}$ $sec^{-1}$) has the order of magnitude of $1.5\times10^4$ dynes $cm^{-1}$. Let $\Delta E$ be 7.5 kcal $mole^{-1}$ (this value increases the reaction rate $10^5$ times). Then $x \sim 0.8$ Å, $y \sim 2.3$ Å, and the elastic energy of the enzyme is $k_e y^2/2 \sim 21$ kcal $mole^{-1}$. Hence the total energy of adsorption expended for elastic deformation is $\sim$30 kcal $mole^{-1}$. This value is not excessive if the adsorption occurs at many points, forming many chemical and nonchemical enzyme-substrate bonds. The observed adsorption energy is the difference between the actual adsorption energy and the elastic energies of the enzyme and substrate.

This elementary calculation was done by assuming that the globule has uniform elastic properties. If the region of substrate binding in an ESC has enhanced stiffness, the energy of elastic deformation of the enzyme will be lower and can even be smaller than that of the substrate molecule.

This is a static model. In reality, the "rack" has to be treated as a dynamic phenomenon, and it can greatly alter the described estimations. Thus, if the oscillations of the enzyme and substrate molecules are in resonance, the necessary mean elastic energy required for acceleration of the reaction is four times smaller than in the static case since the amplitude of the oscillations is periodically doubled.

The "rack" model must be equivalent to the globule theory of Lifshitz (p. 221), according to which even the homogeneous globule is a system with discontinuous free energy levels. Fluctuations of its shell can provide the induced structural fit suggested by Koshland, and the change in the free energy level of the globule due to substrate adsorption becomes equivalent to the storage of elastic deformation energy. It is possible that these ideas will help us to understand the function of enzymes.

None of these notions, however, takes into account an important factor. As was said earlier (p. 371), the substrate molecule enters the hydrophobic, nonpolar cavity in the enzyme molecule. Therefore, the substrate molecule is transferred from aqueous into nonaqueous surroundings. An enzyme is not only a specific reagent, but also a specific reaction medium [88,112]. Perutz wrote [112] that we can ask why a chemical reaction normally requiring powerful organic solvents or strong acids and bases can occur in aqueous solutions near neutral pH if enzymic catalysts are present. Organic solvents have advantages over water, providing as they do a medium with a low dielectric constant in which the strong electric interactions of the reagents can take place. The nonpolar interior of the enzyme provides the living cell with the equivalent of the organic solvents used by chemists. A substrate can be introduced into a medium with a low dielectric constant in which the strong electric interactions of the substrate and the polar groups of the enzyme can take place.

The study of this aspect of enzyme behavior (i.e., of an enzyme as the medium for a reaction) requires on one hand experimental models, and on the other the development of a theory of reaction rates in solutions. There are a few works in which enzymes have been partly modeled by synthetic polyelectrolytes. Let us present the results obtained by Kabanov and co-workers [113-116].

Poly-4-vinylpyridines (PVP) partly alkylated by 2-(2'-chloroethyl)pyridine or by benzylchloride manifest catalytic activity in the hydrolysis of the ester bond, that is, 2.5-3.5 orders higher than the activity of low-molecular-weight analogs. A series of alkylated PVP of the type shown have been studied.

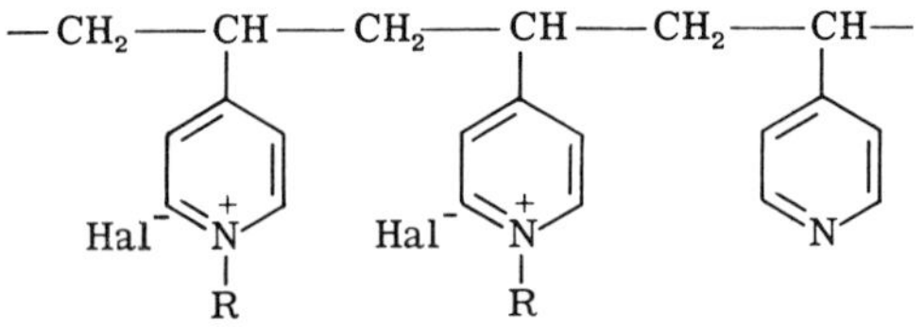


The following substrates have been hydrolyzed: p-nitrophenyl acetate (NPA), p-nitrophenyl trimethyl acetate (NPTMA), p-nitrophenyl benzoate (NPB), p-nitrophenyl cinnamate (NPC). The polymer portions act not as real catalysts but according to the scheme

$$E + S \underset{k_{-1}}{\overset{k_1}{\rightleftharpoons}} ES \xrightarrow{k_2} ES' + P$$

This means that the "catalyst" is changed in the reaction, forming the compound ES'. Hence, "substrate inhibition" has occurred Other alkyl derivatives of PVP act like genuine catalysts--the ES' complex is split. In both cases the kinetics of the hydrolysis could be described by the Michaelis-Menten equation (p. 357). The value $k_2/K_M$ varied from 4.5 to 64.0 liter mole $sec^{-1}$ (depending on the kind and content of the alkyl). On the other hand, the low-molecular-weight analog of PVP (4-ethylpyridine) has $k_2/K_M = 1.6\times10^{-2}$ liter mole $sec^{-1}$. The enhanced catalytic activity of the polymer is explained by the formation of active cavities in the macromolecular coil, made up of hydrophobic substituents. This is confirmed by some interesting facts. Catalytic activity has a pronounced maximum at a definite content $\alpha$ of nonalkylated pyridine rings. The value of $\alpha$ corresponding to the maximum increases with the degree of polymerization. At the same time, the viscosity of the polymer has a sharp minimum at the same value of $\alpha$. Obviously in such cases the polymer acquires the shape of an organized globule with a hydrophobic cavity inside it. A polymer catalyst has high specificity; thus, PV-benzyl-P hydrolyzes NPA 25 times faster than NPTMA and many orders faster than NPB and NPC. Pyridine rings surrounded by N-alkylated ones serve as active sites. The kinetic behavior of polymer catalysts resembles that of $\alpha$-chymotrypsin. The influence of the medium on globularization and hydrolysis have been studied and thermodynamic parameters of the elementary stages of hydrolysis have been estimated.

Kabanov *et al.* [113-116] presented the model of an enzyme acting as the medium for a reaction and made it possible to estimate the corresponding contribution. How can this contribution be evaluated theoretically?

A series of attempts have been made to construct a theory of ionic and ion-dipolar reactions in solutions based on Eyring's theory of absolute reaction rates [2,117-119]. This theory is essentially "gaseous" and in the references quoted here, the conventional vacuum was replaced by a continuous medium having a macroscopic value of the static dielectric permeability $\varepsilon$. All changes in rate constants are expressed through $\varepsilon$, whereas the medium polarizes, becomes reconstructed by changes in the electronic structure of the reagents. Reconstruction of the medium occurs much more slowly than changes in the electronic structure. First calculations of the rates of oxidation-reduction reactions in solutions taking into account the dynamic behavior of the medium were made by Marcus [120]. Later, the rate of electron exchange in polar media was calculated in a more rigorous way by means of quantum mechanics (in the works of Levitch, Dogonadze, Kusnetzov, and others [121-124]). The basic assumption of the theory is that all motions of particles taking part in the reaction, including that of the solvent, can be treated either classically or quantum mechanically. Classical motions possess frequencies $\omega << kT/\hbar$; quantum motions, $\omega >> kT/\hbar$. The activation energy of the process is determined by the height of the potential barrier for the classical subsystem, and the transmission coefficient $\kappa$ (p. 351) is determined by the ability of the quantum subsystem to tunnel through the barrier.

The state of the solvent can be described by a set of harmonic oscillators with coordinates $q_x$ and frequencies $\omega_x$. The dependence of kinetic parameters on the solvent properties can be determined if the complex dielectric permeability of the medium $\varepsilon(k, w)$ is known, where the dependence of $\varepsilon$ on the wave number k describes the spatial change in $\varepsilon$. Thus the energy of reorganization of the medium can be written as

$$E_s = \frac{1}{2} \sum_{x(\text{classical})} \hbar\omega_x (q_{x0}{}^j - q_{x0}{}^i)$$
$$= \frac{1}{8\pi} \sum_k (D_i(k) - D_j(k)) \frac{1}{\pi} \int_{(\text{classical})} \frac{\text{Im}\ \varepsilon(k,\omega)}{\omega|\varepsilon(k,\omega)|} d\omega \qquad (6.81)$$

where $q_{x0}{}^i$, $q_{x0}{}^j$ are the equilibrium values of normal coordinates before and after electronic transitions, $D_i(k)$ and $D_j(k)$ are the Fourier amplitudes of the electric induction vectors of the reagents and reaction products

$$D_{i,j} = \int D_{i,j}(r) e^{-ikr}\ dr \qquad (6.82)$$

In formula (6.81) summation and integration have to be performed over the frequencies of the classical degrees of freedom.

To calculate the activation energy of the simplest processes of electron transfer, the saddle point has to be found at the intersection surface of the terms of initial ($U_i$) and final ($U_j$) states.

$$U_i = I_i + \frac{1}{2} \sum_{(\text{classical})} \hbar\omega_x (q_x - q_{x0}{}^i)^2$$
$$U_j = I_j + \frac{1}{2} \sum_{(\text{classical})} \hbar\omega_x (q_x - q_{x0}{}^j)^2 \qquad (6.83)$$

The coordinate of the saddle point is given by the formula

$$q_x{}^* = \frac{1}{2} (q_{x0}{}^j - q_{x0}{}^i) + \frac{\Delta I}{2E_s} (q_{x0}{}^j - q_{x0}{}^i) \qquad (6.84)$$

where $\Delta I = I_j - I_i$ is the heat of reaction. The activation energy is

$$E^* = \frac{1}{2} \sum_{x(\text{classical})} \hbar\omega_x (q_x{}^* - q_{x0}{}^i) = \frac{(\Delta I + E_s)^2}{4E_s} \qquad (6.85)$$

and the entropy of activation (cf. [124])

$$S^* = k \ln (\hbar\omega_{eff}/kT) \qquad (6.86)$$

where

$$\omega_{eff}^2 = \frac{\Sigma_x \omega_x^3 (q_{x0}^i - q_{x0}^j)^2}{\Sigma_x \omega_x (q_{x0}^i - q_{x0}^j)^2} \tag{6.87}$$

The sums in (6.87) are taken over classical degrees of freedom. The transmission coefficient is

$$\kappa = \begin{cases} 1 & \text{if } |L|^2 \gg \hbar\omega_{eff}(kTE_s)^{1/2} \\ |L|^2 \left( \dfrac{4\pi}{\hbar^2 \omega_{eff}^2 kTE_s} \right)^{1/2} & \text{if } |L|^2 \ll \hbar\omega_{eff}(kTE_s)^{1/2} \end{cases} \tag{6.88}$$

where L is the electron exchange integral.

All of this applies to a polar medium. In the ESC, the reagents are in a nonpolar or partly nonpolar medium. However, the physical nature of the action of the medium on the reaction has to be the same. In the absence of a medium, the relative positions of electron levels of reagents do not change in time. Since electronic transition is practically instantaneous, according to the law of conservation of energy, the change in electron energy has to be compensated by a change in the kinetic energy of the reagents. But because nuclei have such large masses, an abrupt sizable change in their velocity has a low probability. Therefore, electronic transition can occur only if the initial and final levels $E_{red}^0$ and $E_{oxy}^0$ practically coincide. The positions of these levels depend on the dynamic state of the medium. The transition state corresponds to a dipole configuration of the medium and (or) a deformed ESC (enzyme) configuration in which the $E_{red}^0$ and $E_{oxy}^0$ levels coincide. These energies appear as functions of some generalized coordinates q characterizing the state of the medium. In the harmonic approximation, the Hamiltonian of the medium has the form

$$\mathcal{H} = -\frac{1}{2}\sum_x \hbar\omega_x \frac{\partial^2}{\partial q_x^2} + \frac{1}{2}\sum_x \hbar\omega_x (q_x - q_{x0})^2 + I \tag{6.89}$$

where $\omega_x$ and $q_x$ are the frequencies and coordinates of normal vibrations of the medium, and I is the minimum potential (or free) energy of the system. In general, the equilibrium configuration of the medium must be characterized by different conformational states $\alpha$ and different electronic states n of the active site. Taking ECI into account (electron-conformation interactions, p. 94), we have to write $\mathcal{H}$ in the form

$$\mathcal{H} = -\frac{1}{2}\sum_{x} \hbar\omega_x^{\alpha,n} \frac{\partial^2}{\partial q_x^{\,2}} + \frac{1}{2}\sum_{x} \hbar\omega_x^{\alpha,n}(q_x - q_{x0}^{\alpha,n})^2 + I \qquad (6.90)$$

and the probability of an elementary act of reaction is

$$W = \sum_{\alpha\alpha',nn'} \phi_{\alpha n}\, W_{\alpha n,\alpha' n'} \qquad (6.91)$$

where $\phi_{\alpha n}$ is the probability of the conformational state $\alpha$ and electronic state n of the medium before reaction, and $W_{\alpha n,\alpha' n'}$ is the probability of transition into state $\alpha' n'$. The electronic terms (potential energy surfaces) corresponding to different conformational and electronic states are represented schematically by Fig. 6.25 (electronic transition with no change in $\alpha$) and Fig. 6.26 (that with change in $\alpha$).

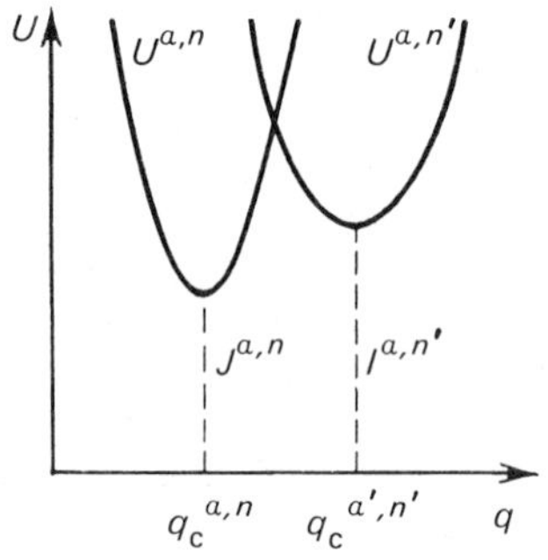


*FIG. 6.25 Electronic transition with no change in α.*

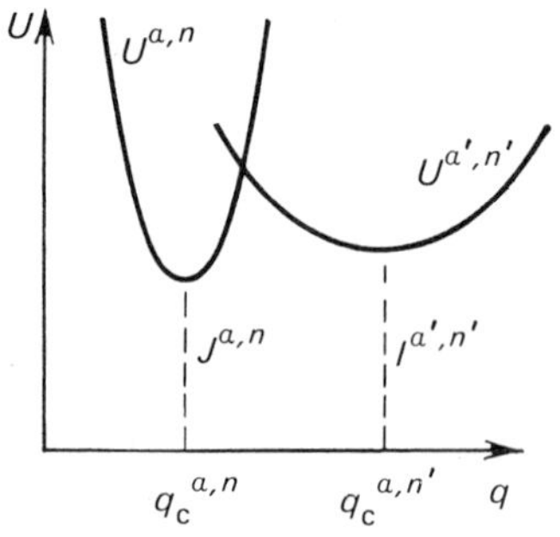


*FIG. 6.26 Electronic transition with change in α.*

The difference between a nonpolar medium and a polar one is that the transfer of an electron or a shift in the electron density can be accompanied by a considerable change in entropy. The frequencies characterizing the fluctuation of polarization do not change in the course of a reaction, but deformational acoustical frequencies can change considerably. Consequently, the reaction rate in nonpolar media can be greatly influenced by the entropy of activation. Because of the low values of acoustical frequencies in nonpolar media, the electronic subsystem always follows the deformational vibrations adiabatically, and the transmission coefficient $\kappa$ must be equal to 1 [125].

This discussion is meant to contribute a program for the development of a physical theory of enzymic catalysis based on an ECI approach (cf. [126]). These notions treat the role of acoustical vibrations (p. 394), of the "rack" (p. 395), and of the so-called complementarity effect from a physical standpoint. Lumry and Biltonen [127] suggest that the shape of conformational

*FIG. 6.27 Complementarity effect. (1) Chemical free energy; (2) conformational free energy; (3) sum of both free energies.*

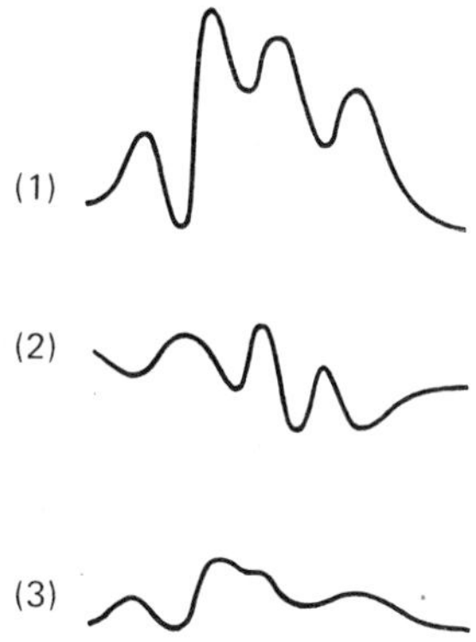


free energy curve along the reaction coordinate is complementary to that of the chemical (electronic) free energy (Fig. 6.27). Consequently, the activation barriers of the total curve are lowered.

Some information about ECI was provided in the studies of the conformational influence on the rate of slow isotope exchange of hydrogen in polynucleotides made by Varshavsky and co-workers [128]. The rate of exchange of hydrogen and tritium at C-8 of the purine ring in polyadenylic acid depends on the electronic density distribution in purine. This distribution is changed by conformational alterations produced by pH and by the formation of a complex with polyuridylic acid. The rate of isotope exchange is rather sensitive here.

The general sense of ECI is that electronic reconstruction in the course of an enzymic reaction is followed by conformational transition of the macromolecule. This situation can be described in the language of the physics of solids (cf., e.g., [129]). The displacement of an electron or of the electron density in a macromolecule produces a deformation of the lattice (i.e., a change in conformation). This deformation can be treated as excitation of the long-wavelength photons and the electron plus macromolecular deformation system becomes like a polaron. Let us call such a system a conformon [129]. The conformon energy is lower than the free electron energy. The reaction barrier is also lowered.

In contrast to a polaron, a conformon is not a real quasiparticle which can be displaced for great distances. Its energy dissipates rapidly because of the lack of homogeneity and periodicity in the globule structure. However, for the occurrence of an enzymic process, a change in conformation (i.e., the excitation of long-wavelength photons) in the range of a few peptide bonds is sufficient. Further conformational reconstruction can occur not as a direct result of ECI but due to the cooperativity of the macromolecule.

It seems natural to apply the conformon concept in treating the semiconductive properties of biopolymers widely quoted in the literature [130]. Evidently, conformational changes have to accompany electron displacement in a biopolymeric system, and therefore they must influence both the preexponential factor $\sigma_0$ and the effective width of the energy slit $\Delta E$ in the expression of conduction

$$\sigma = \sigma_0 \exp(-\Delta E/2RT)$$

However, the existence (and therefore the biological significance) of semiconductive properties in biopolymers is by no means established. The observed phenomena are not reproducible; they can be explained by ionic contaminations and other factors. As has already been said, all we know about the structure and properties of biopolymers contradicts the notion that they are semiconductive (p. 94).

On the other hand, the theory of the ionic conduction of membranes can employ the conformon concept. This will be discussed in "General Biophysics."

The ideas concerning ECI and the conformon can be applied in a direct form to oxidation-reduction enzymes--to cytochromes. Hypotheses have already been formulated explaining oxidative phosphorylation on the basis of ECI [131-133], and the electron exchange in cytochrome c has been calculated quantum mechanically [134].

Experimental information about ECI can be obtained by means of systematic studies of changes in the chemical (electronic) properties of biopolymers due to variations in ligands and cofactors and observation of accompanying conformational changes in the properties of the macromolecule as a whole. Such an investigation was made of the aspartate-aminotransferase (AAT)-ligand (cofactor) systems [135,147,148]. Denaturation in urea solution of an apoenzyme, a normal holoenzyme, and a series of complexes of enzymes with modified coenzymes has been studied. Equilibrium denaturation isotherms for these systems can be divided into two types of curves, with the value of the denaturation midpoint $C_m^*$, equal to ~5 and 7 M of urea. Systems with similar electronic structures, such as those containing C—O—$CH_3$ and C—H groups yield different curves. It is known that the O—$CH_3$ group inhibits the formation of an aldimine bond with AAT (p. 371). On the other hand, curves of the amino form and of holoenzyme reduced by $NaBH_4$ are practically identical, notwithstanding that in the first case the aldimine bond is absent and in the second it is rather strong. These results show directly that local rearrangements of electrons at the active site strongly influence the conformational stability of a protein. The same is attested by myoglobin studies (cf. Chapter 7).

Comparison of the AAT-catalyzed enzymic process with the

reaction in a congruent model system (p. 372) shows that in an enzymic system, the energy levels corresponding to various intermediate forms are equalized. Therefore, according to the Brönsted rule, the activation barriers have to be lowered. This is an expression of the complementarity of chemical (electronic) and conformational energy curves (p. 401). It can be theorized that changes in the conformational stability of a protein on the whole correlate with the conformational free energy of a multistage process. Intermediate forms of the AAT reaction are modeled by complexes of the holoenzyme with inhibitors which stop the reaction at different stages. The denaturation of such complexes has been studied. According to Tanford (p. 229), the absolute free energies of denaturation $\Delta G$ can be derived from the values of $C_m^*$. It has been established that the conformational stability differs at various stages in the process. The forms having the lowest chemical energy in a congruent system have the highest conformational energy. These results agree with the suggestion of complementarity achieved because of ECI [147,148].

## 6.8 Metalloenzymes

ECI is clearly expressed by the properties of metalloenzymes. Metals serve as the cofactors of many enzymes; most enzyme classes include metal-dependent enzymes. A metal ion acts directly on the conformational properties of an enzyme and is a convenient label for studies of the active site. Transient metals are characterized by informative electronic spectra, EPR and NMR spectra, etc. Many enzymes (carboxypeptidase, alkaline phosphatase, carboanhydrase, aldolase, etc.) contain a nontransient metal, Zn, as cofactor. However, Zn can be replaced by transient metals with the retention of enzymic activity. Thus, the peptidase activity of carboxyanhydrase A is preserved if Zn is replaced by Co, Mn, or Ni; esterase activity is preserved if Zn is replaced by Cd, Co, Hg, Mn, Ni, or Pb. The substitution of a metal is a "softer" process than many chemical modifications of proteins.

The action of metal on proteins is due to the fact that they are Lewis acids and catalysts of oxidation-reduction processes. In addition, metal ions support the structural organization of an enzyme.

A Lewis acid is an electron pair acceptor, in contrast to a Brönsted acid, which is a proton donor. According to Brönsted, a base is a proton acceptor. Thus in reaction

$$H_3N: + HA \rightleftharpoons NH_4^+ + A^-$$

$NH_3$ is a Brönsted base.  In the reaction

$$
\begin{array}{ccc}
F & H & \\
| & | & \\
F-B \ + & :N-H & \rightleftharpoons \\
| & | & \\
F & H &
\end{array}
\quad
\begin{array}{cc}
F & H \\
| & | \\
F-B:N-H \\
| & | \\
F & H
\end{array}
$$

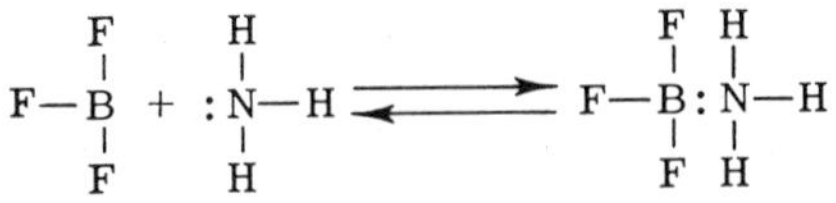

$BH_3$ is a Lewis acid [1].  Metal ions in aqueous solution are acids, and complex ions, such as $Fe(NO)^{2+}$, $Cr(H_2O)_6^{3+}$, and $AlF_6^{3-}$, can be considered acid-base complexes.

Brönsted proposed an empirical relation which characterizes acid-base catalysis [136].  This process amounts to hydrogen transfer.  The catalytic strength of acids is related to their ionization constant.  The chemical equation of acid-base catalysis has the form

$HA + B \rightleftharpoons A^- + HB^+$

where HA, $HB^+$ are the Brönsted acids; B and $A^-$, bases.  The Brönsted equation for an acid is

$$k_{HA} = G_A (K_{HA})^{\alpha} \tag{6.92}$$

and that for a base is

$$k_B = G_B (K_B)^{\beta} \tag{6.93}$$

Here $k_{HA}$ and $k_B$ are second-order catalytic rate constants; $K_{HA}$ and $K_B$ are the corresponding ionization constants.  $G_A$, $G_B$, $0 < \alpha$, and $\beta < 1$ are constants.  Hence

$$\log k_{HA} = \log G_A + \alpha \log K_{HA} \tag{6.94}$$

$$\log k_B = \log G_B + \beta \log K_B \tag{6.95}$$

Figure 6.28 shows the positions of divalent metal ions on a graph of the catalytic constant of the decomposition of acetone dicarboxylate versus the stability constant for malonate metal complexes [137].  The sequence of catalytic strengths of divalent ions in many model reactions is

Ca < Mg < Mn < Fe < Co $\leq$ Zn $\leq$ Ni < Cu

This order, which prevails in nonenzymic catalysis, is disrupted in metalloenzymes [138].  Figure 6.69 shows the relative rates for metalloenzyme-catalyzed reactions and the rates for an idealized model system.  This disturbance of the conventional order can be due to binding of the metal by various functional groups of the protein, probably in this way:

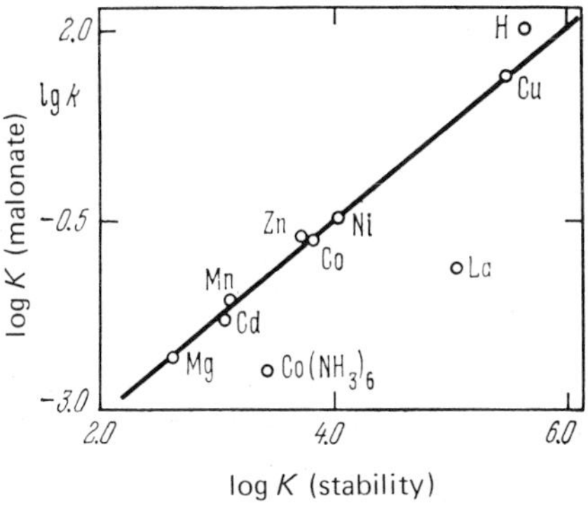


*FIG. 6.28 Correlation of the catalytic constant for the destruction of acetone decarboxylate and the stability constant of malonate complexes of metals.*

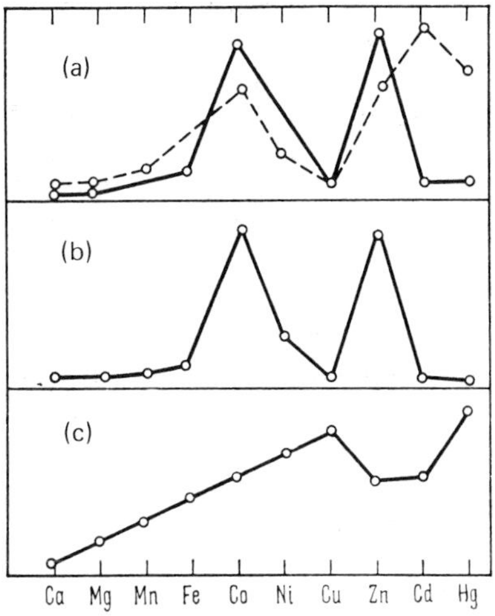


*FIG. 6.29 Rates of reactions catalyzed by metalloenzymes and of a model system: (a) —— carboxypeptidase acting as peptidase; --- carboxypeptidase acting as esterase; (b) carboanhydrase; (c) idealized model system.*

| $—O^-$ | $\equiv N,\ —O^-$ | $\equiv N,\ S^-$ | $—S^-$ |
|---|---|---|---|
| Ca,Mg,Mn | Fe,Co,Ni | Cu,Zn | Cd,Pb,Hg |

True metalloenzymes are characterized by an exact localization of the metal ion at the active site; this position is not changed by the substitution of another metal. In the carboxypeptidase case, the activity of Zn and Co is especially great. The binding constants preserve the usual order for model systems:

Hg > Cu > Ni > Co > Fe > Mn $\geq$ Zn

Consequently, anomalies of catalytic strength are due not to differences in metal binding but to individual features of the metal's physical state at the active site. These features are also expressed by the chemical behavior. Halogens combine with the free Zn ion in the sequence $F^- > Cl^- > Br^- > I^-$. The binding of halogen by Zn in carboxypeptidase [139,140], which inhibits the enzyme, occurs in the reverse sequence $I^- > Br^- > Cl^- > F^-$, and the binding constant of iodine is 1000 times larger than that of fluorine in the case of free Zn. Analysis of reactions with other anions shows that the binding constant is determined by the conformational fit of the anion and the cavity in the protein molecule in the region of the Zn atom [138].

Valley and Williams formulated a statement that agrees with all that has been said about enzymes in the previous sections. Enzymes are catalysts not because they contain some unusual chemical group, but because a specific three-dimensional structure produces the unusual properties of the groups [138]. Also unusual are the spectral properties of metalloenzymes. Table 6.5 presents the spectral characteristics of enzymes containing Cu, nonhemine Fe, and Co in comparison with some model complex compounds [141].

*TABLE 6.5*
*Absorption Bands of Cu, Fe, and Co Complexes and Enzymes*

| Substance | λ, nm (molar absorption, $M^{-1}cm^{-1}$) | Symmetry |
|---|---|---|
| Simple Cu(II) complexes | | |
| $[Cu(NH_3)_4]^{2+}$ | 600(∿20) 660(∿40) 750(∿25) | Tetragonal |
| Bis(3-phenyl-2,4-pentaneditionate) Cu(II) | 490(∿50) 520(∿25) 580(∿50) 650(∿50) | Second order axis |
| $[Cu(H_2O)_2]^{2+}$ | 700(∿5) 790(∿10) 950(∿5) | Tetragonal |
| Protein-Cu complexes | | |
| Cu(II) carboanhydrase | 760(broad) (120) | Unknown low symmetry |
| Cu(II) serum albumin | 570(broad) (90) | Unknown low symmetry |

TABLE 6.5 (cont.)

| Substance | λ, nm (molar absorption, $M^{-1}cm^{-1}$) | Symmetry |
|---|---|---|
| "Inert" Cu- proteins | | |
| Erythrocruorin | 650(broad) (280) | Unknown symmetry, possible tetragonal |
| Cerebrocuprein | 655(broad) (400) | |
| "Active" Cu proteins | | |
| Plastocyanine | 460(500) 597(4400) 770(1600) | Unknown low symmetry |
| Laccase | 450(970) 608(4000) 850(700) | |
| Blue protein of *Pseudomonas aeruginosa* | 500(weak) 625(3000) 725(weak) | |
| Blue protein of *Pseudomonas denitrificans* | 450(weak) 594(3000) 800(1000) | |
| Fe complexes | | |
| $[Fe(CN)_6]^{3-}$ | General rise of absorption | Octahedral |
| $[Fe(H_2O)_5OH]^{2+}$ | with one or two discrete bands | |
| Ferredoxin | 325(5160) 520(5160) 463(4650) | ? |
| Co complexes | | |
| $[Co(H_2O)_6]^{2+}$ | 510(10) 1200(2) | Octahedral |
| $[CoCl_4]^{2-}$ | 685(700) 1700(100) | Tetrahedral |
| Co-carboanhydrase | 510(280) 550(380) 615(300) 640(280) | Nonregular tetrahedron with coordination 5 |
| Co-alkaline phosphatase | 515(265) 555(350) 605(180) 640(230) | |

The special properties of metals in enzymes is also shown by unusual EPR spectra parameters [141]. These data show the specificity of the electron shells of metals in enzymes. Cytochrome spectra also differ from those of model compounds [142]. In many cases, the change in the coordination sphere due to ligand binding transforms the unusual spectra into spectra that are typical for symmetrical complexes.

The oxidation-reduction potentials of metalloenzymes are also unusual. For instance, the potentials of "blue copper enzymes" are much higher than those of all copper complexes except those with strongly disturbed tetragonal symmetry (from 0.3 to 0.4 V in comparison with -0.5 to +0.2 V). The values of these unusual potentials are especially important in the case of cytochromes.

The biological functions of metalloenzymes are related to either electron transfer or hydrolysis. It can be thought that the peculiar electronic state of a metal has a direct relation to these functions. Vallee and Williams suggest that the metals in metalloenzymes are in an entatic (i.e., stressed) state [138, 141]. An enzyme possesses a region with energy which is nearer to the transition state of a monomolecular reaction than to the usual state of stable molecule. Electron transfer with the participation of a complex ion requires a geometry intermediate between those for two different valence states. This means that the activated complex must have a disturbed symmetry. Usually $Cu^{+}$ forms tetrahedral and $Cu^{2+}$ trigonal complexes. In an entatic state the structure of a complex has to be the intermediate one. Apparently precisely such a structure occurs in "blue enzymes." On the other hand, the symmetry of $Fe^{3+}$ and $Fe^{2+}$ complexes is octahedral but the bond lengths are different. Apparently there is some elongation of bonds for $Fe^{3+}$ in cytochrome.

In connection with these ideas we have to recall the work of Pauling [143] in which the suggestion was first made that an enzyme binds especially strongly to a substrate in its transition state. Therefore the activation barrier has to be lowered.

If a metal participates in the transfer of atoms or of some group of atoms, ligand substitution is necessary. Active intermediate compounds in substitution reactions of single chelate compounds of metals presumably possess free coordination bonds or a disturbed coordination sphere. Apparently such structures are realized in Co(II) carboanhydrase, carboxypeptidase, and phosphatase previous to substrate binding. Special activity of $Co^{2+}$ and $Zn^{2+}$ in these enzymes is due to the easier deformation of their complexes in comparison with the complexes of the metals of the transient series.

These peculiarities express the stresses acting on the metal ion because of the multidentate interaction with amino acid residues determined by the conformational structure of the protein. Unusual positions of the ligands form an oriented geometry of the complex. "Enthasis" is produced by both metal and ligands. Metal-ligand interaction in metalloenzymes is similar to the interactions of the active groups His, Ser, Tyr, and -SH of a series of enzymes with protons, which are the simplest cations.

In the active region of a metalloenzyme, the substrate is subjected to the attack of unusually activated groups and consequently the activation energy of the process is lowered [144].

The enthasis concept proposed by Vallee and Williams agrees with the ideas concerning ECI (p. 401). The metal ion changes the enzyme conformation and the conformational state of the enzyme determines the electronic state of the ion. Vallee and

Williams examined the static picture arising because of ECI (the "rack" model) (p. 394).

It can be concluded that a special development of quantum chemistry of coordination compounds is necessary in applications to biophysical problems. In particular, the theory of the ligand field (cf. [145]) has to be applied for the systems with disturbed symmetry.

Metal-activated enzymes, and primarily the ATP-ases activated by ions of alkali and alkaline-earth metals, require special consideration; K- and Na-activated ATP-ase is responsible for the active transport phenomena in biological membranes, Ca- and Mg-activated ATP-ase determines mechanochemical processes in biological contractile systems, particularly in muscle. In both cases the ATP splitting catalyzed by ATP-ase serves as the source of necessary energy. For further details see [146].

Bioinorganic chemistry, that is, the physical chemistry of the metal-containing proteins, is becoming a very important field of knowledge.

## References


1. S. Benson, "The Foundations of Chemical Kinetics." McGraw-Hill, New York, 1960.
2. S. Glasstone, K. Laidler, and H. Eyring, "The Theory of Rate Processes." McGraw-Hill, New York, 1941.
3. M. Volkenstein, "Structure and Physical Properties of Molecules." Ed. Acad. Sci. USSR, Moscow, 1955 (R).
4. L. Landau, Phys. Z. Sowjetun. 1, 88 (1932); 2, 46 (1932).
5. L. Landau and E. Lifshitz, "Quantum Mechanics." Addison-Wesley, Reading, Massachusetts, 1958.
6. C. Zener, Proc. Roy. Soc. **A137**, 696 (1932).
7. M. Temkin, Appendix 1 in N. Semenov, "About Some Problems of Chemical Kinetics and Reactivity." Ed. Acad. Sci. USSR, Moscow, 1954 (R).
8. M. Temkin, Acta Physiochim. USSR 8, 141 (1938).
9. A. Balandin, "Multiplet Theory of Catalysis." Nauka, Moscow, 1970 (R).
10. J. Germain, "Catalyse Hétérogène." Dunod, Paris, 1959.
11. C. Walter, "Enzyme Kinetics." Ronald Press, New York, 1966.
12. J. Romanovsky, N. Stepanova, and D. Tshernavsky, "What is Mathematical Biophysics." Prosveshtshenie, Moscow, 1971 (R).
13. N. Bogoljubov and J. Mitropolsky, "Asymptotic Methods in the Theory of Non-linear Vibrations." Physmathgiz, Moscow, 1958 (R).
14. V. Jakovlev, "Kinetics of Enzymatic Catalysis." Nauka, Moscow, 1965 (R).
15. M. Dixon and E. Webb, "Enzymes." Longmans, Green,

New York, 1958.
16. J. Webb, "Enzyme and Metabolic Inhibitors," Vol. 1, Academic Press, New York, 1963.
17. H. Lineweaver and D. Burk, J. Amer. Chem. Soc. **56**, 658 (1934).
18. H. Bray and K. White, "Kinetics and Thermodynamics in Biochemistry." Churchill, London, 1957.
19. H. Gutfreund, Discuss. Faraday Soc. **20**, 167 (1955).
20. R. Alberty, W. Miller, and H. Fisher, J. Amer. Chem. Soc. **79**, 3973 (1957); J. Phys. Chem. **62**, 154 (1958).
21. R. Lumry, in "The Enzymes" (P. Boyer, H. Lardy , and K. Myrbäck, eds.), Vol. 1, 2nd ed., Chapter 4, Academic Press, New York, 1959.
22. I. Berezin and K. Martinek, J. All-Un. Mendeleev Chem. Soc. **16**, N 4, 411 (1971) (R).
23. L. Blumenfeld, Biophysica **16**, 724 (1971) (R).
24. J. Labedev, J. Tsvetkov, and V. Vojevodsky, Kinet. Catal. **1**, 496 (1960) (R).
25. G. Lichtenstein, Biophysica **11**, 23 (1966) (R).
26. R. Lumry and Sh. Rajender, in "Water Solutions of Proteins and Small Molecules." Wiley (Interscience), New York, 1970.
27. F. Vaslow and D. Doherty, J. Amer. Chem. Soc. **74**, 931 (1952); 75, 928 (1953).
28. B. Belleau and J. Lavoie, Can. J. Biochem. **46**, 1397 (1968)
29. V. Kretovitsh, "Introduction to Enzymology." Nauka, Moscow 1967 (R).
30. E. Goriatshenkova, in "Fundamentals of Molecular Biology. Enzymes," Chapter 2, Nauka, Moscow, 1964 (R).
31. O. Polianovsky, in "Fundamentals of Molecular Biology. Enzymes," Chapter 4, Nauka, Moscow, 1964 (R).
32. D. Nachmansohn, "Chemical and Molecular Basis of Nerve Activity." Academic Press, New York, 1959.
33. F. Westheimer, in "The Enzymes," Vol. 1, 2nd ed. Academic Press, New York, 1959.
34. A. Braunstein, M. Karpeisky, and R. Khomutov, in "Fundamentals of Molecular Biology. Enzymes," Chapter 9, Nauka, Moscow, 1964 (R).
35. A. Braunstein, in "Actual Problems of Modern Biochemistry." Medgiz, Moscow, 1962 (R).
36. C. Blake, L. Johnson, G. Mair, A. North, D. Phillips, and V. Sarma, Proc. Roy. Soc. **B167**, 365, 378 (1967).
37. C. Vernon, Proc. Roy. Soc. **B167**, 389 (1967).
38. N. Sharon, Proc. Roy. Soc. **B167**, 402 (1967).
39. T. Steitz, M. Ludwig, F. Quiocho, and W. Lipcomb, J. Biol. Chem. **242**, 4662 (1967).
40. A. Braunstein and M. Shemiakin, Biochimia **18**, 393 (1953) (R).
41. A. Braunstein, in "The Enzymes," Vol. 2, 2nd ed. Academic

Press, New York, 1960.
42. D. Metzler, M. Ikawa, and E. Snell, J. Amer. Chem. Soc. 26, 648 (1954).
43. E. Snell and W. Jenkins, J. Cell. Comparat. Physiol. 54, 161 (1959).
44. E. Kosower, "Molecular Biochemistry." McGraw-Hill, New York, 1962.
45. D. Auld and T. Bruice, J. Amer. Chem. Soc. **89**, 2098 (1967).
46. J. Thanassi, A. Butler, and T. Bruice, Biochemistry **4**, 1463 (1965).
47. T. French, D. Auld, and T. Bruice, Biochemistry **4**, 77 (1965).
48. T. Bruice and S. Benkovic, J. Amer. Chem. Soc. **86**, 418 (1964).
49. V. Ivanov and M. Karpeisky, Advan. Enzymol. **32**, 21 (1969).
50. J. Tortshinsky and L. Koreneva, Biochimia **28**, 1087 (1963) (R).
51. V. Ivanov, J. Breusov, M. Karpeisky, and O. Polianovsky, Mol. Biol. **1**, 588 (1967) (R).
52. A. Bocharov, V. Ivanov, M. Karpeisky, O. Malaeva, and A. Kuklin, Biochem. Biophys. Res. Commun. **30**, 459 (1968).
53. J. Gass and A. Meister, Biochemistry **9**, 1380 (1970).
54. S. Bresler, Int. Biochem. Congr., 5th, Symp., 1962.
55. E. Smith et al., J. Biol. Chem. **240**, 254 (1965).
56. J. Chevallier et al., Biochem. Biophys. Acta **92**, 521 (1964).
57. A. Braunstein, J. All-Un. Mendeleev Chem. Soc. **8**, N 1, 81 (1963) (R).
58. C. Swain and J. Brown, J. Amer. Chem. Soc. **74**, 2533 (1952).
59. D. Koshland and K. Neet, Ann. Rev. Biochem. **37**, 359 (1968).
60. D. Koshland, J. Theor. Biol. **2**, 85 (1962).
61. D. Koshland, Advan. Enzymol. **22**, 45 (1960).
62. D. Koshland and D. Storm, Proc. Nat. Acad. Sci. U.S. **66**, 445 (1970).
63. F. Karush, J. Amer. Chem. Soc. **72**, 2705 (1950).
64. M. Volkenstein, "Molecules and Life." Plenum Press, New York, 1970.
65. A. Braunstein and M. Karpeisky, J. All-Un. Mendeleev Chem. Soc. **16**, N 4, 362 (1971) (R).
66. F. Vaslow and D. Doherty, J. Amer. Chem. Soc. **75**, 928 (1953).
67. A. Balandin, Dokl. Acad. Sci. USSR **114**, 1008 (1957); Biochimia **83**, 475 (1958) (R).
68. M. Volkenstein, "Enzyme Physics." Plenum Press, New York, 1969.
69. D. Koshland, in "The Enzymes," Vol. 1, 2nd ed., p. 7. Academic Press, New York, 1959.
70. D. Koshland, Proc. Nat. Acad. Sci. U.S. **44**, 98 (1958); Cold Spring Harbor Symp. Quant. Biol. **28**, 473 (1963).

71. D. Koshland, J. Yankeelov, and J. Thoma, Fed. Proc. 21, 1031 (1962).
72. M. Burr and D. Koshland, Proc. Nat. Acad. Sci. U.S. 52, 1017 (1964).
73. D. Koshland, Y. Karkhanis, and H. Latam, J. Amer. Chem. Soc. **86**, 1448 (1964).
74. T. Bruice, A. Brawn, and D. Harris, Proc. Nat. Acad. Sci. U.S. **68**, 658 (1971).
75. S. Grisolia and B. Joyce, Biochem. Biophys. Res. Commun. **1**, 280 (1959).
76. G. Tomkins, K. Yielding, and J. Curban, Proc. Nat. Acad. Sci. U.S. **47**, 270 (1961).
77. H. Schachman, Brookhaven Symp. Theor. Biol. **17**, 91 (1964).
78. M. Citri and N. Garber, Biochem. Biophys. Res. Commun. **4**, 143 (1961).
79. G. Ning Ling, Biopolym. Symp. **1**, 91 (1964).
80. W. Jencks, Ann. Rev. Biochem. **32**, 639 (1963).
81. D. Koshland, in "Horizons in Biochemistry" (M. Kasha and B. Pullman, eds.). Academic Press, New York, 1962.
82. J. Wooten and G. Hess, J. Amer. Chem. Soc. **84**, 440 (1962).
83. B. Laboesse, B. Havsteen, and G. Hess, Proc. Nat. Acad. Sci. U.S. **48**, 2137 (1962).
84. G. Fasman, K. Norland, and A. Pesce, Biopolymers, Symp. **1**, 325 (1964).
85. J. Yankeelov and D. Koshland, J. Biol. Chem. **240**, 1593 (1965).
86. J. Torchinsky, in "Uspekhi Biologitcheskoj Chimii," Vol. 8, Nauka, Moscow, 1967 (R).
87. I. Bolotina, M. Volkenstein, P. Zavodsky, and D. Markovitch Biochimia **31**, 649, 873 (1966) (R); Biochem. Biophys. Acta **132**, 260, 271 (1967); Mol. Biol. **1**, 231 (1967) (R).
88. M. Perutz, Eur. J. Biochem. **8**, 455 (1969).
89. G. Kartha, J. Bello, and D. Harker, Nature (London), **213**, 862 (1967).
90. H. Wykoff, K. Hardman, N. Allewell, T. Inagami, L. Johnson, and F. Richards, J. Biol. Chem. **242**, 3984 (1967).
91. J. Dreuth, J. Jansonius, R. Kockouk, H. Swen, and B. Welters, Nature (London), **218**, 929 (1968).
92. B. Matthews, P. Sigler, R. Headerson, and D. Blow, Nature (London), **214**, 652 (1967); J. Mol. Biol. 35, 143 (1968).
93. D. Blow, J. Birletoft, and B. Hartley, Nature (London), **221**, 337 (1969).
94. G. Reeke, J. Hartsuch, M. Ludwig, F. Quiocho, T. Steits, and W. Lipscomb, Proc. Nat. Acad. Sci. U.S. 58, 2220 (1967).
95. W. Lipscomb et al., Brookhaven Symp. Theor. Biol. 21,
96. 250 (1968).
97. L. Michaelis and H. Davidsohn, Biochem. Z. 35, 386 (1911).
J. Kirkwood and J. Shumaker, Proc. Nat. Acad. Sci. U.S. 38, 863 (1952).

98. J. Kirkwood, Discuss. Faraday Soc. 20, 78 (1955).
99. W. Scheider, Biophys. J. 5, 617 (1965).
100. J. Kirkwood, Symp. Mech. Enzyme Action, Johns Hopkins Univ. Press, Baltimore, Maryland, 1954.
101. M. Volkenstein and S. Fishman, Biophysica **11**, 956 (1966); **12**, 14 (1967) (R).
102. M. Joly, "A Physico-Chemical Approach to the Denaturation of Proteins." Academic Press, New York, 1965.
103. M. Eigen, in "Fast Reactions and Primary Processes in Chemical Kinetics" (Nobel Symp. 5th)(S. Claesson, ed.), Wiley (Interscience), New York, 1967.
104. K. Linderstrøm-Lang and J. Schellmann, in "The Enzymes," Vol. 1, 2nd ed., Chapter 10. Academic Press, New York, 1959.
105. F. Straub and G. Szabolcsi, in "Molecular Biology. lems and Perspectives." Nauka, Moscow, 1964 (R).
106. K. Okunuki, Advan. Enzymol. **23**, 29 (1961).
107. G. Szabolcsi and E. Biszku, Biochim. Biophys. Acta **48**, 335 (1961).
108. N. Slater, "Theory of Unimolecular Reactions." Cornell Univ. Press, Ithaca, New York, 1959.
109. M. Volkenstein, in "Molecular Biology. Problems and Perspectives." Nauka, Moscow, 1964 (R).
110. J. Khurgin, D. Tshernavsky, and S. Schnoll, in "Vibrational Processes in Biological and Chemical Systems." Nauka, Moscow, 1967 (R).
111. M. Volkenstein, L. Gribov, M. Eliashevitch, and B. Stepanov, "Vibrations of Molecules." Nauka, Moscow, 1972 (R).
112. M. Perutz, Proc. Roy. Soc. **B167**, 448 (1967).
113. J. Kirsh, S. Plujnov, T. Shomina, V. Kabanov, and V. Kargin, Vysokomol. Sojedin. **A12**, 186 (1970) (R).
114. N. Vengerova, J. Kirsch, V. Kabanov, and V. Kargin, Dokl. Acad. Sci. USSR **190**, 131 (1970) (R).
115. J. Kirsh, L. Bessmertnaja, V. Tortchilin, I. Papisov, and V. Kabanov, Dokl. Acad. Sci. USSR **191**, 603 (1970) (R).
116. V. Kabanov, J. All-Un. Mendeleev Chem. Soc. **16**, N 4, 446 (1971) (R).
117. G. Scatchard, J. Chem. Phys. **7**, 657 (1939).
118. E. Amis, "Solvent Effects on Reaction Rates and Mechanisms." Academic Press, New York, 1966.
119. E. Caldin, "Fast Reactions in Solution." Oxford Univ. Press, London and New York, 1964.
120. R. Marcus, J. Chem. Phys. **24**, 966 (1956); **26**, 867, 872 (1957); **43**, 697 (1965).
121. V. Levitch, in "Itogi Nauki. Elektrochimia 1965," (VINITI, ed.), Moscow, 1967 (R).
122. R. Dogonadze and A. Kusnetsov, in "Itogi Nauki. Elektrochimia 1967," (VINITI, ed.), Moscow, 1969 (R).

123. R. Dogonadze, A. Kusnetsov, and V. Levitch, Dokl. Acad. Sci. USSR **188**, 383 (1969) (R).
124. V. Levitch, R. Dogonadze, E. German, A. Kusnetsov, and J. Kharkatz, Electrochim. Acta **15**, 353 (1970).
125. M. Volkenstein, R. Dogonadze, A. Madumarov, Z. Urushadze, and J. Kharkatz, Mol. Biol. **6**, 431 (1972) (R).
126. M. Volkenstein, Izv. Acad. Sci. USSR Ser. Biol. N 6, 805 (1971) (R).
127. R. Lumry and R. Biltonen, in "Structure and Stability of Biological Macromolecules," (S. Timasheff and G. Fasman, eds.). Dekker, New York, 1969.
128. R. Maslova, E. Lesnik, and J. Varshavsky, Mol. Biol. **3**, 728 (1969) (R); FEBS Lett. **3**, 211 (1969); Biochem. Biophys. Res. Commun. **34**, 260 (1969).
129. M. Volkenstein, J. Theoret. Biol. **34**, 193 (1972).
130. G. Kemeny and J. Goklany, J. Theoret. Biol. **40**, 107 (1973); **48**, 23 (1974).
131. D. Green, Proc. Nat. Acad. Sci. U.S. **67**, 544 (1970).
132. N. Tshernavskaja, D. Tshernavsky, and A. Grigorov. Preprint N 68. Phys. Inst. Acad. Sci., USSR, Moscow, 1970.
133. L. Blumenfeld and V. Koltover, Mol. Biol. **6**, 161 (1972) (R).
134. L. Blumenfeld, "Problems of Biological Physics." Nauka, Moscow, 1974 (R).
135. V. Ivanov, A. Botcharov, M. Volkenstein, M. Karpeisky, L. Lagutina, and E. Okina, Eur. J. Biochem. **40**, 519 (1973).
136. J. Bronsted and K. Pedersen, Z. Phys. Chem. **108**, 185 (1924)
137. J. Prue, J. Chem. Soc. 2331 (1952).
138. B. Vallee and R. Williams, Chem. in Britain **4**, 397 (1968).
139. S. Lindskod, J. Biol. Chem. **238**, 945 (1963).
140. J. Coleman, Nature (London) **214**, 193 (1967).
141. B. Vallee and R. Williams, Proc. Nat. Acad. Sci. U.S. **59**, 498 (1968).
142. R. Lemberg, A. Ehrenberg, and R. Williams, in "Hemes and Hemeproteins" (B. Chance et al., eds.). Academic Press, New York, 1966.
143. L. Pauling, Chem. Eng. News **24**, 1375 (1946)
144. T. Bruice and S. Benkovic, "Bioorganic Mechanisms." Vol. 1, Benjamin, New York, 1966.
145. C. Ballhausen, "Introduction to Ligand Field Theory." McGraw-Hill, New York, 1962.
146. V. Gorkin, in "Fundamentals of Molecular Biology. Enzymes. Nauka, Moscow, 1964 (R).
147. M. Volkenstein, Pure Appl. Chem. 36, 9 (1973).
148. V. Ivanov, A. Makarov, and M. Volkenstein, Mol. Biol. **8**, 433 (1974) (R).

Chapter 7

# Cooperative Properties of Enzymes

## 7.1 Structure and Properties of Myoglobin and Hemoglobin

This chapter is devoted to those specific properties of enzymes with quaternary structure which are due to interactions of the subunits of a protein molecule.

The function of myoglobin (Mb) and hemoglobin (Hb), which are not enzymes, consists in the reversible binding of molecular oxygen. Mb serves as a depot, storing oxygen for later consumption. Whales, which spend long periods under water, typically contain large amounts of Mb. Hb, the functional protein in erythrocytes, transfers oxygen from the lungs to all organs and tissues and provides the reverse transport of $CO_2$.

However, the study of Mb and Hb gives valuable information about enzyme activity, thereby aiding our understanding of the nature of ECI (electron conformation interactions). The binding of $O_2$ and other ligands by these proteins is quite similar to the binding of substrates by enzymes. $O_2$ penetrates the cavity of an Mb or Hb molecule, but unlike a substrate this oxygen does not undergo chemical transformation. Mb and Hb are considered "honorary enzymes" [1], and are models for a series of enzyme properties.

The proteins Mb and Hb can be obtained in crystalline form comparatively easily, which makes them convenient substances for structural investigations. As we said earlier, both proteins have been studied in detail by X-ray structural analysis with resolution up to 2.8 Å in both oxygenated ($MbO_2$, $HbO_8$) and deoxygenated (Mb, Hb) forms (p. 265). The presence in these proteins of the prosthetic heme group, which has peculiar electronic properties, makes it possible to study Hb and Mb by means of spectroscopy, EPR, magnetic susceptibility, the Mössbauer effect (p. 418), etc.

As we said (p. 216), the Hb molecule contains four subunits, each of which is similar, but not identical, to Mb. Consequently,

Hb possesses cooperative properties due to the so-called heme-heme interaction. Mb lacks these properties.

The binding of $O_2$ and other ligands by Mb and Hb occurs at the heme group. Ligands saturate the sixth coordination valence of the iron atom in heme. The formation of this chemical bond provokes a series of events in the protein molecule.

Heme is ferroprotoporphyrin (Fig. 2.16, page 83). The iron atom in the divalent ($Fe^{2+}$) or ferro state is bound to four nitrogen atoms in the pyrrole groups of the planar porphyrin ring. The fifth coordination bond directed perpendicular to the ring plane links the Fe atom to the imidazole of histidyl; the sixth valence is either free or occupied by a ligand.

Both the oxygenated and the nonoxygenated forms of Hb and Mb contain ferroheme. If oxidized, Hb and Mb form ferri compounds containing the trivalent ($Fe^{3+}$) iron atom. The Fe atom is paramagnetic. Table 7.1 presents some magnetic properties

*TABLE 7.1*
*Magnetic Properties of Hemoglobin*

| Compound | Valence of iron | Sixth ligand | Magnetic moment[a] | Spin | EPR spectrum[b] |
|---|---|---|---|---|---|
| Deoxyhemoglobin (Hb) | 2+ | none | 5.2-5.5 | 2 | none |
| Oxyhemoglobin ($HbO_8$) | 2+ | $O_2$ | 0 | 0 | none |
| Carboxyhemoglobin [$Hb(CO)_4$] | 2+ | CO | 0 | 0 | none |
| Ferrihemoglobin | 3+ | $H_2O$ | 5.6-5.8 | 5/2 | $g_{\parallel}=2, g_{\perp}=6$ |
| Methemoglobin | 3+ | $H_2O$ | 5.6-5.8 | 5/2 | $g_{\parallel}=2, g_{\perp}=6$ |
| Acidic hemoglobin | 3+ | $H_2O$ | 5.6-5.8 | 5/2 | $g_{\parallel}=2, g_{\perp}=6$ |
| Ferri Hb hydroxide | 3+ | $OH^-$ | 4.5-4.7 | 1/2,5/2 | |
| Hydroxyhemoglobin | 3+ | $OH^-$ | 4.5-4.7 | 1/2,5/2 | |
| Basic methemoglobin | 3+ | $OH^-$ | 4.5-4.7 | 1/2,5/2 | |
| Ferri Hb fluoride | 3+ | | 5.8-5.9 | 5/2 | $g_{\parallel}=2, g_{\perp}=6$ |
| Ferri Hb azide | 3+ | | 2.4-2.8 | 1/2 | $g_x=1.72, g_y=2.21$ |
| Ferri Hb cyanide | 3+ | | 2.3-2.5 | 1/2 | $g_z=2.80$ |

[a] In Bohr magnetons.

[b] $g_{\parallel}$ is the factor in the direction of the z axis perpendicular to the heme x,y plane, $g_{\perp}$ the factor in that plane.

of Hb in various states. We see that ferroheme can be in high-spin (Hb) and low-spin [$HbO_8$, $Hb(CO)_4$] states. The ligand binding is expressed by absorption spectra in the visual region (Fig. 7.1). Table 7.2 presents corresponding data.

*FIG. 7.1 Absorption spectra of (1) Hb and (2) $HbO_8$.*

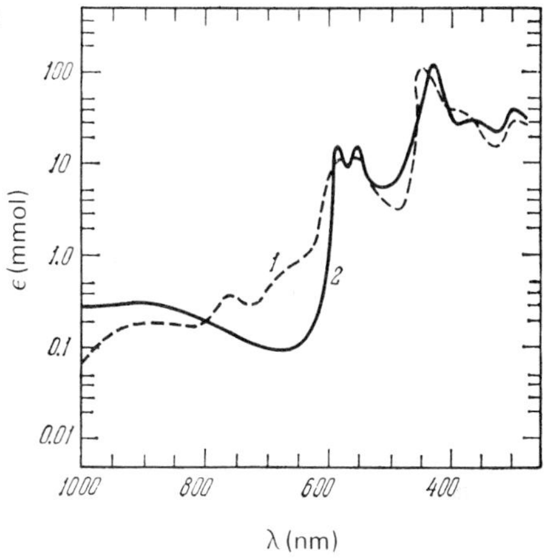


*TABLE 7.2*
*Absorption Spectra of Some Heme Compounds[a]*

| | α Band | | β Band | | γ Band (Soret band) | |
|---|---|---|---|---|---|---|
| Compound | λ (nm) | $\varepsilon \times 10^{-3}$ | λ (nm) | $\varepsilon \times 10^{-3}$ | λ (nm) | $\varepsilon \times 10^{-3}$ |
| Heme | 565 | 6.1 | | | 390 | 39.6 |
| CO··heme | 562 | 14.6 | 530 | 11.9 | 406.5 | 147 |
| Hb | 555 | 13.5 | | | 430 | 119 |
| $Hb(O_2)_4$ | 577 | 14.6 | 542 | 13.8 | 412 | 135 |
| $Hb(CO)_4$ | 569 | 13.4 | 539 | 13.4 | 419 | 191 |
| Mb | 555 | 12.0 | | | 435 | 114 |
| $MbO_2$ | 582 | 13.1 | 544 | 12.7 | 417 | 119 |
| MbCO | 578 | 12.3 | 541 | 14.1 | 423 | 185 |

[a]Data for Hb and its derivatives [3] are for human hemoglobin; those for Mb and its derivatives [4] are for sperm whale myoglobin.

The interpretation of the magnetic and spectral properties of heme is based on quantum-mechanical analysis. The external electrons of the Fe atom have the configuration $3d^6 4s^2$ for $Fe^{2+}$ and $3d^5 4s^2$ for $Fe^{3+}$ (Fig. 7.2). Theoretical calculations apply the ligand-field theory [5]. In the molecular orbital approximation, ligands are considered together with the central ion and the MO of the system are expressed as linear combinations of the orbitals of Fe and the ligands. Corresponding calculations for Fe-porphyrin complexes have been made by many authors [6-8]. The electron density distribution and the set of energy levels have been calculated; however, the obtained interpretation of spectra cannot be considered completely established. Further details are presented in [2,9].

An important problem in quantum biochemistry is the treatment of the stressed 5-coordinated heme in Hb and Mb (cf. [131]). All natural iron compounds contain 2.2% of the isotope $^{57}Fe$; this makes it possible to investigate Hb and Mb by means of Mössbauer spectroscopy (cf. [10]). The beam of monochromatic γ rays radiated by γ-radioactive $^{57}Fe$ (obtained from $^{57}Co$) act on the sample. If the source moves with low velocity relative to the γ-absorbing sample, the frequency of the γ quantum will be slightly changed because of the Doppler effect. At some frequency value the $^{57}Fe$ nuclei in the sample will absorb the γ quanta resonantly, making the transition into the excited state. The resonance frequency of a $^{57}Fe$ nucleus in the heme group is very susceptible to the influence of neighboring atoms and groups. Therefore, it is possible with the Mössbauer effect to obtain valuable information about the electronic structure of the heme group in various states of Hb and Mb (cf. [2,11]).

The primary structure of the Hb and Mb in a series of animal species is known, as is the primary structure of many mutants of human hemoglobin (Section 2.5). The deciphering of the structure of Mb and Hb and elucidation of the conformational changes that occur during their oxygenation are very important. For these proteins, the relation between structure and properties has been studied in special detail.

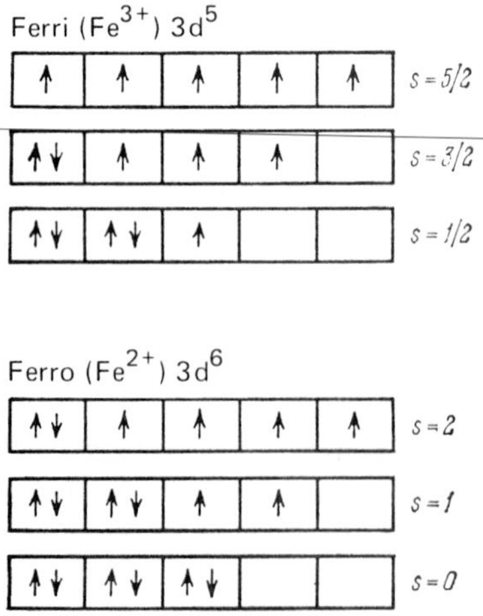


*FIG. 7.2 Electronic structure of $Fe^{3+}$ and $Fe^{2+}$.*

*FIG. 7.3 Curves of the saturation by $O_2$ of (1) Mb; (2) Hb.*

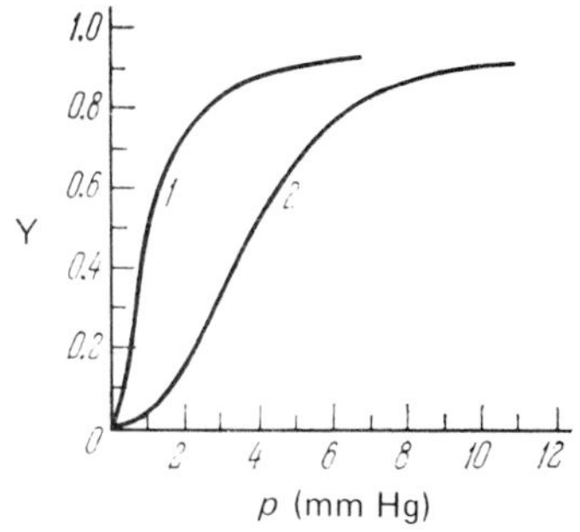


The curve of the saturation of Mb by molecular oxygen Y(p), where p is the partial oxygen pressure, lacks singularities and is similar to Langmuir's isotherm

$$Y = Kp/(1+Kp) \tag{7.1}$$

On the other hand, the curve for Hb has an S-like shape and can be expressed by Hill's equation [12]

$$Y = Kp^n/(1+Kp^n) \tag{7.2}$$

where the parameter $n \cong 2.8$. Both curves are shown in Fig. 7.3 (cf. p. 34).

The physiological significance of the S-shaped Y(p) curve for Hb consists in the decrease in its affinity for oxygen when oxygen is split from Hb. The changes of the partial pressure of $O_2$ in tissues are smaller than those in the lungs. Should Hb be characterized by a hyperbolic Y(p) curve like that of Mb, only a small portion of the transferred $O_2$ would be split from the Hb in tissues. An organism would suffocate even in an atmosphere of pure oxygen. The effectiveness of respirative transport is also enhanced by 2,3-diphosphoglycerate (DPG) and because of the Bohr effect; the first factor lowers the affinity of Hb for the $O_2$ in tissues, and the second factor increases this affinity.

The Bohr effect, which is peculiar to Hb but not to Mb, is the dependence of oxygen affinity on the pH of the medium. It is minimal close to pH 6 and maximal at pH 9. In other words, small concentrations of protons enhance the binding of $O_2$, and conversely, small concentrations of $O_2$ enhance the binding of protons. By increasing the pH of venous blood, and therefore increasing its capacity for bicarbonate absorption, the Bohr effect provides the main mechanism for the transport of $CO_2$ from tissues to the lungs. The subsequent mechanism depends directly on the interaction of Hb with $CO_2$. The pH dependence of Hb's affinity for $O_2$ (i.e., the Bohr effect), expressed by log $p_{1/2}$ ($p_{1/2}$ is the partial pressure corresponding to half-saturation of Hb by oxygen), is shown in Fig. 7.4 [13].

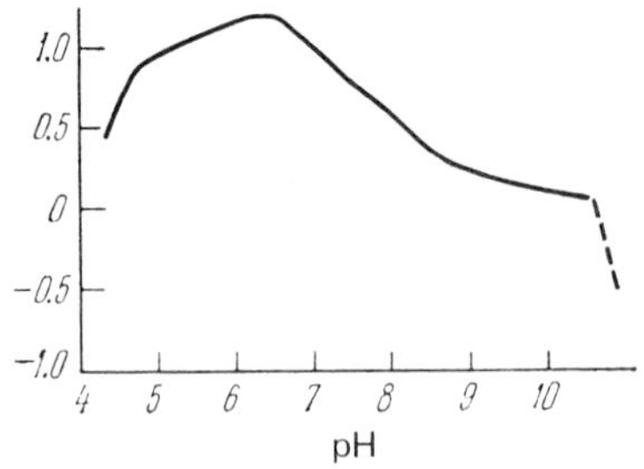


*FIG. 7.4 Dependence of the affinity of Hb for $O_2$ on pH.*

The difference of parameter n from 1 in Hill's equation and the corresponding S-shaped saturation curve express the heme-heme interaction (i.e., the interaction of subunits) and hence, the cooperativity of oxygen binding. This problem was first examined by Adair [14], who suggested that ligand binding occurs in four stages, each of which is characterized by its own equilibrium constant $K_i$

$$HbX_i + X \underset{k_{-(i+1)}}{\overset{k_{i+1}}{\rightleftharpoons}} HbX_{i+1} \qquad (i = 0, 1, 2, 3) \tag{7.3}$$

Partial saturation by a ligand is

$$Y = \frac{K_1X+2K_1K_2X^2+3K_1K_2K_3X^3+4K_1K_2K_3K_4X^4}{4(1+K_1X+K_1K_2X^2+K_1K_2K_3X^3+K_1K_2K_3K_4X^4)} \tag{7.4}$$

where X is the ligand concentration.

Other more accurate expressions of Y were proposed by Pauling [15], Wyman [16], Roughton [17], and Allen et al. [18]. Guidotti [19] took into account the dissociation of Hb and $HbO_8$ into dimers. Experiment gives the Hill expression (7.2), which can coincide with an expression of the type of (7.4) at proper values of the constants.

Detailed reviews of the physical and chemical properties of Hb and Mb are given in [20], [21] and [132]. We present here some thermodynamic characteristics (Tables 7.3 [20] and 7.4 [21]).

*TABLE 7.3*
*Equilibrium of Mb and $O_2$ at pH 7, 20°C[a]*

| | $p_{1/2}$ (mm Hg) | $K \times 10^{-6}$ ($M^{-1}$) | $\Delta \log p_{1/2}/\Delta pH$ | $\Delta H$ (kcal $mole^{-1}$) |
|---|---|---|---|---|
| Horse Mb | 0.65 | 0.85 | 0.09 | -20.6 |
| Human Mb | 0.72 | 0.77 | 0.03 | -13.1 |

[a] The smallness of $\Delta \log p_{1/2}/\Delta pH$ shows the absence of the Bohr effect.

TABLE 7.4
*Equilibria of Single α and β Chains and of Hb with $O_2$, CO, and Ethylisocyanide (EIC) at pH 7, 20°C*[a]

| | $O_2$ | | CO | | EIC | |
|---|---|---|---|---|---|---|
| | $K \times 10^{-5}$ ($M^{-1}$) | ΔG | $K \times 10^{-7}$ ($M^{-1}$) | ΔG | $K \times 10^{-4}$ ($M^{-1}$) | ΔG |
| α Chain | 12 | -8.2 | 30 | -11.4 | 45 | -7.65 |
| β Chain | 14 | -8.25 | 48 | -11.7 | 42 | -7.6 |
| Hb | 0.56 | -6.4 | 1.4 | -9.6 | 2 | -5.8 |

[a]In each case, the Gibbs free energy ΔG is given in kilocalories per mole of $O_2$.

According to estimates made by Roughton and others [22], the heats of reaction for various stages of interaction with $O_2$ (7.3) are equal to (sheep Hb) $\Delta H_1 = -15.7 \pm 0.8$, $\Delta H_2 = -11.4 \pm 2.5$, $\Delta H_3 = -7.8 \pm 3.7$, $\Delta H_4 = -8.7 \pm 3.3$ kcal $mole^{-1}$. The mean value is

$$\Delta H = \frac{1}{4}(\Delta H_1+\Delta H_2+\Delta H_3+\Delta H_4) = -10.9 \pm 0.3 \text{ (kcal mole}^{-1}\text{)}$$

The events that occur in an Hb molecule during oxygenation were investigated by Perutz by X-ray structural analysis [23] (cf. p. 216). The $O_2$ molecule is bound by the Fe atom of the heme group. In oxyHb the Fe atom is situated in the plane of the porphyrin ring, in its center (with exactness to 0.05 Å). In high-spin deoxyHb the Fe atom is shifted from this plane by ∿0.8 Å in the direction of the imidazole ring of His-F8. The coordination number of Fe in this state is 5. Oxygenation transfers Fe into the low-spin state and increases the number of ligands by one. In this state the Fe atom is shifted back into the ring plane by 0.8 Å. These changes provoke changes in the contacts between the porphyrin ring and the closely packed amino acid residues in the globule. In other words, there occurs reorganization of the protein globule due to ECI.

Sixty atoms of protein have direct contact with the porphyrin ring at the van der Waals contact distances [24]. Introduction of even the smallest ligand, $OH^-$, with a radius of 1.5 Å, produces a change in the conformation of the β subunits in deoxyHb. In these subunits the γ-$CH_3$ group belonging to the Val-E11(67) residue is 2.5 Å away from the $OH^-$, a distance that is less than the sum of the van der Walls radii. Hence, in Hb β globules there is no room for even the smallest ligand, and oxygenation increases the distance between the ring and Val-E11(67) by approximately 1 Å. Such displacement does not occur in the α subunits, however, since in them the width of the "pocket" is sufficient for incorporation of the ligand.

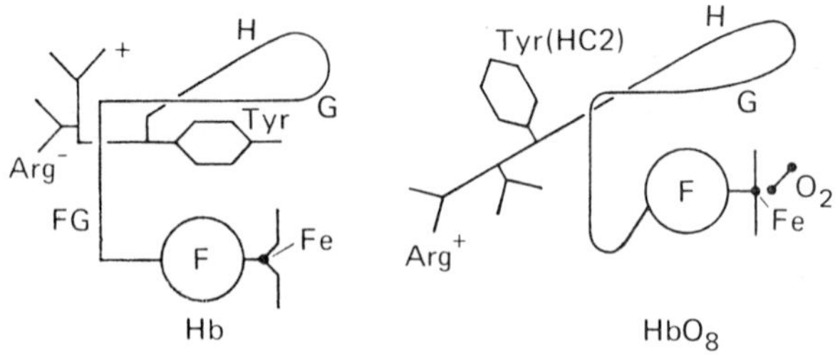


*FIG. 7.5 Scheme of the shift of Tyr-HC2(140) upon oxygenation of Hb.*

In oxyHb, the C-terminal ends of all four chains have total freedom of rotation, and the next to last tyrosyls have partial freedom. They spend only a small amount of time bound between the F and H helices. In deoxyHg, however, every C-terminal end is fixed twice by the salt bridges: the α-carboxyl of Arg-HC3(141)$\alpha_1$ is bound by the α-$NH_2$ of Val-A1(1)$\alpha_2$, and the guanidine group of Arg-141 by Asp-H9(126)$\alpha_2$; α-carboxyl of His-HC3(146)$\beta_1$ is bound by ε-$NH_2$ of Lys-C5(40)$\alpha_2$, and its imidazole by Asp-FG1(94)$\beta_1$. All four next-to-last Tyr are fixed in the cavities between the F and H helices by the van der Waals and hydrogen bonds. Oxygenation of Hb provokes displacement of the F helix, and breakage of the salt bridges. Therefore, Tyr-HC2(140) is shifted from the "pocket" between the F and H helices. Oxygenation diminishes the width of the "pockets" for Tyr-HC2 by 1.3 Å in α chains and by 2 Å in the β chain. This reconstruction is shown schematically in Fig. 7.5 and is confirmed by data obtained by means of corresponding chemical modifications of protein. In heme-heme interactions, these events are brought about by breakage of the salt bridges connecting the subunits.

Transformations in single subunits and the breakage of the salt bridges alter the quaternary structure of Hb. In $\alpha_1\beta_1$ and $\alpha_2\beta_2$ contacts, shifts of 1 Å occur; in $\alpha_1\beta_2$ and $\alpha_2\beta_1$, 7 Å shifts ensue! These latter contacts are especially significant for heme-heme interactions. We have to emphasize that nearly all the residues in $\alpha_1\beta_2$ contacts are invariant for vertebrate animals. On the other hand, broad variations are observed in the residues forming an $\alpha_1\beta_1$ contact. Mutant substitutions in the $\alpha_1\beta_2$ contact considerably decrease the heme-heme interaction.

In the $\alpha_1\beta_2$ contact, the CD region of one chain enters the FG region of another one. Deoxygenation disturbs this close joining of the chains. In oxyHb the projection formed by the side chain of Thr-C3(38)α enters the groove in the chain at Val-FG5(98)β. In deoxyHb this groove is occupied by the side chain of Thr-C6(41)α (i.e., by the chain projected from the next turn in the C helix). On the other hand, the hydrogen bond connecting Asp-Gl(94)α with Asn-G4(102)β is replaced in oxyHb by the hydrogen bond between Tyr-C7(42)α and Asp-Gl(99)β.

Thus, Hb oxygenation provokes a series of conformational events, which are triggered by the displacement of Fe into the ring plane and by a corresponding 0.75- to 0.95-Å shift in the ring's proximal His. This shift produces a change in the tertiary structure because the Fe atom is rigidly bound by His-F8 and the ring contacts 60 globin atoms. The F helix is displaced toward the center of the molecule, pushing Tyr-HC2(140) out of the cavity between the F and H helices. The expelled Tyr pulls with it Arg-HC3(141), thereby breaking the salt bridges with the α chain. The situation in the β chain is different. Before the ligand approaches the Fe atom, it must "open" a cavity near the heme group. The result of the subsequent ligand binding is to shift the Fe toward the ring plane. The F helix is displaced toward the center of the molecule and pushes Tyr-HC2(145) out of its cavity. This residue removes His-HC3(146) and breaks the salt bridge connecting it with Asp-FG1(94).

Perutz found that the structure of deoxyHb is stabilized by a cofactor, 2,3-diphosphoglycerate (DPG), which forms additional salt bridges between the β chains. DPG is removed from a molecule by oxygenation. What is the sequence of events in the four-chain Hb molecule? Perutz proposed a reliable hypothesis, according to which the binding of $O_2$ by every subunit transforms it into the oxy conformation but the quaternary structure remains in the deoxy form until two $O_2$ molecules have been bound. In other words, every subunit can exist in two alternative conformations, but the quaternary structure can be "erroneous." Oxygenation presumably begins in the α chains, since they, unlike β chains, have sufficient room for ligand intercalation. The general scheme of the proces (O, oxy form; D, deoxy form) is

$$(\alpha_1{}^D\alpha_2{}^D\beta_1{}^D\beta_2{}^D)^D \rightarrow (\alpha_1{}^O\alpha_2{}^D\beta_1{}^D\beta_2{}^D)^D \rightarrow (\alpha_1{}^O\alpha_2{}^O\beta_1{}^D\beta_2{}^D)^D$$

$$\xrightarrow{\searrow \text{DPG}} (\alpha_1{}^O\alpha_2{}^O\beta_1{}^D\beta_2{}^D)^O \rightarrow (\alpha_1{}^O\alpha_2{}^O\beta_1{}^O\beta_2{}^D)^O \rightarrow (\alpha_1{}^O\alpha_2{}^O\beta_1{}^O\beta_2{}^O)^O$$

A pictorial presentation of this scheme appears in Fig. 7.6. The essence of the mechanism is that because of its specific construction the heme group reinforces the slight change in the atomic radius of Fe due to the transition from the high-spin to the low-spin state and transforms this change into a large displacement of the heme-bound His. This displacement is followed by the events described earlier.

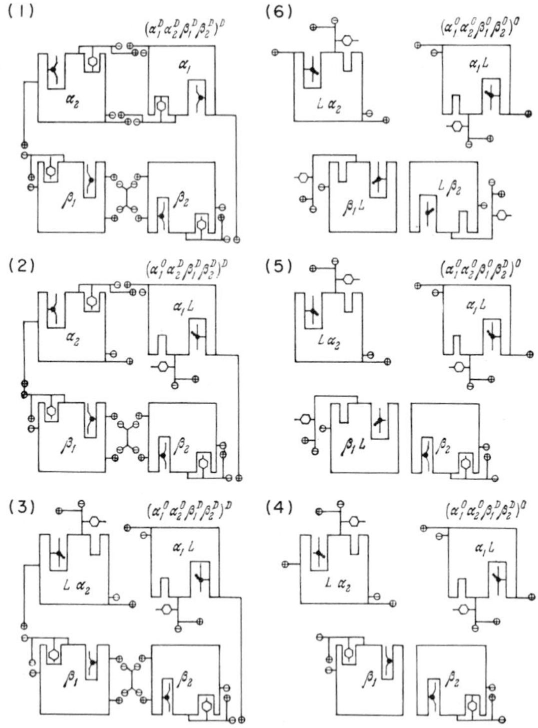


*FIG. 7.6 Scheme of Perutz's hypothesis regarding the oxygenation of Hb. (1) Hb with salt bridges intact and one DPG molecule between β chains; (2) $HbO_2$; (3) $HbO_4$. Oxygenation of the α chains occurs during stages 1-2 and 2-3. (4) $HbO_4$ with changed conformation. Changes in conformation occur during stages 3-4. (5) $HbO_6$; (6) $HbO_8$.*

The subunit interaction energy is approximately 12 kcal $mole^{-1}$, a value that agrees well with the energy of the six salt bridges forming the $\alpha_1\alpha_2$, $\alpha_1\beta_2$, and $\alpha_2\beta_1$ contacts. (The energy of one salt bridge is 1-2 kcal $mole^{-1}$.) Actually, enzymic separation of the four C-terminal residues destabilizes the quaternary structure of deoxyHb and inhibits heme-heme interaction.

Hb is tetrameric; its splitting into dimers can occur only at very low concentrations. The salt bridges have to be broken and the quaternary structure must take on the form corresponding to oxyHb. All the cooperative effects observed during ligand binding are due to tetrameric deoxy structure. However, another model has been suggested, in which the fundamental functional unit determining cooperative properties is the dimer [21].

The Bohr effect can be interpreted in molecular terms on

the basis of Perutz's investigations. The structural changes produced by oxyHb⇄deoxyHb transitions change the neighborhood of the three pairs of weak bases in such a way that they tend to bind protons if $O_2$ is split off. In oxyHb, the $\alpha NH_2$ groups of Val-1α and imidazoles of His-146β are free, but in deoxyHb they become drawn together with carboxyl groups. The imidazole of His-122α is close to guanidine in oxyHb, but close to the carboxyl in deoxyHb. Deoxygenation results in rotation of the $\alpha_1$ chain in relation to the $\beta_2$ chain by 13.5°. The $\varepsilon$-$NH_2$ group of Lys-C5(40)$\alpha_1$ is shifted by 7 Å into the region of the $\alpha_1\beta_2$ contact and forms a salt bridge with the C-terminal carboxyl of $\beta_2$. This bridge fixes Tyr-145$\beta_2$ in the cavity between F$\beta_2$ and H$\beta_2$ and makes possible the formation of a bridge between the imidazole of His-146$\beta_2$ and carboxyl of Asp-94$\beta_2$. These events are shown schematically in Fig. 7.7. His-146$\beta_2$ participates in the Bohr effect because of the change in the pK of the imidazole due to the changed surroundings.

The α-$NH_2$ groups of Val-1α have normal pK values in oxyHb but the pK values in deoxyHb are elevated as a result of binding with α-carboxyls of another α chain. The C-terminal arginyls of every α chain also participate in the Bohr effect, forming in deoxyHb two bridges with another α chain (Fig. 7.8).

Further details are contained in Perutz's paper [23], in which the acid Bohr effect is also examined. Discussing these

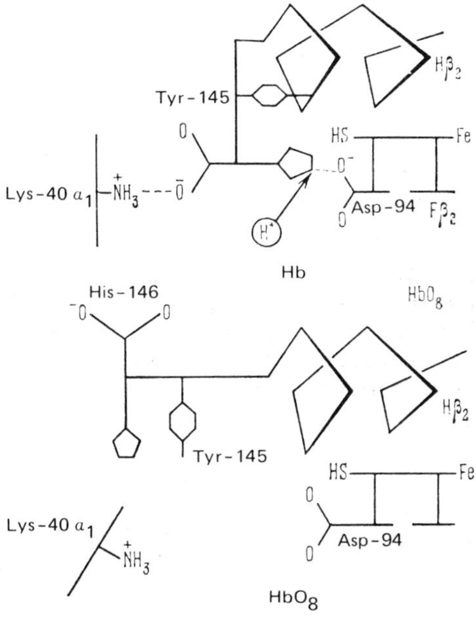


*FIG. 7.7 Scheme of conformational changes in* $\beta_2$ *chain upon oxygenation.*

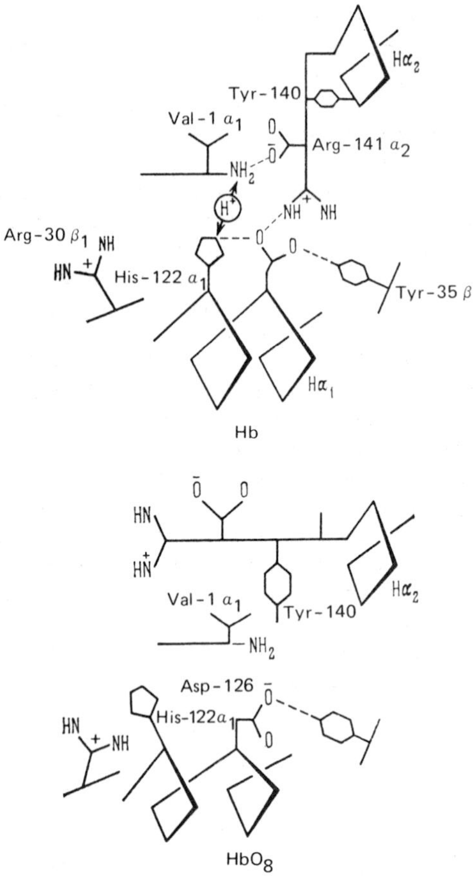


*FIG. 7.8 Scheme of conformational changes in α chains.*

conformational phenomena, Perutz states quite correctly that a protein is a dynamic system. Both tertiary and quaternary structures of Hb oscillate continuously between oxy and deoxy conformations. The presence of a ligand induces not the exclusion of a deoxy conformation but a shift in the conformational equilibrium [23]. The phenomena examined by Perutz can be considered a clear expression of ECI (p. 401). The shift in electron density in the heme group provokes conformational reconstruction of the globule.

Atanasov studied the conformational properties of myoglobin (Mb) by a series of physical methods. The presence of conformers of Mb was established and the general results of these works correlate with the described model ([25,26]; cf. also [27]).

Physical understanding of heme-heme interactions and related events requires not only a molecular picture but phenomenological treatment, too. This can be done on the basis of the theory of coupled functions developed by Wyman (next section) and of a general theory of the cooperative kinematic behavior of proteins (Section 7.5).

## 7.2 Phenomenological Theory of the Equilibrium Properties of Hemoglobin

Let us consider the theory of coupled functions suggested by Wyman [28]. A macromolecule M possesses q ligand binding sites. The total concentration of bound ligand is

$$X = M_0 \sum_{i=0}^{q} iK_i x^i \tag{7.5}$$

where $M_0$ is the concentration of free macromolecules, x is the ligand activity, and $K_i$ is the apparent macroscopic equilibrium constant of the reaction

$$M+iX \rightleftharpoons MX_i$$

Evidently $K_0 = 1$. The total concentration of macromolecules in all ligand complexes is

$$M = M_0 \sum_{i=0}^{q} K_i x^i \tag{7.6}$$

The amount of ligand bound by 1 mole of macromolecules is

$$\overline{X} = \frac{X}{M} = \frac{\Sigma_i iK_i x^i}{\Sigma_i K_i x^i} = \frac{d \ln \Sigma_i K_i x^i}{d \ln x} \tag{7.7}$$

If two different ligands are bound and there are r sites for ligand Y, then

$$M = M_0 \sum_{i=1}^{q} \sum_{j=1}^{r} K_{ij} x^i y^j \tag{7.8}$$

where $K_{ij}$ is the equilibrium constant for the reaction

$$M+iX+jY \rightleftharpoons MX_iY_j$$

We obtain the saturation functions

$$\overline{X} = \frac{\partial \ln \Sigma\Sigma K_{ij} x^i y^j}{\partial \ln x}, \qquad \overline{Y} = \frac{\partial \ln \Sigma\Sigma K_{ij} x^i y^j}{\partial \ln y} \tag{7.9}$$

We get Wyman's fundamental equation, coupling the functions

$\bar{X}$ and $\bar{Y}$

$$\left( \frac{\partial \ln \bar{X}}{\partial \ln y} \right)_x = \left( \frac{\partial \ln \bar{Y}}{\partial \ln x} \right)_y \tag{7.10}$$

If the number of ligands is n, the number of equations required is $n(n - 1)/2$. We can write

$$d\bar{X} = \left( \frac{\partial \bar{X}}{\partial \ln x} \right)_y d \ln x + \left( \frac{\partial \bar{Y}}{\partial \ln y} \right)_x d \ln y \tag{7.11}$$

$$d\bar{Y} = \left( \frac{\partial \bar{Y}}{\partial \ln x} \right)_y d \ln x + \left( \frac{\partial \bar{Y}}{\partial \ln y} \right)_x d \ln y \tag{7.12}$$

From (7.10)-(7.12), a series of differential relations follows, in particular

$$\left( \frac{\partial \bar{X}}{\partial \bar{Y}} \right)_x = - \left( \frac{\partial \ln y}{\partial \ln x} \right)_Y \tag{7.13}$$

and

$$\left( \frac{\partial \ln x}{\partial \bar{Y}} \right)_X = \left( \frac{\partial \ln y}{\partial \bar{X}} \right)_Y \tag{7.14}$$

This formalism was also developed by Wyman for systems containing various types of macromolecules and for systems where macromolecules polymerize or participate in chemical reactions. The action of ligands on the respective equilibrium constants can be elucidated. We limit ourselves here to the case of one kind of macromolecule and two kinds of ligands.

For an equilibrium system at constant temperature and pressure, thermodynamics gives

$$\left( \frac{\partial \mu_1}{\partial n_2} \right)_{n_1} = \left( \frac{\partial \mu_2}{\partial n_1} \right)_{n_2} \tag{7.15}$$

where $\mu_1$, $\mu_2$ are the chemical potentials and $n_1$, $n_2$ are the amounts of ligand. If we substitute

$$d\mu_1 = RT \, d \ln x, \qquad d\mu_2 = RT \, d \ln y$$

and replace $n_1$, $n_2$ by $\bar{X}$ and $\bar{Y}$, we again get the fundamental equation (7.10).

Equation (7.5) or (7.7) describes the equilibrium between macromolecule and ligands in terms of q equilibrium constants $K_i$. We can write an expression for the difference between the chemical potential of the ligand when it is in equilibrium with the macromolecule and when it is in standard state in the form

$$\mu_x - \mu_0 = RT \ln x = -RT \ln K_x = \Delta G_x \tag{7.16}$$

where $\Delta G_x$ is the change in free energy due to interaction with the macromolecule. If

$$(\partial \bar{X}/\partial \ln x) > q$$

the interaction is positive or stabilizing.

It is convenient to express the equilibrium curve of the ligand not by $\bar{X}$ or by the partial saturation $\bar{x} = \bar{X}/q$, but by the value of $\ln[\bar{x}/(1-\bar{x})]$ as a function of ln x, and to introduce the parameter n, defined as

$$n = \frac{d \ln[\bar{x}/(1-\bar{x})]}{d \ln x} = \frac{1}{x(1-x)} \frac{d\bar{x}}{d \ln x} \tag{7.17}$$

If all sites are identical and independent, the graph is rectilinear with n = 1, as follows from the law of mass action. If there are strong stabilizing interactions, the curve is rectilinear in a broad range with the center at $x = \frac{1}{2}$ with $n > 1$. This corresponds to the Hill equation (7.2), which can be rewritten in the form

$$\bar{x} = Kx^n/(1+Kx^n) \qquad \text{or} \qquad \ln[\bar{x}/(1-\bar{x})] = \ln K + n \ln x$$

Therefore, Wyman calls the graph of $\ln[\bar{x}/(1-\bar{x})]$ versus ln x Hill graph. Hill's parameter n is related to the mean free energy of the interaction of sites. It follows from (7.16) and (7.17) that

$$\frac{d\Delta G_x}{d\bar{x}} = RT \frac{c \ln x}{d\bar{x}} = \frac{RT}{n\bar{x}(1-\bar{x})} \equiv \Delta G_x' \tag{7.18}$$

If the sites are identical and independent, $n = 1$ and $\Delta G_x' = \Delta G_x'^{(1)}$. The difference

$$\Delta G_x' - \Delta G_x'^{(1)} \equiv \Delta G_{xx}$$

expresses the stabilization energy per site. Assuming that it is positive at $n > 1$, we obtain

$$\Delta G_{xx} = \frac{RT}{\bar{x}(1-\bar{x})} \left(1 - \frac{1}{n}\right) \tag{7.19}$$

where n cannot be larger than q, the total number of sites. n = 1 also in the case of independent sites that are not identical. The integral

$$\int_0^1 \Delta G_{xx}\, d\bar{x} = RT \int_{x=0}^{\infty} (n-1)\, d \ln x \tag{7.20}$$

expresses the minimum value of the total interaction energy per site at total saturation of the macromolecule. The

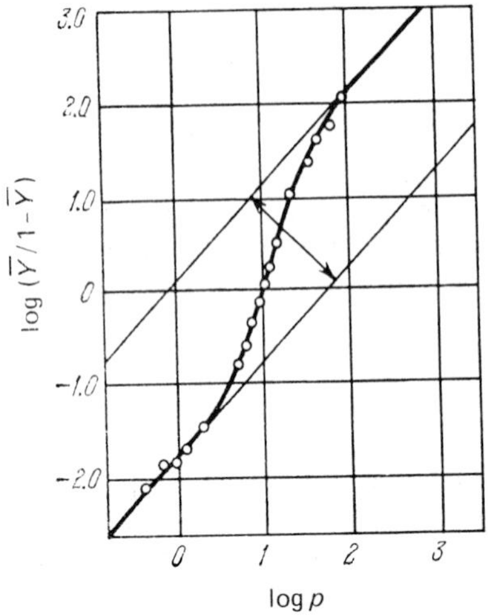


*Fig. 7.9 Hill's graph for horse hemoglobin saturation by oxygen.*

integral does not become infinity if n tends to 1 when $\Delta G_{xx}$ tends to 0 or ∞, that is, x = 0, or $\overline{x}$ = 1.

Let us apply Wyman's method to hemoglobin. Figure 7.9 shows the Hill graph for the saturation of horse Hb by oxygen (0.6 M phosphate buffer, pH 7.0, 19°C). The total free energy of the interaction is 2.6 kcal mole$^{-1}$, n = 2.95 ± 0.05 at $\overline{Y} = \frac{1}{2}$. At very large and very small saturations n → 1.

Hb can bind $O_2$ and CO. At total saturation the distribution coefficient R does not depend on the partial pressures of both gases (Haldane's first law)

$$[Hb(CO)_4]/[Hb(O_2)_4] = R(pCO/pO_2) \tag{7.21}$$

According to Haldane's second law, the total saturation of Hb, interacting with a mixture of $O_2$ and CO, is a function of $pO_2$ + RpCO. The second law follows from the first. Let X = $O_2$, Y = CO. Let $\sigma_x = \overline{X}/x$ and $\sigma_y = \overline{Y}/y$. According to (7.21) $\sigma_x = \sigma_y = \sigma$. The coupling equation (7.13) gives

$$\left(\frac{\partial\sigma}{\partial x}\right)_y = \left(\frac{\partial\sigma}{\partial Ry}\right)_x \tag{7.22}$$

The integral of this equation is

$$\sigma = \phi(x+Ry)$$

Total saturation is

$$\overline{X}+\overline{Y} = (x+Ry)\sigma = f(x+Ry) \tag{7.23}$$

Conversely, the first law follows from the second.

Coupling equations describe the Bohr effect. Two ligands, $O_2$ and $H^+$, are involved. Equation (7.13) gives ($\overline{H}^+$ and $\overline{Y}$ are the amounts of bound $H^+$ and $O_2$ per heme equivalent, $a_H$ is the proton activity, p is the partial pressure of $O_2$)

$$\left(\frac{\partial \bar{H}^+}{\partial \bar{Y}}\right)_{a_H} = -\left(\frac{\partial \ln p}{\partial \ln a_H}\right)_{\bar{Y}} \tag{7.24}$$

and Eq. (7.14) gives

$$\left(\frac{\partial \ln a_H}{\partial \bar{Y}}\right)_{\bar{H}^+} = \left(\frac{\partial \ln p}{\partial \bar{H}^+}\right)_{\bar{Y}} \tag{7.25}$$

Assume that the protons enter the system along with an anion $A^-$ of a strong acid whose activity is a, and the amount per heme amount is $\bar{C}$. If we identify $\bar{H}^+$ and C, then (7.25) can be rewritten in the form

$$\left(\frac{\partial \ln a}{\partial \bar{Y}}\right)_{\bar{C}} = \left(\frac{\partial \ln p}{\partial \bar{C}}\right)_{\bar{Y}} \tag{7.26}$$

Let us introduce the relation

$$a = a_H a_A = a_H \gamma_A \bar{C}$$

where $a_A$ is the activity of the anion and $\gamma_A$ is the part of the anion remaining in solution. Equation (7.26) assumes the form

$$\left(\frac{\partial \ln a_H}{\partial \bar{Y}}\right)_{\bar{C}} + \left(\frac{\partial \ln \gamma_A}{\partial \bar{Y}}\right)_{\bar{C}} = \left(\frac{\partial \ln p}{\partial \bar{C}}\right)_{\bar{Y}} \tag{7.27}$$

or

$$\left(\frac{\partial \bar{C}}{\partial \bar{Y}}\right)_{a_H} = -\left(\frac{\partial \ln p}{\partial \ln a_H}\right)_{\bar{Y}} + \left(\frac{\partial \ln \gamma_A}{\partial \bar{Y}}\right)\left(\frac{\partial \bar{C}}{\partial \ln a_H}\right)_{\bar{Y}} \tag{7.28}$$

The last multiplier on the right-hand side of (7.28) gives the slope of the titration curve of Hb at a degree of oxygenation equal to $\bar{Y}$. Equations (7.27) and (7.28) coincide with (7.24) and (7.25) if $\gamma_A$ does not change when $\bar{Y}$ is changed (i.e., the bound part of the anion does not depend on oxygenation). Experiment shows that in fact the Bohr effect, when measured as change in affinity for $O_2$ due to pH change, coincides with the same effect when measured as proton liberation due to oxygenation. Experiment gives

$$\left(\frac{\partial \bar{C}}{\partial \mathrm{pH}}\right)_Y = 3; \quad \left|\left(\frac{\partial \log p}{\partial \mathrm{pH}}\right)_{\bar{Y}}\right| = 0.6; \quad \left|\left(\frac{\partial \log \gamma_A}{\partial Y}\right)_{\bar{C}}\right| < 0.01$$

The method of coupled functions shows the general thermodynamic sense of the cooperative behavior of Hb, of the competitive interactions of ligands, and of the Bohr effect.

Wyman generalized his theory and introduced the so-called coupling potential $\Lambda$ [29], which corresponds to any thermodynamic potential in the following way. Let F be any continuous function of r variables $n_i$; its first and second derivatives are also continuous. We have

$$\mu_i = \partial F/\partial n_i \qquad (i = 1,2,\ldots,r) \tag{7.29}$$

The coupling potential is given by the relations

$$n_i = \partial\Lambda/\partial\mu_i \tag{7.30}$$

For instance, if $F = n_1n_2+n_2^3$, then $\mu_1 = n_2$, $\mu_2 = n_1+3n_2^2$, or $n_1 = \mu_2-3\mu_1^2$. Correspondingly, $\Lambda = \mu_1\mu_2-\mu_1^3$, which does not coincide with $F = \mu_1\mu_2-2\mu_1^3$. The general expression for the coupling potential is

$$\Lambda = \Lambda_\mu + \sum_{i=1}^{r} \Lambda_{\bar{\mu}_i} \tag{7.31}$$

where $\Lambda_\mu$ is the function of all $\mu$, and $\Lambda_{\bar{\mu}_i}$ is the function of all $\mu$ except $\mu_i$. This expression shows that the binding of one component of the system is coupled with that of another one. Thus, the components form a coupling group. If one of the components is a polyfunctional macromolecule, its ability to bind each of the ligands can be called the coupling function. It is also possible to speak not about the coupling of components but about the coupling of functions, and to use the notion of coupled functions.

If we are concerned about one ligand X bound by a macromolecule, then the potential $\Lambda$ can be expressed in the form

$$\Lambda = RT\ \ln(1+K_1x+K_2x^2+\cdots+K_qx^q) \tag{7.32}$$

where x is the activity of X, $K_i$ are constants, and q is the number of binding sites. This polynomial can be factored; that is,

$$\Lambda = RT\ \ln[(1+k_1x)(1+k_2x)\cdots(1+k_qx)] \tag{7.33}$$

if all q sites are independent. Assuming that in Hb there are two acid sites per heme group, and that these acid sites are responsible for the Bohr effect and are independently ionized, we can obtain the polynomial for Hb. It has to be expressed as the product of two polynomials, the first one in the fourth power of p ($O_2$ activity) and the eighth power of $H^+$ ($H^+$ activity), and the second in the (q-8)th power of $H^+$, where q is the total number of ionizable groups in the molecule. The first polynomial must have the form

$$N_1 = (1+\alpha_0 H^+)^4(1-\beta_0 H^+)^4+(1+\alpha_0 H^+)^3(1+\beta_0 H^+)^3(1+\alpha_1 H^+)(1+\beta_1 H^+)K_1 p$$
$$+\cdots+(1+\alpha_1 H^+)^4(1+\beta_1 H^+)^4 K_4 p^4 \quad (7.34)$$

Here $1/\alpha_0$ and $1/\beta_0$ are the ionization constants of the two sites in the deoxy form; $1/\alpha_1$ and $1/\beta_1$ are the same constants for the oxy form; $K_i$ are the equilibrium constants of sequential stages in the reaction at $H^+ = 0$. If $p \to 0$ (i.e., for the deoxy form), only the first term need be taken into account; at large p, only the last one. This enables us to estimate the constants of the group coupled with oxygen. Should the ionization of every pair of such groups not be independent, we would obtain multipliers of the type $(1+\alpha' H^+ + \beta' H^{+2})$ instead of $(1+\alpha_0 H^+)(1+\beta_0 H^+)$ and $(1+\alpha_1 H^+)(1+\beta_1 H^+)$. The titration curves ov Hb and $Hb(O_2)_4$ show that the second polynomial $N_2$ of the (q-8)th power of $H^+$ can be factored.

Further development of the theory and its application to proteins containing a very large number of functional sites (erythrocruorins) are presented in [30].

## 7.3 The Faraday Effect

In 1845 Faraday discovered that when a strong magnetic field is set up parallel to a beam of polarized light passing through glass, the plane of polarization of the light is rotated by an angle $\phi$ proportional to the thickness $\ell$ of the layer of glass and to the field intensity H

$$\phi = V\ell H \quad (7.35)$$

where V is the constant of magnetic rotation, the Verdet constant. Faraday wrote that he had succeeded in magnetizing the light ray and in illuminating the magnetic lines of force [31]. Actually, the magnetic field acts not on the light but on the substance, making the electron shells of the molecules asymmetric, thereby inducing circular birefringence in the substance [cf. p. 281, (5.47)]:

$$\phi = (\pi/\lambda)(n_L - n_D)\ell \quad (7.36)$$

As has already been said, the optical properties of molecules are determined by their polarizability. Polarizability, in turn, depends on the intensity of the magnetic field. In the first approximation the components of a molecule's polarizability tensor can be written as

$$a_{\sigma\tau} = a_{\sigma\tau}^{0} + \sum_{\rho} a_{\sigma\tau,\rho} H_{\rho} + \cdots \quad (7.37)$$

Here $\sigma,\tau,\rho = \xi,\eta,\zeta$ are the coordinates in the system fixed in the molecule; $H_\rho$ are the components of the magnetic intensity vector in the same system. Since H = rot A (A is the vector potential) is an axial vector, the values $a_{\sigma\tau,\rho}$ are antisymmetric in relation to permutation of the indexes $\sigma,\tau$:

$$a_{\xi\eta,\zeta} = \left(\frac{\partial a_{\xi\eta}}{\partial H_\zeta}\right)_{H=0} = \left(\frac{\partial a_{\xi\eta}}{\partial[(\partial A_\eta/\partial\xi)-(\partial A_\xi/\partial\eta)]}\right)_{H=0} = -a_{\eta\xi,\zeta}$$

It follows from the law of conservation of energy that the tensor $a_{\sigma\tau}$ (7.37) is Hermitian. Hence, the values $a_{\sigma\tau,\rho}$ are imaginary. If the field H is directed along the z axis of the coordinate system fixed in space, the material equations are

$$\begin{aligned} D_x &= \varepsilon E_x - i\varepsilon' E_y \\ D_y &= i\varepsilon' E_x + \varepsilon E_y \\ D_z &= \varepsilon E_z \end{aligned} \qquad (7.38)$$

where $\varepsilon' = i\varepsilon_{xy} = 4\pi i N_1 a_{xy}$ (for a gas, $N_1$ is the number of molecules in 1 $cm^3$). Since $a_{xy}$ is an imaginary quantity, arising because of the presence of imaginary asymmetric polarizability components depending on $H_z$, $\varepsilon'$ is substantial and proportional to $H_z$. If the plane wave passes along the z axis, we have

$$D_x = n^2 E_x, \quad D_y = n^2 E_y, \quad D_z = 0 \qquad (7.39)$$

It then follows from (7.38) and (7.39) that

$$\begin{aligned} (\varepsilon-n^2)E_x - i\varepsilon' E_y &= 0 \\ i\varepsilon' E_x + (\varepsilon-n^2)E_y &= 0 \end{aligned} \qquad (7.40)$$

We obtain two values for the refractive index

$$n^2 = \varepsilon \pm \varepsilon' \qquad (7.41)$$

The solution with the plus sign corresponds to a wave with left circular polarization, that with the minus sign to a wave with right polarization (p. 280). The circular birefringence is

$$\Delta n = n_L - n_D \cong \varepsilon'/n \qquad (7.42)$$

where $n = \frac{1}{2}(n_L+n_D)$. Comparing (7.35), (7.36), and (7.42), we find

$$V = \pi\varepsilon'/\lambda n H \qquad (7.43)$$

This is the phenomenological theory of the Faraday effect [32,33]. In classical electron theory the Faraday effect is connected with the Zeeman effect. In the absence of a magnetic field, an electron harmonic oscillator oscillates with circular frequency $\omega_0$. In the presence of a field directed parallel to the light ray, a spectral line with frequency $\omega_0$ splits into two lines with right and left circular polarization. The splitting value is $2|\omega_H|$ where $\omega_H$ is the Larmor precession frequency (cf. pp. 324,330)

$$|\omega_H| = eH/2mc$$

where e, m are the charge and mass of the electron, respectively, and c is the velocity of light. The circular birefringence is expressed as

$$\Delta n = \left(\frac{\partial n}{\partial \omega}\right)_{H=0} 2\omega_H = -\frac{eH}{mc}\left(\frac{\partial n}{\partial \omega}\right)_{H=0}$$

or

$$\Delta n = \lambda^2\left(\frac{\partial n}{\partial \lambda}\right)_{H=0} \frac{eH}{2\pi mc^2} \tag{7.44}$$

We obtain

$$\phi = \frac{eH}{2mc^2}\lambda\left(\frac{\partial n}{\partial \lambda}\right)_{H=0} \ell$$

and

$$V = \frac{e\lambda}{2mc^2}\left(\frac{\partial n}{\partial \lambda}\right)_{H=0} \tag{7.45}$$

This is Becquerel's formula, which in the case of diamagnetic substances is in rough agreement with experiment.

The only true theory of the effect has to be a quantum-mechanical one. Under the action of a magnetic field several phenomena occur. First, the ground energy level of an electronic system is split. Second, the resultant right wave has a different transition probability from the resultant left wave. Third, the excited state also splits. If the molecules are paramagnetic, the population of the sublevels created by the magnetic field changes (due to the orientation of the magnetic moments).

According to the general principles of molecular optics (Section 5.5), the magnetic circular dichroism (MCD) in a substance's absorption bands corresponds to the magnetic optical rotation (MOR). It is convenient to demonstrate these phenomena in absorption (Fig. 7.10 [34]). Splitting of the ground state gives $\omega_D = \omega_0 + m_G H$, $\omega_L = \omega_0 - m_G H$; the difference in transition probabilities can be expressed as $\varepsilon_D = \varepsilon_m(1-bH)$, $\varepsilon_L = \varepsilon_m(1+bH)$; the different sublevel populations are 1 and $1-2m_G H/kT$ ($m_G$ represents the splitting). We obtain [34]

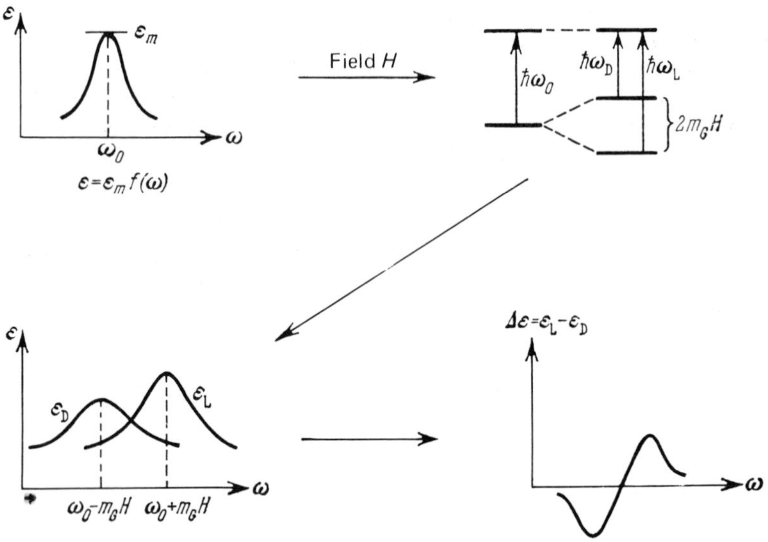


*FIG. 7.10 Scheme explaining the origin of magnetic circular dichroism.*

$$\varepsilon_L = \varepsilon_m(1+bH)f(\omega_0-m_GH)$$

$$\varepsilon_D = \varepsilon_m(1-bH)f(\omega_0+m_GH)(1-2m_GH/kT)$$

and

$$\Delta\varepsilon = 2\varepsilon_mH\{-m_G\frac{df(\omega)}{d\omega} + (b+\frac{m_G}{kT})f(\omega)\} = A+B+\frac{C}{kT} \tag{7.46}$$

If the splitting of the excited state, characterized by $m_E$, is also taken into account, then

$$\Delta\varepsilon = 2\varepsilon_mH\{-(m_G+m_E)\frac{df(\omega)}{d\omega} + (b+\frac{m_G}{kT})f(\omega)\} = A'+B+\frac{C}{kT} \tag{7.47}$$

The magnetic circular dichroism is expressed as a combination of the absorption curve B+C/kT and of its first derivative (A or A'). The same is true of the MOR, where A represents the derivative of the dispersion curve n(ω) and B and C represent the curve itself. Evidently the Becquerel formula corresponds only to A [35]. For a diamagnetic substance (C = 0) in which both components are of equal intensity (b = 0) the MCD has the shape shown in Fig. 7.11. On the other hand, if A term is negligibly small, the MCD curve corresponds to B+C/kT and is symmetrical (Fig. 7.12).

A rigorous quantum-mechanical theory of the MOR (and hence

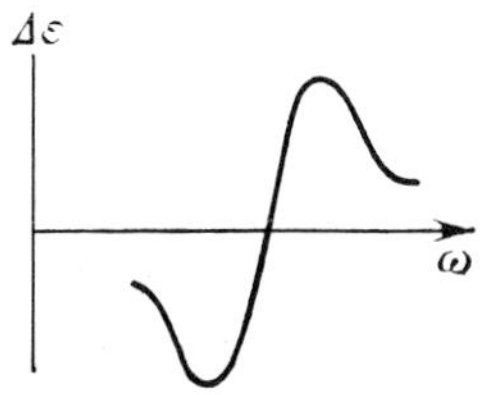


*FIG. 7.11 MCD curve for term A.*

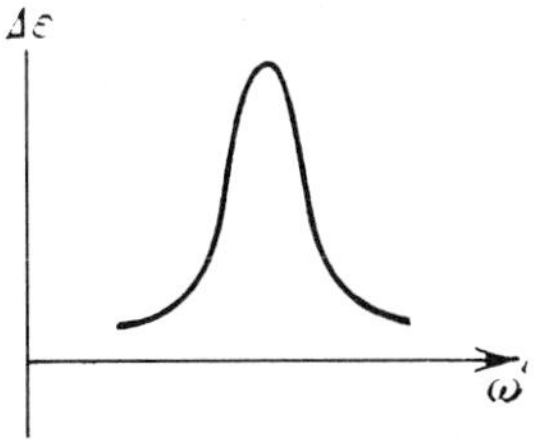


*FIG. 7.12 MCD curve for terms B and C.*

of the MCD) has been developed on the basis of perturbation theory [36-39]. Far from the absorption band, the specific rotation due to electronic transition from the ground level 0 to the excited one j is expressed as

$$\phi(0\to j) = -\frac{4\pi}{\hbar c} N_1\left\{ \frac{2\omega_{j0}\omega^2 A_j}{\hbar(\omega_{j0}^2-\omega^2)^2} + \frac{\omega^2}{\omega_{j0}^2-\omega^2}\left(B_j + \frac{C_j}{kT}\right)\right\}H_z \tag{7.48}$$

where

$$A_j = [(j|m_z|j)-(0|m_z|0)]\ \mathrm{Im}[(0|p_x|j)(j|p_y|0)]$$

$$B_j = \mathrm{Im}\Big\{\sum_{s\neq 0}\frac{(s|m_z|0)}{\hbar\omega_{s0}}[(0|p_x|j)(j|p_y|s)-(0|p_y|j)(j|p_x|s)]$$
$$+\sum_{s\neq j}\frac{(j|m_z|0)}{\hbar\omega_{sj}}[(0|p_x|j)(s|p_y|0)-(0|P_y|j)(s|p_x|0)]\Big\} \tag{7.49}$$

$$C_j = (0|m_z|0)\ \mathrm{Im}[(0|p_x|j)(j|p_y|0)]$$

Here $\omega$ is the frequency of the incident light; $(j|m_z|j)$, etc. are the matrix elements of the components of the magnetic moment along the z axis; $(0|p_x|j)$, $(s|p_y|0)$, etc., are the matrix elements of the components of the electric dipole moment; Im indicates that only the imaginary part of the expression need be taken. For diamagnetic molecules $(0|m_z|0) = 0$; hence, C = 0 and the A term is determined only by the magnetic moment of the excited state. The contribution of C has the form in (7.49) if $(0|m_z|0)H_z \ll kT$.

In the absorption region, the anomalous dispersion of magnetic rotation (ADMR) is observed, described by the formula

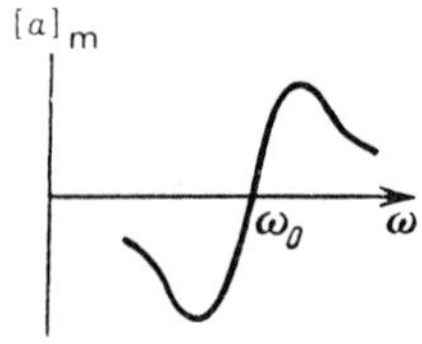


*FIG. 7.13 ADMR curve for term A.*

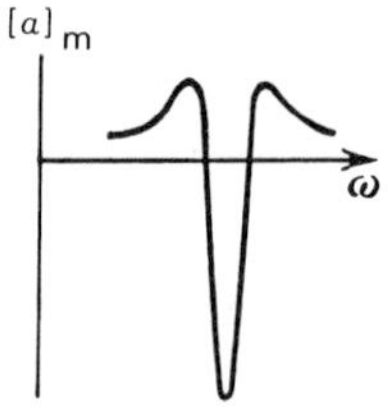


*FIG. 7.14 ADMR curve for terms B and C.*

$$\phi(0\to j) = -\frac{4\pi}{\hbar c} N_1 \left\{ \frac{2\omega_{j0}\omega^2[(\omega_{j0}{}^2-\omega^2)^2-\omega^2\Gamma_{j0}{}^2]}{[(\omega_{j0}{}^2-\omega^2)^2+\omega^2\Gamma_{j0}{}^2]^2} A_j + \frac{\omega^2(\omega_{j0}{}^2-\omega^2)}{(\omega_{j0}{}^2-\omega^2)^2+\omega^2\Gamma_{j0}{}^2}\left(B_j + \frac{C_j}{kT}\right)\right\} H_z \quad (7.50)$$

where $A_j$, $B_j$, $C_j$ have the values assigned earlier (7.49) and $\Gamma_{j0}$ coincides approximately with the bandwidth of the $0 \to j$ band. The shape of the ADMR curve for A is shown in Fig. 7.13 and for B and C in Fig. 7.14. ADMR of the A type was first observed in sodium vapor by Macaluso and Corbino in 1898 ([40], cf. also [32,41]). Such a symmetrical ADMR curve is called the Macaluso-Corbino effect. Simple classical considerations show that this effect must be very susceptible to the positions and intensities of the spectral bands. Term A, as we said, is due to the longitudinal Zeeman effect and, as shown in Fig. 7.15, expresses a differential interference effect inside the absorption band.

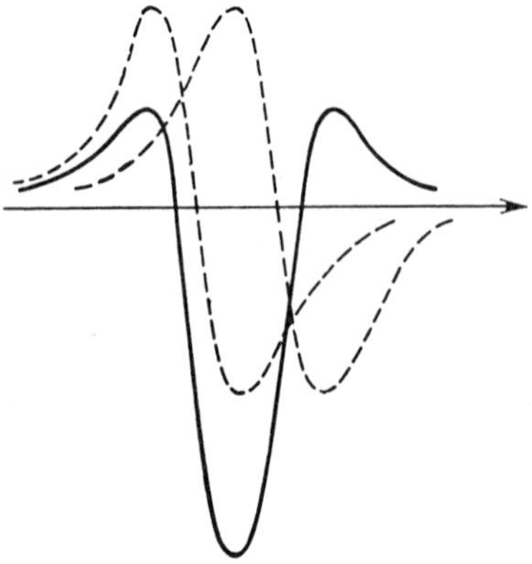

*FIG. 7.15 Scheme explaining the origin of ADMR (term A).*

Direct observation of the Zeeman effect in molecular spectra is practically impossible, since the bands are broad and the splitting $2\omega_H$ is very small. On the other hand, symmetrical ADMR can be expressed very clearly, particularly in the case of heme-containing proteins (cf. next section).

The MCD, characterized by the ellipticity $\theta$ (p. 282), is expressed by the formula

$$\theta(0\to j) = -\frac{4\pi}{\hbar c} N_1 \left\{ \frac{4\omega_{j0}\omega^3(\omega_{j0}^2-\omega^2)^2\Gamma_{j0}}{\hbar[(\omega_{j0}^2-\omega^2)^2+\omega^2\Gamma_{j0}^2]^2} A_j + \frac{\omega^3\Gamma_{j0}}{(\omega_{j0}^2-\omega^2)^2+\omega^2\Gamma_{j0}^2}\left(B_j + \frac{C_j}{kT}\right)\right\} H_z \tag{7.51}$$

It is easy to see that an asymmetrical MCD curve corresponds to term A (Fig. 7.11) and a symmetrical one to terms B and C (Fig. 7.12).

For an isotropic medium, we obtain the values of A, B, and C, which are independent of the orientation of the spatially fixed axis, by averaging over all orientations of the molecule

$$\begin{aligned} A_j &= \frac{1}{6}[(j|\mathbf{m}|j)-(0|\mathbf{m}|0)]\ \mathrm{Im}[(0|\mathbf{p}|j)(j|\mathbf{p}|0)] \\ B_j &= \frac{1}{3}\mathrm{Im}\left\{\sum_{s\neq 0}\frac{(s|\mathbf{m}|0)}{\hbar\omega_{s0}}(0|\mathbf{p}|j)(j|\mathbf{p}|s) + \sum_{s\neq j}\frac{(j|\mathbf{m}|s)}{\hbar\omega_{sj}}(0|\mathbf{p}|j)(s|\mathbf{p}|0)\right. \\ C_j &= \frac{1}{6}(0|\mathbf{m}|0)\ \mathrm{Im}[(0|\mathbf{p}|j)(j|\mathbf{p}|0)] \end{aligned} \tag{7.52}$$

where **m** and **p** are the vectors of the magnetic and electric moments of the molecule. The interrelation of $\phi$ and $\theta$ is given by the Kramers-Kronig expressions (p. 287). MOR, ADMR, and MCD differ from the corresponding quantities for natural optical activity (DOA, ADOA, and CD; Section 5.5). The Faraday effect has nothing to do with natural optical activity--it is observed in both chiral and nonchiral substances. This distinction is most obvious in the behavior of the secondary ray reflected from a mirror. If such a ray passes through an optically active substance, the rotation is compensated and vanishes, but the Faraday effect is doubled. Thus, the Faraday effect provides new information about molecular structure and important data for the interpretation of spectra.

ADMR can be observed experimentally in substances that have high magnetic rotation but comparatively small absorption, such as porphyrin compounds and rare-earth compounds. If the magnetic rotation is small, measurement of the ADMR is rather difficult. The Faraday effect is linear in relation to the magnetic field; therefore, an increase in H can improve the conditions for observation. A series of recent works report the use of superconducting magnets to set up magnetic fields with intensities of (and higher than) 50,000 G [42]. With conventional techniques, fields of only 20,000-30,000 G can be obtained. The description of a corresponding differential device is given in [43].

Observation of MCD is not limited by bands with high rotation. It is made with circular dichrograph with a magnetic additional device [42].

Detailed experimental data on the ADMR and MCD of many substances are contained in [39] and [44]. Broad examination of these effects began only recently, particularly in the work of Djerassi [45-47].

## 7.4 Magnetic Optical Rotation of Heme-Containing Proteins

Some characteristic MCD values for various organic compounds are the following: porphyrins, ∿100; annulenes, ∿3; purine, 0.2; cyclohexanone, 0.00002 [48]. Porphyrins are peculiar because of their high symmetry. This makes the Faraday effect a powerful method with which to investigate heme-containing proteins, cobamide enzymes, and chlorophyll and its derivatives. We have seen that ADMR and MCD possess specific features which make it possible to obtain information inaccessible by absorption spectra. Transitions corresponding to very weak absorption bands can give pronounced effects in ADMR and MCD. These effects are characterized by high band resolution and are very susceptible to small structural changes.

The first qualitative data on the ADMR of heme-containing proteins were obtained by Shashoua [49]. Systematic quantitative investigation of hemoproteins was done by Sharonov and co-authors [50-57].

Figure 7.16 shows absorption spectra and ADMR curves of metmyoglobin containing $Fe^{3+}$. Strongly overlapped 500-, 540-, 580-, and 630-nm absorption bands are well resolved in magnetic rotation. The 540- and 580-nm bands apparently correspond to $\pi\ \pi^*$ transitions in the porphyrin ring, the 500- and 630-nm bands to charge transfer transitions. Addition of a strong ligand, such as azide, transfers Fe into the low-spin state. This is expressed by an increase in the intensity of the $\beta$, and especially of the $\alpha$, absorption bands, and by an increase in the DMR. Similar changes are observed if metMb is transformed into hydroxyMb by an increase in the pH of the medium. This example demonstrates the possibilities of the method. In the absorption spectrum of CN-metMb, no $\alpha$ band is observed, whereas

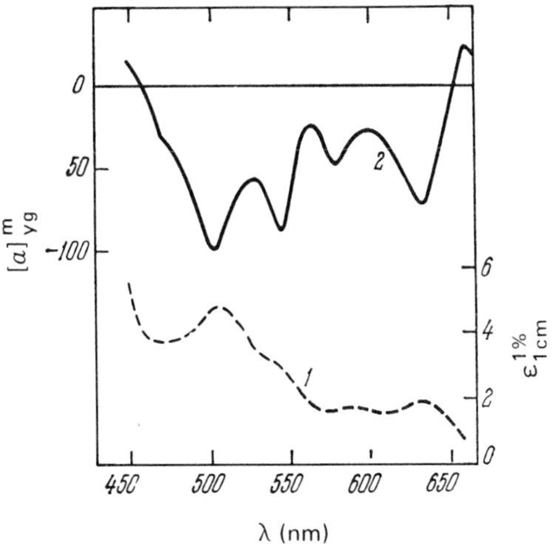


FIG. 7.16 (1) Absorption and (2) ADMR curves for metmyoglobin.

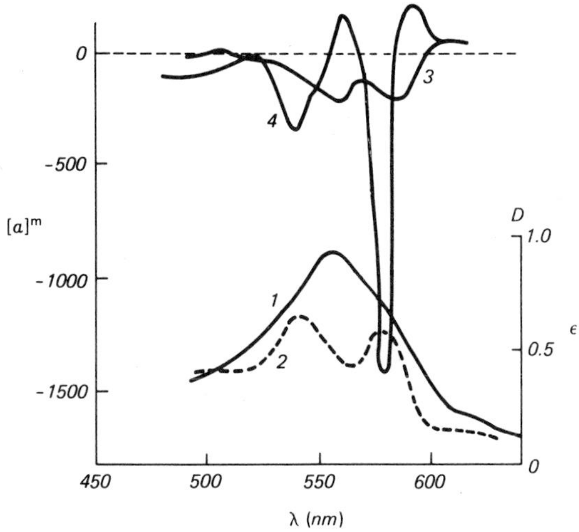


FIG. 7.17 (1) Mb and (2) $MbO_2$ absorption curves, (3) Mb and (4) $MbO_2$ DMR curves.

in the DMR curve it gives a strong effect.

The addition of one electron to Fe of the heme group with preservation of the low-spin state increases the DMR in the α band many times, but the absorption by only 20–30%. Figure 7.17 shows absorption and DMR curves of oxyMb and deoxyMb. The DMR in the α band is very susceptible to changes in the electron density of the heme group. The minimum depth of the DMR curve

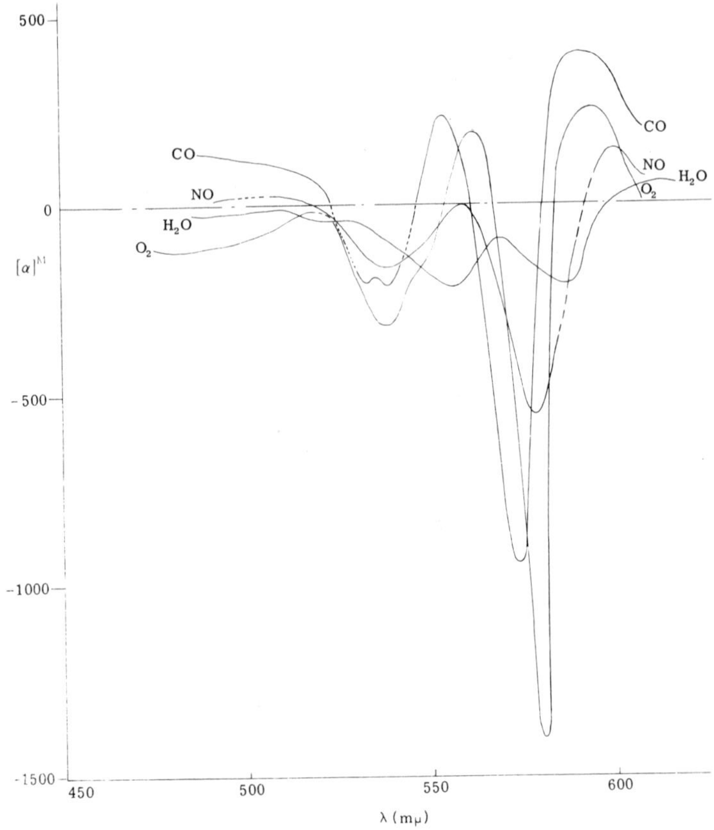


*FIG. 7.18 DMR curves for Mb complexes.*

of $MbO_2$ is approximately 40 times larger than that of metMb. These differences are actually determined by the changes in the electron density of the heme group due to different ligands. This is proved by correlation of the minimum depth with the isomeric shift in the Mössbauer effect, which measures the electron density in the Fe atom [51,58]. Figure 7.18 shows several curves for Mb complexes, Fig. 7.19 the DMR Mössbauer effect correlation for Mb complexes with different ligands.

The observed correlation of the DMR with conformational stability of Mb complexes, studied by means of urea denaturation, is a clear and direct expression of ECI [59].

The α band of deoxyMb, which is hardly observable in absorption spectra (p. 417), Table 7.2) but very obvious in DMR, is very susceptible to heme-heme interaction. In the DMR of Mb, the α and β bands give approximately equal effects (Fig. 7.17). In Hb, whose absorption spectrum is practically identical with that of Mb, the effect on the β band is the same as

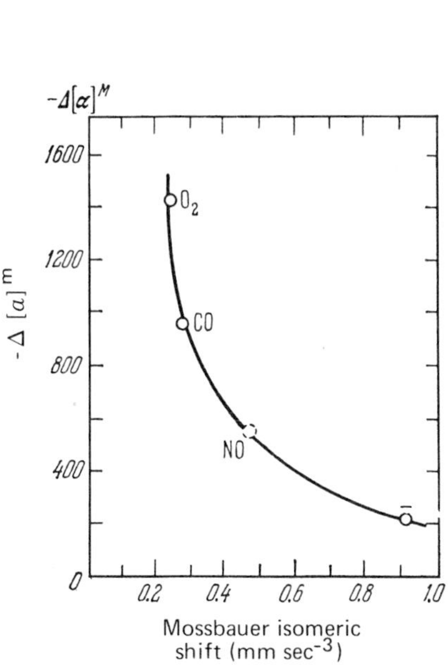


FIG. 7.19 Correlation of the DMR with the Mössbauer effect for Mb complexes with various ligands.

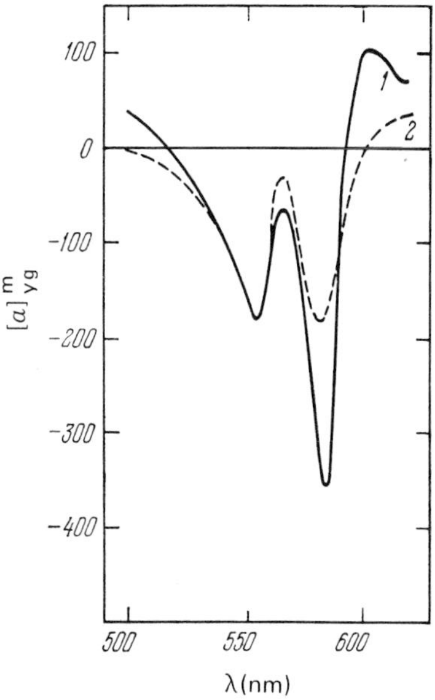


FIG. 7.20 DMR curves of (1) Hb and (2) isolated α and β chains.

in Mb, but the effect on the α band is twice as large (Fig. 7.20) [50]. Single α and β subunits of Hb give the same picture as Mb (Fig. 7.21). This must also be considered as an expression of ECI. The electronic state of the heme group, expressed as the DMR, appears to depend on the quaternary structure (i.e., on the conformational state of the protein moiety). The DMR method is an excellent technique with which to study the dissociation of Hb into subunits. With this method it has been shown that in the pH range from 10 to 11 tetramers dissociate into noncooperative dimers αβ, and the dependence of the dissociation constant $K_{4,2}$ on the pH has been determined. This dependence, in turn, shows that the dissociation is accompanied by the ionization of four amino acid residues (two Tyr and apparently two Lys). Extrapolation toward the neutral pH region has shown that the constant $K_{4,2}$ of Hb is no larger than $10^{-8}$ (i.e., at least two orders of magnitude less than $K_{4,2}$ for $HbO_8$). If the pH increases, the minimum corresponding to the β band remains unchanged at first but the α minimum diminishes in the course of dissociation. Both minima become equal at total dissociation. Upon further increase in the pH, the β minimum increses nearly 30 times and the α minimum vanishes (Fig. 7.21) [55]. This effect is due to denaturation of the chains. The

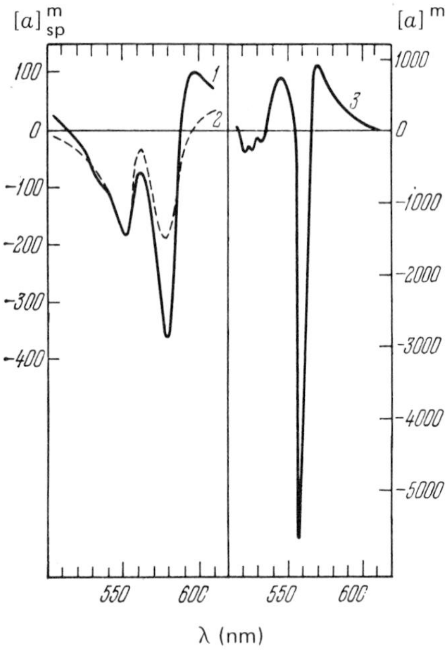


*FIG. 7.21 DMR of (human) deoxyHb (1) at pH 7.0 in 0.2 M phosphate buffer and (2) at pH 11.2 in 0.2 M phosphate-glycine buffer, at 20°C. (3) DMR of denatured Hb in 0.2 M NaCl + NaOH at pH 13.0 (Hb concentration 0.11%).*

DMR method is very susceptible to denaturation and makes it possible to detect 0.5% of denatured Hb molecules.

It has to be emphasized that the natural optical rotation (ADOA) and CD of Mb and Hb in heme absorption bands (induced optical activity, p. 302) are relatively small (cf. [60]).

The DMR method is also very effective in studies of another heme-containing protein, cytochrome *c*. This protein takes part in electron transfer in oxidative phosphorylation and the Fe atom of the heme group becomes functionally oxidized and reduced (p. 82). The causes of the different behavior of the heme group in Mb and Hb on the one hand and in cytochrome *c* on the other are related to the difference in protein structure close to the heme group. In cytochrome *c* the sixth valence of Fe is occupied by methionyl of the protein chain; in Mb and Hb it is free and able to bind the ligand. If cytochrome *c* becomes modified as a result of interaction with bromoacetic acid (pH 7.0, room temperature), the Met becomes alkylated and its bond with Fe becomes weaker. Such a carboxymethylated ferrocytochrome *c* can bind CO and $O_2$ without oxidation, like Mb and Hb. Native cytochrome *c* lacks this ability. The DMR of native cytochrome *c* and its absorption spectrum are shown in Fig. 7.22,

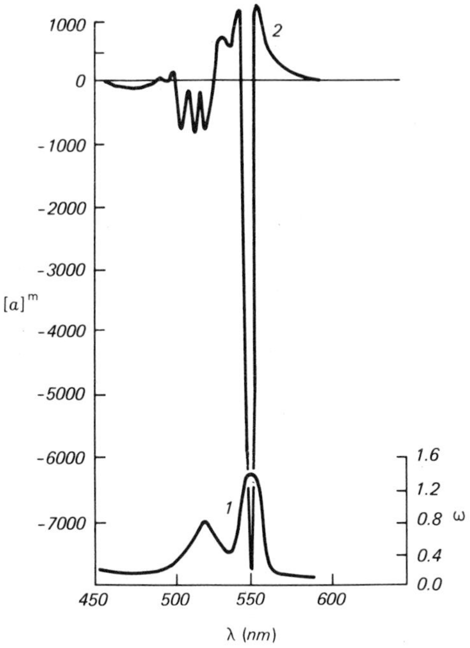


*FIG. 7.22 (1) Absorption curve and (2) DMR curve of cytochrome c.*

its MCD curve in Fig. 7.23. The advantages of DMR and MCD are clearly demonstrated in this case; the absorption spectrum contains only two bands; the magnetic spectra include up to five bands which are susceptible to structural changes. Figure 7.24 shows the DMR curves of carboxymethylated cytochrome *c* and of

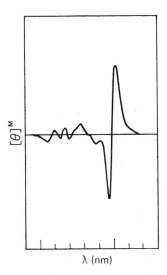

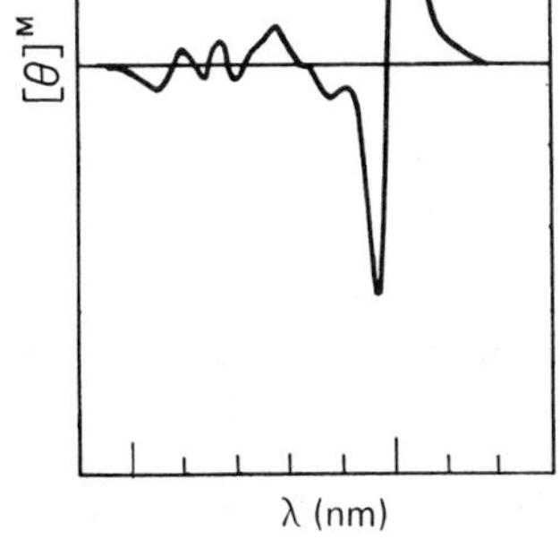


*FIG. 7.23 MCD curve of native cytochrome c.*

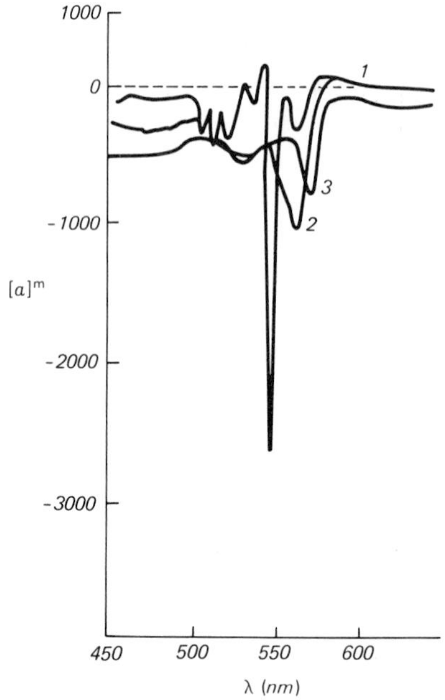


*FIG. 7.24 MCD curves of (1) carboxymethylated cytochrome c and its complexes with (2) CO and (3) $O_2$.*

its CO and $O_2$ complexes. The picture is very similar to that of Mb (Fig. 7.18). Carboxymethylated cytochrome *c* is myoglobin-like [54,56].

Interpretation of the magnetic spectra of heme-containing proteins requires detailed quantum-mechanical calculations with the aim of applying the theoretical results presented on p. 439. Such calculations have been made for low-molecular-weight porphyrin complexes (cf. [62,63]). However, not only are more exact calculations required, but the role of protein has to be taken into account.

## 7.5 Allosteric Enzymes

Catalytic enzyme action determines the course of all biochemical reactions--the nature of the products and the reaction rate. Biochemical reactions *in vivo* have to be precisely regulated and self-consistent, providing optimum chemical behavior of cell and organism. The most general considerations show that enzymes have to play not only an organizing but also a regulatory role. Regulation requires channels for forward and reverse communication, transmitting the proper information. Information transmission in cells is the transmission of chemical signals (i.e., of molecules and ions). The substances must function in the cell, the reactivity of which depends on the action of molecular signals. Since enzymes constitute the source of these signals, they have to be controlled by chemical feedback. We conjecture that there exist regulatory monoenzyme and polyenzyme systems. As we said earlier, an enzyme can be considered a transformer of signals--the input signal (i.e., substrate) is transformed into an output signal (i.e., product). If the output signal affects the function of the transformer, feedback occurs (either positive or negative, depending on whether the product activates or inhibits the enzyme).

Such systems do actually function in cells. Umbarger discovered the presence of successive enzymic reactions in which the end metabolite influences the activity of the enzyme, which catalyzes the first reaction in the sequence [64]. Inhibition has been detected, the kinetics of which is like competitive kinetics, while the structure of the inhibitor, called in this case allosteric (from ancient Greek *allos*, another; *steric*, spatial) is different from that of the substrate. Umbarger found that threonine-deaminase of *E. coli*, whose substrate is threonine, is inhibited by isoleucine, the end product of the subsequent reaction chain. Many allosteric enzymes (ASE) are known at present, similarly regulated according to the feedback principle. A list of 24 such systems was published in 1965 [65]; many more are now known. Detailed reviews of more recent data have been published by Bulgarian scientists ([66], cf. also [67]). Figure 7.25 shows an allosteric feedback scheme for the system of CTP synthesis where the ATC is aspartate transcarbamylase and CTP is the allosteric effector [68].

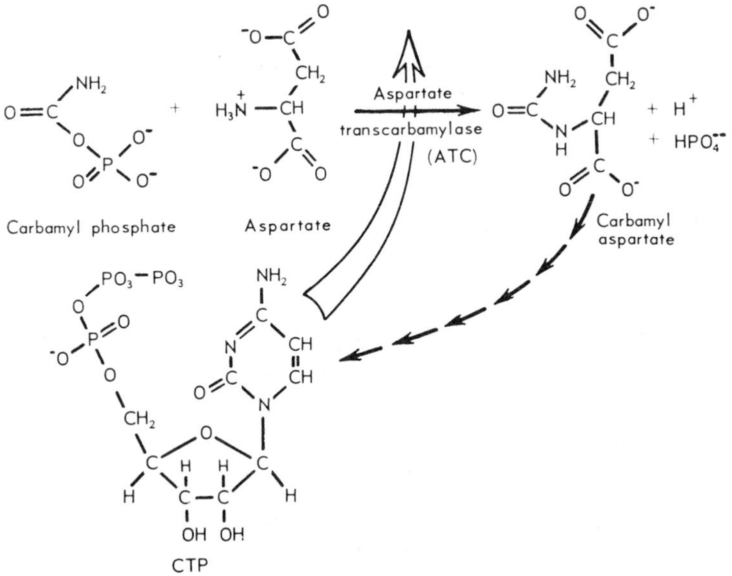


*FIG. 7.25 Allosteric feedback scheme.*

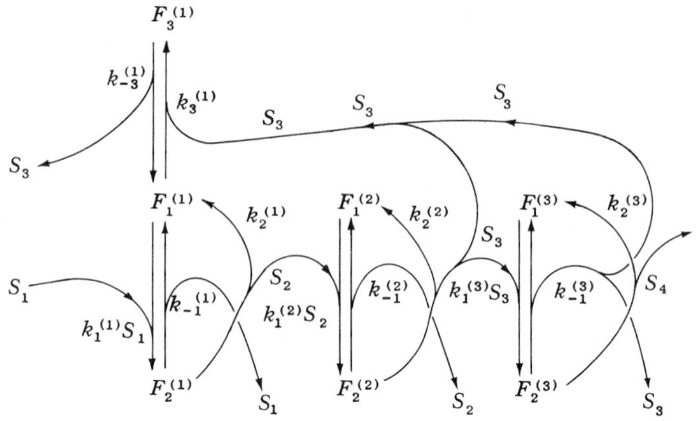

FIG. 7.26 *Allosteric inhibition scheme.*

We present an illustrative calculation of stationary-state feedback in an elementary allosteric system containing three enzymic reactions, the catalyst of the first of which is ASE [33,62,70]. The scheme of the reactions is shown in Fig. 7.26. In stationary-state conditions the reaction rates $S_1 \to S_2$, $S_2 \to S_3$, and $S_3 \to S_4$ are

$$v_1 = \frac{v_{m1} S_1}{S_1 + K_1 (1 + K_I S_3)}, \qquad v_2 = \frac{v_{m2} S_2}{S_2 + K_2}, \qquad v_3 = \frac{v_{m3} S_3}{S_3 + K_3} \tag{7.53}$$

where $v_{m1}$, $v_{m2}$, $v_{m3}$ are the maximal reaction rates; $K_1$, $K_2$, $K_3$ are the Michaelis constants; and $K_I$ is the inhibition constant (cf. p. 360). The stationarity condition is $v_1 = v_3$; hence

$$S_3^2 + \frac{1}{K_I} \left[1 + \frac{S_1}{K_1} \left(1 - \frac{v_{m1}}{v_{m3}}\right)\right] S_3 - \frac{v_{m1} K_3 S_1}{v_{m3} K_1 K_I} = 0 \tag{7.54}$$

The stationary rate $v_1 = v_3$ is smaller than in the system without feedback (i.e., at $K_I = 0$). Regulation (i.e., the maintenance of $S_3$ at a stationary level) is effected because an increase in $S_3$ provokes an unlimited increase in $S_1$, since

$$S_1 = v_{m3} K_1 S_3 \frac{1 + K_I S_3}{v_{m1} K_3 + (v_{m1} - v_{m3}) S_3}$$

On the other hand, in the absence of feecback the limiting value of $S_1$ does not depend on $S_3$

$$\left(S_1\right)_{\substack{S_3\to\infty \\ K_I\to 0}} = \frac{v_{m3}}{v_{m1}-v_{m3}} K_1$$

The rate of $S_3$ formation is determined by the consumption of this substance in the subsequent reactions, that is, by the rate $V_3$ of transformation $S_3 \to S_4$. If $v_{m3}$ is large ($v_{m3} >> v_{m3}$), then according to Eq. (7.54) $S_3$ is small and the first enzyme is not inhibited. Conversely, if $v_{m3}$ is small (i.e., $v_{m3} << v_{m1}$), $S_3$ is large. If $S_3 >> K_3$, then

$$S_3 \cong (S_1/K_1K_I)(v_{m1}/v_{m3})$$

and

$$v_1 \cong \frac{v_{m1}v_{m3}S_1}{(v_{m1}+v_{m3})S_1+v_{m3}K_1} \cong v_{m3} << \frac{v_{m1}S_1}{S_1+K_1}$$

The maximum rate of the first reaction becomes not $v_{m1}$ but a much smaller value, $v_{m3}$. If $v_{m3} \to 0$, we get $v_1 = v_3 \to 0$ (i.e., total inhibition of the process).

The difference between the allosteric effector and the substrate of the ASE suggests that the effector is bound by an enzyme site other than the active one, where the substrate is bound and transformed. Allosteric inhibition has to be considered a result of a change in the conformation of the entire ASE molecule. This transformation also affects the active site, which is probably rather far from the effector binding site, and disturbs its catalytic properties [71]. Such a model correlates with Koshland's theory (p. 381), and if it is correct, allosterism has to be considered an important confirmation of this theory.

However, the actual situation is more complicated. Studies have established two fundamental features of ASE. First, all the ASE that have been studied have quaternary structure (i.e., their molecules contain several globular subunits). Second, the stationary-state kinetics of corresponding reactions differs from Michaelis-Menten kinetics; the v(S) curves have singularities, mainly bends. Figure 7.27 shows the sigmoidal v(S) curve for aspartate transcarbamylase [68]. In the presence of allosteric inhibitor, CTP the singularity vanishes. Inhibition kinetics is also different from elementary kinetics. As was shown by Changeux [72], the kinetics of the transformation of threonine by the ASE threonine deaminase can be expressed by the formula

$$\log[v/(v_m-v)] = n \log S - \log K \qquad (7.55)$$

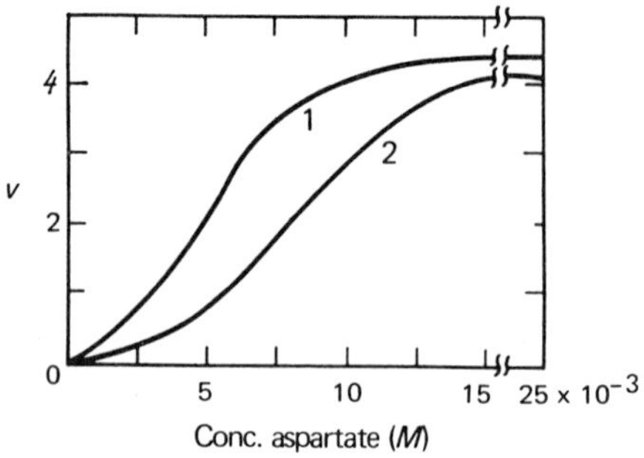


*FIG. 7.27 Curve of v versus S for aspartate transcarbamylase: (1) control; (2) $2\times10^{-4}$ M CTP.*

and the kinetics of inhibition by allosteric effector by the formula

$$\log[v/(v_0-v)] = \log K' - n' \log I \tag{7.56}$$

Here v is the stationary rate, $v_m$ is the maximum rate, $v_0$ is the rate at I = 0, and K and K' are constants. In elementary kinetics n and n' must be equal to 1; Changeux found n = 1.37 and n' = 1.86.

These ASE properties are similar to those of hemoglobin, which has quaternary structure and an S-shaped curve of saturation by oxygen. As was shown in Section 7.1, this singularity is due to cooperative interaction of the subunits. We can conclude on the same ground that the peculiarities of ASE kinetics are due to cooperative interaction of enzyme subunits. In this sense Hb is not only an "honorary enzyme" (p. 415) but also an "honorary allosteric enzyme." Hb is the model of ASE and its inclusion in the ASE list [65] was not accidental. Treating the above-mentioned results from the standpoint of cooperativity enables us to interpret formulas (7.55) and (7.56) [70].

Let us examine the elementary model of an enzyme whose molecule contains two interacting identical subunits [39,70]. Each subunit contains an active site. The scheme of a stationary-state process is presented in Fig. 7.28. The enzyme molecule can be in three states: $F_{00}$, in which both active sites are free; $F_{10} = F_{01}$, in which one site is free, and the other is occupied by substrate; and $F_{11}$, in which both sites are occupied by substrate. The steady-state kinetics equations are

$$\begin{aligned} \dot{F}_{00} &= -2k_1SF_{00}+2(k_{-1}+k_2)F_{10} = 0 \\ 2\dot{F}_{10} &= 2k_1SF_{00}-2(k_{-1}+k_2+k_3S)F_{10}+2(k_{-3}+k_4)F_{11} = 0 \\ \dot{F}_{11} &= 2k_3SF_{10}-2(k_{-3}+k_4)F_{11} = 0 \end{aligned} \tag{7.57}$$

*FIG. 7.28 Scheme of a reaction with an enzyme that contains two subunits.*

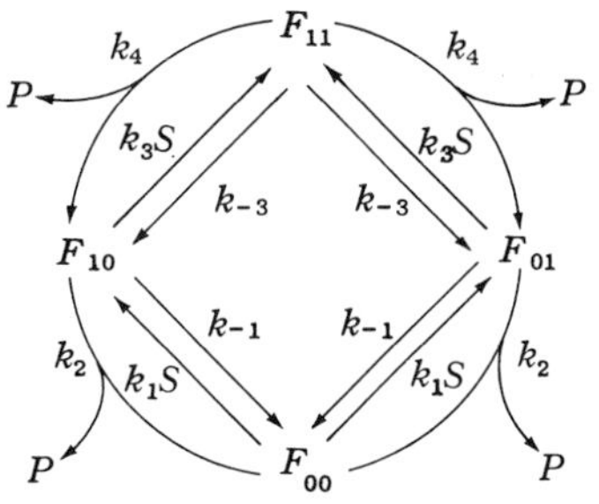


The total enzyme concentration is constant

$$E = F_{00}+2F_{10}+F_{11} \tag{7.58}$$

According to the scheme Fig. 7.28 the reaction rate is equal to

$$v = 2k_2F_{10}+2k_4F_{11} \tag{7.59}$$

Solving Eqs. (7.57) and (7.58), we find

$$v = 2k_2ES \frac{\beta K+\alpha S}{\beta K^2+2\beta KS+S^2} \tag{7.60}$$

where $\alpha = k_4/k_2$, $\beta = K'/K$, $K = (k_{-1}+k_2)/k_1$, and $K' = (k_{-3}+k_4)/k_3$. Cooperativity (i.e., the interaction of two active sites) is expressed by the difference from unity of $\alpha$ and $\beta$. Indeed, if $\alpha = \beta = 1$,

$$v = 2k_2ES \frac{K+S}{(K+S)^2} = \frac{v_mS}{K+S} \tag{7.61}$$

that is, the kinetics coincides with Michaelis-Menten kinetics. On the other hand, if this condition is not obeyed, the v(S) curve can have a bend or a maximum, or both. Analysis of formula (7.60) shows that a bend is possible

(1) if $\alpha < 0.5$ and $\beta$ has any value;
(2) if $1 > \alpha > 0.5$ and $\alpha/(2\alpha-1) > \beta > \alpha^2/(2\alpha-1)$; and
(3) if $\alpha > 1$ and $\alpha^2/(2\alpha-1) > \beta > \alpha/(2\alpha-1)$.

A maximum can appear only if $\alpha < 0.5$ and at any value of $\beta$.

Assume that each of the two subunits simultaneously includes a site which binds the inhibitor I, an allosteric effector. Correspondingly, we shall have nine enzyme states: $F^{00}_{00}$, $F^{00}_{10} = F^{00}_{01}$, $F^{00}_{11}$, $F^{20}_{00} = F^{02}_{00}$, $F^{20}_{01} = F^{02}_{10}$, $F^{22}_{00}$, $F^{00}_{11}$, where 0 corresponds to the free site, 1 to the site occupied by S, and 2 to the site occupied by I; the sites for S are indicated by subscripts, those for I by superscripts. Assuming for simplicity that states $F_{01}^{20}$ and $F_{10}^{02}$ are not attainable and that the substrate-product transformation occurs only in the $F_{11}^{00}$ state, we get [70]

$$\frac{v_0}{v_m - v_0} = \frac{zS^2}{x_1 + y_1 S}, \qquad \frac{v}{v_0 - v} = \frac{x_1 + y_1 S + zS^2}{x_2 I + x_3 I^2 + y_2 SI + y_3 SI^2}$$

where $x_i$, $y_i$, z are the combinations of rate constants. These formulas agree with (7.55), (7.56) if $2 > n, n' > 1$.

The treatment of such enzymic processes requires mathematical algorithms simplifying the solution of the complex systems of linear equations. They will be discussed in the next section.

The model discussed here represents direct cooperativity, that is, the difference in rate constants for the states in which the substrate occupies one or two sites. The equilibrium function for saturation by substrate for the two-site system under consideration is expressed as

$$Y = \frac{2F_{10} + 2F_{11}}{2(F_{00} + 2F_{10} + F_{11})} = \frac{\beta KS + S^2}{\beta K^2 + 2\beta KS + S^2} \qquad (7.62)$$

which is the formula of Langmuir's isotherm at $\beta = 1$. At $\beta \neq 1$ the Y(S) curve can have a bend but not a maximum.

An indirect cooperativity model was suggested by Monod et al., in order to treat ASE properties [65]. The enzyme molecule is an oligomer containing two or more identical subunits (protomers) having equivalent spatial positions. Therefore, the molecule has symmetry elements. It can be built isologously or homologously; in the latter case an unlimited oligomer is possible (Fig. 7.29). There is an active site corresponding to every ligand (substrate or allosteric effector) on every protomer.

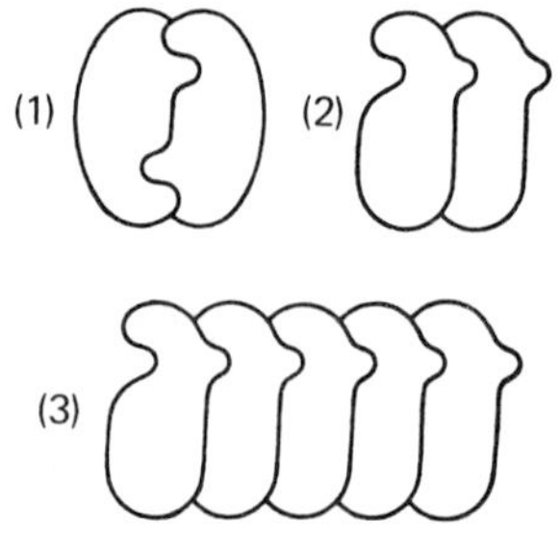


*FIG. 7.29 Possible structures of (1) an isologous and (2) a heterologous dimer; (3) possible structure of a heterologous polymer.*

It is hypothesized that the protomer as a whole can be in two or more conformational states, preserving its symmetry. The affinity of stereospecific sites for ligands is altered if the state of the oligomer is changed. Such a system is a cooperative one.

Without loss of generality let us consider a dimer which can be in two different states R and T. In every state the dimer can bind 0, 1, or 2 molecules of ligand S. We have six states $R_{00}$, $R_{10}$ (includes two states), $R_{11}$, $T_{00}$, $T_{10}$ (two states), $T_{11}$. The equilibrium conditions are

$$T_{00} = LR_{00}$$

$$R_{10} = 2R_{00}\frac{S}{K_R}, \quad R_{11} = \frac{1}{2}R_{10}\frac{S}{K_R} = R_{00}\frac{S^2}{K_R{}^2}$$

$$T_{10} = 2T_{00}\frac{S}{K_T}, \quad T_{11} = \frac{1}{2}T_{10}\frac{S}{K_T} = T_{00}\frac{S^2}{K_T{}^2}$$

The total enzyme concentration is

$$E = R_{00}+R_{10}+R_{11}+T_{00}+T_{10}+T_{11}$$

Here $K_R$ and $K_T$ are the dissociation constants for S in states R and T; L is the equilibrium constant for the $R \rightleftarrows T$ transition if S is absent. The dissociation constants in states $R_{10}$ and $R_{11}$ (and correspondingly in $T_{10}$ and $T_{11}$) are assumed to be equal. The function of saturation of the protein by substrate is

$$Y = \frac{R_{10}+2R_{11}+T_{10}+2T_{11}}{2(R_{00}+R_{10}+R_{11}+T_{00}+T_{10}+T_{11})}$$

$$= \left(K_R\frac{1+Lc}{1+Lc^2}S+S^2\right)\Big/\left(K_R\frac{1+L}{1+Lc^2}+2K_R\frac{1+Lc}{1+Lc^2}S+S^2\right) \tag{7.63}$$

where $c = K_R/K_T$. The Y(S) curve has a point of inflection (a bend). At $c = 1$ or at $L \to 0$ or $L \to \infty$ the cooperativity vanishes and Y becomes Langmuir's isotherm

$$Y = S/(K_R+S) \tag{7.64}$$

The expression for the rate of substrate transformation obtained from equilibrium conditions has the form

$$v = 2Ek\frac{1+\kappa Lc^2}{1+Lc^2}\left(K_R\frac{1+\kappa Lc}{1+\kappa Lc^2}S+S^2\right)\Big/\left(K_R\frac{1+L}{1+Lc^2}+2K_R\frac{1+Lc}{1+Lc^2}S+S^2\right) \tag{7.65}$$

where k is the rate constant for states $R_{10}$ and $R_{11}$ and $\kappa k$ is that for states $T_{10}$ and $T_{11}$.

Generalizing Eq. (7.63) for a system containing n protomers, we obtain

$$Y = \frac{Lcx(1+cx)^{n-1} + x(1+x)^{n-1}}{L(1+cx)^n + (1+x)^n} \tag{7.66}$$

where $x = S/K_R$. This equation was applied to horse Hb, for which n was taken to be equal to 4 [65]. Comparison of formula (7.66) with the experimental oxygenation curve showed that L = 9054 and c = 0.014. It has to be emphasized, however, that only the Hb subunit pairs are identical, $\alpha^2\beta^2$ (p. 216), and a more rigorous theory is required.

These calculations concern the homotropic cooperative effect expressed by a shift in the $R \rightleftarrows T$ equilibrium due to the action of one ligand S. If $c \ll 1$ and L and $\kappa$ are not too large, formulas (7.63) and (7.65) for a dimer take on the form

$$\bar{Y} \cong x \frac{1+x}{L+(1+x)^2} \tag{7.67}$$

$$v \cong 2Ekx \frac{1+x}{L+(1+x)^2} \tag{7.68}$$

Assume that along with the substrate S the inhibitor I and activator A act upon a dimer, and that each of the two protomers contains three active sites--for S, I, and A. Assuming for simplicity that the dimer has an affinity for I only in the R state and to A only in the T state, we obtain the saturation function for S at $c \ll 1$

$$\bar{\bar{Y}} \cong x \frac{1+x}{L'+(1+x)^2} \tag{7.69}$$

where

$$L' \cong L \frac{(1+I/K_I)^2}{(1+A/K_A)^2} \tag{7.70}$$

Heterotropic interaction with effectors influences Y(S). I increases and A decreases the cooperativity.

The Monod-Wyman-Changeux model was generalized by Kotani [73], who examined the continuous distribution of protein conformations and independent conformational changes in every subunit (cf. also the work of Whitehead [74]).

In the case of positive cooperativity the models of direct

and indirect cooperativity can give equivalent results [33]. Unlike the Monod-Wyman-Changeux model, the direct cooperativity model can also describe negative cooperativity (i.e., a decrease in the affinity for ligand in the course of saturation of active sites).

Koshland et al., [75] examined direct cooperativity in tetrameric proteins. If the tetramer is a tetrahedron, then the pair interactions of the four subunits are equal. If it is a square, then every subunit interacts only with two neighboring ones, but not with the diagonal one. Let us assume that every subunit can be in two conformations A and B, and that only B can bind the ligand. The saturation function for the "square" model is

$$Y = \frac{K_{AB}^2 x' + (K_{AB}^4 + 2K_{AB}^2 K_{BB}) x'^2 + 3K_{AB}^2 K_{BB}^2 x'^3 + K_{BB}^4 x'^4}{1 + 4K_{AB}^2 x' + (2K_{AB}^4 + 4K_{AB}^2 K_{BB}) x'^2 + 4K_{AB}^2 K_{BB}^2 x'^3 + K_{BB}^4 x'^4} \qquad (7.71)$$

where $K_{AA} = 1$; $K_{AB}$ and $K_{BB}$ are the interaction constants of the subunits in corresponding conformations. The effective ligand concentration x' involves as multipliers both the affinity constant of the ligand and the equilibrium constant of the two subunit conformations. If $K_{BB}/K_{AB} < 1$, Y(x') has the shape of a curve with an intermediate plateau; if $K_{BB}/K_{AB} > 1$, the curve has a sigmoidal shape. See also [76,77].

The presence in ASE of quaternary structure determines the possibility of their dissociation under the action of substrates and effectors. This has actually been observed in a series of enzymes. Dissociation provides allosteric interactions, due to the shift in equilibrium of oligomeric ASE forms under the action of allosteric effectors. These effects are expressed in cooperative kinetics. Corresponding theory was developed by Kurganov (cf. [67,78-80]). The quaternary structure of ASE was directly observed by means of electron microscopy [81].

The singularities in the v(S) curve can be due in principle not to cooperative interactions of subunits but to relaxational conformational properties of the enzyme. Assume that an enzyme molecule, having already transformed the substrate into a product, remains in an active conformationally changed state. If the time of relaxation toward the initial nonexcited state is larger than the time between successive collisions of enzyme and substrate molecules, or if these times are of the same order, the stationary-state kinetics would imitate the cooperative kinetics. Such a model was proposed by Rabin [82] (see also the later work [83]); the scheme of the process is presented in Fig. 7.30. Here $F_0$ is the free enzyme molecule in its initial conformation, $F_1$ is the nonactive enzyme-substrate complex, $F_2$ is the active complex, $F_3$ is the free enzyme with

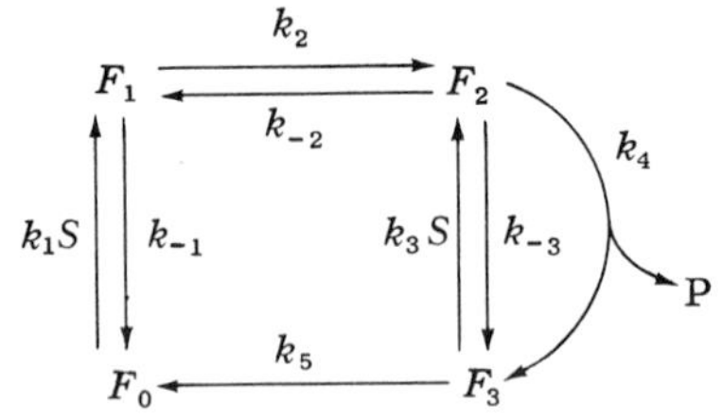


*FIG. 7.30 Reaction scheme according to Rabin.*

active conformation. The kinetic equations are

$$\dot{F}_0 = -k_1SF_0+k_{-1}F_1+k_5F_3$$
$$\dot{F}_1 = k_1SF_0-(k_{-1}+k_2)F_1+k_{-2}F_2 \qquad (7.72)$$
$$\dot{F}_2 = k_2F_1-(k_{-2}+k_{-3}+k_4)F_2+k_3SF_3$$
$$\dot{F}_3 = (k_{-3}+k_4)F_2-(k_3S+k_5)F_3$$

The reaction rate is $v = k_4F_2$. The stationary-state solution under the condition

$$E = F_0+F_1+F_2+F_3 = \text{const}$$

has the form

$$v = \{k_1k_2k_4S(k_3S+k_5)E\}/\{k_1(k_{-2}+k_2)(k_3S+k_5)S$$
$$+[k_{-1}k_{-2}k_3+(k_{-3}+k_4)k_5]S+(k_{-1}+k_2)(k_{-3}+k_4)k_5+k_{-1}k_{-2}k_5\} \qquad (7.73)$$

This function has the same form as (7.60); that is

$$v = (AS^2+BS)/(CS^2+DS+K) \qquad (7.74)$$

and can therefore possess singularities. "Cooperativity" vanishes if $k_3 = 0$. In this case v does not depend on $k_5$. If $k_5 = 0$, "cooperativity" is absent too, and v does not depend on $k_3$. v becomes zero at both $k_2 = 0$ and $k_4 = 0$.

Obviously the appearance in this case of a rate curve does not depend on the presence of quaternary structure. On the other hand, the pronounced changes in the quaternary structure of hemoglobin produced by oxygenation (p. 424) demonstrate the direct role of subunit interaction in the cooperative behavior of proteins. This calculation has only an illustrative meaning. The nonequilibrium enzymic processes require special investigation.

Allosteric effects, cooperative properties of oligomeric proteins, which play a regulatory role, are of great importance in many biological processes. It can be thought that functional proteins in membranes and contractile proteins have allosteric properties. Allosterism is the most important feedback mechanism at the molecular level.

## 7.6 The Kinetics of Complex Enzymic Reactions

Solution of kinetic problems for complex chemical reactions, particularly enzymic ones, requires simplifying algorithms, even in the stationary case. Structural methods for the analysis of complex reactions have been developed by Schwab [84], Horiuti [85], Christiansen [86], and Semenov [87]. King and Altman [88] suggested an effective algorithm for the treatment of stationary-state enzymic reactions which has been used to solve a series of concrete problems [89-91]. Employing this algorithm, Cleland [92] suggested a nomenclature for many-substrate reactions and a method for representing them graphically. However, the methods of King and Altman and of Kleland are practically inapplicable in complicated cases. The best algorithm known at present is based on the application of graph theory. A graph is a topological pattern formed by the knot points or intersections and the lines connecting them [93-95]. The theory of nondirectional graphs was first applied for calculation of reaction rates in the works of Temkin [96]. The application of directed graphs to enzymic reactions was proposed in [97]. A directed graph is a set of points connected by directed lines (arrows) [93]. Such graphs can be applied to solve various problems involving branched and unbranched flows of matter, charges, or information. Graph theory is very effective in electro- and radiotechnics [98-100].

Let n+1 enzyme complexes $EX_i$, including free enzyme, i.e., $i = 0,1,\ldots,n$) be formed in an enzymic reaction. Let us correlate a point on the graph of i, j, etc., with every complex and two branches with opposite directions with every reversible stage of the interaction. If a stage is irreversible, we have to consider only one directed branch. Every branch can be characterized by its value--by the probability of stage $w_{ij}$ equal to the rate constant $k_{ij}$ or $k_{ij}$ multiplied by the ligand concentration in the stages of enzyme-ligand interaction. The rate of stage $v_{ij}$ along the branch $i \rightarrow j$ is equal to

$$v_{ij} = [EX_i]w_{ij} \tag{7.75}$$

The continuity condition has to be obeyed at every stage; that is,

$$[EX_i] \sum_{j \neq i} w_{ij} = \sum_{j \neq i} [EX_j]w_{ji} \tag{7.76}$$

This condition is equivalent to the condition of the steady state.

Let us examine as an example the reaction of one substrate S with one modifier M. (Such a reaction was investigated by King and Altman [88].) The chemical equations have the form

$$E + S \underset{k_2}{\overset{k_1}{\rightleftharpoons}} ES \xrightarrow{k_\beta} E + P$$

$$E + M \underset{k_8}{\overset{k_7}{\rightleftharpoons}} EM$$

$$ES + M \underset{k_4}{\overset{k_3}{\rightleftharpoons}} EMS \xrightarrow{k_\alpha} EM + P$$

$$EM + S \underset{k_6}{\overset{k_5}{\rightleftharpoons}} EMS$$

The graph of a stationary system is shown in Fig. 7.31. Determination of the stationary-state rate is made according to simple rules without cumbersome calculations. Let us define some notions.

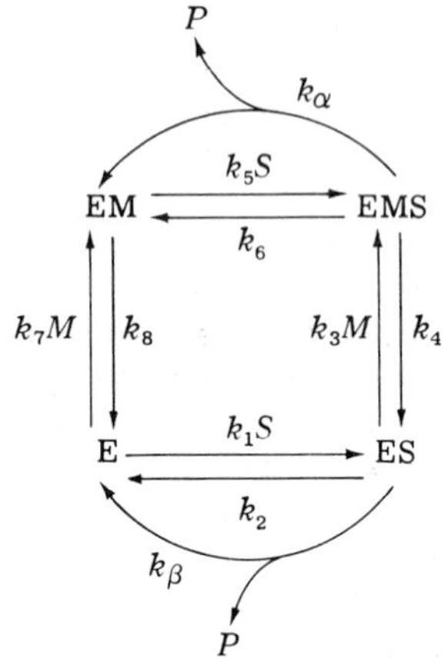


*FIG. 7.31 Graph of the reaction of E with S and M.*

The system is a sequence of branches directed toward one side. Thus, in our case there are four paths from point ES to point EM. Their values P are the products of the branches along the path. We have

$$P^{(1)} = k_2k_7[M], \quad P^{(2)} = k_\beta k_7[M], \quad P^{(3)} = k_3[M]k_6, \quad P^{(4)} = k_3[M]k_\alpha$$

The tree of the graph directed toward point i is the set of branches connecting all points on the graph without forming a closed cycle and directed toward point i. Thus, there are ten trees directed toward the vertex (Fig. 7.32). The cycle is a continuous sequence of branches directed to one side.

The determinant of the point i is the sum of the values

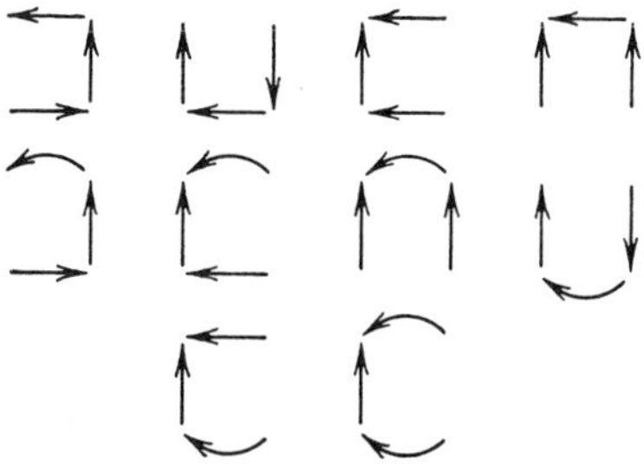

*FIG. 7.32 The trees of the graph.*

of all trees directed toward point i. The stationary-state rate of an enzymic reaction is expressed by means of the graph determinants

$$v = E(\Sigma_i k_i D_i / \Sigma_i D_i) \quad (7.77)$$

where E is the total enzyme concentration, $D_i$ is the determinant of point i, and $k_i$ is the rate constant of product flow from this point. The summation in the denominator has to be performed over all points; that in the numerator over the points corresponding to product formation. This rule coincides with the one suggested by King and Altman [88], but the graph method permits us to simplify the problem, reducing the number of trees. The method of directed graphs also has many advantages over Temkin's method of nondirected graphs [96], since it makes it possible to obtain quantitative solutions by using structural rules without solving equations.

The calculation of v is simplified first by means of the junction of $i \to j$ branches if there are several (N) branches of this kind. They can be replaced by one branch with the value

$$w_{ij} = \sum_{m=1}^{N} w_{ij}^{(m)} \quad (7.78)$$

Thus, in our case the summary branch EMS → EM has the value $k_\alpha + k_6$, and branch ES → E has the value $k_\beta + k_2$. We obtain four instead of ten trees (Fig. 7.33).

The base determinants are different for different paths. Let us examine $D_i$. We select some auxiliary point j and examine all paths $j \to i$ with values $P_{ji}^{(1)}, P_{ji}^{(2)}, \ldots$ . If we contract the path into a point, the graph will be simplified. Let us denote the determinant of the graph obtained by such a junction as $D_{ji}^{(1)}, D_{ji}^{(2)}, \ldots$ for every one of the paths $P_{ji}^{(1)}$, $P_{ji}^{(2)}, \ldots$ contracted into one point. Then the determinant $D_i$ of the point i becomes

$$D_i = \sum_m P_{ji}^{(m)} D_{ji}^{(m)} \quad (7.79)$$

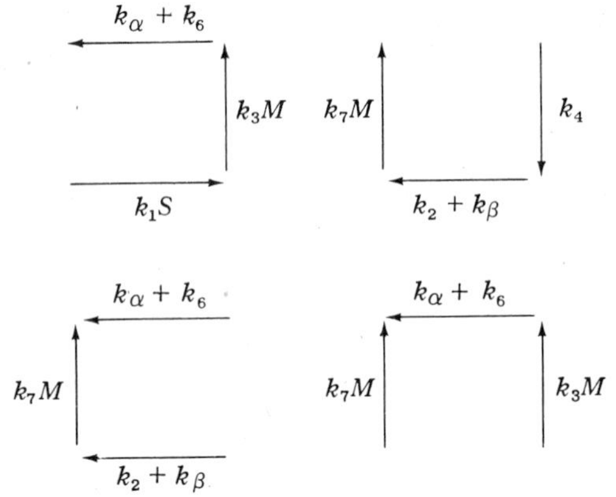


*Fig. 7.33 The four basic trees of the graph.*

Calculate, for example, $D_{EM}$ on our graph. We select ES as an auxiliary point. Taking into account the addition of parallel branches, we find two paths ES → E → EM and ES → EMS → EM with values

$$P^{(1)} = (k_\beta+k_2)k_7M \quad \text{and} \quad P^{(2)} = k_3M(k_\alpha+k_6)$$

After the contraction of these paths into points, we obtain the graphs with determinants

$$D^{(1)} = k_\alpha+k_6+k_4, \quad D^{(2)} = k_7M+k_1S$$

Let us explain the derivation of these branches and trees graphically (Fig. 7.34). The dotted arrows show the paths from ES toward EM, the solid arrows the determinants of the subgraphs obtained after the contraction of paths into points (the trees directed toward contracted points). Each such subgraph contains two parallel branches whose values have to be added. We obtain

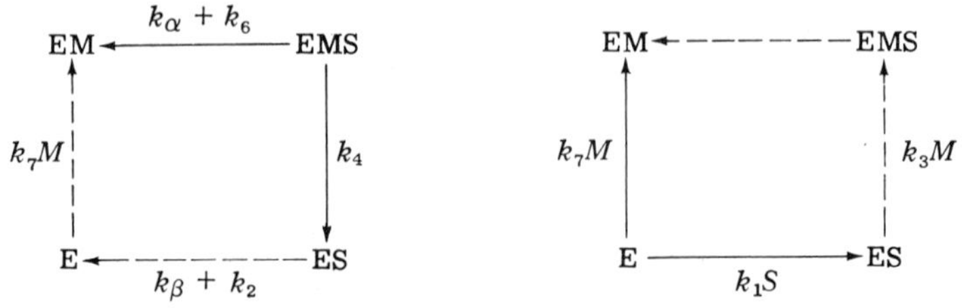


*FIG. 7.34 Establishment of paths and trees.*

$$D_{EM} = P^{(1)}D^{(1)} + P^{(2)}D^{(2)}$$

$$= (k_2+k_\beta)k_7M(k_\alpha+k_6+k_4)+k_3M(k_\alpha+k_6)(k_7M+k_1S)$$

Expansion in the paths calculating $D_{EM}$ leads to the determination of only two terms instead of ten according to the King and Altman rule. Other determinants are evaluated in the same way. We finally obtain

$$v = E\,\frac{k_\beta D_{ES}+k_\alpha D_{EMS}}{D_E+D_{ES}+D_{EM}+D_{EMS}} \tag{7.80}$$

where

$$D_E = (k_\alpha+k_5)k_7(k_\beta+k_2+k_3M)+k_4(k_\beta+k_2)(k_6S+k_7)$$

$$D_{ES} = k_1k_7S(k_\alpha+k_4+k_5)+k_4k_6S(k_8M+k_1S)$$

$$D_{EM} = (k_2+k_\beta)k_7M(k_\alpha+k_4+k_5)+k_3M(k_\alpha+k_5)(k_7M+k_1S)$$

$$D_{EMS} = (k_8Mk_6S(k_\beta+k_2+k_3M)+k_1Sk_3M(k_8S+k_7)$$

Further improvement of some details of the method suggested in [97] has been done by Fromm [101].

In a multistage enzymic reaction some stages can proceed at a much faster rate than others. In the corresponding subgraphs of a stationary-state system, a quasi-equilibrium will be established. This permits us to simplify the calculation of the rate [102]. The subgraph of the fast equilibrium (the part of the graph containing only rapid stages and separated from other parts by slow stages) can be isolated. If the rate of formation of ESC is considerably larger than that of product formation, there exists only one fast equilibrium subgraph containing the points that represent the complexes. It becomes possible to neglect slow stages, since the trees that involve these stages have small values. The reaction rate is then expressed by the equilibrium constants and rate constants of the isomerization of enzyme-substrate into enzyme-product complexes and vice versa. If the isomerization stage is the limiting one, the complex concentration is determined only by the equilibrium constants. It is possible to use the diagrammatic method [103,104], which is a further simplification of the graph method. The diagrammatic method was developed in relation to reactions with one and several outputs (i.e., with several complexes from which the product is formed). The rules simplifying the calculations have been worked out in detail, and the method has been successfully applied to a series of complicated reactions, particularly those in which isoenzymes participate. This method has independent significance. Mechanisms of fast equilibrium can be regenerated only from the experimental dependence of the reaction rate on ligand concentrations [105,106].

The graph method permits us to generalize the different ASE models [65,74,75], all of which can be represented by one graph which can be uniquely regenerated from the dependence of the initial stationary-state rate on ligand concentrations.

Let us examine the binding of one ligand by several enzyme sites. In the simplest case two identical sites, the graph of fast quasi-equilibrium substrate binding with slow subsequent transformation into product has the form

$$E \xrightarrow{(2/1)K_1S} ES^{\uparrow k_1} \xrightarrow{(1/2)K_2S} ES_2^{\uparrow 2k_2} \tag{7.81}$$

$K_1$, $K_2$ are the quasi-equilibrium constants for the fast stages; $k_1$, $k_2$ are the rate constants for the slow stages. The coefficients 2/1 and 1/2 mean that in the first stage one of the two free sites can be occupied and one occupied site can be liberated, and that in the second stage one free site can be occupied and one of two occupied sites can be liberated. Applying this notation, we get

$$v = 2E \frac{k_1K_1S+k_2K_1K_2S^2}{1+2K_1S+K_1K_2S^2} \tag{7.82}$$

a formula equivalent to (7.60). The graph can be uniquely and totally regenerated from the experimental function v(S). (The dependence of the shape of the curve on the values of the constants was discussed on p. 451).

The graph for an enzyme with four identical sites contains 16 points. However, the highest power of S in the expression for v is not larger than 4. Therefore, the equivalent graph with the minimum number of branches is expressed by a four-stage chain of substrate binding

$$E \xrightarrow{(4/1)K_1S} ES^{\uparrow k_1} \xrightarrow{(3/2)K_2S} ES_2^{\uparrow 2k_2} \xrightarrow{(2/3)K_3S}$$

$$ES_3^{\uparrow 3k_3} \xrightarrow{(1/4)K_4S} ES_4^{\uparrow 4k_4} \tag{7.83}$$

The rate expression is

$$v = 4E \frac{k_1K_1S+3k_2K_1K_2S^2+3k_3K_1K_2K_3S^3+k_4K_1K_2K_3K_4S^4}{1+4K_1S+6K_1K_2S^2+4K_1K_2K_3S^3+K_1K_2K_3K_4S^4} \tag{7.84}$$

The various models of a tetrameric enzyme discussed in [13] (tetrahedron, square, etc.) are described by this equation with various values of the constants and can be represented by a unique graph (7.83).

In the general case of n identical sites we have

$$E \xrightarrow{(n/1)K_1S} ES^{\uparrow k_1} \xrightarrow{[(n-1)/2]K_2S} ES^{\uparrow 2k_2} \rightarrow \cdots$$

$$\rightarrow ES_{n-1}^{\uparrow (n-1)k_{n-1}} \xrightarrow{(1/n)K_nS} ES_n^{\uparrow nk_n} \tag{7.85}$$

The Monod-Wyman-Changeux model can be represented by such a graph with the definite law of increase of $K_i$ constants. Therefore, this model is equivalent to the direct cooperativity model of Koshland and coauthors in the case of positive cooperativity. Assume that an enzyme can exist in N+1 conformational states $E_0$, $E_1$, $\cdots$, $E_N$. The equilibrium constants of the transformations $E_0 \rightleftharpoons E_1$, $E_4 \rightleftharpoons E_2$, etc., are $L_1$, $L_2$, etc. The ligand binding constant of one site in state $E_0$ is K, in $E_1$, $\phi_1 K$, in $E_i$, $\phi_i K$. The rate constants of slow stages are equal in all states $E_i$. The graph has the form shown in Fig. 7.35. For this tree of stages, v can be found by determining the paths from point $E_0$ toward all points on the graph (Fig. 7.35). Comparing the coefficients at powers of S in the rate equation for graphs (7.85) and Fig. 7.35, we obtain the relation of the constants $K_i$ to the parameters $L_i$ and $\phi_i$ [107]

$$\frac{K_{i+1}}{K_i} = \frac{(\Sigma_{m=0}^{N} a_m^{i-1})(\Sigma_{m=0}^{N} a_m^{i+1})}{(\Sigma_{m=0}^{N} a_m^{i})^2} \tag{7.86}$$

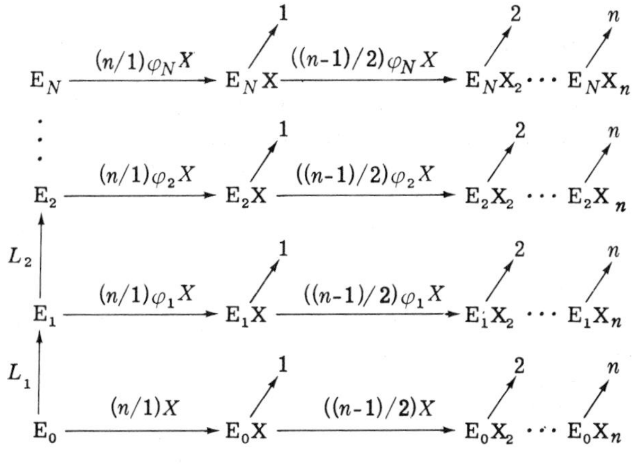


*FIG. 7.35 Graph of a system with indirect cooperativity.*

where $a_m = \phi_m \Pi_{j=}^{m} L_j$ for $m = 1,2,\cdots,n$, $a_0 = 1$. Equation (7.86) is convenient for application of the Cauchy-Buniakovsky inequality, which gives

$$K_1 \leq K_2 \leq K_3 \leq \cdots \leq K_n$$

If only two conformations $E_0$ and $E_i$ exist, Eq. (7.86) becomes

$$\frac{K_i}{K_n} = \frac{L_1\phi_1{}^{i}+1}{L_1\phi_1{}^{i-1}+1} \quad (7.87)$$

If only one state binds the ligand, $\phi_1 = 0$ and (7.87) transforms into

$$K_1/K_n = 1/(L_1+1) < 1, \quad K_{i\neq 1}/K_n = 1$$

This case is equivalent to a smaller affinity for the first ligand molecule than for the successive ones

$$K_1 < K_2 = K_3 = \cdots = K_n$$

If $K_1 = K$, $K_{i\neq 1} = \alpha K$, then

$$v = nkE \frac{x(1+x)^{n-1}}{\alpha-1+(1+x)^n} \quad (7.88)$$

where $x = \alpha KX$ and $\alpha$ is the cooperativity parameter. If the cooperativity is positive, $\alpha > 1$ and Eq. (7.88) is identical with Eq. (7.66) in the Monod-Wyman-Changeux theory at $c = 0$. If $\alpha < 1$, the cooperativity is negative (cf. [108]) and Eq. (7.88) cannot be derived from the Monod-Wyman-Changeux model where $\alpha - 1$ is a positive allosteric constant. Equation (7.88) is a general one.

Thus, the models of quasi-equilibrium reactions for enzymes with any number of sites can be represented by a unique graph. The parameters of this graph characterize the fitness of the model. The possibility of regenerating the graph from the experimental curve in a unique way depends, of course, on the exactness of the experiment (cf. [109]).

Direct methods for slow stage determination, and hence for the establishment of fast equilibrium subgraphs, are provided by relaxation (cf. next section) and pre-stationary-state kinetics (cf. [110]). On the other hand, some criteria for the applicability of the quasi-equilibrium approximation can be obtained by analyzing the curve of the initial stationary rate and of the saturation function.

For some enzymes, curves with intermediate plateaus have been observed (cf., e.g., [111]). The Monod-Wyman-Changeux model cannot give such curves. On the other hand, they can be satisfactorily explained as a result of the combination of positive and negative cooperativity.

Saturation curves cannot have intermediate plateaus. The curvature of these curves has limitations from both above and below (cf. [112]). The number of points of inflexion of saturation curves is limited by the number of sites: there are no bends at $n = 1$; at $n = 2$, only one bend is possible; at $n = 3$, two are possible; at $n = 4$, three. This can readily be shown by determination of the number of positive roots of a polynomial with the help of Descartes's method [113]. In contrast to saturation curves, the rate curve $v(S)$ can possess extremum points. Let us examine the conditions for the appearance of intermediate maxima for a tetrameric enzyme. For simplicity, we put all constants $K_i$ equal; that is,

$$v = 4E \frac{k_1\sigma+3k_2\sigma^2+3k_3\sigma^3+k_4\sigma^4}{(1+\sigma)^4} \tag{7.89}$$

where $\sigma = KS$. Let us find $v' \equiv dv/d\sigma$. Equation $v' = 0$ has the form

$$\sigma^6(4k_4-3k_3)+6\sigma^5(2k_4-k_2)+3\sigma^4[4k_4-k_1+2(3k_3-2k_2)]$$
$$+4\sigma^3[2(3k_3-k_1)+k_4]+3\sigma^2[2(2k_2-k_1)+3k_3]+6\sigma k_2+k_1 = 0 \tag{7.90}$$

According to the Descartes rule [113], this equation has a maximum number of positive roots if the signs of the coefficients alternate, that is, we have -+-+-+-, or $k_4 < \frac{3}{4}k_3$, $k_4 > \frac{1}{2}k_2$, $4k_2-k_1 < 2(2k_2-3k_3)$, $k_4 > 2(k_1-3k_3)$, $3k_3 < 2(k_1-2k_2)$. Hence, Eq. (7.90) cannot have more than five positive roots. However, it is easy to show that this system of inequalities is insoluble and in fact there can be only three changes of sign. Therefore, $v(\sigma)$ (7.89) cannot have more than three extrema--two intermediate maxima and one minimum. If the constants $k_i$ change monotonically with $i$, only one maximum can appear in the curve $v(\sigma)$ at any number of sites.

Further development of the stationary-state kinetics of cooperative enzymes both in the presence and in the absence of a detailed equilibrium, and analyses of various experimental curves, appear in works by Goldstein and others [107,114,115], Kurganov and others [78-80,116,117], and Magarshak and others [102-106].

Graph theory can also be effectively applied in studies of the pre-stationary-state kinetics of enzymic reactions [118]. In this case the time derivatives of the concentrations of the enzymic complexes are not zero. We have

$$\frac{d[EX]}{dt} = \sum_{j\neq i}^{n} w_{ji}[EX_j] - [EX_i]\sum_{j\neq i}^{n} w_{ij} \tag{7.91}$$

It is assumed that in a pre-stationary-state regime the time dependence of the ligand concentrations can be neglected, since their changes are small in comparison with $[EX_i]$. Therefore, the $w_{ij}$ in Eq. (7.91) are constant.

Using the integral Laplace-Carson transform, Eq. (7.91) can be transformed into a system of algebraic equations that can be solved by the graph method [33,116,118].

Let $[EX_i]/[E] = x_i(t)$, where [E] is the total enzyme concentration. The transform is

$$x_i^*(z) = z\int_0^\infty e^{-zt}x_i(t)\,dt \quad \text{or} \quad x_i(t) \longleftrightarrow x_i^*(z) \tag{7.92}$$

$$dx_i/dt \longleftrightarrow zx_i^* - zx_i(0) \tag{7.93}$$

The complex variable z is the transformed time t. Initial concentration values are given in the form

$$x_i(0) = \delta_{i0}$$

The transformed system (7.91) has the form

$$\begin{aligned} -z &= -x_0^*(w_{00}+z)+x_1^*w_{10}+x_2^*w_{20}+ \cdots +x_n^*w_{n0} \\ 0 &= x_0^*w_{01}-x_1^*(w_{11}+z)+x_2^*w_{21}+ \cdots +x_n^*w_{n1} \\ &\vdots \\ 0 &= x_0^*w_{0n}+x_1^*w_{1n}+x_2^*w_{2n}+ \cdots -x_n^*(w_{nn}+z) \end{aligned} \tag{7.94}$$

where $w_{ii} = \Sigma_{j\neq i} w_{ij}$. The condition of constancy of the total enzyme concentration in the transformed form is

$$\sum_{i=0}^{n} x_i^* = 1$$

Hence, the first equation in system (7.94) can be rewritten as

$$0 = -x_0^*w_{00}+x_1^*(w_{10}+z)+ \cdots +x_n^*(w_{n0}+z)$$

and the whole system as

$$\sum_{j=0}^{n}{}' x_j^*w_{ji}^* = x_i^*\sum_{j=0}^{n}{}' w_{ij}^*, \quad i = 0,1,\cdots,n \tag{7.95}$$

where

$$w_{ij}^{*} = \begin{cases} w_{ij}, & j \neq 0 \\ w_{i0}+z, & j = 0 \end{cases}$$

We have obtained a system of equations whose form is similar to that of stationary-state kinetics. In contrast to the stationary-state graph, the pre-stationary-state one contains additional branches with the value z directed from every point on the graph toward the point corresponding to the free enzyme. Calculation by the graph method gives the transform of the reaction rate $v^*$; the original v can be obtained from $v^*$ by standard methods of the operational calculus. If the product is formed irreversibly

$$v^*(z) = \sum_i k_i D_i^* / \sum_i D_i^* \tag{7.96}$$

where $D_i^*$ are the determinants of the graph of the pre-stationary-state reaction system. Otherwise,

$$v^* = \sum_i k_i x_i^*$$

where $x_i^* = D_i^*/\Sigma_j D_j^*$. The functions $x_i^*$ have the form

$$x_i^*(z) = \frac{\delta_{i0}z^n + z^{n-1}b_{n-1}^{(i)} + \cdots + b_0^{(i)}}{z^n + z^{n-1}a_{n-1} + \cdots + a_0} \tag{7.97}$$

The power of the polynomial in the denominator for the graph with n+1 points is n, since n is the number of z branches. The highest power $z^n$ arises as the value of the tree formed by the z branches directed toward the point 0. This tree is unique. The value of each of its branches is z; therefore, the coefficient at z in both the numerator and the denominator of (7.97) is 1. Coefficients a and b can be expressed by means of $w_{ij}$. The stationary-state solution can be had from (7.97) if $t \to \infty$, or $z \to 0$. Hence

$$x_i(\infty) = x_i^*(0) = b_0^{(i)}/a_0 = D_i/\sum_i D_i \tag{7.98}$$

Every addend in the numerator and denominator of (7.97) is the sum of trees having n branches (including z branches). Therefore, $a_{n-1}$ and $b_{n-1}^{(i)}$ are the sums of some graph branches without z branches; $a_{n-2}$ and $b_{n-2}^{(i)}$ are the sums of pair products of branches; etc. The equalities are valid

$$a_m = \sum_{i=0}^{n} b_m^{(i)}, \quad m = 0,1,\cdots,n$$

Let us examine as an example the case of a cooperative enzyme with two identical sites. In the case of stationarity in absence of quasi-equilibria we obtain the simplified graph shown in Fig. 7.36 instead of (7.81). We have

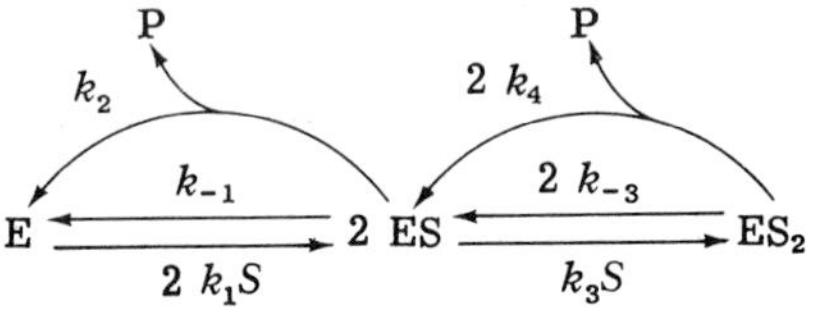


*FIG. 7.36 Graph of a two-center system (stationary-state kinetics).*

$$D_1 = (k_2+k_{-1})2(k_4+k_{-3}), \quad D_2 = 2k_1S2(k_4+k_{-3}), \quad D_3 = 2k_1Sk_3S$$

and

$$v = E\,\frac{2k_2D_2+2k_4D_4}{D_1+D_2+D_3} = 2E\,\frac{k_2K'S+k_4S^2}{KK'+2K'S+S^2} \tag{7.99}$$

where $K = (k_{-1}+k_2)/k_1$, $K' = (k_{-3}+k_4)/k_3$. We obtain the already known expression [cf. (7.60),(7.82)]. In pre-stationary-state conditions the graph has the form shown in Fig. 7.37. We have

$$D_1{}^* = (k_2+k_{-1}+z)2(k_4+k_{-3})+z(k_{-1}+k_2+z)+zk_3S$$

$$D_2{}^* = 2k_1S2(k_{-3}+k_4)+z2k_1S, \quad D_3{}^* = 2k_1Sk_3S$$

and

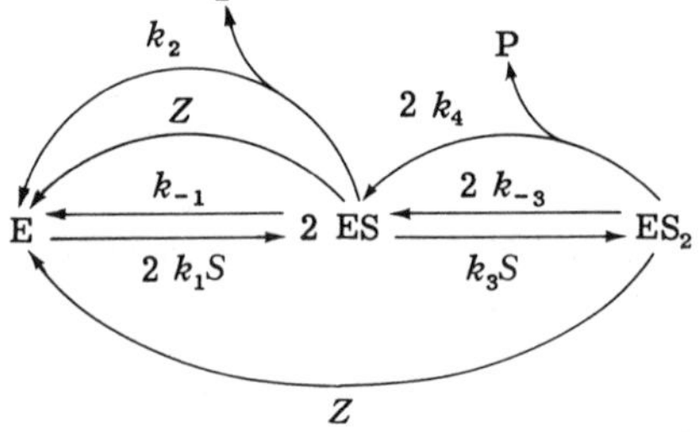


*FIG. 7.37 Graph of a two-center system (pre-stationary-state kinetics).*

$$v^*(z) = E\frac{2k_2D_2{}^*+2k_4D_3{}^*}{D_1{}^*+D_2{}^*+D_3{}^*} = 2k_1k_2ES\frac{z+C}{z^2+2Az+B} \tag{7.100}$$

where

$$2A = 2k_4+2k_{-3}+k_{-1}+k_2+k_3S+2k_1S$$

$$B = 2(k_2+k_{-1})(k_4+k_{-3})+4k_1S(k_4+k_{-3})+2k_1k_3S^2$$

$$C = 2(k_{-3}+k_4)+2Sk_4k_3/k_2$$

If $t \to \infty$, $z \to 0$, and

$$v^*(0) = 2k_1k_2ES(C/B) = v$$

The original of Eq. (7.100) gives

$$v(t) = v_{stat} - 2k_1k_2SE\left[\frac{z_1+C}{z_1(z_2-z_1)}e^{z_1t} + \frac{z_2+C}{z_2(z_1-z_2)}e^{z_2t}\right] \tag{7.101}$$

where $z_{1,2} = -A\pm(A^2-B)^{1/2}$ are the roots of the corresponding characteristic equation.

If a complex reaction obeys the detailed equilibrium condition, the roots are substantial and negative [119]. Such reactions are represented by graphs with unclosed cycles (i.e., by stage trees), though the reaction scheme can include closed cycles. Consequently all reactions whose graphs are represented by trees of stages cannot produce oscillations in the pre-stationary-state. If the detailed equilibrium is not obeyed, oscillations can arise in the system. The Laplace-Carson transform of the reaction rate is a meromorphic function of the complex variable z. The roots of the characteristic polynomial lie in the left half-plane of complex numbers for all enzymic reactions. Therefore, the enzymic system is stable in the pre-stationary-state regime and only damped oscillations can be formed. They can arise only if the graph contains a cycle of at least three stages.

Oscillating enzyme reactions and other biological oscillating processes will be treated in "General Biophysics." For further details concerning the application of graph theory in stationary- and pre-stationary-state kinetics of enzymic reactions see the papers already cited, Appendix I of [33], and Goldstein's thesis [115].

The directed graph method functions as a convenient algorithm for the kinetics of enzymes. On the other hand, it demonstrates the deep analogy between the processes in complicated electronic networks and those in enzymic reactions.

Both systems use signals related by similar functional interrelations. In electronic chains, voltages and currents serve as signals; in enzymic reactions, the concentrations and rates of stages behave this way. The mass action law is the analog of Ohm's law. However, Ohm's law requires a voltage difference at the ends of a two-pole device, and the mass action law requires the concentration of an enzymic complex (the voltage analog) only at the input of the two-pole device (the graph branch). This distinction determines the inapplicability of the graphical rules worked out for electric circuits to enzymic reaction in a direct way and makes difficult the direct electric modeling of reactions [120,121].

## 7.7 Chemical Relaxation

The mechanism of enzyme action can be effectively studied by means of the chemical relaxation methods developed by Eigen [122]. The system is shifted from its equilibrium or steady state by rapid changes in an external parameter, and the kinetics of its approach to a new equilibrium or steady state is studied. Relaxation times $\tau$ up to $10^{-10}$ sec can be measured. Instantaneous and stationary-state perturbation methods such as concentration and temperature jump, or perturbation by an electric field or ultrasound, are used. The relaxation times are measured spectrophotometrically. Reviews of experimental methods and theory are given in [123-126].

Relaxational methods make it possible to determine the entire set of rate constants of a complex process. The main advantage is that at small perturbations of the equilibrium or stationary state of a liquid reagent mixture, nonlinear kinetic equations become linearized.

Examine the one-stage reaction (M is the ligand)

$$F_0 + M \underset{k_{-1}}{\overset{k_1}{\rightleftharpoons}} F_1$$

The kinetic equation has the form

$$\dot{F}_0 = -k_1 F_0 M + k_{-1} F_1 \tag{7.102}$$

In the steady state $\dot{F}_0 = \dot{F}_1 = \dot{M} = 0$. If there is a deviation from the steady state, characterized by $\overline{F}_0$, $\overline{F}_1$, $\overline{M}$, then

$$F_0 = \overline{F}_0 + \Delta F_0, \quad M = \overline{M} + \Delta M, \quad F_1 = \overline{F}_1 + \Delta F_1$$

We get

$$d\Delta F_0/dt = -k_1 \overline{F}_0 \Delta M - k_1 \overline{M} \Delta F_0 - k_1 \Delta F_0 \Delta M + k_{-1} \Delta F_1 \tag{7.103}$$

If the perturbation is small, $k_1 \Delta F_0 \Delta M$ can be omitted. From the law of conservation of mass we get $\Delta M = \Delta F_0 = -\Delta F_1$. Hence

$$d\Delta M/dt = -(k_1F_0+k_1M+k_{-1})\Delta M \tag{7.104}$$

or

$$\tau(d\Delta M/dt) + \Delta M = 0 \tag{7.105}$$

where

$$\tau^{-1} = k_1(F_0+M)+k_{-1}$$

The solution of (7.105) is

$$\Delta M = (\Delta M)_0 e^{-t/\tau} \tag{7.106}$$

where $(\Delta M)_0$ is the $\Delta M$ value at $t = 0$. The measurement of $\tau$ at various concentrations of $F_0$ and M permits us to determine $k_1$ and $k_{-1}$.

Consider now more complicated models of enzymic reactions. Let the enzyme interact with some modifying but not reacting ligand M. For a two stage modification

$$F_0 + M \underset{k_{-1}}{\overset{k_1}{\rightleftharpoons}} F_1 \underset{k_{-2}}{\overset{k_2}{\rightleftharpoons}} F_2$$

where $F_1$ and $F_2$ are the two enzyme-modifier complexes. The study of stationary-state kinetics does not enable us to determine the four rate constants.

The kinetic equations are

$$\dot{M} = -k_1F_0M + k_{-1}F_1$$

$$\dot{F}_1 = k_1F_0M - (k_{-1}+k_2)F_1 + k_{-2}F_2 \tag{7.107}$$

Introducing small deviations from the stationary-state concentrations

$$M = \bar{M}+x_1, \quad F_0 = \bar{F}_0+x_1, \quad F_1 = \bar{F}_1+x_2, \quad F_2 = \bar{F}_2-(x_1+x_2)$$

we obtain two linear equations if the second-order terms are omitted

$$\dot{x}_1 = \alpha_{11}x_1 + \alpha_{12}x_2, \qquad \dot{x}_2 = \alpha_{21}x_1 + \alpha_{22}x_2 \tag{7.108}$$

where

$$\alpha_{11} = -k_1(\bar{F}_0+\bar{M}), \qquad \alpha_{12} = k_{-1}$$

$$\alpha_{21} = k_1(\bar{F}_0+\bar{M})-k_{-2}, \quad \alpha_{22} = -(k_{-1}+k_2+k_{-2})$$

We look for a solution of system (7.108) in the form

$$x_1 = x_{10}\, e^{-t/\tau}, \quad x_2 = x_{20}\, e^{-t/\tau} \tag{7.109}$$

The characteristic equation has the form

$$\begin{vmatrix} \alpha_{11}+\tau^{-1} & \alpha_{12} \\ \alpha_{21} & \alpha_{22}+\tau^{-1} \end{vmatrix} = 0 \tag{7.110}$$

and we get two substantial values of $\tau$

$$\tau_{1,2}^{-1} = -\frac{\alpha_{11}+\alpha_{22}}{2}\left[1 \pm \left(1 - \frac{4(\alpha_{11}\alpha_{22}-\alpha_{12}\alpha_{21})}{(\alpha_{11}+\alpha_{22})^2}\right)^{1/2}\right] \tag{7.111}$$

or

$$\tau_1^{-1}+\tau_2^{-1} = -(\alpha_{11}+\alpha_{22}) = k_1(F_0+M)+k_{-1}+k_2+k_{-2}$$

$$\tau_1^{-1}\tau_2^{-1} = \alpha_{11}\alpha_{22}-\alpha_{12}\alpha_{21} = k_1(k_2+k_{-2})(F_0+M)+k_{-1}k_{-2}$$

These two combinations of $\tau_1$ and $\tau_2$ give all four rate constants. The method has been applied for the determination of the rate constants of the reaction of α-chymotrypsin with proflavin [126].

In general, we obtain a system of n linearized kinetic equations with variables $x_1, x_2, \cdots, x_n$ and correspondingly the characteristic nth order equation, giving n values of $\tau$. It is rather difficult to visualize the general solution but it becomes simplified if the various $\tau$ values differ greatly. Eigen considered as an example the simple enzymic reaction

$$F_0+S \underset{k_{-1}}{\overset{k_1}{\rightleftharpoons}} F_1 \underset{k_{-2}}{\overset{k_2}{\rightleftharpoons}} F_2 \underset{k_{-3}}{\overset{k_3}{\rightleftharpoons}} F_0+P$$

where $F_1$ is the enzyme-substrate and $F_2$ the enzyme-product complex. Assume that stages 1 and 3 are fast and that stage $F_1 \rightleftharpoons F_2$ is slow and therefore limits the $S \rightarrow P$ transformation. For slow stages the linearized equations have the same form as (7.108), where

$$x_1 = \Delta S, \quad -x_1 = \Delta F_1, \quad x_2 = \Delta P, \quad -x_2 = \Delta F_2, \quad x_1+x_2 = \Delta F_0$$

and

$$\alpha_{11} = -[k_1(F_0+S)+k_{-1}], \qquad \alpha_{21} = -k_{-3}P,$$

$$\alpha_{12} = -k_1 S, \qquad \alpha_{22} = -[k_{-3}(F_0+P)+k_3]$$

The solution gives

$$\tau_1^{-1}+\tau_2^{-1} = (k_1+k_{-3})F_0+(k_1+k_{-3}K)S+k_{-1}+k_3$$

$$\tau_1^{-1}\tau_2^{-1} = k_1k_{-3}F_0^2+k_1k_{-3}(1+K)F_0S+(k_1k_3+k_{-1}k_{-3})F_0$$
$$+(k_1k_3+k_{-1}k_{-3}K)S+k_{-1}k_3 \qquad (7.112)$$

where K is the equilibrium constant for $S \rightleftarrows P$. The rate equation for the slow stage is

$$\frac{d(S+F_1)}{dt} = -\frac{d(P+F_2)}{dt} = -k_2F_1+k_{-2}F_2 \qquad (7.113)$$

corresponding to a relaxation time

$$\tau_3^{-1} = \frac{k_2}{1+(K_1/F_0)f_2} + \frac{k_{-2}}{1+(K_2/F_0)f_1} \qquad (7.114)$$

where $K_1$, $K_2$ are the equilibrium constants for binding reactions (i.e., for the formation of $F_1$ from $F_0$ and S, and $F_2$ from $F_0$ and P).

$$f_1 = \frac{K_1+F_0+S+PK_1/K_2}{K_1+F_0+S+P}$$
$$f_2 = \frac{K_2+F_0+S+PK_2/K_1}{K_2+F_0+S+P} \qquad (7.115)$$

Relaxational kinetics makes it possible to determine all the constants.

If all stages have rates of similar orders of magnitude, the situation becomes more complicated. In such cases it appears expedient to consider averaged values of $\tau$, which also give valuable information about the rate constants [125,127].

At small enzyme concentrations, we are concerned with the stationary-state concentrations of intermediate complexes. The relaxational spectrum comes down to one $\tau$ value depending on S in a definite way. Relaxational methods make it possible to obtain detailed information about intermediate complexes and the corresponding rate constants.

Let us consider now the relaxational spectrum of an allosteric protein, also studied by Eigen [125]. According to the Monod-Wyman-Changeux theory (p. 452), an allosteric enzyme can be characterized by three parameters: the binding constants of the ligand in the R and T states, and the parameter describing the conformational transitions $R \rightleftarrows T$. Consequently, the system must have at least three time constants. Assume that the $R \rightleftarrows T$ transition is the slowest stage. The binding by R subunits is characterized by one observable time constant $\tau_1$ corresponding to the reaction of the free substrate molecules with the free R sites: $\phi_R = 4R_0+3R_1+2R_2+R_3$ (for a tetramer). The value $\tau_1^{-1}$ has to be the function of the sum of S and $\phi_R$. The same is true of the T state of a tetramer, characterized by $\tau_2$. Both equilibria are coupled by S--when the T states interact with the substrate, the change in its concentration influences the more rapid binding by the R state. The relaxational spectrum of an nmeric enzyme was calculated by Eigen suggesting the constancy of S concentration (buffer) or the rapid establishment of equilibrium with the R state in comparison with the T state; that is, $n\tau_T^{-1} << \tau_R^{-1}$. We have

$$\phi_R = nR_0+(n-1)R_1+(n-2)R_2+\ldots+R_{n-1} \tag{7.116}$$

and

$$\tau_{1R}^{-1} = k_A^R(S+\phi_R)+k_D^R, \quad \tau_{2R}^{-1} = 2(k_A^R S+k_D^R), \ldots, \quad \tau_{nR}^{-1} = n(k_A^R S+k_D^R) \tag{7.117}$$

and

$$\tau_{1T}^{-1} = k_A^T(S+f_R\phi_T)+k_D^T, \quad \tau_{2T}^{-1} = 2(k_A^T S+k_D^T), \ldots, \quad \tau_{nT}^{-1} = n(k_A^T S+k_D^T) \tag{7.118}$$

where $k_A$ and $k_D$ are the rate constants for ligand binding and dissociation; $\phi_T$ is analogous to $\phi_R$; and

$$f_R = \frac{K_R+S}{K_R+S+\phi_R}$$

where $K_R$ is the binding constant for the R state.

The R $\rightleftharpoons$ T transformation is characterized by one relaxation time $\tau_{conf}$ depending in general on S. At small S there is no binding and the system is in the $T_0$ and $R_0$ states. At very large S the sites are saturated and the system is in the $T_4$ and $R_4$ states. In these two cases $\tau_{conf}$ does not depend on S. If $T_0$ is more stable than $R_0$ but $R_4$ is, conversely, more stable than $T_4$, a sigmoidal transition occurs between the two values of $\tau_{conf}$. At S = const we get

$$\sum R_i \rightleftharpoons \sum T_i$$

$$d\Sigma R_i/dt = -k_0^{RT}R_0+k_0^{TR}T_0-k_1^{RT}R_1+k_1^{TR}T_1- \cdots \tag{7.119}$$

Replacing $\Delta R_0$ with

$$\frac{\Delta\Sigma R_i}{1+(\Delta R_1/\Delta R_0)+(\Delta R_2/\Delta R_1)(\Delta R_1/\Delta R_0)+ \cdots}$$

and performing analogous substitutions for all $\Delta R_i$ and $\Delta T_i$, we find

$$\tau_{conf}^{-1} = \frac{k_0^{RT}+k_1^{RT}(\Delta R_1/\Delta R_0)+k_2^{RT}(\Delta R_2/\Delta R_1)(\Delta R_1/\Delta R_0)+\cdots}{1+(\Delta R_1/\Delta R_0)+(\Delta R_2/\Delta R_1)(\Delta R_1/\Delta R_0)+\cdots}$$

$$+\frac{k_0^{TR}+k_1^{TR}(\Delta T_1/\Delta T_0)+\cdots}{1+(\Delta T_1/\Delta T_0)+\cdots} \qquad (7.120)$$

The calculation at $\Delta S = 0$ gives the approximate value

$$\tau_{conf}^{-1} = \left[\frac{k_0^{TR}}{(1+S/K_T)^n} + \frac{k_0^{RT}}{(1+S/K_R)^n}\right] f\left(\frac{S}{K_T}\right) \qquad (7.121)$$

where

$$f\left(\frac{S}{K_T}\right) = \sum_{i=0}^{n} \alpha_i^T \frac{n!}{(n-i)!i!}\left(\frac{S}{K_T}\right)^i$$

$\alpha_i^T = k_i^{TR}/k_0^{RT}$. At $\alpha_i^T \to 1$ the function $f(S/K_T)$ becomes $(1+S/K_T)^n$.

We find in limiting cases:

$$\tau^{-1} = k_0^{TR}+k_0^{RT}, \quad \text{if} \quad S \ll K_T, \quad S \ll K_R$$

$$\tau^{-1} = k_0^{TR}+k_n^{RT}, \quad \text{if} \quad S \ll K_T, \quad S \gg K_R$$

$$\tau^{-1} = k_n^{TR}+k_n^{RT}, \quad \text{if} \quad S \gg K_T, \quad S \gg K_R$$

The presence of the first two relaxational effects, for which $\tau^{-1}$ increases linearly with concentration, gives detailed information about the binding in both states. The third value of $\tau$ characterizes the conformational transition. If $\tau_{conf}$ transfers from one constant value to the other, only two conformations are present. The sharpness of the transition characterizes cooperativity. Kirschner investigated D-glyceraldehyde-3-phosphate dehydrogenase (GAPD) by means of the relaxation method and actually found these peculiarities of the relaxational spectrum to be in agreement with the Monod-Wyman-Changeux theory [125]. These results do not, however, exclude the alternative model suggested by Koshland and coauthors (p. 455).

Recently Schwartz suggested a theory of chemical relaxation of cooperative conformational transitions in linear biopolymers [128]. The relaxation in helix-coil transitions in polypeptides was investigated on the basis of the linear Ising model and a theory for both long and short chains was developed. It was shown that the largest relaxation time controls the conformational change.

The relaxational spectrum obviously gives the most direct information about ECI, providing separate studies of rapid and slow motions in ESC. It would be of special interest to study hemoglobin oxygenation with the aim of proving the dynamic model proposed by Perutz (p. 424). A significant development in relaxational studies will probably occur soon. Effective methods of inducing rapid perturbations and of registering relaxational processes rapidly are being used [123-127,129,130], but there are undoubtedly many possibilities for improvement here, as well.

References


1. M. Perutz, Proc. Roy. Soc. **B167**, 448 (1967).
2. M. Weissbluth, Physics of hemoglobin, in "Structure and Bonding," Vol. 2, Springer-Verlag, Berlin and New York, 1967.
3. A. Rossi-Fanelli, E. Antonini, and A. Caputo, Biochim. Biophys. Acta 35, 93 (1959).
4. T. Samejima and J. Yang, J. Mol. Biol. 8, 863 (1964).
5. C. Ballhausen, "Introduction to Ligand Field Theory." McGraw-Hill, New York, 1962.
6. K. Ohno, Y. Tanable, and F. Tasaki, Theor. Chim. Acta 1, 378 (1963).
7. A. Veillard and B. Pullmann, J. Theor. Biol. 8, 317 (1965).
8. M. Zerner, M. Gouterman, and H. Kobayashi, Theor. Chim. Acta 6, 363 (1966).
9. G. Gurinovitch, A. Sevtchenko, and K. Soloviev, "Spektroscopy of Chlorophyll and Related Compounds." Nauka i Technika, Minsk, 1968 (R).
10. C. Johnson, Phys. Today **24**, 35 (1971).
11. G. Lang, T. Asakura, and T. Yonetani, Biochim. Biophys. Acta **214**, 381 (1970).
12. A. Hill, J. Physiol. **40**, IV (1910).
13. E. Antonini, A. Rossi-Fanelli, and A. Caputo, Arch. Biochem. Biphys. **97**, 343 (1962).
14. G. Adair, J. Biol. Chem. **63**, 529 (1925); Proc. Roy. Soc. **A109**, 292 (1925).
15. L. Pauling, Proc. Nat. Acad. Sci. U.S. **21**, 186 (1935).
16. J. Wyman, Advan. Protein Chem. **4**, 407 (1948).
17. F. Roughton, in "Haemoglobin" (F. Roughton and J. Kendrew, eds.). Butterworths, London and Washington, D.C., 1949.
18. D. Allen, K. Guthe, and J. Wyman, J. Biol. Chem. **187**, 393 (1950).

19. G. Guidotti, J. Biol. Chem. **242**, 3694 (1967).
20. A. Rossi-Fanelli, E. Antonini, and A. Caputo, Advan. Protein Chem. **19**, 73 (1964).
21. E. Antonini and M. Brunelli, Ann. Rev. Biochem. **39**, 977 (1970).
22. F. Roughton, A. Otis, and R. Lyster, Proc. Roy. Soc. **B144**, 29 (1955).
23. M. Perutz, Nature (London) **228**, 726 (1970).
24. M. Perutz, H. Muirhead, J. Cox, and L. Gooman, Nature (London) **219**, 139 (1968).
25. B. Atanasov, Nature (London) **233**, 560 (1971).
26. B. Atanasov, Mol. Biol. **4**, 51, 348 (1970) (R).
27. E. Saburova, D. Markovitch, and M. Volkenstein, Mol. Biol. **5**, 461 (1971) (R).
28. J. Wyman, Advan. Protein Chem. **19**, 223 (1964); Quart. Rev. Biophys. **1**, 35 (1968).
29. J. Wyman, J. Mol. Biol. **11**, 631 (1965).
30. J. Wyman, J. Mol. Biol. **39**, 523 (1969).
31. M. Faraday, Phil. Trans. 1 (1846).
32. M. Volkenstein, "Molecular Optics." Gostekhizdat, Moscow, 1951 (R).
33. M. Volkenstein, "Enzyme Physics." Plenum Press, New York, 1969.
34. B. Briat and C. Djerassi, Nature (London) **217**, 918 (1968).
35. Y. Sharonov, Opt. Spectrosc. **29**, 463 (1970) (R).
36. R. Serber, Phys. Rev. **41**, 489 (1932).
37. I. Tobias and W. Kautzmann, J. Chem. Phys. **35**, 538 (1961).
38. M. Groenewege, Mol. Phys. **5**, 541 (1962).
39. A. Buckingham and P. Stephens, Ann. Rev. Phys. Chem. **17**, 399 (1967).
40. D. Macaluso and O. Corbino, Nuovo Cimento **8**, 257 (1898); **9**, 381 (1899).
41. R. Wood, "Physical Optics." Dover, New York, 1968.
42. D. Schooley, E. Bunnenberg, and C. Djerassi, Proc. Nat. Acad. Sci. U.S. **53**, 579 (1965).
43. Y. Sharonov, Opt. Spectrosc. **25**, 930 (1968) (R).
44. P. Shatz and A. McCaffry, Quart. Rev. Phys. Chem. **23**, 552 (1969).
45. W. Voelter, G. Barth, R. Records, E. Bunnenberg, and C. Djerassi, J. Amer. Chem. Soc. **91**, 6165 (1969).
46. W. Voelter, R. Records, E. Bunnenberg, and C. Djerassi, J. Amer. Chem. Soc. **90**, 6163 (1968).
47. G. Barth, R. Records, E. Bunnenberg, C. Djerassi, and W. Voelter, J. Amer. Chem. Soc. **93**, 254 (1971).
48. C. Djerassi, Lecture at the Symp. Chem. Natural Products, Riga (1970).

49. V. Schashoua, J. Amer. Chem. Soc. 86, 2109 (1964); 87, 4044 (1965); Nature (London) 203, 972 (1964); Biochemistry 3, 1719 (1964); Arch. Biochem. Biophys. 111, 550 (1965).
50. M. Volkenstein, Y. Sharonov, and A. Shemelin, Nature (London) 209, 709 (1966); Mol. Biol 1, 467 (1967) (R).
51. B. Atanasov, M. Volkenstein, Y. Sharonov, and A. Shemelin, Mol. Biol. 1, 477 (1967) (R).
52. M. Volkenstein, L. Govshovitchus, Y. Sharonov, and A. Shemelin, Mol. Biol. 1, 854 (1967) (R).
53. M. Volkenstein, Y. Sharonov, and A. Shemelin, Mol. Biol. 2, 864 (1968) (R).
54. B. Atanasov, M. Volkenstein, and Y. Sharonov, Mol. Biol. 3, 518, 696, 804 (1969) (R).
55. N. Sharonova, Y. Sharonov, and M. V. Volkenstein, Biochem. Biophys. Acta 271, 65 (1972).
56. A. Arutiunian and Y. Sharonov, Mol. Biol. 7, 587 (1973) (R).
57. Y. Sharonov and N. Sharonova, FEBS Lett. 27, 221 (1972).
58. G. Lang and W. Marshall, Proc. Phys. Soc. 87, 3 (1966).
59. M. Volkenstein, Izv. Acad. Sci. USSR Ser. Biol. N 6, 805 (1971) (R).
60. T. Li and B. Johnson, Biochemistry 8, 3638 (1969).
61. P. Stephens, W. Suetaaka, and P. Schatz, J. Chem Phys. 44, 4592 (1966).
62. M. Malley, G. Feher, and O. Manzerall, J. Mol. Spectrosc. 26, 320 (1968).
63. A. McHugh, M. Gouterman, and C. Weiss, Theor. Chim. Acta 24, 346 (1972).
64. H. Umbarger, Science 123, 848 (1956); 145, 674 (1964).
65. J. Monod, J. Wyman, and J.-P. Changeux, J. Mol. Biol. 12, 88 (1965).
66. N. Tomova, M. Setchenska, G. Detchev, N. Krysteva, and L. Dimitrieva, "Enzymatic Regulation of Cellular Metabolism according to the Feed-Back Principle." Ed. Bulgarian Acad. Sci., Sofia, 1968 (R).
67. "Itogi Nauki. Biological Chemistry 1969. Allosteric Regulation of Enzyme Action." Ed. VINITI, Moscow, 1971 (R).
68. J. Gerhart and A. Pardee, Cold Spring Harbor Symp. Quant. Biol. 28, 329 (1963).
69. J. Webb, "Enzyme and Metabolic Inhibitors," Vol. 1. Academic Press, New York, 1963.
70. M. Volkenstein, in "Molecular Biophysics," Nauka, Moscow, 1965 (R).
71. J. Monod, J.-P. Changeux, and F. Jacob, J. Mol. Biol. 6, 306 (1963).
72. J.-P. Changeux, Cold Spring Harbor Symp. Quant. Biol. 28, 313 (1963).

73. M. Kotani, Progr. Theor. Phys. Suppl. extra number 233 (1968).
74. E. Whitehead, Biochemistry 9, 1440 (1970).
75. D. Koshland, G. Nemethy, and D. Filmer, Biochemistry 5, 365 (1966).
76. J. Teipel and D. Koshland, Biochemistry 3, 4656 (1969).
77. D. Koshland, in "Current Topics in Cellular Regulation" (B. Horecker and E. Stadtman, eds.), Vol. 1. Academic Press, New York, 1969.
78. B. Kurganov, Mol. Biol. 1, 17 (1967); 2, 166, 430 (1968) (R).
79. B. Kurganov and V. Jakovlev, Mol. Biol. 4, 781 (1970) (R).
80. B. Kurganov and O. Polianovsky, J. All-Un. Mendeleev Chem. Soc. 16, N 4, 421 (1971) (R).
81. N. Kiselev, J. All-Un. Mendeleev Chem. Soc. 16, N 4, 431 (1971) (R).
82. B. Rabin, Biochem. J. 102, 226 (1967).
83. N. Sidorenko and V. Deshtsherevsky, Biophysica 15, 785 (1970) (R).
84. G. Schwab, Z. Phys. Chem. 88, 141 (1930).
85. I. Horiuti, J. Res. Inst. Catal. Hokkaido Univ. 5, I (1957).
86. J. Christiansen, Advan. Catal. 5, 311 (1953).
87. N. Semenov, J. Phys. Chem. 17, 183 (1943); 27, 187 (1953) (R).
88. E. King and C. Altman, J. Phys. Chem. 60, 1375 (1956).
89. J. Wong and C. Hanes, Can. J. Biochem. Physiol. 40, 763 (1962); Arch. Biochem. Biophys. 135, 50 (1969).
90. A. Schulz and D. Fisher, Can. J. Biochem. 47, 889 (1969).
91. R. Hurst, Can. J. Biochem. 47, 643, 941 (1969).
92. W. Cleland, Biochim. Biophys. Acta 67, 104, 173, 183 (1963).
93. K. Berge, "Theory of Graphs and Its Applications." Wiley, New York, 1962.
94. L. Bessonov, "Fundamentals of Graph Theory." Fizmatgiz, Moscow, 1964 (R).
95. F. Harary (ed.), "Graph Theory and Theoretical Physics" (Proc. NATO Symp.) Academic Press, New York, 1967.
96. M. Temkin, Dokl. Acad. Sci. USSR 152, 156 (1963); 165, 615 (1965).
97. M. Volkenstein and B. Goldstein, Dokl. Acad. Sci. USSR 170, 963 (1966); Biochimia 31, 541, 679 (1966) (R); Biochim. Biophys. Acta 115, 471, 478 (1966).
98. S. Mason and H. Zimmerman, "Electronic Circuits, Signals, and Systems." MIT Press, Cambridge, Massachusetts, 1960.
99. L. Robichau, M. Boiver, and J. Robert, "Directional Graphs."

100. P. Ionkin and A. Sokolov, Electritchestvo N 5, 57 (1964); N 8, 26 (1964) (R).
101. H. Fromm, Biochem. Biophys. Res. Commun. 40, 692 (1970).
102. B. Goldstein, Y. Magarshak, and M. Volkenstein, Dokl. Acad. Sci. USSR 191, 1172 (1970) (R).
103. M. Volkenstein and Y. Magarshak, Biophysica 15, 777, 949 (1970); Dokl. Acad. Sci. USSR 192, 665 (1970) (R).
104. M. Volkenstein, Y. Magarshak, and V. Stefanov, Dokl. Acad. Sci. USSR 197, 958, 1193 (1971) (R).
105. M. Volkenstein, Y. Magarshak, and V. Stefanov, Biophysica 17, 379 (1972) (R).
106. M. Volkenstein and Y. Magarshak, Biophysica 18 (1973) (R).
107. B. Goldstein and M. Volkenstein, Mol. Biol. 5, 555 (1971) (R).
108. A. Conway and D. Koshland, Biochemistry 7, 4011 (1968).
109. A. Cornish-Bowden and D. Koshland, Biochemistry 9, 3325 (1970).
110. J. Darvey, J. Theoret. Biol. 19, 125 (1968).
111. A. Levitzki and D. Koshland, Proc. Nat. Acad. Sci. U.S. 62, 1121 (1969).
112. G. Weber and S. Anderson, Biochemistry 4, 1942 (1965).
113. J. Botts, Trans. Faraday Soc. 54, 593 (1958).
114. B. Goldstein and M. Volkenstein, Dokl. Acad. Sci. USSR 178, 386 (1968) (R).
115. B. Goldstein, Method of Graphs in Kinetics of Enzymatic Reactions. Thesis, Institute of Biophysics, Acad. Sci. USSR (1971) (R).
116. B. Kurganov and V. Jakovlev, Mol. Biol. 6, 113 (1972) (R).
117. Z. Kagan, D. Khashimov, and B. Kurganov, Biochemia 35, 937 (1970) (R).
118. M. Volkenstein, B. Goldstein, and V. Stefanov, Mol. Biol. 1, 52 (1967) (R).
119. J. Wei and Ch. Preter, Advan. Catalysis 13, (1962).
120. A. Deri and M. Wolleman, Acta Biochem. Biophys. Acad. Sci. Hung. 5, 177 (1970).
121. C. Schugurensky and J. Olavarria, Acta Physiol. Latino-Amer. 19, 153 (1969).
122. E. Eigen, Discuss. Faraday Soc. 17, 194 (1954).
123. M. Eigen and L. de Mayer, Tech. Org. Chem. 8, N 2, 895 (1963).
124. G. Czerlinsky, "Chemical Relaxation." Dekker, New York, 1966; in "Theoretical and Experimental Biophysics" (A. Cole, ed.), Vol. 2, p. 69. Dekker, New York, 1969.
125. M. Eigen and G. Hammes, Advan. Enzymol. 25, I (1963).
126. G. Hammes, Advan. Protein Chem. 23, I (1968).

127. M. Eigen, in "Fast Reactions and Primary Processes in Chemical Kinetics" (Nobel Symp. 5) (S. Claesson, ed.). Wiley (Interscience), New York, 1968; Quart. Rev. Biophys. **1**, 3 (1968).
128. G. Schwartz, J. Theoret. Biol. **36**, 569 (1972).
129. B. Havsteen, J. Biol. Chem. **242**, 769 (1967).
130. G. Schwartz, Biopolymers **5**, 321 (1967).
131. I. Golovanov, V. Sobolev, and M. Volkenstein, Dokl. Acad. Sci. USSR 218, 478 (1974) (R).
132. Y. Sharanov and N. Sharonova, Mol. Biol. 9, 145 (1975) (R).

# Chapter 8

# The Physics of Nucleic Acids

## 8.1 Molecular Biology

Molecular biology investigates the molecular nature of the fundamental phenomena of life, primarily those of heredity and variability, which are determined by the structure and properties of nucleic acids. The rise of molecular biology is directly related to the discovery of the biological significance of these acids. Molecular biology as a scientific discipline originated with the discovery, by Avery et al. [1] in 1944, of the transformation of bacteria by DNA (p. 486). This discipline strives to explain biological phenomena through physics and chemistry, a procedure that entails the incorporation of biology into one unified natural science. Molecular biology studies not only heredity and variability but all life phenomena: enzymic catalysis, membrane transport, mechanochemical phenomena, etc. This undertaking constitutes an authentic atomic-molecular approach to biological problems.

Although it is, of course, impossible to draw a line between molecular biology and biochemistry, the identification of these two fields of knowledge (cf., e.g., [2]) is not expedient. Biochemistry studies any chemical reaction in living organisms, the reactions of all substances having biological functions, but does not always solve the problems related to fundamental life phenomena. Biochemistry has limitless applications in medicine, pharmacology, and agriculture, whereas practical applications of molecular biology are only beginning. In contrast to classical biochemistry, molecular biology interlocks with physics and chemistry and is unique in that it examines new aspects of these studies and provides a new formulation of problems.

The notion of "molecular biochemistry" [3] is obviously a tautology and therefore makes no sense. Chemistry and biochemistry deal only with molecules--"nonmolecular biochemistry" does not exist.

Molecular biophysics cannot be separated from molecular biology. The distinction here does not concern final goals, but the formulation of problems (p. 36). The history of molecular biology has been presented in a very elegant form by Watson [4]. Some other books that serve as an introduction to this field include [5-11].

The unity of fundamental life phenomena follows from the Darwinian theory of evolution, as well as from cell theory, which was developed by Schleiden and Schwann in 1839. The existence of unicellular organisms and the formation of a multicellular organism from one cell--a zygote--shows that the single cell possesses the properties of a living body. The mechanism of heredity and variability responsible for biological evolution is contained in the cell. Biology has localized this mechanism with increasing precision. The zygote arising as a result of the union of an oocyte and a sperm obtains hereditary properties from both cells. Since the sperm consists mainly of the nuclear substance, not the whole cell but only its nucleus is responsible for heredity (Heckel, 1868). Cytology and genetics have shown that the apparatus of heredity is concentrated in the chromosomes, which are located in the nucleus.

The next step was the discovery of the gene and the determination of the positions of the genes in the chromosomes (Morgan). Finally, molecular biology has shown that the genes are the regions in DNA molecules which form the chromosomes. Specification of the material carrier of heredity can be represented by the following historical scheme:

Organism → cell → nucleus → chromosomes → genes → DNA

It is clear that Mendel's great discovery (1865) constitutes the basis of molecular biology. Mendel's laws, as he fully understood, posit the existence in the cell of "material elements" responsible for the transmission of heredity traits [12]. In other words, the existence of genes follows directly from Mendel's laws. What is more, it follows from two quite evident and well-known facts: that parental traits absent in the first generation of offspring are frequently quite obvious in the second and subsequent generations; and that an organism develops from one fertilized oocyte.

However, determination of the molecular structure of genes does not complete the foundation of molecular biology. This requires the discovery of the molecular mechanism of gene functionality. At every cell division, the chromosomes duplicate themselves. Consequently, the genes (i.e., DNA) must have the ability to reproduce themselves (i.e., to reduplicate). The existence of mutations indicates that the reduplication of genes is covariant [13]; that is, mutational changes that arose before or in the course of copying are evident in the gene copies. That DNA molecules reduplicate (p. 535) does not explain the function of the gene in terms of chemistry and molecular physics. As early as 1909 the English physician A. E. Garrod suggested that the wild type gene is responsible for the presence in an organism of a specific enzyme. Homozygous individuals with the corresponding mutant gene lack this enzyme. In this way Garrod explained such hereditary (i.e., inborn) metabolic disturbances as phenylketonuria, a serious hereditary disease stemming from an organism's inability to destroy phenylalanine (cf. [14]). Consequently, the function of a gene is the synthesis of a definite protein. Attempts to introduce molecular ideas into genetics had been made earlier as well; the first such attempt was apparently suggested by the Russian physicist Kolly as early as 1893 (cf. [13]). The biosynthetic function of the gene has since become increasingly clear. Koltsov formulated a hypothesis on this point, but he thought that the substance of a gene is protein (the significance of nucleic acids was not yet known) [15]. Molecular biology stated that nucleic acids are responsible for the biosynthesis of protein chains. The fundamental thesis of molecular biology, which was formulated by Beadle, is: one gene--one enzyme (cf., e.g., [16]). Now this thesis has to be formulated in a more precise way: one gene--one protein chain.

Thus heredity at the molecular level means the DNA-programmed reproduction of the synthesis of definite proteins in subsequent generations. Gene mutations amount to changes in this program (i.e., to changes in the DNA structure). The molecular nature or mutations and their correlation with the laws of physics were discovered by Timofeev-Ressovsky, Delbrück, and Zimmer (cf. [17-18]). On the other hand, the so-called modifications (i.e., nonhereditary changes) are due to alterations in the protein (enzyme) structure and function in ontogeny and do not concern the genetic program (i.e., the structure of DNA).

The development of molecular biology has been marked by great discoveries made in a comparatively short time. In 1953 Watson, Crick, and Wilkins discovered the secondary structure of DNA by X-ray structural analysis (p. 489). Watson has written a vivid description of the history of this discovery [19]. The structure of DNA, the double helix, explains its capacity for covariant reduplication. A general understanding of the biosynthetic function of DNA made it possible to formulate the genetic code problem (Gamow, 1954, cf. Chapter 9). Later, details of the biosynthetic process, as well as of the participation of other nucleic acids (messenger and transfer RNA's) and nucleoproteins (ribosomes), were elucidated. These successes in molecular biology are closely related to developments in genetics, physics, and chemistry. The genetics of bacteria and viruses makes it possible to study the transmission of hereditary traits at the chemical level. Of great importance in this area was the discovery of the sexual reproduction of bacteria and their hybridization (Lederberg and Tatum [20]). The phenomena arising at bacterial conjugation were investigated and described by Jacob and Vollman [21]. Kornberg performed DNA reduplication *in vitro* (p. 535). Successes with polynucleotide synthesis made it possible to decipher the genetic code (Chapter 9). Khorana performed gene synthesis (Chapter 9). We see that a series of problems in heredity and variability has not only been stated but solved, enabling us to turn to more complicated problems that are far from solution: cell differentiation; embryogenesis; and carcinogenesis. Understanding the molecular foundations of the evolutionary process [5], we are only beginning to comprehend the molecular processes of ontogeny and phylogeny.

Molecular biology relates biology to physics and chemistry and totally excludes the possibility of a vitalistic treatment of life phenomena. The development of molecular biology means a revolution in the natural sciences, the reconstruction of a scientific Weltanschauung.

Let us consider now the biological function of nucleic acids. The proofs of the genetic role of DNA are indisputable. DNA is localized in chromosomes. The DNA content of diploid (somatic) cells in different tissues in individuals of a given species is practically constant. The DNA content in haploid (sexual) cells is half this amount [22]. In the course of mitosis (i.e., at chromosome duplication), a cell's DNA content doubles.

Schrödinger raised the question of the causes of the formidable stability of the hereditary substance, which is made up of light atoms, in a series of generations [18]. Indeed DNA has great metabolic stability. Adenine labeled by $^{32}P$ is not incorporated in the DNA of nondividing cells. If cells divide in a medium that contains labeled atoms, these atoms are incorporated in the DNA being formed, but are later retained in the DNA and do not participate in metabolism. DNA's high stability is a result of its peculiar secondary structure (Section 8.2). The question asked by Schrödinger has been answered. On the other hand, the participation of DNA in mutagenesis has been established. Short-wavelength radiation has been found to induce changes in DNA and chemical agents have produced changes in the nitrogen bases (p. 604).

DNA determines the transformation of bacteria, discovered by Griffith in 1928 [23]. He found that mutant strains of pneumococci (*Diplococcus pneumoniae*) differ in the character of their polysaccharide cell walls. These differences can be detected by immunological methods and are visually manifest in the shape of colonies. Strain S forms "smooth" shiny colonies (i.e., the cells have strong capsule-like walls). Colonies of strain R, which consists of cells that have lost their ability to produce the polysaccharide that is essential to such walls, are "furry." If a living culture of type R and a dead culture of type S pneumococci are injected into a mouse, the mouse becomes infected with pneumonia; a culture of the bacteria that multiplies within the mouse generates a living virulent type S pneumococci, from which it follows that dead strain S bacteria contain a factor that can transform strain R into strain S. Similar phenomena were observed in a number of other bacterial species, and it has since been established that the transforming factor is DNA [1]. Pure samples of DNA extracted from specific strains produce analogous hereditary changes in other strains.

Deoxyribonuclease (DNase), an enzyme that splits DNA, stops transformation. DNA may carry several genetic indicators at once: resistance to antibiotics; the ability to form capsules; etc.

When penicillin is added to an initial pneumococcus culture containing penicillin-resistant mutants, all the cells perish except the mutant ones, which form culture P after multiplication. When DNA extracted from this culture is added to the initial culture, many more P mutants form within it than would form in the absence of transforming factor. If the number of spontaneous P mutants is 1 in $10^7$ cells, then as result of transformation it is $10^4$ to $10^5$ times greater.

Bacteriophages (i.e., viruses multiplying in bacteria) consist of DNA surrounded by a protein envelope. By using labeled atoms $^{35}S$ and $^{32}P$, it was established that no ($^{35}S$-labeled) protein penetrates a phage-infected bacterial cell, but that ($^{32}P$-labeled DNA) does penetrate such a cell. The phage particle "injects" its DNA into the cell. Multiplication of the phage particles in the cell shows that DNA organizes both the synthesis of its copies and the synthesis of the protein envelopes. This means that DNA is the genetic material of the phage.

In many virus particles the role of DNA is played by RNA. It has been shown that RNA extracted from tobacco mosaic virus (TMV) has infectious activity [24,25]; that is, viral RNA molecules organize the synthesis of proteins in the new virus particles when they are introduced into plant cells. Viral RNA plays a genetic role. Mixing TMV RNA with its protein reconstitutes the particles possessing the properties of the initial virus. Hybrid viruses have also been obtained by combining the RNA of one strain of TMV with the protein of another strain. The protein of the progeny of the hybrid virus contains amino acid residues that are characteristic of the protein of the RNA donor virus, but are absent from the protein of the second strain. These results again manifest the genetic role of viral RNA [26].

Another proof of the genetic role of phage RNA is the transduction phenomenon. Some phages are able to transduce genetic material from the donor bacteria in which the phage multiplies to the recipient bacteria infected by this phage. Transduction is the transfer of the bacterial genetic traits; that is, phage transduces both its own genetic material and part of the genetic material of the donor bacteria. This material is DNA [27-28].

This is also shown by the phenomenon of lysogeny. So-called moderate phages can remain in bacterial cells in the form of harmless prophage. The cells multiply quite normally for several generations and the prophage, too, is regenerated. If such a cell is acted on by an inductor, for example, by ultraviolet radiation, it begins to produce phage and dies. Lysogeny is a hereditary property of bacteria that is acquired during infection with the phage, whose DNA unites with the DNA of the bacterial cell.

These facts and many others demonstrate the proof of the statement that genes are DNA. Hereditary information (i.e., the program of protein synthesis) is stored in DNA molecules.

Ribonucleic acids are also essential participants in protein synthesis. As was said earlier, the RNA of a series of viruses plays a role similar to that of DNA. High-molecular-weight, or messenger, RNA (mRNA) transmits genetic information from the chromosomal DNA toward cell organelles, where protein synthesis takes place. These organelles are ribosomes, composed of ribosomal RNA (rRNA) and proteins. Low-molecular-weight transfer RNAs (tRNA) are substances that interact directly with the amino acids which assemble in the protein chain. All three kinds of RNA, m-, r-, and t-, are synthesized on DNA (with the participation of enzymes); they further organize protein synthesis (Chapter 9) [4-11].

A series of physical problems arises in connection with the function of nucleic acids. It is necessary to determine the mechanism of covariant reduplication, the mechanism of the template synthesis of RNA in DNA and of polypeptide chains in complexes of mRNA with ribosomes. The genetic code problem of the correlation between the primary structure of DNA (and, hence, of mRNA) and that of the protein chain has to be solved. It is not sufficient to describe the chemical pattern of protein synthesis; the physical and physicochemical mechanisms and conditions of the synthesis must be established, particularly the enzyme action mechanisms. Biophysics investigates the thermodynamics and kinetics of these reduplicative and biosynthetic processes, as well as the thermodynamics and kinetics of the denaturation of nucleic acids. The physics of mutagenesis--spontaneous, chemical, and radiational--is also an important field in biophysics.

## 8.2 The Structure of Nucleic Acids

The interrelation between molecular structure and function is not always evident. We have seen that the problem of establishing such an interrelation in proteins is far from solution. The situation for nucleic acids, particularly DNA, is quite different. At least one of the most important functions, the reduplication of DNA, could be qualitatively explained immediately after the discovery of the secondary structure of DNA.

The secondary structure of DNA was determined by X-ray diffraction in work done by Franklin, Crick, Watson, and Wilkins [29-32]. Oriented fibers of the Li and Na salts of DNA yielded X-ray diffraction patterns containing up to 100 reflections (Fig. 5.8, p. 267). The crosslike positions of the reflections show that the structure is helical. The deciphering of roentgenograms and the application of molecular models revealed the double-helical structure of DNA, which can be crystallized in two different helical forms. At low relative humidity (less than 70%) DNA is crystallized in the A form, monoclinic with lattice parameters a = 40.4 Å, b = 22.07 Å, β = 97.1°. If the humidity is high, the hexagonal B form, characterized by a = 46 Å, comes about. Both forms of Na-DNA from calf thymus are shown in Fig. 8.1. Figure 8.2 shows the molecular model of the B form. The C form, which is obtainable at low humidity, was discovered later [33]. Geometric characteristics of the A, B, and C forms of DNA are presented in Table 8.1 [34]. They differ in the projection values of the residues on the helix axis, the angles of rotation between neighboring base pairs, and the inclination of these pairs relative to the helix axis.

*TABLE 8.1*
*Geometry of DNA in Different Forms*

| DNA, salt | Humidity (%) | Number of nucleotides per helix turn | Step of helix (Å) | Translation per nucleotide (Å) | Rotation per nucleotide | $\phi^a$ | $\psi^a$ |
|---|---|---|---|---|---|---|---|
| A-form, Na | 75 | 11 | 28.15 | 2.55 | 32.7 | 20 | 0 |
| B-form, Na | 92 | 10 | 33.6 | 3.36 | 36 | -- | -- |
| B-form, Li | 66 | 10 | 33.7 | 3.37 | 36 | 2 | 5 |
| C-form, Li | 66 | 9.3 | 31.0 | 3.32 | 39 | 6 | 10 |

[a] $\phi$ is the angle between the perpendicular to the helix axis and the plane of the bases; $\psi$ is the dihedral angle between the planes of the base pairs.

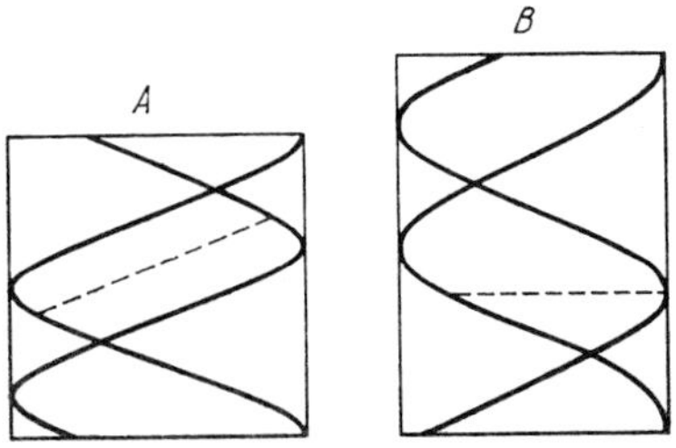


*FIG. 8.1 A and B forms of DNA.*

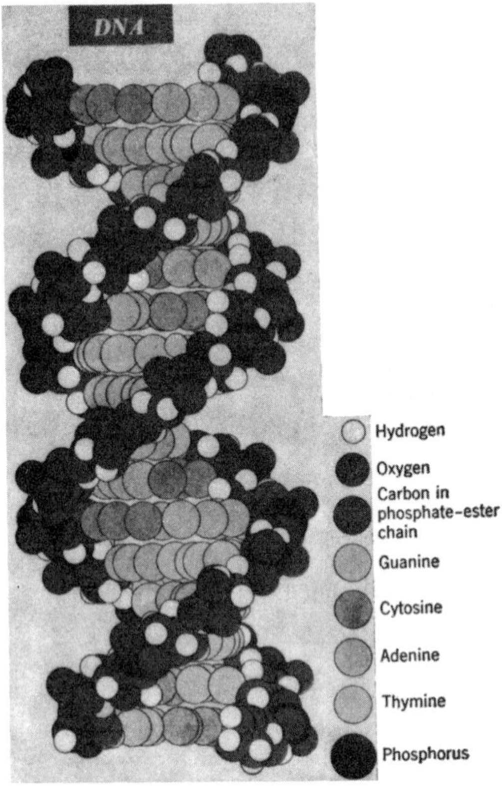


*FIG. 8.2 Molecular model of the B form of DNA.*

DNA of the T2 phage has a peculiar T form [35]. Instead of cytosine, this DNA contains oxymethylcytosine (70% of all oxymethylcytosines are glucosylated and 5% are diglucosylated). At low humidity phage DNA has the T form with a step of helix of 27.2-28.2 Å and 8.0-8.3 base pairs per turn.

The conformation of a polynucleotide chain depends on the six angles of rotation (Fig. 8.3), which are counted off in the following way: if we look along the bond around which rotation occurs, then the most remote bond rotates clockwise in relation to the nearest one (Fig. 8.4). The angle corresponding to the cis position of the bonds is taken as zero. Comparison of the structures of polynucleotides and nucleotides makes it possible to determine the value of the angles that occur in nature. Calculation of the most advantageous conformations is also of great importance. The conformation of a polynucleotide chain is essentially dependent on the conformation of its ribose (deoxyribose). In nucleoside and nucleotide crystals, as well as in polynucleotides, ribose occurs in four conformations, which differ as

*FIG. 8.3 Rotation angles in a polynucleotide chain.*

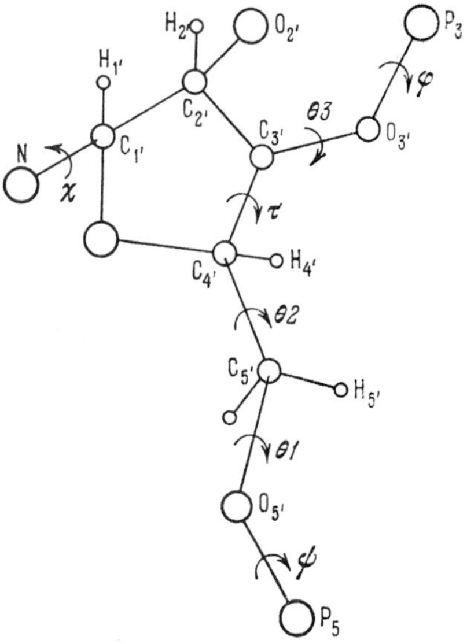


*FIG. 8.4 Direction of rotation.*

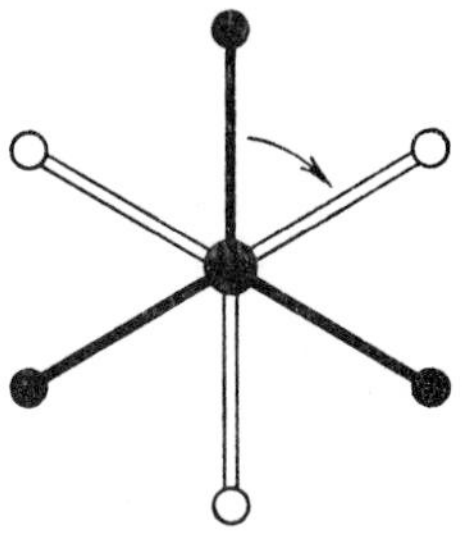

regards the kind of atom--$C_2$' or $C_3$'--of the five-membered ring that is shifted out of the plane of the other four atoms. The conformations are $C_2$'-endo; $C_3$'-endo; $C_2$'-exo; and $C_3$'-exo. If the $C_2$' or $C_3$' atom is on the same side of the plane as the $C_5$' atom, the displacement is called an exo displacement. In the opposite case we refer to an endo displacement. $C_2$'-endo and $C_3$'-endo conformations are the most frequent ones. The $C_2$'-endo conformation of ribose occurs in the B and C forms of DNA. In the DNA A form, and in double-helical RNA and poly-A, ribose occurs in the $C_3$'-endo conformation. The $C_2$' and $C_3$' atoms are displaced no more than 0.6 Å.

There are two principal conformations (syn and anti) of rotation around the glycoside bond, connecting the $N_9$ atom of purine (or $N_1$ of pyrimidine) with the $C_1$ atom of ribose. The anti conformation corresponds to the maximally elongated form of the nucleotide, and it occurs in the majority of nucleotides and nucleosides under investigation and in polynucleotides. The syn conformation has been observed in deoxyguanosine crystals and in some minor nucleotides occurring in tRNA. The sterically allowed regions of the angle $\chi$ (Fig. 8.3) are different from the $C_2$'-endo and $C_3$'-endo conformations in sugar.

There are three energetically favorable conformations formed by rotation around the $C_4$'-$C_5$' bond. They correspond to $\theta 2$ angle values of 60°, 180°, and 300°. Only one conformation $\theta 1 \simeq 180°$ occurs for the rotation around $C_5$'-$O_5$'. This corresponds to the most elongated conformation of the polynucleotide chain. All $\theta 3$ values in the range 210°-260° are sterically allowed. The conformations relative to the $O_3$'-P-$O_5$' bonds in all polynucleotides are gauche-gauche or gauche-trans; the angles $\phi$ and $\psi$ have the values (60°,60°) and (300°,300°) for the gauche-gauche conformation, and (180°,300°) and (60°, 180°) for the gauche-trans conformation. In polynucleotides only the gauche-300°, but not the gauche-60°, conformation occurs. The most elongated trans-trans conformation does not occur in polynucleotides.

According to Sundaralingam's calculations, the six favorable polynucleotide conformations are characterized by the angles listed in Table 8.2, which also presents the rotation angle $\tau$ around the $C_3$'-$C_4$' bond. Just this angle effects the displacement of the $C_3$' and $C_2$' atoms of ribose from the plane.

The conventional DNA double helix consists of two intertwined chains whose nitrogen bases are pairwise hydrogen bound. The adenine (A) of one chain is bound to the thymine (T) of the other, and guanine (G) to cytosine (C). The schemes of these pairs (Watson-Crick pairs) are shown in Fig. 8.5. Therefore,

*TABLE 8.2*
*Conformations of Polynucleotides and Nucleic Acids*

| Conformation | $\psi$° | $\theta 1$° | $\theta 2$° | $\tau$° | $\theta 3$° | $\phi$° |
|---|---|---|---|---|---|---|
| 1 | 285 | 170 | 60 | 80 | 210 | 280 |
| 2 | 285 | 170 | 60 | 150 | 210 | 280 |
| 3 | 285 | 170 | 175 | 150 | 210 | 280 |
| 4 | 285 | 170 | 175 | 80 | 210 | 280 |
| 5 | 285 | 170 | 175 | 80 | 210 | 210 |
| 6 | 285 | 170 | 175 | 150 | 210 | 210 |
| DNA-B | 281 | 212 | 58 | 130 | 147 | 282 |
| DNA-C | 315 | 143 | 48 | 168 | 211 | 212 |
| DNA-A | 283 | 167 | 67 | 76 | 221 | 279 |
| RNA-11 | 282 | 165 | 74 | 95 | 216 | 273 |
| Poly-A | 285 | 165 | 69 | 83 | 216 | 293 |
| RNA-10 | 257 | 188 | 88 | 80 | 203 | 285 |
| A-2'p5'-U | 313 | 170 | 57(U)<br>45(A) | 84(U)<br>184(A) | 244 | 232 |

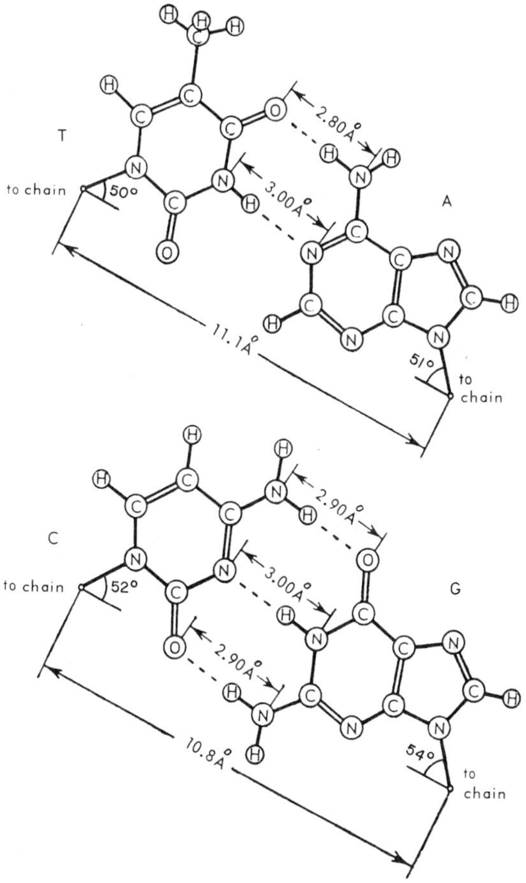


*FIG. 8.5 A-T and G-C pairs (Watson-Crick).*

two DNA chains are complementary--a unique correlation exists between their nucleotides. This correlation explains the sense of Chargaff's rules (p. 73).

The peculiar structure of double-helical DNA is characterized by broad and narrow grooves, or furrows, on the surface of the macromolecule.

Evidently, there are two possible arrangements of the double helix--in parallel or antiparallel chains (Fig. 8.6). The second model agrees better with X-ray diffraction data. The proof of the antiparallel arrangement was obtained biochemically [39]. The reduplicative synthesis of DNA *in vitro* is possible (Section 8.8). This occurs in a mixture of all four types of nucleoside triphosphates (NTP), the necessary enzyme (DNA-polymerase), some ions, and native DNA as initiator

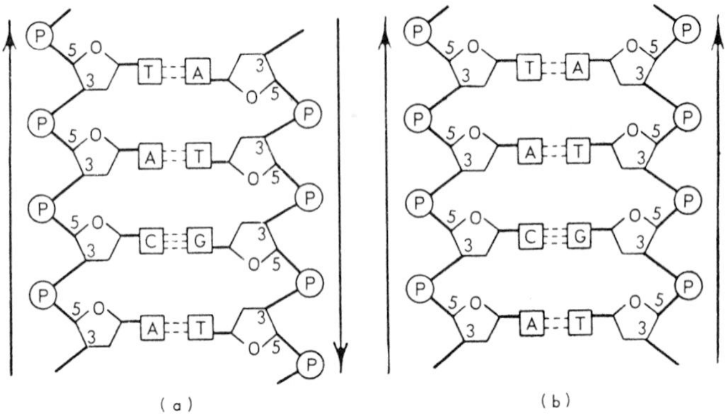

*FIG. 8.6 Double helices formed by (a) antiparallel and (b) parallel chains.*

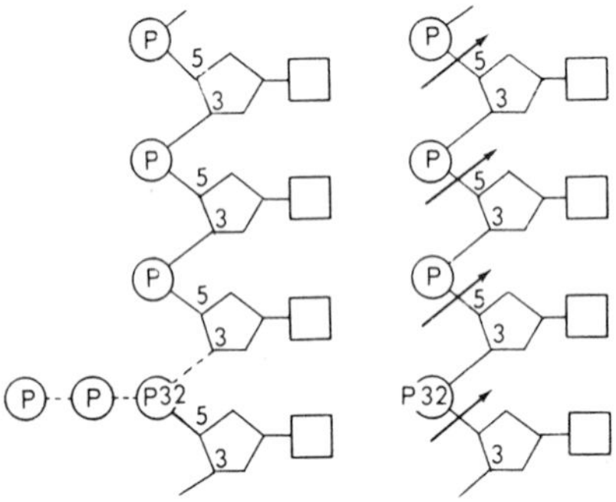

*FIG. 8.7 Synthesis and splitting of a $^{32}P$-labeled DNA chain.*

of the template synthesis of identical DNA. In the experiments of Josse one of the four NTP contained $^{32}P$ in the phosphate connected with deoxyribose in the 5' position (Fig. 8.7). With the help of enzymes, the DNA obtained was split into 3'-deoxyribonucleotides in such a way that the labeled P atom occurred in the 3' position of the neighboring nucleotide (Fig. 8.7). Four different 3'-deoxyribonucleotides were obtained and separated by paper electrophoresis. By determining the radioactivity of the monomers obtained, it was possible to establish the extent to which $^{32}P$ was linked to the other nucleotides. The experiments were performed with all four NTP; therefore, all 16 possible combinations were investigated.

The results are shown in Table 8.3. The rule of Chargaff is, of course, obeyed. Identical Greek letters indicate the combinations whose content must be equal if the chains are antiparallel; identical Latin letters indicate those that must be equal if the chains are parallel. Indeed, in the first case (Fig. 8.6) the contents of the APC and GPT combinations must

*TABLE 8.3*
*Relative Content of Various Combinations of Neighboring Nucleotides in a DNA Preparation*[a]

| Labeled triphosphate | Isolated 3'-deoxyribonucleotides | | | |
|---|---|---|---|---|
| | TP | AP | CP | GP |
| D ATP-$^{32}$P | a<br>TPA<br>0.053 | b<br>APA<br>0.089<br>α | c<br>CPA<br>0.080<br>β | d<br>GPA<br>0.064<br>γ |
| D TTP-$^{32}$P | b<br>TPT<br>0.087<br>α | a<br>APT<br>0.073 | d<br>CPT<br>0.067<br>δ | c<br>GPT<br>0.056<br>ε |
| D GPT-$^{32}$P | e<br>TPG<br>0.076<br>β | f<br>APG<br>0.072<br>δ | g<br>CPG<br>0.016 | h<br>GPG<br>0.050<br>φ |
| D CPT-$^{32}$P | f<br>TPC<br>0.067<br>γ | e<br>APC<br>0.052<br>ε | h<br>CPC<br>0.054<br>φ | g<br>GPC<br>0.044 |
| Total | 0.283 | 0.286 | 0.217 | 0.214 |

[a]The preparation was obtained by using native calf thymus DNA as the DNA initiator; from [39].

be equal, for example, whereas in the second case APC and TPG combinations must be identical, since the bonds with $^{32}$P that are broken by the enzyme are distributed differently in these two cases. We see that the chains are antiparallel.

The double-helical structure explains qualitatively the reduplication of DNA and its covariance. According to Crick's and Watson's ideas, the double helix separates because of the breakage of hydrogen bonds and unwinding of the helix. Each liberated chain serves as a template for the formation of a new chain. Monomers bind to the chain, forming the Watson-Crick pairs A-T, T-A, G-C, and C-G. Simultaneously, polycondensation of nucleotides occurs and we obtain two new double helices identical with the initial one. If an unsuitable nucleotide is incorporated, then at the subsequent duplication it occupies its position lawfully and reduplicates also. Such a semiconservative model (the new double helix contains one old and one new chain) was actually confirmed by experiment (Section 8.8).

The structure of DNA can be observed with an electron microscope. Figures 8.8 and 8.9 show electron micrographs of native DNA. Native DNA molecules with lengths up to 49 μm have been extracted from the T2 phage; the length of DNA from *E. coli* reaches 400 μm, corresponding to molecular weights of the order of $10^9$. The entire DNA of a phage particle consists of one macromolecule.

Low-angle X-ray scattering yields results that are in good agreement with the double-helical structure [40]. Solutions of native DNA are very viscous. However, this does not mean that the entire DNA macromolecule is a stiff "stick." Data obtained by means of light scattering, sedimentation, viscosimetry, and dynamic birefringence show that the double helix of native DNA is folded into a loose coil [41]. The intrinsic viscosity $[\eta]$ of DNA is approximately proportional to the first power of the molecular weight MW, corresponding to the loose coil. The dependence of $[\eta]$ on M has the form (in 0.15 M NaCl) [42]

$$[\eta] = 6.9 \times 10^{-4} M^{0.7} \qquad (2.10^6 < MW < 130 \times 10^6)$$
$$[\eta] = 1.05 \times 10^{-7} M^{1.32} \qquad (MW < 2 \times 10^6) \qquad (8.1)$$

Investigation of the flow birefringence permits us to determine the thermodynamic flexibility of DNA. The flexibility can be characterized by the so-called persistent length of the approximately rectilinear region. At 0.15 M NaCl, the persistent

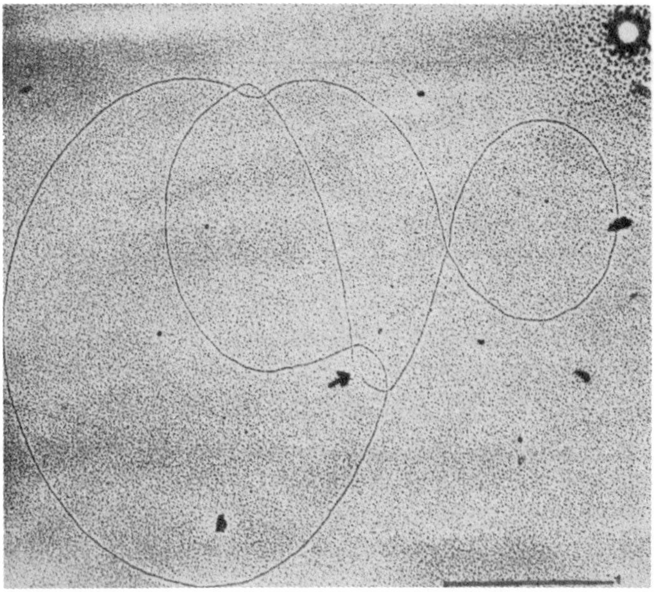

*FIG. 8.8 Electron micrograph of the cyclic DNA double helix of E. coli. Magnification 35,000.*

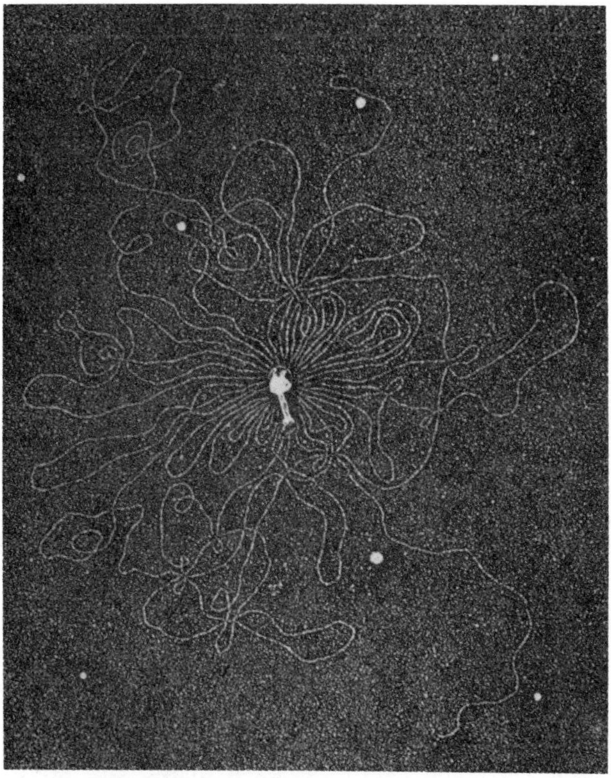

*FIG. 8.9 Electron micrograph of DNA of the T2 phage. Magnification 64,000.*

length is equal to 500 Å, at 0.0014 M, 800 Å (i.e., the flexibility decreases with increasing ionic strength) [43]. Approximately 150 monomer units correspond to the 500-Å length. Apparently the native DNA macromolecule cannot be considered a zigzag, consisting of rigorously rectilinear regions. It has rather a "wormlike" structure, with a continuously changing curvature. The persistent length is a conventional value which expresses the length of the rectilinear region if the wormlike chain is replaced by the zigzag one.

Doty showed that the density of DNA (of the order of 1.7 g $cm^{-3}$) increases linearly with increasing G+C content (MW of A+T = 247, that of G+C = 273). On the other hand, these pairs occupy approximately the same volume of the double helix. An increase in density equal to 0.00103 g $cm^{-3}$ corresponds to a 1% increase in the G+C content. These results were obtained by density gradient sedimentation (p. 139).

Heating, changes in the pH, etc., cause denaturation of DNA, that is, the transition double helix → two coils (Section 8.4).

The helicity of DNA is clearly expressed by its spectral properties, namely by the hypochromism effect in the absorption region of nitrogen bases at 260 nm (p. 277-280). The absorption intensity of double-helical nucleic acid is considerably less than that of the denatured form. The degree of helicity of DNA and RNA can be readily determined by hypochromism.

The circular dichroism (CD) of oriented and nonoriented films of DNA at various values of humidity and salt concentration has been investigated in detail [34]. The CD spectra of Na and Li DNA salts at relative humidity (r.h.) 92% coincide with those of solutions and correspond to the DNA B form. At r.h. in the range 75-66% films with 3% $Na^+$ yield a CD spectrum identical with that of the A form solution. Li-DNA at low r.h. gives the spectrum typical of the C form. Figure 8.10 shows the CD spectra of Na-DNA (from calf thymus) in solution and in a film at r.h. 92%. Hence, film CD makes possible the estimation of secondary DNA structure. A CD theory for films that simultaneously possess linear dichroism (LD) was developed in [34]. Analysis of CD and LD permits us to determine the orientations of optical transitions in DNA. Results have been obtained showing the possible existence of the n $\pi^*$ transition with $\lambda < 240$ nm, probably with $\lambda = 225$ nm. Conformational transitions of DNA in solution are observed by means of CD spectra. Double-helical regions of RNA and the hybrid double helices DNA-RNA are in the A form; DNA of the heads of bacteriophages are in the more compact C form; DNA in solution, in the B form. The transition into the C form occurs under the action of monovalent cations, which neutralize the phosphate groups. This transition is facilitated in methanol-water mixtures. Ethanol provokes the cooperative B → A transition [44].

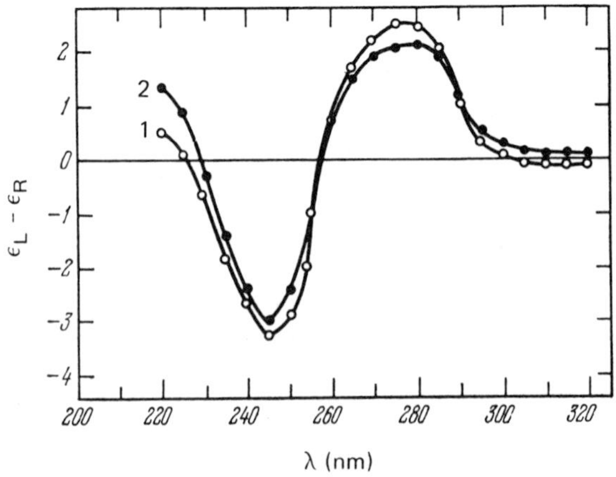


*FIG. 8.10 Comparison of CD spectra of Na-DNA in (1) solution and (2) a film.*

The double-helical regions of nucleic acids are modeled by the synthetic polynucleotides. At 0.1 M NaCl, poly A forms a double helix with poly U; the maximum hypochromism (i.e., maximum helicity) is observed at the 1:1 content of the mixture [45]. If bivalent cations are present, at $1.2\times10^{-2}$ M $MgCl_2$ the maximum hypochromism corresponds to the ratio poly U:poly A = 2:1. This means the formation of a triple helix. Apparently the Mg ions compensate the charges of the phosphate groups and, hence, diminish the electrostatic repulsion of the chains.

Investigation of mixtures of poly U and of the copolymers poly AU of various composition made it possible to determine the nature of the defective regions in the double helix [45]. A is complementary to U; therefore A and U are bound together; similar binding of U and U is impossible. If poly U interacts with poly AU, two possibilities exist, as shown in Fig. 8.11. Either (a) there are no hydrogen bonds at those points where U is positioned opposite U, but the helix remains a helix; or (b) the unpaired groups are pushed out of the helix and loops are formed. These two cases differ as regards the ratio of poly U and poly AU of equal polymerization degree in the system having maximum hypochromism (i.e., in the double helix). For the helix fragments shown in Fig. 8.9 in (a) the ratio poly AU:poly U is 10:10, in (b) 6:10. Experiment confirms only the second possibility ([45], cf. [6]). Examination of the structure of type (b) with the help of a molecular model shows that the loops do not disturb the organization of the double helix. Only a slight rapproachement of the phosphate groups occurs--their distance decreases from 7 to 6 Å. It can be thought that in native DNA, too, the accidental incorporation of an unsuitable nucleotide (i.e., a mutation) results in the formation of a loop.

An investigation of double-helical poly-deoxy IC is reported in [46]. The X-ray diffraction patterns of the Na salt at r.h. 75% and the CD spectra both show an unusual conformation of the helix (probably a left one) different from all known

*FIG. 8.11 Possible secondary structures of the complexes formed by poly U and poly AU.*

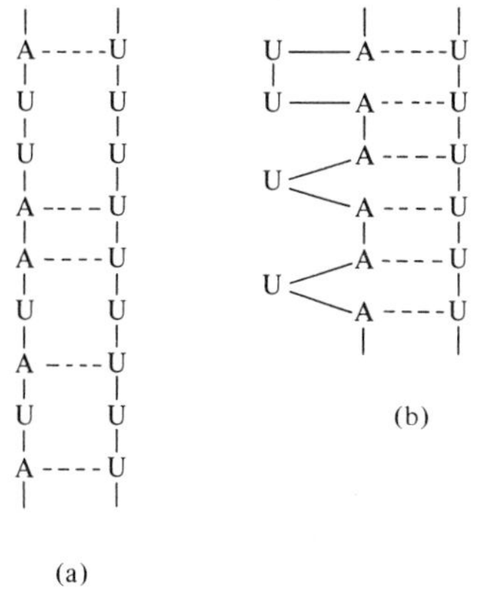

double helices and having eight monomers per turn. We cannot, however, exclude the possibility that the helix remains a right one, but in another conformation. The solution of such problems requires the development of a more perfect CD theory for polynucleotides. Later it was shown that the conclusion about the left helix was erroneous [193].

Viral RNA and rRNA do not crystallize. In solution these types of RNA have much smaller viscosities and higher sedimentation rates than DNA.

The denaturability of RNA, expressed particularly as an increase in light absorption at 260 nm, the greater compactness of the coils ($[\eta]$ is approximately proportional to $MW^{0.5}$), and the CD spectra manifest the partly helical structure of native RNA. The resistance of RNA to denaturation increases with its G+C content. Doty and co-workers suggested a "starfish" structure for RNA, with the "rays" formed by defective double-helical regions (cf. [6,47]). This structure satisfactorily explains the observed high percentage of helical reactions--up to 77% in rRNA and as high as 88% in TMV RNA. Spirin described the structure of high-molecular-weight RNA with the help of the model shown in Fig. 8.12 [48]. The structure consists of a single chain forming a compact coil at high ionic strength, a compact "stick-like" structure at low ionic strength, and a statistical coil at denaturation. Such a structure agrees with the results of investigations of RNA in solution. There are no major differences between the models of Doty and Spirin. The structure of low-

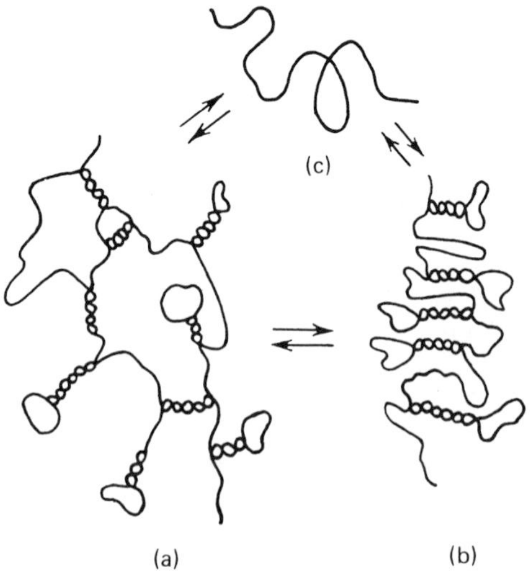


*FIG. 8.12 Spirin's model of RNA structure at (a) high ionic strength; (b) low ionic strength of solution; and (c) denatured RNA.*

molecular-weight tRNA, which is also partly double-helical, is described in Section 9.4.

Recently, the formation of a tertiary structure of double-stranded DNA in solution was discovered in the presence of polyethylene glycol (PEG) [49]. The conditions for its formation are room temperature; 0.25-0.5 M NaCl; $MW_{PEG}$ = 20,000; $C_{PEG}$ = 80 mg $ml^{-1}$; $MW_{DNA} = 1.4\times10^6$; $C_{DNA}$ = 10 g $ml^{-1}$. The DNA takes on the form of a torus, a form corresponding to a decrease in the intensity of the 260-nm band and the appearance of an intensely negative CD band with a maximum in the 260 - 270-nm range. The toruses can be observed in an electron microscope; their diameters lie in the 800 - 1400-Å range. The specific reversible conformational transition of DNA that is thus established may have some biological significance.

## 8.3 Intramolecular Interactions in the Double Helix

According to the original suggestion, the double-helical structure of DNA is due to hydrogen bonds fixing the Watson-Crick pairs (Fig. 8.5). The determining role of three hydrogen bonds in G-C and of two hydrogen bonds in A-T seem to be quite natural. This suggestion agrees with the results of X-ray analysis. However, actual determination of the nature of the interactions in the double helix requires special theoretical and experimental investigation. The quantitative data characterizing these interactions can be obtained by quantum-mechanical calculations and physical studies of simple monomeric models.

Free nitrogen bases form the hydrogen-bonded purine-pyrimidine complexes in the solid state. This was first established by Hoogsteen [50], who obtained such a complex by means of the common crystallization of 9-methyladenine with 1-methylthymine (MA-MT). The $N_1$ atom in T and $N_9$ atom in A were blocked by the methyl groups, preventing the formation of additional hydrogen bonds. X-ray analysis of the MA-MT crystal showed that the structure of the complex differs from the Watson-Crick structure (WCS) (Fig. 8.13; cf. Fig. 8.5). The $N_1$ atom in MT forms a hydrogen bond not with MA but with the imidazole $N_7$. A similar Hoogsteen structure (HS) was established for the complex formed by 9-ethyladenine with 1-methyluracil (EA-MU) [51]. It is possible that the HS is more stable than the WCS already in the solution. Another possibility is that the type of complex occurring in the crystal is determined by the crystalline packing. Infrared and NMR spectroscopy of solutions show that purine-pyrimidine complexes actually are formed, but these techniques do not enable us to determine the complexes' structure [52-57]. It has to be noted that hydrogen bonds are formed in solutions in the A-U, A-T, and G-C pairs, but not in the G-U, A-G, and A-C pairs. Consequently, crystals with noncomplementary pairs could not be obtained from solution.

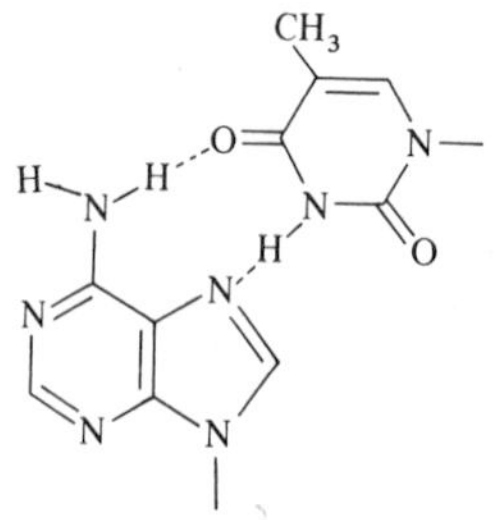

*FIG. 8.13 Structure of the MA-MT pair.*

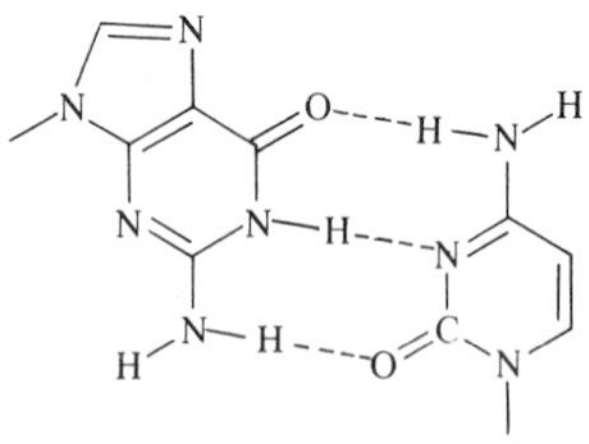

*FIG. 8.14 Structure of the EG-MC pair.*

A WCS with three hydrogen bonds was detected in the crystalline complex of 9-ethylguanine with 1-methylcytosine (EG-MC) (Fig. 8.14) [58]. The complex of 9-ethylguanine with 1-methyl-5-fluorocytosine has a similar structure [59]. The structure of the complex of 9-ethylguanine with 1-methyl-5-bromocytosine is characterized by somewhat different crystalline packing [59]. The system deoxyguanosine-5-Br-deoxycytidine, which is closer to DNA, also contains three hydrogen bonds connecting the bases [60].

The strong mutagenic substance bromouracil (methylated) was investigated in pairs with ethyladenine [61] and methyladenine [62]. It is interesting that different results were obtained in these two cases. In the first pair the hydrogen bonds $O_2$-$N_6$ and $N_3$-$N_7$ form; in the second, one $O_4$-$N_6$ and $N_3$-$N_7$. These results show that the nitrogen bases are able not only to tautomerize but also to form various types of hydrogen bonds. Further details are contained in [63] and [64].

Quantum-mechanical calculations of the interactions in purine-pyrimidine pairs that took into account the monopole-monopole forces, the forces of the interaction of monopoles with induced dipoles, and dispersion forces have been performed [65,66] (cf. also [6]). The calculations showed a greater stability of the HS in comparison with that of WCS for the A-T and A-U pairs and the stability of WCS for the G-C pair.

Obviously the structure of a complementary base pair in a double helix cannot be determined solely from the properties of an isolated pair. Along with the "horizontal" interactions in the pair there must exist the "vertical" interactions of the neighboring bases. The structure optimal for the double helix as a whole must be stabilized. On the other hand, the possibility of the formation of various hydrogen bonds is probably of importance for mutagenesis (p. 603).

Theoretical calculations of interactions in the double helix have been reported in a series of works, starting with that of De Voe and Tinoco ([67]; cf. [6]), who sought to elucidate the role of the vertical interactions of the parallel (or quasi-parallel) nitrogen bases ("stacking"). It is known that strong dispersion interactions occur between planar π-electron cycles that are parallel to each other (p. 181). Therefore, dye molecules in solution can form polymers (the so-called Scheibe polymers) that have peculiar optical properties. These properties are related to the exciton transfer of energy. In DNA and in ordered double-helical polynucleotides the fact that the packing of bases is sufficiently dense is expressed, in particular, by hypochromism (Section 5.4).

De Voe and Tinoco calculated the energies of the van der Waals dipole-dipole, induced, and dispersion horizontal and vertical interactions. The dipole moments of A and T are directed nearly antiparallel in the A-T pair and their interaction amounts to repulsion. On the other hand, the dipole-dipole interaction in the G-C pair is an attractive one. The summary energies of the van der Waals interactions in A-T and G-D pairs (horizontal interactions) calculated with the dielectric constant equal to 1 are correspondingly 0.2 and -3.9 kcal $mole^{-1}$. In a real medium they must be several times smaller, since the effective dielectric constant is of the order of 2-5. Consequently, the single pair is stabilized mainly by the hydrogen bonds. More rigorous calculation, taking into account the interaction of monopoles, but not of dipoles, gives much larger energies of attraction, -5.5 kcal $mole^{-1}$ and -19.5 kcal $mole^{-1}$, respectively, for the Watson-Crick A-T and G-C pairs [65].

In subsequent works, side by side with the "monopole" calculations of the van der Waals interactions [68,69], the repulsion forces estimated on the basis of Kitaigorodsky's atom-atom potentials (p. 110), are considered [70]. Poltev and Sukhorukov developed a semiempirical method for calculating the energies of the horizontal and vertical interactions, using the approximation of the atom-atom potentials. The sum of the energies of the electrostatic (E), polarization (P), and dispersion (D) interactions, and the repulsion energy at small distances, have been determined. π-electronic charges have been calculated according to the method of Hückel, σ-electronic charges according to Del Re [71]. The parameters found in the work [72] were used in all these calculations. The polarization energy is estimated as the sum of the polarization energies of all atoms of the base in the electric field of other atoms. The dispersion energy and repulsion energy are calculated by means of the modified Buckingham function (p. 110)

$$U = Be^{-qr} - Ar^{-6} \tag{8.2}$$

where A, B, and q are semiempirical parameters, characterizing the interaction in a given pair. Parameters A are found from the approximate London formula, q from the graph drawn on the basis of data obtained for the inert gases. Parameter B is expressed as

$$B_{mn} = 6A_{mn}(e^{q_{mn}\rho_{mn}}/q_{mn}\rho_{mn}{}^{7}) \tag{8.3}$$

where $\rho_{mn}$ is the position of the minimum of the curve of U versus the interatomic distance. It is suggested that $\rho_{mn} = \rho_m+\rho_n$ where $\rho_m$ and $\rho_n$ are the equilibrium atomic radii. Calculation of the interaction of the layers in a graphite crystal for various interlayer distances yielded the ρ value for carbon in the $sp^2$ hybridization state, equal to 1.86 Å. Using this value, the values of ρ for other atoms have been estimated. For hydrogen participating in the hydrogen bond ρ = 0.4 Å, three hydrogen atoms of the methyl group were considered as one atom with ρ = 1.9 Å [73].

The criteria of effectivity of such a method are given by the calculations of the energies of a series of molecular crystals and their comparison with the experimental sublimation heats. The calculated value for naphthalene was 18.1 kcal $mole^{-1}$ (experimental values 17.3, 17.03); for anthracene 24.4 (23.54, 23.35, 24.4, 23.9); for p-benzoquinone 14.1 (15.0); for pyrazine 14.4 (13.5); for imidazole 20.2 (20.4).

In Table 8.4 the interaction energies obtained in a series of works are compared.

Table 8.5 shows the results of Poltev and Sukhorukov's calculations of interaction energies of bases in triplets (cf. also the calculations of Claverie [70]). According to these data, the double helix regions with minimum stability must contain A and T alternating in both chains, and the most stable one must have alternating G and C. Both tables show that the nucleotide sequence greatly influences the interactions.

Further improvement of these calculations is important, both for the determination of probabilities of the various point mutations and for the estimation of codon-anticodon interactions (Chapter 9). It has to be emphasized, however, that the biological function of nucleic acids is probably due mainly to kinetic and not to thermodynamic factors.

## 8.4 The Thermodynamics of Helix-Coil Transitions

Denaturation of nucleic acids amounts to the destruction of the double helix. Apparently only tRNAs have a fixed tertiary structure in the cell. The separation of the two chains of the double helix and their subsequent folding into random coils occur upon the heating of a native DNA solution at certain pH and ionic strength values. Viscosity and optical activity decrease markedly and hypochromism vanishes (i.e., the absorption intensity in the 260-nm region increases; p. 498 [75]) Separation of the chains is directly proved by the centrifugation of DNA containing $^{15}N$ in a CsCl density gradient (p. 139).

TABLE 8.4

*Interactions of the Neighboring Base Pairs in the Double Helix*[a]

| Neighbor-ing base pairs | a | b | c | d | e | f |
|---|---|---|---|---|---|---|
| ↑G-C / G-C↓ | 22.8 | 15.42 | 7.20 | 1.26 | 8.72 | 9.5 |
| ↑C-G / G-C↓ | 39.6 | 22.66 | 10.96 | 4.99 | 13.50 | 15.0 |
| ↑G-C / C-G↓ | 11.4 | 16.08 | 15.18 | 9.80 | 16.01 | 15.5 |
| ↑A-T / A-T↓ | 1.0 | 14.98 | 8.32 | 4.08 | 9.76 | 12.8 |
| ↑T-A / A-T↓ | 10.4 | 10.26 | 7.28 | 2.52 | 8.64 | 11.7 |
| ↑A-T / T-A↓ | 3.2 | 13.28 | 7.22 | 4.48 | 7.97 | 11.9 |
| ↑A-T / G-C↓ | 14.8 | 20.32 | 8.04 | 3.50 | 9.63 | 12.3 |
| ↑G-C / A-T↓ | 22.4 | 13.90 | 11.36 | 6.04 | 12.69 | 13.6 |
| ↑A-T / C-G↓ | 15.6 | 14.88 | 11.16 | 7.25 | 12.05 | 13.8 |
| ↑C-G / A-T↓ | 14.4 | 14.24 | 7.52 | 2.41 | 9.09 | 12.1 |

[a]Negative energy in kcal $mole^{-1}$. The arrows indicate the direction of the sugar-phosphate chain from the $C_3$' atom of deoxyribose towards the $C_5$' atom. Column headings denote: a, the approximation of molecular point dipoles [67]; b, approximation of monopoles and molecular polarizability [68]; c, approximation of monopoles and bond polarizabilities [69]; d, the same as c plus the repulsion energy [70]; e, the same as c plus the exchange interaction energy [74]; f, approximation of Poltev and Sukhorukov [73] without polarization energy.

Cells of *E. coli* grown in a medium with $^{15}N$ are transferred into a medium with the usual $^{14}N$. During cell division, reduplicated double helices containing one $^{15}N$ chain and one $^{14}N$ chain are formed. Before denaturation only one density peak is observed, 1.717 g $cm^{-3}$, corresponding to $^{15}N$-$^{14}N$ double helices. After denaturation, two peaks, 1.740 and 1.724 g $cm^{-3}$, arise, corresponding to single-thread coils containing $^{15}N$ and

### TABLE 8.5
#### Negative Energies of the Base Interactions in Triplets[a]

| Triplet | Energy | Triplet | Energy | Triplet | Energy | Triplet | Energy |
|---|---|---|---|---|---|---|---|
| ▲G-C / C-G / G-C▼ | 42.0 | ▲G-C / C-G / C-G▼ | 38.4 | ▲G-C / A-T / C-G▼ | 23.2 | ▲A-T / A-T / A-T▼ | 21.7 |
| ▲C-G / G-C / A-T▼ | 40.5 | ▲A-T / G-C / A-T▼ | 38.3 | ▲G-C / A-T / A-T▼ | 22.5 | ▲C-G / A-T / G-C▼ | 21.7 |
| ▲C-G / G-C / T-A▼ | 40.3 | ▲C-G / C-G / G-C▼ | 38.3 | ▲G-C / A-T / G-C▼ | 22.4 | ▲C-G / A-T / A-T▼ | 21.5 |
| ▲T-A / G-C / C-G▼ | 40.1 | ▲G-C / G-C / A-T▼ | 36.8 | ▲A-T / A-T / G-C▼ | 22.3 | ▲T-A / T-A / A-T▼ | 21.3 |
| ▲A-T / G-C / C-G▼ | 40.1 | ▲G-C / G-C / T-A▼ | 36.7 | ▲C-G / A-T / C-G▼ | 22.1 | ▲T-A / A-T / G-C▼ | 21.3 |
| ▲T-A / G-C / T-A▼ | 38.6 | ▲A-T / G-C / G-C▼ | 36.3 | ▲G-C / A-T / C-G▼ | 22.1 | ▲A-T / T-A / T-A▼ | 21.1 |
| ▲A-T / G-C / T-A▼ | 38.5 | ▲T-A / G-C / G-C▼ | 36.2 | ▲T-A / A-T / C-G▼ | 21.9 | ▲C-G / A-T / T-A▼ | 21.0 |
| ▲T-A / G-C / A-T▼ | 38.4 | ▲G-C / G-C / G-C▼ | 34.4 | ▲A-T / A-T / G-C▼ | 21.7 | ▲A-T / T-A / A-T▼ | 20.7 |

[a] In kilocalories per mole of base pairs.

[14]N.  The density increases because the coils are more compact
than the helix [76].  Direct determinations of the MW of DNA
show that it decreases by one half after denaturation [75,77].
The formation of a coil by denaturation is directly observed
in the electron microscope (Fig. 8.15).

As in the case of polyamino acids (Section 4.5), the transition can be considered as the melting of the helix.  The simplest model for investigations of these processes is a synthetic
homopolynucleotide containing complementary pairs of only one
kind, for example, poly A-poly U.  Such a double helix melts
at 65°C at 0.15 M NaCl and pH 7.0.  The absorption intensity
at 260 nm increases by 34% and the specific rotation $[\alpha]_D$ becomes 275° smaller.

As could be anticipated, the melting point $T_m$ of DNA

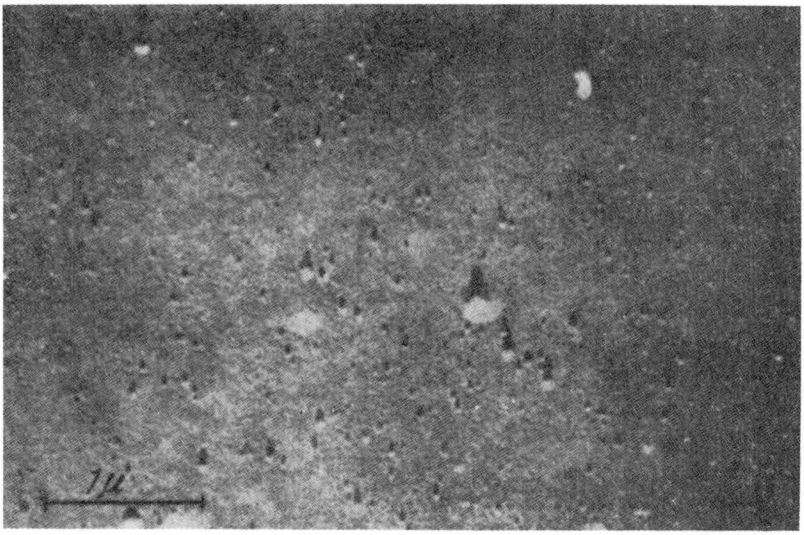


*FIG. 8.15 Electron micrograph of denatured DNA.*

increases with the relative content of G+C: these nucleotides are more strongly bound than A+T (Section 8.4). The dependence of $T_m$ on the G+C content is linear [78]. Extrapolation of this straight line gives the limiting values: $T_m$ = 69°C for poly-d-(A-T) and 110°C for poly-d(G-C), which agree with experimental values obtained for the corresponding synthetic polynucleotides: 65°C and 104°C. The $T_m$ of DNA increases with the ionic strength of the solution, approximately proportionally to the logarithm of the cation concentration.

The melting theory of the simplest models, homopolynucleotides, has been developed in a series of works [79-82] (cf. also the monographs [6,83,84].

In the transition stage the molecule consists of alternating helical and nonordered regions. If we denote the number of broken pairs by $N_1$, the number of bound pairs by $N_2$, and the number of helical regions (equal to the number of nonordered regions) by n, then the free energy of the system is equal to

$$G(N_1,N_2,n) = N_1G_1+N_2G_2+nG_s-TS_0 \tag{8.4}$$

where $G_1$ and $G_2$ are the free energies of the totally melted and totally helical molecules per base pair, $G_s$ the energy required for the formation of a melted region between two helical ones. Formation of a helical region containing $\nu$ pairs occurs as a result of the breakage of $\nu$ transverse and $\nu+1$ longitudinal bonds. Finally, $S_0$ is the mixing entropy of the helical and nonhelical regions

$$S_0 = R \ln [N_1!/n!(N_1-n)!][N_2!/n!(N_2-n)!] \tag{8.5}$$

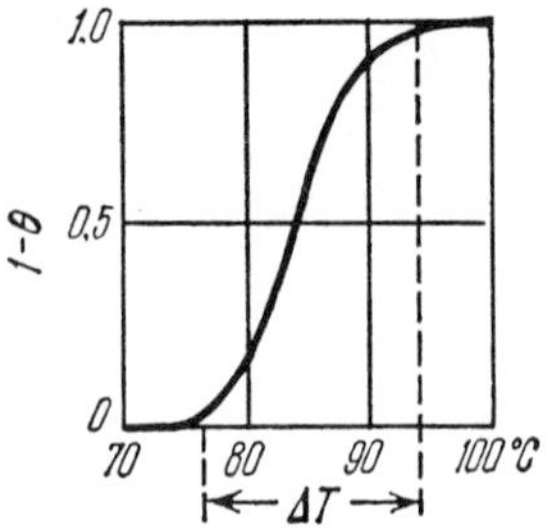


*FIG. 8.16 Helix-coil transition of DNA.*

G has a minimum obeying the condition

$$\left(\frac{N_1}{n} - 1\right)\left(\frac{N_2}{n} - 1\right) = \frac{1}{\sigma} \tag{8.6}$$

where $\sigma = \exp(-G_s/RT)$ is the cooperativity factor (cf. p. 196).

Equilibrium $N_1$, $N_2$, and n values are determined by the condition obtained via minimization of G at a constant value of $N = N_1+N_2$

$$\frac{1-(n/N_2)}{1-(n/N_1)} = s \equiv \exp\frac{G_1-G_2}{RT} \tag{8.7}$$

The melting curve, that is, the dependence of the fraction of nonordered pairs $1-\theta = N_1/N$ ($\theta$ is the degree of helicity) on T, is the sharper the smaller $\sigma$ is. At $\sigma = 1$ there is no cooperativity; at $\sigma = 0$ there is total cooperativity. The temperature range of transition $\Delta T$ (Fig. 8.16) is determined by the condition

$$\Delta T = \frac{1}{|d\theta/dT|_{max}} \tag{8.8}$$

and the calculation based on Ising's model (p. 194) gives

$$\Delta T = 4\sigma^{1/2}(RT_m^2/\Delta H) \tag{8.9}$$

where $\Delta H$ is the difference in enthalpy of the helical and nonhelical molecules per pair of bases. A more rigorous theory must take into account that the nonordered region forms a closed loop, which can be nonsymmetrical [81,82]. The improved calculation gives

$$\Delta T = \frac{16\pi^{1/3}}{3}(2\sigma)^{2/3}\frac{RT_m^2}{\Delta H} \tag{8.10}$$

From the experimental values of $\Delta T$ for synthetic homopolymers the value $\sigma \sim 10^{-4}$–$10^{-5}$ has been derived, that is, $G_s \sim 7$ kcal

$mole^{-1}$. The degree of cooperativity is very large.

The detailed analysis of the situation for heteropolymers performed by Lazurkin, Frank-Kamenetskij, and their co-workers made it possible to formulate a theory applicable to DNA and its complexes with small molecules, and resulted in the creation of a method for investigating the structure of nucleic acids [85].

The consideration of heterogeneity requires taking account of two factors: additional reinforcement or weakening of the double helix due to ligands, and the different stability of the A-T and G-C pairs. These factors must be treated in differing ways: the ligands redistribute along the helix in the melting process, and the primary structure remains unchanged.

The influence of reversible ligand binding on the melting of homopolymers has been examined [82,85,86]. Assume that $m_1$ ligand molecules are bound by the helical regions of a polymer and $m_2$ by the nonhelical ones. Instead of (8.4) we have

$$G(N_1,N_2,n,m_1,m_2)$$

$$= N_1G_1+N_2G_2+nG_s+m_1\psi_1+m_2\psi_2-TS_0$$

$$-RT \ln \frac{N_1!}{m_1!(N_1-m_1)!} \frac{N_2!}{m_2!(N_2-m_2)!} \qquad (8.11)$$

where $\psi_1$ and $\psi_2$ are the free energies of the ligand bound by regions 1 and 2. Equation (8.11) corresponds to the case of noninteracting ligands. The minimum condition $\partial G/\partial n = 0$ coincides with (8.6); that is, the average length of helical regions at a given $\theta$ does not depend on the presence of ligands. On the other hand, instead of (8.7) we obtain from $\partial G/\partial N_1 = 0$ that

$$\frac{1-n/N_2}{1-n/N_1} = s^* \equiv s \frac{1-c_1}{1-c_2} \qquad (8.12)$$

where $c_1 = m_1/N_1$, $c_2 = m_2/N_2$. These ligand concentrations can be expressed in terms of the ligand concentration in solution $c_0$ and of the binding constants $K_1$ and $K_2$

$$K_1 = \frac{c_1}{c_0(1-c_1)}, \qquad K_2 = \frac{c_2}{c_0(1-c_2)} \qquad (8.13)$$

The melting curve of the homopolymer without ligands is

$$\theta = f(s) \qquad (8.14)$$

We can put $\theta = 0.5$ at $s_m = 1$. If ligands are present

$$\theta = f(s^*) \tag{8.15}$$

and $\theta = 0.5$ at $s^* = 1$; that is, at $s_m = (1-c_2)/(1-c_1)$. The shift in melting point as compared with that of the pure polymer $T_0$ is

$$\delta\left(\frac{1}{T_m}\right) = \frac{1}{T_0} - \frac{1}{T_m} = \frac{R}{\Delta H}\ln\left(\frac{1-c_1}{1-c_2}\right) \tag{8.16}$$

The change in the melting range

$$\delta\left(\frac{\Delta T}{T_m^2}\right) = \frac{\Delta T}{T_m^2} - \frac{\Delta_0 T}{T_0^2} = -\frac{R}{\Delta H}\left[\frac{\partial}{\partial\theta}\ln\left(\frac{1-c_1}{1-c_2}\right)\right]_{\theta=0.5} \tag{8.17}$$

If there is a surplus of ligands in the solution, then

$$\delta\left(\frac{1}{T_m}\right) = \frac{R}{\Delta H}\ln\left(\frac{1+K_1C}{1+K_2C}\right) \tag{8.18}$$

where C is the total concentration of ligands in solution. The value of $\delta(\Delta T/T_m^2)$ is zero in this case. On the other hand, if the ligands are strongly bound by the polymer in the entire transition range, that is, at $K_1P \gg 1$ and $K_2P \gg 1$ (P is the concentration of the phosphate binding groups in a polynucleotide)

$$\delta T_m = 2\left(\frac{p-1}{p+1}\right)\frac{RT_0^2}{\Delta H}c \tag{8.19}$$

$$\delta\,\Delta T = 4\left(\frac{p-1}{p+1}\right)^2\frac{RT_0^2}{\Delta H}c \tag{8.20}$$

where $p = K_2/K_1$ and $c = 2C/p$. Formulas (8.19) and (8.20) are valid for small c. The theory describes also the general case [85]. Thus, at $p > 1$ ligands act as "clips," stabilizing the double helix.

These expressions are confirmed by experiment. Equation (8.18) gives a satisfactory description of the pH influence on the melting curve. With this equation the value $\Delta H = 10$–$11$ kcal $\text{mole}^{-1}$ was found at high ionic strengths and neutral pH [87]. Equations (8.19) and (8.20) agree well with the data obtained in studies of the complexes of DNA (from T2 phage) with acridine dyes and actinomycin at low ionic strengths [88,89]. The theory was also applied successfully in studies of DNA complexes with ribonuclease [85].

Metal ions (of the alkaline metals, and $Ag^+$, $Cu^+$) are bound by DNA and stabilize its structure [90–93]. The ions $Ag^+$ and $Cu^+$ can redistribute along the molecule and are bound mainly by G-C pairs. Experiments with metal ions made it possible to determine a $\Delta H$ of 8.1 ± 1 kcal mole$^{-1}$ in 0.01 M NaCl and 11.0 ± 2 kcal mole$^{-1}$ in 0.1 M NaCl [93]. Determinations of $\Delta H$ based on this theory agree well with results of direct microcalorimetric investigations, which gave $\Delta H \cong 9$ kcal mole$^{-1}$ [94].

Let us consider now the melting of a double-helical heteropolymer. In principle such melting can occur with or without the formation of loops. The general theory for a monohelical heteropolymer (say, a polyamino acid) was developed in [95-97]. The theory for the double helix assumes a random distribution of A-T and G-C pairs. The linear dependence of $T_m$ on [G-C] (p. 507) agrees with that suggestion.

Peculiar to heteropolymers is the relatively small number of microstates corresponding to a given energy. It is not likely that two different distributions of coils and helices with the same values of $N_1$, $N_2$, and n have equal energy, since they will almost certainly contain different numbers of G-C pairs in the melted regions. Therefore the entropy of mixing cannot be essential for the melting of heteropolymers. On the other hand, a new energy factor must be involved, because a decrease in the mean length of the melted regions entails a reduction in their content of the more stable G-C pairs. The competition between this factor and the factor due to disadvantageous junctions of helical and nonhelical regions must result in the alternation of helical and nonhelical regions of definite mean length at a given value of $\theta$.

An approximate solution to the problem of heteropolymer melting with a random sequence of pairs was given in [85]. Let us divide a molecule into equal segments, each containing $\lambda$ pairs. If $\lambda$ is sufficiently large, then the distribution of G-C pairs over these segments will be Gaussian:

$$P_\lambda(x) = \frac{1}{(2\pi)^{1/2}\sigma_\lambda} \exp\left[-\frac{(x-x_0)^2}{2\sigma_\lambda^2}\right] \qquad (8.21)$$

where $\sigma_\lambda^2 = x_0(1-x_0)/\lambda$, x is the content of the G-C pairs. The total number of G-C pairs in the melted regions is minimal if all segments with a G-C concentration smaller than some limiting value $x_\lambda$ are melted and all segments with $x > x_\lambda$ are helical. Therefore, the $x_\lambda$ value is determined by

$$\int_0^{x_\lambda} P_\lambda(x)\, dx = N_1/N \qquad (8.22)$$

The average concentration $\bar{x}_1$ of G-C pairs in the melted regions is

$$\bar{x}_1 = \left[\int_0^{x_\lambda} xP_\lambda(x)\, dx\right] / \left[\int_0^{x_\lambda} P_\lambda(x)\, dx\right] = (N/N_1) \int_0^{x_\lambda} xP_\lambda(x)\, dx \qquad (8.23)$$

The probability of formation of a melted region of k sequential segments is

$$\left(1 - \frac{N_1}{N}\right)\left(\frac{N_1}{N}\right)^{k-1} \tag{8.24}$$

since the G-C contents of the segments are not correlated.

The mean number of segments in the melted region becomes

$$\bar{k} = \frac{1}{1-(N_1/N)} \tag{8.25}$$

The mean number of nucleotides in this region is

$$\bar{\nu}_1 = \lambda\bar{k} \tag{8.26}$$

The number of melted regions in the whole molecule is

$$n = \frac{N_1}{\lambda\bar{k}} = N_1 \frac{1-(N_1/N)}{\lambda} \tag{8.27}$$

The free energy of a polymer is equal to

$$G(N_1,N_2,n) = N_1[\bar{x}_1 G_1{}^{GC} + (1-\bar{x}_1)G_1{}^{AT}] + N_2[\bar{x}_2 G_2{}^{GC} + (1-\bar{x}_2)G_2{}^{AT}] + nG_s \tag{8.28}$$

Introducing the conditions $N_1+N_2 = N$, $\bar{x}_1 N_1+\bar{x}_2 N_2 = x_0 N$ and denoting $\Delta G_{GC} = G_1{}^{GC}-G_2{}^{GC}$, $\Delta G_{AT} = G_1{}^{AT}-G_2{}^{AT}$, we obtain (omitting the constant terms and using the variable $\lambda$ instead of n)

$$G(N_1,\lambda) = N_1\bar{x}_1\Delta G_{GC} + N_1(1-\bar{x}_1)\Delta G_{AT} + \frac{N_1}{\lambda}\left(1 - \frac{N_1}{N}\right)G_s \tag{8.29}$$

where $\bar{x}_1$ and $x_\lambda$ are defined by Eqs. (8.23) and (8.22). The equilibrium values of $N_1$ and $\lambda$ are found from the corresponding conditions of minimum G. Examination of the G expression in the transition region (i.e., close to $N_1/N = \frac{1}{2}$) is done via expansion in a series of the powers of $\varepsilon = N_1/N-\frac{1}{2}$. Using Eq. (8.29), and again neglecting the constant members, we obtain

$$G(N_1,\lambda) = N\left\{\varepsilon\,\Delta G - \frac{1}{(2\pi)^{1/2}}\left[\frac{x_0(1-x_0)}{\lambda}\right]^{1/2}(\Delta G_{GC}-\Delta G_{AT})(1-\pi\varepsilon^2) + \frac{G_s}{4\lambda}(1-4\varepsilon^2)\right\} \tag{8.30}$$

where

$$\Delta G = x_0\Delta G_{GC} + (1-x_0)\Delta G_{AT} \tag{8.31}$$

From the condition $\partial G/\partial\lambda = 0$ (at $\varepsilon = 0$) we find

$$\lambda_0 = \frac{\pi}{2}\left(\frac{G_s}{\Delta G_{GC}-\Delta G_{AT}}\right)^2 \frac{1}{x_0(1-x_0)} \tag{8.32}$$

and since

$$\Delta G_{GC}-\Delta G_{AT} = \frac{(T_{GC}-T_{AT})\Delta H_{AT}}{T_{AT}} \tag{8.33}$$

where $T_{GC}$ and $T_{AT}$ are the melting temperatures for the corresponding polymers, we get

$$\bar{\nu}_1 = \pi\left[\frac{T_{AT}}{\Delta H_{AT}}\,\frac{G_s}{(T_{GC}-T_{AT})}\right]^2 \frac{1}{x_0(1-x_0)} \tag{8.34}$$

Equating the derivative in $\varepsilon$ to zero at constant $\lambda$ and differentiating the equation obtained in T, we find

$$\Delta T = 2\,\frac{\pi-2}{\pi}\,\frac{\Delta H_{AT}}{T_{AT}}\,\frac{(T_{GC}-T_{AT})^2}{G_s}\,x_0(1-x_0) \tag{8.35}$$

The rigorous solution of the problem differs from the one obtained only by numerical factors [98].

The formation of loops during melting is taken into account by an additional term in G, corresponding to the entropy of loops [85].

Numerical calculations by means of computers, using as independent parameters the fraction of G-C pairs $x_0$, homopolymer melting points $T_{AT}$ and $T_{GC}$, $\Delta H_{AT}$, and $\sigma$ made it possible to determine the melting curves, the dependence of the mean length of the helical region on the helicity $\theta$, and the dependence of the width of the transition range on $x_0$ at various values of these parameters [85,95,99]. In Figs. 8.17-8.19 are shown the curves obtained at $T_{AT}$ = 340°K, $T_{GC}$ = 380°K, $\Delta H_{AT}$ = 7 kcal mole$^{-1}$, $\sigma = 5\times10^{-4}$, and $N = 2\times10^4-10^6$. The linear dependence of $T_m$ on $x_0$ is obtained directly

$$T_m = T_{AT} + (T_{GC}-T_{AT})x_0 \tag{8.36}$$

These calculations were performed without taking the formation of loops into account. The loops lead to a decrease in $\Delta T$ [85, 100].

The shape of the melting curve agrees with experiment practically independently of the model. On the other hand, $\Delta T$ depends essentially on the model and the parameter values, all of which (except $\sigma$) can be found by independent experiments. Perfect agreement with experiment for a series of different types

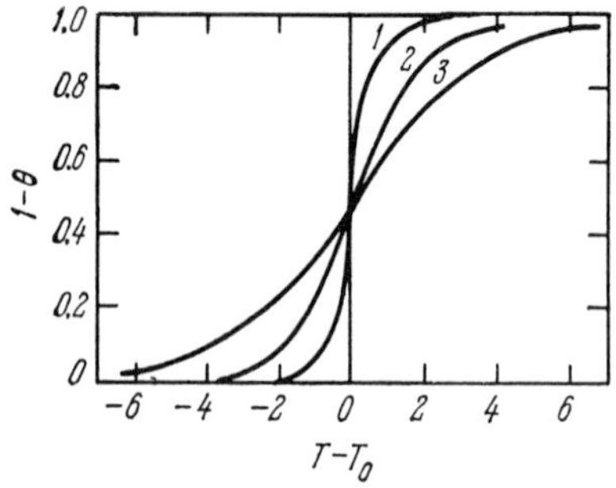


FIG. 8.17 Dependence of θ on T for a polyheteronucleotide. (1) $x_0 = 0$, $T_m = 340°K$, $N = 2\times10^4$; (2) $x_0 = 0.1$, $T_m = 344°K$, $N = 4\times10^4$; (3) $x_0 = 0.5$, $T_m = 360°K$, $N = 6\times10^4$.

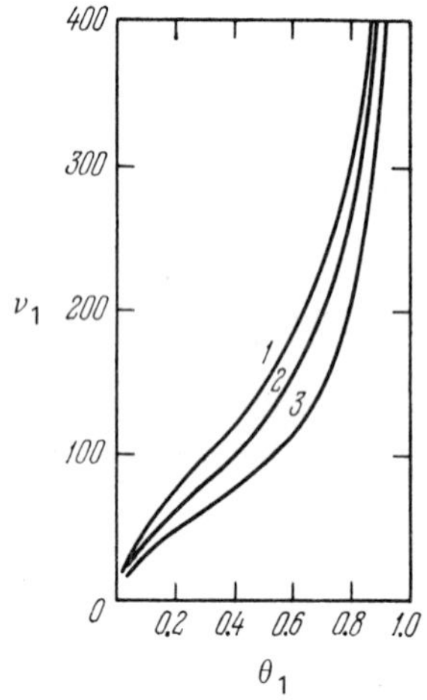


FIG. 8.18 Dependence of the average number of base pairs in helical regions on $\theta_1$. Curves 1-3 correspond to the same parameters as in Fig. 8.17.

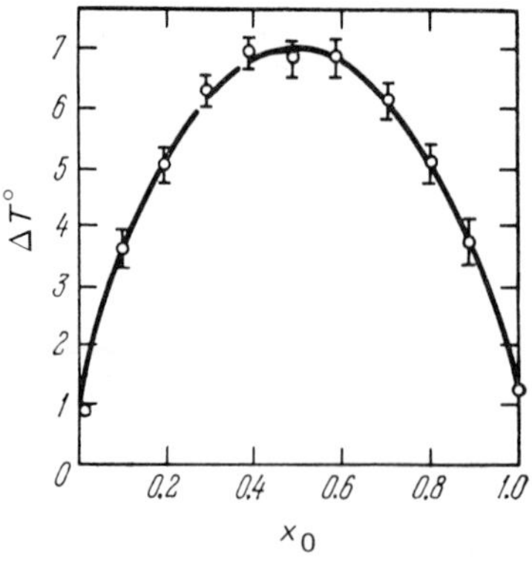


Fig. 8.19 Dependence of the melting range on the concentration of G-C pairs.

of DNA was obtained at $\sigma = 5\times10^{-5}$ [85]. Hence, the experiment permits us to find $\sigma$.

Determination of the optical density, and therefore of the degree of helicity $\theta$, and determination of the DNA size in solution by means of measurements of the intrinsic viscosity in the process of thermal denaturation made it possible to determine the mean lengths of the helical regions in an independent way [100]. In agreement with theory in the range $\theta = 0.8\text{–}0.9$ the length of helical segments $\bar{\nu}$ varies from 1000 to 2500 pairs; in the transition range (i.e., at $\theta = 0.5$), $\bar{\nu} = 400\text{–}500$ pairs.

$T_m$ decreases and $\Delta T$ increases with decreasing chain length [102]. The theory of this phenomenon was developed in [103] and confirmed experimentally [104].

These results make it possible to detect the defects in the secondary structure of DNA on the basis of the measurements of the melting curves and to determine their concentration. Ultraviolet radiation causes broadening of the $\theta(T)$ curves due to local denaturation [105]. The presence of defects can be proved by kinetic experiments [106]; their number is estimated via calibration of the thermodynamic method using fragmented DNA [104].

The possibility of establishing the character of the nucleotide distribution in a chain by investigating the helix-coil transition is especially interesting. It has been found that the DNA in a series of moderate phages contains regions of differing G-C pair concentration [107]. This is expressed by steps in the melting curve and by the results of centrifugation of fragmented DNA in a density gradient. Such block heterogeneity can be investigated by parallel examination of melting curves and of the dependence of the intrinsic viscosity on T in the transition range [107].

Let the block sequence have the form

$$\cdots \vdash \overset{x_0^{(i)}}{\text{———}} + \overset{x_0^{(i+1)}}{\text{——————}} + \overset{x_0^{(i+2)}}{\text{———}} \dashv \cdots$$

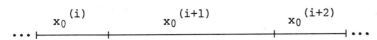

G-C pairs are randomly distributed inside a block. Fragments of DNA, called molecules, usually contain $n_0 = 1000\text{–}50{,}000$ nucleotide pairs. In this region of values the $\theta(T)$ curves do not depend on $n_0$. On the other hand, the curves of the dependence of the intrinsic viscosity $[\eta]$ on T depend on $n_0/n_B$ where $n_B$ is the mean block length. If $n_B \ll n_0$, every molecule in the solution contains a large number of blocks with varying G-C pair content. Therefore, every molecule that is melted will disintegrate into the helical and nonordered regions without the complete separation of the chains up to the final stages of melting. The $[\eta](T)$ curves will be shifted toward lower temperatures in comparison with the $\theta(T)$ curves because the formation of a small number of short melted regions disposed far from one another diminishes the dimensions of the molecule. This occurs due to the smallness of the nonordered regions and to the rapprochement of the helical regions as well as because of the appearance of the new sharp breaks diminishing the size of the statistical segment.

If $n_B >> n_0$, the molecules differ considerably in their G-C content. $\Delta T$ for every kind of molecule is smaller than the total $\Delta T$ of their mixture. In the melting range most molecules will be either totally helical or totally melted. Consequently

$$[\eta] = [\eta]_1\theta_1 + [\eta]_2(1-\theta_1) \quad (8.37)$$

and the fraction of melted base pairs is

$$\theta_1 = g = \frac{[\eta]_2-[\eta]}{[\eta]_2-[\eta]_1} \quad (8.38)$$

Figure 8.20 shows the functions $g(T)$ and $\theta_1(T)$ for various DNA types; Fig. 8.21 shows the $g(\theta)$ curve [85,101]. The curves for higher organisms are similar to the theoretical ones; for *E. coli* and phage T2 this is not valid. Hence, block heterogeneity is a property of the DNA higher organisms. These results enable us to estimate the lower limit of block sizes from the condition $n_B >> n_0$. Molecular weights of DNA from higher organisms have been of the order of $10^9$; the mean block weight is undoubtedly higher than $10^7$.

Thus, careful investigation of the melting curves of DNA yields valuable information about its structure. This method may also be useful in RNA studies.

Lifshitz studied theoretically the dependence of the form of the DNA melting curve on the sequence of the pairs A-T and G-C [108]. He showed that in an ideal case the investigation of this curve can yield valuable information about the DNA

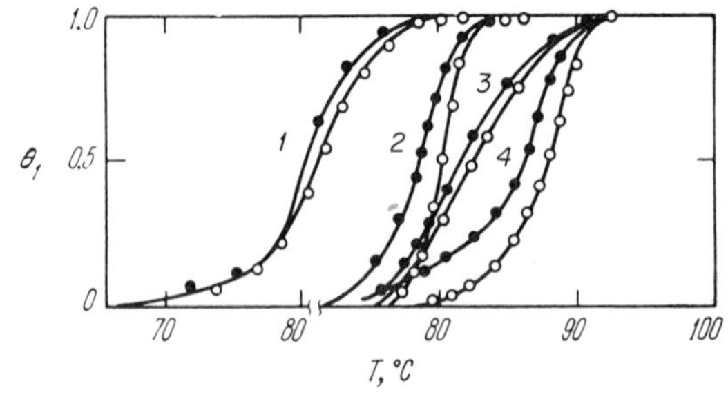


*FIG. 8.20 Dependence of the relative changes in [η] (closed circles) and in the optical density (open circles) on T at 0.075M. (1) Herring sperm DNA; (2) T2 phage DNA; (3) calf thymus DNA; (4) E. coli DNA.*

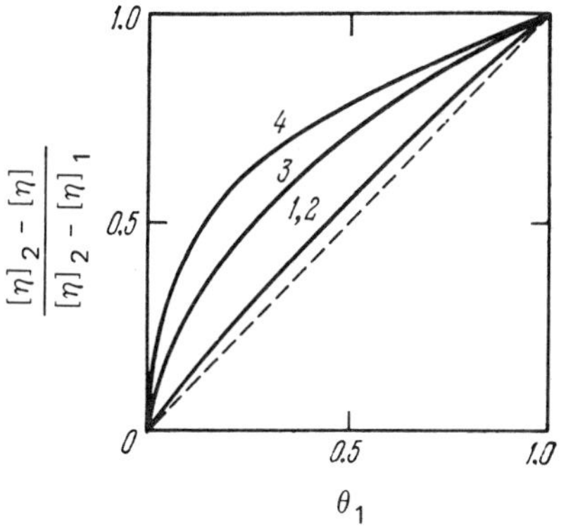


*FIG. 8.21 Dependence of the relative changes in g on the fraction of melted base pairs. (1) Herring sperm DNA; (2) calf thymus DNA; (3) E. coli DNA; (4) T2 phage DNA.*

primary structure. However, the precision of the current experiments is not sufficient for such investigation.

The thermodynamic parameters of the denaturation of DNA were determined microcalorimetrically by Privalov [94,109,110]. Heat of denaturation depends strongly on the pH. If the pH increases from 7.0 to 9.7, $T_m$ decreases from 84.8°C to 66.3°C, $\Delta H$ from 9650 to 7140 cal $mole^{-1}$, and $\Delta S$ from 27 to 21 eu. If the pH decreases from 5.4 to 3.2, $T_m$ decreases from 84°C to 55°C, $\Delta H$ from 9400 to 4000 cal $mole^{-1}$, and $\Delta S$ from 25.6 to 12.4 eu. $T_m$ changes considerably with ionic strength, whereas $\Delta H$ depends neither on it nor on the temperature. The difference in the free energies of denatured and native DNA at 37°C decreases from 1250 to 620 cal $mole^{-1}$ if the pH increases from 7.0 to 9.7, and from 1250 to 220 cal $mole^{-1}$ if the pH decreases from 5.4 to 3.2. At pH 7.0 $\Delta G$ decreases from 1250 to 710 cal $mole^{-1}$ if the pNa increases from 0.84 to 2.04. The free energy of stabilization of the native DNA structure depends linearly on NaCl activity. The logarithmic dependence of $\Delta G$ on the ionic strength at constant pH shows that the mechanism of stabilization of the helix by monovalent ions is an entropical one. Apparently the number of counterions bound by a macromolecule is determined by the linear density of the charges on the chain and does not depend on the counterion concentration in the solution. Denaturation diminishes the density of the charges on the chain, counterions disperse into the solution and a change in entropy occurs.

A detailed investigation of the thermodynamics of helix-coil transitions was conducted with the simplest models, oligoriboadenylic acids [111,112]. At low pH oligomer A forms double helices starting with a tetranucleotide. The pairing model suggested in [111] is shown in Fig. 8.22 for polymerization degree N = 3. In general, the number of different double helices with n bound pairs at given N is $(N-n+1)^2$. The transitions have been studied

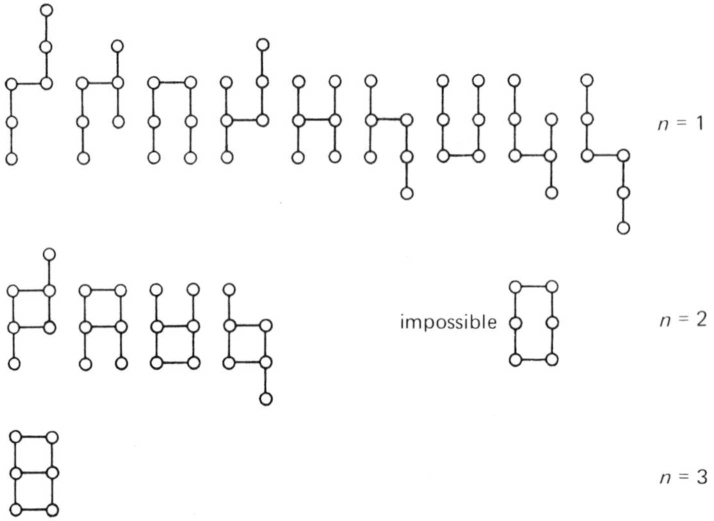


*FIG. 8.22 Models of trinucleotide pairing.*

by means of the hypochromic effect at 265 nm for N from 6 to 10. Formation of the double helix is characterized by two constants: that of nucleation β and of helix growth s. The constant $s = \exp(-\Delta H/RT+\Delta S/R)$. The equilibrium constant of the transition two chains $\rightleftharpoons$ double helix is equal to

$$K = \beta \sum_{n=1}^{N} (N-n+1)^2 s^n \equiv \beta\, L(s)$$

$$= \beta\, s[N^2-(2N^2+2N-1)s + (N+1)^2 s^2 - s^{N+1} - s^{N+2}]/(1-s)^3 \qquad (8.39)$$

The fraction of paired nucleotides is [113]

$$\theta = \frac{s[1+4\gamma L-(1+8\gamma L)^{1/2}]L'}{4\gamma N L^2} \qquad (8.40)$$

where $L' = dL/ds$, $\gamma = \beta c$, c is the total molar concentration of both the free and the bound chains. At $s \gg 1$ (strong dilution) $L(s) \rightarrow s^N$ and

$$\theta = \frac{1+4\gamma s^N-(1+8\gamma s^N)^{1/2}}{4\gamma s^N} \qquad (8.41)$$

At the midpoint of the transition $\theta = 0.5$ and $\gamma s^N = 1$. The sharpness of transition at this point is

$$\left(\frac{d\theta}{d\ln s}\right)_m = \frac{N}{6} \qquad (8.42)$$

Hence

$$\left(\frac{d\theta}{dT}\right)_m = \frac{N\ \Delta H}{6RT_m^{\ 2}} \tag{8.43}$$

that is, $\Delta H$ can be determined from the slope of the $\theta(T)$ curve at $T_m$. The values of $T_m$ increase with c. If $\theta$ = const, $\gamma s^N$ = const. We have $(\partial \ln s/\partial \ln \gamma)_{\theta,N} = -N^{-1}$ and

$$\left(\frac{\partial T}{\partial \ln \gamma}\right)_{\theta,N} = -\frac{RT^2}{N\ \Delta H} \tag{8.44}$$

Integrating this equation from c to c' at constant $\beta$ we find

$$\frac{1}{T'} = \frac{1}{T} + \frac{R}{N\ \Delta H} \ln \frac{c'}{c} \tag{8.45}$$

The dependence of $T_m$ on the chain length follows from $(\gamma s^N)_m = 1$

$$\frac{1}{T_m} = \frac{1}{T_c} + \frac{R \ln \gamma}{N\ \Delta H} \tag{8.46}$$

Careful measurements [112] have given $\Delta H$ = 5.85 kcal mole$^{-1}$ at N = 6; 6.7 at N = 7; 7.5 at N = 8; 8.1 at N = 9; and 8.9 kcal mole$^{-1}$ at N = 10. The statistical distribution of various n values, the presence of vertical interactions in the single chains, giving $\Delta S_{st}$ = 26 eu and $\Delta H_{st}$ = 6.1, 5.4, 4.4, 4.1, 4.0 kcal mole$^{-1}$ at N = 6,7,8,9,10, respectively, as well as the protonation of the chains must, however, be taken into account. Single chains are 10% protonated at pH 4.2; double-helical oligomers 80% [114]. A 70% change in protonation gives $\Delta H$ = 2.2 kcal mole$^{-1}$. Consideration of all these corrections at $\beta = 10^{-3}$ liter mole$^{-1}$ gives, for N = 6,7,8,9,10, $\Delta H$ values 10.0, 10.2, 10.2, 10.6, 11.4 kcal mole$^{-1}$, respectively. Further analysis has shown, however, that the hypothesis that $\beta$ is not dependent on T is not borne out by experiment, which shows that $\beta s$ is independent of T (i.e., $\Delta H_\beta \cong -\Delta H_s$). This gives a $\Delta H$ value independent of N and equal to 11.8 ± 0.6 kcal mole$^{-1}$. The $\beta s$ value is $3\times10^{-4}$ liter mole$^{-1}$.

Evdokimov and Varshavsky investigated DNA melting in light and heavy water over a broad range of pH: from 2 to 12 [115]. The dependence of $T_m$ on the pH has a bell-like shape--a plateau at 80-85°C (T2 phage DNA) at pH from 4.5 to 8.5, and a sharp decrease in $T_m$ at lower and higher pH (ionic strength 0.18-0.25). At neutral pH the $T_m$ values for $H_2O$ and $D_2O$ coincide. Consequently the stability of DNA cannot be attributed mainly to hydrogen bands. In the acid region H-DNA is more stable; in the alkaline one, D-DNA is more stable. The dependence of $T_m$ of H-DNA and D-DNA on the ionic strength is also different at low values of ionic strength (∿0.01). This effect can be explained by different hydration (it is lower in $D_2O$). Values of $\Delta H$ and $\Delta S$ have been determined from experimental data: at 37% G-C pairs, $\Delta H = 8.0$, $\Delta S = 23.0$; at 40%, $\Delta H = 8.7$, $\Delta S = 25.4$; at 48%, $\Delta H = 9.5$ kcal mole$^{-1}$, $\Delta S = 27.4$ eu. The values for $H_2O$ and $D_2O$ are practically identical.

The acid denaturation of DNA due to protonation of the nitrogen bases was investigated by means of calorimetry and optical methods [116]; the appearance of an intermediate DNA conformation differing from both the native and the denatured ones was established. Reviews of the theory of helix-coil transitions in DNA are given in [117].

## 8.5 The Kinetics of DNA Denaturation

The kinetics of the unwinding of the DNA double helix was first examined by Kuhn [118] (cf. also [6]), who showed that if the unwinding (occurring after the breakage of the interchain bonds) occurs as a result of rotational Brownian motion, then this process requires a time $\tau$ much larger than the observed one. Thus, for DNA with MW $3\times10^6$, the 450 turns of the helix necessary for total separation of the chains require 150 days. But the observed $\tau$ for DNA with MW $10^8$ is of the order of 1 min. Kuhn assumed that the unwinding occurred at the ends of the helix, but he did not take into account the coiling of the liberated chains. If the inevitable entanglement is neglected, then $\tau$ for the unwinding of the coiling chains will be much smaller than that calculated by Kuhn [6,119]. Kuhn examined the separation of chains resulting from the combination of rotational and translational thermal motion and obtained, for DNA with MW $3\times10^6$, $\tau = 1$ min. This value is also too large.

Longuet-Higgins and Zimm considered the rotational momentum due to the increase in entropy of the separated chains [120]

$$P = \frac{T\,\Delta S}{\phi} = \frac{57.3}{36}\,T\,\Delta S \qquad (8.47)$$

where $\phi = 36°/57.3°$ is the angle of rotation (expressed in radians) of the chain end which is required for liberation of one pair of bases. The rate of unwinding is $\omega = P/K$, where K is the friction coefficient. For helices with MW of the order of $10^6$, $\tau$ happens to be of the order of seconds. This estimate does not consider the irreversibility of unwinding and entanglement of the liberated chains.

Fixman examined the simultaneous action of the diffusional mechanism proposed by Kuhn and the entropical one [121]. Crothers also studied the combination of these mechanisms but considered the possibility of simultaneous melting at different regions of the molecule [122]. The work of Fong is especially interesting [123]. The entanglement of the free chains can be avoided if we assume that breakage occurs at the midpoint of the helix. The two halves of the helix fluctuate independently; half the time they are unwinding and half the time rewinding. During the initial stages of the process, however, secondary winding is impossible because of the close packing of the helix. The kinetic energy of winding transforms into potential energy enhancing the unwinding. If many turns have been performed, secondary winding has become possible. Therefore, the two halves of the helix unwind and rewind without a change in radius of the helix. Winding cannot proceed for an indefinitely long time; therefore, this random motion amounts mainly to unwinding. The number of turns $N(t)$ at $t > 0$ is therefore smaller than or equal to the initial number of helix turns $N_0$. The problem is to estimate $\tau$ using the condition

$$\int_0^\tau \omega(t)\, dt = -2\pi N_0 \tag{8.48}$$

where $\omega(t)$ is the relative angular velocity of the two halves of the helix. The upper and lower limits of $\tau$ can be estimated. The mean value of $\omega(t)$ in some time interval $\bar{\omega}(t)$ is a monotonic decreasing function of t. The lower limit of $\tau$ can be found by replacing $\bar{\omega}(t)$ with the initial value $\bar{\omega}(o)$. Let the initial angular velocity be $\omega_0$. If both halves of the helix unwind, then $\omega(o) = 2\omega_0$. Since at $t = 0$, DNA is unwinding half the time, $\bar{\omega}(o) = 0.5\ \omega(o) = \omega_0$. This value can be found from the condition

$$\frac{1}{2} I\, \omega_0{}^2 = \frac{1}{2}\, kT \tag{8.49}$$

where I is the moment of inertia of half the helix relative to its axis. If $N_0 = 2.10^4$ (MW $= 1.2\times10^8$), $T = 20°C$, and $\omega_0 = 3\times10$ rad/sec$^{-1}$, the lower limit of $\tau$ is

$$\tau_\ell = 2\pi N_0/\bar{\omega}(o) = 4.2\times10^{-4} \text{ sec}$$

The upper limit $\tau_u$ corresponds to random rewinding and unwinding without advantages for the latter. Assume that the DNA molecule is in water at 20°C and pressure equal to 1 atm. The mean angular momentum of rotation P of half the DNA molecules is $1.34\times10^{-22}$ erg rad$^{-1}$, the momentum of the water molecule $p = 1.9\times10^{-18}$ erg rad$^{-1}$. Multiplying P by the radius of the helix, we find the order of magnitude of the mean angular momentum resulting from collisions between the DNA and $H_2O$ molecules, $\ell = 1.7\times10^{-25}$, $m = P/\ell = 7.8\times10^2$. Hence, the lowering of P to zero requires a large number of collisions equal to $m^2$. Calculation gives $2\times10^{15}$ collisions sec$^{-1}$ and the time required for $m^2$ collisions, that is, for decreasing the angular momentum from P to zero, is equal to $3\times10^{-10}$ sec. The mean time between two sequential zeros of the fluctuating angular momentum $\lambda$ is twice as large, due to the symmetry of fluctuations in time. During the time $\lambda$ the angular velocity does not change signs and both $\lambda$ and the number of turns n during $\lambda$ can be considered as a step in the Brownian motion. We get

$$n = \omega_0\lambda = 2.9\times10^{-2} \text{ turns}$$

and the time $\tau_u$ for $10^4$ turns

$$\tau_u = (10^4/n)^2\lambda = 73 \text{ sec}$$

It can be shown that at $N >> 1$, $\tau$ must be close to the upper limit. This value agrees with experiment.

Such a process is more effective than that of Kuhn, since during the time $\lambda$ the value of the angular velocity, but not its direction, fluctuates, and the required turns occur one after another.

Consider the collective motion of the unwound part of the DNA. Unwinding starts in many regions of the helix, the fragmentation of which can only diminish $\tau$. In Fong's model the rotation of the liberated ends of the chains can be neglected, since they must be rather short. Calculation shows that $\tau_u \sim MW^3$ and $\tau_\ell \sim MW^{3/2}$. The real $\tau$ lies between these values.

Fong's theory is rather rough, but it gives the correct order of magnitude of $\tau$, which must not be changed by rigorous solution of the stochastic problem. The unwinding of DNA with simultaneous reduplication will be discussed later (p. 540).

Apparently there exists the possibility of an internal disordering of the double helix without separation of the chains. The denaturation of circular double-helical DNA without unwinding has been demonstrated [124]. The time of denaturation of phage DNA (of the order of 25 sec) depends on the MW. For a series of phage DNAs it is proportional to $MW^2$. The square dependence follows directly from the proportionality of $\tau$ to the number of turns (i.e., to MW) and to the viscous resistance (i.e., also to MW). Massey and Zimm [125] investigated the denaturation of DNA by relaxational methods (p. 470). They established that $\tau$ depends on many factors--the stage of the helix-coil transition, the ionic strength and viscosity, the concentration of DNA, and the MW, as well as on the number and distribution of breaks. They state that

$$\tau = \alpha(1+c[\eta])\eta MW\psi \tag{8.50}$$

where $0 < \psi < 1$ is the factor characterizing the transition stage, c is the concentration, $[\eta]$ is the intrinsic viscosity of the DNA, $\eta$ is the viscosity of the solvent, and $\alpha$ is a constant depending on the ionic strength. At small c, $\tau \sim MW$. The increase in $\tau$ during transition shows that the resistance of the medium increases when the molecule unwinds. The effective radius of the double helix increases simultaneously because of loop formation. This is not taken into account in Fong's theory [123].

The same relaxational techniques were applied to the study of DNA denaturation [126] in an investigation that showed the presence of three successive processes. At moderate temperature perturbation (from 6° to 18°C) an "instantaneous" response appears ($\tau$ < 20 msec), consisting in a rapid structural disorganization of the helix without the separation of chains. Disorganization must start in regions that have a surplus of A-T pairs. The "instantaneous" response is followed by a smooth kinetic curve which the authors call the fast effect. The time of this effect is proportional to $MW^{2.3}$. Later small perturbations in the transition range are expressed by a very slow kinetic component. This slow effect, characterized by a large activation energy (∿100 kcal $mole^{-1}$), is practically independent of MW. This can be treated as a nucleation phenomenon in the cooperative transition (i.e., the destruction of helical regions separating the nonordered ones).

We see that the kinetics of the melting of native DNA corresponds to the spectrum of relaxation times. A real theory of melting has not yet been formulated and it meets with great difficulty in studies of the kinetic relationship between the approach to internal equilibrium and unwinding. This has not been considered in previous calculations, where unwinding is treated as the limiting stage. In reality the process can be limited by the "fusing" of the disorganized regions.

Varshavsky and Evdokimov investigated the unwinding of DNA by applying the "flash heating" method. The DNA solution was heated by 5°-20°C (depending on the ionic strength) for 0.5 sec. Kinetic curves show the presence of two or even three stages of structural transition. In the first rapid stage, nearly the whole hypochromic effect arises. The half-period of the total transition varies from several seconds to several tens of seconds, depending on conditions. The rate constant obeys the Arrhenius equation. Unwinding occurs with the maximum rate at the extreme pH values; the dependence of the activation energy on the pH is bell-shaped. The maximum value $E_a$ = 170 kcal $mole^{-1}$ corresponds to pH 7.5 (ionic strength 0.18-0.25); it drops to 20-25 kcal $mole^{-1}$ at pH 3 and pH 10.5. The authors interpret the first stage as the formation of immovable loops and the entire unwinding occurs at the second stage. The time of unwinding increases with MW. The results obtained show that the character of the dependence of $\tau$ on MW is determined by the ionic strength and by the number and distribution of breakages [127-129].

Lazurkin and co-workers developed the kinetic formaldehyde method (KFM), based on studies of the kinetics of unwinding, to determine defects in the secondary structure of DNA [85,130].

If the solution contains a substance reacting with locally disordered nucleotides and hindering the formation of Watson-Crick pairs, the reaction will proceed up to total disordering.

If the binding time of such a reagent is much larger than the unwinding time, the reaction itself will limit the rate of unwinding. The rate constant of the reaction of native DNA with the reagent is $wK = k$, where $w$ is the probability of thermal denaturation for any pair of bases in the molecule and $K$ is the rate constant of the reaction of the separated nucleotides with this reagent. If some bases of the native DNA have already reacted and formed the locally denatured region, the probability of heat denaturation of the neighboring nucleotides is $w' \gg w$. Reaction of these pairs proceeds at a much greater rate constant $k' = w'K$. The formation of a locally denatured segment requires some free energy $G_s$ (p. 507). Therefore, $k/k' = w/w' \sim \exp(-G_s/RT) = \sigma \approx 10^{-4}$–$10^{-5}$. Since $k \ll k'$ at the start of the process, the DNA unwinding should proceed slowly and be accelerated by the accumulation of centers of unwinding. Thereafter, due to the merging of denatured sections, the process must be decelerated. The kinetic curve should be S-shaped. The presence of defects in a double helix would increase the initial slope of this curve. Let us determine its shape.

Let $w(x,t)$ be the probability that at time $t$ a section with $x$ pairs belongs completely to the helix region. Let a DNA molecule consisting of $N$ nucleotide pairs be at some stage of unwinding. The degree of helicity at time $t$ is

$$\theta = W(x,t) \tag{8.51}$$

and at time $t + dt$

$$W(x,t + dt) = W(x + 2k'dt,t)[1 - (x + y - 1)k\, dt] \tag{8.52}$$

Indeed, for the total helicity of a section containing $x$ elements at time $t + dt$ it is necessary that at time $t$ each end of the section have a "reserve" no less than $k'\, dt$ in length and that no new unwinding center arise in this section. The initial size (the number of pairs) of an unwinding center is denoted by $y$.

Expression (8.52) leads to the differential equation

$$\frac{\partial W}{\partial t} = 2k' \frac{\partial W}{\partial x} - (x + y - 1)kW \tag{8.53}$$

Let us solve it with an initial condition

$$W(x,0) = f(x)$$

If there are no defects, then $f(x) = 1$. If defects are present and the lengths of the helical sections between them are distributed randomly, then $f(x) = \exp[-c(x-1)]$, where $c = n^{-1}$ is the concentration of defects and $n$ is the mean number of pairs in the helical section. Solution of Eq. (8.53) with the stated initial condition gives [106]

$$\theta = f(1 + 2k't) \exp(-kyt - kk't^2) \tag{8.54}$$

In the case of random distribution of defects we have

$$\theta = \exp[-(ky + 2k'c)t - kk't^2] \quad (8.55)$$

or

$$-(\ln \theta)/t = ky + 2k'c + kk't \quad (8.56)$$

Curves (8.55) and (8.56) are shown in Figs. 8.23 and 8.24. The intersection of the straight line with the ordinate permits us to determine c.

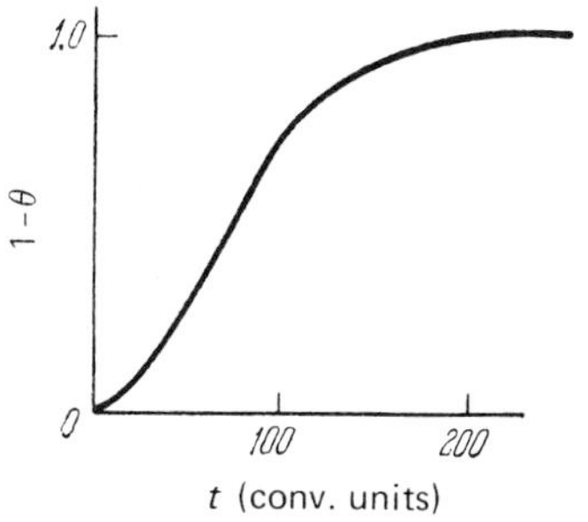


*FIG. 8.23 Kinetic curve for DNA with defects.*

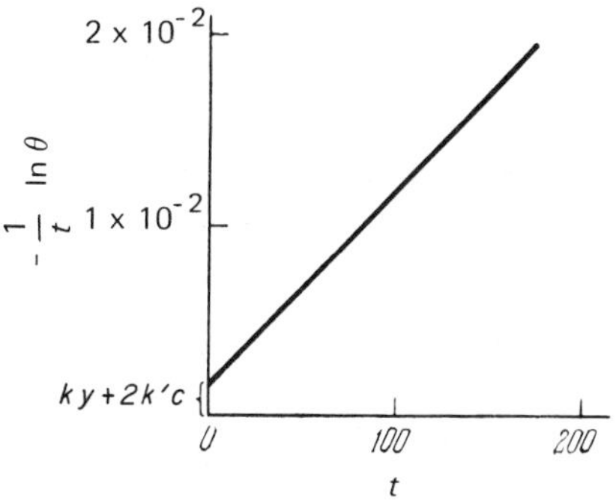


*FIG. 8.24 Kinetic curve transformed according to (8.56).*

Formaldehyde is an unwinding agent. Examination of experimental θ(t) curves (which agree with the theoretical curves) has made it possible to show that ultraviolet irradiation of DNA increased c, and to find its values. The values of k' (of the order of 2 $min^{-1}$) have been estimated. The minimal concentration of defects found by the kinetic formaldehyde method (KFM) is one defect per $10^4$ nucleotide pairs. The method is very sensitive [106].

The KFM theory was improved with the aim of treating variations in k and k' at different sites in the molecule [131]. For a homopolymer having "weak points," the dependence of (1/t)ln θ on t is nonrectilinear. The block heterogeneity of DNA is successfully incorporated in the theory.

KFM has been effectively applied in studies of DNA with breakages due to the action of pancreatic DNase, of DNA irradiated by short-wavelength radiation, etc. Native, intact DNA appears to be practically devoid of defects or "weak points." This disproves the zigzag model of native DNA with its sharp breaks, and it can be suggested that the wormlike model (p. 497) in which chain flexibility is due to small turns in monomers is valid. KFM is applicable in studies of the complexes of DNA with RNA-polymerase (p. 566), as well as in studies of the fine structure of DNA [131,132].

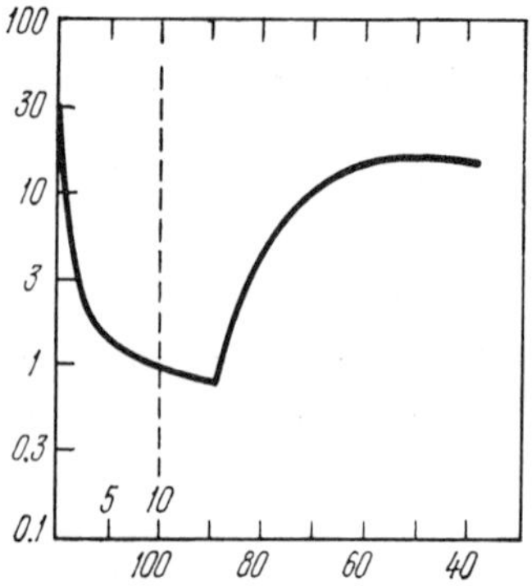


*FIG. 8.25 Renaturation of DNA. The numbers 5 and 10 mean the time of annealing of the sample at 100°C.*

Denatured, coiled single DNA chains are able to renature (to form double-helical DNA again). The renaturation of DNA was discovered and investigated by Marmur and Doty and their co-workers [133,134]. When denatured DNA is slowly cooled ("annealed"), its biological activity and hypochromic effect partly regenerate. This is shown in Fig. 8.25. The regeneration is incomplete because some of the DNA has been destroyed. Renaturation also occurs after the rapid cooling of denatured DNA if the sample has been heated to a temperature somewhat lower than $T_m$. DNA renaturation is clearly expressed as density changes measured by centrifugation in CsCl. For example, the density of native DNA is 1.704; that of denatured DNA that is "chilled" in this state is 1.716; and that of renatured DNA is 1.700 g $cm^{-3}$.

Evidently denaturation is strongly temperature dependent. Renaturation requires a temperature that is high enough for uncoiling and at the same time low enough for stabilization of the double helix. Therefore, an optimal renaturation temperature exists. In fact, DNA from *Diplococcus pneumoniae* renatures 2.5 times faster at 67°C than at 80°C and 50°C.

Renaturation of phage DNA proceeds readily; in the case of bacterial DNA, renaturation can be effected with some difficulty; but it is practically impossible to effect renaturation of the DNA of higher organisms. Should all the DNA samples obtained from calf thymus, *E. coli* cells, and the T2 phage have MW $10^7$, then the corresponding number of DNA molecules per cell (or per phage particle) would be $10^6$, $10^2$, and 10. At equal weight concentrations these three samples would contain different concentrations of complementary chains, and the probability of the meeting of two such chains formed by the denaturation of one molecule in a sample obtained from mammalian cells would be very small.

Renaturation makes it possible to obtain hybrid double helices made up of DNA chains of different origins. DNA hybrids extracted from different strains of *E. coli* and interbacterial hybrids have been obtained [134]. The hybridization of DNA reveals the evolutionary-genetic relations of species.

## 8.6 Interactions of DNA with Small Molecules

Investigation of the interaction of DNA (and RNA) with small molecules is important for knowledge of DNA structure and its possible changes. The small molecules in a series of compounds affect the biological function of DNA as mutagens (e.g., acridine dyes, p. 528) and as inhibitors of transcription (actinomycin and other antibiotics). It has been established that this influence is due to the ability of antibiotics to form slowly dissociating complexes with DNA [135-137]. The inhibition of transcription (p. 566) is due to a hindering of the unwinding of DNA [89] and to the practical irreversibility of complex formation. Acridine dyes (AD) that have almost the same binding constant with DNA as actinomycin at low ionic strength, and an increase in the $T_m$ of DNA by approximately the same amount [88], do not influence transcription [138]. The dissociation time of the DNA-proflavin complex is $\sim 10^{-3}$ sec, that of the DNA-actinomycin complex is $\sim 260$ sec.

The binding of actinomycin and AD occurs stepwise, via a series of intermediate forms. The formation of complexes alters the DNA conformation. X-ray analysis has shown that the DNA in a proflavin complex has an increased pitch of the helix and is partly unwound [139]. Experiments with supercoiled cyclic DNA have shown that AD and actinomycin unwind the DNA double helix [140,141]. In such DNA the right double helix is folded, forming a helix of a higher order. The unwinding of supercoiled DNA changes the geometry of the entire macromolecule because of the topological features of the ring structure. It has been stated that a critical ligand concentration, corresponding to total uncoiling of the superhelix, exists, and that addition of more AD produces reverse winding of the DNA and the formation of a left superhelix. These facts have been observed by sedimentation methods.

The spatial model of the complex must take into account the unwinding of DNA and its features. Lerman proposed the "intercalation" model, according to which the AD molecules lie between the pairs of DNA bases and interact with them "vertically" (stacking) [142]. Intercalation requires unwinding of the sugar-phosphate frameworks of both DNA chains to form sufficient free space between the base pairs for the inclusion of an AD molecule. The van der Waals "width" of this molecule is 3.4 Å. Lerman's model explains unwinding, but detailed quantitative analysis shows that the increase in the pitch of the helix corresponding to this model is too large. The modified intercalation model suggests a turn of the base pairs by 24° [126]. Therefore, the sugar-phosphate framework becomes unwound by 12°, in agreement with experiment (cf. [143]).

However, the unwinding of DNA can also be explained by the "external binding" model. Nonplanar molecules lacking conjugated bonds (some steroids) also cause unwinding. Some angles of unwinding of base pairs in DNA complexes are the following: with ethidium bromide, 12°; with proflavin, 8.4° ±2.4°; with daunomycin, 5.2° ±1.4°; with nogalomycin, 8.1° ±2.3°; with actinomycin, 11.4° ±3.0° [141].

The change in the pitch of the DNA helix is provoked also by metal ions and by some proteins, including histones [144]. However, the alteration in the helix pitch in complexes with AD and actinomycin is larger. This is quite natural, since additional geometric conditions are required for the packing of a bulky organic cation. We present the formulas.

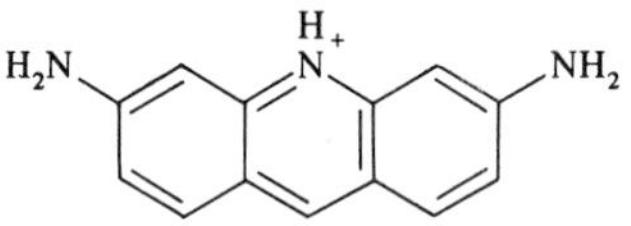


Proflavin

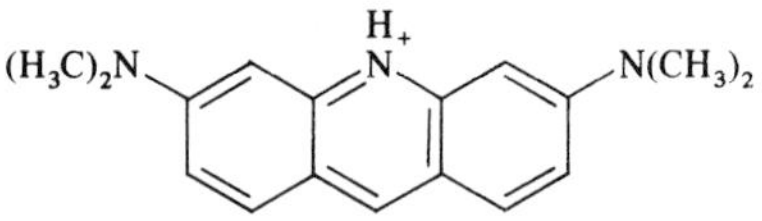


Acridine orange

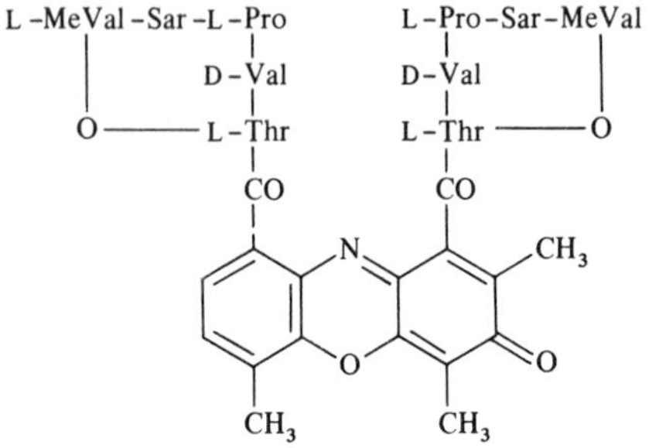


Actinomycin $C_1$
(Sar-sarcolysine)

The external binding models were developed by Gursky [145], who theorized that the AD or actinomycin is located in a narrow groove in the surface of the DNA double helix. The proflavin molecule is submerged in the narrow groove in such a way that the ring nitrogen and amino group, being positively charged, are located on the outer surface of the DNA. They interact with oxygen atoms in the phosphate groups by means of electrostatic and hydrogen bonds. Diethyl and dimethyl amino groups in the 2 and 8 positions form only electrostatic bonds. These interactions "tighten" the phosphate groups in the two chains and provoke unwinding of the helix. The chromophoric ring of actinomycin forms hydrogen bonds with amino group 2 of guanine, and amides of the peptide rings form them with the phosphate groups. Peptide rings are located in the narrow groove of the DNA on different sides of the chromophoric ring. They are connected by a second-order symmetry axis coinciding with a similar axis in the DNA and directed perpendicular to the long axis of the chromophore. This model is partly confirmed by X-ray analysis of the complex of actinomycin with deoxyguanosine [146]. Another external binding model, which was proposed earlier [147], was not confirmed in this way.

Both models suggested by Lerman and Gursky agree with a series of facts [145,148-150]. However, they meet with difficulty in the explanation of some experimental data [145,150,151]. A single model cannot yet be chosen.

Crothers described the possibility of obtaining the crystalline complex of actinomycin with a short oligomer containing five base pairs [152]. An oligomer has only one binding site for an antibiotic. Investigation of such complexes by X-ray analysis and NMR should enable us to determine the structures of DNA complexes with AD and actinomycin.

Investigation of the binding of small molecules by a DNA macromolecule requires the development of a theory of one-dimensional adsorption on a polymer. If the binding sites are independent, then the adsorption isotherm is Langmuir's isotherm

$$y = \frac{Kan}{1 + Ka} \tag{8.57}$$

where y is the amount of bound ligand per monomer, K is the binding constant, n is the number of binding sites, and a is the ligand activity. Such a dependence on a [153] is rather an exclusive case. If adsorption occurs in a cooperative way, then the adsorbed molecules form a "nonideal gas." In the model of Crothers [154] every ligand molecule occupies m successive sites, and we have a "nonideal gas with repulsion." This model explains the experimental curves of AD adsorption on DNA using one binding constant.

Crother's model was generalized [155] to take into account the dependence of the macroscopic binding constant on the distance between neighboring ligands. Analysis of the experimental adsorption isotherms enables us to obtain information about the precise shape of this function.

Crothers also calculated the adsorption isotherms for heteropolymers with various sequences of nucleotides [154]. The adsorption on a nonuniform line and surface has been investigated [156].

The exact solution of the problem for a heteropolymer with two kinds of binding sites (A-T and G-C pairs) is given in [157]. The great partition function has the form

$$Z = \sum_{q=0}^{n} a^q Z_q \tag{8.58}$$

where $Z_q$ is the partition function for a polymer that has bound q ligand molecules, and n is the maximal number of molecules adsorbed. For a homopolymer

$$Z_q = \frac{(N-mq+q)!}{(N-mq)!q!} \exp \frac{-\Delta Gq}{RT} \tag{8.59}$$

where N is the number of the binding sites (nucleotide pairs) in the polymer, m is the number of sites occupied by one ligand molecule, and $\Delta G = -RT \ln K$. The preexponential factor is the number of ways to distribute sections of length m on a one-dimensional lattice.

Assume that the ligand binding is characterized by binding constants $K_A$ and $K_G$, depending on the kind of pair (A-T or G-C), which are $\ell$ links away from the left edge of the ligand. We ascribe to the ligand some recognizing group in a definite position. Let us number the sites from left to right. Site j is occupied by ligand if the left edge of the ligand is at site j, j-1,$\cdots$,j-m+1. Each such situation contributes to the partition function $b_j^1$, $b_j^2$, $\cdots$, $b_j^m$. Contribution $b_j^0$ corresponds to the members of the sum at unoccupied site j. We have

$$\begin{aligned}
b_j^0 &= 1\, b_{j-1}^0 + 0\, b_{j-1}^1 + \cdots + 1\, b_{j-1}^m \\
b_j^1 &= s_{j+\ell}\, b_{j-1}^0 + 0\, b_{j-1}^1 + \cdots + s_{j+\ell}\, b_{j-1}^m \\
&\vdots \\
b_j^m &= 0\, b_{j-1}^0 + 0\, b_{j-1}^1 + \cdots + 1\, b_{j-1}^{m-1} + 0\, b_{j-1}^m
\end{aligned} \tag{8.60}$$

where $s_j = s_A = K_A a$ if site j is an A-T pair, and $s_j = s_G = K_G a$ if it is a G-C pair. Sum (8.58) assumes the form

$$Z_N = b_{N-m+1}^0 + b_{N-m+1}^1 + \cdots + b_{N-m+1}^m \tag{8.61}$$

From (8.60) and (8.61) it follows that

$$Z_N = Z_{N-1} + s_{N-m+\ell}\, Z_{N-m} \tag{8.62}$$

We obtain the number of adsorbed molecules per pair of bases

$$y = N^{-1}\left[s_A \frac{\partial}{\partial s_A} (\ln Z_N) + s_G \frac{\partial}{\partial s_G} (\ln Z_N)\right] \tag{8.63}$$

where y is an integral parameter that does not depend strongly on the sequence of the A-T and G-C pairs. For a random sequence

$$y = s_A \frac{\partial}{\partial s_A} \left\langle N^{-1} \ln Z_N \right\rangle + s_G \frac{\partial}{\partial s_G} \left\langle N^{-1} \ln Z_N \right\rangle \tag{8.64}$$

The parentheses here mean the averaging over all possible sequences. We have

$$N^{-1} \ln Z_N = N^{-1} \sum_{j=1}^{N-m+1} \ln(1+w_j) \tag{8.65}$$

where $w_j = b_j^{1}(b_j^{0}+b_j^{2}+\cdots+b_j^{m})^{-1}$. On the other hand,

$$w_j = s_{j+\ell} / \prod_{j=1}^{\ell-1} (1+w_{j-i}) \qquad \text{if} \quad m+1 \le j \le N-m+1$$

$$w_j = s_{j+\ell} / \prod_{j=1}^{\ell-1} (1+w) \qquad \text{if} \quad 2 \le j \le m \tag{8.66}$$

$$w_1 = s_\ell$$

The averaging of $\ln Z_N$ is performed with the help of the function [158]

$$C_N(t) = N^{-1} \left\langle \sum_{j=1}^{N-m+1} \ln(1+w_j) \right\rangle \tag{8.67}$$

If $m = 2$, we get [157,158]

$$\frac{N}{N-1} C_N(t) = x_A C_{N-1} \frac{t+s_A}{t} + x_G C_{N-1} \frac{t+s_G}{t} - C_{N-1}(1) + \frac{2}{N-1} C_2(t) + \frac{N-2}{N-1} \ln t \tag{8.68}$$

where

$$C_2(t) = \frac{1}{2} [x_A \ln(t+s_A) + x_G \ln(t+s_G)]$$

and $x_A$, $x_G$ are the probabilities of the appearance of A-T and G-C pairs.

If $N \to \infty$, (8.68) gives

$$C_\infty(t) = x_A C_\infty[(t+s_A)/t] + x_G C_\infty[(t+s_G)/t] - C_\infty(1) + \ln t \tag{8.69}$$

Equations (8.63), (8.64), (8.68), and (8.69) provide a rigorous solution at $m = 2$. If $m > 2$, $C_\infty(1)$ can be estimated. If the ligands are bound by only one kind of site ($s_A = 0$), the calculation gives

$$\left(1 + \frac{y}{x_G - y[1+x_G(m-1)]}\right)^{x_G(m-1)} \frac{y}{x_G - y[1+x_G(m-1)]} = s_G \tag{8.70}$$

If $x_G$ is small, then this isotherm becomes the Langmuir isotherm $y/(x_G-y) = x_G a$. If $x_G \to 1$, isotherm (8.70) transforms into that of a homopolymer. The agreement of formula (8.70) with that obtained by Crothers improves the larger m is. The calculations are done by computers. However, if we want to determine the fundamental parameters from the experimental isotherm, it is sufficient to examine the asymptotic functions at

$a \to 0$ and $a \to \infty$. We get

$$\lim_{y \to 0} \frac{y}{a} = \frac{\langle Z_1 \rangle}{NZ_0}, \quad \lim_{y \to 0} \frac{\partial (y/a)}{\partial y} = \frac{\langle 2Z_2 - Z_1^2/Z_0 \rangle}{\langle Z_1 \rangle} \tag{8.71}$$

where $Z_1$ and $Z_2$ are the partition functions of the polymer that has adsorbed one and two ligands. Averaging must be performed over all possible sequences of A-T and G-C pairs. The partition function $Z_2$ can be expressed as

$$Z_2/Z_0 = \frac{1}{2}\left((Z_1^2/Z_0) - \sum_{\xi=1}^{\nu} \sum_{\eta=1}^{\nu} \phi_{\xi\eta} K_\xi K_\eta\right) \tag{8.72}$$

where $\nu$ is the number of different types of complexes formed by the ligand molecules with the polymer, and $K_\xi$, $K_\eta$ are the binding constants. $Z_1^2$ takes into consideration the number of possible microstates for two adsorbed ligands. $\phi_{\xi\eta}$ is the total number of microstates with the constants $K_\xi$, $K_\eta$.

Assume that $\nu = 2$ and there are two constants $K_A$ and $K_G$. It follows from (8.71) and (8.72) that

$$\lim_{y \to 0} \frac{\partial (y/a)}{\partial y} = - \frac{\langle \phi_{AA} \rangle K_A^2 + 2 \langle \phi_{AG} \rangle K_A K_G + \langle \phi_{GG} \rangle K_G^2}{\langle Z_1 \rangle / Z_0} \tag{8.73}$$

Let us number sequentially all possible ways of binding one ligand molecule, denoting every binding site by $i = 1, 2, \ldots, N-m+1$. We have

$$Z_1 = (N - m + 1)(K_A x_A + K_G x_G) Z_0 \tag{8.74}$$

The average values of the quantities $\phi_{\xi\eta}$ are

$$\langle \phi_{\xi\eta} \rangle = \sum_{i=j}^{N-m+1} \langle \phi_{\xi\eta}^{\,i} \rangle \tag{8.75}$$

where $\langle \phi_{\xi\eta}^{\,i} \rangle$ is the average number of sites with the constant $K_\eta$ nonaccessible for the second ligand because of the adsorption of the first ligand at site i with constant $K_\xi$. We have

$$\langle \phi_{\xi\eta} \rangle = x_\xi (N - m + 1) \delta_{\xi\eta} + x_\xi x_\eta (m-1)(2N - 3m + 2) \tag{8.76}$$

where $\delta_{\xi\eta}$ is the Kronecker symbol. We obtain finally

$$\begin{gathered} \lim_{y \to 0} \frac{y}{a} = \frac{N - m + 1}{N} (x_A K_A + x_G K_G) \\ \lim \frac{\partial (y/a)}{\partial y} = - \frac{x_A K_A^2 + x_G K_G^2}{x_A K_A + x_G K_G} - \left(2 - \frac{m}{N-m+1}\right)(m-1)(x_A K_A + x_G K_G) \end{gathered} \tag{8.77}$$

These formulas are simplified at $N \to \infty$:

$$\lim_{y\to 0} \frac{y}{a} = x_A K_A + x_G K_G$$

$$\lim_{y\to 0} \frac{\partial (y/a)}{\partial y} = - \frac{x_A K_A^2 + x_G K_G^2}{x_A K_A + x_G K_G} - 2(m-1)(x_A K_A + x_G K_G) \tag{8.78}$$

If the concentration of free ligand is very large, then maximally dense packing of adsorbed molecules on the polymer will occur. The values of $y_{max}$ can be determined by using the geometric properties of such packing.

If K are nonzero for all possible positions of the ligand on the polymer, then $y_{max} = m^{-1}$.

Assume that ligand is bound by only one type of pair. Let the ligand molecule have a site recognizing the G–C pair at a distance $\ell \leq [m/2]$ from the left edge of this molecule. Here $[m/2]$ is an integer, $(m/2)-1 < [m/2] \leq m/2$. Let us bind ligand molecules starting from the left in such a way that the recognizing site of every subsequent ligand coincides with the nearest G–C pair, which is at a distance equal to or larger than $\ell$ from the right edge of the previous ligand. This gives a dense packing. If the G–C pairs occur at random, their mean fraction directly occupied by bound molecules is equal to $y[1+x_G(m-1)]$. Only those G–C pairs can remain unoccupied which are at distances from the right edges of the bound molecules of ligands equal to $1,2,\ldots,\ell-1$ pairs. The probability $w_i$ of finding i unoccupied pairs is

$$w_i = W^{-1}\, x_G (1-x_G)^i\,, \qquad 1 \leq i \leq \ell-2$$

$$w_i = W^{-1}\, (1-x_G)^{\ell-1}\,, \qquad i = \ell-1 \tag{8.79}$$

where

$$W = \sum_{i=1}^{\ell-1} w_i$$

The total fraction of unoccupied G–C pairs is $y_{max} \Sigma_{i=1}^{\ell-1} i x_G w_i$. Since every G–C pair can be either occupied or free, we get

$$y_{max} = x_G \{1 + x_G(m-1) + (1-x_G)[1-(1-x_G)^{\ell-1}]\}^{-1} \tag{8.80}$$

if $[m/2] < 1 \leq m$, we obtain

$$y_{max} = x_G \{1 + x_G(m-1) + (1-x_G)[1-(1-x_G)^{m-\ell}]\}^{-1} \tag{8.81}$$

If $\ell = 1$ or $\ell = m$, then $y_{max} = X_G/[1+x_G(m-1)]$, in agreement with the approximate formula (8.57). In general, $y_{max}$ depends on the position of the recognizing site on the ligand molecule.

In the case of a homopolymer or of ligand adsorption independent of the kind of monomer, the adsorption isotherm can be obtained in an analytical form [155,156]. According to [155] we have

$$\left(1 + \frac{y}{1-ym}\right)^{m-1} \frac{y}{1-ym} = Ka \qquad (8.82)$$

However, this formula does not consider the interaction of ligand molecules--direct interaction or interaction resulting from alteration of the polymer structure due to ligand adsorption. Combinatorial calculation of such cooperative binding has been performed and is reported in [157].

The theory presented permits us to interpret experimental results. Antibiotics of the actinomycin type (p. 528) are adsorbed by DNA and poly(G-C) but not by poly(A-T) [158-160]. The typical adsorption isotherm of actinomycin by thymus DNA is shown in Fig. 8.26. From the initial slope and intersection of the curve with the ordinate axis both K and m (i.e., the size of the DNA section occupied by one ligand molecule) can be determined. The theoretical adsorption curve calculated according to formula (8.70) is shown by the solid line in Fig. 8.26. Curve (8.70) coincides with the experimental one in the region of small y at m = 5 and K = $5\times10^6$ liter $mole^{-1}$ of pairs of bases. The same estimate of m is valid for DNA from T-even phages, having a nearly random sequence of nucleotides. Müller and Crothers [161] obtained a similar estimate of m using the data concerning DNA with a nonrandom sequence. This shows that the mean number of G-C pairs occupied by one actinomycin molecule is near this value for a random polymer. The value m = 5 is in agreement with the van der Waals dimensions of the actinomycin molecule.

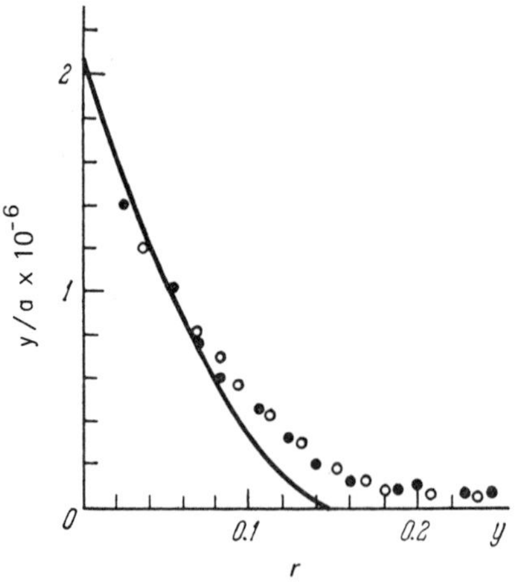


*FIG. 8.26 Adsorption isotherm of actinomycin by DNA.*

The adsorption isotherm of proflavin on DNA agrees with the theoretical calculation considering the interaction of adsorbed molecules. At small y the value of m is 3, and $K = 16\times10^6$ liter $mole^{-1}$. In the region $0.2 \leq y \leq 0.33$ m decreases to 2, and K to $1.6\times10^6$ liter $mole^{-1}$.

## 8.7 Reduplication of DNA

In principle three reduplication mechanisms are possible [162]: the conservative mechanism, in which the initial double helix is preserved and a new daughter helix is formed; the semiconservative; and the dispersive one, in which the initial material is uniformly distributed between the four chains of two daughter double helices (cf. [6]). Meselson and Stahl investigated the reduplication of DNA during the division of *E. coli* cells with the help of labeled atoms and density gradient sedimentation [76]. A cell population labeled by $^{15}N$ was obtained. During the interphase period the cells were transferred into a medium that contained only $^{14}N$ but not $^{15}N$. The cells duplicated in this medium. DNA was extracted from the initial population, its "children," and its "grandchildren," and its density and radioactivity was examined. The initial DNA had the highest density, being completely labeled with $^{15}N$. The DNA of the children was half labeled (i.e., its density was equal to the arithmetic mean between the densities of $^{15}N$ DNA and $^{14}N$ DNA). Finally, the DNA of the grandchildren was separated in a density gradient into two zones, one corresponding to $^{14}N$ DNA and the other to DNA labeled by 50% $^{15}N$. This is exactly the pattern which must be obtained with the semiconservative mechanism.

With the help of radioautography, Taylor showed that the reduplication of a chromosome as a whole also proceeds in a semiconservative manner; he devised a model relating the reduplication of chromosomes to DNA reduplication [6,163].

The same method of radioautography was used to study the reduplication of cyclic double helical DNA--of the sole chromosome of *E. coli*. The inclusion of the label $^3H$ proceeds sequentially from one nucleotide to the neighboring one. The reduplication starts at some definite locus [6,164].

Kornberg performed reduplicational synthesis of DNA *in vitro* [165,166]. The incubation mixture contained the triphosphates of all four deoxynucleosides (TPP, CTP, GTP, ATP), ions of $Mg^{2+}$ ($6.10^{-3}$ M), DNA-polymerizing enzyme extracted from *E. coli* and native DNA from calf thymus as an initiator, pH 7.5. DNA reduplicates in such a complete system. The reaction does not proceed if any one of the components of the mixture is absent, or if the initiator DNA is treated first with deoxyribonuclease, an enzyme that destroys DNA. Using nucleotides labeled by $^{32}P$ it has been shown that DNA synthesis is a polycondensation.

As each nucleotide is included in the chain, one molecule of pyrophosphate $H_4P_2O_7$ is released. According to the mass action law, the reaction is retarded if a surplus of pyrophosphate is added. If its amount is 100 times larger than the total amount of deoxynucleoside triphosphates, the reaction is slowed down by 50%. Later, using an especially well-purified enzymic system, Kornberg effected the reduplication of DNA from $\phi$X174 phage. The DNA obtained had an infective action; when it was introduced into a cell, phage particles were synthesized [167]. With the help of polymerase double helices of synthetic polynucleotides can be synthesized:

```
-A-T-A-T-A-T-A-T-              -G-G-G-G-G-G-G-
                      and
-T-A-T-A-T-A-T-A-              -C-C-C-C-C-C-C-
```

It has since been established that the polymerase used by Kornberg was rather a reparase, a slow-acting enzyme linking the polynucleotide blocks. Genuine polymerase works *in vivo*. We do not yet know much about these enzymes.

Statistical-thermodynamic investigation of DNA reduplication is described in [168,169] (cf. [6]). The change in the free energy due to the formation of two double helices from an initial double helix and nucleoside triphosphates can be expressed as

$$\Delta G = -N(2E_1+\bar{E}_2) - T\,\Delta S \tag{8.83}$$

where N is the number of nucleotides in the chain, $E_1$ is the energy released when one nucleotide becomes included in the chain, and $E_2$ is the mean gain in energy due to interactions which stabilize the Watson-Crick pairs. $E_2$ depends on the relative content of G-C pairs $\alpha$

$$\bar{E}_2 = \alpha\, E_{GC} + (1-\alpha)E_{AT} \tag{8.84}$$

The change in entropy is determined by the replacement of the nucleoside triphosphate (NTP) with pyrophosphate (PP) in solution, and by the loss of degrees of freedom of rotational and translational motion by the nucleotide being incorporated into the double helix. This last factor can be written as k ln q. We obtain

$$\Delta S = 2Nk\,\ln[q(\bar{x}_{NTP}/x_{PP})] \tag{8.85}$$

where $x_{PP}$ is the concentration of PP in solution, $\bar{x}_{NTP}$ is the mean concentration of NTP, equal to $(x_G^{\alpha}x_C^{\alpha}x_A^{1-\alpha}x_T^{1-\alpha})^{1/2}$ where $x_G$, etc., are the respective NTP concentrations. If they are identical, then

$$\bar{x}_{NTP} = x_A = x_T = x_G = x_C = x_{NTP}/4$$

where $x_{NTP}$ is the total concentration of all four NTP. The reduplication condition is

$$2E_1 + \bar{E}_2 + 2kT \ln[q(\bar{x}_{NTP}/x_{PP})] \geq 0 \qquad (8.86)$$

The values of $E_1$ and $E_2$ are of the order of 4 and 5 kcal mole$^{-1}$. q is equal to $Z_1'Z_2/Z_1Z_2'$, where $Z_1$ and $Z_1'$ are the respective partition functions of NTP in solution and of NTP bound to the template by hydrogen bonds, $Z_2$ and $Z_2'$ are the respective partition functions of PP in solution and of PP included in template-bound NTP. The estimate is $q \sim 10^{-2}$. At 300°K condition (8.86) is obeyed up to $x_{PP}/x_{NTP} \sim 400$. In cells, as in Kornberg's system, this ratio is much smaller. Condition (8.86) is valid and the reduplication is obviously limited not by thermodynamic but by kinetic causes or by the shortage of NTP.

Similar consideration of denaturation leads to the value of the difference in free energies of the separated chains and the double helix, which is equal to

$$\Delta G' = N\bar{E}_2 - 2NkT \ln q' \qquad (8.87)$$

where q' characterizes the change in the chain entropy per nucleotide. The denaturation condition is $\Delta G' \leq 0$. q' can be considered as equal to the number of conformations of a single chain. One link of the chain consists of five single bonds; two to three conformations can arise due to rotation around every bond. q' is of the order of 10-100. From Eq. (8.87) it follows that the denaturation temperature is

$$T_m = \frac{\bar{E}_2}{2k \ln q'} = \frac{E_{AT}}{2k \ln q'} + \alpha \frac{E_{GC} - E_{AT}}{2k \ln q'} \qquad (8.88)$$

The experimental dependence of $T_m$ on α [171]

$$T_m = 342°K + 41\alpha°K$$

can be obtained if $E_{AT}$ = 4.7 kcal mole$^{-1}$, $E_{GC}$ = 5.3 kcal mole$^{-1}$, and q' = 32. These calculations are very rough; they do not take into consideration the role of the counterions. Of course, these calculations have no rigorous quantitative significance. However, they permit us to estimate the orders of magnitude of ΔH and ΔS.

The fate of the double helix depends on the relationship between ΔG and ΔG'. If ΔG < ΔG' and $\Delta G \leq 0$, then reduplication becomes possible. If ΔG' < ΔG and $\Delta G' \leq 0$, then denaturation occurs. If $\Delta G = \Delta G' \leq 0$, both processes are possible, and finally, at ΔG > 0 and ΔG' > 0 the double helix remains intact.

The statistical mechanics of reduplication [6,168,169] is based on the Ising model (pp. 28,123). The first theory was that unwinding of the helix occurred at both its ends. NTP are adsorbed on the sites of the liberated chains. In the new chain, a nucleotide bond arises if nucleotides that are suitable for the formation of Watson-Crick pairs have been adsorbed on any two neighboring sites of the template. Study of the partition function obtained on the basis of this theory reveals the dependence of the reduplication degree on NTP concentration. Because of the cooperativity condition (the requirement of a suitable neighborhood for the adsorbed NTP) reduplication must proceed at some critical value of $x_{NTP}/x_{PP}$ according to the all-or-none principle, like a phase transition. This is valid, of course, only at large N; if N is of the order of $10^2$, only an S-shaped curve can be obtained. The same theory provides a rough description of the helix-coil transition.

However, a series of data shows that NTP is not adsorbed on the liberated chains in a random fashion. The synthesis of a new chain occurs by means of the sequential binding of NTP by the 3' nucleoside end of the growing chain [170,171]. Therefore the process is "antiparallel"; reduplication occurs at the first end of one chain and at the second end of another [172]. This process is also a cooperative one because the next NTP can be bound only if the previous site has already been occupied by NTP. The partition function has the form [6,169],

$$Z = \sum_{r=0}^{N-1} \sum_{i=0}^{r} \sum_{p=0}^{N-1-r} \sum_{\ell=0}^{p} b_1^{r+p} b_2^{i+\ell} + b_1^N \sum_{r'=0}^{N} \sum_{r''=0}^{N} b_2^{r'+r''} \tag{8.89}$$

Here $b_1 = \exp(-\bar{E}_2/kT+2 \ln q)$, $b_2 = a_3/a_2$, $a_2 = \exp(-F_2/kT)$, $a_3 = \exp[-(F_1-F_3)/kT]$, $F_2 = -kT \ln q$, $F_1 = E_2-kT \ln(x_{NTP}Z_1'/Z_1)$, $F_3 = -E_1-kT \ln(x_{PP}Z_2'/Z_2)$. The first term in (8.89) corresponds to the states of the nonseparated chains, which are connected by at least one hydrogen bond. As an arbitrary intermediate state we take the state in which r bonds between the bases have been broken from one end and a new chain containing i nucleotides has grown. The second term refers to a state in which all the hydrogen bonds are broken and synthesis proceeds on single DNA chains. As an intermediate state, we take the state in which a new chain of r' nucleotides has grown on one chain, and a new chain of r" nucleotides has grown on the other. The complicated summation can be simplified. At moderate temperatures, when denaturation is thermodynamically unfavorable, $b_1 < 1$. Reduplication takes place only when $b_1b_2^2 \geq 1$. In this region we may neglect terms of the order of $b_1^N$ and $b_1^Nb_2^N$. Then

$$Z = \frac{1}{(1-b_1)^2(1-b_1b_2)^2} + \frac{(b_1b_2^2)^N b_2^2}{(1-b_2)^2} \tag{8.90}$$

The average number of bound nucleotides is

$$2r = \frac{\partial \ln Z}{\partial \ln b_2} \tag{8.91}$$

If $b_1b_2^2 < 1$, then $2\bar{r} \cong 2b_1b_2/(1-b_1b_2)$, and if $b_2$ is small, $\bar{r}$ is also small. If $b_1b_2^2 > 1$, then $2\bar{r} = 2N-2/(b_r-1)$, that is, $\bar{r} \cong N$, and reduplication proceeds completely. The condition of reduplication is $b_1b_2^2 = 1$, which coincides with (8.86). At the transition point $\bar{r}$ is equal not to 0.5N but to 0.2N. This is due to the role of intermediate states during an incomplete separation of the chains when the synthesis of the new chain is less favorable, since the breakage at a given base pair may lead to the growth of only a single chain. Just as before, the transition is very sharp at large values of N.

The statistical-thermodynamic theory does not contradict experiment but it is not directly confirmable, since the influence of $x_{PP}$ on DNA synthesis *in vitro* has not been investigated. At the same time the theory is very rough, because it does not consider the counterions and pH of the medium.

Several facts show that reduplication actually requires the unwinding of the initial double helix. Replicative synthesis proceeds via the participation of DNA-polymerase. The molecule of this enzyme displaces apparently along the double helix, unwinding it, and synthesizing the new chains (the "zipper" model). Bacteriophage $\phi$X174 contains not double-helical but one-thread DNA. Sinsheimer showed that the multiplication of this phage proceeds via the replicative stage when phage DNA becomes double-helical DNA [173]. Hence, in this case, too, only the double helix reduplicates. This is also true of viral RNA reduplicating via an intermediate two-chain structure [174]. The one-chain DNA of phage $\phi$X174 is the cyclic one. In the above-mentioned experiments of Kornberg (p. 535) this DNA served as initiator. DNA-polymerase forms a new chain that is not closed in a ring. But only ring DNA is biologically active (i.e., infective). Kornberg used another enzyme, ligase (Lehman-Oliver preparation), for the closing of the new chain. New chains have been built not with thymine but with bromouracil. To extract synthetic DNA, the preparation was incubated briefly with DNase in such a way that approximately half of the molecules underwent one ring breakage. The resulting mixture contains equal amounts of template rings, synthetic rings, and template and synthetic open DNAs, as well as unbroken double rings. Synthetic rings containing heavy bromouracil were extracted by density gradient sedimentation. These rings are able to infect *E. coli* cells. It was shown earlier by Sinsheimer that the substitution of one nucleotide in 5500 nucleotides of phage $\phi$X174 DNA resulted in the loss of infectivity. Hence, Kornberg's synthesis was an absolutely exact one. Synthetic double rings (the replicative form) have also been obtained [167].

The kinetics of DNA reduplication is of greater interest than the thermodynamics. The thermodynamic conditions for reduplication undoubtedly occur in a dividing cell. The study of kinetics can yield information about polymerase action, which has not yet been adequately studied.

Unwinding of the DNA double helix that occurs simultaneously with reduplication (and with mRNA synthesis, cf. p. 566) must proceed differently from simple denaturation. The gain in free energy due to the formation of new chains creates considerable rotational momentum. This problem is examined in [120,175-177]. Phillips's work [177] is based on the idea that the enzyme being displaced along the double helix creates local conditions that disturb the balance of the chain's interactions. It can be assumed that the enzyme hinders the formation of the hydrophobic bonds between the bases. The model is shown in Fig. 8.27. However, the quantitative theory of the process has not been developed.

An elementary model for the study of the kinetics of DNA reduplication is a one-dimensional template on which NTP are adsorbed. Irreversible polycondensation proceeds from one end. Analysis of such a model, considering the necessity of the activation energy for NTP adsorption and desorption (the breakage of the hydrogen bonds with water at adsorption, and that with template at desorption) was first reported in [178] (cf. [6]). Rough estimates show that NTP diffusion from a solution occurs at a higher rate than the processes on a template, and the rate of synthesis is limited not by sorption and desorption, but by the enzymic process. In accordance with experiment, theory shows that synthesis must occur without any lag period.

Another cooperative model, suggesting that the combination of nucleotides occurs at every site on the template where two or more neighboring sites are occupied by adsorbed nucleotides, contradicts the experiment, since it gives a considerable lag period.

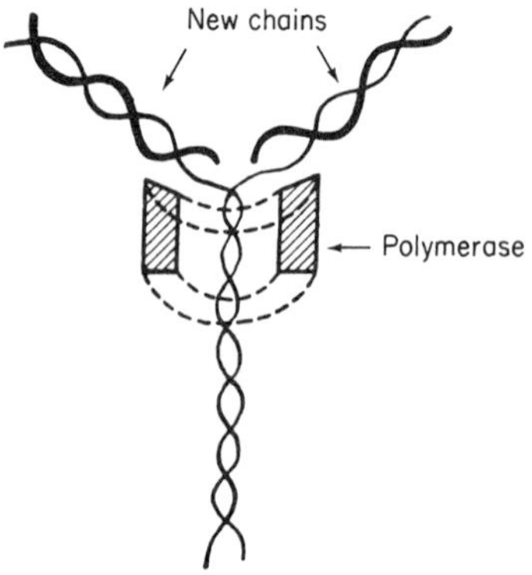


*FIG. 8.27 Model of the unwinding of the DNA double helix occurring simultaneously with reduplication.*

A more rigorous examination of linear template synthesis has since been performed [179-184]. Let us consider the work reported in [184]. We have a one-dimensional template with m sites. The growth centrum (polymerase) enters the first site and leaves the mth one. The synthesis is irreversible, that is, the process of centrum displacement from the (i-1)th towards the ith site proceeds far from equilibrium. The rate constant $k_p$ is the same for every site; the rate constant of initiation $k_0$ is much smaller than $k_p$. The motion of the growth centrum from the mth site occurs rapidly and does not limit the process. The problem amounts to an examination of irreversible wandering in a one-dimensional system. The kinetic equations have the form

$$\begin{aligned} \dot{N}_1 &= k_0 - k_r N_1 \\ \dot{N}_i &= k_r N_{i-1} - k_r N_i \qquad i = 2,\ldots,m-1 \\ \dot{N}_m &= k_r N_{m-1} \end{aligned} \tag{8.92}$$

where $N_i$ is the probability of the presence of enzyme at the ith site. We look for the mean numerical polymerization degree

$$\langle i(t)\rangle = \sum_{i=1}^{m} i\, N_i(t) \Big/ \sum_{i=1}^{m} N_i(t) \tag{8.93}$$

Under the initial conditions $N_1(0) = \cdots = N_m(0) = 0$. The solution of system (8.92) with the help of the Laplace-Carson transform (cf. p. 466) has the form

$$\begin{aligned} N_i(t) &= \frac{k_0}{k_r} \sum_{\ell=1}^{\infty} f(\ell), \qquad i = 1,\ldots,m-1 \\ N_m(t) &= k_0 t \sum_{\ell=m-2}^{\infty} f(\ell) - (m-1)\frac{k_0}{k_r} \sum_{\ell=m-1}^{\infty} f(\ell) \end{aligned} \tag{8.94}$$

where

$$f(\ell) = (k_r t)^{\ell} \exp(-k_r t)/\ell!$$

We get

$$\begin{aligned} 2k_r t\langle i\rangle = {} & \Big[2k_r t \sum_{\ell=0}^{m-3} f(\ell) + (k_r t)^2 \sum_{\ell=0}^{m-4} f(\ell) \\ & + 2mk_r t \sum_{m-2}^{\infty} f(\ell) - m(m-1) \sum_{m-1}^{\infty} f(\ell)\Big] \end{aligned} \tag{8.95}$$

At $k_r t \ll 1$, $N_i \to 0$, $N_m \to 0$, $\langle i \rangle \to 1$, that is, there is no synthesis. At $k_r t \gg 1$, $N_i \to k_0/k_r$, $N_m \to k_0 t$, $\langle i \rangle \to m$, the synthesis proceeds completely. These results coincide with those obtained in [179,180].

The rate constant $k_r = wk_t$, where w is the probability of the presence of an adsorbed monomer at a site before the growth centrum, and $k_t$ is the constant of its incorporation into the growing chain. Since the process is not limited by diffusion [178,184], w possesses an equilibrium value and is expressed by the Langmuir isotherm

$$w = qa/(1+qa) \tag{8.96}$$

where $q = \exp(-E/kT)$, $a = \exp(\mu/kT)$, E is the sorption energy, a is the absolute activity, and $\mu$ is the chemical potential of NTP. Since $\mu = \mu_0 + kT \ln(\gamma c)$, we have $a = Ac$, $A = \gamma \exp(\mu_0/kT)$, where $\gamma$ is the coefficient of activity of NTP and $\mu_0$ is the standard chemical potential. At $w \ll 1$, $qa \ll 1$ and

$$w = qa = Ac \exp(-E/kT) \tag{8.97}$$

$$k_p = k_t Ac \exp(-E/kT) \tag{8.98}$$

For the transition from a small to a large value of $\langle i \rangle$ a sharp increase in c is required. It is doubtful that such an increase occurs during cell division. On the other hand, the dependence of $\langle i \rangle$ on E is S-shaped, and a small change in E is sufficient for complete synthesis. This change can be due to cooperativity, that is, to the interaction of the neighboring nucleotides on the template (stacking). At small values of $\theta$ Eqs. (8.97) and (8.98) are valid in this case as well.

This treatment is equivalent to the one based on the theory of absolute reaction rates (p. 347). The reaction rate of the imer, which incorporates two nucleotides into the chain simultaneously, can be written as

$$v = (kT/h)w \tag{8.99}$$

Experimental values for the rate of synthesis of DNA are, for *E. coli*, $1.8\times10^3$; for mouse lymphoma cells, 200 monomers link$^{-1}$ sec$^{-1}$. This corresponds to $w = 2.8\times10^{-10}$ and $3.1\times10^{-11}$.

The free energy of activation $G^*$ can be expressed through the equilibrium constant of the activated complex of the i-mer with two nucleotides

$$G^* = -RT \ln K^*$$

where

$$K^* = w/[i]c$$

[i] and c are dimensionless, and c expresses the number of NTP molecules in the cell. Estimates of $G^*$ give for *E. coli* 13.0 kcal mole$^{-1}$, for mice 9.3 kcal mole$^{-1}$. It has to be emphasized

that this energy is related only to NTP adsorption but not to its incorporation into the chain. The theory ignores a series of facts: the heterogeneity of the template, the cooperativity of the process, and the solvent influence. It gives, however, a reasonable semiquantitative description of template synthesis and can also be applied for the synthesis of RNA on DNA and for the synthesis of protein on mRNA.

Enzymic synthesis of polynucleotides can also proceed without a template. In this case the polymer can be obtained after a long latent period decreasing with the increase in enzyme concentration [185]. An example is poly-(A-T) synthesis. If an oligomer is introduced as an initiator, then the lag period decreases rapidly with its length. Apparently even short oligomers can serve as templates for a growing polymer if it slides down from the template [186,187]. This was proved in a work [188] devoted to poly A synthesis on oligomers. The theory of such a synthesis based on the sliding of the new chain was developed in [189]. Figure 8.28 explains the model. We look for the time τ necessary for the formation of a chain

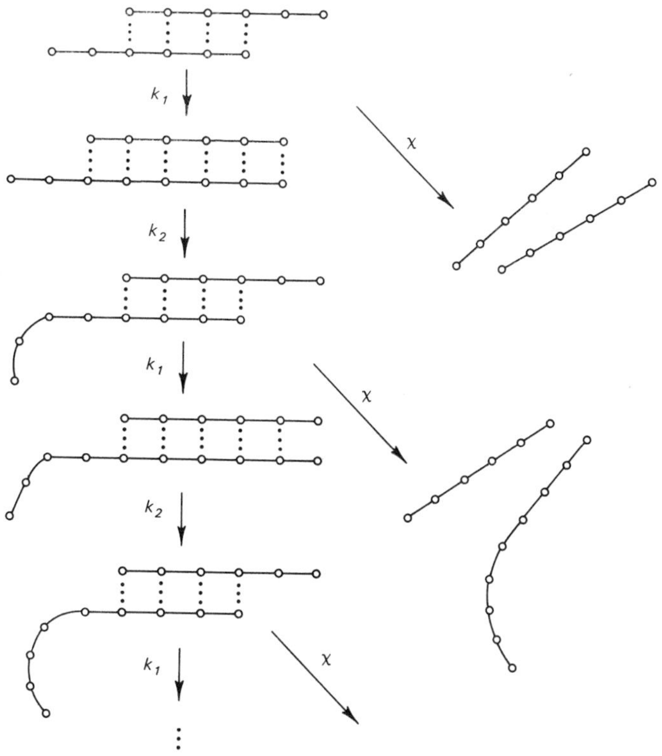


*FIG. 8.28 Scheme of polynucleotide synthesis without template.*

containing N links on a template having n < N links. It is necessary also to take into account the possibility of the complete removal of the new chain from the template, stopping the synthesis. Let the growing chain attain the size of template n at the moment t = 0 and be shifted by one pair of bonds. The system can be transmitted from this state into state i, corresponding to the further growth and sliding of the chain. The rate constant of this process k is expressed as

$$\frac{1}{k} = \frac{1}{k_1} + \frac{1}{k_2} \qquad (8.100)$$

where $k_1$ characterizes the chain growth, $k_2$ the sliding. Besides that, the system can be transmitted into the state i' of the separation of the chain from the template characterized by the rate constant $\chi$. Let us denote the number of subsystems in state i by $m_i(t)$, in state i' by $m_{i'}(t)$. We have the kinetic equations

$$\begin{aligned} \dot{m}_0 &= -(k+\chi)m \\ \dot{m}_i &= km_{i-1} - (k+\chi)m_i \\ \dot{m}_{i'} &= \chi m_{i-1} \end{aligned} \qquad (8.101)$$

$$m_i(0) = m_{i'}(0) = 0$$

The solutions of these equations are

$$\begin{aligned} m_i(t) &= m_0(0)\,\frac{(kt)^i}{i!}\exp[-(k+\chi)t], \quad i \geq 0 \\ M_{i'}(t) &= m_0(0)\chi \int_0^t \frac{(kt)^{i-1}}{(i-1)!}\exp[-(k-\chi)t]\,dt, \quad i \geq 1 \end{aligned} \qquad (8.102)$$

The probability of finding the system in state i is

$$w_i = Am_i \qquad (8.103)$$

The proportionality coefficient can be found from the normalization condition

$$\sum_{i=0}^{\infty} (w_i + w_{i'+1}) = 1$$

States i and i'+1 correspond to chains that have the lengths n+1. In state i the chain is bound to the template, in state i'+1 the chain is free. We get $A = m_0(0)^{-1}$. The average length of the growing chain is

$$\bar{\ell} = \sum_{i=0}^{\infty} i(w_i + w_{i'+1})$$

$$= \sum_{i=0}^{\infty} i \frac{(kt)^i}{i!} \exp[-(k+\chi)t]$$

$$+ \chi \int_0^t \sum_{i=0}^{\infty} i \frac{(kt)^i}{i!} \exp[-(k+\chi)t]\ dt$$

$$= \frac{k}{\chi}[1-\exp(-\chi t)] \tag{8.104}$$

and at $t \to \infty$, $\bar{\ell} \to k/\chi$.

$\bar{\ell}$ becomes equal to N-n during the time

$$t_{N-n} = \frac{1}{\chi} \ln \frac{1}{1-\chi/k(N-n)} \tag{8.105}$$

The average time of growth of a chain with length n is

$$t_n = \frac{n}{k_1} + \frac{1}{k_2} \tag{8.106}$$

The time $\tau_N$ is equal to the sum of $t_n$ and $t_{N-1}$. If $N >> n$, we get

$$\tau_N = \frac{1}{\chi} \ln \frac{1}{1-(\chi\ N/k)} \tag{8.107}$$

The observed strong dependence of i on n is due to a decrease in $\chi$ with increasing n. Experiment shows that $\chi$ decreases especially fast in the range between n = 4 and n = 5. This means that n = 4 is the critical size of the template at which the new chain can apparently form a double helix with the template chain. The parameter estimates are $N/k \sim 1$ min, $k_1 \sim 30N$ $\text{min}^{-1}$, $k_2 \sim N\ \text{min}^{-1}$.

The new chain is always synthesized *in vitro* in the 5' → 3' direction. On the other hand, there are data showing that both chains are reduplicated *in vivo*, one in the 5' → 3' direction and the other in the 3' → 5' direction. Okazaki and co-workers proposed a solution to this contradiction, theorizing that reduplicative synthesis occurred discontinuously. The short regions of the new DNA are synthesized on one or on both chains at the reduplication point, always in the 5' → 3' direction (Fig. 8.29). Later these regions combine, forming new chains. If this model is correct, the last replicated regions of one or both daughter chains can be separated. The selective and temporal inhibition of the enzyme catalyzing the formation of phosphodiester bonds between the DNA chains must result in the accumulation of the new short chains. These predictions have

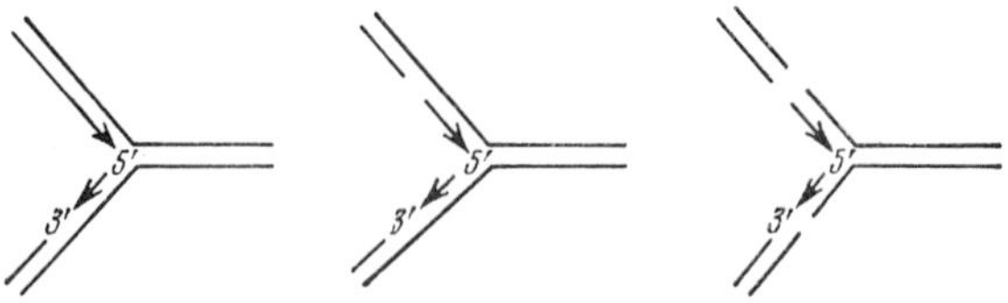

*FIG. 8.29 Possible reduplication schemes according to Okazaki.*

been confirmed experimentally. It has been shown that $^3$H-labeled thymidine is incorporated in short chains [190,191] that later combined into long ones.

Werner investigated reduplication by using labeled thymine, and showed that bacteria use thymine for reduplication, and thymidine for reparative synthesis [192]. Werner suggests another model: DNA reduplicates continuously in the fork via the simultaneous elongation of both new chains. The preferential use of thymine rather than thymidine shows that the process of reduplication is different from reparative synthesis. Werner suggests that the short chains found by Okazaki appear because of the action of specific nucleases. However, this scheme is not supported by experiment.

The precise mechanism of DNA reduplication is unknown at present. Reduplication *in vivo* starts at a definite stage of cell development with the participation of the regulatory factors apparently being related to the cell membrane. The problems of regulation at the molecular level will be discussed in "General Biophysics."

## References


1. O. Avery, C. McLeod, and M. McCarty, J. Exp. Med. **79**, 137 (1944).
2. E. Kreps, Vest. Acad. Sci. USSR N 10, 84 (1970) (R).
3. E. Kosower, "Molecular Biochemistry." McGraw-Hill, New York, 1962.
4. J. Watson, "Molecular Biology of the Gene." Benjamin, New York, 1965.
5. C. Anfinsen, "The Molecular Basis of Evolution." Wiley, New York, 1959.
6. M. Volkenstein, "Molecules and Life." Plenum Press, New York, 1970.
7. S. Bresler, "Introduction to Molecular Biology." Academic Press, New York, 1970.
8. G. Haggis, D. Michie, A. Muir, K. Roberts, and P. Walker, "Introduction to Molecular Biology." Wiley, New York, 1964.
9. W. Braun, "Bacterial Genetics." Saunders, Philadelphia, Pennsylvania, 1965.

10. J. Kendrew, "The Thread of Life." Bell and Sons, London, 1966.
11. W. Hayes, "The Genetics of Bacteria and Their Viruses." Wiley, New York, 1964.
12. G. Mendel, "Versuche über Pflanzenhybriden. Ostwald's Klassiker der exakten Wissenschaften." Leipzig, 1940.
13. N. Timofeev-Resovsky, Cytologia 2, 45 (1960) (R).
14. V. Efroimson, "Introduction to Medical Genetics." Medgiz, Moscow, 1964 (R).
15. N. Koltsov, "Organization of the Cell," Biomedgiz, Moscow, 1936; Bull. Moscow Soc. Invest. of Nature 70, N 4, 75 (1965) (R).
16. J. Beadle, in "The Chemical Basis of Heredity" (W. McElroy and B. Glass, eds.). Johns Hopkins Press, Baltimore, Maryland, 1957.
17. M. Delbrück, N. Timofeev-Resovsky, and K. Zimmer, Nachr. Biolog. Ges. Wiss. Göttingen 1, 189 (1935).
18. E. Schrödinger, "What Is Life?" Cambridge Univ. Press, London and New York, 1945.
19. J. Watson, "Double Helix." Atheneum, New York, 1968.
20. J. Lederberg and E. Tatum, Cold Spring Harbor Symp. Quant. Biol. II, 113 (1946).
21. F. Jacob and E. Wollman, "Sexuality and the Genetics of Bacteria." Academic Press, New York, 1961.
22. J. Davidson, "Biochemistry of Nucleic Acids." Wiley, New York, 1961.
23. F. Griffith, J. Hyg. Cambridge 27, 113 (1928).
24. H. Fraenkel-Conrat, J. Amer. Chem. Soc. 78, 882 (1956).
25. A. Gierer and G. Schramm, Nature (London) 177, 702 (1956).
26. H. Fraenkel-Conrath, V. Singer, and R. Williams, in "The Chemical Basis of Heredity" (W. McElroy and B. Glass, eds.). Johns Hopkins Press, Baltimore, Maryland, 1957.
27. F. Hartman, in "The Chemical Basis of Heredity" (W. McElroy and B. Glass, eds.). Johns Hopkins Press, Baltimore, Maryland, 1957.
28. G. Stent, "Molecular Biology of Bacterial Viruses." Freeman, San Francisco, California, 1963.
29. R. Franklin and R. Gosling, Nature (London) 171, 156, 740 (1953).
30. J. Watson and F. Crick, Nature (London) 171, 737, 964 (1953).
31. M. Wilkins, A. Stokes, and H. Wilson, Nature (London) **172**, 759 (1953).
32. F. Crick, Sci. Amer., October (1954).
33. D. Marvin, M. Spencer, and M. Wilkins, J. Mol. Biol. 3, 547 (1961).
34. M. Tunis-Schneider and N. Maestre, J. Mol. Biol. 52, 521 (1970).
35. T. Mokulskaja, Biophysica 11, 528 (1966) (R).

36. M. Sandaralingam, Biopolymers 7, 821 (1969).
37. A. Lakshminarayanam and V. Sasisekharan, Biopolymers 8, 475, 489, 505 (1969).
38. S. Arnott, Prog. Biophys. and Mol. Biol. 21, 265 (1970).
39. J. Josse, in Symp. 1, Int. Biochem. Congr., 5th
40. D. Luzzati, A. Nicolaieff, and F. Masson, J. Mol. Biol. 3, 185 (1961).
41. P. Doty, in Biophysical science. A study program. Rev. Mod. Phys. 31, N 1 (1959).
42. J. Eigner and P. Doty, J. Mol. Biol. 12, 549 (1965).
43. L. Shtshagina, D. Richter, E. Frisman, and V. Vorobiev, Mol. Biol. 3, 221 (1969) (R).
44. V. Ivanov et al., Biopolymers 12, 89 (1973): Studia Biophys. 40, 1 (1973).
45. J. Fresco and B. Alberts, Proc. Nat. Acad. Sci. U.S. 46, 311 (1960).
46. Y. Mitsui et al., Nature (London) 228, 1166 (1970).
47. J. Fresco, B. Alberts, and P. Doty, Nature (London) 188, 98 (1960).
48. A. Spirin, "Some Problems of Macromolecular Structure of Ribonucleic Acids." Ed. Acad. Sci. USSR, Moscow, 1963 (R).
49. Y. Evdokimov, N. Akimenko, N. Glukhova, A. Tikhonenko, and J. Varshavsky, Mol. Biol. 7, 131 (1973) (R).
50. K. Hoogsteen, Acta Crystallogr. 12, 822 (1959); 16, 907 (1963).
51. F. Mathews and A. Rich, J. Mol. Biol. 8, 89 (1964).
52. R. Hamlin, R. Lord, and A. Rich, Science 148, 1734 (1965).
53. H. Miles, Nature (London) 195, 459 (1962).
54. Y. Kyogoku, T. Shimanouchi, M. Tsuboi, and I. Watanabe, Nature (London) 195, 460 (1962).
55. Y. Kyogoku, R. Lord, and A. Rich, Science 154, 518 (1966).
56. L. Katz and S. Penman, J. Mol. Biol. 15, 220 (1966).
57. P. Ts'o, in "Molecular Associations in Biology" (B. Pullman, ed.). Academic Press, New York, 1968.
58. E. O'Brien, J. Mol. Biol. 7, 107 (1963); 22, 377 (1966), Acta Crystallogr. 23, 92 (1967).
59. H. Sobell, K. Tomita, and A. Rich, Proc. Nat. Acad. Sci. U.S. 49, 885 (1963).
60. A. Hashemeyer and H. Sobell, Nature (London) 202, 969 (1964); Acta Crystallogr. 19, 125 (1965).
61. L. Katz, K. Tomita, and A. Rich, J. Mol. Biol. 13, 340 (1965).
62. Y. Baklagina, M. Volkenstein, and Y. Kondrashev, J. Struct. Chem. (USSR) 7, 399 (1966) (R).
63. Y. Baklagina, Mol. Biol. 2, 635 (1968) (R).
64. K. Hoogsteen, in "Molecular Associations in Biology" (B. Pullman, ed.). Academic Press, New York, 1968.
65. B. Pullman, P. Claverie, and J. Caillet, Proc. Nat. Acad. Sci. U.S. 55, 904 (1966); J. Mol. Biol. 22, 373 (1966).

66. H. Nash and D. Bradly, J. Chem. Phys. **45**, 1380 (1966).
67. H. DeVoe and I. Tinoco, J. Mol. Biol. **4**, 500 (1962).
68. P. Claverie, B. Pullman, and J. Caillet, J. Theor. Biol. **12**, 419 (1966).
69. R. Rein, P. Claverie, and M. Pollack, Int. J. Quant. Chem. **2**, 129 (1968).
70. P. Claverie, J. Chim. Phys. **65**, 57 (1968), in "Molecular Associations in Biology" (B. Pullman, ed.). Academic Press, New York, 1968; Studia Biophys. **24/25**, 161 (1970).
71. G. Del Re, J. Chem. Soc. 4031 (1958).
72. H. Berthod and A. Pullman, J. Chem. Phys. **62**, 942 (1965).
73. V. Poltev and B. Sukhorukov, Piophysica **12**, 763 (1967); **13**, 941 (1968); J. Struct. Chem. **9**, 298 (1968) (R); Studia Biophys. **24/25**, 179 (1970).
74. R. Rein et al., Ann. N.Y. Acad. Sci. **153**, 805 (1969).
75. J. Marmur, R. Round, and C. Shildkraut, "Progress in Nucleic Acid Research," Mol. Biol. 1 (1963).
76. M. Meselson and F. Stahl, Proc. Nat. Acad. Sci. U.S. **44**, 671 (1958).
77. C. Thomas and K. Berns, J. Mol. Biol. **3**, 277 (1961).
78. J. Marmur and P. Doty, Nature (London) **183**, 1429 (1959).
79. B. Zimm, J. Chem. Phys. **33**, 1349 (1960).
80. S. Lifson and B. Zimm, Biopolymers **1**, 15 (1963).
81. D. Crothers and B. Zimm, J. Mol. Biol. **9**, 1 (1964).
82. M. Frank-Kamenetsky, Mol. Biol. **2**, 408 (1968) (R).
83. T. Birstein and O. Ptitsyn, "Conformations of Macromolecules." Wiley (Interscience), New York, 1966.
84. D. Poland and H. Scheraga, "Theory of Helix-Coil Transitions in Biopolymers." Academic Press, New York, 1970.
85. Y. Lazurkin, M. Frank-Kamenetsky, and E. Trifonov, Biopolymers **9**, 1253 (1970).
86. M. Frank-Kamenetsky, Dokl. Acad. Sci. USSR **157**, 187 (1964); Vysokomol. Sojedin. **7**, 354 (1965) (R).
87. B. Sukhorukov, Y. Moshkovsky, T. Birstein, and V. Lystsov, Biophysics **8**, 294 (1963) (R).
88. V. Permogorov, M. Frank-Kamenetsky, L. Serdiukova, and Y. Lazurkin, Vysokomol. Sojedin. **7**, 362 (1965) (R).
89. V. Permogorov, A. Prozorov, M. Shemiakin, Y. Lazurkin, and R. Khesin, in "Molecular Biophysics," Nauka, Moscow, 1965 (R).
90. M. Daune, C. Dekker, and H. Schachman, Biopolymers **4**, 51 (1966).
91. R. Jensen and M. Davidson, Biopolymers **4**, 17 (1966).
92. L. Mintshenkova and V. Ivanov, Biopolymers **5**, 615 (1967).
93. A. Poletaev, V. Ivanov, L. Mintshenkova, and A. Shtshelkina, Mol. Biol. **3**, 303 (1969) (R).
94. P. Privalov, Mol. Biol. **3**, 690 (1969) (R).
95. A. Vedenov, A. Dykhne, A. Frank-Kamenetsky, and M. Frank-Kamenetsky, Mol. Biol. **1**, 313 (1967) (R).

96. G. Lehman and J. McTague, J. Chem. Phys. 49, 3170 (1968).
97. T. Birstein and N. Namoradze, Dokl. Acad. Sci. Georgian SSR 44, 4567 (1966) (R).
98. A. Vedenov and A. Dykhne, J. Exp. Theor. Phys. 55, 357 (1968) (R).
99. M. Frank-Kamenetsky and A. Frank-Kamenetsky, Mol. Biol. 3, 375 (1969) (R).
100. D. Crothers, Biopolymers 6, 1391 (1968).
101. A. Shugalyi, M. Frank-Kamenetsky, and Y. Lazurkin, Mol. Biol. 3, 133 (1969); 4, 275 (1970) (R).
102. D. Crothers, N. Kallenbach, and B. Zimm, J. Mol. Biol. 11, 802 (1965).
103. M. Frank-Kamenetsky and A. Frank-Kamenetsky, Mol. Biol. 2, 778 (1968) (R).
104. I. Berestetskaja, Y. Kasaganov, Y. Lazurkin, and E. Trifonov, Mol. Biol. 4, 137 (1970) (R).
105. J. Marmur et al., J. Cell. Compart. Physiol. Suppl. I, 58, 33 (1961).
106. E. Trifonov, N. Shafranovskaja, M. Frank-Kamenetsky, and Y. Lazurkin, Mol. Biol. 2, 887 (1968) (R).
107. S. Falkow and D. Cowie, J. Bacteriol. 96, 777 (1968).
108. I. Lifshitz, J. Exp. Theor. Phys. 65, 1100 (1973) (R).
109. P. Privalov, Thesis, Inst. of Biol. Phys. Acad. Sci. USSR (1970).
110. P. Privalov, O. Ptitsyn, and T. Birstein, Biopolymers 8, 559 (1969).
111. J. Applequist and V. Damle, J. Amer. Chem. Soc. 87, 1450 (1965).
112. M. Eigen and D. Pörschke, J. Mol. Biol. 53, 123 (1970).
113. J. Applequist and V. Damle, J. Chem. Phys. 39, 2719 (1963).
114. D. Holcomb and S. Timasheff, Biopolymers 6, 513 (1968).
115. Y. Evdokimov and J. Varshavsky, Biophysica 11, 7 (1966) (R); FEBS, 3rd Meeting, Warsaw (1966); in "Structure and Function of Cellular Nucleus," Nauka, Moscow, 1967 (R).
116. A. Platonov, Y. Evdokimov, and J. Protasevitsh, Mol. Biol. 10, 321 (1976) (R).
117. A. Vedenov, A. Dykhne, and M. Frank-Kamenetsky, Usp. Phys. Nauk 109, N 3 (1971) (R).
118. W. Kuhn, Experientia 13, 301 (1957).
119. M. Volkenstein, N. Godjaev, and Y. Gottlieb, Biophysica 7, 16 (1962) (R).
120. H. Longuet-Higgins and B. Zimm, J. Mol. Biol. 2, 1 (1960).
121. M. Fixman, J. Mol. Biol. 6, 39 (1963).
122. D. Crothers, J. Mol. Biol. 9, 712 (1964).
123. P. Fong, Proc. Nat. Acad. Sci. U.S. 52, 939 (1964).
124. J. Vinograd, J. Lebowitz, and R. Watson, J. Mol. Biol. 33, 173 (1968).
125. H. Massie and B. Zimm, Biopolymers 7, 475 (1969).
126. H. Spatz and D. Crothers, J. Mol. Biol. 42, 191 (1969).

127. Y. Evdokimov and J. Varshavsky, Dokl. Acad. Sci. USSR **170**, 1205 (1966) (R).
128. Y. Evdokimov, K. Knorre, and J. Varshavsky, Mol. Biol. 3, 163 (1969).
129. Y. Evdokimov, N. Akimenko, Y. Kaurov, S. Ogurtsov, S. Tulkes, and J. Varshavsky, Studia Biophys. 24/25, 83 (1970).
130. E. Trifonov, M. Frank-Kamenetsky, and Y. Lazurkin, Mol. Biol. 1, 164 (1967) (R).
131. Y. Bannikov and E. Trifonov, Mol. Biol. 4, 734 (1970) (R).
132. J. Marmur and P. Doty, J. Mol. Biol. 3, 585 (1961).
133. C. Schildkraut, J. Marmur, and P. Doty, J. Mol. Biol. 4, 430 (1962).
134. J. Marmur, R. Round, and C . Schildkraut, "Progress in Nucleic Acid Research," Mol. Biol. 1 (1963).
135. W. Müller and D. Crothers, J. Mol. Biol. 35, 251 (1968).
136. W. Behr, K. Honikel, and G. Hartmann, Eur. J. Biochem. 9, 82 (1969).
137. M. Tsuboi, K. Matsuo, and P. Ts'o, J. Mol. Biol. 15, 256 (1966).
138. J. Richardson, J. Mol. Biol. 21, 83 (1966).
139. D. Neville and D. Davies, J. Mol. Biol. 17, 57 (1966).
140. L. Crawford and M. Waring, J. Mol. Biol. 25, 23 (1967).
141. M. Waring, J. Mol. Biol. 54, 247 (1970).
142. L. Lerman, J. Mol. Biol. 3, 18 (1961).
143. W. Fuller and M. Waring, Ber. Bunsenges, Phys. Chem. 68, 805 (1964).
144. B. Richards and J. Pardon, Exp. Cell. Res. 62, 184 (1970).
145. G. Gursky, Biophysica 11, 737 (1966); Mol. Biol. 3, 749 (1969) (R); Studia Biophys. 24/25, 265 (1970).
146. H. Sobell, S. Jain, T. Sakore, and C. Nordman, Nature (London) New Biol. 231, 200 (1971).
147. L. Hamilton, W. Fuller, and A. Rich, Nature (London) 198, 538 (1963).
148. A. Blake and A. Peacocke, Biopolymers 6, 1225 (1968).
149. M. Waring, Nature (London) 219, 1320 (1968).
150. A. Peacocke, Studia Biophys. 24/25, 213 (1970).
151. G. Löber, Studia Biophys. 24/25, 447 (1970).
152. D. Crothers, Studia Biophys. 24/25, 449 (1970).
153. G. Scatchard, Ann. N.Y. Acad. Sci. 52, 660 (1949).
154. D. Crothers, Biopolymers 6, 574 (1968).
155. A. Zasedatelev, G. Gursky, and M. Volkenstein, Mol. Biol. 5, 245 (1971) (R).
156. Y. Chismadjev and V. Markin, Elektrochimia 3, 127 (1968); 4, 3 (1969) (R).
157. G. Gursky, A. Zasedatelev and M. Volkenstein, Mol. Biol. 6, 479 (1972) (R).
158. R. Wells and J. Larson, J. Mol. Biol. 49, 319 (1970).
159. A. Cerami, E. Reich, D. Ward, and J. Goldberg, Proc. Nat.

Acad. Sci. U.S. 57, 1036 (1967).
160. J. Goldberg, M. Rabinowitz, and E. Reich, Proc. Nat. Acad. Sci. U.S. **48**, 2094 (1962).
161. W. Müller and D. Crothers, J. Mol. Biol. 35, 251 (1968).
162. M. Delbrück and G. Stent, in "The Chemical Basis of Heredity" (W. McElroy and B. Glass, eds.). Johns Hopkins Press, Baltimore, Maryland, 1957.
163. J. Taylor, in "Molecular Genetics" (J. Taylor, ed.), Part I. Academic Press, New York, 1963.
164. J. Cairns, Cold Spring Harbor Symp. Quant. Biol. 28, 43 (1963).
165. A. Kornberg, Biophysical science. A study program. Rev. Mod. Phys. 31, N 1 (1959).
166. A. Kornberg, "Enzymatic Synthesis of DNA." Wiley, New York, 1961.
167. A. Kornberg, Sci. Amer., October (1968).
168. M. Volkenstein, Dokl. Acad. Sci. USSR **130**, 889 (1960) (R); in Symp. 1, Int. Biochem. Congr., 5th.
169. M. Volkenstein and A. Eliashevitch, Dokl. Acad. Sci. USSR **131**, 538 (1960); **132**, 565 (1960); **136**, 1216 (1961); Biophysica 6, 513 (1961) (R).
170. J. Marmur and P. Doty, J. Mol. Biol. 3, 585 (1961).
171. J. Adler, Proc. Nat. Acad. Sci. U.S. **44**, 651 (1958).
172. K. Atwood, Science **132**, 617 (1960).
173. R. Sinsheimer, J. Chem. Phys. **58**, 986 (1961); J. Mol. Biol. **4**, 142 (1962).
174. C. Weissman et al., Proc. Nat. Acad. Sci. U.S. **49**, 407 (1963); Science **142**, 1188 (1963).
175. C. Levinthal and H. Crane, Proc. Nat. Acad. Sci. U.S. **42**, 436 (1956).
176. S. Erhan, J. Theoret. Biol. 23, 339 (1969).
177. A. Phillips, J. Theoret. Biol. **24**, 273 (1969).
178. M. Volkenstein, N. Godjaev, Y. Gottlieb, and O. Ptitsyn, Biophysica **8**, 3 (1963) (R).
179. R. Simha, J. Zimmerman, and J. Moacanin, J. Chem. Phys. 39, 1239 (1963).
180. J. Zimmerman, and J. Moacanin, J. Theoret. Biol. 9, 156 (1965); **13**, 106 (1966).
181. A. Pipkin and J. Gibbs, Biopolymers **4**, 3 (1966).
182. C. MacDonald, J. Gibbs, and A. Pipkin, Biopolymers **6**, 1 (1968).
183. M. Garrick, J. Theoret. Biol. 17, 19 (1967).
184. J. Maniloff, J. Theoret. Biol. 23, 441 (1969).
185. H. Schachman, J. Adler, C. Radding, I. Lehman, and A. Kornberg, J. Biol. Chem. 235, 3242 (1960).
186. M. Chamberlin and P. Berg, Proc. Nat. Acad. Sci. U.S. 48, 81 (1962).
187. A. Kornberg, L. Bertsch, J. Jackson, and M. Khorana, Proc. Nat. Acad. Sci. U.S. **51**, 315 (1964).

188. A. Falashi, J. Adler, and M. Khorana, J. Biol. Chem. 238, 3080 (1963).
189. M. Volkenstein and S. Fishman, Biopolymers 4, 77 (1966).
190. R. Okazaki, T. Okazaki, K. Sugimoto, R. Kainuma, A. Sugino, and N. Iwatsuki, Cold Spring Harbor Symp. Quant. Biol. 33, 129 (1968).
191. R. Okazaki, K. Sugimoto, Y. Imae, and A. Sugino, Nature (London) 228, 223 (1970).
192. R. Werner, Nature (london) 230, 570 (1971).
193. S. Arnott et al., J. Mol. Biol. 88, 523 (1974).

# Chapter 9

# Biosynthesis of Protein

## 9.1 The Problem of the Genetic Code

The gene (or the cistron, cf. [1,2]) is the part of the DNA macromolecule responsible for the synthesis of the protein chain. Is the role of DNA an instructive or an enabling one? Many data testify in favor of the first possibility. Protein synthesis is programmed by the DNA molecule, which contains information about the primary structure of protein chains, that is, genetic information. This information is "written" in the primary structure of DNA (i.e., in the nucleotide sequence). In RNA-type virus particles the genetic program is stored in the RNA molecules.

The problem of the genetic code consists in determining the correlation between the nucleotide sequence in DNA (and viral RNA) and the sequence of amino acid residues in the protein chain. This is a physical problem involving, first, the correlation of the informational content of DNA and protein; second, the discovery of the quantitative relationships between nucleotides and amino acids, which are ultimately determined by the molecular interactions that occur during protein synthesis; and third, the physical meaning of the genetic code, that is, the nonaccidental character of this relationship. It must be emphasized that the problem of the genetic code was first formulated by a physicist, Gamow [3], and that many physicists have taken part in solving this problem. However, the code was deciphered by biological and chemical methods.

The formulation of the code problem (i.e., determining the molecular structure and properties of matter) is a typical problem in physics. The chemical properties of atoms (as shown by the structure of Mendeleev's periodic chart) are coded by the number of electrons in each atom, in accordance with the principles of quantum mechanics. The same is true of atomic spectra. Knowing the sequence of the quantum levels, we possess the code for chemistry and optics. The system of elementary particles and their transformations, one of the most urgent physical problems, also requires the establishment of a code. The obligatory character of the quantitative relationships is typical for each code problem.

Gamow's first attempt to decipher the genetic code [3,4] was made immediately after the discovery of the double-helical structure of DNA. Although his hypothesis was erroneous, this initial effort by Gamow deserves our attention because the series of important questions raised in this work has played a major role in attracting the attention of physicists to molecular biology.

The protein "text" is "written" by means of a 20-letter alphabet; that of DNA (or RNA) by a 4-letter one. It follows from elementary considerations that the code ratio (i.e., the number of nucleotides coding one amino acid) cannot be less than three. The amount of information per amino acid is $i_a = \log_2 20 = 4.322$ bits and the information per nucleotide is $i_n = \log_2 4 = 2.000$ bits. Hence, the smallest number of nucleotides corresponding to one amino acid is $i_a/i_n = 2.161$.

An amino acid cannot be coded by a nonintegral number of nucleotides. Hence, the code ratio cannot be less than three. Indeed, the number of combinations of $4^2 = 16$, which is less than 20, the number of amino acids. The number of combinations of $4^3 = 64$. Three nucleotides in one codon (i.e., in the aggregate of nucleotides coding one amino acid) is more than sufficient.

Gamow hypothesized that the protein chain is synthesized directly on the DNA double helix, while every amino acid is located in a groove between four nucleotides, two of which belong to one chain of the helix, and two to the other chain. One nucleotide in the first chain forms a Watson-Crick pair with one nucleotide in the second chain.

The "diamond code" of Gamow provides just 20 "letters." Every letter is a rhombus consisting of four nucleotides. The number of combinations of $4^4$ is 256. But there are constraints, since the shorter diagonal of the rhombus must connect A with T or G with C. If the right and left forms of nonsymmetrical rhombuses are considered to be identical, for instance,

$$\begin{array}{ccc} & G & \\ A & & T \\ & A & \end{array} \quad \equiv \quad \begin{array}{ccc} & G & \\ T & & A \\ & A & \end{array}$$

then 20 different rhombuses will be obtained.

The diamond code is an overlapping one. Since every rhombus contains nucleotides from three successive pairs, two nucleotides on one side of each rhombus are shared with its two neighboring rhombuses. Therefore, the neighboring amino acids must be correlated. A given residue can be followed not by any one of the 20 possible residues, but only by certain of them (cf. [1]).

Gamow attempted to verify his code by comparing the possible combinations of rhombuses with the known primary structures of insulin and adrenocorticotropin. This effort met with a series of irreconcilable contradictions. Further investigations showed that no overlapping codes agree with experiment. The presence of overlaps may be expressed by the correlations of the neighboring amino acid residues. In other words, some residue pair combinations must be forbidden. Analysis of the primary structures of proteins has shown that there are no such correlations; any residue can be the neighbor of any other one, although different residues are present with differing frequencies [4,5]. It is possible, however, to form overlapping nucleotide codes which allow any sequence of amino acids [6].

Assume that a codon contains four nucleotides, and that the last nucleotide in a given codon is the first one in the next codon. The code ratio is three. If all coding quartets start, e.g., with G or U (RNA), they must also end with G and U. The total number of such codons is 64 ($=2\times16\times2$). These codons are distributed among the 20 amino acids in such a way that every amino acid possesses at least one quartet starting with G and one starting with U. Since every quartet must also end with G or U, any amino acid sequences become possible.

Therefore, the lack of correlation in polypeptide chains cannot be considered a decisive argument against overlapping codes, but the lack of correlation makes it impossible to decipher the code theoretically.

A more essential argument is that the mutational replacement of two neighboring amino acid residues in protein has never been observed. Yet the mutation of a nucleotide shared by two neighboring amino acids must result in such a double replacement. Wall argued against this statement [6]. However, since no experimental proof of overlapping has ever been found, we can consider the genetic code as a nonoverlapping one, as has in fact been shown by its complete deciphering.

Studies of mutations have shown that the code is colinear; that is, the codons in a nucleic acid and the corresponding amino acid residues in a protein are arranged in the same linear sequence. A series of proofs of this statement is presented in the monograph by Yčas [5]. Other attempts to decipher the code theoretically are also described there; of course, these attempts are now of historical interest only (cf. also [1] and [7]).

The code ratio has been found experimentally as a result of genetic investigation performed by Crick and co-workers [8]. Cistron B of the rII region of bacteriophage T4 (a parasite of *E. coli*) was investigated. A detailed analysis of the genetic properties of the phage T4-*E.coli* system had been done earlier by Benzer [9]. Among the point mutations of rII are deletions of nucleotides from and additions of nucleotides to the DNA chain. Such mutations can be provoked by acridine dyes.

The wild type of phage w multiplies on strains B and K12(λ) of *E. coli*. The mutant phages r multiply only on strain B, forming plaques with sharp edges. The FCO mutants induced by proflavin belong to type r. They are able to revert to the wild type w. Genetic analysis shows that such reversions occur not as a result of the reverse mutation r → w, but because a second suppressing mutation appears close to the first (w → r) one. Suppressors belong to the same phenotype r as the suppressed mutants. Each of these mutations separately leads to loss of the ability to synthesize the corresponding protein, but the combination of the two mutations in one cistron restores this ability. Eighty r mutants, including their double and triple combinations (suppressors of suppressors and suppressors of suppressors of suppressors) were studied. All suppressors belong to two classes: + (addition of a nucleotide) and − (deletion). If the initial mutation r is +, then its suppressor is −, and vice versa. Combinations +−, −+, +++, −−−, give the wild phenotype but combinations ++, −−, ++++, −−−−, etc., do not.

These facts can be explained if we assume that the code is triplet, nonoverlapping, and read sequentially starting from some definite nucleotide. Let us represent the DNA chain by the letter sequence ABCABCABC··· (Fig. 9.1). Reading the code starting from a given nucleotide is equivalent to the superposition on this sequence of a frame with holes. If one of the letters is deleted (−) or added (+), the whole sequence beginning at the mutation site becomes wrong. This means that normal wild type protein cannot be synthesized (Fig. 9.1b). If the suppressing mutation is added (−+ or +−), the sequence is disturbed only in the region between the two mutations

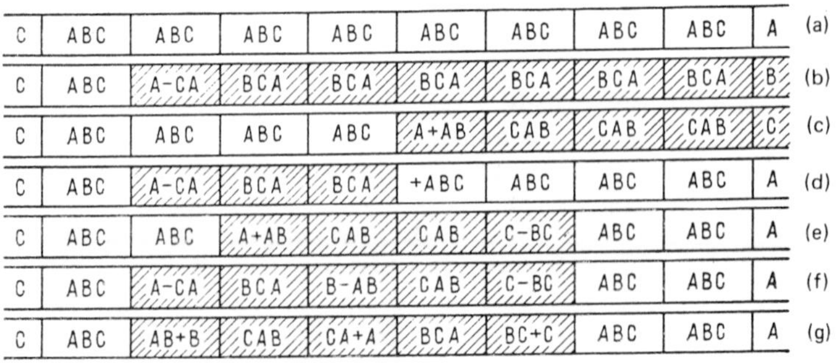


*FIG. 9.1 DNA text: (a) wild type; (b) mutant (deletion); (c) mutant (addition); (d,e) double mutants of type +−; (f) triple mutant −−−; (g) triple mutant +++.*

(Fig. 9.1d,e). If the incorrect region is not too large, then the synthesized protein can maintain its function and reversion will be observed. Figure 9.1f,g shows reversions due to three mutations of the same type, +++ or ---. It is easy to understand why two, four, and five mutations of the same type do not produce reversions.

These results show that the code ratio is three or a multiple of three. It is natural to hypothesize that it equals three. Nevertheless, Wall showed that the results obtained by Crick and co-workers also agree with his own overlapping quaternary code [5,6]. (Wall's code, however, is unacceptable for other reasons.)

Many details about these "frame shift mutations" are described in [8] and [1,5]. At first no direct proofs of the protein-synthesizing role of the rII cistron region were obtained. Later, Terzagi and co-workers [10] investigated similar mutations in the phage T4 cistron producing lysozyme. It was found that the frame shifts really disturbed the protein text. The double reverted mutant differs from the wild type by a sequence of six residues:

| | |
|---|---|
| w type | •••Thr-Lys-Ser-Pro-Ser-Leu-Asn-Ala••• |
| revertant (pseudo-w type) | •••Thr-Lys-Val-His-His-Leu-Met-Ala••• |

Thus a series of physical questions was answered by these biological investigations. The code ratio is established, the collinearity of the code is proved once more, it is shown that the code is read starting from a definite nucleotide, and that there are no commas (i.e., material units) separating one codon from another. It follows from these studies that a linear sequence of three nucleotides forms the codon. The total number of codons is 64.

These results led to inquiries about the degeneracy of the code. Since 64 is more than 20, we have to ask how many different codons correspond to a given amino acid? We have to emphasize the physical formulation of the problem and the physical sense of the word "degeneration," in contrast to its conventional biological sense.

The frame shift mutations investigations testify in favor of a degenerate code. This follows clearly from [10]. If the code is nondegenerate, then both codons for His in the lysozyme revertant have to be identical, say, ABC. If the shift is due to the addition of a nucleotide at the left side and the codon for Pro is BCA, then the relationship between the wild type and the revertant can be expressed in the following way:

··· Pro-Ser ···

A BCA BCA

ABC ABC

··· His-His ···

But this means that the codon for Ser is also BCA. The contradiction can be avoided only by the presence of different codons for His.

Another proof of code degeneracy is that, even with considerable variations in the DNA composition of various organisms (G + C varies from 0.25 to 0.75, p. 74), the mean composition of proteins does not change markedly. On the other hand, a series of facts shows the code's universality. It can be suggested that a considerable part of DNA nucleotides is nonfunctional, but this hypothesis cannot be coordinated with the uniformity of DNA composition for a given biological species. On the other hand, these facts are explicable if the code is degenerate.

Other indirect data pointing to a degenerate code were obtained in studies of both spontaneous and chemically induced mutations. Wittmann [11] studied mutations produced in TMV protein by nitrite. Under the action of nitrous acid, an amino group is replaced by the hydroxyl one

$$-NH_2 + HNO_2 \rightarrow -OH + N_2 + H_2O$$

Therefore nitrite transforms C into U, G into xanthine, and A into hypoxanthine. After reduplication, xanthine and hypoxanthine are replaced by G. Hence, replacement of the bases consists of C → U and A → G. The 64 triplet codons can be represented by eight octets. These octets are shown in Fig. 9.2, in which the arrows indicate point mutations. The substitutions observed in TMV protein are disposed in these same octets. Wittman found that at least Ser and Ile must be coded by several triplets (Fig. 9.3); that is, the code is degenerate. At the same time, these data show that most substitutions in protein occur due to the replacement of only one nucleotide in the codon.

There are 190 (20×19÷2) possible replacements of one amino acid by another. Some are the result of the replacement of one nucleotide in the triplet codon (allowed substitutions), others require the change of two or three nucleotides. Eck analyzed the large number of known mutations in protein, assuming that the allowed substitutions must be uniformly distributed [12]. According to his estimate, the number of allowed substitutions is 75 ± 20. More detailed statistical analysis gives a value close to 70 [13]. This figure shows the degeneration of the

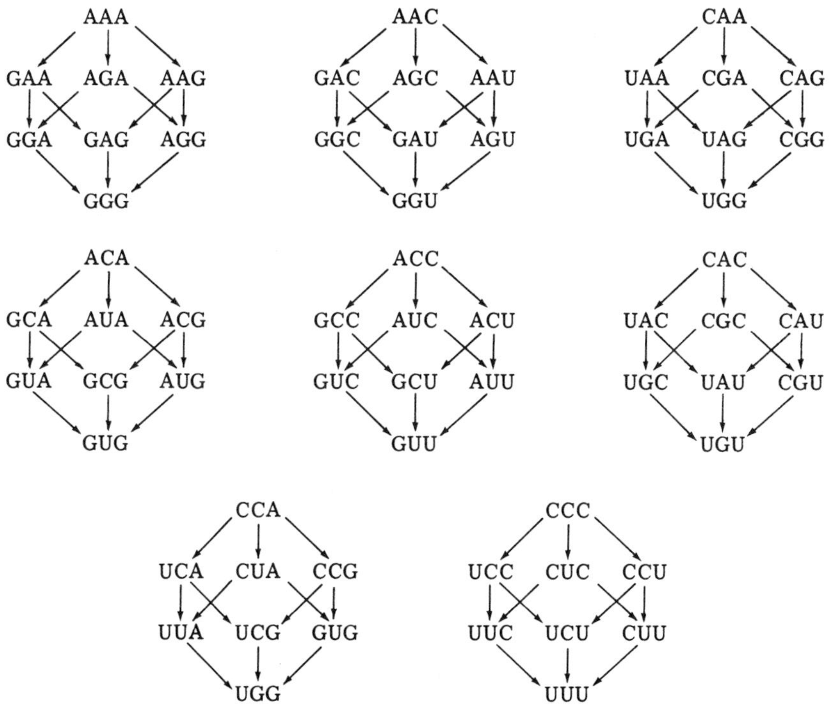


*FIG. 9.2 Classification of codons in octets. The arrows show the point mutations.*

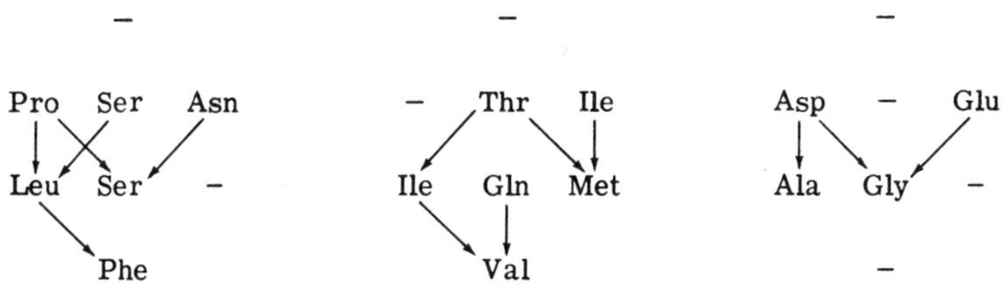


*FIG. 9.3 Correlation between amino acid residues and octets based on the nitrite mutations of TMV.*

the degeneration of the code, but does not give the degree of degeneration (i.e., the number of codons corresponding to one amino acid). However, statistical analysis shows that along with the 75 allowed substitutions there are approximately 100 substitutions due to a change in two nucleotides in a codon, and only 12 due to a change in three nucleotides. Such a distribution agrees with both double (40 codons) and triple (60 codons) uniform degeneration [13].

Thus, the genetic code is triplet, nonoverlapping, lacking

commas, and degenerate. Complete deciphering of the code required direct biochemical experiments that became possible only after the discovery of the mechanism of protein synthesis.

## 9.2 Biosynthesis of Protein

The molecular biology of protein synthesis is presented in [14] (cf. also [2]). Only a short account is given here; the remainder of the chapter will be devoted to physical problems related to protein synthesis.

The information contained in DNA is transmitted to mRNA in the transcription process. Protein is synthesized on mRNA, that is, the informational chain is DNA → RNA → protein.

The first experimental proof of the existence of mRNA was obtained by Belozersky, Spirin, and co-workers [15,16]. They showed that the composition of bacterial RNA correlated with the composition of bacterial DNA, and concluded that there were at least two types of RNA, one (the major fraction) differing in composition from DNA, and the second having a composition similar to DNA. It has since been discovered that the first fraction is ribosomal RNA (rRNA), and the second is messenger RNA (mRNA). The template function of RNA was revealed in experiments with plant viruses. It has been stated that the RNA extracted from TMV manifested infective activity--molecules of this viral RNA that were introduced into plant cells organized the synthesis of new TMV particles there (i.e., synthesis of their proteins) [17,18]. Fraenkel-Conrath regenerated infective TMV particles by mixing RNA of TMV with its protein, thereby obtaining hybrid viruses combining RNA from one TMV strain with protein from another [19].

Volkin and Astrakhan studied the events that take place within an *E. coli* cell after infection with phage T2. During phage multiplication some RNA that has an increased metabolic activity is synthesized. RNA incorporates $^{32}P$ rapidly. The composition of the synthesized RNA is similar to that of the phage DNA [20,21].

Later mRNA was also discovered in uninfected bacterial cells [22]. Its composition was analogous to that of the bacterial DNA. It was shown that mRNA is synthesized directly on DNA, as on a template. The presence of a special enzyme, RNA-polymerase, was discovered [23]. Later, the hybridization of DNA with its complementary mRNA was established [24-26].

The statement that this mRNA transfers genetic information from DNA to protein can be criticized (cf. [5]). It has been stated firmly only that a fraction of cellular RNA codes protein. However, all experimental data are in accord with the transcription theory and there are no grounds to doubt it correctness.

Thus mRNA is the carrier of information from DNA to

ribosomes (i.e., to the protein-synthesizing cellular device). mRNA provides the program for biosynthesis, but biosynthesis requires amino acids and the proper thermodynamic conditions.

Amino acids are present in the cell in the free form. Their direct polycondensation is an endergonic process; it is accompanied by increase in the free energy of approximately 3 kcal $mole^{-1}$ during peptide bond formation. The polycondensation of amino acids in a cell is coupled with an exergonic reaction of ATP dephosphorylation (cf. p. 87). Amino acid enters the biosynthesis reaction in an activated form [26-28]:

amino acid + ATP + enzyme ⇌ aminoacyladenylate + enzyme + PP

Aminoacyladenylate contains an activated form of the amino acid. Its structure is

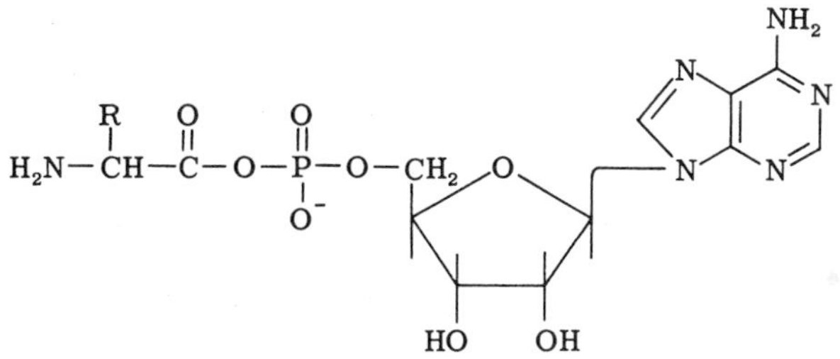


The enzyme aminoacyl-tRNA-synthetase is a bifunctional one. It catalyzes the reaction in which aminoacyladenylate is formed and the transfer of an activated amino acid to the end of the adaptor molecule, tRNA. The scheme of this reaction is

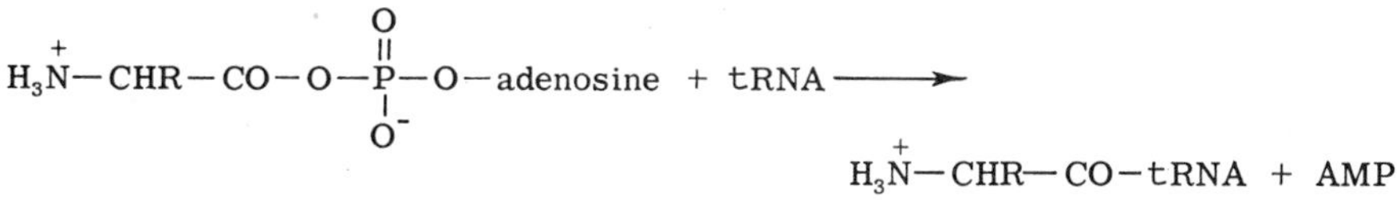


The energy required for biosynthesis is stored in the chemical bond linking the amino acid with tRNA.

For every amino acid there exists at least one specific tRNA. Hence, the number of different tRNA's must be no less than 20 and, correspondingly, no less than 20 different aminoacylsynthetases have to be active. Actually the number of tRNA's is more than 20. The number of enzymes is not yet known.

Biosynthesis of proteins occurs on ribosomes. Ribosomes are nucleoprotein particles consisting of proteins and ribosomal RNA. A growing *E. coli* cell contains approximately 15,000 ribosomes with MW ∿ $3\times10^6$. They form up to 25% of the whole mass of the cell. Ribosomes provide the appropriate interaction of mRNA with AA-tRNA. In this sense the ribosome is similar to an enzyme.

The ribosome binds the 5' end of mRNA, coding the start of the polypeptide chain, its N end. The ribosome moves along the mRNA chain, "reading the text" in such a way that the codons in an mRNA-ribosome complex sequentially bind the AA-tRNA. tRNA, which carries an amino acid, interacts in a complementary way with the mRNA codon by the anticodon of tRNA. The tRNA specific for a given amino acid contains an anticodon, a triplet complementary to the mRNA codon. In a given ribosome, in the given mRNA locus the sequential binding of two AA-tRNA's is effected by the nth and (n+1)th mRNA codons. This permits the formation of a peptide bond connecting the nth and (n+1)th amino acids; the nth tRNA becomes separated, and the ribosome carrying the (n+1)th tRNA with the growing polypeptide chain is displaced along the mRNA thread by one codon. In its new position the ribosome interacts with the (n+1)th and (n+2)th mRNA codons. This process is shown schematically in Fig. 9.4. Not one but a series of ribosomes are displaced along one mRNA chain, and at every ribosome the polypeptide chain grows. The mRNA-ribosomes system, which can be compared with beads on a thread, is called a polysome. Thus, a series of identical protein chains is synthesized on one mRNA chain. The general scheme of protein synthesis is shown in Fig. 9.5.

Formation of the peptide bond requires the participation of macroergic GTP molecules. Consequently, protein is synthesized *de novo* from amino acids and not via the assembling of previously made polypeptide blocks. The sequence of events in biosynthesis is rather complicated but it is characterized by unique principles.

Four types of nucleic acids participate in biosynthesis: DNA, mRNA, rRNA, and tRNA. Their functions are different but their structural features have much in common. In biosynthesis the fundamental biological and biophysical principle, namely, the principle of structural fit, or of the complementarity of

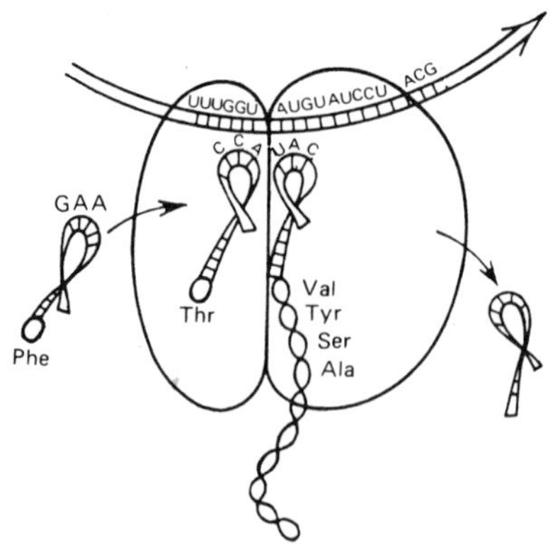


*FIG. 9.4 Scheme of the binding of AA-tRNA and mRNA with the participation of ribosome.*

*FIG. 9.5 General scheme of protein synthesis.*

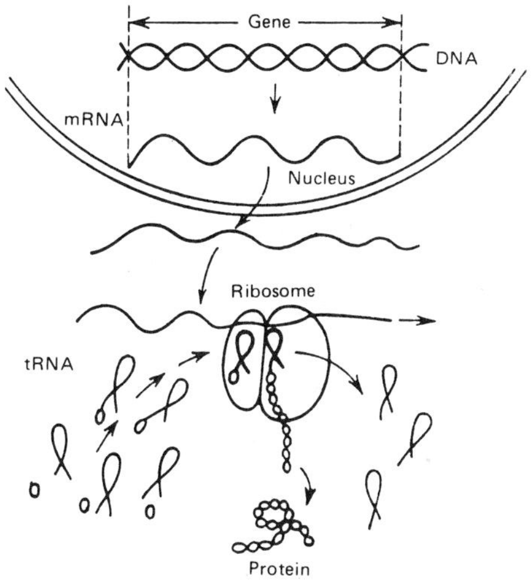


interacting molecules is again evident. The two chains in the DNA double helix are complementary, as are the DNA and mRNA chains, and the anticodons of tRNA and codons of mRNA. Complementarity exists for DNA and tRNA, and for DNA and rRNA, since these RNA's are also synthesized on DNA as on a template. In this sense DNA contains two kinds of genes: structural genes responsible for the synthesis of mRNA and of the corresponding proteins; and genes responsible for the synthesis of tRNA and rRNA.

Interactions between complementary molecules are weak interactions. However, the complicated interplay of these weak forces leads to the formation of strong chemical (peptide) bonds.

The biosynthesis of protein is template synthesis. We meet here two types of "reading" of the linear template. The syntheses of mRNA, tRNA, and rRNA on DNA, like DNA reduplication, occur via the direct growth of a new chain along the initial DNA chain with the participation of an enzyme, RNA-polymerase, whose molecule is displaced along the initial template. The synthesis of protein is more complicated, since it involves the participation of tRNA, and the ribosomes displace along the template. The protein chain itself has no complementary relationship with mRNA. Here occurs not the transcription of the nucleotide text of DNA into that of mRNA, but the translation of the nucleotide text into the amino acid one.

All these processes require the expenditure of energy, first, for the formation of chemical bonds (internucleotide and peptide bonds). On the other hand, the displacement of RNA-polymerase along the DNA chain and the displacement of ribosomes along the mRNA chain are mechanochemical processes due to the transformation of chemical into mechanical energy.

The macroergic nucleoside triphosphates serve as energy sources for all these processes.

The investigation of protein biosynthesis advances a series of physical problems. It is necessary to develop the theory of template synthesis, to understand the dynamic and structural features of RNA-polymerase, of the ribosomes and enzymes taking part in biosynthesis. It is necessary to investigate the structural basis of the bioenergetical processes determining biosynthesis.

New problems arise in connection with the discovery of DNA synthesis on RNA. The "central dogma" of molecular biology is expressed by the scheme

$$\text{DNA} \rightarrow \text{RNA} \rightarrow \text{protein} \qquad \text{(A)}$$

Actually it is not a dogma but a law of nature. However, the transfer of information from RNA to DNA has been discovered in some oncogenic viruses (cf. [29-32]). This finding implies the existence of RNA-dependent DNA-polymerase. DNA extracted from cells infected by virus hybridizes with viral RNA. Apparently several enzymes, "reversal transcriptases," act in the system. More recently, RNA-dependent DNA-polymerase was also discovered in uninfected cells of mice and man [33]. This enzyme can copy synthetic RNA-RNA and RNA-DNA templates. As early as 1963 Gershenson found that DNA-type virus responsible for nuclear polyhedrose in insects formed mRNA in the host cells. Injection of this mRNA into healthy insects infects them (cf. [34]). All these facts do not violate the dogma, but they require some supplementing. Instead of scheme A, the scheme

$$\text{DNA} \rightleftarrows \text{RNA} \rightarrow \text{protein} \qquad \text{(B)}$$

or scheme

$$\text{RNA}' \rightarrow \text{DNA}' \rightarrow \text{RNA} \rightarrow \text{protein} \qquad \text{(C)}$$

has to be used. These phenomena have not yet been sufficiently studied.

## 9.3 Transcription

The synthesis of RNA on a DNA template occurs with the help of RNA-polymerase. Most investigations have been performed using the RNA-polymerase from *E. coli* or *Azotobacter vinelandii*. These enzymes are specific molecular engines; they have a complicated quaternary structure and perform a series of functions: recognition of the initial locus of synthesis and specific binding of DNA at this locus; initiation of RNA synthesis (i.e., formation of the first internucleotide bond); direct synthesis of RNA; termination (i.e., breakage) of the synthesized RNA chain at the end of a gene or linear sequence of genes (if polycistronic RNA is being synthesized); and removal of the enzyme from the template.

At low ionic strengths RNA-polymerase is a dimeric enzyme with MW less than $10^6$ and sedimentation constant 23 S. At ionic strengths higher than 0.1 M the enzyme dissociates reversibly into two monomers: 21 S $\rightleftharpoons$ 2×13 S. It has been shown that 13 S is the enzymically active form [35]; little is known about the 21 S monomer (cf. [14]).

The monomeric 13 S enzyme in its turn consists of subunits: two large subunits $\beta$ and $\beta'$ with MW 150,000 and 165,000, respectively; two identical smaller subunits $\alpha$ with MW ~40,000; and the so-called $\sigma$ factor, with MW 80,000. Consequently, the 13 S RNA-polymerase can be represented symbolically as $\sigma\alpha_2\beta\beta'$ [36].

The role of the subunits is not yet known, but it is known that the $\sigma$ factor provides the specific "reading" of a definite system of genes. If the $\sigma$ factor is removed, the effectiveness of transcription of the so-called pre-early genes of the phage T4 is lowered by two orders of magnitude. On the other hand, the influence of $\sigma$ factor on a template that is not specific for this enzyme (thymus DNA) is small. The $\sigma$ factor does not participate in the direct synthesis of RNA, and it goes away into solution when the polymerase starts its motion along the template. The $\sigma$ factor unites with the fundamental part $\alpha_2\beta\beta'$ after the enzyme separates from the template. It has to be emphasized that the complicated polymerase structure is related to the diversity of the functions performed by this enzyme and to the necessity of regulating the transcription process. In those cases when only RNA synthesis must occur, the RNA-polymerase can be arranged in a much simpler way. Thus, for instance, the specific RNA-polymerase of the T7 phage contains only one polypeptide chain with MW ~100,000. Termination of *in vivo* synthesis is brought about with the help of the so-called $\rho$ factor, which is a separate protein not normally connected with RNA-polymerase. This factor recognizes the end of the chain and forces the polymerase to stop the RNA synthesis [37].

The RNA-polymerases discussed here are more active with double-stranded than with denatured DNA. They use only the double-helical template *in vivo*, unwinding it at the point of growth of the RNA chain [38]. The rates of RNA synthesis *in vivo* and in optimal *in vitro* conditions are similar: 20-30 nucleotides/sec. Only one DNA strand is read *in vivo*, and on a specific nondestroyed template *in vitro*. However, for different gene sequences (different operons) this strand is not always the same. Therefore, the start of a sequence has to give information about the DNA strand which must be read by the polymerase. In connection with this requirement we have to mention Szybalski's hypothesis [39], according to which the start of each gene sequence is characterized by a nonuniform distribution of purine and pyrimidine bases between the DNA strands: one strand contains more pyrimidines and the other more purines. Experiments *in vitro* with artificial polynucleotides tend to confirm this hypothesis. RNA-polymerase has been found to favor the pyrimidine-rich strand. However, the deciphering of the sequences in the regions of the polymerase binding did not confirm the Szybalski hypothesis [149].

X-ray structural analysis of polynucleotides in which one strand contained only purines and the other only pyrimidines showed that the spatial structure of these polynucleotides differs from that of the B form of DNA [40]. It is therefore possible that RNA-polymerase recognizes the start of "reading" by the spatial structure.

The application of physical methods to the study of the transcription mechanism is hindered by the fact that RNA-polymerase very rarely "sits" on DNA. Therefore, the conformational transformations of the template that accompany RNA synthesis affect only an insignificant part of the long DNA molecule. Therefore, studies of the complexes formed by RNA-polymerase with the shortest DNA sections that manifest all specific functions except termination are of great interest. Such complexes can be obtained by means of the action of deoxyribonuclease on the complex of DNA and polymerase. The section of DNA which binds polymerase and contains less than three turns of the helix (∿30 nucleotide pairs) remains intact in such experiments. It was found that such a complex can perform the fundamental polymerase function, the synthesis of short RNA molecules containing 15 - 20 nucleotides [41]. Work with these complexes

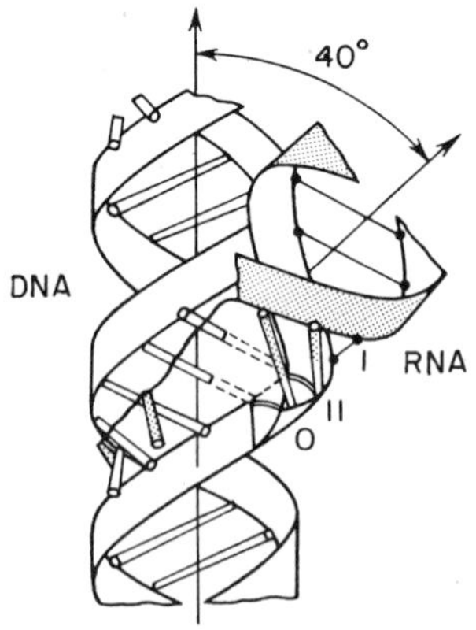


*FIG. 9.6 Model of RNA-polymerase action. The structure of the two-helical DNA template and of the growing RNA end.*

is only beginning, but we can hope that the use of various physical methods, including X-ray analysis, will enable us to elucidate what occurs during the displacement of RNA-polymerase along DNA, which proceeds at the rate of 20 nucleotides/sec.

Florentiev and Ivanov suggested a hypothetical model of RNA-polymerase action [42]. When an RNA chain grows (Fig. 9.6), binding of the 3'-hydroxyl of ribose to the 5'-phosphate group of the attacking nucleotide occurs. The sequence of bases in RNA is determined by that in DNA [43].

It is known that splitting of the phosphoester bond proceeds via an intermediate state in which the five covalent bonds of the P atom are in a trigonal bipyramidal configuration. Therefore, the nucleotide addition reaction can be represented by a two-stage scheme (Fig. 9.7). The formation of the intermediate system (II) requires energy, whereas the separation of pyrophosphate liberates the free energy. The model takes into account the overlapping of both stages. It is theorized that the A form (not the conventional B form) of the DNA template participates in the unwinding of DNA and the formation of a

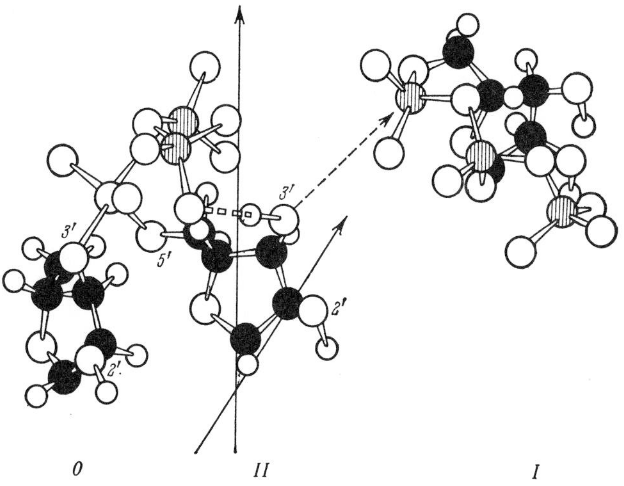


*FIG. 9.7 Two-stage scheme of nucleotide binding. At the left the conformation of the two last ribonucleotides at the site of RNA growth is shown; at the right, the newly bound nucleoside triphosphate. 0 is the nucleotide which has formed a covalent bond with II, but has not yet liberated the pyrophosphate.*

hybrid DNA-RNA double helix. This means that the A form is stabilized in the template region adjoining the system's active site. This hypothesis is in accord with a series of facts [42]. The local transition B → A during polymerase activity can be induced by removal of the water molecules from the enzyme-bound DNA region, and favors nucleotide polycondensation. The study of molecular models shows that there are no steric hindrances to the removal of the base of one DNA chain from the glycoside groove in the A form of DNA after breakage of the hydrogen bonds with the base of the complementary chain. The conformation of the second chain is not changed here.

The overlapping of the stages requires the simultaneous participation of two nucleotides. One is in state I, and the other in state II (Fig. 9.7); both are paired with DNA nucleotides. The model shows that the axis of the hybrid helix being formed is inclined to that of the initial DNA double helix at an angle of 40°.

Consequently, at every step the new base is "extracted" from the DNA template, and the deoxyribonucleotide whose complementary ribonucleotide has lost pyrophosphate turns back into DNA with the regeneration of hydrogen bonds. The axis of the hybrid helix moves along the DNA in a screwlike way, with simultaneous displacement of the enzyme.

It is known that in the absence of substrates RNA-polymerase is only weakly bound to DNA. On the other hand, during RNA synthesis a nondissociating DNA-enzyme complex is formed [43]. According to the model of Florentiev and Ivanov, this complex is stabilized by the bonds between the pyrophosphate groups of two interacting ribonucleoside triphosphates and the cationic groups of polymerase subunits. The bases of these triphosphates form Watson-Crick pairs with the DNA bases. Nucleotides I and II form an intermediate hybrid region with nucleotide 0. At any moment two ribonucleoside triphosphates are functioning; therefore, the model explains strong enzyme binding. The model presents the conformations of the sugar-phosphate part of nucleotides 0 and II and introduces well argumented suggestions about the conformation of nucleotide I. When the enzyme is displaced along DNA, conformational changes that enhance the reaction occur. The model takes into account the subunit structure of polymerase.

The energy aspect of the process, which has been examined qualitatively, amounts to the coupling of exergonic and endergonic steps between two successive reaction cycles. The model is similar to that of transaminase action (p. 374) in the sense that the enzymes provide the proper conditions at every stage of the multistage process. These conditions are brought about by structural rearrangements during the preceding stage. In essence, the model takes into consideration electron-conformation interactions (p. 400).

An alternative model was simultaneously suggested by Riley [44]; this model assumed that transcription occurs without unwinding of the DNA double helix. RNA grows in the broad groove of DNA because of a specific stereochemical interaction between ribonucleotides and the DNA base pairs. At the intermediate stages triplets are formed by two DNA bases and one RNA base. This model does not take into account the work of the enzyme.

Experimental data that appeared soon after the work of Florentiev and Ivanov [42] and Riley [44] confirmed the unwinding model. Using the formaldehyde method (p. 523), Kosaganov and co-workers investigated the T2 phage DNA complex with the RNA-polymerase from *E. coli* in the presence of all four nucleoside triphosphates [38]. The formaldehyde method makes it possible to detect defects in DNA in the ratio of 1 per 10,000 base pairs. In the complete system the concentration of defects in DNA was found to equal $6 \pm 2 \times 10^{-4}$, which corresponded to an average distance of 1600 ± 500 base pairs between two defects. In the initial DNA and in the DNA + polymerase complex in the absence of NTP, the concentration of defects is less than $1 \times 10^{-4}$ and the distance between defects is more than $10^4$ base pairs. Inactivation of polymerase by heating also decreases the number of defects to $10^{-4}$. Later, using the modified formaldehyde method, the same authors observed the unwinding of DNA which had already reached the stage of its binding with enzyme. The study of the previously mentioned (p. 568) complexes of RNA-polymerase with short DNA fragments, made with the help of CD methods, showed that during the mRNA synthesis the A form of either DNA or the hybrid DNA-RNA arises[150].

## 9.4 Transfer RNA's

Transfer RNA's were discovered in 1957 [45,46]. They make up approximately 10% of the total RNA content of a cell. The complete primary structure of a series of tRNA's was determined. This was possible because of the presence in tRNA of the minor nitrogen bases, methylated adenosine, guanine, etc. The minor nucleotides serve as easily identifiable labels when tRNA is split under the action of nuclease. Transfer RNA is split in blocks, the oligonucleotides obtained are fractionated, and the analysis of the chain ends and determination of the oligonucleotide structure is performed. This method was first applied to the study of Ala-tRNA from yeast by Holley and co-workers [47,48]. In the works of Bayev and co-workers which resulted in the determination of the primary structure of Val-tRNA from yeast [49], optical methods (spectrophotometry and CD) were used to identify the oligonucleotides [50,51]. A detailed account of these studies is given in [52]. By the beginning of 1975 the primary structures of more than 60 tRNA's of varying origins had been established.

The secondary structure of tRNA can be visualized if the possibility of complementary base pairing is considered and it is assumed that the nonpairing bases form loops. Then all the known tRNA's can be represented by a "cloverleaf" scheme having four or five double-helical regions (Fig. 9.8). The degree of helicity determined by various methods is in agreement with this scheme. The cloverleaf represents the projection of the unfolded tRNA molecule on a plane and reproduces only the topology of the molecule, the positions of the double-helical regions.

The problem of determining the secondary structure of RNA on the basis of the known primary structure was solved by Tumanian [53-55]. The algorithm involves two stages. First, the trigonal matrix of all possible pairs is formed according to the simple rule: the matrix element at the intersection of the ith column and jth row is equal to 1 or 0 if the ith and jth nucleotides are, respectively, complementary and noncomplementary. If the difference in free energy of the G-C and A-T pairs has to be taken into account, the corresponding statistical weights are used instead of 1.

More than five years later the first part of Tumanian's algorithm was reproduced by Tinoco and co-workers [56]. After the matrix is constructed, the problem can be solved by sorting out (as in [56]) all the possible variations of pair sequences (of course, of those allowed topologically), taking into consideration the free energy of various kinds of loops.

Tumanian's solution also contains a second stage, the "maximal hairpin" algorithm. The maximal hairpin is a secondary structure with one double-helical region that has the maximum negative free energy, corresponding in the first approximation to the maximum number of pairs. The validity of the algorithm is proved by the method of complete mathematical induction. Then a rather limited sorting out of the secondary structures with different numbers of maximal hairpins has to be performed.

This algorithm provides the complete solution to the problem of determining the secondary structure of RNA. Of course, the correct parameterization is required--the correct values of the free energy of formation of the various pairs and loops, etc.

Estimates of the free energy of formation of the loops (p. 499) and double-helical regions were offered in [56]. The initiation of base pairing in a one-chain RNA molecule involves loop formation. The free energy of initiation is $-RT \ln \gamma_m$, where $\gamma_m$ is the probability of loop formation from m unpaired links. $\gamma_m$ increases with decreasing m until the formation of the loop becomes hindered by steric factors. If $T_m^{\infty}$ is the temperature corresponding to the equilibrium of equal amounts of double-helical and one-chain regions (i.e., the melting point of the double helix), then for the helix formation $\Delta G(\text{helix}) = 0$, and

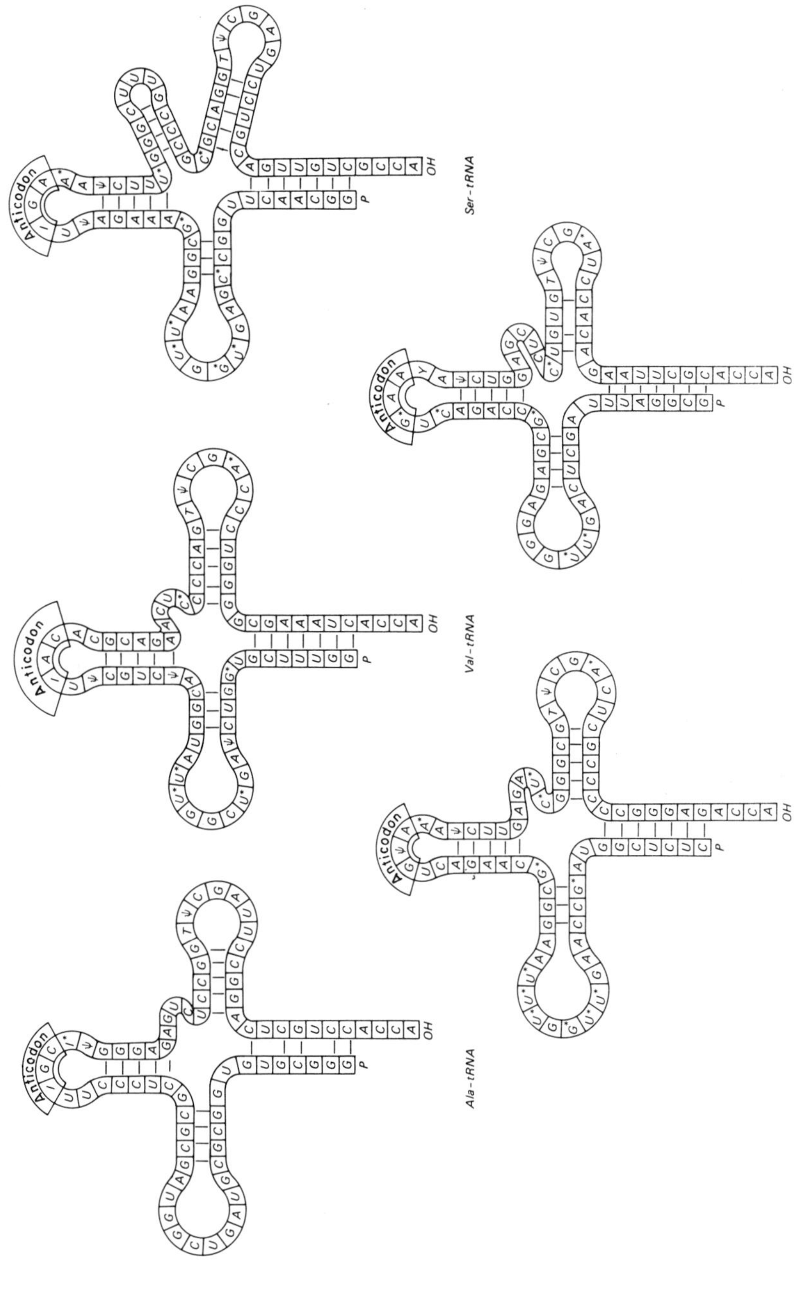


FIG. 9.8 Secondary structures of some tRNA's. ψ, $A^*$, $U^*$, $G^*$, $I^*$ are minor nucleotides.

$$\Delta S(\text{helix}) = \Delta H(\text{helix})/T \tag{9.1}$$

At temperature T for a short double helix of RNA

$$\Delta G(\text{helix}) = \Delta H(1 - T/T_m^{\infty}) \tag{9.2}$$

and the total free energy of the double helix formation

$$\Delta G = \Delta H(1 - T/T_m^{\infty}) - RT \ln \gamma_m \tag{9.3}$$

The structure with the lowest ΔG value is the most stable one. Formation of the A-U pair gives -1.2 kcal $\text{mole}^{-1}$. Let us represent this value by the "stability number" β, equal to +1. Then the following values of β can be obtained from the data on the melting points and thermodynamic characteristics of melting: (1) A-U pairs, +1; (2) G-C pairs, +2; (3) G-U, 0; (4) hairpin loops, from -5 to -7; (5) internal loops, from -4 to -7; (6) one-sided loops, from -2 to -6. With the help of these values the ΔG values corresponding to three possible structures of the sequence of 55 bases from the RNA of virus R17 (shown in Fig. 9.9) were calculated [56]. The most stable structure is II. These calculations were based on preliminary data; they do not take into consideration factors pertaining to the medium. These results of Tinoco et al. [56] were improved later [151]. Two neighboring base pairs were considered instead of using a definite free energy value for each base pair, and therefore the interactions of the pairs were taken into account. The necessary thermodynamic data were obtained from the oligonucleotide studies. The estimated free energies of loops were also changed.

The causes of the formation of a cloverleaf tRNA structure can be visualized in the "tRNA game" described by Eigen [57]. Given are a random sequence of N digits of four classes: A, U, G, C; and a tetrahedral die, each face of which represents one of the four letters. There are two players. By throwing the die in turn and replacing a position in the sequence with the digit obtained, each player tries to approach the double-stranded structure with the greatest number of A-U or G-C pairs. The game is over whenever a player announces a "complete" structure. The winner is the player who then has the greatest number of points. The G-C pair is worth two points, the A-U pair one point. The constraint is that pairs can be formed only beginning with the minimal sequences 2 G-C, or 1 G-C and 2 A-U, or 4 A-U pairs (cooperativity). Every loop in the structure must contain at least 5 unpaired units. The result of the game is in fact a cloverleaf. A hairpin with one loop has the largest number of bases which are allowed to pair, but the cloverleaf is more flexible in that it gives many more combinations than the hairpin. For 80 nucleotides the optimum is around three or four leaves. Eigen writes that "nature apparently played such a game long ago."

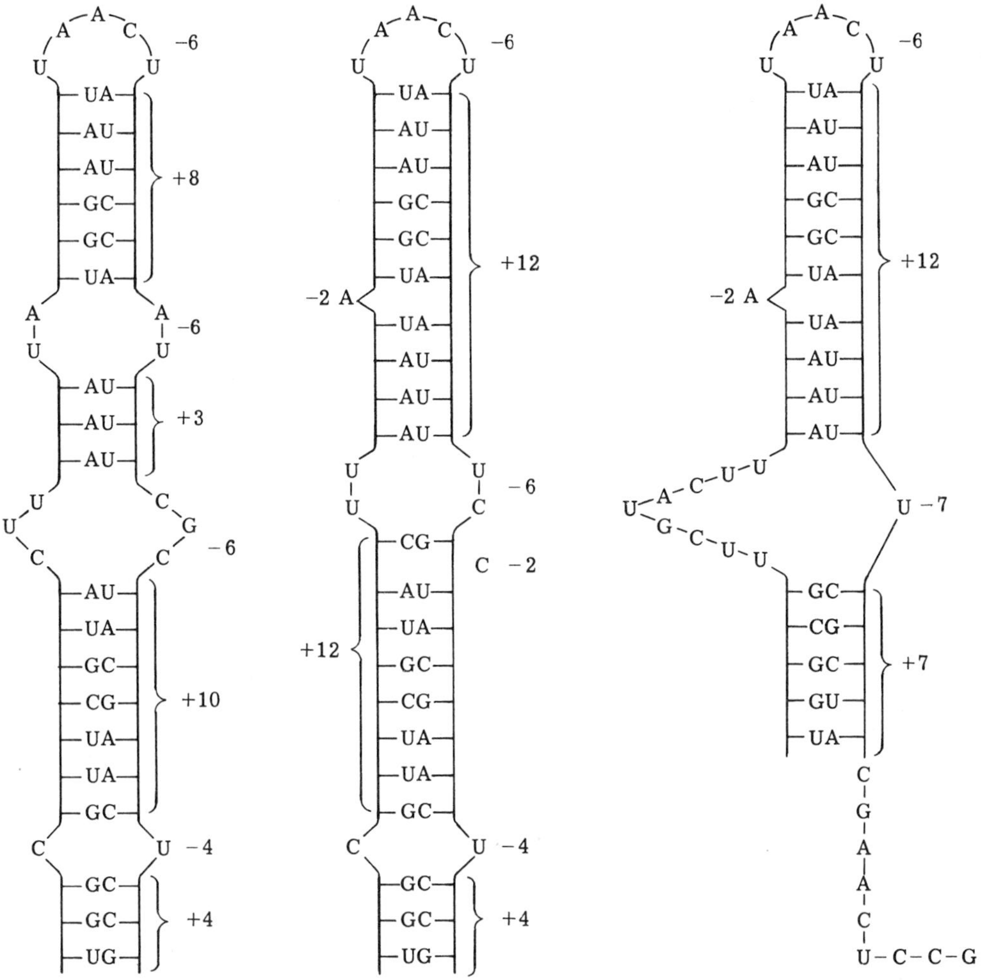

Sum +3
ΔG = −3.6 kcal
(I)

Sum +8
ΔG = −9.6 kcal
(II)

Sum +4
ΔG = −4.8 kcal
(III)

*Fig. 9.9 Three possible structures of a fragment of viral RNA.*

The spatial structure of tRNA is characterized by the compactness of the molecules in the native state. The positions of the helical regions are fixed, probably as a result of the interaction of the nonhelical regions. In this sense the tRNA molecule is similar to a protein globule. For maintenance of the spatial structure Mg ions are required. Thermally denatured tRNA can be renatured. These features of the structure were discovered by hydrodynamic methods and low-angle X-ray scattering. (Cf. in particular [58,59].)

tRNA has been obtained in a crystalline form and its crystal structure has been subjected to X-ray analysis [60-62]; the 12 Å resolution gave molecular dimensions of 80×25×35 $Å^3$ for formylmethionine tRNA from *E. coli*. High resolution (2.3 Å) was used in studies of crystals of Phe-tRNA from yeast. The molecular dimensions are 80×33×28 $Å^3$, and the dimensions of an orthorhombic unit cell are $a = 33.2$, $b = 56.1$, and $c = 161$ Å. The diffraction pattern along the a axis (but not that along the b axis) is typical of that of double-helical nucleic acids. Apparently the length of the double-helical regions of the tRNA molecule correspond to not less than half a turn of the helix (i.e., they contain more than 4-7 base pairs). These results agree with the cloverleaf structure [63].

tRNA crystallizes in various elementary cells; different crystalline forms can be obtained under changed conditions of crystallization.

Later Rich and co-workers deciphered the complete spatial structure of Phe-tRNA using X-ray diffraction with 4 Å resolution [152]. The tertiary structure contains two helical segments forming an angle near 90°; thus the molecule is similar to the letter Γ. The anticodon hairpin is connected in a helical system with the hairpin, containing the dehydrouridyl loop. This structure is shown in Fig. 9.10.

The compact structure of tRNA is stabilized by the horizontal and vertical interactions of the bases. Their packing is also due to hydrophobic interactions and Coulombic repulsion of the charged phosphate groups. Investigations of tritium exchange in the reaction of formaldehyde with tRNA showed the existence in the tRNA structure of a stable two-stranded framework (48 slowly exchangeable protons, which agrees quantitatively with the cloverleaf model), surrounded by the bases, forming the more labile tertiary structure [64].

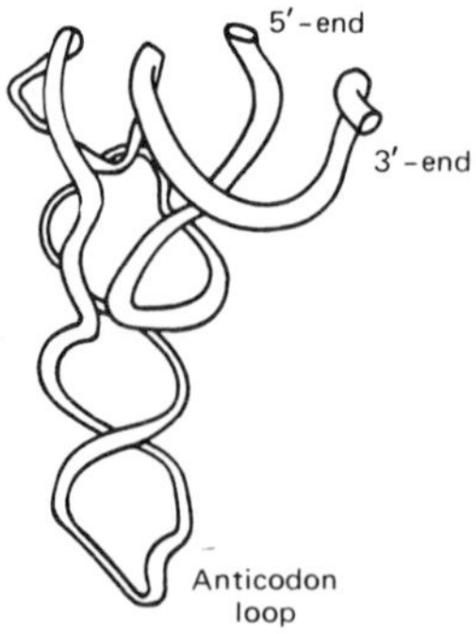


*FIG. 9.10 Spatial structure of Phe-tRNA.*

The Mg ions neutralize the negative charges of the phosphate groups and provide the maximum stabilization of the compact molecular structure, which has a specific optical activity [65,66]. The CD spectrum of tRNA contains a pronounced positive maximum at 264 nm ($\Delta\varepsilon \sim 5.5$) and minima at 296 nm ($\Delta\varepsilon = -0.35$), 233 nm ($\Delta\varepsilon = -1.0$), and 210 nm ($\Delta\varepsilon \sim -5.0$). In the short-wave region another intense positive band is noted at 188 nm ($\Delta\varepsilon \sim 18$). The amplitude of this band is very susceptible to structural changes [67,68]. The amplitudes of the 264-nm, 210-nm and 296-nm bands also "feel" conformational changes in tRNA [67-69]. If the molecules are partly disordered (low ionic strength, increased temperature, absence of Mg ions, presence of an organic solvent), the 296-nm band vanishes, the amplitude of the positive band decreases by 20-25%, and that of the 210-nm band by 50-55% [67, 70]. The loss of ordered structure is accompanied by a shift of the positive band toward the long-wave end [71]. It has been shown [70,72] that this effect, like the vanishing of the 296-nm band, is due to changes in the mutual orientation of the pentoses and bases, and of the bases. Investigation of the temperature dependence of the CD spectra of tRNA showed several gradual changes in molecular conformation in the melting process and again confirmed a secondary structure scheme like the cloverleaf (cf. [70-73]).

The ordering of the structure of the tRNA molecule is accompanied by the appearance of hypochromism (up to 22%) [64], an increase in the number of slowly exchangeable protons [74], a decrease in the number of bases accessible to formaldehyde [64] and to various nucleases [75], an increase in the sedimentation coefficient, and a decrease in the intrinsic viscosity. Complete stabilization by Ca and Mg ions occurs at a ratio of their concentration to the phosphate concentration equal to 1:2 [70].

The melting of tRNA molecules in the absence of Mg ions possesses a noncooperative character; the width of the melting range $\Delta T$ is $\sim 55°C$ [76]. Addition of Mg ions makes the process a cooperative one ($\Delta T \sim 18°C$) and increases the melting point to 75-80°C [76]. Theoretical analysis of the melting curves again confirms the cloverleaf structure [64,77].

When tRNA is heated, structural changes are observed at 30-40°C (i.e., before melting starts). These changes were observed upon applying hydrodynamic methods [78], and when the luminescence polarization of tRNA-bound dye molecules was investigated. Apparently some degrees of freedom arise in tRNA molecules before melting, allowing separate displacements of the molecule fragments. Results of luminescence polarization studies show that tRNA molecules in solution have ellipsoidal shapes, with the axes ratio 1:4 [79].

Investigations of energy migration between the luminescent base, located in the anticodon loop of Phe-tRNA molecules, and the acridine-type chromophores covalently bound at the acceptor end, made it possible to estimate the distance between them [80]. This distance happens to be more than 40 Å. The results of investigations of the energy migration between molecules of acridine dyes adsorbed by a tRNA molecule favor a compact tertiary structure [81].

The anticodon triplet is located in the "upper leaf" of the cloverleaf. As a rule it contains the minor nucleotide, inosine I. Abraham suggested a model for Phe-tRNA containing as many vertical interactions as the cloverleaf, but arranged in one continuous double helix with defects. The main argument of this work is the necessity for joint packing of two tRNA's, bound by neighboring mRNA codons, in a ribosome [82]. At present, this model lacks any structural confirmation.

The natural way to elucidate the structural and functional properties of tRNA consists in enzymic splitting of tRNA and in studies of the properties of the fragments thus obtained. Bayev and co-workers [83] showed that the halves of a yeast Val-$tRNA_1$ molecule, obtained by splitting the phosphodiester bond in the anticodon, are biologically inactive. When mixed, they resume acceptor activity (they can bind valine). On the other hand, splitting this bond irreversibly destroys the anticodon's ability to interact with the codon [84]. It has since been stated that the regeneration of activity is preceded by self-assembly of the fragments; the aggregated tRNA molecule can even lose a part of the polynucleotide chain [85,86]. During self-assembly the 3' and 5' halves aggregate in an equimolar ratio and the self-assembly takes less than 2 minutes. It is quite specific; the 3' and 5' halves of Val-tRNA form only a homologous complex in the presence of the 5' halves of other tRNA's. Apparently every tRNA has its own nucleotide sequence. The halves have no affinity for Valyl-tRNA-synthetase. Bayev suggests that the structural Val-tRNA regions contained in the halves do not participate directly in recognizing Valyl-tRNA-synthetase, but the helical parts of the anticodon and acceptor branches are necessary to an active tRNA conformation.

The spatial structure of tRNA is also investigated by indirect methods. Acridine dyes are adsorbed by the double-helical regions of the nucleic acids by pairs between neighboring A-U pairs (or A-T in DNA. By investigating the quenching of luminescence due to energy migration between the adsorbed molecules, the relative positions of the double pairs $(A\text{-}U)_2$ can be established (cf. [73]).

Isoacceptor tRNA's differ in their primary structure. Obviously this heterogeneity is due to more than the degeneracy of the genetic code (cf, Section 9.6). Species variations are apparent in the structure of tRNA. It has been established that the minor bases have a secondary, not a genetic origin;

tRNA's have been methylated by the action of an enzyme (methylase).

## 9.5 Ribosomes

Ribosomes are directly observed by means of the electron microscope as roundish granules with a diameter of 150-300 Å. The physical characteristics of *E. coli* ribosomes are presented in Table 9.1 [87].

*TABLE 9.1*
*Physical Properties of E. coli Ribosomes*

| | |
|---|---|
| Sedimentation coefficient $S_{20°,w}$ (Svedbergs) | 69.1-70.5 |
| Intrinsic viscosity [η] ($cm^3\ g^{-1}$) | 6.1-6.8 |
| Coefficient of translational diffusion $D_{20°,w}$ ($\times 10^7\ cm^2\ sec^{-1}$) | 1.83 |
| Partial specific volume ($cm^3\ g^{-1}$) | 0.64-0.60 |
| Molecular weight | $3\times10^6$ |
| Dimensions in the dried state (Å) | 200×170×170 |
| Volume ($Å^3$) | $3\times10^6$ |
| Dimensions in aqueous medium (Å) | 300×300×200 |
| Volume ($Å^3$) | $7\text{-}10\times10^6$ |
| Probable amount of retained water w ($g\ g^{-1}$) | 0.9 |

The ribosomes of eucaryotes (cells of higher organisms which contain nuclei) are characterized by a larger $S_{20°,w}$ value (80 S instead of 70 S), MW = $4.1\text{-}4.7\times10^6$, and larger size (dry volume $5\times10^6$). The ribosomes of mitochondria and chloroplasts of eucaryotes belong to the 70 S and not to the 80 S type.

70 S ribosomes contain two subunits, 30 S and 50 S. They dissociate into subunits upon a decrease in the concentration of bivalent ions ($Mg^{2+}$) or an increase in the monovalent ion concentration. The MW of 50 S particles is twice that of 30 S particles ($1.8\times10^6$ and $0.7\text{-}1.0\times10^6$), respectively. The shapes and sizes of the 70 S, 30 S, and 50 S particles are shown in Fig. 9.11. Similarly, 80 S ribosomes are formed by 60 S and 40 S particles. The presence of Mg ions is necessary for the maintenance of the ribosome structure [88,89]. There are two

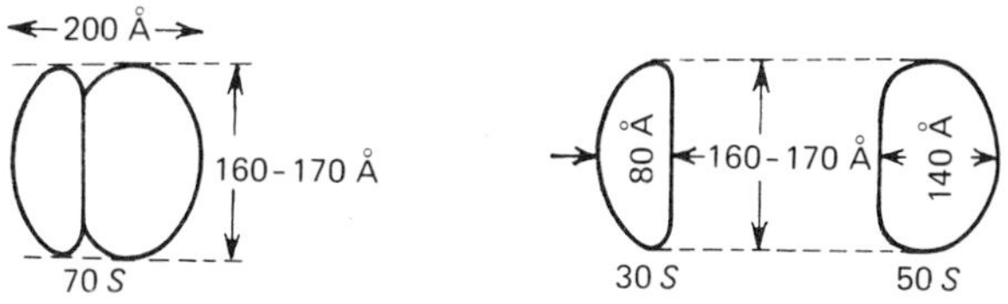


*FIG. 9.11 Scheme of 70 S, 30 S, and 50 S particles.*

critical levels of the $Mg^{2+}$ content. If it is lower than the first one, the ribosomes dissociate into subunits; if it is lower than the second, the subunits themselves lose their compact structure. The first level is 0.5 μmol of $Mg^{2+}$ per micromole of phosphate for 70 S particles and 0.3-0.1 for 80 S particles (i.e., 2500 - 3000 atoms of Mg per ribosome). The second level is 0.3 for 70 S and 0.15 for 80 S (i.e., less than 1000 atoms of Mg per ribosome) [88].

Ribosomal RNA forms about 65% of the dry weight of a ribosome, proteins about 35%. rRNA's can be divided into three classes: high-molecular-weight, 23 S - 28 S (MW > $1\times10^6$) and 16 S - 18 S (MW < $1\times10^6$), and low-molecular-weight 5 S (MW ∿ 40,000). In the native state these components of rRNA are continuous chains [90]. The nucleotide composition of various rRNA components is different, and it does not obey Chargaff's rules. Investigations of the structural transformations of ribosomes lead to hypotheses about the structure of rRNA in solution and about its interactions with proteins [87,91]. Apparently the secondary structure of rRNA can be characterized by many short double-helical regions connected by intermediate single-stranded regions, forming a flexible "stick." The protein molecules interact with the nonhelical rRNA regions, not destroying its secondary structure. The ribonucleoprotein system obtained in this way is folded into the compact structure of a ribosomal subunit. The 70 S ribosome contains approximately 65 polypeptide chains with an average MW of 65,000; the number of such chains in the 80 S ribosome is twice as great. In 30 S particles there are approximately 19 - 20 kinds of proteins; in 50 S this number is more than 50. These proteins are partly globular, and they contain α-helical regions.

Ribosomes are subject to three types of structural transformation: reversible dissociation into two subunits; unfolding of subunits; and disassembling of subunits. As we said earlier, dissociation can be produced by a decrease in the Mg ion concentration. Electron microscopy shows that the associating subunits interact via definite regions on their surfaces. The role of Mg (and Ca) ions probably consists in the screening of negative charges of phosphate and carboxyl groups. The interaction of subunits in a ribosome has not yet been studied in detail. There are data showing the existence in the cell of a pool of free subunits, in equilibrium with the nonfunctioning ribosomes, in which the bond of the subunits is not sufficiently stable. This bond is stabilized by interaction with the components of a protein-synthesizing system, particularly with tRNA [92]. Spirin and Gavrilova emphasized the role of the labile association of two nonidentical subunits in a ribosome [87].

The unfolding of subunits due to the removal of Mg ions and to an increased concentration of monovalent ions was discovered in studies of *E. coli* ribosomes [93-95]. The unfolding occurs stepwise, via discrete stages, each stage occurring in a narrow range of ionic surroundings. These jump-like transitions testify in favor of the cooperativity of the processes. Finally, a loose, nonordered ribonucleoproteide is formed.

The disassembling of ribosomal particles consists in the separation of protein from rRNA due to an increase in the concentration of monovalent ions. These processes also have a cooperative and multistage character. Of great interest is the possibility of reverse self-assembly of ribosomal particles due to appropriate changes in the medium. The reconstructed particles are biologically functional [96,97]. According to Spirin, these phenomena are due to the specific role of rRNA, which serves "as a framework or template for the specific accommodation of the molecules of ribosomal protein" [87]. The self-assembling is based on recognition of an RNA region by a molecule of the definite protein.

The complete disassembling of ribosomes and their reconstruction from the proteins obtained and rRNA was effected by Nomura and co-workers [98-100]. The role of RNA is specific but not absolutely so; functional 30 S subunits can be obtained from the 16 S rRNA of one bacterial species, and from the ribosomal 30 S proteins of another. Nineteen proteins have been extracted. It has been established that the susceptibility of the 30 S particle to streptomycin (hindering the correct "reading" of the mRNA "text") is determined by only one protein, denoted as P10. It turned out that particles devoid of P10 "read the text" in a more precise way in the presence of substances that hinder this reading, which include not only antibiotics, but also $C_2H_5OH$ and $Mg^{2+}$ in high concentration. The absence of any one of the 19 proteins influences a series of functions of 30 S particles. Several of their different functions are changed simultaneously in the absence of any one of the 19 proteins. Consequently, all the proteins act in an interrelated way. This coordination is also necessary for self-assembling.

The kinetics of self-assembling corresponds to a first-order reaction. This implies the occurrence of a slow reconstruction of some intermediate complex which limits the process. The strong temperature dependence of the reaction rate shows that the reconstruction requires $\sim$40 kcal $mole^{-1}$ of ribosome. The self-assembling of a ribosome from all components *in vitro* is completed in 5 min.

Genetic analysis of *E. coli* mutants characterized by various defects in the ribosomal self-assembly process shows that self-assembling *in vivo* proceeds in a manner similar to that *in vitro*. The molecular mechanisms of these interesting and important phenomena have not yet been studied.

Investigation of ribosome functioning is a current problem. The presence of special proteins (transfer factors) is necessary for ribosome work (cf. [87]). Three fractions of transfer factors are known: G factor (depending on GTP), $T_c$ (stable factor), and $T_n$ (nonstable factor) [101]. In the presence of G factor, GTPase activity appears. Transfer factors unite with a ribosome if bivalent cations are present. The structure and role of the transfer factors have not been studied sufficiently. The presence of protein factors affecting initiation and termination of polypeptide chain synthesis has also been established (cf. [87]).

For the complex building of ribosomes with mRNA or artificial polynucleotides an $Mg^{++}$ concentration no lower than 0.005 M is required. Every ribosome combines with only one polynucleotide chain. Binding occurs in the 30 S subunit. In the triple complex ribosome mRNA-aminoacyl-tRNA the last component is bound in a strictly selective way: the tRNA anticodon combines with the mRNA codon. Formation of the strong codon-anticodon pair also occurs in the 30 S subunit. In the absence of template, deacylated tRNA or aminoacyl-tRNA binds with the 50 S subunit at some peptidyl-tRNA binding region.

The translation process starts by initiation (i.e., by synthesis of the first peptide bond). Two aminoacyl-tRNA's must be simultaneously involved in the initiation. One of them happens to be initiating. It has been established that this is always N-formylmethionyl-tRNA, which has an enhanced affinity for the peptidyl-tRNA binding region of the ribosome localized on the 50 S subunit [101]. AUG or GUG, located at the 5' end of the polynucleotide, serves as an initiating codon of mRNA. Apparently, this codon is previously bound by the aminoacyl-tRNA binding region of the 30 S subunit. The general scheme of the initiation process is (cf. [87])

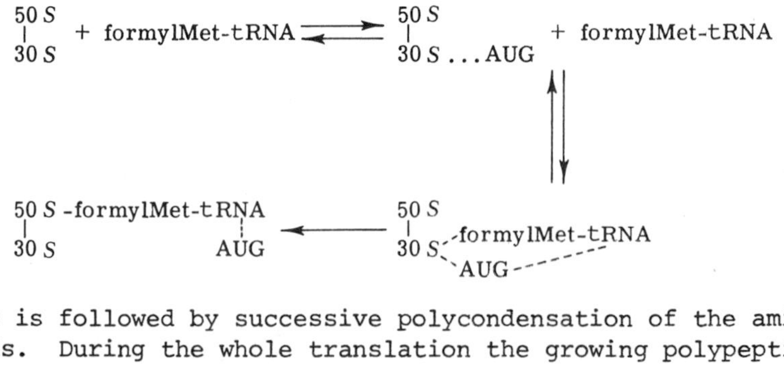


This is followed by successive polycondensation of the amino acids. During the whole translation the growing polypeptide is retained by the ribosome. The addition of each successive aminoacyl occurs at the C end of the polypeptide tRNA which has brought the last added aminoacyl, remains bound to it. The next aminoacyl is bound via the replacement of tRNA by

*FIG. 9.12 Probable scheme of the ribosome working cycle.*

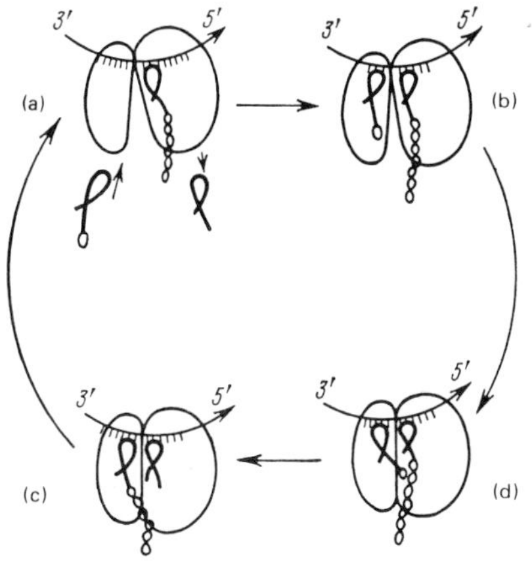


aminoacyl-tRNA. The general scheme of translation (cf. Fig. 9.12) is

peptidyl(n)-tRNA'+aminoacyl-tRNA" → peptidyl(n+1)-tRNA"+tRNA'

This cycle is reproduced many times. Its definitive description was provided by Watson [2,102]. Assume that at some arbitrary moment the peptidyl-tRNA is at the peptidyl-tRNA binding site of the 50 S particle, and at the 30 S particle is the aminoacyl-tRNA (Fig. 9.12b). Also on the 50 S particle is the peptidyl transferase center, providing peptidyl transfer. In Fig. 9.12c the aminoacylic end of aminoacyl-tRNA is close to the esterified carboxyl of the peptidyl; later this carboxyl is transferred to the amino group of aminoacyl-tRNA and deacylated tRNA remains in the 50 S particle (Fig. 9.12d). Then translocation occurs; the tRNA residue of the molecule of newly formed peptidyl-tRNA transfers from the 30 S to the 50 S particle, along with the bound mRNA codon, and supplants the deacylated tRNA from the peptidyl-tRNA binding site (Fig. 9.12a). Then the new aminoacyl-tRNA approaches the new codon located in the 30 S particle and the cycle begins again. The result of each cycle is the formation of one peptide bond and the displacement of mRNA relative to the ribosome by one codon. It can be suggested that translocation occurs as a result of a conformational change in the ribosome, the required energy comes from GTP. A mechanochemical process occurs: the work of mRNA displacement occurs on account of chemical energy.

As has been said, one mRNA molecule combines with several ribosomes, forming a polysome. The sizes of the polysomes depend on the lengths of the mRNA chains. At maximum use of mRNA one ribosome corresponds to every 80 nucleotides of mRNA. During the synthesis of proteins containing approximately 150 amino acids, the polysome contains 4-6 ribosomes; during synthesis of longer protein chains, 12-20 and more ribosomes. The presence of polysomes explains the relatively low content of mRNA in the cell (1-2% of the total amount of RNA), since one mRNA chain can assist in the synthesis of many polypeptide chains. For further description of translation see Section 9.8.

## 9.6 The Genetic Code

The code was deciphered by synthetic, not analytic, methods. The analytic path is closed for the present, because the primary structure of the template (DNA or mRNA) corresponding to the synthesized protein chain with the known sequence of amino acid residues has not been established.

The path to the solution of the problem began with the work reported by Nirenberg and Mattheai at the Fifth International Congress of Biochemistry in Moscow in 1961 [103]: If synthetic polyribonucleotides are introduced into a cell-free system, amino acids are incorporated into the polypeptide chain. The cell-free system contains ribosomes; the set of tRNA, ATP, and GTP; and all the necessary enzymes; but it does not contain DNA and mRNA. The system was prepared from broken *E. coli* cells. The ribosomal and supernatant fractions were extracted by centrifugation. The ribosomes were washed out, and the supernatant liquid containing the acylating enzymes and tRNA's was dialyzed using a special buffer which included mercaptoethanol to stabilize the system. To the mixture of these two purified fractions were added the ATP-generating system and 20 amino acids. RNA or synthetic polynucleotides were introduced into the cell-free system, and the incorporation of $^{14}C$-labeled amino acid into the fraction that was insoluble in trichloroacetic acid (i.e., into the polypeptides) was investigated. It turned out that much more label was incorporated in the presence of RNA than in its absence. The incorporation is suppressed by the RNA-destroying enzyme (ribonuclease) as well as by puromycin and chloramphenicol. It was found that poly-U stimulates the incorporation of $^{14}C$-phenylalanine. If poly-U was absent, there were 44 radioactive counts in 1 mg of polypeptide, while in the presence of poly-U 39,800 counts were observed.

Thus in these experiments the biological system was "fooled," that is, instead of natural mRNA, a synthetic poly-U template interacted with ribosomes. Poly-U codes phenylalanine. If the code is a triplet code, then Phe corresponds to the UUU codon. Nirenberg and Matthaei found that poly-C stimulated the incorporation of proline. Thus, the CCC codon was established for Pro.

Work with copolymers that have regular sequences (e.g.,

•••AUAUAU•••) proved impossible since these polymers form double-helical structures which do not have template activity [104]. Poly-G was also inactive. Polynucleotides with random double-helical structures do not bind ribosomes. Further, the action of copolymers of a known composition but with a random nucleotide sequence have also been investigated (cf. [1,5,7]). For such polymers the frequency of occurrence of triplets having a definite composition can be readily calculated. Copolymer AU in a 1:5 ratio stimulates the incorporation of Phe, Leu, Ile, and, in smaller amounts, Asn and Lys. If 100 is the probability of a 3U triplet, the probability of 1A2U is 20 (3U/1A2U=5); that of a 2A1U is 4 (3U/2A1U=25); and that of a 3A is 0.8 (3U/3A=125). Comparing these values with the degree of incorporation (that for Phe is taken as 100), we can establish the composition of codons for named amino acids, as is shown in Table 9.2. Thus, these triplets were established: 3U for Phe; 1A2U for Tyr; 2A1U and 1A2U for Ile; 1A2U for Leu, 2A1U for Asn; 3A and 2A1U for Lys. The last triplet has since been found to be erroneous; the others have been confirmed.

*TABLE 9.2*
*Determination of Triplets in Experiments with Poly-AU (1:5)[a]*

| Amino acid | Calculated probabilities of triplets | | | | Sum of probabilities | Incorporation of amino acid (conventional units) |
|---|---|---|---|---|---|---|
| | 3A | 2A1U | 1A2U | 3U | | |
| Asn | -- | 4 | -- | -- | 4 | 6.6 |
| Ile | -- | 4 | 20 | -- | 24 | 20 |
| Leu | -- | -- | 20 | -- | 20 | 15 |
| Lys | 0.8 | 4 | -- | -- | 4.8 | 3.1 |
| Tyr | -- | -- | 20 | -- | 20 | 25 |
| Phe | -- | -- | -- | 100 | 100 | 100 |

[a]From [3].

However, these experiments do not completely decipher the code. It is not known which of the three codons AUU, UAU, or UUA codes Tyr, etc. A detailed history of the code investigation was set forth by Yčas [5]. The complete deciphering of the code was presented in the works of Nirenberg, in which trinucleotides of known structure were applied instead of polynucleotides [105]. Trinucleotide-tRNA-amino acid complexes are formed. Polypeptide synthesis does not occur, but since the trinucleotide imitates the codon, the complex formation enables us to read it. All tRNA's that bind $^{14}C$-labeled amino acids must be examined. Thus, all 64 triplets were investigated and which amino acids bound with them was established. The definitive deciphering, which confirmed these results, was performed by Khorana [106,107]. Oligodeoxyribonucleotides were synthesized that contained from 8 to 12 links in the chain with known sequence. They contained repeated triplets (e.g., $(TTC)_4$). Such an oligomer was used as a template for the *in vitro* synthesis of a DNA-like polymer in a system which contained the nucleoside triphosphates ATP, CTP, GTP, TTP; the necessary amount of ions; and DNA-polymerase (cf. p. 537). A DNA-like double helix was synthesized in this way. Both strands of the helix contained complementary triplets repeated n times. Every strand was later used as the template for polyribonucleotide synthesis with the help of RNA-polymerase. In this way Khorana obtained two chains that imitated mRNA with a known sequence of the repetitive triplets. These chains were introduced into a cell-free system and the incorporation of amino acids in the polypeptide fraction was determined according to Nirenberg's method. These elegant experiments made it possible to check six codons in one multistage synthesis according to the scheme shown.

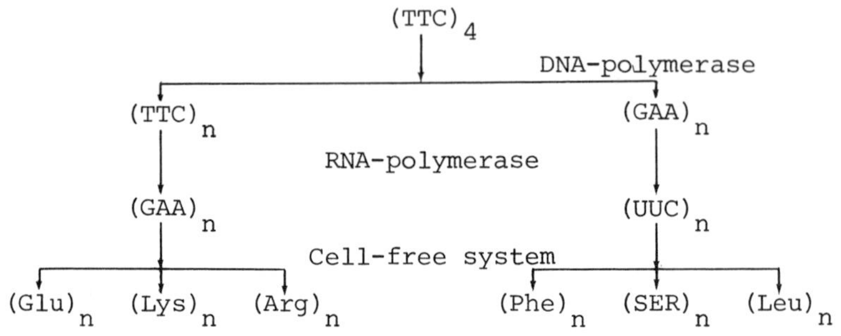


The polymer $(GAA)_n$ contains the codons GAA, AAG, and AGA, and the polymer $(UUC)_n$ the codons UUC, UCU, and CUU. Since the functionality of homopolymers $(AAA)_n$, $(CCC)_n$, $(GGG)_n$, and $(UUU)_n$ has already been established, checking the remaining 60 codons requires 10 multistage syntheses.

Tables 9.3 and 9.4 present the codon-amino acid and the reverse amino acid-codon dictionaries established by Nirenberg and co-workers and by Khorana and co-workers.

The codons GUG (Val) and AUG (Met code the designated amino acids in the middle of the chain. On the other hand, they serve as initiators of chain synthesis, coding formylmethionine at the N end of the chain

*TABLE 9.3*
*Codon-Amino Acid Dictionary*

| | | | | | | | | | | | |
|---|---|---|---|---|---|---|---|---|---|---|---|
| 1 | AAA | Lys | 17 | CAA | Gln | 33 | GAA | Glu | 49 | UAA | -- |
| 2 | AAC | Asn | 18 | CAC | His | 34 | GAC | Asp | 50 | UAC | Tyr |
| 3 | AAG | Lys | 19 | CAG | Gln | 35 | GAG | Glu | 51 | UAG | -- |
| 4 | AAU | Asn | 20 | CAU | His | 36 | GAU | Asp | 52 | UAU | Tyr |
| 5 | ACA | Thr | 21 | CCA | Pro | 37 | GCA | Ala | 53 | UCA | Ser |
| 6 | ACC | Thr | 22 | CCC | Pro | 38 | GCC | Ala | 54 | UCC | Ser |
| 7 | ACG | Thr | 23 | CCG | Pro | 39 | GCG | Ala | 55 | UCG | Ser |
| 8 | ACU | Thr | 24 | CCU | Pro | 40 | GCU | Ala | 56 | UCU | Ser |
| 9 | AGA | Arg | 25 | CGA | Arg | 41 | GGA | Gly | 57 | UGA | -- |
| 10 | AGC | Ser | 26 | CGC | Arg | 42 | GGC | Gly | 58 | UGC | Cys |
| 11 | AGG | Arg | 27 | CGC | Arg | 43 | GGG | Gly | 59 | UGG | Trp |
| 12 | AGU | Ser | 28 | CGU | Arg | 44 | GGU | Gly | 60 | UGU | Cys |
| 13 | AUA | Ile | 29 | CUA | Leu | 45 | GUA | Val | 61 | UUA | Leu |
| 14 | AUC | Ile | 30 | CUC | Leu | 46 | GUC | Val | 62 | UUC | Phe |
| 15 | AUG | Met | 31 | CUG | Leu | 47 | GUG | Val | 63 | UUG | Leu |
| 16 | AUU | Ile | 32 | CUU | Leu | 48 | GUU | Val | 64 | UUU | Phe |

*TABLE 9.4*
*Amino Acid-Codon Dictionary*[a]

| | | | |
|---|---|---|---|
| 1 | Ala | GCA, GCC, GCG, GCU | (4) |
| 2 | Arg | AGA, AGG, CGA, CGC, CGG, CGU | (6) |
| 3 | Asn | AAC, AAU | (2) |
| 4 | Asp | GAC, GAU | (2) |
| 5 | Cys | UGC, UGU | (2) |
| 6 | Gln | CAA, CAG | (2) |
| 7 | Glu | GAA, GAG | (2) |
| 8 | Gly | GGA, GGC, GGG, GGU | (4) |
| 9 | His | CAC, CAU | (2) |
| 10 | Ile | AUA, AUC, AUU | (3) |
| 11 | Leu | CUA, CUC, CUG, CUU, UUA, UUG | (6) |
| 12 | Lys | AAA, AAG | (2) |
| 13 | Met | AUG | (1) |
| 14 | Phe | UUC, UUU | (2) |
| 15 | Pro | CCA, CCC, CCG, CCU | (4) |
| 16 | Ser | AGC, AGU, UCA, UCC, UCG, UCU | (6) |
| 17 | Thr | ACA, ACC, ACG, ACU | (4) |
| 18 | Trp | UGG | (1) |
| 19 | Tyr | UAC, UAU | (2) |
| 20 | Val | GUA, GUC, GUG, GUU | (4) |

[a]The last column shows the degeneracy, the number of codons corresponding to the given amino acid.

$$H_3C-S-CH_2-CH_2-\overset{H}{C}\begin{matrix} \diagup NH-CO-H \\ \diagdown COOH \end{matrix}$$

If the artificial template does not contain these codons, then chains that start at an arbitrary link, and therefore have different N ends are synthesized. Conversely, in the presence of GUG or AUG the standard formylmethionine N ends are formed. Khorana showed that poly-UG formed the chain formylMet-(Cys-Val)$_n$. However, natural proteins do not contain formylmethionine. In a cell-free system of *E. coli* chains are formed with N ends: formylMet-Ala-formylMet-Ser but not formylMet-Met. The natural proteins of *E. coli* usually have Met, Ala, Ser, at the N ends. It can be theorized that two enzymes act in living systems: one splits formylMet from the chain, the other splits the formyl group from Met (cf. [107], the paper of Zinder et al.).

The UAA ("ochre"), UAG ("amber"), and UGA ("opal") codons do not code any amino acid; the protein chain is broken by each of these codons. These are the "senseless" codons. Khorana found that protein synthesis did not proceed on (GAUA)$_n$ and (GUAA)$_n$ templates. These templates contain many nonfunctioning triplets.

The established code is confirmed by many facts (cf. [5, 7]).

Even before the code was deciphered, Sueoka investigated the correlation of the composition of complete bacterial protein with the composition of bacterial DNA [108]. He found a linear dependence of the content of 16 amino acids on the GC content in DNA. The amino acids can be divided into three groups. In the first group the amino acid content increases with the GC content; the correlation coefficient b is large. In the second group the amino acid content is practically independent of the GC content and b is close to zero. In the third group the amino acid content decreases with an increase in GC and b is negative. These results can be well explained by the code dictionary; this is shown in Table 9.5 [1,109].

Is the code universal? Do the same codons act in the same way in different organisms? The answer to this question is positive. Poly-U stimulates the incorporation of Phe in the polypeptide chain in cell-free systems obtained from the cells of mammals and of algae. The same is true of the other synthetic polynucleotides (cf. [5]). The trinucleotide technique of Nirenberg was applied in cell-free systems from cells of the amphibia *Xenopus laevis* and of the guinea pig with the same results [110]. Apparently only the relative participation of the various codons belonging to the same amino acid is altered, but the code dictionary remains the same as in the case of *E. coli*.

The multiplication of phages and viruses in cells can be

*TABLE 9.5*
*Correlation of Amino Acid Content of Protein with GC Content of DNA*

| Amino acid residue | Correlation coefficient b | Codons | Number of G+C in codons | Mean fraction of G+C in codons |
|---|---|---|---|---|
| | | Group 1, b > 0 | | |
| Ala | 0.164 | GCA, GCC, GCG, GCU | 2, 3, 3, 2 | 0.83 |
| Arg | 0.089 | AGA, AGG, CGA, CGC, CGG, CGU | 1, 2, 2, 3, 3, 2 | 0.72 |
| Gly | 0.051 | GGA, GGC, GGG, GGU | 2, 3, 3, 2 | 0.83 |
| Pro | 0.024 | CCA, CCC, CCG, CCU | 2, 3, 3, 2 | 0.83 |
| Average | 0.082 | | 2.39 | 0.79 |
| | | Group 2, b ≃ 0 | | |
| Val | 0.008 | GUA, GUC, GUG, GUU | 1, 2, 2, 1 | 0.50 |
| Thr | 0.000 | ACA, ACC, ACG, ACU | 1, 2, 2, 1 | 0.50 |
| Leu | -0.006 | CUA, CUC, CUG, CUU, UAA, UUG | 1, 2, 2, 1, 0, 1 | 0.39 |
| His | -0.010 | CAC, CAU | 2, 1 | 0.50 |
| Ser | -0.017 | AGC, AGU, UCA, UCC, UCG, UCU | 2, 1, 1, 2, 2, 1 | 0.50 |
| Average | -0.005 | | 1.41 | 0.47 |
| | | Group 3, b < 0 | | |
| Met | -0.024 | AUG | 1 | 0.33 |
| Phe | -0.040 | UUC, UUU | 1, 0 | 0.17 |
| Tyr | -0.047 | UAC, UAU | 1, 0 | 0.17 |
| Gln } | -0.052 | GAA, GAG | 1, 2 | 0.50 |
| Glu } | | CAA, CAG | 1, 2 | 0.50 |
| Asn } | -0.053 | GAC, GAU | 2, 1 | 0.50 |
| Asp } | | AAC, AAU | 1, 0 | 0.17 |
| Lys | -0.084 | AAA, AAG | 0, 1 | 0.17 |
| Ile | -0.098 | AUA, AUC, AUU | 0, 1, 0 | 0.11 |
| Average | -0.057 | | 0.83 | 0.28 |

considered as proof of the code's universality. Viral DNA or RNA uses the cell's biosynthetic apparatus and synthesizes its own proteins. The protein is coded by the viral nucleic acid but the "cell-free system" is quite another one.

The methods developed by Khorana permitted researchers to solve the problem of gene synthesis. Khorana performed the chemical synthesis of the deoxynucleotide sequence complementary to the known sequence of ribonucleotides in the yeast Ala-tRNA [111].

Double-helical DNA was obtained (i.e., the gene that codes the synthesis of this tRNA). Synthesis of the first DNA chain was performed by the addition of oligonucleotides to the growing chain with the participation of enzymes, DNA ligases, and ATP as the energy source. The second chain was obtained in reduplicative synthesis (cf. p. 586). The functionality of the synthesized gene has not yet been verified, but that verification seems beyond question.

The discovery of the genetic code is a great achievement of molecular biology, biochemistry, and biophysics. Only 12 years elapsed between the formulation of the problem and its complete solution.

New problems have arisen because of the deciphering of the genetic code. It must be investigated whether the genetic dictionary has real physical, molecular significance, or whether the correlation between the codons and amino acids is quite accidental. What can be said about the evolution of the code in this respect? What factors influence the "reading" of the code, the processes of transcription and translation? What disturbs the code, and in what manner? What are the physicochemical mechanisms of mutations? With these questions we approach the general problem of biological regulation at the molecular level, which will be treated in "General Biophysics."

The code is a translation program. It is important to understand the conditions necessary for the realization of this program. Molecular mechanisms are responsible for the kinetics of codon-anticodon interaction, as well as for the action of ribosomes and the enzymes. At present we know almost nothing about these mechanisms and we have no physical theory for the reading of the code (cf. Section 9.8).

## 9.7 The Physical Sense of the Genetic Code

The primary structure of protein is genetically coded. However, the biological functions are determined by the spatial protein structure. The primary and spatial structures must be connected in a unique way (Section 4.9). Therefore, the spatial structure and biological function of protein are also coded genetically. On the other hand, natural selection involves not the primary but the spatial structure i.e., the biological behavior.

Various mutational substitutions influence protein structure in various ways. From the facts presented in Section 4.6 it follows logically that mutations which considerably alter the hydrophobicity of an amino acid residue must influence the biological properties of the protein more than mutations which produce only small changes in hydrophobicity. The first type of mutation is more threatening to the existence of a species than the second. It can be theorized that the genetic code is formed by nature in such a way that it provides advantages for mutations of the first type. These features of the code have to be elucidated.

This formulation of the problem was presented in [112], where it was shown that the code is very reliable as regards the unfavorable replacements of polar amino acids by nonpolar ones, and vice versa. A meaningful correlation does exist between codons and amino acids. Cf. also [113].

First, the character of mutational substitutions of amino acids has to be established. We have already seen that the internal amino acid residues in hemoglobin and myoglobin are variable but always remain nonpolar (p. 217). Margoliash turned his attention to the constancy of the positions of polar (basic) and nonpolar residues within the cytochrome c of five types of vertebrates [114]. He asked whether this is an evolutionary constancy, a result of the pressure of selection on the structural features important for the function of cytochrome, or even in part a result of genetic variability? Analysis of the genetic dictionary enables us to answer this question [112,115].

It has been noted already that the sequence of residues near the active site of hemoglobin obeys the same regularity. Most natural mutants of human hemoglobin are characterized by maintenance of the amino acid class; the most frequent substitution is that of a polar residue by another polar one. In many other cases most mutations retain the amino acid class.

This situation can have two alternative explanations. First, the "wrong" mutations, those that change the amino acid class, can be so dangerous for protein function that they are eliminated by natural selection. These mutations are lethal and therefore not observable. Second, the genetic code itself is arranged in such a way that it provides advantages for the "right" mutations. Let us consider the second possibility.

As we have seen (p. 215), the hydrophobicity of an amino acid residue can be estimated quantitatively, according to Tanford, by the change of free energy $\Delta G$ when an amino acid is transferred from $C_2H_5OH$ into water (Table 4.11). Calculate the mean difference of the $\Delta G$ values upon arbitrary replacement of any residue by any other one, applying the data in Table 4.11. The different frequencies of the residues in proteins are not taken into account. The mean difference over all 20 amino acids is $\Delta\Delta G = 1280$. The mean differences over amino acids of the first and second classes, introduced conventionally, over the hydrophobic (first 10) and hydrophilic (second 10) amino acids are equal to 805 and 392, respectively.

Let us present the dictionary of codons written as xyz by the square table in Fig. 9.13. The hydrophobic residues are shaded in the figure. If the second site of the codon is U (y = U), the residue is always hydrophobic.

| x \ y | A | C | G | ∩ | z |
|---|---|---|---|---|---|
| A | Lys | Thr | Arg | Ile | A |
| | Asn | Thr | Ser | Ile | C |
| | Lys | Thr | Arg | Met | G |
| | Asn | Thr | Ser | Ile | U |
| C | Gln | Pro | Arg | Leu | A |
| | His | Pro | Arg | Leu | C |
| | Gln | Pro | Arg | Leu | G |
| | His | Pro | Arg | Leu | U |
| G | Glu | Ala | Gly | Val | A |
| | Asp | Ala | Gly | Val | C |
| | Glu | Ala | Gly | Val | G |
| | Asp | Ala | Gly | Val | U |
| U | - | Ser | - | Leu | A |
| | Tyr | Ser | Cys | Phe | C |
| | - | Ser | Trp | Leu | G |
| | Tyr | Ser | Cys | Phe | U |

*FIG. 9.13 Table of the genetic code.*

If only one of the nucleotides xyz is replaced, we obtain the following distribution of "right" and "wrong" mutations.

| Substitution | | |
|---|---|---|
| in x | in y | in z |
| 120 right<br>54 wrong | 74 right<br>102 wrong | 156 right<br>20 wrong |

There are a total of 350 (i.e., 66.5%) right substitutions, and 176 (33.5%) wrong ones. The probability of a right mutation is twice as great as that of a wrong one.

Let us compute the mean difference in $\Delta G$ for the substitution of one nucleotide in a codon. We get: Substitution in x, 1000; in y, 1280; and in z, 340. If averaged over all three nucleotides, we get 870, which is considerably less than 1280 for the accidental substitution of an amino acid. The mean difference in hydrophobicity of the initial and substituting amino acid for 70 mutants of human hemoglobin is 834; the mean difference for six cytochrome *c*'s (of vertebrates and invertebrates) is 900; that for mutants of tryptophan synthetase A, 1030 [116]. The respective fractions of "right" substitutions in these three cases are 57.0%, 56.3%, and 61.3%. Analysis of 423 substitutions in six homologous proteins of various species (cytochrome *c*, hemoglobin, insulins A and B, and ferredoxin [117]) gives a mean $\Delta G$ value of 772. All these data agree with the code table.

The most dangerous substitution is that of the middle nucleotide y; it leads to the greatest changes in hydrophobicity. The least dangerous is the substitution of z.

This regularity expresses the physical sense of the code dictionary. The advantage of "right" mutations is determined by the protein structure in an aqueous medium, and therefore by the peculiar physical properties of water. Consequently, the correlation between codons and amino acids is dictated by the physics of water. The code has high but not complete reliability in relation to the "wrong" mutations.

Yčas states that both too large and too small mutation frequencies are disadvantageous for evolution; there must be an intermediate optimum frequency [5]. If the mutability determined by the code is less than optimal, the arguments presented here lose some weight. However, we know nothing about the optimum mutability, and have no grounds to think that the code does not correlate with it. The reliability of the code is an undoubted fact.

The molecular mechanism determining the correlation of the codons and amino acids is located in the acylating enzyme which provides the binding of the proper amino acid to the proper tRNA (i.e., the correlation between the anticodon of tRNA and aminoacyladenylate). The enzyme is apparently susceptible to both the amino acid and the structure of tRNA as a whole.

Rumer stated that 16 xy doublets can be grouped in two octets: the first one contains the xy doublets that define the coded residue uniquely and independently on n; the second contains xy's that code one residue if z = purine and another if z = pyrimidine [118]. In Table 9.6 we present a table of

*TABLE 9.6*
*Systematics of Codons*

| First octet | | | | | Second octet | | | | | |
|---|---|---|---|---|---|---|---|---|---|---|
| z = A, G, U, C | | | | | z = U, C | | | z = A, G | | |
| x | y | residue | group | n | residue | x | y | residue | group | n |
| C | C | Pro | 1 | 6 | Phe | U | U | Leu | 1 | 4 |
| C | U | Leu | 1 | 5 | Ile | A | U | Ile z=A<br>Met z=G | 1 | 4 |
| G | U | Val | 1 | 5 | Cys | U | G | -- z=A<br>Trp z=G | 2 1 | 5 |
| C | G | Arg | 2 | 6 | Tyr | U | A | -- | 1 | 4 |
| G | C | Ala | 2 | 6 | His | C | A | Gln | 1 2 | 5 |
| A | C | Thr | 2 | 5 | Asn | A | A | Lys | 2 1 | 4 |
| U | C | Ser | 2 | 5 | Asp | G | A | Glu | 2 | 5 |
| G | G | Gly | 2 | 6 | Ser | A | G | Arg | 2 | 5 |

the two octets (cf. [119]).

The compositions of the xy's in the first and second octet have a marked difference. In the first octet A is met only once; in the second, this is true of C. In the first octet, for both x and y, (G+C)/(A+U) = 3; in the second (G+C)/(A+U) = $\frac{1}{3}$. The last column on each side of the table lists the numbers n of hydrogen bonds connecting the xy nucleotides of the codon and the complementary nucleotides of the anticodon. For the first octet the value of n (which can be called the degree of complementarity) is either 6 or 5, and for the second octet either 5 or 4. It can be thought that at n = 6 for xy, the $z \leftrightarrow z'$ codon-anticodon interaction is not significant, since the $xy \leftrightarrow x'y'$ binding is strong enough and provides the necessary complementarity. Therefore, the codons of the first octet are indifferent to z. In this case there are 16 codon-anticodon combinations which correspond to the incorporation of the same amino acid at any z and z' at fixed xy, and hence at x'y'. If n = 5 in the first octet, then it is possible that the $z \leftrightarrow z'$ interaction plays a role in providing complementarity, and the number of permissible codon-anticodon combinations can be less than 16 and more than 4. This does not mean, of course, that all possible combinations can be found in nature. It would be interesting to determine experimentally the number of different anticodons (i.e., different tRNA's) corresponding to a given amino acid and the number of corresponding codons. It is possible that the experiment would actually show the presence of a larger number of codons and anticodons for those amino acids of the first octet for which n = 6, and a smaller number of codons and anticodons for those for which n = 5. The effectivity of mutational substitutions in the codons of mRNA for n = 6 and n = 5 may differ.

In the second octet the role of the last nucleotide is essential. At n = 5 the maximum number of codon-anticodon combinations corresponding to the same amino acid is 8 (if z' is not fixed). At n = 4 the $z \leftrightarrow z'$ bonds are fixed uniquely and this number becomes 2.

Crick formulated the properties of the third nucleotide of the anticodon in his hypothesis of "wobbles" [120]. The complementary codon-anticodon pairing in the first two nucleotides of the anticodon is unique and standard. The $z \leftrightarrow z'$ binding, however, is not unique. Analysis of the possible structure of pairs leads to the results presented in Table 9.7. These results are confirmed by experiment. The nonuniqueness of the $z \leftrightarrow z'$ pairing has a structural explanation. The minor nucleotide inosine, I, that appears in the anticodon is especially polyfunctional.

Figure 9.14 represents a table of doublets showing the codons of the first and second octets and the distribution of hydrophobic amino acids (shaded).

*TABLE 9.7*
*Properties of the Third Anticodon Nucleotide*

| Anticodon nucleotide z' | Codon nucleotides that bind with z' |
|---|---|
| U | A<br>G |
| C | G |
| A | U |
| G | U<br>C |
| I | U<br>C<br>A |

The correlation of codons and amino acids is directly related to the evolution of the code. If the contemporary code is optimal in relation to mutations which disturb the spatial protein structure, then it can be thought that the code appeared as a result of biochemical evolution. At present, however, we have no serious arguments supporting the code's optimality except the fact of its reliability in relation to the "wrong" mutations. Woese stated that it is not the errors in translation themselves which are essential but their biological significance [121]. Woese suggested that the code of the primitive cell was nonunique; it could not distinguish between

| x \ y | A | C | G | U |
|---|---|---|---|---|
| A | Lys<br>Asn<br>Lys<br>Asn | Trp | Arg<br>Ser<br>Arg<br>Ser | Ile<br>Ile<br>Met<br>Ile |
| C | Gln<br>His<br>Gln<br>His | Pro | Arg | Leu |
| G | Glu<br>Asp<br>Glu<br>Asp | Ala | Gly | Val |
| U | –<br>Tyr<br>–<br>Tyr | Ser | –<br>Cys<br>Trp<br>Cys | Leu<br>Phe<br>Leu<br>Phe |

*FIG. 9.14 Doublet table.*

individual amino acids, but only between their "functional" or "nonfunctional" groups. Later the code was perfected in such a way that errors of translation and the influence of these errors became minimal. A series of other speculative ideas has been put forth (cf. [5,7]). Crick criticized these hypotheses and emphasized the impossibility of testing them experimentally [122]. Today the important features of the code are established, a reasonable correlation of codons and amino acids has been found, but nothing definite can be said about the origin of the code and its evolution. This problem will also be discussed in "General Biophysics."

## 9.8 Translation

The fundamental physical problems related to protein synthesis on a polysome consist in the determination of the mechanisms by which the codon recognizes the anticodon, those of peptide bond formation, and those of the displacement of ribosomes along mRNA.

Recognition of the anticodon by the codon is very precise, the errors being hundredths of a percent [123]. The exact pairing cannot be determined by thermodynamic factors, by the differences in the energies of various pairs. Using quantum-mechanical calculations of the horizontal and vertical interactions between the triplets of the codon and anticodon, Ninio concluded that the recognition is nonunique [124]. The nonuniqueness of recognition increases also because of the "wobbles." The interaction energy for the erroneous anticodons does not differ much from that of the right ones. The interaction energies of GUG with CAC and CCC differ by only 2.1 kcal $mole^{-1}$. Attempting to overcome this difficulty, Ninio suggested the hypothesis of "superfluous triplets:" the anticodons which lead to considerable nonuniqueness (ACG, ACC, ICC) simply do not exist.

This hypothesis contradicts the general ideas of molecular biophysics. Translation is not an equilibrium state, but a kinetic process occurring with the participation of an enzyme, which in this case is the ribosome. Translation provides the precise recognition and optimal rate of "reading" of mRNA. Therefore, not only the thermodynamics, but also the kinetics, of recognition are essential: various tRNA anticodons can interact with the codons of mRNA at differing rates determined by the structure and properties both of the tRNA molecule as a whole and of the ribosomes.

Woese proposed hypothetical models of recognition that take into account the properties of ribosomes [125]. Recognition cannot be considered a simple codon-anticodon interaction. The codon is read twice, by the ribosome and by tRNA. The first reading is an approximate one--the ribosome does not distinguish between all the letters of the codon, but because of its allosteric properties it allows the interaction of only some definite tRNA with the codon. Hence, the model suggests the existence of various classes of tRNA, distinguished by the ribosome. Not only the anticodon structure but that of the tRNA molecule as a whole is important. "Key" links of tRNA are located close to the anticodon.

The tRNA structures have to be classified from a physical point of view. A conformational change is brought about by the action of a codon on a ribosome. Experiment shows that the state of the ribosome strongly affects the precision of translation. Such an influence is produced by mutations of the ribosomal proteins [126]. The determining role of the ribosome conformations follows also from the studies of *in vitro* translation of mRNA which contains halogenated bases. In such cases errors of reading the first nucleotide of the codon occur [7].

Woese suggested that tRNA also performed conformational transitions when it interacted with a ribosome and then with mRNA, and proposed a model of this transition. Translation is an allosteric process.

It has been shown experimentally that translation depends on mRNA conformation. In [127] RNA of the R17 phage was incubated in the presence of $Mg^{2+}$ and then introduced as a template in a cell-free system of *E. coli*. Incorporation of Phe and His and the amount of protein synthesized happened to differ from those in the case when the RNA was not incubated with $Mg^{2+}$. These alterations cannot be explained as due to degradation of RNA; they vanish if RNA is treated properly. In this situation three proteins are synthesized. The succession of their synthesis depends on the conformational state of the RNA.

Spirin proposed a visual model of the work of ribosome [87,128]. His hypothesis starts by localization of the aminoacyl-tRNA binding site and that for the peptidyl-tRNA on two different subunits. The mRNA binding center and tRNA binding sites are localized on internal surfaces of two subunits which are in contact. The 70 S ribosome is periodically transformed from the closed into the open state and vice versa. If the two subunits become closed up, the peptidyl-tRNA and aminoacyl-tRNA are drawn together and a peptide bond is formed. If the system becomes open, the new aminoacyl-tRNA enters the ribosome and the dezcylated tRNA leaves it. "The periodic closing and opening of the ribosome subunits is the driving mechanism providing all spatial displacements of tRNA and mRNA in the translation process" [87]. The periodic alteration of the quaternary structure of the ribosome is transformed into translational motion of the mRNA chain (cf. Fig. 9.12). The closing of the subunits is induced by the entrance of aminoacyl-tRNA; the opening requires the energy of GTP splitting (cf. below).

Spirin's hypothesis gives concrete expression to the conformational changes in ribosome. The hypothesis gives a visual explanation of a series of facts. However, a quantitative physical theory of the molecular mechanism of the ribosome work has not yet been devised. Sufficient data are not available; the spatial structures of ribosome and tRNA are not yet sufficiently known; and the kinetics of biosynthesis has not been sufficiently studied.

Experiment shows that an active ribosomal complex actually performs periodic conformational transitions during translation. According to [129], in a synchronized system that contains poly-U, subunits of *E. coli* ribosomes, and purified transfer factors, the active complex is assembled in stages:

$$\text{poly-U} + 30\ \text{S} \rightleftharpoons [\text{poly} - 30\ \text{S}]$$

$$[\text{poly-U} - 30\ \text{S}] + \text{HO} - \text{tRNA}^{\text{Phe}} \rightleftharpoons [30\ \text{S}]$$

$$[30\ \text{S}] + 50\ \text{S} \longrightarrow [60\ \text{S}]$$

$$[60\ \text{S}] + [\text{Phe-tRNA} - \text{T}_n + \text{GTP}] \longrightarrow [70\ \text{S}] + \text{T}_n + \text{GDP} + \text{P}$$

where [ ] denotes the active form, $T_n$ is the transfer factor, and P is phosphate. The transformations were investigated by means of sedimentation and electrophoresis. It was established that an active complex undergoes a cycle of compression and expansion when every amino acid is added. The binding of aminoacyl-tRNA catalyzed by the transfer factor T produces the stable, compact 70 S conformation; translation by means of the G factor transforms it into the less stable 60 S form. The formation of a peptide bond is preceded by translocation. These facts agree with the model of chain initiation without an initiating factor. The deacylated Phe-tRNA bound by ribosome, produces a translocation with participation of a G factor and GTP immediately after the 60 S-70 S transition due to the binding of Phe-tRNA.

Gavrilova and Spirin showed that a ribosome can also work without the participation of the macroergic substance GTP [130]. It was shown that p-chloromercuribenzoate (PCMB) activates the ability of *E. coli* ribosomes to synthesize polyphenylalanine in the presence of poly-U and Phe-tRNA when the transfer factors and GTP are absent. PCMB modifies 30 S subunits. As in normal synthesis, translation is suppressed by some antibiotics and SH compounds. Apparently the capacity for translation is peculiar to the structural organization of the ribosome itself (cf. also [131]). PCMB blocks one or several SH groups, producing some structural changes which permit the ribosome to work without catalytic factors and GTP. Blocking of SH groups reveals the potential ability of ribosomes to perform "nonenzymic" translocation.

The kinetics of protein synthesis on polysomes has been

studied in a series of works. The time necessary for the synthesis of a protein chain of conventional size (400 residues) is of the order of 30 sec. The protein-synthesizing ribosomes move along mRNA from the 5' to the 3' end. The mean distance $\delta$ between two neighboring ribosomes on a polysome is equal to 90 nucleotides of mRNA, according to the data of sedimentation analysis and electron microscopy. A ribosome covers this distance in $\tau \cong 3$ sec, corresponding to a linear velocity of the order of $10^{-6}$ cm $sec^{-1}$. The kinetics is studied using the measurements of the rate of incorporation of a $^{14}C$-labeled amino acid (cf., e.g., [132]).

The theory of the kinetics of translational synthesis on a polysome was presented in [133]. The dependence of the amount of incorporated amino acid on time for polysomes with various numbers of ribosomes n has been examined. The theoretical curves agree with the experimental ones obtained in [132]. However, this theory does not take into account the degradation of mRNA due to the action of exonuclease, which has not been studied up to the present time. The lifetime of mRNA is approximately 40 sec.

A theory has been developed which takes into account mRNA degradation [134-136]. Let us discuss the work of Singh [135]. It is suggested that ribosomes are bound at the 5' end of mRNA; exonuclease acts also starting from this end. The ribosome protects the end of mRNA from degradation. When the ribosome is displaced, the 5' end either degrades or binds a new ribosome. The degraded end cannot bind a ribosome, but the ribosomes, which were bound earlier and are in advance, continue to move and to synthesize protein.

For a random process the probability of the existence of the 5' end during the time t is exp(-kt). The number of protein molecules synthesized on polysomes $P_1, P_2, \ldots, P_{n-1}$ which contain 1, 2,...,n-1 ribosomes, during the time $(n-1)\tau$ is equal to

$$(1-e^{-k})+2e^{-k}(1-e^{-k})+\cdots+(n-1)e^{-(n-2)k}(1-e^{-k}) \tag{9.4}$$

In this expression the time is calculated in units of $\tau$. The number of the protein molecules synthesized by the $P_n$ polysome during $t \to \infty$ is

$$e^{-(n-1)k}+e^{-nk}+\cdots+(n-1)(1-e^{-k})(e^{-(n-1)k}+e^{-nk}+\cdots) \tag{9.5}$$

Summing (9.4) and (9.5) we get the total number of protein molecules synthesized on one mRNA chain during infinite time

$$N = (1-e^{-k})^{-1} \tag{9.6}$$

For a stable mRNA, $k = 0$ and $N \to \infty$; for a very unstable one, $k \to \infty$ and $N \to 1$.

Information about the degradation of mRNA is based on the

observed rate of decay of labeled chains. It may be thought that the decay is due to accidental inactivation of the 5' end, followed by a stepwise degradation with the rate determined by the ribosome displacement. If the mRNA chains of length $(n - 1)\delta$ are labeled, then the number of units of length $\delta$ which remain intact at time n - 1 is equal to

$$M_0 = (1-e^{-k})[e^{-k}+2e^{-2k}+3e^{-3k}+\cdots+(n-2)e^{-(n-2)k}+(n-1)e^{-(n-1)k}]$$

$$= [e^{-k}(1-e^{-(n-1)k})]/(1-e^{-k}) \quad (9.7)$$

The number of intact units at any subsequent moment t is equal to

$$M_t = [e^{-(t+1)k}(1-e^{-(n+t-1)k})]/(1-e^{-k})$$

and the number of units f that survive until t is

$$f = M_t/M_0 = [e^{-kt}(1-e^{-(n+t-1)k})]/(1-e^{-(n-1)k}) \quad (9.8)$$

if $n \gg 1$, $f = e^{-kt}$.

The actual situation depends on the relative rates of transcription and translation. According to [137], the first process is much faster. On the other hand, Stent suggests that both these rates are of the same order [138]. The correlation between the lengths of the labeled mRNA and polysomes and the absence of free labeled mRNA testifies in favor of the second possibility.

Thus, the theory permits us to calculate the distribution of mRNA dimensions. Sedimentational analysis of labeled bacterial mRNA gives a broad range of radioactivity from 4 S to 80 S with a peak in the range of 8 S-12 S. Singh's theory gives a similar distribution.

When the distribution of the enzymic activity of synthesized protein is calculated, it can be thought that only polysomes with ribosomes at the end of the chain contain the protein. In the steady state the amount of protein bound by a polysome that has r ribosomes is equal to $e^{-kr}$, and

$$P_n = (e^{-(n-1)k})/(1-e^{-k}) \quad (9.9)$$

For a protein made up of several chains synthesized by one polycistronic mRNA the calculation becomes more complicated but is also possible [135].

The experimental data were analyzed in detail [136]. The observed sedimentation curve of fast-labeled bacterial mRNA does not necessarily express the size distribution of intact mRNA. The 8 S-12 S peak depends on the rate of translation, the lifetime of the mRNA, and on the duration of labeling. The model predicts that the peak must shift toward larger S when the duration of labeling increases, achieving a stationary-state value determined by the rate of translation and the lifetime of RNA. The theoretical curves agree well with the experimental ones.

This stochastic theory does not exhaust the problem. The molecular mechanisms of the work of ribosomes, of the action of macroergs (GTP), and of the work of exonucleases have to be established. For further details concerning the translation mechanism see [14].

## 9.9 Mutations

Changes in the hereditary program, mutations, occur either spontaneously or through the influence of powerful external factors: chemical or radiational action on chromosomes. Chromosomal mutations which are observed microscopically, i.e., rearrangements of chromosomes, imply alterations in the supramolecular structures; point, or gene mutations in DNA mean a change in the molecular primary structure. The problems of mutagenesis have been touched on many times earlier (pp. 484, 557,560, etc.).

Several mutationally changed proteins have already been investigated, in particular, mutants of TMV protein and of *E. coli* alkaline phosphatase and tryptophan synthetase, etc. [107]. In many cases, direct correlation between changes in the nucleotides and changes in the amino acid residues could be established. It is reasonable to distinguish three types of point mutations: (a) mutations that change the sense of a codon, missense mutations which produce the substitution of an amino acid residue in protein; (b) mutations that transform a normal codon into a senseless one, UAA, UAG, AGA (nonsense mutations) and break the protein chain; (c) deletions and additions, i.e., frame shift mutations (p. 598).

Spontaneous gene mutations are due to errors which arise because of the thermal motion of atoms and molecules in the process of DNA reduplication. Obviously, errors in transcription and translation are not inherited (excluding the possibilities connected with RNA-dependent DNA-polymerase, p. 566, which have not yet been studied).

Reduplication of DNA is a stochastic process; "noise" (p. 24) cannot be entirely excluded.

The incorporation of a nucleotide which is not complementary to that of the template must apparently amount to "loop" formation (p. 499). In subsequent generation the loop vanishes because of semiconservative synthesis but the primary structure of the DNA remains altered. The main difficulty in the physicochemical treatment of such errors is connected with the necessity to elucidate the relative roles of thermodynamic and kinetic factors.

It follows from elementary considerations that replacement of the A-T pair by G-C is thermodynamically advantageous. The thermodynamic probability of such a substitution can be computed [139]. Were the mutational processes determined only by thermodynamics, the relative G-C content in DNA would increase in the course of evolution. This conclusion contradicts experiment; the G-C content of higher organisms is stabilized at the 40-45% level. The calculated thermodynamic probabilities of the formation of various kinds of loops (from $10^{-2}$ to $10^{-4}$) are much larger than the experimental ones (in bacteria the number of mutations per gene per generation is no larger than $10^{-5}$-$10^{-7}$).

Both the thermodynamics and kinetics of template synthesis are important for reduplication. The appearance in a chain of an erroneous nucleotide resulting from addition, substitution, or deletion, is also determined by the rate of the process, i.e., by the behavior of DNA-polymerase, which catalyzes DNA synthesis. If the synthesis proceeds in a discontinuous way, according to Okazaki (p. 545), then the kinetic conditions are important both in the replication of the DNA sections and in their unification into a common chain.

The results of the mutational "noise" give the material for natural selection at the molecular and organismic levels and therefore are under the powerful action of selection. Selection has formed the contemporary code, whose degeneracy has strongly diminished the danger of errors (i.e., the probability of protein mutation). The features of the code in the sense of its reliability in relation to the most dangerous mutations were discussed in Section 9.7.

Evidently the evolutionary formation of a "G-C organism," though advantageous from the thermodynamic point of view, is impossible, since triplets that do not contain A and T (i.e., A and U in mRNA) code only Pro, Arg, Ala, and Gly (i.e., only one fifth of all amino acids). Natural selection, the biological factor, constrains the mutations compatible with life, limits the role of thermodynamic and kinetic factors.

The incorporation of an erroneous nucleotide in the growing chain of DNA is due, first, to its binding by the given template locus. The formation of a non-Watson-Crick pair can, in turn, be determined by the tautomery of the nucleotide and by its ability to form unconventional hydrogen bonds (p. 501). Freese considered tautomery the principal cause of mutations [140].

Certain derivatives of nitrogen bases have high mutagenic activity (chemical mutagenesis). Such are in particular 5-bromouracil (BU) and 2-aminopurine (AP). In the *in vitro* synthesis of DNA, BU is incorporated instead of T. Apparently BU can form a pair with A if it is present in the conventional keto form. In the rarer enol form BU imitates C and forms a

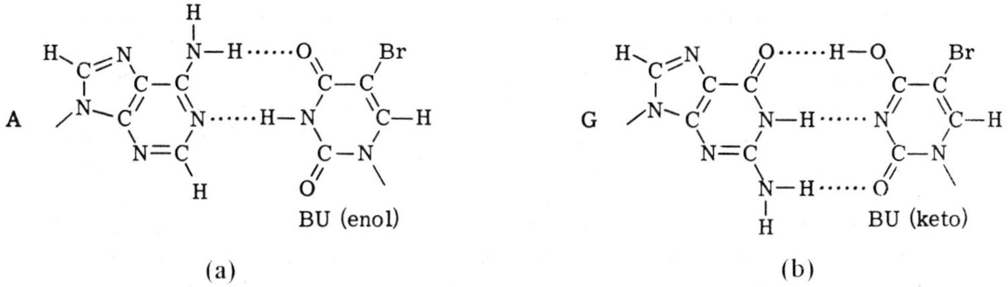


*FIG. 9.15 BU-A and BU-G pairs.*

pair with G (Fig. 9.15b). In the first case an error of incorporation occurs during reduplication. In the second case an error of reduplication occurs--the DNA chain, which already contained BU, forms at this place not a BU-A pair (which would improve the error of incorporation: A-BU → A, BU → A-T, BU-A → A-T, BU, A → A-T, BU-A, A-T, etc.), but a BU-G pair because of the tautomeric transformation of BU [135]. The probability of such a transformation is increased in the presence of the bromine atom.

The importance of the quantum-chemical analysis of tautomerism in nitrogen bases is obvious. Along with tautomerism an essential role in mutagenesis is played by the formation of non-Watson-Crick pairs due to different localization of hydrogen bonds (p. 501).

All that we know about DNA obliges us to think about the cooperative nature of mutagenesis. The nitrogen bases interact both horizontally and vertically (stacking). Any error of reduplication implies an alteration in these interactions. Therefore, the probability of a mutation at a given DNA locus must depend on the neighboring base pairs. This idea has been offered as an explanation of the "hot points" that were observed in studies of the mutability of the rII locus of phage T4 [141,142]. The main difficulty in treating the rates of direct and reverse mutations is that which of the degerate codons is located at the given point of DNA is not known. However, determination of the amino acid content of a series of revertants of "amber" mutants (i.e., of mutants that lead to the nonsense codon UAG) in the protein of the head of phage T4 has shown that substitutions of the base pairs in various loci of cistron have different probabilities [143]. The CAA → UAA mutants in T4 rII are induced by $NH_2OH$ with rates that change up to 20 times, depending on the DNA locus [144]. Similar facts were discovered in the ultraviolet reversion of these mutants [145]. All these facts can be explained in other ways--by the dependence of the mutation probabilities inside a cistron on the direction of reduplication and transcription of the gene; by the proximity of the controlling elements; etc. However, Koch obtained direct proof of the influence of neighboring pairs on mutagenesis [146].

The chain-breaking amber (UAG) and ochre (UAA) codons can be mutually transformed by substitutions of the third nucleotide. The mutation rates of the first two nucleotides of UAA and UAG can be measured in the T4 rII system. AP as a mutagen produces transitions (i.e., the replacement of purine by purine or of pyrimidine by pyrimidine) but not transversions (i.e., the replacement of purine by pyrimidine and vice versa) [147]. Koch identified the genotypes of revertants arising in T4 rII under the action of AP. The scheme of these mutations is

| | |
|---|---|
| CAA (Gln) | CAG (Gln) |
| ↑ | ↑ |
| UAA (Ochre) | UAG (Amber) |
| ↓ | ↓ |
| UGA (Opal) | UGG (Trp) |

It was found that the transition frequencies of the first two nucleotides of the codon do not depend on the nature of the third, but increase strongly if an A-T pair is replaced by G-C.

The mechanism of this cooperativity has not been investigated. The physics of spontaneous and chemical mutagenesis is not yet developed. This development requires detailed deciphering of the structure and action of DNA-polymerase. Valuable information on the molecular mechanism of mutations is presented in [147] and [148].

Along with the analogs of nitrogen bases, several other substances act as strong mutagens: nitrous acid (p. 560); diethylsulfate $(C_2H_5)_2SO_4$, which produces ethylation of G and its subsequent removal from DNA; $NH_2OH$, which interacts with C and forms a substance imitating T. Especially strong mutagens have been discovered: NN'-nitrosonitroguanidine, acridine derivatives of nitrogen mustard, etc.

A series of substances, particularly antibiotics, influence the translation of the code. These are not mutagens in the real sense of the word. The study of their action is important in connection with the problems of molecular regulation of biological processes. These problems will be discussed in "General Biophysics."

We cannot draw a boundary between molecular biophysics and investigation of the general biophysical problem at the molecular and supramolecular level. The presentation in this book is limited to the most important properties of proteins and nucleic acids. In "General Biophysics" we shall again discuss biopolymers, particularly the proteins which are antibodies, from the point of view of the general concept of molecular "recognition." The molecular regulatory processes are directly related to this concept. Consequently, molecular regulation of protein biosynthesis will be considered.

## References


1. M. Volkenstein, "Molecules and Life." Plenum Press, New York, 1970.
2. J. Watson, "Molecular Biology of the Gene," Benjamin, New York, 1965.
3. J. Gamow, Nature (London) **173**, 318 (1954): Kgl. Dansk. Videnskab. Selskab. Biol. Medd. **22**, 1 (1954).
4. G. Gamow, A. Rich, and M. Yčas, Advan. Biol. Med. Phys. **4**, 23 (1956).
5. M. Yčas, "The Biological Code," North-Holland Publ., Amsterdam, 1969.
6. R. Wall, Nature (London) **193**, 1268 (1962).
7. C. Woese, "The Genetic Code." Harper, New York, 1967.
8. F. H. C. Crick, L. Barnett, S. Brenner, and R. Watts-Tobin, Nature (London) **192**, 1227 (1961).
9. S. Benzer, Proc. Nat. Acad. Sci. U.S. **45**, 1607 (1959); **47**, 403 (1961).
10. E. Terzaghi, Y. Okada, G. Streisinger, J. Emrich, M. Inoue, and A. Tsugita, Proc. Nat. Acad. Sci. U.S. **56**, 500 (1966).
11. H. Wittmann, Z. Vererbungs. **93**, 491 (1962).
12. R. Eck, J. Theor. Biol. **2**, 139 (1962).
13. N. Luchnik, Works of Inst. of Biol., Acad. Sci. USSR, Urals Dept., N37, Sverdlovsk (1963) (R).
14. L. Kiselev, V. Nikiforov, O. Astaurova, B. Gottikh, and A. Krajevsky, "Molecular Basis of Biosynthesis of Proteins." Nauka, Moscow, 1971 (R).
15. A. Spirin, A. Belozersky, N. Shugajeva, and V. Vanjushin, Biochimia **22**, 744 (1957) (R).
16. A. Belozersky and A. Spirin, Nature (London) **182**, 111 (1958); Progr. Nucleic Acids Res. Mol. Biol. **3**, 147 (1960).
17. H. Fraenkel-Conrath, J. Amer. Chem. Soc. **78**, 147 (1960).
18. A. Gierer and G. Schramm, Nature (London) **177**, 702 (1956).
19. H. Fraenkel-Conrath, V. Singer, and R. Williams, in "The Chemical Basis of Heredity" (W. McElroy and B. Glass, eds.). Johns Hopkins Press, Baltimore, Maryland, 1957.
20. E. Volkin and L. Astrachan, Virology **2**, 149 433 (1956).
21. E. Volkin, in "Molecular Genetics" (J. Taylor, ed.), Part I. Academic Press, New York, 1963.
22. F. Gros, W. Gilbert, H. Hiatt, G. Attardi, P. Spahr, and J. Watson, Cold Spring Harbor Sump. Quant. Biol. **26**, 111 (1961).
23. J. Hurwitz and J. August, Progr. Nucleic Acids Res. Mol. Biol. **1**, 59 (1963).
24. B. Hall and S. Spiegelman, Proc. Nat. Acad. Sci. U.S. **47**, 137 (1961).
25. S. Spiegelman, Sci. Amer. **210**(5), 48 (1964).
26. M. Hoagland, E. Keller, and P. Zamecnik, J. Biol. Chem. **218**, 345 (1956).

27. A. Maister, Biophysical science. A study program, Rev. Mod. Phys. 31(1) (1959).
28. M. Hoagland, in Symp. 8, Int. Congr. Biochem. 4th (1958).
29. H. Temin and S. Mizutani, Nature (London) 226, 1211 (1970).
30. S. Mizutani, D. Boettiger, and H. Temin, Nature (London) 228, 424 (1970).
31. S. Spiegelman et al., Nature (London) 228, 430 (1970).
32. E. Skolnik, E. Rands, S. Aaronson, and G. Todaro, Proc. Nat. Acad. Sci. U.S. 67, 1789 (1970).
33. E. Skolnik, S. Aaronson, G. Todaro, and W. Parks, Nature (London) 229, 318 (1971).
34. S. Gershenson et al., "Investigation of the Possibility of Transfer of Genetic Information from RNA to DNA in Reproduction of the Nuclear Polyhedrose Viruses." Naukova Dumka, Kiev, 1971 (R).
35. D. Anthony and D. Goldthwait, Biochem. Biophys. Acta 205, 156 (1970).
36. A. Travers and R. Burgess, Nature (London) 222, 537 (1969).
37. J. Roberts, Proc. Int. Repetit Colloq., lst, Florence, p. 208, Milan (November 1969).
38. Y. Kosaganov, M. Zarudnaja, Y. Lazurkin, M. Frank-Kamenetsky, R. Bibilashvili, and L. Savotshkina, Nature (London), New Biol. 231, 212 (1971).
39. W. Szybalski et al., J. Cell. Physiol. Suppl. 1, 74, 33 (1969).
40. R. Langridge, J. Cell. Physiol. Suppl. 1, 74, 3 (1969).
41. W. Rüger, Biochem. Biophys. Acta 238, 202 (1971).
42. V. Florentiev and V. Ivanov, Nature (London) 228, 519 (1970).
43. E. Geiduschek, E. Brody, and D. Wilson, in "Molecular Associations in Biology" (B. Pullman, ed.). Academic Press, New York, 1968.
44. P. Riley, Nature (London) 228, 522 (1970).
45. M. Hoagland, P. Zamecnik, and M. Stephenson, Biochem. Biophys. Acta 24, 215 (1957).
46. K. Ogata and H. Nohara, Biochem. Biophys. Acta 25, 659 (1957).
47. R. Holley, G. Everett, J. Madison, and A. Zamir, J. Biol. Chem. 240, 2122 (1965).
48. R. Holley, Sci. Amer.
49. A. Bajev, T. Venkstern, A. Mirsabekov, A. Krutilina, L. Axelrod, and V. Axelrod, Mol. Biol. 1, 754 (1967) (R).
50. G. Zavilgelsky, T. Venkstern, and A. Bajev, Dokl. Acad. Sci. USSR 166, 978 (1966) (R).
51. G. Zavilgelsky and L. Li, Mol. Biol. 1, 323 (1967) (R).
52. T. Venkstern, "Primary Structure of Transfer Nucleic Acids." Nauka, Moscow, 1970 (R).
53. V. Tumanian, L. Sotnikova, and A. Kholopov, Dokl. Acad.

Sci. USSR **166**, 1465 (1966) (R).
54. V. Tumanian, in "Nucleic Acids." Medicina, Moscow, 1966 (R).
55. V. Tumanian and L. Sotnikova, Biophysica 12, 5 (1967) (R).
56. I. Tinoco, O. Uhlenbeck, and M. Levine, Nature (London) 230, 362 (1971).
57. M. Eigen, Naturwissenschaften 58, 465 (1971).
58. J. Fresco, A. Adams, R. Ascione, D. Henley, and T. Lindahl, Cold Spring Harbor Symp. Quart. Biol. 31, 527 (1966).
59. J. Fresco, T. Lindahl, and D. Henley, Fed. Proc. 27, 796 (1968).
60. S. Kim and A. Rich, Science **162**, 1381 (1968); **166**, 1621 (1969).
61. A. Hampel et al., Science **162**, 1384 (1968).
62. J. Fresco, R. Blake, and R. Langridge, Nature (London) 220, 1285 (1968).
63. S. Kin, G. Quigley, F. Suddath, and A. Rich, Proc. Nat. Acad. Sci. U.S. **68**, 841 (1971).
64. A. Rosenfeld, C. Stevens, and M. Priatz, Biochemistry 9, 4971 (1970).
65. M. Lanborg, P. Zamecnik, T. Li, J. Kagi, and B. Vallee, Biochemistry **4**, 63 (1965).
66. T. Samejima and J. Yang, J. Biol. Chem. **240**, 2094 (1965).
67. P. Sarin, P. Zamecnik, P. Bergquist, and J. Scott, Proc. Nat. Acad. Sci. U.S. 55, 579 (1966).
68. F. Wolfe, K. Oikawa, and C. Kay, Biochemistry 7, 3361 (1968).
69. G. Zavilgelsky, A. Mirzabekov, A. Poletajev, and A. Bajev, Mol. Biol. 6, 231 (1972) (R).
70. A. Poletajev, Mol. Biol. **7**, 84 (1973) (R).
71. H. Hashizume and K. Imahory, J. Biochem. (Tokyo) 61, 738 (1967).
72. V. Ivanov, L. Mintshenkova, A. Shtshelkina, and A. Poletajev, Biopolymers (1973).
73. G. Melcher, D. Paulin, and W. Guschelbauer, Biochemie 53, 43 (1971).
74. S. Englander and J. Englander, Proc. Nat. Acad. Sci. U.S. **53**, 370 (1965).
75. S. Nishimura and G. Novelli, Biochem. Biophys. Acta 80, 574 (1964).
76. S. Nishimura, F. Harada, V. Narushima, and T. Seno, Biochem. Biophys. Acta **142**, 133 (1967).
77. N. Kalenbach, J. Mol. Biol. 37, 445 (1968).
78. D. Henley, T. Lindahl, and J. Fresco, Proc. Nat. Acad. Sci. U.S. 55, 191 (1966).
79. A. Surovaja, O. Borisova, T. Yulajeva, V. Sheinker, and L. Kiselev, FEBS Lett. 8, 201 (1970).
80. R. Beardsley and C. Kantor, Proc. Nat. Acad. Sci. U.S. 65, 39 (1970).

81. S. Trubitsyn, A. Surovaja, and O. Borisova, Mol. Biol. 5, 419 (1971) (R).
82. D. Abraham, J. Theor. Biol. 30, 83 (1971).
83. A. Bajev, I. Fodor, A. Mirzabekov, V. Axelrod, and L. Kazarinova, Mol. Biol. 1, 859 (1967) (R).
84. A. Mirzabekov, D. Grünberger, and A. Bajev, Biochem. Biophys Acta 166, 68 (1968) (R).
85. A. Mirzabekov, L. Kazarinova, and A. Bajev, Mol. Biol. 3, 879 (1969) (R).
86. A. Mirzabekov, L. Kazarinova, D. Lastit, and A. Bajev, Mol. Biol. 3, 909 (1969) (R).
87. A. Spirin and L. Gavrilova, "The Ribosome." Nauka, Moscow, 1971 (R).
88. A. Tissières and J. Watson, Nature (London) 182, 778 (1958).
89. M. Hamilton and M. Petermann, J. Biol. Chem. 234, 1441 (1959).
90. A. Spirin, J. Mol. Biol. 2, 436 (1960); Biochimia 26, 511 (1961) (R).
91. A. Spirin, "Some Problems of Macromolecular Structure of Nucleic Acids." Ed. Acad. Sci. USSR, Moscow, 1963 (R).
92. D. Schlessinger, G. Mangiarotti, and D. Apirion, Proc. Nat. Acad. Sci. U.S. 58, 1782 (1967).
93. A. Spirin, N. Kiselev, R. Shakulov, and A. Bogdanov, Biochimia 28, 920 (1963) (R).
94. L. Gavrilova, M. Lerman, and A. Spirin, Izv. Acad. Sci. USSR Ser. Biol. 826 (1966) (R).
95. L. Gavrilova, D. Ivanov, and A. Spirin, J. Mol. Biol. 16, 473 (1966).
96. A. Spirin and N. Belitsyna, J. Mol. Biol. 15, 282 (1966).
97. A. Spirin, M. Lerman, L. Gavrilova, and N. Belitsyna, Biochimia 31, 424 (1966) (R).
98. K. Nosokawa, R. Fujimura, and M. Nomura, Proc. Nat. Acad. Sci. U.S. 55, 198 (1966).
99. P. Traub and M. Nomura, Proc. Nat. Acad. Sci. U.S. 59, 777 (1968); J. Mol. Biol. 40, 391 (1969).
100. M. Nomura, Sci. Amer. October (1969).
101. M. Bretscher and K. Marcker, Nature (London) 211, 380 (1966).
102. J. Watson, Bull. Soc. Chem. Biol. 46, 1399 (1964).
103. M. Nirenberg and G. Mattheai, Symp. 1, Int. Biochem. Cong., 5th (1961).
104. H. Khorana, Fed. Proc. 24, 1473 (1965).
105. M. Nirenberg et al., Proc. Nat. Acad. Sci. U.S. 53, 1161 (1965).
106. D. Jones, S. Nishimura, and H. Khorana, J. Mol. Biol. 16, 454 (1966).
107. The genetic code, Cold Spring Harbor Symp. Quant. Biol. 31, (1966).

108. N. Sueoka, Cold Spring Harbor Symp. Quant. Biol. 26, 35 (1961); Proc. Nat. Acad. Sci. U.S. **48**, 582 (1963).
109. M. Volkenstein, Biophysica 8, 394 (1963) (R).
110. R. Marshall, T. Caskey, and M. Nirenberg, Science **155**, 180 (1967).
111. N. Gupta et al., Proc. Nat. Acad. Sci. U.S. **60**, 1338 (1968).
112. M. Volkenstein, Genetika N 2 54 (1965); N 4, 119 (1966) (R); Biochim. Biophys. Acta **119**, 421 (1966).
113. C. Woese, D. Dugre, W. Saxinger, and S. Dugre, Proc. Nat. Acad. Sci. U.S. **55**, 966 (1966).
114. E. Margoliash, Can. J. Biochem. **42**, 745 (1964).
115. M. Volkenstein, "Enzyme Physics." Plenum Press, New York, 1969.
116. C. Yanofsky, Cold Spring Harbor Symp. Quant. Biol. **31**, 151 (1966).
117. M. Dayhoff and R. Eck, Atlas of Protein Sequence and Structure 1967-1968. Nat. Biomed. Res. Foundation, Silver Spring, Maryland (1968).
118. G. Rumer, Dokl. Acad. Sci. USSR **167**, 1394 (1966); **183**, 225 (1968) (R).
119. M. Volkenstein and G. Rumer, Biophysica **12**, 10 (1967) (R).
120. F. Crick, J. Mol. Biol. **19**, 548 (1966).
121. C. Woese, Proc. Nat. Acad. Sci. U.S. **54**, 1546 (1965).
122. F. Crick, Cold Spring Harbor Symp. Quant. Biol. **31**, 3 (1966).
123. W. Szer and S. Ochoa, J. Mol. Biol. 8, 323 (1964).
124. J. Ninio, J. Mol. Biol. 56, 63 (1971).
125. S. Woese, J. Theor. Biol. 26, 83 (1970); Nature (London) **226**, 817 (1970).
126. R. Rosset and L. Gorini, J. Mol. Biol. **39**, 95 (1969).
127. H. Fukami and K. Imahori, Proc. Nat. Acad. Sci. U.S. **68**, 570 (1971).
128. A. Spirin, Dokl. Acad. Sci. USSR **179**, 1467 (1968) (R).
129. M. Schreier and H. Noll, Nature (London) 227, 128 (1970); Proc. Nat. Acad. Sci. U.S. **68**, 805 (1971).
130. L. Gavrilova and V. Smolianinov, Mol. Biol. **5**, 883 (1971) (R).
131. L. Gavrilova and A. Spirin, Mol. Biol. 6, 311 (1972) (R).
132. H. Noll, T. Staehelin, and F. Wettstein, Nature (London) **198**, 632 (1963).
133. M. Volkenstein and S. Fishman, Dokl. Acad. Sci. USSR **160**, 1407 (1965) (R).
134. R. Parker and T. Linkoln, J. Theor. Biol. **15**, 218 (1967).
135. U. Singh, J. Theor. Biol. **25**, 444 (1969).
136. U. Singh and R. Gupta, J. Theor. Biol. **30**, 603 (1971).
137. O. Maaloe and N. Kjeldgaard, in "Control of Macromoleculae Synthesis" (O. Maaloe and N. Kheldgaard, eds.). Benjamin, New York, 1966.

138. G. Stent, in "Mendel Centennial Symp." Roy. Soc., London, 1965.
139. M. Volkenstein and A. Eliashevitch, Dokl. Acad. Sci. USSR **136**, 1216 (1961) (R).
140. E. Freese, Symp. 1, Int. Biochem. Congr., 5th, (1961).
141. S. Benzer and E. Freese, Proc. Nat. Acad. Sci. U.S. **44**, 112 (1958).
142. S. Benzer, Proc. Nat. Acad. Sci. U.S. **47**, 403 (1961).
143. A. Stretton, S. Kaplan, and S. Brenner, Cold Spring Harbor Symp. Quant. Biol. **31**, 173 (1966).
144. S. Brenner, A. Stretton, and S. Kaplan, Nature (London) **206**, 994 (1965).
145. F. Sherman, J. Stewart, M. Cravens, F. Thomas, and N. Shipman, Genetics **61**, 55 (1969).
146. P. Koch, Proc. Nat. Acad. Sci. U.S. **68**, 773 (1971).
147. J. Drake, "The Molecular Basis of Mutation." Holden Day, San Francisco, California, 1970.
148. E. Freese, in "Molecular Genetics," (J. Taylor, ed.), Part 1. Academic Press, New York, 1963.
149. T. Sekiya and H. Khorana, Proc. Nat. Acad. Sci. U.S. **71**, 2978 (1974).
150. R. Bibilashvili, V. Ivanov, L. Mintshenkova, and L. Savotchkina, Biochim. Biophys. Acta **259**, 35 (1972).
151. I. Tinoco et al., Nature (London), New Biol. **246**, 40 (1973).
152. S. Kim et al., Proc. Nat. Acad. Sci. U.S. **69**, 3746 (1972); Science **281**, 3322 (1973).

# Selected Additional References


Chapter 2

1. R. Bonnett, Neovitamin $B_{12}$ (cyano-13-epicobalamin, Phil. Trans. Roy. Soc. London Ser. B **273** (N 924), 295-301 (1976).
2. P. Mitchell, Possible molecular mechanisms of the protonmotive function of cytochrome systems, J. Theor. Biol. **62**, 327-367 (1976).
3. "Electron transfer chains and oxidative phosphorylation" (Proceedings of the international symposium on electron transfer and oxidative phosphorylation, Selva di Fasano, Italy, September 15-18, 1975), North Holland, Amsterdam 1975.
4. D. Dolphin, ed., "The Porphyrins, Vol. 2, Organic Chemistry, Part B." Academic Press, New York, 1977.
5. V. Prelog, Chirality in chemistry, Science, **193**, 17-24 (1976)

Chapter 3

1. M. D. Frank-Kamenetskii, A. V. Lukashin, and A. V. Vologodskii, Statistical mechanics and topology of polymer chains, Nature **258**, 398-402 (1975).
2. E. Suzuki and B. Robson, Relationship between helix-coil transition parameters for synthetic polypeptides and helix conformation parameters for globular proteins. A simple model. J. Mol. Biol. **107**, 357-367 (1976).
3. P. J. Flory, Spatial configuration of macromolecular chains, Science **188**, 1268-1276 (1975).

Chapter 4

1. E. Adman, K. D. Watenpaugh, and L. H. Jensen, NH S hydrogen bonds in *Peptococcus aerogenes* ferredoxin, *Clostridium pasteurianum* rubredoxin, and chromatium high potential iron proteins, Proc. Nat. Acad. Sci. U.S. **72**, 4854-4858 (1975).
2. A. Yu. Grosberg, Theoretical model of the adsorption of homopolymer chain on the homogeneous surface, Biophysics **21**, 603-609 (1976) (R).
3. A. Yu Grosberg and B. D. Liberol, About the nature of elasticity of a polymer globule, Biophysica **21**, 610-614 (1976) (R).
4. J. N. Israelachvili and D. Taber, Van der Waals forces: theory and experiment, Progr. Surf. and Membrane Sci. **7**, 1-55 (1973).
5. S. Tanaka and H. A. Scheraga, Statistical mechanical treatment of protein conformation: I. Conformational properties of amino acids in proteins, Macromolecules 9(1), 142-182 (1977).
6. B. Robson and E. Suzuki, Conformational properties of amino residues in globular proteins, J. Mol. Biol. **107**, 327-356 (1976).
7. M. J. E. Sternberg and J. M. Thornton, On the conformation of proteins: The handedness of the β-strand-α-helix-β-strand unit, J. Mol. Biol. **105**, 367-382 (1976).
8. C. Chothia, The nature of the accessible and buried surfaces in proteins, J. Mol. Biol. **105**, 1-14 (1976).
9. M. G. Rossmann and P. Argos, Exploring structural homology of proteins, J. Mol. Biol. **105**, 75-95 (1976).

Chapter 5

1. D. T. Clark and W. J. Feast, Application of electron spectroscopy for chemical applications (ESCA) to studies of structure and bonding in polymeric systems, J. Macromol. Sci. Rev. Macromol. Chem. **13**, 191-286 (1975).
2. Su-yun Chung and G. Holzwarth, Circular dichroism of flow-oriented DNA, J. Mol. Biol. **92**, 449-466 (1975).
3. S. K. Zavriev, M. V. Volkenstein, V. I. Ivanov, and L. E. Minchenkova, Conformational transition in the B-family of DNA as revealed by anisotropic circular-dichroism method, Mol. Biol. **11**, 2 (1977) (R).
4. T. G. Sprio, Resonance Raman spectroscopic studies of heme proteins, Biochim. Biophys. Acta **416**, 169-189 (1975).
5. R. M. Stephens, Structural investigations of peptides and proteins. Infrared and Raman spectroscopy, in "Amino-Acids, Peptides and Proteins, Vol. 5, Rev. Lit. Publ. 1972," pp. 223-228, London (1974).
6. K. A. Hartman, R. C. Lord, and G. J. Thomas, Structural

studies of nucleic acids and polynucleotides by infrared and Raman spectroscopy, in "Physical and Chemical Properties of Nucleic Acids," Vol. 2, London-New York (1973).

7. F. C. Phillips, A. Wlodawer, M. M. Gevitz, and K. O. Hodgson, Application of synchrotron radiation to protein crystallography: preliminary results. Proc. Nat. Acad. Sci. U.S. **73**(1), 128-132 (1976).
8. J. R. Wasson and D. R. Lorenz, Nuclear magnetic resonance spectroscopy, Anal. Chem. **48**(5), 246-261 (1976).
9. I. V. Duditch, V. P. Timofeev, M. V. Volkenstein, and A. Yu. Misharin, ESR measurement of rotational correlation time of macromolecules with a covalently bonded spin label, Mol. Biol. USSR **11**(3), 686-695 (1977).
10. M. R. Wilcott, III and R. E. Davis, Determination of molecular conformation in solution, Science **190**, 850-857 (1975).
11. J. Bordas, I. H. Munro, and A. M. Glazer, Small-angle scattering experiments on biological materials using synchrotron radiation, Nature **262**, 541-545 (1976).

Chapter 6

1. R. E. Dickerson and R. Timkovich, Cytochromes *c*, in "The Enzymes" (P. Boyer, ed.) 3rd ed., Vol. 11, pp. 397-547, Academic Press, New York 1975.
2. L. A. Blumfeld, The physical aspects of enzyme functioning, J. Theor. Biol. **58**, 269-284 (1976).
3. W. Lovenberg, ed., "Iron-sulfur proteins," Vol. 2, Molecular Properties (1973), Vol. 3, Structure and Metabolic Mechanisms (1977), Academic Press, New York.
4. H. Gutfreund, "Enzymes: Physical Properties," Wiley (Interscience), New York 1975. Reprint with corrections of 1972 edition.
5. A. Warshel and M. Levitt, Theoretical studies of enzymic reactions: dielectric, electrostatic and steric stabilization of the carbonium ion in the reaction of lysozyme, J. Mol. Biol. **103**, 227-249 (1976).

Chapter 7

1. Yu. A. Sharonov and N. A. Sharonova, Structure and function of hemoglobin, Mol. Biol. **9**(1), 119-140 (1975) (R).
2. J. M. Baldwin, Structure and function of haemoglobin, Progr. Biophys. Mol. Biol. **29**(3), 225-320 (1975).
3. M. Weisbluth, "Hemoglobin, Cooperativity and Electronic Properties," London, Chapman and Hall, 1974.
4. P. J. Stephens, Magnetic circular dichroism, Ann. Rev. Phys. Chem. **25**, 201-232 (1974).
5. M. R. F. Wikström, H. J. Harmon, N. J. Ingledew, and B. Chance, A re-evaluation of the spectral, potentiometric

and energy-linked properties of cytochrome *c* oxidase in mitochondria, FEBS Letters **65**(3), 259-278 (1976).
6. A. Cornish-Bowden, The physiological significance of negative co-operativity, J. Theor. Biol. **51**(1), 233-235 (1975).
7. K. Dalziel, Kinetics of control enzymes in "Rate Control in Biological Processes," pp. 21-48, Cambridge, 1973.

Chapter 8

1. A. A. Ahrem, D. Yu. Lando, and U. I. Krot, Investigation of nucleoproteins melting: I. Theory of helix-coil transition of DNA in the presence of proteins with cooperative character of interaction under conditions of reversible binding, Mol. Biol. USSR **10**(6), 1332-1340 (1976).
2. V. E. Khutorsky and V. I. Poltev, Theoretical conformational analysis of two-stranded helical polynucleotides, Biophysics USSR **21**, 201-207 (1976).
3. V. B. Zhurkin, Yu. P. Lysov, and V. I. Ivanov, Computer analysis of conformational possibilities of double-helical DNA, FEBS Letters **59**, 44-47 (1975).
4. V. I. Ivanov, Yu. P. Lysov, G. G. Malenkov, L. E. Minchenkova E. E. Miniat, and V. B. Zhurkin, Conformational possibilities of double-helical DNA, Studia biophysica **55**, 5-13 (1976).
5. W. Fiers et al., Complete nucleotide sequence of bacteriophage MS2 RNA: primary and secondary structure of the replicase gene, Nature **260**, 500-507 (1976).
6. B. G. Barrell, G. M. Air, and C. A. Hutchison, III, Overlapping genes in bacteriophage X174, Nature **264**, 34-41 (1976).
7. F. Sanger et al., Nature **265**, 687-695 (1977).

Chapter 9

1. G. V. Gursky, V. G. Tumanyan, A. S. Zasedatelev, A. L. Zhuse, S. L. Grokhovsky, and B. P. Gottikh, A model for the binding of lac repressor to the lac operator, Mol. Biol. Repts. **2**(5), 427-434 (1976).
2. M. V. Volkenstein, About transversions and transitions, Mol. Biol. **10**(3), 498-500 (1976) (R).
3. M. V. Volkenstein, Probabilities of transversions and transitions, Mol. Biol. **10**(4), 737-741 (1976) (R).
4. Tse-Fei Wong, A co-evolution theory of the genetic code, Proc. Nat. Acad. Sci. U.S. **72**, 1909-1912 (1975).
5. J. D. Robertus et al., Structure of yeast phenylalanine tRNA at 3Å resolution, Nature **250**, 546-551 (1974).
6. S. H. Kim et al., Three-dimensional tertiary structure of yeast phenylalanine transfer RNA, Science **185**, 435-440 (1974).
7. G. W. Moore, M. Goodman, C. Callahan, R. Holmquist, and

H. Moise, Stochastic versus augmented maximum parsimony method for estimating superimposed mutations in the divergent evolution of protein sequences. Methods tested on cytochrome *c* amino acid sequences, J. Mol. Biol. 105, 15-37 (1976).

# Index

## U



## V



## W



## X



## Z



A
B 7
C 8
D 9
E 0
F 1
G 2
H 3
I 4
J 5